T0188955

Fundamentals of Statistical Hydrology

Mauro Naghettini
Editor

Fundamentals of Statistical Hydrology

 Springer

Editor
Mauro Naghettini
Federal University of Minas Gerais
Belo Horizonte, Minas Gerais
Brazil

ISBN 978-3-319-82855-8   ISBN 978-3-319-43561-9   (eBook)
DOI 10.1007/978-3-319-43561-9

Printed on acid-free paper

This Springer imprint is published by Springer Nature
The registered company is Springer International Publishing AG
The registered company address is: Gewerbestrasse 11, 6330 Cham, Switzerland

To
Ana Luisa and Sandra.

Preface

This book has been written for civil and environmental engineering students and professionals. It covers the fundamentals of probability theory and the statistical methods necessary for the reader to explore, interpret, model, and quantify the uncertainties that are inherent to hydrologic phenomena and so be able to make informed decisions related to planning, designing, operating, and managing complex water resource systems.

Fundamentals of Statistical Hydrology is an introductory book and emphasizes the applications of probability models and statistical methods to water resource engineering problems. Rather than providing rigorous proofs for the numerous mathematical statements that are given throughout the book, it provides essential reading on the principles and foundations of probability and statistics, and a more detailed account of the many applications of the theory in interpreting and modeling the randomness of hydrologic variables.

After a brief introduction to the context of Statistical Hydrology in the first chapter, Chap. 2 gives an overview of the graphical examination and the summary statistics of hydrological data. The next chapter is devoted to describing the concepts of probability theory that are essential to modeling hydrologic random variables. Chapters 4 and 5 describe the probability models of discrete and continuous random variables, respectively, with an emphasis on those that are currently most commonly employed in Statistical Hydrology. Chapters 6 and 7 provide the statistical background for estimating model parameters and quantiles and for testing statistical hypotheses, respectively. Chapter 8 deals with the at-site frequency analysis of hydrologic data, in which the knowledge acquired in previous chapters is put together in choosing and fitting appropriate models and evaluating the uncertainty in model predictions. In Chaps. 9 and 10, an account is given of how to establish relationships between two or more variables, and of the way in which such relationships are used for transferring information between sites by regionalization. The last two chapters provide an overview of Bayesian methods and of the modeling tools for nonstationary hydrologic variables, as a gateway to the more advanced methods of Statistical Hydrology that the interested reader should consider. Throughout the book a wide range of worked-out examples are discussed as a means to illustrate the application of theoretical concepts to real-world practical cases. At the end of each chapter, a list of homework exercises, which both illustrate and extend the material given in the chapter, is provided.

The book is primarily intended for teaching, but can also be useful for practitioners as an essential text on the foundations of probability and statistics, and as a summary of the probability distributions widely encountered in water resource literature, and their application in the frequency analysis of hydrologic variables. This book has evolved from the text "Hidrologia Estatística," written in Portuguese and published in 2007 by the Brazilian Geological Survey CPRM (Serviço Geológico do Brasil), which has been extensively used in Brazilian universities as a reference text for teaching Statistical Hydrology. *Fundamentals of Statistical Hydrology* incorporates new material, within a revised logical sequence, and provides new examples, with actual data retrieved from hydrologic data banks across different geographical regions of the world. The worked-out examples offer solutions based on MS-Excel functions, but also refer to solutions using the R software environment and other free software, as applicable, with their respective Internet links for downloading. The book also has 11 appendices containing a brief review of basic mathematical concepts, statistical tables, the hydrological data used in the examples and exercises, and a collection of solutions to a few exercises and examples using R. As such, we believe this book is suitable for a one-semester course for first-year graduate students.

I take this opportunity to acknowledge and thank Artur Tiago Silva, from the Instituto Superior Técnico of the University of Lisbon, in Portugal, and my former colleagues Eber José de Andrade Pinto, Veber Costa, and Wilson Fernandes, from the Universidade Federal de Minas Gerais, Belo Horizonte, Brazil, for their valuable contributions authoring or coauthoring some chapters of this book and reviewing parts of the manuscript. I also wish to thank Paul Davis for his careful revision of the English. Finally, I wish to thank my family, whose support, patience, and tolerance were essential for the completion of this book.

Belo Horizonte, Brazil   Mauro Naghettini
11th May 2016

Contents

List of Contributors

Veber Costa Universidade Federal de Minas Gerais, Belo Horizonte, Minas Gerais, Brazil

Wilson Fernandes Universidade Federal de Minas Gerais, Belo Horizonte, Minas Gerais, Brazil

Mauro Naghettini Universidade Federal de Minas Gerais, Belo Horizonte, Minas Gerais, Brazil

Eber José de Andrade Pinto CPRM Serviço Geológico do Brasil, Universidade Federal de Minas Gerais, Belo Horizonte, Minas Gerais, Brazil

Artur Tiago Silva CERIS, Instituto Superior Técnico, Universidade de Lisboa, Lisbon, Portugal

Chapter 1
Introduction to Statistical Hydrology

Mauro Naghettini

1.1 The Role of Probabilistic Reasoning in Science and Engineering

Uncertainty is a word largely used to characterize a general condition where vague, imperfect, imprecise, or incomplete knowledge of a reality prevents the exact description of its actual and future states. As in everyday life, uncertainties are inescapable in science and engineering problems. Scientists and engineers are asked to comprehend complex phenomena, to ponder competing alternatives, and to make rational decisions on the basis of uncertain quantities. Uncertainties can arise from many sources (Morgan and Henrion 1990): (1) random and/or systematic errors in measurements of a quantity; (2) linguistic imprecision derived from qualitative reasoning; (3) quantity variability over time and space; (4) inherent randomness; (5) unpredictability (chaotic behavior) of dynamical systems; (6) disagreement or different opinions among experts about a particular quantity; and (7) approximation uncertainty arising from a simplified model of the real-world system. In such a context, quantity may refer either to an empirically measurable variable or to a model parameter.

Another possible categorization of uncertainty sources groups them into the aleatory and the epistemic types (Ang and Tang 2007). The former type includes the sources of uncertainties associated with natural and inherent randomness, such as sources (3), (4), (5) and part of (1) previously mentioned, and are said to be irreducible. The epistemic type encompasses all other sources and the corresponding uncertainties are said to be reducible, in the sense that the imperfect knowledge we have about the real world, as materialized by our imprecise and/or inaccurate measurement techniques and our simplified models (or statements), is

M. Naghettini (✉)
Universidade Federal de Minas Gerais Belo Horizonte, Minas Gerais, Brazil
e-mail: mauronag@superig.com.br

© Springer International Publishing Switzerland 2017
M. Naghettini (ed.), *Fundamentals of Statistical Hydrology*,
DOI 10.1007/978-3-319-43561-9_1

subject to further improvement. Such a categorization is debatable since in real-world applications, where both aleatory and epistemic uncertainties coexist, the distinction between them is not straightforward and, in some cases, depends on the modeling choices (Kiureghian and Ditlevsen 2009).

No matter what the sources of uncertainties are, they need to be assessed and combined, in a systematic and logical way, within the framework of a sound scientific approach. The best-known and most widely used mathematical formalism to quantify and combine uncertainties is embodied in the probability theory (Lindley 2000; Morgan and Henrion 1990). According to this theory, uncertainty is quantified by a real number between 0 (impossibility) and 1 (certainty). This real number is named probability. However, this role of probability in dealing with scientific and technical problems, albeit logical and clear, has not remained undisputed over the history of science and engineering (Yevjevich 1972; Koutsoyiannis 2008).

Since early times, one of the main functions of Science has been to predict future events from the knowledge acquired from the observation of past events. In the realm of determinism, a philosophical idea that has deeply influenced scientific thought, such predictions are made possible by inferring cause–effect relations between events from observed regularities. These strictly deterministic causal relations are then synthesized into "laws of nature," which are utilized to make predictions (Moyal 1949). This line of thought is demonstrated in the laws of classical Newtonian mechanics, according to which, given the boundary and initial conditions, and the differential equations that govern the evolution of a system, any of its future states can, in principle, be precisely determined. As an intrinsically time-symmetrical action, the specification of the state of the system at an instant t determines also its states before t.

Adherence to the principles of strict determinism has led the eminent French scientist Pierre Simon Laplace (1749–1827) to state in his *Essai philosophique sur les probabilités* (Laplace 1814) that:

> An intelligence which, for one given instant, would know all the forces by which nature is animated and the respective situation of the entities which compose it, if besides it were sufficiently vast to submit all these data to mathematical analysis, would encompass in the same formula the movements of the largest bodies in the universe and those of the lightest atom; for it, nothing would be uncertain and the future, as the past, would be present to its eyes (Laplace 1814, pp. 3–4).

Such a hypothetical powerful entity, well equipped with the attributes of being capable (1) of knowing the state of the whole universe, with perfect resolution and accuracy; (2) of knowing all the governing equations of the universe; (3) of an infinite instantaneous power of calculation; and (4) of not interfering with the functioning of the universe, later became known as Laplace's *demon*.

The counterpart of determinism, in mathematical logic, is expressed through deduction, a systematic method of deriving conclusions (or theorems) from a set of premises (or axioms), through the use of formal arguments (syllogisms): the conclusions cannot be false when the premises are true. The principle implied by deduction is that any mathematical statement can be proved from the given set of axioms. Besides deduction, another process of reasoning in mathematical logic is induction, by which a conclusion does not follow necessarily from the premises and actually goes beyond the information they contain. Instead of proving an argument is true or false, induction offers a highly probable conclusion, being subject, however, to occasional errors. When the use of deduction is not possible, as in decision making with incomplete information, induction can be helpful.

Causal determinism dominated scientific thought until the late nineteenth century. Random events, which occur in ways that are uncertain, were believed to be the mere product of human ignorance or insufficient knowledge of the initial conditions. As argued by Koutsoyiannis (2008), in the turn of the nineteenth century and in the first half of the twentieth century, the deterministic view which held that uncertainty could in principle be completely disregarded turned out to be deceptive, as it suffered serious setbacks in four major scientific areas.

(a) Statistical physics: a branch of physics that uses methods of probability and statistics to deal with large populations of elementary particles, where we cannot keep track of causal relations, but only observe statistical regularities. In particular, statistical mechanics has been very successful in explaining phenomenological results of thermodynamics from a probability-based perspective of the underlying kinetic properties of atoms and molecules.

(b) Complex and chaotic nonlinear dynamics: a field of mathematics whose main objects of study are dynamical systems, which are governed by complex nonlinear equations that are very sensitive to the initial conditions. Examples of such systems include the weather and climate system. Even in the case of a completely deterministic model of a system, with no random components, very small differences in the initial conditions can result in highly diverging outcomes. This characteristic of nonlinear dynamical system models makes them unpredictable in the long term, in spite of their deterministic nature.

(c) Quantum physics: an important field of physics dealing with the behavior of particles, as packets of energy, at the atomic and subatomic scales. The serious implications of quantum physics for determinism became clear in 1926, with Heisenberg's principle of uncertainty on the disturbances of states by observation. According to this principle, it is impossible to measure simultaneously the position and the momentum of a particle, as the more precisely we measure one variable the less precisely we can predict the other. Hence, in contrast with determinism, quantum physics is inherently uncertain and thus probabilistic in nature.

(d) Incompleteness theorems: two theorems in the domain of pure mathematical logic, proven in 1931 by the mathematician Kurt Gödel. They are generally interpreted as demonstrating that finding a complete and consistent set of axioms for all mathematics is impossible, thus revealing the inherent limitations of mathematical logic. An axiomatic set is complete if it does not contain a contradiction, whereas it is consistent if, for every statement, either itself or its denial can be derived from the system's axioms (Enderton 2002). Koutsoyiannis (2008) notes that Gödel's theorems imply that deductive reasoning has limitations and uncertainty cannot be completely disregarded. This conclusion strengthens the role of probabilistic reasoning, as extended logic, and opens up space for induction.

These appear to be compelling arguments in favor of the systematic use of probability and statistical methods instead of the pure deterministic rationale in science and engineering. However, it is not too hard to perceive a somewhat recalcitrant intellectual resistance to these arguments in some of today's scientists, engineers, and educators. This may be related to traditional science education, which unfortunately keeps teaching many disciplines with the same old perspective of depicting nature as fully explicable, in principle, by the laws of science, thus reinforcing the obsolete ideas of causal determinism. This book is intended to recognize from its very beginning that uncertainties are always present in natural phenomena, with a particular focus on those of the water cycle, and are best described and accounted for by the methods of probability and statistics.

1.2 Hydrologic Processes

Hydrology is a geoscience that deals with the natural phenomena that determine the occurrence, circulation, and distribution of the waters of the Earth, their biological, chemical, and physical properties, and their interaction with the environment, including their relation to living beings (WMO and UNESCO 2012). Hydrologic phenomena are at the origin of the different fluxes and storages of water throughout the several stages that compose the hydrologic cycle (or water cycle), from the atmosphere to the Earth and back to the atmosphere: evaporation from land or sea or inland water, evapotranspiration by plants, condensation to form clouds, precipitation, interception, infiltration, percolation, runoff, storage in the soil or in bodies of water, and back to evaporation. The continuous sequences of magnitudes of flow rates and volumes associated with these natural phenomena are referred to as hydrologic processes and can show great variability both in time and space. Seasonal, inter-annual, and quasi-periodic fluctuations of the global and/or regional climate contribute to such variability. Vegetation, topographic and geomorphic features, geology, soil properties, land use, antecedent soil moisture, the temporal and areal distribution of precipitation are among the factors that greatly increase the variability of hydrologic processes.

Applied Hydrology (or Engineering Hydrology) utilizes the scientific principles of Hydrology, together with the knowledge borrowed from other academic disciplines, to plan, design, operate and manage complex water resources systems. These are systems designed to redistribute, in space and time, the water that is available to a region in order to meet societal needs (Plate 1993), by considering both the quantity and quality aspects. Fulfillment of these objectives requires the reliable estimation of the time/space variability of hydrologic processes, including precipitation, runoff, groundwater flows, evaporation and evapotranspiration rates,

surface and subsurface water volumes, water-quality-related quantities, river sediment loads, soil erosion losses, etc.

Hydrologic processes often relate flows (or concentrations or mass or volumes)
of water (or dissolved oxygen or soil or energy) to their chronological times of
occurrence, though, in general, such a correspondence can be done also in space, or
both in time and space. The geographical scales used to study hydrologic processes
are diverse and go from the global to the most frequently used scale of the
catchment. For instance, the continuous time evolution of the discharges flowing
through the outlet section of a drainage basin is an example of a hydrologic process,
which, in this case, represents the time-varying synthesis of a very complex and
dynamical interaction of the many hydrologic phenomena operating in, over, and
under the surface of that particular catchment. Hydrologic processes can be monitored at discrete times, according to certain measurement standards, and then form
samples of hydrologic data. These are key elements for hydrologic analysis and
decision making concerning water resource projects.

In order to solve water-resource-related problems, hydrologists make use of
models, which are simplified representations of reality. Models can be generally
categorized as physical, analog, or mathematical. The former two are hardly used
by today's hydrologists who most often choose mathematical models, given their
vast possibility of applications and the widespread use of computers. Chow
et al. (1988) introduce the concept of a hydrologic system *as a structure or volume
in space, surrounded by a boundary, that accepts water and other inputs, operates
on them internally, and produces them as outputs*. A mathematical model, in this
perspective of system analysis, may be viewed as a collection of mathematical
functions, organized in a logical structure, linking the input and the output. Inputs
can be, for example, precipitation data or flood flow data, whereas outputs can be a
sequence of simulated flows or a probability-based summary of flood flows. The
model (or the system in mathematical form) is composed of a set of equations
containing parameters.

Chow et al. (1988) used the concept of hydrologic system analysis to classify the
mathematical models with respect to randomness, spatial variation, and time
dependence. If randomness is of concern, models can be either deterministic or
stochastic. A model is deterministic if it does not consider randomness: as an
inherent principle, both its input and output do not contain uncertainties and,
being of a causative nature, a given input always produces the same output. In
contrast, a stochastic model has several outputs produced by the same input and it
allows the quantification of their respective likelihoods.

It is rather intuitive to realize that hydrologic processes are random in nature. For
instance, next Monday's volume of rainfall at a specified location cannot be
forecast with absolute certainty. As being a pivotal process in the water cycle,
any other derived process, such as streamflow, for example, not only will inherit the
uncertainties from rainfall but also from other intervening processes. Another fact
that shows the ever-present randomness built in hydrologic processes refers to the
impossibility of establishing a functional cause–effect relation among variables
related to them. Let us take as an example one of the most relevant characteristics of

a flood event, the peak discharge, in a given catchment. Hydrology textbooks teach us that flood peak discharge can be conceivably affected by many factors: the time-space distribution of rainfall, the storm duration and its pathway over the catchment, the initial abstractions, the infiltration rates, and the antecedent soil moisture, to name a few. These factors are also interdependent and highly variable both in time and space. Any attempt to predict the next flood peak discharge from its relation to a finite number of intervening factors will result in error as there will always be a residual portion of the total dispersion of flood peak values that is left unexplained.

Although random in nature, hydrologic processes do embed seasonal and periodic regularities, and other identifiable deterministic signals and controls. Hydrologic processes are definitely susceptible to be studied by mass, energy and momentum conservation laws, thermodynamic laws, and to be modeled by other conceptual and/or empirical relations extensively used in modern physical hydrology. Deterministic and stochastic modeling approaches should be combined and wisely used to provide the most effective tools for decision making for water resource system analysis. There are, however, other views on how deterministic controlling factors should be included in the modeling of hydrologic processes. The reader is referred to Koutsoyiannis (2008) for details on a different view of stochastic modeling of hydrologic processes.

An example of a coordinated combination of deterministic and stochastic models comes from the challenging undertaking of operational hydrologic forecasting. A promising setup for advancing solutions to such a complex problem is to start from a probability-informed ensemble of numerical weather predictions, accounting for possibly different physical parametrizations and initial conditions. These probabilistic predictions of future meteorological states are then used as inputs to a physically plausible deterministic hydrologic model to transform them into sequences of streamflow. Then, probability theory can be used to account for and combine uncertainties arising from the different possible sources, namely, from meteorological elements, model structure, initial conditions, parameters, and states. Seo et al. (2014) recognize the technical feasibility of similar setups and point out the current and expected advances to fully operationalize them.

The combined use of deterministic and stochastic approaches to model hydrologic processes is not new but, unfortunately, has produced a philosophical misconception in past times. Starting from the principle that a hydrologic process is composed of a signal, the deterministic part, and a noise, the stochastic part, the underlying idea is that the ratio of the signal, explained by physical laws, to the unexplained noise will continuously increase with time (Yevjevich 1974). In the limit, this means that the process will be entirely explained by a causal deterministic relation at the end of the experience and that the probabilistic modeling approach was regarded only as temporary, thus reflecting an undesirable property of the phenomenon being modeled. This is definitely a philosophical journey back to the Laplacian view of nature. No matter how greatly our scientific physical knowledge about hydrologic processes increases and our technological resources to measure and observe them advance, uncertainties will remain and will need to be interpreted and accounted for by the theory of probability.

The theory of probability is the general branch of mathematics dealing with random phenomena. The related area of statistics concerns the collection, organization, and description of a limited set of empirical data of an observable phenomenon, followed by the mathematical procedures for inferring probability statements regarding its possible occurrences. Another area related to probability theory is stochastics, which concerns the study and modeling of random processes, generally referred to as stochastic processes, with particular emphasis on the statistical (or correlative) dependence properties of their sequential occurrences in time (this notion can also be extended to space). In analogy to the classification of mathematical models, as in Chow et al. (1988), general stochastic processes are categorized into *purely random processes*, when there is no statistical dependence between their sequential occurrences in time, and (simply) stochastic processes when there is. For purely random processes, the chronological order of their occurrences is not important, only their values matter. For stochastic processes, both order and values are important.

The set {theory of probability-statistics-stochastics} forms an ample theoretical body of knowledge, sharing common principles and analytical tools, that has a gamut of applications in hydrology and engineering hydrology. Notwithstanding the common theoretical foundations among the components of this set, it is a relatively frequent practice to group the hydrologic applications of the subset {theory of probability-statistics} into the academic discipline of Statistical Hydrology, concerning purely random processes and the possible association (or covariation) among them, whereas the study of stochastic processes is left to Stochastic Hydrology. This is an introductory text on Statistical Hydrology. Its main objective is to present the foundations of probability theory and statistics, as used (1) to describe, summarize and interpret randomness in variables associated with hydrologic processes, and (2) to formulate or prescribe probabilistic models (and estimate parameters and quantities related thereto) that concern those variables. .

1.3   Hydrologic Variables

A hydrologic process at a given location evolves continuously in time t. Such a continuous variation in time of a specific hydrologic process is referred to as a basic hydrologic variable (Yevjevich 1972) and is denoted as $x(t)$. Examples of basic hydrologic variables are instantaneous river discharge, instantaneous sediment concentration at a river cross section and instantaneous rainfall intensity at a site. For varying t, the statistical dependence between pairs of $x(t)$ and $x(t + \Delta t)$ will depend largely on the length of the time interval Δt. For short intervals, say 3 h or 1 day, the dependence will be very strong, but will tend to decrease as Δt increases.

For a time interval of one full year, $x(t)$ and $x(t + \Delta t)$ will be, in most cases, statistically independent (or time uncorrelated).

A basic hydrologic variable, representing the continuous time evolution of a specific hydrologic process, is not a practical setup for most applications of Statistical Hydrology. In fact, the so-called derived hydrologic variables, as extracted from $x(t)$, are more useful in practice (Yevjevich 1972). The derived hydrologic variables are, in general, aggregated total, mean, maximum, and minimum values of $x(t)$ over a specific time period, such as 1 day, 1 month, one season, or 1 year. Derived variables can also include highest/lowest values or volumes above/below a given threshold, the annual number of days with zero values of $x(t)$, and so forth. Examples of derived hydrologic variables are the annual number of consecutive days with no precipitation, the annual maximum rainfall intensity of 30-min duration at a site, the mean daily discharges of a catchment, and the daily evaporation volume (or depth) from a lake. As with basic variables, the time period used to single out the derived hydrologic variables, such as 1 day, 1 month, or 1 year, also affects the time interval separating their sequential values and, of course, the statistical dependence between them. Similarly to basic hydrologic variables, hourly or daily derived variables are strongly dependent, whereas yearly derived variables are, in most cases, time uncorrelated.

Hydrologic variables are measured at instants of time (or at discrete instants of time or still in discrete time intervals), through a number of specialized instruments and techniques, at site-specific installations called gauging stations, according to international standard procedures (WMO 1994). For instance, the mean daily water level (or gauge height) at a river cross section is calculated by averaging the instantaneous measurements throughout the day, as recorded by different types of sensors (Sauer and Turnipseed 2010), or, in the case of a very large catchment, by averaging staff gauge readings taken at fixed hours of the day. Analogously, the variation of daily evaporation from a reservoir, throughout the year, can be estimated from the daily records of pan evaporation, with readings taken at a fixed hour of the day (WMO 1994).

Hydrologic variables are random and the likelihoods of specific events associated with them are usually summarized by a probability distribution function. A sample is the set of empirical data of a derived hydrologic variable, recorded at appropriate time intervals to make them time-uncorrelated. The sample contains a finite number of independent observations recorded throughout the period of record. Clearly, the sample will not contain all possible occurrences of that particular hydrologic variable. These will be contained in the notional set of the population. The population of a hydrologic random variable would be a collection, infinite in some cases, of all of its possible occurrences if we had the opportunity to sample them. The main goal of Statistical Hydrology is to extract sufficient elements from the data to conclude, for example, with which probability, between the extremes of 0 and 1, the hydrologic variable of interest will exceed a given reference value, which has not yet been observed or sampled, within a given time horizon. In other words, the implicit idea is to draw conclusions on the population probabilistic behavior of the random variable, based on the information provided by the available sample.

According to the characteristics of their outcomes, random variables can be classified into qualitative (categorical) or quantitative (numerical) types. Qualitative random variables are those whose outcomes cannot be expressed by a number but by an attribute or a quality. They can be further classified in either ordinal or nominal, with respect to the respective possibility of their attributes (or qualities) be ordered or not, in a single way. The storage level of a reservoir, selected among the possible states of being {A: excessively high; B: high; C: medium; D: low; and E: excessively low} is an example of a categorical ordinal hydrologic random variable. On the other hand, the sky condition singled out among the possibilities of {sunny; rainy; and cloudy}, as reported in old-time weather reports, is an example of a categorical nominal random variable, as its possible outcomes are neither numerical nor susceptible to be ordered.

Quantitative random variables are those whose outcomes are expressed by integer or real numbers, receiving the respective type name of discrete or continuous. The number of consecutive dry days in a year at a given location is totally comprised of the subset of integer numbers given by {0, 1, 2, 3, ..., 366}. On the other hand, the annual maximum daily rainfall depth at the same location is a continuous numerical random variable because the set of its possible outcomes belongs to the subset of nonnegative real numbers. Numerical random variables can also be classified into the limited (bounded) or unlimited (unbounded) types. The former type includes the variables whose outcomes are upper-bounded, lower-bounded or double-bounded, either by some natural constraint or by the way they are measured. The random variable dissolved oxygen concentration in a lake is lower-bounded by zero and bounded from above by the oxygen dissolution capacity of the water body, which depends strongly but not only on the water temperature. In an analogous way, the wind direction at a site, measured by an anemometer or a wind vane, is usually reported in azimuth angles from 0 to 360°. The possible outcomes of unbounded continuous random variables are all real numbers. Most hydrologic continuous random variables are nonnegative and thus lower-bounded at 0.

Univariate and multivariate analyses are yet other possible types of formalisms involving hydrologic random variables. The univariate type refers to the analysis of a single random quantity or attribute, as in the previous examples, and the multivariate type involves more than one random quantity. In general terms, multivariate analysis describes the joint (and conditional) covariation of two or more random variables observed simultaneously. This and other topics briefly mentioned in this introduction are detailed in later chapters. As it is the most frequent case in applications of Statistical Hydrology, we intend to focus almost exclusively on hydrologic numerical random variables throughout this text.

1.4 Hydrologic Series

Hydrologic variables have their variation recorded by hydrologic time series, which contains observations (or measurements) organized in a sequential chronological order. It is again worthwhile noting that such a form of organization in time can also be replaced by space (distance or length) or, in some other cases, even extended to encompass time and space. Because of practical restrictions imposed by observational or data processing procedures, sequential records of hydrologic variables are usually separated by intervals of time (or distance). In most cases, these time intervals separating contiguous records, usually of 1-day but also of shorter durations, are equal throughout the series. In some other cases, particularly for water-quality variables, there are records taken at irregular time intervals.

As a hypothetical example, consider a large catchment, with a drainage area of some thousands square kilometers. In such a case, the time series formed by the mean daily discharges is thought to be acceptably representative of streamflow variability. However, for a much smaller catchment, with an area of some dozens square kilometers and time of concentration of the order of some hours, the series of mean daily discharges is insufficient to capture the streamflow variability, particularly in the course of a day and during a flood event. In this case, the time series formed by consecutive records of mean hourly discharges would be more suitable.

A hydrologic time series is called a historical series if it includes all available observations organized at regular time intervals, such as the series of mean daily discharges or less often a series of mean hourly discharges, chronologically ordered along the period of record. In a historical series, sequential values are time-correlated and, therefore, unsuitable to be treated by conventional methods of Statistical Hydrology. An exception is made for the statistical analysis through flow-duration curves, described in Chap. 2, in which the data time dependence is broken up by rearranging the records according to their magnitude values.

The reduced series, made of some characteristic values abstracted or derived from the records of historical series, are of more general use in Statistical Hydrology. Typical reduced series are composed of annual mean, maximum, and minimum values as extracted from the historical series. If for instance, a reduced series of annual maxima is extracted from a historical series of daily mean discharges, its elements will be referred to as annual maximum daily mean discharges. A series of monthly mean values, for the consecutive months of each year, may also be considered a reduced series, but their sequential records can still exhibit time dependence. In some cases, there may be interest in analyzing hydrologic data for a particular season or month within a year, such as the summer mean flow or the April rainfall depth. As they are selected on a year-to-year basis, the elements of such series are generally considered to be independent.

The particular reduced series containing extreme hydrologic events, such as maximum and minimum values that have occurred during a time period, is called an extreme-value series. If the time period is a year, extreme-value series are annual; otherwise, they are said to be non-annual, as extremes can be selected within a season or within a varying time interval (Chow 1964). An extreme-value series of maxima is formed by selecting the maximum values directly from instantaneous records, retained in archives of paper charts and files from electronic data loggers and data collection platforms, when available. In some cases, maximum flow values are derived from crest gauge heights or high-water marks (Sauer and Turnipseed 2010). When annual maxima are selected from instantaneous flow records they are referred to as annual peak discharges. When instantaneous records are not available, the series of annual maxima selected from the historical series is a possible acceptable option for flood analysis. In general, the water-year (or hydrological year), spanning from the 1st day of the wet season (e.g.: October 1st of a given year) to the last day of the dry season (e.g.: September 30th of the following year), is generally used in place of the calendar year. Because dry-season low flows vary at a much slower pace than high flows, extreme-value series for minima can be filled in by values extracted from the historical series.

Among the non-annual extreme-value series, there is the partial-duration series in which only the independent extreme values that are higher, in the case of maxima, or lower, in the case of minima, than a specified reference threshold are selected from the records. In a given year, there may be more than one extreme value, say 3 or 4, whereas in another year there may be none. Attention must be paid to the selection of consecutive elements in a partial duration series to ensure that they are not dependent or do not refer to the same hydrologic episode. For maxima, partial duration series are also referred to as peaks-over-threshold (POT) series, whereas for minima and particularly for the statistical analysis of dry spells, Pacheco et al. (2006) suggest the term pits-under-threshold (PUT). These concepts are revisited in following chapters of this book.

Figure 1.1 depicts the time plot of the annual peak discharges series, from October 1st, 1941 to September 30th, 2014, with a missing-data period from October, 1st 1942 to September, 30th 1943, of the Lehigh River at Stoddartsville, located in the American state of Pennsylvania, as summarized from the records provided by the streamflow gauging station USGS 01447500, owned and operated by the US Geological Survey. The Lehigh River is a tributary of the Delaware River. Its catchment at Stoddartsville has a drainage area of 237.5 km^2 and flow data are reported to be not significantly affected by upstream diversions or by reservoir regulation. In the USA, the water-year is a 12-month period starting in October 1st of any given year through September 30th of the following year. Available data at the gauging station USGS 01447500 can be retrieved by accessing the URL http://waterdata.usgs.gov/pa/nwis/inventory/?site_no=01447500 and include: daily series of water temperature and mean discharges, daily, monthly, and annual statistics for the daily series, annual peak streamflow series (Fig. 1.1), field measurements, field and laboratory water-quality samples, water-year summaries, and an archive of instantaneous data.

Figure 1.1 shows two extraordinary flood events: one in May 1942, with a peak discharge of 445 m^3/s, and the other, in August 1955, with an even bigger peak flow of 903 m^3/s. Note that both peak discharges are many times greater than the average peak flow of 86.8 m^3/s, as calculated for the remaining elements of the series. The

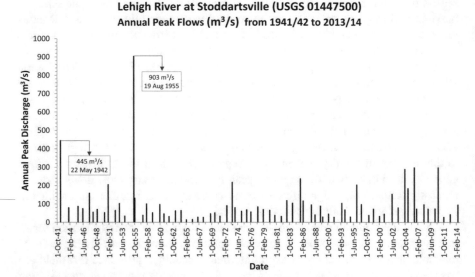

Fig. 1.1 Annual peak discharges in m^3/s of the Lehigh River at Stoddartsville (PA, USA) for water-years 1941/42 to 2013/14

first outstanding flood peak, on May 22nd, 1942, was the culmination of 3 weeks of frequent heavy rain over the entire catchment of the Delaware River (Mangan 1942). However, the record-breaking flood peak discharge, on August 19th 1955, is more than twice the previous flood peak record of 1942 and was due to a rare combination of very heavy storms associated with the passing of two consecutive hurricanes over the northeastern USA: hurricane Connie (August 12–13) and hurricane Diane (August 18–19). The 1955 flood waters ravaged everything in their path, with catastrophic effects, including loss of human lives and very heavy damage to properties and to the regional infrastructure (USGS 1956). Figure 1.1 and related facts illustrate some fascinating and complex aspects of flood data analysis.

Data in hydrologic reduced series may show an apparent gradual upward or downward trend with time, or a sudden change (or shift or jump) at a specific point in time, or even a quasi-periodical behavior with time. These time effects may be due to climate change, or natural climate fluctuations, such as the effects of ENSO (El Niño Southern Oscillation) and NAO (North Atlantic Oscillation), or natural slow changes in the catchment and/or river channel characteristics, and/or human-induced changes in the catchment, such as reservoir regulation, flow diversions, land-use modification, urban development, and so forth. Each of these changes, when properly and reliably detected and identified, can possibly introduce a nonstationarity in the hydrologic reduced series as the serial statistical properties and related probability distribution function will also change with time. For instance, if a small catchment, in near-pristine natural condition, is subject to a sudden and extensive process of urban development, the hydrologic series of mean

hourly discharges observed at its outlet river section will certainly behave differently from that time forth, as a result of significant changes in pervious area and
other factors controlling the rainfall-runoff transformation. For this hypothetical
example, it is likely that a reduced series extracted from the series of mean hourly
discharges will fall in the category of nonstationary. Had the catchment remained in
its natural condition, its reduced series would remain stationary. Conventional
methods of Statistical Hydrology require hydrologic series to be stationary. In
recent years, however, a number of statistical methods and models have been
developed for nonstationary series and processes. These methods and models are
reviewed in Chap. 12.

Conventional methods of Statistical Hydrology also require that a reduced
hydrologic time series must be homogeneous, which means that all of its data
must (1) refer to the same phenomenon in observation under identical conditions;
(2) be unaffected by eventual nonstationarities; and (3) originate from the same
population. Sources of heterogeneities in hydrologic series include: damming a
river at an upstream cross section; significant upstream flow diversions; extensive
land-use modification; climate change and natural climate oscillations; catastrophic
floods; earthquakes and other natural disasters (Haan 1977); man-made disasters,
such as dam failure; and occurrence of floods associated with distinct
flood-producing mechanisms, such as snowmelt and heavy rainfall (Bobée and
Ashkar 1991).

It is clear that the notion of heterogeneity, as applied particularly to time series,
encompasses that of nonstationarity: a nonstationary series is nonhomogeneous
with respect to time, although a nonhomogeneous (or heterogeneous) series is not
necessarily nonstationary. In the situation arising from upstream flow regulation by
a large reservoir, it is possible to transform regulated flows into *naturalized* flows,
by applying the water budget equation to inflows and outflows to the reservoir, with
the previous knowledge of its operating rules. This procedure can provide a single,
long and homogeneous series formed partly by true natural discharges and partly by
naturalized discharges. However, in other situations, such as in the case of an
apparent monotonic trend possibly due to gradual land-use change, it is very
difficult, in general, to make calculations to reconstruct a long homogeneous series
of discharges. In cases like this, it is usually a better choice to work with the most
recent homogeneous subseries, as it reflects approximately the current conditions
on which any future prediction must be based. Statistical tests concerning the
detection of nonstationarities and nonhomogeneities in reduced hydrologic series
are described in Chap. 7 of this book.

Finally, hydrologic series should be representative of the variability expected for
the hydrologic variable of interest. Representativeness is neither a statistical
requirement nor an objective index, but a desirable quality of a sample of empirical
data. To return to the annual peak discharges of the Lehigh River, observed at
Stoddartsville (Fig. 1.1), let us suppose we have to predict the flooding behavior at
this site for the next 50 years. In addition, let us suppose also that our available flood
data sample had started only in 1956/57 instead of 1941/42. Since the two largest
and outstanding flood peaks are no longer in our data sample, our predictions would

result in a severe underestimation of peak values and, if they were eventually used to design a dam spillway, they might have caused a major disaster. In brief, our hypothetical data sample is not representative of the flooding behavior at Stoddartsville. However, representativeness even of our actual flood peak data sample, from 1941/42 to 2013/14, cannot be warranted by any objective statistical criterion.

1.5 Population and Sample

The finite (or infinite) set of all possible outcomes (or realizations) of a random variable is known as the population. In most situations, with particular emphasis on hydrology, what is actually known is a subset of the population, containing a limited number of observations (data points) of the random variable, which is termed sample. Assuming the sample is representative and forms a homogeneous (and, therefore, stationary) hydrologic series, one can say that the main goal of Statistical Hydrology is to draw valid conclusions on the population's probabilistic behavior, from that finite set of empirical data. Usually, the population probabilistic behavior is summarized by the probability distribution function of the random variable of interest. In order to make such an inference, we need to resort to mathematical models that adequately describe such probabilistic behavior. Otherwise, our conclusions would be restricted by the range and empirical frequencies of the data points contained in the available sample. For instance, if we consider the sample of flood peaks depicted in Fig. 1.1, a further look at the data points reveals that the minimum and maximum values are 14 and 903 m^3/s, respectively. On the sole basis of the sample, one would conclude that the probability of having flood peaks smaller than 14 or greater than 903 m^3/s, in any given year, is zero. If one is asked about next year's flood peak, a possible but insufficient answer would be that it will probably be in the range of 14 and 903 m^3/s which is clearly an unsatisfactory statement.

The underlying reasoning of Statistical Hydrology begins with the proposal of a plausible mathematical model for the probability distribution function of the population. This is a deductive way of reasoning, by which a general mathematical synthesis is assumed to be valid for any case. Such a mathematical function contains parameters, which are variable quantities that need to be estimated from the sample data to fully specify the model and adjust it to that particular case. Once the parameters have been estimated and the model has been found capable of fitting the sample data, the now adjusted probability distribution function allows us to make a variety of probabilistic statements on any future value of the random variable, including on those not yet observed. This is an inductive way of reasoning where the general mathematical synthesis of some phenomena is particularized to a specific case. In summary, deduction and induction are needed to understand the probabilistic behavior of the population.

The easiest way of sampling from a population is termed simple random sampling and is the one which is most often used in Statistical Hydrology. One can think about simple random sampling in hydrology by devising an abstract plan according to which the data points (sample elements) are drawn from the population of the hydrologic variable, one by one, in a random and independent way. Each drawn element, such as the 1955 flood peak discharge in the Lehigh River at Stoddartsville, is the result of many causal, interdependent, and dynamic factors at the origin of that particular event. Such a sampling plan means that for a sample made of the elements $\{x_1, x_2, \ldots, x_N\}$, any one of them has been randomly drawn from the population, among a large number of equally probable choices. Element x_1, for example, had the same chance of being drawn as x_{25} or any other x_i had, including one or more than one repeated occurrences of x_1 itself. The latter possibility, called sampling with replacement, logically implies the N sample elements are statistically independent among themselves. The combination of the attributes of equal probability and collective statistical independence defines a simple random sample (SRS). A homogeneous and representative SRS is, in general, the first step for a successful application of Statistical Hydrology.

1.6   Quality of Hydrologic Data

Quantifying hydrologic variables, their variability, and their possible statistical association (covariation) requires the systematic collection of data, which develop in time and space. Longer samples of accurate hydrologic data, collected at many sites over a catchment or geographic area, are at the origin of effective solutions to the diverse problems of water resource system analysis. The hydrologic series comprehend rainfall, streamflow, groundwater flow, evaporation, sediment transport, and water-quality data observed at site-specific installations, known as gauging stations, in appropriately defined time intervals, according to standard procedures (WMO 1994). The group of gauging stations within a state (a province, a region, or a country) is known as the hydrometric network (Mishra and Coulibaly 2009), whose spatial density and maintenance are essential for the quality and practical value of hydrologic data. Extensive hydrometric networks are usually maintained and operated by national and/or provincial (federal and/or state) agencies. In some countries, private companies, from the energy, mining, and water industry sectors maintain and operate smaller hydrometric networks for meeting specific demands.

Hydrologic data are not error-free. Errors can possibly arise from any of the four phases of sensing, transmitting, recording and processing data (Yevjevich 1972). Errors in hydrologic data may be either random or systematic. The former are inherent to the act of measuring, carrying with them the unavoidable imprecision of readings and measurements which will scatter around the true (and unknown) value. For instance, if in a given day of a given dry season, 10 discharge measurements were made, using the same reliable current meter, operated by the same team of skilled personnel, one would have 10 close but different results, thus showing the essence of random errors in hydrologic data.

Systematic errors are the result of uncalibrated or defective instruments, repeatedly wrong readings, inappropriate measuring techniques, or any other inaccuracies coming from (or conveyed by) any of the phases of sensing, transmitting, recording and processing hydrologic data. A simple example of a systematic error refers to eventual changes that may happen around a rainfall gauge, such as a tree growing or the construction of a nearby building. They can possibly affect the prevalent wind speed and direction, and thus the rain measurements, resulting in systematically positive (or negative) differences as compared to previous readings and to other rainfall gauges nearby. The incorrect extension of the upper end of a rating curve is also a potential source of systematic errors in flow data.

Some methods of Statistical Hydrology can be used to detect and correct hydrologic data errors. However, most methods of Statistical Hydrology to be described in this book assume hydrologic data errors have been previously detected and corrected. In order to detect and modify incorrect data one should have access to the field, laboratory, and raw measurements, and proceed by scrutinizing them through rigorous consistency analysis. Statistical Hydrology deals most of the time with sampling variability or sampling uncertainty or even sampling *errors*, as related to samples of hydrologic variables. The notion of sampling error is best described by giving a hypothetical example. If five samples of a given hydrologic variable, each one with the same number of elements, are used to estimate the true mean value, they would yield five different estimates. The differences among them are part of the sampling variability around the true and unknown population mean. This, in principle, would be known only if all the population had been sampled. Sampling the whole population, in the case of natural processes such as those pertaining to the water cycle, is clearly impossible, thus disclosing the utility of Statistical Hydrology. In fact, as already mentioned, the essence of Statistical Hydrology is to draw valid conclusions on the population's probabilistic behavior, taking into account the uncertainty due to the presence and magnitude of sampling errors. In this context, it should be clear now that the longer the hydrologic series and the better the quality of their data, the more reliable will be our statistical inferences on the population's probabilistic behavior.

Exercises

1. List the main factors that make the hydrologic processes of rainfall and runoff random.
2. Give an example of a controlled field experiment involving hydrologic processes that can be successfully approximated by a deterministic model.
3. Give an example of a controlled laboratory experiment involving hydrologic processes that can be successfully approximated by a deterministic model.

Table 1.1 Annual maximum daily rainfall (mm) at gauge P. N. Paraopeba (Brazil)

Water year	Max daily rainfall (mm)	Water year	Max daily rainfall (mm)	Water year	Max daily rainfall (mm)	Water year	Max daily rainfall (mm)
41/42	68.8	56/57	69.3	71/72	70.3	86/87	109
42/43	Missing	57/58	54.3	72/73	81.3	87/88	88
43/44	Missing	58/59	36	73/74	85.3	88/89	99.6
44/45	67.3	59/60	64.2	74/75	58.4	89/90	74
45/46	Missing	60/61	83.4	75/76	66.3	90/91	94
46/47	70.2	61/62	64.2	77/77	91.3	91/92	99.2
47/48	113.2	62/63	76.4	78/78	72.8	92/93	101.6
48/49	79.2	63/64	159.4	79/79	100	93/94	76.6
49/50	61.2	64/65	62.1	79/80	78.4	94/95	84.8
50/51	66.4	65/66	78.3	80/81	61.8	95/96	114.4
51/52	65.1	66/67	74.3	81/82	83.4	96/97	Missing
52/53	115	67/68	41	82/83	93.4	97/98	95.8
53/54	67.3	68/69	101.6	83/84	99	98/99	65.4
54/55	102.2	69/70	85.6	84/85	133	99/00	114.8
55/56	54.4	70/71	51.4	85/86	101		

Table 1.2 Annual maximum mean daily flow (m³/s) at gauge P. N. Paraopeba (Brazil)

Water year	Max daily flow (m³/s)	Water year	Max daily flow (m³/s)	Water year	Max daily flow (m³/s)	Water year	Max daily flow (m³/s)
38/39	576	53/54	295	68/69	478	86/87	549
39/40	414	54/55	498	69/70	340	87/88	601
40/41	472	55/56	470	70/71	246	88/89	288
41/42	458	56/57	774	71/72	568	89/90	481
42/43	684	57/58	388	72/73	520	90/91	927
43/44	408	58/59	408	73/74	449	91/92	827
44/45	371	59/60	448	74/75	357	92/93	424
45/46	333	60/61	822	75/76	276	93/94	603
46/47	570	61/62	414	77/78	736	94/95	633
47/48	502	62/63	515	78/79	822	95/96	695
48/49	810	63/64	748	79/80	550	97/98	296
49/50	366	64/65	570	82/83	698	98/99	427
50/51	690	65/66	726	83/84	585		
51/52	570	66/67	580	84/85	1017		
52/53	288	67/68	450	85/86	437		

4. Give three examples of discrete and three of continuous random variables, associated with the rainfall phenomenon.

5. Tables 1.1 and 1.2 refer respectively to the annual maxima of daily rainfall depth (mm) at the rainfall gauging station 19440004 and of mean daily discharge of the Paraopeba river (m³/s) recorded at the gauging station 40800001,

both located in Ponte Nova do Paraopeba, in southcentral Brazil. The water-year from October, 1st to September 30th has been used to select values for both reduced series. The drainage area of the Paraopeba river catchment at Ponte Nova do Paraopeba is 5680 km^2. Make a scatter plot of concurrent data, with rainfall maxima in abscissa and flow maxima in ordinate, using the arithmetic scale, first, and, then, the logarithmic scale in both axes. Provide hydrologic arguments to explain why the points show scatter in the charts? List a number of unaccounted physical factors and hydrologic variables that can possibly help to find further explanation of the variation of annual maximum discharge, in addition to that provided by rainfall maxima. By adding as many influential factors and variables as possible to a multivariate relation, do you expect to fully explain the variation of annual maximum discharge?

6. Regarding the data in Tables 1.1 and 1.2, discuss their possible attributes of randomness and independence as required by simple random sampling. What is the best way to deal with missing data as in Table 1.1?

7. List and discuss possible random and systematic errors that may exist in flow measurements.

8. List and discuss possible random and systematic errors that may exist in rainfall gauging.

9. List and discuss possible sources of nonhomogeneities (heterogeneities) that may exist in flow data.

10. List and discuss possible sources of nonhomogeneities (heterogeneities) that may exist in rainfall data.

11. Access the URL http://waterdata.usgs.gov/pa/nwis/inventory/?site_no=01447500 and download the historical series of mean daily discharges and the reduced series of annual peak discharges, of the Lehigh River at Stoddartsville, both for the period of record. Extract from the downloaded daily flows the reduced series of annual maximum mean daily discharges, on the water-year basis. Plot on the same chart the two reduced series of annual peak discharges and annual maximum mean daily discharges, against time. Explain what causes their values to be different. Describe the possible disadvantages and drawbacks of using the reduced series of annual maximum mean daily discharges for designing flood control hydraulic structures?

12. Using the historical series of mean daily discharges as downloaded in Exercise 11, make a time plot of daily flows from January 1st, 2010 to December 31st, 2014. Compare the different results you find in maximum, mean, and minimum daily mean flows on the basis of water-year and calendar-year. Which time period, between water-year and calendar-year, would be recommended for selecting independent sequential data points in reduced series of annual maxima and annual minima?

13. Take the sample of 72 annual peak discharges as downloaded in Exercise 11 and separate it into 6 time-nonoverlapping sub-samples of 12 elements each. Calculate the mean value for each sub-series and for the complete series. Are all values estimates of the true mean annual flood peak discharge? Which one is more reliable? Why?

Table 1.3 Annual maximum mean daily discharges (m^3/s) of the Shokotsu River at Utsutsu Bridge (Japan Meteorological Agency), in Hokkaido, Japan

Date (m/d/y)	Q (m^3/s)	Date (m/d/y)	Q (m^3/s)	Date (m/d/y)	Q (m^3/s)	Date (m/d/y)	Q (m^3/s)
08/19/1956	425	11/01/1971	545	04/25/1985	169	04/13/1999	246
05/22/1957	415	04/17/1972	246	04/21/1986	255	09/03/2000	676
04/22/1958	499	08/19/1973	356	04/22/1987	382	09/11/2001	1180
04/14/1959	222	04/23/1974	190	10/30/1988	409	08/22/2002	275
04/26/1960	253	08/24/1975	468	04/09/1989	303	04/18/2003	239
04/05/1961	269	04/15/1976	168	09/04/1990	595	04/21/2004	287
04/10/1962	621	04/16/1977	257	09/07/1991	440	04/29/2005	318
08/26/1964	209	09/18/1978	225	09/12/1992	745	10/08/2006	1230
05/08/1965	265	10/20/1979	528	05/08/1993	157	05/03/2007	224
08/13/1966	652	04/21/1980	231	09/21/1994	750	04/10/2008	93.0
04/22/1967	183	08/06/1981	652	04/23/1995	238	07/20/2009	412
10/01/1968	89.0	05/02/1982	179	04/27/1996	335	05/04/2010	268
05/27/1969	107	09/14/1983	211	04/29/1997	167	04/17/2011	249
04/21/1970	300	04/28/1984	193	09/17/1998	1160	11/10/2012	502

Courtesy: Dr. S Oya, Swing Corporation, for data retrieval

14. Consider the annual maximum mean daily discharges of the Shokotsu River at Utsutsu Bridge, in Hokkaido, Japan, as listed in Table 1.3. This catchment, of 1198 km^2 drainage area, has no significant flow regulation or diversions upstream, which is a rare case in Japan. In the Japanese island of Hokkaido, the water-year coincides with the calendar year. Make a time plot of these annual maximum discharges and discuss their possible attributes of representativeness, independence, stationarity, and homogeneity.

15. Suppose a large dam reservoir is located downstream of the confluence of two rivers. Each river catchment has a large drainage area and is monitored by rainfall and flow gauging stations at its outlet and upstream. On the basis of this hypothetical scenario, is it possible to conceive a multivariate model to predict inflows to the reservoir? Which difficulties do you expect to face in conceiving and implementing such a model?

References

Ang AH-S, Tang WH (2007) Probability concepts in engineering: emphasis on applications to civil and environmental engineering, 2nd edn. Wiley, New York

Bobée B, Ashkar F (1991) The gamma family and derived distributions applied in hydrology. Water Resources Publications, Littleton, CO

Chow VT (1964) Section 8-I statistical and probability analysis of hydrological data. Part I frequency analysis. In: Chow VT (ed) Handbook of applied hydrology. McGraw-Hill, New York

Chow VT, Maidment DR, Mays LW (1988) Applied hydrology. McGraw-Hill Book Company, New York

Enderton HB (2002) A mathematical introduction to logic, 2nd edn. Harcourt Academic Press, San Diego

Haan CT (1977) Statistical methods in hydrology. Iowa University Press, Ames, IA

Kiureghian AD, Ditlevsen O (2009) Aleatory or epistemic? Does it matter? Struct Saf 31:105–112

Koutsoyiannis D (2008) Probability and statistics for geophysical processes. National Technical University of Athens, Athens. https://www.itia.ntua.gr/docinfo/1322. Accessed 14 Aug 2014

Laplace PS (1814) Essai philosophique sur les probabilités. Courcier Imprimeur–Libraire pour les Mathématiques, Paris. https://ia802702.us.archive.org/20/items/essaiphilosophi00laplgoog/essaiphilosophi00laplgoog.pdf. Accessed 14 Nov 2015

Lindley DV (2000) The philosophy of statistics. J R Stat Soc D 49(3):293–337

Mangan JW (1942) The floods of 1942 in the Delaware and Lackawanna river basins. Pennsylvania Department of Forests and Waters, Harrisburg. https://ia600403.us.archive.org/28/items/floodsofmay1942i00penn/floodsofmay1942i00penn.pdf. Accessed 14 Nov 2015

Mishra AK, Coulibaly P (2009) Developments in hydrometric network design: a review. Rev Geophys 47:RG2001:1–24

Morgan MG, Henrion M (1990) Uncertainty—a guide to dealing with uncertainty in quantitative risk and policy analysis. Cambridge University Press, Cambridge

Moyal JE (1949) Causality, determinism and probability. Philosophy 24(91):310–317

Pacheco A, Gottschalk L, Krasovskaia I (2006) Regionalization of low flow in Costa Rica. In: Climate variability and change—hydrological impacts, vol 308. IAHS Publ., pp 111–116

Plate EJ (1993) Sustainable development of water resources, a challenge to science and engineering. Water Int 18(2):84–94

Sauer VB, Turnipseed DP (2010) Stage measurement at gaging stations. US Geol Surv Tech Meth Book 3:A7

Seo D, Liu Y, Moradkhani H, Weerts A (2014) Ensemble prediction and data assimilation for operational hydrology. J Hydrol. doi:10.1016/j.jhydrol.2014.11.035

USGS (1956) Floods of August 1955 in the northeastern states. USGS Circular 377, Washington. http://pubs.usgs.gov/circ/1956/0377/report.pdf. Accessed 11 Nov 2015

WMO and UNESCO (2012) International glossary of hydrology. WMO No. 385, World Meteorological Organization, Geneva

WMO (1994) Guide to hydrological practices. WMO No. 168, vol I, 5th edn. World Meteorological Organization, Geneva

Yevjevich V (1974) Determinism and stochasticity in hydrology. J Hydrol 22:225–238

Yevjevich V (1972) Probability and statistics in hydrology. Water Resources Publications, Littleton, CO

Chapter 2
Preliminary Analysis of Hydrologic Data

Mauro Naghettini

2.1 Graphical Representation of Hydrologic Data

Hydrologic data are usually presented in tabular form (as in Table 1.3), which does not readily depict the essence of the empirical variability pattern that may exist therein. Such a desirable depiction can be easily conveyed through the graphical representation of hydrologic data. In this section, a non-exhaustive selection of various types of charts for graphically displaying discrete and continuous hydrologic random variables is presented and exemplified. The reader interested in more details on graphing empirical data is referred to Tufte (2007), Cleveland (1985) and Helsel and Hirsch (1992).

2.1.1 Bar Chart

The varying number of occurrences of a discrete hydrologic variable can be well summarized by a bar chart, which displays the possible integer values on the abscissa axis, while the number of occurrences, as corresponding to each possibility, is drawn as a vertical bar and read on the ordinate axis. Figure 2.1 illustrates an example of a bar chart representing the number of years in the record, from 0 to 34, for each one of which a corresponding annual number of flood occurrences, from 0 to 9, has been observed for the Magra River at the Calamazza gauging station, located in northwestern Italy. A single flood occurrence at this site is counted for every time flow increases above and then decreases below the threshold of 300 m^3/s. This threshold corresponds to a specific water level, above which the

M. Naghettini (✉)
Universidade Federal de Minas Gerais Belo Horizonte, Minas Gerais, Brazil
e-mail: mauronag@superig.com.br

© Springer International Publishing Switzerland 2017
M. Naghettini (ed.), *Fundamentals of Statistical Hydrology*,
DOI 10.1007/978-3-319-43561-9_2

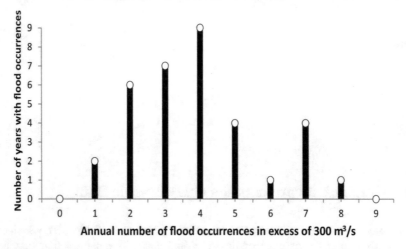

Fig. 2.1 Example of a bar chart for the number of years in the record with a specific count of flood occurrences per year, for the Magra River at Calamazza (adapted from Kottegoda and Rosso 1997)

river starts posing a threat to lives and properties along the reach's flood plain (Kottegoda and Rosso 1997). The flow data sample at Calamazza spans from 1939 to 1972. A further look at Fig. 2.1 suggests an approximately symmetrical distribution of the number of years in the record with a specific count of floods per year, centered around 4 floods every year.

2.1.2 Dot Diagram

Dot diagrams are useful tools for depicting the shape of the empirical frequency distribution of a continuous random variable, when only small samples, with typical sizes of 25–30, are available. This is a quite common situation in hydrological analysis, due to the usually limited periods of records and data samples of short lengths.

For creating dot diagrams, data are first ranked in ascending order of their values and then plotted on a single horizontal axis. As an example, Table 2.1 lists the annual mean daily discharges of the Paraopeba River at the Ponte Nova do Paraopeba gauging station, located in southeastern Brazil, for calendar years 1938–1963. In the table's second column, flows are listed according to the chronological years of occurrence, whereas the third column gives the discharges as ranked in ascending order. These ranked flow data were then plotted on the dot diagram shown in Fig. 2.2, where one can readily see a distribution of the sample points slightly skewed to the right of the discharge 85.7 m^3/s, the midpoint value

Table 2.1 Annual mean daily discharges of the Paraopeba River at the Ponte Nova do Paraopeba gauging station (Brazil), from 1938 to 1963

Calendar year	Annual mean daily flow (m^3/s)	Ranked flows (m^3/s)	Rank order
1938	104.3	43.6	1
1939	97.9	49.4	2
1940	89.2	50.1	3
1941	92.7	57	4
1942	98	59.9	5
1943	141.7	60.6	6
1944	81.1	68.2	7
1945	97.3	68.7	8
1946	72	72	9
1947	93.9	80.2	10
1948	83.8	81.1	11
1949	122.8	83.2	12
1950	87.6	83.8	13
1951	101	87.6	14
1952	97.8	89.2	15
1953	59.9	92.7	16
1954	49.4	93.9	17
1955	57	97.3	18
1956	68.2	97.8	19
1957	83.2	97.9	20
1958	60.6	98	21
1959	50.1	101	22
1960	68.7	104.3	23
1961	117.1	117.1	24
1962	80.2	122.8	25
1963	43.6	141.7	26

Paraopeba River at Ponte Nova do Paraopeba (Brazil)
Dot Diagram of Annual Mean Daily Flows (m^3/s)

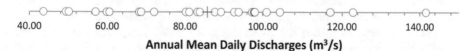

Annual Mean Daily Discharges (m^3/s)

Fig. 2.2 Dot diagram of annual mean daily discharges of the Paraopeba River at the Ponte Nova do Paraopeba gauging station (Brazil), from 1938 to 1963

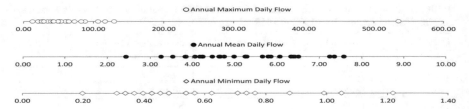

Fig. 2.3 Dot diagrams of annual minimum, mean, and maximum daily discharges of the Lehigh River at the Stoddartsville gauging station (PA, USA), from 1943/44 to 1972/73

between the 13th and 14th ranked discharges, marked with a cross on the chart. Another noticeable fact is the occurrence of the sample's wettest year in 1943, with an annual mean daily flow distinctly higher than those observed in other years, which partially explains the distribution's moderate asymmetry to the right.

Other examples of dot diagrams are shown in Fig. 2.3, for the annual mean, minimum, and maximum daily discharges, as abstracted (or reduced) from the historical series of mean daily flows of the Lehigh River recorded at the Stoddartsville gauging station (http://waterdata.usgs.gov/pa/nwis/inventory/?site_ no=01447500), for the water years 1943/44 to 1972/73. One can readily notice the remarkable differences among these plots: annual maximum daily flows are strongly skewed to the right, largely due to the record-breaking flood of 1955, whereas mean flows are almost symmetrically distributed around their respective midpoints, and minimum flows are slightly right-skewed. In the charts of Fig. 2.3, only the first 30 annual flows from the entire 72-year period of available records were utilized to illustrate the recommended application range of dot diagrams. Had the entire record been utilized, the data points would appear unduly close to each other and too concentrated around the center of the diagram.

2.1.3 Histogram

Histograms are graphical representations employed for displaying how data are distributed along the range of values contained in a sample of medium to large size when it becomes convenient to group data into classes or subsets in order to identify the data's patterns of variability in an easier fashion. For hydrologic variables, usually extracted from hydrologic reduced series, samples may be arbitrarily categorized as small, if $N \leq 30$, medium, if $30 < N \leq 70$ or large, if $N > 70$, where N denotes the sample size. The reduced time series given in Table 2.2 are the mean annual daily discharges abstracted from the historical series of mean daily discharges of the Paraopeba River at Ponte Nova do Paraopeba, from January 1st, 1938 to December 31st, 1999. The sample formed by the annual mean flows {104.3, 97.9, …, 57.3}, with $N = 62$, is considered as medium-sized and is used to exemplify how to make a histogram. It is worth reminding ourselves that data in a hydrologic

Table 2.2 Annual mean daily flows of the Paraopeba River at Ponte Nova do Paraopeba, from 1938 to 1999

Calendar year	Annual mean flow (m^3/s)	Calendar year	Annual mean flow (m^3/s)
1938	104.3	1969	62.6
1939	97.9	1970	61.2
1940	89.2	1971	46.8
1941	92.7	1972	79
1942	98	1973	96.3
1943	141.7	1974	77.6
1944	81.1	1975	69.3
1945	97.3	1976	67.2
1946	72	1977	72.4
1947	93.9	1978	78
1948	83.8	1979	141.8
1949	122.8	1980	100.7
1950	87.6	1981	87.4
1951	101	1982	100.2
1952	97.8	1983	166.9
1953	59.9	1984	74.8
1954	49.4	1985	133.4
1955	57	1986	85.1
1956	68.2	1987	78.9
1957	83.2	1988	76.4
1958	60.6	1989	64.2
1959	50.1	1990	53.1
1960	68.7	1991	112.2
1961	117.1	1992	110.8
1962	80.2	1993	82.2
1963	43.6	1994	88.1
1964	66.8	1995	80.9
1965	118.4	1996	89.8
1966	110.4	1997	114.9
1967	99.1	1998	63.6
1968	71.6	1999	57.3

sample are extracted from a reduced series of uncorrelated elements that, in turn, are abstracted from the instantaneous records or from the historical series of daily values, often with one single value per year to warrant statistical independence.

In order to create a histogram, one first needs to group the sample data into classes or bins, defined by numerical intervals of either fixed or variable width, and, then, count the number of occurrences, or the absolute frequency, for each class. The number of classes (or number of bins), here denoted by NC, depends on the sample size N and is a key element to histogram plotting. In effect, too few classes will not allow a detailed visual inspection of the sample characteristics, whereas too many will result in excessively large fluctuations of the corresponding frequencies. Kottegoda and Rosso (1997) suggest that NC may be approximated by the nearest integer to $\sqrt{N}$, with a minimum value of 5 and a maximum of 25, positing that histograms for sample sizes of less than 25 are not informative. An alternative is the Sturges' rule, proposed by Sturges (1926), who suggested that the number of bins should be given by

$$NC = 1 + 3.3\log_{10}N \qquad (2.1)$$

The Sturges' rule was derived on the assumption of approximately symmetrical data distribution, which is not the general case for hydrologic samples. For sample sizes lower than 30 and/or asymmetrical data, the Eq. (2.1) does not usually provide the best results. For a description of a more elaborate data-driven method for determining the optimal bin-width and the corresponding number of bins, the reader is referred to Shimazaki and Shinomoto (2007). Matlab, Excel, and R programs for optimizing bin-width, through the Shimazaki–Shinomoto method, are available for downloading from the URL http://176.32.89.45/~hideaki/res/histogram.html#Matlab.

To show how to calculate a frequency table, which is the first step of histogram plotting, let us take the example of the flows listed in Table 2.2, which forms a sample of $N = 62$. By applying the $\sqrt{N}$ and Sturges criteria, NC must be a number between 7 and 8. Let us fix $NC = 7$ and remind ourselves that the lower limit for the first class must be less than or equal to the sample minimum value which is 43.6 m³/s, whereas the upper limit of the seventh class must be greater than the sample maximum of 166.9 m³/s. Since the range R, between the sample maximum and minimum values, is $R = 123.3$, and $NC = 7$, the assumed fixed class-width (or bin-width) may be taken as $CW = 20$ m³/s, which is the nearest multiple-of-ten to 17.61, the ratio between R and NC. Table 2.3 summarizes the results for (a) the absolute frequencies, given by the number of flows within each class; (b) the relative frequencies, calculated by dividing the corresponding absolute frequencies by $N = 62$; and (c) the cumulative relative frequencies.

Table 2.3 Frequency table of annual mean flows of the Paraopeba River at Ponte Nova do Paraopeba—Records from 1938 to 1999

Class j	Class interval (m³/s)	Absolute frequency f_j	Relative frequency fr_j	Cumulative frequency $F = \sum_j fr_j$
1	(30,50]	3	0.0484	0.0484
2	(50,70]	15	0.2419	0.2903
3	(70,90]	21	0.3387	0.6290
4	(90,110]	12	0.1935	0.8226
5	(110,130]	7	0.1129	0.9355
6	(130,150]	3	0.0484	0.9839
7	(150,170]	1	0.0161	1
Sum		62	1	

With the results of Table 2.3, making the histogram shown in Fig. 2.4 is a rather straightforward operation. A histogram is actually a simple bar chart, with the class intervals on the horizontal axis and the absolute (and relative) frequencies on the vertical axis. The salient features of the histogram, shown in Fig. 2.4, are: (a) the largest concentration of sample data on the third class interval, which probably contains the central values around which data are spread out; (b) a moderately asymmetric frequency distribution, as indicated by the larger distance from the 3rd to the last bin, compared to that from the 3rd to the first bin; and (c) a single occurrence of a large value in the 7th class interval. It is important to stress, however, that the histogram is very sensitive to the number of bins, to bin-width, and to the bin initial and ending points as well. To return to the example of Fig. 2.4, note that the two last bins have absolute frequencies respectively equal to 3 and 1, which could be combined into a single class interval of width 40 m³/s, absolute frequency of 4, and ending points of 130 and 170 m³/s. This would significantly change the overall shape of the resulting histogram. As an aside, application of the Shimazaki–Shinomoto method to data of Table 2.2 results in a histogram with an optimized number of bins of only 4.

Histograms of annual mean, minimum, and maximum flows usually differ much among themselves. The panels (a), (b), and (c) of Fig. 2.5 show histograms of annual mean, minimum, and maximum daily discharges, respectively, of the Lehigh River at Stoddartsville for the water years 1943/44 to 2014/15. The histogram of annual mean flows shows a slight asymmetry to the right, which is a bit more pronounced if one consider the annual minimum flows histogram. However, in respect of the annual maximum discharges, the corresponding histogram, in

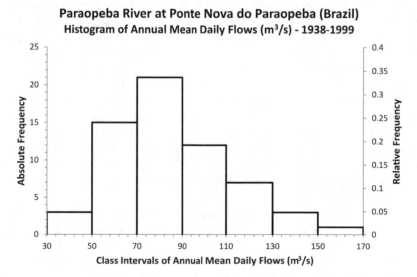

Fig. 2.4 Histogram of annual mean daily flows of the Paraopeba River at Ponte Nova do Paraopeba—Records from 1938 to 1999

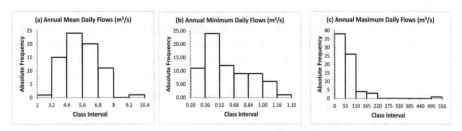

Fig. 2.5 Histograms of annual mean, minimum, and maximum daily flows of the Lehigh River at Stoddartsville—Records from 1943/44 to 2014/15

panel (c) of Fig. 2.5, is much more skewed to the right, thus illustrating a prominent characteristic of frequency distributions of extreme high flows. As a matter of fact, ordinary floods occur more frequently and usually cluster around the center of the histogram. On less frequent occasions, however, extraordinary floods, such as the 1955 extreme flood in the Lehigh River, add a couple of occurrences at bins far distant from the central ones, giving the usual overall shape of a histogram of maximum flows. This characteristic pattern of variability is used in later chapters to formulate and prescribe mathematical models for the probability distributions of hydrologic maxima.

2.1.4 Frequency Polygon

The frequency polygon is another chart based on the frequency table and is also useful to diagnose the overall pattern of the empirical data variability. Such a polygon is formed by joining the midpoints of the topsides of the histogram bars, after adding one bin on both sides of the diagram. The frequency polygon, based on the relative frequencies of the histogram of Fig. 2.5, is depicted in Fig. 2.6. Note that, as the frequency polygon should start and end at zero frequency and have the total area equal to that of the histogram, its initial and end points are respectively located at the midpoints of its first and last bins. Thus, the frequency polygon of Fig. 2.6 starts at the abscissa equal to 20 m^3/s and ends at 180 m^3/s, both with a relative frequency equal to zero. For a single-peaked frequency polygon, the abscissa that corresponds to the largest ordinate is termed mode and corresponds to the most frequent value in that particular sample. In the case of Fig. 2.4, the sample mode is 80 m^3/s.

Frequency polygons are usually created on the basis of relative frequencies instead of absolute frequencies. The relative frequencies, plotted on the vertical axis, should certainly be bounded by 0 and 1. As the sample size increases, the number of bins also increases and the bin-width decreases. This has the overall effect of gradually smoothing the shape of the frequency polygon, turning it into a frequency curve instead. In the limiting and hypothetical case of an infinite sample, such a frequency curve would become the population's probability density function, whose formal definition is one of the topics discussed in Chap. 3.

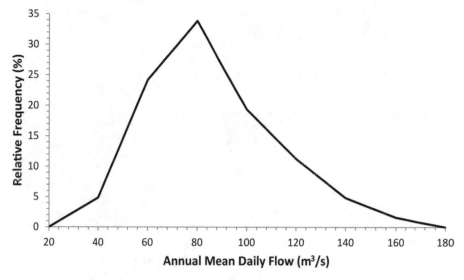

Fig. 2.6 Frequency polygon of annual mean daily flows of the Paraopeba River at Ponte Nova do Paraopeba—Records from 1938 to 1999

2.1.5   Cumulative Relative Frequency Diagram

The cumulative relative frequency diagram is made by joining, through straight lines, the pairs formed by the upper limit of each bin interval and the cumulative relative frequency up to that point, as read from a frequency table. On the vertical axis, the diagram gives the frequency that a random variable is equal to or less than a value read on the horizontal axis. As an alternative and more practical method of drawing it, the cumulative relative frequency diagram can also be created without previously making a frequency table. This can be carried out through the following steps: (1) rank data in ascending order of values; (2) associate sorted data with their respective ranking orders m, with $1 \leq m \leq N$; and (3) associate ranked data with their corresponding non-exceedance frequencies, as calculated by the ratio m/N. Besides being more expeditious and practical, this alternative method has the additional advantage of not depending on the previous definition of the number of class intervals, which always entails an element of subjectivity. Both methods have been used in plotting the cumulative relative frequency diagrams shown in Fig. 2.7, for the annual mean daily flows of the Paraopeba River at Ponte Nova do Paraopeba. The diagram in dashed line corresponds to plotting cumulative frequencies as calculated in the 5th column of Table 2.3, whereas the diagram in continuous line results from ranking data the alternative way.

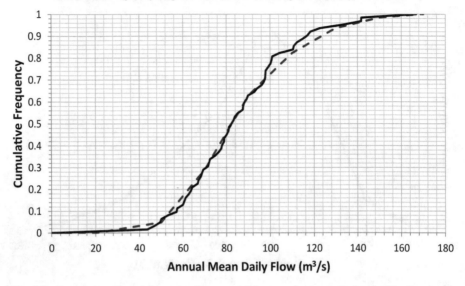

Fig. 2.7 Cumulative relative frequency diagrams of the annual mean daily flows of the Paraopeba River at Ponte Nova do Paraopeba—Records from 1938 to 1999. Dashed line: diagram with frequencies from Table 2.3. Continuous line: diagram for ranked data

The cumulative frequency diagram allows the immediate identification of the second quartile, denoted by Q_2, which corresponds to the value associated with the non-exceedance frequency of 0.5. The first quartile Q_1 and the third quartile Q_3 correspond respectively to the frequencies of 0.25 and 0.75. For the continuous-line diagram of Fig. 2.7, $Q_2 = 82.7$, $Q_1 = 67.95$; and $Q_3 = 99.38$ m^3/s. The Inter-Quartile Range, or IQR, is given by the difference between Q_3 and Q_1 and is commonly utilized as a criterion to identify outliers, which are defined as points (or elements) that deviate substantially from the pattern of variation shown by the other elements of the sample. According to this criterion, a high outlier is any sample element that is larger than $(Q_3 + 1.5IQR)$ and, analogously, a low outlier is any sample element that is smaller than $(Q_1 - 1.5IQR)$. An outlier might be the result either of a gross observational error or of an extraordinary single event. If, after a close look at data, the former case is confirmed, the outlier must simply be removed from the sample. However, in the case of a rare and extraordinary occurrence, removing the outlier from the sample would be an incorrect decision since it would make the sample less representative of the variation pattern of the random quantity in question. To return to the example of Fig. 2.7, according to the IQR criterion, the 1983 annual mean flow of 166.9 m^3/s is a high outlier.

Analogously to quartiles, one can make reference to deciles, for cumulative frequencies of 0.1, 0.2, 0.4, ..., 0.9, to percentiles for frequencies of 0.01, 0.02,

0.03, ..., 0.99, and more generically to quantiles. It is worth noting that by transposing the horizontal and vertical axes of a cumulative frequency diagram, one would get the quantile plot or Q-plot. Similarly to the frequency polygon, as the sample size increases, the cumulative frequency diagram is smoothened and turns itself into a cumulative frequency distribution curve. In the limiting and hypothetical case of an infinite sample, such a frequency curve would become the population's cumulative probability distribution function. These concepts are further discussed in Chap. 3.

2.1.6 Duration Curves

The duration curve is a variation of the cumulative relative frequency diagram, obtained by replacing the non-exceedance frequency by the fraction of a specific time period, during which a given value, now read on the horizontal axis, has been equaled or exceeded. In engineering hydrology, duration curves, in general, and flow duration curves (FDCs), in particular, are used with much success to graphically synthesize the variability of a hydrologic quantity, especially daily flows ranked according to their values. FDCs are also very frequently used for water resources planning and management, and, in some countries, as a means to calculate maximum flow abstractions related to water users' rights.

In general, for a flow gauging station with N days of records, a flow duration curve can be created through the following sequential steps: (1) rank flows Q in descending order of values; (2) assign to each sorted flow Q_m its respective ranking order m; (3) relate to each ranked flow Q_m its respective empirical frequency of being equaled or exceeded $P(Q \geq Q_m)$, which can be estimated by the ratio (m/N); and (4) plot the ranked flows on the vertical axis, as matched by their respective percent frequencies $100(m/N)$, on the horizontal axis. Flow duration curves can be plotted on a yearly basis, with $N = 365$ (or 366), for any given year, when it is termed an Annual Flow Duration Curve (AFDC). When it is calculated over a larger-than-a-year period or over all daily flow records available in the sample, it is referred to as just an FDC. In order to give an example of AFDC and FDC charting, let us take the case of all daily flows of the Lehigh River recorded at the gauging station of Stoddartsville (http://waterdata.usgs.gov/pa/nwis/inventory/?site_no=01447500). The annual hydrographs for all water-years in the available records, from 1943/44 to 2014/2015, are plotted on the background of Fig. 2.8, whereas the hydrograph for the water-year 1982/83, considered as a typical one, is highlighted in the foreground. Note that the vertical axis in Fig. 2.8 is displayed on a logarithmic scale so that to allow all flow data be plotted on a single chart.

Figure 2.9 depicts AFDCs for four particular years and the FDC for the period-of-record at the Stoddartsville gauging station. The 1964–1965 AFDC corresponds to the year with the smallest annual mean daily flow and is representative of dry conditions all year round. In contrast, the 2010–2011 AFDC reflects the year-round

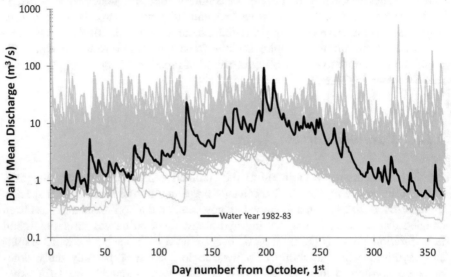

Fig. 2.8 Annual hydrographs of the Lehigh River at Stoddartsville for the period-of-record (1943/44 to 2014/15)

wettest conditions in the period-of-record. The AFDC for 1954–1955 depicts the flow behavior for the year, throughout which discharges vary the most, while for 1989–1990 they vary the least. Finally, the FDC, comprehending all 26,298 daily flows recorded from October 1st, 1943 to September 30th, 2015, synthesize all possible variations of daily flows throughout the period-of-record. One might ask why, in the case of AFDCs and FDCs, daily flows are employed with no concerns regarding the unavoidable statistical dependence that exist among their time-consecutive values. In fact, the chronological order of daily flows does not matter for AFDCs and FDCs, since it is expected to be entirely disrupted by ranking data according to their magnitudes, which are the only important attributes in this type of analysis.

2.2 Numerical Summaries and Descriptive Statistics

The essential features of the histogram (or frequency polygon) shape can be summarized through the sample descriptive statistics. These are simple and concise numerical summaries of the empirical frequency distribution of the random variable. They have the advantage, over the graphical representation of data, of

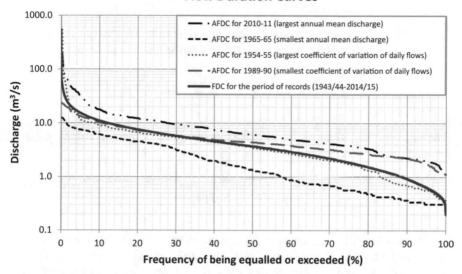

Fig. 2.9   AFDCs and FDC of the Lehigh River at Stoddartsville

providing the sample numerical information to later infer the population probabilistic behavior. Descriptive statistics may be grouped into (1) measures of central tendency; (2) measures of dispersion; and (3) measures of asymmetry and tail weight.

2.2.1   Measures of Central Tendency

Hydrologic data usually cluster around a central value, as in the dot diagrams of Figs. 2.2 and 2.3. The sample central value can be located by one of the measures of central tendency, among which the most often used are the mean, the median, and the mode. The right choice among the three depends on the intended use of the central value location.

2.2.1.1   Mean

For a sample of size N consisting of the data points $\{ x_1, x_2, \ldots, x_N \}$, the arithmetic mean is estimated as

$$\bar{x} = \frac{x_1 + x_2 + \ldots + x_N}{N} = \frac{1}{N} \sum_{i=1}^{N} x_i \tag{2.2}$$

If from the N data points of variable X, N_1 are equal to x_1, N_2 are equal to x_2 and so forth up to the kth sample value, then, the arithmetic mean is given by

$$\bar{x} = \frac{N_1 x_1 + N_2 x_2 + \ldots + N_k x_k}{N} = \frac{1}{N} \sum_{i=1}^{k} N_i x_i \tag{2.3}$$

Analogously, if f_i denotes the relative frequency of a generic datum point x_i, Eq. (2.3) can be rewritten as

$$\bar{x} = \sum_{i=1}^{k} f_i x_i \tag{2.4}$$

The sample arithmetic mean is the most used measure of central tendency and has an important significance as an estimate of the population mean μ. As mentioned in Sect. 2.1.4, in the limiting case of an infinite sample of a continuous random variable X or, in other terms, in the case of a frequency polygon becoming a probability density function, the population mean μ would be located on the horizontal axis exactly at the coordinate of the centroid of the area enclosed by the probability curve and the abscissa axis.

Alternatives to the sample arithmetic mean, but still using the same implied idea, are two other measures of central tendency which can be useful in some special cases. They are the harmonic mean, denoted by $\bar{x}_h$, and the geometric mean $\bar{x}_g$. The harmonic mean is the reciprocal of the arithmetic mean of the reciprocals of the sample data points. It is formally defined as

$$\bar{x}_h = \frac{1}{(1/N)[(1/x_1) + (1/x_2) + \ldots + (1/x_N)]} \tag{2.5}$$

Typically, the harmonic mean gives a better notion of a mean, in situations involving rates of variation. For example, if a floating device traverses the first half of a river reach with the velocity of 0.4 m/s and the other half at 0.60 m/s, then the arithmetic mean speed would be $\bar{x} = 0.50\,\text{m/s}$ and the harmonic mean would be $\bar{x}_h = 0.48$ m/s. The latter is actually the true mean velocity at which the floating device crosses the entire river reach.

On the other hand, the geometric mean $\bar{x}_g$ is more meaningful for estimating the mean of a variable whose sample points either can vary throughout orders of magnitude and are best described by their logarithms, such as the fecal coliform concentrations in a water body, or refer to proportional effects rather than additive, such as the percentage growth of the human population. The geometric mean can be applied only to a variable of the same sign and it is consistently smaller than or equal to the arithmetic mean. For a sample of size N, the geometric mean is given by

$$\bar{x}_g = \sqrt[N]{x_1.x_2.\;\cdots\;.x_N} = \prod_{i=1}^{N}(x_i)^{1/N} = \exp\left(\frac{1}{N}\sum_{i=1}^{N}\ln x_i\right) \qquad (2.6)$$

and it can be easily proved that the geometric mean is equal to the antilogarithm of the arithmetic mean of the logarithms of data elements x_i.

2.2.1.2 Median

The arithmetic mean, by virtue of taking into account all sample elements, has the disadvantage of being possibly affected by outliers. The median, denoted by x_{md}, is another measure of central tendency that is considered resistant (or robust) to the effects that eventual sample outliers can produce. The median is defined as the value that separates the sample into two subsamples, each with 0.5 cumulative frequency and, thus, is equivalent to the second quartile Q_2. If sample data are ranked in ascending order such that $\{x_{(1)} \leq x_{(2)} \leq \;\cdots\; \leq x_{(N)}\}$, then the median is given by

$$x_{md} = x_{\left(\frac{N+1}{2}\right)} \text{ if } N \text{ is an odd number or } x_{md} = \frac{x_{\left(\frac{N}{2}\right)} + x_{\left(\frac{N}{2}+1\right)}}{2} \text{ if } N \text{ is even} \qquad (2.7)$$

2.2.1.3 Mode

The mode x_{mo} is the value that occurs most frequently and is usually taken from the frequency polygon (in Fig. 2.6, $x_{mo} = 80$ m^3/s). In the limiting case of an infinite sample of a continuous random variable and the frequency polygon turning into a probability density function, the mode will be located at the abscissa value corresponding to the point where the derivative of the function is zero, observing however that, for multimodal density functions, more than one such point can occur. For skewed frequency polygons, where the ranges of sample values to the right and to the left of the mode greatly differ, the measures of central tendency show peculiar characteristics. When the right range is much larger than the left one, the polygon is positively asymmetric (or skewed), a case in which $x_{mo} < x_{md} < \bar{x}$. For the opposite case, the polygon is said to be negatively asymmetric (or skewed) and $\bar{x} < x_{md} < x_{mo}$. When both ranges are approximately equivalent, the polygon is symmetric and the three measures of central tendency are equal or very close to each other.

2.2.2   *Measures of Dispersion*

The degree of variability of sample points, around their central value, is given by
the measures of dispersion. Among these, the simplest and most intuitive is the
range R, given by $R = x_{(N)} - x_{(1)}$, where $x_{(N)}$ and $x_{(1)}$ denote, respectively, the Nth
and the 1st sample elements, ranked in ascending order of values. Note that the
range R depends only on the maximum and minimum sample points, which might
be strong outliers and thus make R an exaggerated measure of scatter. A measure
that is more resistant (or robust) to the presence of outliers is given by the Inter-
quartile Range (*IQR*) as defined by $IQR = Q_3 - Q_1$.

Both measures already mentioned, based on only two sample characteristic
values though easy to calculate, are not representative of the overall degree of
scatter because they actually ignore the remaining sample elements. Such problems
can be overcome through the use of other dispersion measures based on the mean
deviation of sample points to their central value. The main dispersion measures in
this category are the mean absolute deviation and the standard deviation.

2.2.2.1   Mean Absolute Deviation

The mean absolute deviation, denoted by d, is the arithmetic mean of the absolute
deviations from the sample mean. For a sample consisting of the elements $\{x_1,$
$x_2, \ldots, x_N\}$, d is defined as

$$d = \frac{|x_1 - \bar{x}| + |x_2 - \bar{x}| + \ \cdots \ |x_N - \bar{x}|}{N} = \frac{1}{N} \sum_{i=1}^{N} |x_i - \bar{x}| \qquad (2.8)$$

The mean absolute deviation linearly weights both small and large deviations from
the sample mean, which is seen as simpler and more intuitive, as compared to other
measures of dispersion. However, in spite of being a rather natural measure of
dispersion, the mean absolute deviation of a sample leads to a biased estimation of
the population equivalent measure, which is an undesirable attribute which will be
explained in later chapters.

2.2.2.2   Standard Deviation

An alternative to replacing the mean absolute deviation as a dispersion measure is
to square the deviations from the sample mean, which would give more weight to
the large deviations. For a data set $\{x_1, x_2, \ldots, x_N\}$, the so-called uncorrected
variance, here denoted by s_u^2, is defined as the arithmetic mean of the squared
deviations from the sample mean. Formally,

$$s_u^2 = \frac{(x_i - \bar{x})^2 + (x_2 - \bar{x})^2 + \ldots + (x_N - \bar{x})^2}{N} = \frac{1}{N} \sum_{i=1}^{N} (x_i - \bar{x})^2 \qquad (2.9)$$

Analogously to the population mean μ, the population variance, represented by σ^2, can be unbiasedly estimated from a sample through the corrected variance equation, given by

$$s^2 = \frac{1}{N-1} \sum_{i=1}^{N} (x_i - \bar{x})^2 \qquad (2.10)$$

The term unbiasedness is freely used here to indicate that, by averaging the estimates of a notional large set of samples, there will be no difference between the population variance σ^2 and the average sample variance s^2. Equation (2.9) yields a biased estimate of σ^2, whereas Eq. (2.10) is unbiased; to adjust the uncorrected variance s_u^2 for bias, one needs to multiply it by the factor $N/(N-1)$. Another way to interpret Eq. (2.10) is to state that there has been a reduction of 1 degree of freedom, from the N original ones, as a result of the previous estimation of the population mean μ by the sample mean $\bar{x}$. The terms bias and degrees of freedom will be formally defined and further explained in Chap. 6.

The variance is expressed in terms of the squared units of the original variable. To preserve the variable's original units, the sample standard deviation s is defined as the positive square root of the sample unbiased variance s^2, and given by

$$s = \sqrt{\frac{(x_i - \bar{x})^2 + (x_2 - \bar{x})^2 + \ldots + (x_N - \bar{x})^2}{N-1}} = \sqrt{\frac{1}{(N-1)} \sum_{i=1}^{N} (x_i - \bar{x})^2}$$

$$(2.11)$$

Unlike the mean absolute deviation, the standard deviation stresses the largest (positive and negative) deviations from the sample mean and is the most used measure of dispersion. Expansion of the right-hand side of Eq. (2.11) can facilitate the calculations to obtain the standard deviation, as follows:

$$s = \sqrt{\frac{1}{(N-1)} \left(\sum_{i=1}^{N} x_i^2 - 2\bar{x} \sum_{i=1}^{N} x_i + N\bar{x}^2 \right)} = \sqrt{\frac{1}{(N-1)} \sum_{i=1}^{N} x_i^2 - \frac{N}{(N-1)} \bar{x}^2}$$

$$(2.12)$$

On comparing the degrees of variability or dispersion among two or more different samples, one should employ the sample coefficient of variation CV, given by the ratio between the standard deviation s and the mean $\bar{x}$. The coefficient CV is a dimensionless positive number and should be applied, as is, only to samples of positive numbers with non-zero means. If data are negative, CV must be calculated with the absolute value of the sample mean.

2.2.3 Measures of Asymmetry and Tail Weight

Additional descriptive statistics are needed to fully characterize the shape of a histogram or of a frequency polygon. These are given by the measures of asymmetry and tail weight, among which the most important are the coefficients of skewness and kurtosis, respectively.

2.2.3.1 Coefficient of Skewness

For a sample $\{x_1, x_2, \ldots, x_N\}$, the coefficient of skewness g is a dimensionless number given by

$$g = \frac{N}{(N-1)(N-2)} \frac{\sum_{i=1}^{N} (x_i - \bar{x})^3}{s^3} \tag{2.13}$$

In Eq. (2.13), the first ratio term of its right-hand side represents the necessary correction to make g an unbiased estimate of the population coefficient of skewness γ. The second ratio term is dimensionless and measures the cumulative contributions of the cubic deviations from the sample mean, positive and negative, as scaled by the standard deviation raised to the power 3.

Positive and negative deviations, after raised to the power 3, will keep their signs, but will result in much larger numbers. The imbalance, or balance, of these cubic deviations, as they are summed throughout the data set, will determine the sign and magnitude of the coefficient of skewness g. If g is positive, the histogram (or frequency polygon) is skewed to the right, as in Figs. 2.5 and 2.6. In this case, one can notice, from the charts and previous calculations, that the sample mode is smaller than the median, which, in turn, is smaller than the mean. The opposite situation would arise if g is negative. When there is a balance between positive and negative cubic deviations, the sample coefficient of skewness will be equal to zero (or close to zero) and all three measures of central tendency will converge to a single value. The coefficient of skewness is a bounded number since it can be proved that $|g| \leq \sqrt{N-2}$.

In general, samples of hydrologic maxima, such as annual peak discharges or annual maximum daily rainfall depths, have positive coefficients of skewness. As already mentioned in this chapter, such a general statement is particularly true for flood data samples, for which ordinary frequent floods are usually clustered around the mode of flood discharges, while extraordinary floods can show great deviations from it. Just a few occurrences of such large floods are needed to determine the usual right-skewed shape of a frequency polygon of flood flows. It appears, then, logical to prescribe positively skewed probability density functions to model flood data samples. It should be recognized, however, that as the sample coefficient of skewness is very sensitive to the presence of extreme-flood data in small-sized samples, the pursuit of close matching the model and sample skewness cannot be per se an unequivocal argument in favor of a specific probability density function.

2.2.3.2 Coefficient of Kurtosis

A measure of the tail weight (or heaviness of tails) of a frequency curve, in the case of a sample, or of a probability density function, in the case of a mathematical model of a continuous random variable, is given by the coefficient of kurtosis. For a sample, this dimensionless coefficient is calculated by

$$k = \frac{N(N+1)}{(N-1)(N-2)(N-3)} \frac{\sum_{i=1}^{N}(x_i - \bar{x})^4}{s^4} - 3\left[\frac{(N-1)^2}{(N-2)(N-3)} - 1\right] \quad (2.14)$$

The interpretation of the coefficient of kurtosis, as a frequency distribution shape descriptor, has been much debated by statisticians. The classical notion used to be that the coefficient k measures both *peakedness* (or flatness) and tail weight. Westfall (2014) contends that the interpretation of k as a measure of *peakedness* is incorrect and asserts that it reflects only the notion of tail extremity, meaning the presence of outliers, in the case of a sample, or the propensity to produce outliers, in the case of a probability distribution. However, the classical interpretation of kurtosis as a measure of *peakedness* and tail weight remains valid if applied to unimodal symmetric distributions.

For being a coefficient based on the sum of deviations from the mean, raised to the power 4, it is evident that to yield reliable estimates of kurtosis, the sample size must be sufficiently large, of the order of $N = 200$ as suggested by Kottegoda and Rosso (1997). The coefficient of kurtosis is more relevant if used to compare symmetric unimodal distributions, as a relative index for the distribution tail heaviness. In fact, as k indicates how clustered around the sample mean the data points are, it also reflects the presence of infrequent points, located in the lower and upper tails of the distribution.

It is common practice to subtract 3 from the right-hand side of Eq. (2.14) to establish the coefficient of excess kurtosis k_e, relative to a perfectly symmetric unimodal distribution with $k = 3$. In this context, if $k_e = 0$, the distribution is said mesokurtic; if $k_e < 0$, it is platykurtic; and if $k_e > 0$, it is leptokurtic. Figure 2.10 illustrates the three cases. Note in Fig. 2.10 that the leptokurtic distribution is sharply peaked at its center, but, for values much smaller or much larger than its mode, the corresponding frequencies decrease at a lower rate when compared to

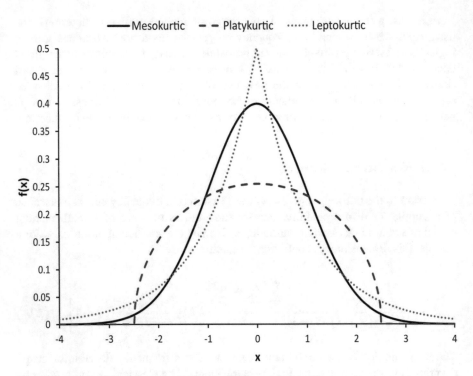

Fig. 2.10 Types of frequency distributions with respect to kurtosis

that of the mesokurtic distribution, thus revealing its relative heavier tails. The opposite line of reasoning applies to the platykurtic distribution.

In dealing with reduced hydrologic time series, which usually produce samples of sizes smaller than 80, the most useful descriptive statistics to describe the shape of a frequency distribution are (1) the mean, median, and mode, for central tendency; (2) the inter-quartile range, variance, standard deviation, and coefficient of variation, for dispersion; and (3) the coefficient of skewness for asymmetry. Table 2.4 gives a summary of these and other descriptive statistics for the annual mean flows of the Paraopeba River at Ponte Nova do Paraopeba, listed in Table 2.2. Results from Table 2.4 show that the mode is lower than the median, which is lower than the mean, thus indicating a positively skewed frequency distribution. This is confirmed by the sample coefficient of skewness of 0.808. Although the sample has only 62 elements, the sample coefficient of excess kurtosis of $k_e = 0{,}918$ suggests a leptokurtic distribution, relative to a perfectly symmetric and unimodal mesokurtic distribution with $k_e = 0$.

Table 2.4 Descriptive statistics for the sample of annual mean daily flows of the Paraopeba River at Ponte Nova do Paraopeba—Records from 1938 to 1999

Statistic	Notation	Estimate	Unit	Calculation
Mean	$\bar{x}$	86.105	m³/s	Eq. (2.2)
Mode	x_{mo}	80	m³/s	Frequency polygon
Median	x_{md}	82.7	m³/s	Eq. (2.7)
Harmonic mean	$\bar{x}_h$	79.482	m³/s	Eq. (2.5)
Geometric mean	$\bar{x}_g$	82.726	m³/s	Eq. (2.6)
Range	R	123.3	m³/s	(Maximum–minimum)
First quartile	Q_1	67.95	m³/s	Eq. (2.7) → 1st half-sample
Third quartile	Q_3	99.38	m³/s	Eq. (2.7) → 2nd half-sample
Inter-quartile range	IQR	31.43	m³/s	(Q_3-Q_1)
Mean absolute deviation	d	19.380	m³/s	Eq. (2.8)
Variance	s^2	623.008	(m³/s)²	Eq. (2.10)
Standard deviation	s	24.960	m³/s	Eq. (2.11)
Coefficient of variation	CV	0.290	Dimensionless	$s/\bar{x}$
Coefficient of skewness	g	0.808	Dimensionless	Eq. (2.13)
Coefficient of kurtosis	k	3.918	Dimensionless	Eq. (2.14)
Excess kurtosis	k_e	0.918	Dimensionless	$(k-3)$

2.3   Exploratory Methods

Tukey (1977) coined the term *EDA—Exploratory Data Analysis—*to identify an approach which utilizes a vast collection of graphical and quantitative techniques to describe and interpret a data set, without the previous concern of formulating assumptions and mathematical models for the random quantity being studied. EDA is based on the idea that data reveal by themselves their underlying structure and model. Among the many techniques proposed in the EDA approach, we highlight the two most commonly used: the box plot and the stem-and-leaf chart.

2.3.1   Box Plot

The box plot consists of a rectangle aligned with the sample's first (Q_1) and third (Q_3) quartiles, containing the median (Q_2) on its inside, as illustrated in the example chart of Fig. 2.11, for the annual mean daily flows of the Paraopeba River at Ponte Nova do Paraopeba. From the rectangle's upper side, aligned with Q_1, a line is extended up to the sample point whose magnitude does not exceed ($Q_3 + 1.5IQR$), the high-outlier detection bound. In the same way, another line is extended from the rectangle's lower side, aligned with Q_3, down to the sample point whose magnitude is not less than ($Q_1 - 1.5IQR$), the low-outlier detection bound. These extended lines are called *whiskers*. Any sample element lying outside the *whiskers* is

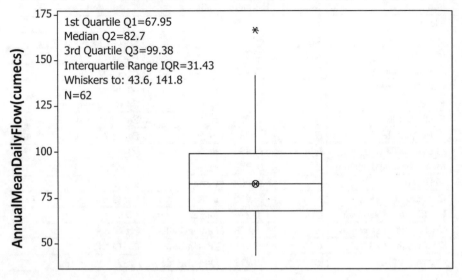

Fig. 2.11 Box plot of annual mean daily discharges (in m³/s or cumecs) of the Paraopeba River at Ponte Nova do Paraopeba—Records from 1938 to 1999

considered an outlier and pinpointed, as the 1983 mean flow of 166.9 m³/s marked with an asterisk in the box plot of Fig. 2.11.

A box plot is very useful for data analysis as it allows a broad view of sample data, showing on a single chart the measures of central tendency and dispersion, in addition to pinpointing possible outliers and providing a graphical outline of skewness. The central value is given by the median. Data dispersion is graphically summarized in the box defined by Q_1 and Q_3. The whiskers provide the range of non-outlier sample points. Possible outliers are directly identified in the plot. A clue on the skewness is given by the asymmetry between the lengths of the lines departing from the rectangle to the sample maximum and minimum values. Box plots can also be drawn horizontally and are particularly helpful for comparing two or more data samples in a single chart.

2.3.2   The Stem-and-Leaf Diagram

For samples of moderate to large sizes, the histogram and the frequency polygon are effective for illustrating the shape of a frequency curve. For small to moderate sample sizes, an interesting alternative is given by the stem-and-leaf diagram. This simple chart groups data in a singular and convenient way, so that all information contained in the data set is graphically displayed. The stem-and-leaf diagram also allows the

Cumulative
Frequency

Cumulative Frequency		Stem	Leaf
0		2	
0		3	
0		3	
1		4	36
3		4	68 94
5		5	01 31
8		5	70 73 99
13	Q1	6	06 12 26 36 42
18		6	68 72 82 87 93
22		7	16 20 24 48
27		7	64 76 80 89 90
(6)	Q2	8	02 09 11 22 32 38
29		8	51 74 76 81 92 98
23		9	27 39
21	Q3	9	63 73 78 79 80 91
15		10	02 07 10 43
11		10	
11		11	04 08 22 49
7		11	71 84
5		12	28
4		12	
4		13	34
3		13	
3		14	17 18
1		14	
1		15	
1		16	
1		16	69
0		17	

Fig. 2.12 Stem-and-leaf diagram of annual mean daily flows of the Paraopeba River at Ponte Nova do Paraopeba—Records from 1938 to 1999

location of extreme points and possible data gaps in the sample. To give an example of a stem-and-leaf diagram consider again the sample of annual mean daily flows of the Paraopeba River, listed in Table 2.2. First, the 62 flows in the sample are ranked in ascending order, ranging from the minimum of 43.6 m³/s to the maximum of 166.9 m³/s. There is no fixed rule in grouping data on a stem-and-leaf diagram since it depends largely on the data set specifics. The key idea is to separate each sample datum into two parts: the first, called stem, is placed on the left of a vertical line on the chart, and the second, the leaf, stays on the right, as shown in Fig. 2.12.

The stem indicates the initial numerical digit (or digits) of each sample point while the leaf gives its remaining digits. In the example of Fig. 2.12, the sample minimum of 43.6 is shown in the fourth row from top to bottom, with a stem of 4 and a leaf of 36, whereas the maximum is located in the penultimate row, with a stem of 16 and a leaf of 69. Note that, in the example of Fig. 2.12, the stems correspond to tens and/or hundreds, while the leaves are assigned to units multiplied by 10. A stem with many leaves corresponds to a large number of occurrences, as in stems numbered 8 in Fig. 2.12. In addition, leaf absolute frequencies can be cumulated, from the top down and from the bottom up to the row that contains the median. Cumulative frequencies are annotated on the left of stems, except for the median row, which shows only its respective absolute frequency. As a complement, the rows containing the first and third quartiles may be highlighted as well.

The stem-and-leaf diagram, after being rotated 90° to the left around its center, resembles a histogram but without any loss of information resulting from grouping sample data. With the stem-and-leaf diagram, it is possible to visualize the location of the median, the overall and inter-quartile ranges, the data scatter, the asymmetry (symmetry) with which the sample points are distributed, the data gaps, and possible outliers. In Fig. 2.12, for convenience, the stems were taken as double digits to enhance the way the leaves are distributed. If desired, the first of a double-digit stem can be marked with a minus sign ($-$) to identify the leaves that vary from 0 to 4, while the second digits are marked with a plus sign ($+$) for leaves going from 5 to 9. Also, there may be cases where the leaves can possibly be rounded off to the nearest integer, for more clarity.

2.4 Associating Data Samples of Different Variables

In the preceding sections, the main techniques for organizing, summarizing, and charting data from a sample of a single hydrologic variable were presented. It is relatively common, however, to be interested in analyzing the simultaneous behavior of two (or more) variables, looking for possible statistical dependence between them. In this section, we will look at the simplest case of two variables X and Y, whose data are concurrent or abstracted over the same time interval, and organized in pairs denoted by $\{(x_1,y_1), (x_2,y_2), \ldots, (x_N, y_N)\}$, where the subscript refers to a time index. What follows is a succinct introduction to the topics of correlation and regression, which are detailed in Chap. 9. In this introduction, the focus is on charting scatterplots and quantile–quantile plots for two variables X and Y.

2.4.1 Scatterplot

A scatterplot is a Cartesian-coordinate chart on which the pairs $\{(x_1,y_1), (x_2,y_2), \ldots, (x_N, y_N)\}$, of concurrent data from variables X and Y are plotted. To illustrate how to create a scatterplot and explore its possibilities, let us take as an example the variables: $X =$ annual total rainfall depth, in mm, and $Y =$ annual mean daily discharges, in m^3/s, for the Paraopeba River catchment, of 5680-km^2 drainage area. A sample of variable Y can be reduced from the historical time series of mean daily discharges from the gauging station coded 40800001. Similarly, a

Table 2.5 Annual mean flows and annual total rainfall depths for water-years 1941/42 to 1998/99—Gauging stations: 4080001 for discharges and 01944004 for rainfall

Water year	Annual rainfall depth (mm)	Annual mean daily flow (m³/s)	Water year	Annual rainfall depth (mm)	Annual mean daily flow (m³/s)
1941/42	1249	91.9	1970/71	1013	34.5
1942/43	1319	145	1971/72	1531	80.0
1943/44	1191	90.6	1972/73	1487	97.3
1944/45	1440	89.9	1973/74	1395	86.8
1945/46	1251	79.0	1974/75	1090	67.6
1946/47	1507	90.0	1975/76	1311	54.6
1947/48	1363	72.6	1976/77	1291	88.1
1948/49	1814	135	1977/78	1273	73.6
1949/50	1322	82.7	1978/79	2027	134
1950/51	1338	112	1979/80	1697	104
1951/52	1327	95.3	1980/81	1341	80.7
1952/53	1301	59.5	1981/82	1764	109
1953/54	1138	53.0	1982/83	1786	148
1954/55	1121	52.6	1983/84	1728	92.9
1955/56	1454	62.3	1984/85	1880	134
1956/57	1648	85.6	1985/86	1429	88.2
1957/58	1294	67.8	1986/87	1412	79.4
1958/59	883	52.5	1987/88	1606	79.5
1959/60	1601	64.6	1988/89	1290	58.3
1960/61	1487	122	1989/90	1451	64.7
1961/62	1347	64.8	1990/91	1447	105
1962/63	1250	63.5	1991/92	1581	99.5
1963/64	1298	54.2	1992/93	1642	95.7
1964/65	1673	113	1993/94	1341	86.1
1965/66	1452	110	1994/95	1359	71.8
1966/67	1169	102	1995/96	1503	86.2
1967/68	1189	74.2	1996/97	1927	127
1968/69	1220	56.4	1997/98	1236	66.3
1969/70	1306	72.6	1998/99	1163	59.0

sample of variable X can be extracted from the series of daily rainfall depths observed at the gauging station 01944004, assuming these are good estimates of the areal mean rainfall over the catchment. The samples of concurrent data on X and Y, as abstracted on the water-year from October to September, are listed in Table 2.5. Figs. 2.13 and 2.14 show different possibilities for the scatterplot between X and Y: the first with marginal histograms and the second with marginal box plots for both variables.

Looking at the scatterplots of Figs. 2.13 and 2.14, one can readily see that the higher the annual rainfall depth, the higher the annual mean discharge, thus indicating a positive association between X and Y. However, one can also note

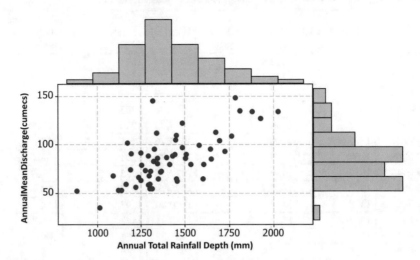

Fig. 2.13 Example of a scatterplot with marginal histograms

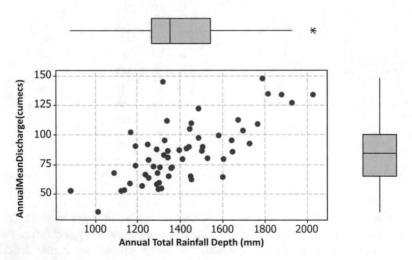

Fig. 2.14 Example of a scatterplot with marginal box plots

the pairs $\{(x_1,y_1), (x_2,y_2), \ldots, (x_N, y_N)\}$ are considerably scattered, thus showing that the randomness carried by Y cannot be fully explained by the variation of X. Additional concurrent observations of other relevant variables, such as, for instance, pan evaporation data as an estimate of potential evapotranspiration, could certainly reduce the degree of scatter. However, for a large catchment of 5680 km², with a substantial space-time variation of climate characteristics, soil properties, vegetation, and rainfall, the inclusion of more and more additional variables would never be enough to fully explain the variation of annual flows. Furthermore, the marginal histograms and box plots show that annual rainfalls are more scattered and more positively skewed than annual flows.

The degree of linear association of a set of N pairs $\{(x_1,y_1), (x_2,y_2), \ldots, (x_N, y_N)\}$ of concurrent data from variables X and Y can be quantified by the sample correlation coefficient, denoted by $r_{X,Y}$, and given by

$$r_{X,Y} = \frac{s_{X,Y}}{s_X\, s_Y} = \frac{1}{N-1}\, \frac{\displaystyle\sum_{i=1}^{N}(x_i - \bar{x})(y_i - \bar{y})}{s_X\, s_Y} \qquad (2.15)$$

This dimensionless coefficient is the result of scaling the sample covariance, which is represented in Eq. (2.15) by $s_{X,Y}$, by the product $s_X\, s_Y$ of the standard deviations of the variables. The correlation coefficient must satisfy the condition $-1 \le r_{X,Y} \le 1$ and it reflects the degree of linear association between variables X and Y: in the extreme cases, a value of 1 or -1 indicates perfectly linear positive or negative association, respectively, and a value of 0 means no linear association.

Figure 2.15a illustrates the case of a positive partial association, when Y increases as X increases, whereas Fig. 2.15b, c show, respectively, a negative partial association and a no linear association. Figure 2.15c further shows that a zero correlation coefficient does not necessarily imply there is no other form of association or dependence between the variables. In the particular case depicted in Fig. 2.15c, the dependence between X and Y indeed exists, but it is not of the linear form. Finally, it is worth noting that an eventual strong association between two variables does not imply a cause–effect relation. There are cases, such as the association between annual flows and rainfall depths, when one variable causes the other to vary. However, in other cases, even with a high correlation coefficient, such a physical relation between variables may not be evident or even plausible, or may not exist at all.

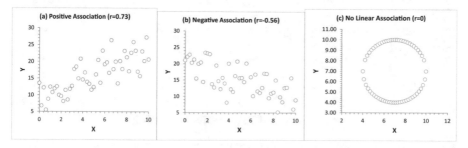

Fig. 2.15 Types of association between two variables

2.4.2 Empirical Quantile–Quantile Diagram (Empirical Q–Q Plot)

The original quantile–quantile diagram, or Q–Q plot, is a chart for comparing two probability distributions through their corresponding quantiles. It can be easily adapted to empirical data and turn into a useful chart for previewing a possible association between variables X and Y, through the comparison of their quantiles. Unlike the scatterplot of simultaneous or concurrent data from the variables, the empirical Q–Q plot links ranked data, or ranked quantiles, from the set $\{x_1, x_2, \ldots, x_N\}$ to ranked quantiles from $\{y_1, y_2, \ldots, y_N\}$, both samples being assumed to have the same size N. To create a Q–Q plot, one needs: (1) rank data from X and then from Y, both in ascending order of values; (2) assign to each sorted datum its respective rank order m, with $1 \leq m \leq N$; (3) associate with each order m the pair of ranked data $[x_{(m)}, y_{(m)}]$; and (4) plot on a Cartesian-coordinate chart the X and Y paired-data of equal rank orders. Figure 2.16 is an example of a Q–Q plot for the data given in Table 2.5.

Contrasting it to the scatterplot, the singular feature of a Q–Q plot is that it readily allows the comparison of high and low data points of X with their homologous points in the Y sample. As a hypothetical case, if the frequency distributions of both samples, after being conveniently scaled by their respective means and standard deviations, were identical, then all pairs on the Q–Q plot would be located exactly on the straight line $y = x$. In actual cases, however, if the frequency

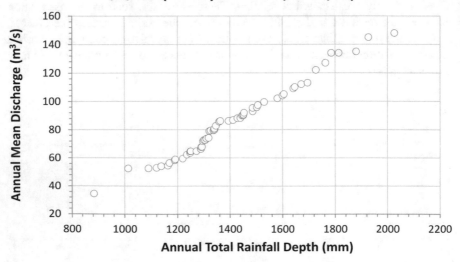

Fig. 2.16 Quantile–quantile diagram (Q–Q plot) for annual mean flows and annual total rainfalls for the Paraopeba River catchment

distributions of X and Y were very similar to each other, their respective quantiles will nearly lie on a straight line, but not necessarily on $y = x$. The empirical Q–Q plot is useful to see how X and Y paired-data deviate from a notional straight line that would link both variables, had their frequency distributions be identical. That is the idea behind empirical Q–Q plots.

Exercises

1. Table 2.6 gives the partial duration series of independent peak discharges over the 17,000-CFS threshold of the Greenbrier River at Alderson (WV, USA), recorded at the gauging station USGS 03183500 from 1896 to 1967. For the years not listed in the table, no discharges greater than 17,000 CFS were observed. Prepare a bar chart, similar to the one depicted in Fig. 2.1, for the

Table 2.6 Independent peak discharges of the Greenbrier River at Alderson (West Virginia, USA) that exceeded the threshold 17,000 cubic feet per second (CFS)

Year	Flow (CFS)	Year	Flow (CFS)	Year	Flow (CFS)	Year	Flow (CFS)
1896	28,800	1915	34,000	1935	20,800	1954	29,700
1897	27,600		40,800	1936	19,400		18,800
	54,000	1916	27,200		20,800	1955	32,000
	40,900		24,400		27,100		28,000
1898	17,100	1917	17,300		58,600		44,400
	18,600		43,000		28,300		26,200
	52,500		28,000	1937	21,200	1956	18,200
1899	25,300	1918	17,900		22,300	1957	23,900
	20,000		77,500		36,600		28,900
	23,800		24,000		26,400		22,000
	48,900	1919	28,600	1938	21,200	1958	21,800
1900	17,100		24,800		32,800		23,900
1901	56,800		49,000		22,300		22,200
	21,100	1920	38,000	1939	40,200		17,500
	20,400		20,700		41,600		26,700
	19,300		33,500		21,200	1959	17,200
	20,000	1922	21,500		17,200		23,900
1902	36,700		20,100		19,400	1960	17,800
	43,800		22,200	1940	29,900		35,500
1903	25,300	1923	19,500		21,500		32,500
	29,600	1924	26,500		19,400	1961	25,000
	33,500		20,400		18,700		21,800
	34,400		36,200	1942	35,300		31,400
	48,900		17,900	1943	33,600		17,200

(continued)

Table 2.6 (continued)

Year	Flow (CFS)	Year	Flow (CFS)	Year	Flow (CFS)	Year	Flow (CFS)
1904	25,700	1926	20,700		17,200	1962	34,700
	25,700		17,600		36,200		20,100
1905	29,600	1927	17,900		21,200		21,500
	37,600		24,000	1944	25,200		17,800
1906	18,200		40,200		17,200		23,200
	26,000		18,800	1945	17,900		35,500
1907	17,500		19,500		19,000	1963	22,700
	52,500	1928	18,000	1946	43,600		34,800
1908	17,800	1929	22,800	1947	20,000		47,200
	23,000		32,700		24,400		26,100
	31,500		23,800	1948	35,200		30,400
	52,500		20,000		23,500	1964	19,100
	26,800	1930	36,600		40,300		39,600
	27,600	1932	50,100	1949	18,500		22,800
	31,500		17,600		37,100	1965	22,000
1909	20,000		31,500		26,300		28,400
1910	45,900		27,500		23,200		19,800
1911	43,800		21,900	1950	31,500		18,600
	20,000	1933	26,400	1951	25,600	1966	26,400
	23,800	1934	32,300		27,800	1967	54,500
	18,900		20,500		26,700		39,900
	18,900		27,900		18,500		20,900
	35,500	1935	19,400		19,800		
	27,200		49,600		29,300		
	20,000		22,300	1952	17,800		
	21,100		17,900		19,100		
1913	21,800		24,800		27,600		
	64,000		20,100	1953	47,100		
	20,000		24,800		20,100		

number of years in the record with different annual frequencies of floods. Consider a peak flow over 17,000 CFS as a flood.

2. Consider the flow data listed in Table 2.5. These are the annual mean flows of the Paraopeba River at Ponte Nova do Paraopeba (Brazil), abstracted on the water-year basis, from October to September. For the flow data in that table, use MS Excel, or other similar software, to make the following charts: (a) dot diagram; (b) histogram; (c) relative frequency polygon; and (d) cumulative frequency polygon.

3. Compare the graphs created with the solution to Exercise 2 with those shown in Sect. 2.1 of this chapter. The former are abstracted on the water-year basis, whereas the latter are on the calendar year, but both concern the mean annual discharges. Discuss the relevance and the adequacy of using one or the other time basis for abstracting mean annual flows.

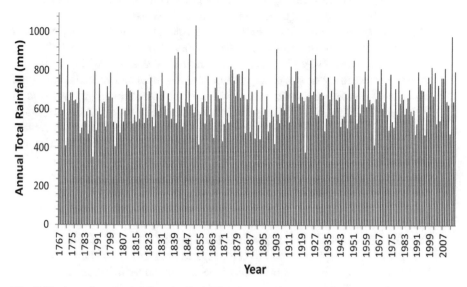

Fig. 2.17 Annual total rainfall at the Radcliffe Meteorological Station, in Oxford

4. The Australian Bureau of Meteorology selected a number of streamflow gauging stations throughout the Australian territory, which are considered *hydrologic reference stations*, given their high-quality data and being located in unregulated catchments with minimal land use change, among other criteria. Flow data and other information for those reference stations can be retrieved from the URL http://www.bom.gov.au/water/hrs/index.shtml. In particular, the 2677-km^2 Avoca River catchment at Coonooer, in the state of Victoria, has been monitored since August 1966, through gauging station coded 408200. The regional water-year extends from March to February, but the Avoca River is intermittent and ceases to flow at an uncertain time of any given year. Retrieve all daily flow data available for the gauging station 408200 and plot the annual hydrographs and the AFDCs for water-years 1969/70 and 2009/10. Plot the FDC for the period of record. Download data of the annual number of cease-to-flow days and plot the corresponding histogram.

5. One of the longest time series of rainfall has been recorded since 1767, at the Radcliffe Meteorological Station, in Oxford (England). Figure 2.17 depicts the annual total rainfall depths, measured at the Radcliffe station, from 1767 to 2014. The daily rainfall series has undergone a recent analysis aiming to correct nonhomogeneities due mainly to changing instrumentation over the period of records (Burt and Howden 2011). Retrieve the annual total rainfall data, from 1767 to the latest year, by accessing the URL http://www.geog.ox.ac.uk/research/climate/rms/rain.html. Plot the histogram and the relative frequency polygon for that data set. Comment on the asymmetry (or symmetry) shown by these graphs.

6. Use the rainfall data set, as retrieved in Exercise 5, and prepare a complete numerical summary of it, calculating all pertinent measures of central tendency, dispersion, asymmetry, and tail weight. Interpret your results considering the histogram and frequency polygon from Exercise 5.

7. The Dore River, located in the French Department of Puy-de-Dôme, has been gauged at the station of Saint-Gervais-sous-Meymont (code K2871910/Hydro/EauFrance), since 1919. The catchment area is 800 km^2 and flows are not severely affected by upstream regulation and diversions. The variable *annual minimum 7-day mean flow*, denoted by Q_7, is the lowest average discharge of 7 consecutive days in a given year and is commonly used for the statistical analysis of low flows. The yearly values of Q_7 from 1920 to 2014, abstracted from the daily flows for the Dore River at Saint-Gervais-sous-Meymont, are listed in Table 2.7. For the listed Q_7 flow data, use MS Excel, or other similar software, to make the following charts: (a) histogram; (b) relative frequency polygon; and (c) cumulative frequency polygon. Calculate the sample measures of central tendency, dispersion and asymmetry. Interpret your graphs and results.

8. The first one-third of a river reach is travelled by a floating device with the velocity of 0.3 m/s, the second at 0.5 m/s, and the third at 0.60 m/s. Show the harmonic mean is more meaningful of the true mean velocity, if compared to the arithmetic mean. .

9. The population of a town increases geometrically with time. According to the 1980 demographic census, the town population was 150,000 inhabitants, whereas the 2000 census counted 205,000 people. An engineer wants to check the operative conditions of the local water supply system at an intermediate point in time. To do it he/she needs an estimate of the 1990 population to calculate the *per capita* water consumption. Determine the central value needed by the engineer. Justify your answer.

10. A random variable can pass through linear and nonlinear transformations. An example of a linear transformation of X is to change it to the standard variable Z by applying the relation $z_i = (x_i - \bar{x})/s_x$ to its sample points. In such transformation, X is centered, by subtracting the arithmetic mean, and scaled by the standard deviation. Go back to the data of Exercise 2, transform X into Z, calculate $\bar{z}$, s_Z, g_Z and k_Z, and compare these statistics to those of X. Which conclusions can be drawn from such transformed variable? Can the conclusions be extended to the untransformed variable? What are the potential uses of variable transformation in data analysis?

11. An example of nonlinear transformations is given by the logarithmic conversion of X to Z by means of $z_i = \log_{10} x_i$ or $z_i = \ln x_i$. Solve Exercise 10 using the logarithmic transformation.

12. A family of possible transformation of variable X can be summarized by the Box–Cox formula $z_i = (x_i^\lambda - 1)/\lambda$, for $\lambda \neq 0$ or $z_i = \ln x_i$, for $\lambda = 0$ (Box and Cox 1964). The right choice of λ can potentially transform asymmetric data into symmetric. Employ the Box–Cox formula with $\lambda = -1, -0.5, 0, +0.5, +1$, and $+2$ to transform the data listed in Table 2.2. Calculate the coefficients of

Table 2.7 Q_7 flows for the Dore River at Saint-Gervais-sous-Meymont, 1920–2014

Calendar year	Q_7 (m³/s)	Calendar year	Q_7 (m³/s)	Calendar year	Q_7 (m³/s)
1920	1.06	1952	0.32	1984	1.16
1921	0.74	1953	0.81	1985	0.45
1922	1.50	1954	1.79	1986	0.74
1923	0.50	1955	1.27	1987	1.38
1924	0.89	1956	4.87	1988	1.28
1925	1.77	1957	0.87	1989	0.97
1926	0.96	1958	2.34	1990	0.85
1927	3.31	1959	1.05	1991	1.00
1928	0.69	1960	1.12	1992	1.57
1929	1.07	1961	0.70	1993	1.46
1930	2.73	1962	0.32	1994	1.32
1931	1.79	1963	2.24	1995	Missing
1932	2.04	1964	0.86	1996	1.62
1933	0.73	1965	1.70	1997	0.97
1934	0.20	1966	1.05	1998	0.91
1935	0.97	1967	1.20	1999	1.24
1936	2.66	1968	1.86	2000	1.79
1937	0.83	1969	3.03	2001	1.68
1938	1.17	1970	1.09	2002	1.69
1939	1.47	1971	1.58	2003	Missing
1940	1.23	1972	1.77	2004	1.49
1941	2.01	1973	2.00	2005	Missing
1942	0.86	1974	0.74	2006	Missing
1943	0.63	1975	0.88	2007	Missing
1944	0.93	1976	0.82	2008	2.31
1945	0.72	1977	3.13	2009	0.81
1946	1.31	1978	0.75	2010	1.61
1947	0.79	1979	1.00	2011	0.88
1948	1.47	1980	1.86	2012	1.52
1949	0.08	1981	2.00	2013	2.54
1950	0.19	1982	1.61	2014	3.11
1951	2.31	1983	0.49		

skewness and kurtosis and check which value of λ makes data approximately symmetric. Draw the frequency polygon for the transformed data and compare it with that of Fig. 2.6.

13. To create a cumulative frequency diagram according to one of the procedures outlined in Sect. 2.1.5, one needs to estimate the empirical non-exceedance frequency, by sorting data and using the rank order m. In the example shown in Sect. 2.1.5, we used the ratio m/N to estimate the non-exceedance frequency. However, this is a poor estimate because it implies a zero chance of any future occurrence being larger than the sample maximum. To overcome this

drawback, a number of *plotting-position* formulas have been introduced in statistical and hydrologic literature. One of these is the Weibull plotting-position formula, which calculates the non-exceedance frequency of the m-ranked sample point as $m/(N + 1)$. Redraw the diagram of Fig. 2.7 with the Weibull formula.

14. Draw a box plot for the annual total rainfall data measured at the Radcliffe Meteorological Station, in Oxford, as retrieved in Exercise 5.
15. Draw a stem-and-leaf diagram for the total annual rainfall listed in Table 2.5.
16. Interpret the Q–Q Plot of Fig. 2.16.
17. Table 2.8 lists data of Total Dissolved Solids (TDS) concentration and discharge (Q) concurrently measured at the USGS/NASQAN stream quality station 4208000, of the Cuyahoga River at Independence (Ohio, USA), as compiled by Helsel and Hirsch (1992). In the table, 'Mo' indicates the month when the measures were made and "T" the corresponding dates, as expressed in "decimal time" and listed as (date-1000). Discharges are in CFS and TDS in mg/l.

   (a) Draw a simple scatterplot for Q and TDS;
   (b) Draw scatterplots with marginal histograms and box plots for Q and TDS;
   (c) Calculate the sample correlation coefficient for Q and TDS;
   (d) Provide a physical explanation for the sign of the correlation coefficient between Q and TDS; and
   (e) Create and interpret the Q–Q plot for Q and TDS.

Table 2.8 Discharges and total dissolved solids concentrations of the Cuyahoga River at Independence, Ohio

Mo	T	TDS	Q	Mo	T	TDS	Q	Mo	T	TDS	Q	Mo	T	TDS	Q
1	74.04	490	458	2	78.12	680	533	10	79.79	410	542	7	81.54	560	444
2	74.12	540	469	3	78.21	250	4930	11	79.87	470	499	8	81.62	370	595
4	74.29	220	4630	4	78.29	250	3810	12	79.96	370	741	9	81.71	460	295
7	74.54	390	321	5	78.37	450	469	1	80.04	410	569	10	81.79	390	542
10	74.79	450	541	6	78.46	500	473	2	80.12	540	360	12	81.96	330	1500
1	75.04	230	1640	7	78.54	510	593	3	80.21	550	513	3	82.21	350	1080
4	75.29	360	1060	8	78.62	490	500	4	80.29	220	3910	5	82.37	480	334
7	75.54	460	264	9	78.71	700	266	5	80.37	460	364	6	82.46	390	423
10	75.79	430	665	10	78.79	420	495	6	80.46	390	472	8	82.62	500	216
1	76.04	430	680	11	78.87	710	245	7	80.54	550	245	11	82.87	410	366
4	76.29	620	650	12	78.96	430	736	8	80.62	320	1500	2	83.12	470	750
8	76.62	460	490	1	79.04	410	508	9	80.71	570	224	5	83.37	280	1260
10	76.79	450	380	2	79.12	700	578	10	80.79	480	342	8	83.62	510	223
1	77.04	580	325	3	79.21	260	4590	12	80.96	520	732	11	83.87	470	462
4	77.29	350	1020	4	79.29	260	4670	1	81.04	620	240	2	84.12	310	7640
7	77.54	440	460	5	79.37	500	503	2	81.12	520	472	5	84.37	230	2340
10	77.79	530	583	6	79.46	450	469	3	81.21	430	679	7	84.54	470	239
11	77.87	380	777	7	79.54	500	314	4	81.29	400	1080	11	84.87	330	1400
12	77.96	440	1230	8	79.62	620	432	5	81.37	430	920	3	85.21	320	3070
1	78.04	430	565	9	79.71	670	279	6	81.46	490	488	5	85.37	500	244

Data from Helsel and Hirsch 1992

References

Box GEP, Cox DR (1964) An analysis of transformations. J R Stat Soc Series B 26:211–252

Burt TP, Howden NJK (2011) A homogeneous daily rainfall record for the Radcliffe Observatory, Oxford, from the 1820s. Water Resour Res 47, W09701

Cleveland WS (1985) The elements of graphing data. Wadsworth Advanced Books and Software, Monterrey, CA

Helsel DR, Hirsch RM (1992) Statistical methods in water resources. USGS Techniques of Water-Resources Investigations, Book 4, hydrologic analysis and interpretation, Chapter A-3. United States Geological Survey, Reston, VA. http://pubs.usgs.gov/twri/twri4a3/pdf/twri4a3-new.pdf. Accessed 1 Feb 2016

Kottegoda NT, Rosso R (1997) Statistics, probability, and reliability for civil and environmental engineers. McGraw-Hill, New York

Shimazaki H, Shinomoto S (2007) A method for selecting the bin size of a time histogram. Neural Comput 19:1503–1527

Sturges HA (1926) The choice of a class interval. J Am Stat Assoc 21:65–66

Tufte ER (2007) The visual display of quantitative information. Graphics Press LLC, Cheshire, CO

Tukey JW (1977) Exploratory data analysis. Addison Wesley, Reading, MA

Westfall PH (2014) Kurtosis as peakedness, 1905-2014. Am Stat 68(3):191–195

Chapter 3
Elementary Probability Theory

Mauro Naghettini

3.1 Random Events

Probability theory deals with experiments, whose results or outcomes cannot be predicted with certainty. They are referred to as random experiments. Even though the outcomes of a random experiment cannot be anticipated, it is possible to assemble the set of all of its possible results. This set is termed sample space, is usually denoted by S (or Ω) and contains the collection of all possible sample points, as related to distinguishable events and their outcomes. Suppose, for instance, a random experiment is conducted to count the annual number of rainy days, denoted by y, at some gauging station; a rainy day is any day with total rainfall depth greater than 0.3 mm. For this example, the sample space is the finite set of integer numbers, contained into $S = S_D = \{y = 0,\ 1,\ 2,\ \ldots, 366\}$, with $S_D \subset N_0$. In contrast, if the experiment concerned the streamflow monitoring at a gauging station, with flows denoted by x, the corresponding sample space would be the set $S \equiv S_C = \{x \in \mathbf{R}_+\}$ of non-negative real numbers.

Any subset of the sample space S is termed an event. In the sample space S_C, of flows X, we could distinguish the flows that lie below a given threshold x_0 and group them into the event $A = \{x \in \mathbf{R}_+ | 0 \leq x < x_0\}$, such that A is contained in S_C. In this case, the complement event A^c of A, will consist of all elements of S_C that are not included in A. In other terms, $A^C = \{x \in \mathbf{R}_+ | x \geq x_0\}$, whose occurrence implies the denial of event A. Analogously to the sample space S_D, of the annual number of rainy days Y, it would be feasible to categorize as "dry years" those for which $y < 30$ days and group them into the event $B = \{y \in \mathbf{N} | y < 30\}$. In this case, the complement of B will be given by the elements of the finite subset $B^c = \{y \in \mathbf{N} | 30 \leq y \leq 366\}$ of "wet years." For the preceding examples, events

M. Naghettini (✉)
Universidade Federal de Minas Gerais Belo Horizonte, Minas Gerais, Brazil
e-mail: mauronag@superig.com.br

© Springer International Publishing Switzerland 2017
M. Naghettini (ed.), *Fundamentals of Statistical Hydrology*,
DOI 10.1007/978-3-319-43561-9_3

A and A^c (or B and B^c) are considered disjoint or mutually exclusive events, since the occurrence of one of them determines the non-occurrence of the other. In other words, none of the elements contained in one event will be in the other.

The events of a sample space can be related to each other through the operations of intersection and union of sets. If two non-mutually exclusive events A_1 and A_2 have elements in common, the subset formed by those common sample points is named intersection and is represented by $A_1 \cap A_2$. Contrarily, if events A_1 are A_2 are disjoint, then $A_1 \cap A_2 = \varnothing$, where $\varnothing$ denotes the null set. The subset containing all elements of A_1 and A_2, including those that are common to both, is the union, which is represented by $A_1 \cup A_2$. The intersection is linked to the Boolean logical operator "AND," or conjunction, implying the joint occurrence of A_1 and A_2, whereas the union refers to "AND/OR" or conjunction/disjunction, relating the occurrences of A_1 or A_2 or both. In respect of the sample space S_C, from the example of flows X, consider the hypothetical events, defined as $A_1 = \{x \in \mathbf{R}_+ | 0 \leq x \leq 60 \text{ m}^3/\text{s}\}$, $A_2 = \{x \in \mathbf{R}_+ | 30 \text{m}^3/\text{s} \leq x \leq 80 \text{ m}^3/\text{s}\}$, and $A_3 = \{x \in \mathbf{R}_+ | x \geq 50 \text{ m}^3/\text{s}\}$. For these, one can write the following statements:

- $A_1 \cap A_2 = \{x \in \mathbf{R}_+ | 30 \text{m}^3/\text{s} \leq x \leq 60 \text{ m}^3/\text{s}\}$
- $A_2 \cap A_3 = \{x \in \mathbf{R}_+ | 50 \text{m}^3/\text{s} \leq x \leq 80 \text{ m}^3/\text{s}\}$
- $A_1 \cap A_3 = \{x \in \mathbf{R}_+ | 50 \text{m}^3/\text{s} \leq x \leq 60 \text{ m}^3/\text{s}\}$
- $A_1 \cup A_2 = \{x \in \mathbf{R}_+ | 0 \text{m}^3/\text{s} \leq x \leq 80 \text{ m}^3/\text{s}\}$
- $A_2 \cup A_3 = \{x \in \mathbf{R}_+ | 30 \text{m}^3/\text{s} \leq x \leq \infty\}$
- $A_1 \cup A_3 = \{x \in \mathbf{R}_+ | x \geq 0\} \equiv S_C$

The operations of intersection and union can be extended to more than two events and have the associative and distributive properties of sets, which are, respectively, analogous to the rules affecting addition and multiplication of numbers. The following compound events are examples of the associative property of set algebra: $(A_1 \cup A_2) \cup A_3 = A_1 \cup (A_2 \cup A_3)$ and $(A_1 \cap A_2) \cap A_3 = A_1 \cap (A_2 \cap A_3)$. The set operations $(A_1 \cup A_2) \cap A_3 = (A_1 \cap A_3) \cup (A_2 \cap A_3)$ and $(A_1 \cap A_2) \cup A_3 = (A_1 \cup A_3) \cap (A_2 \cup A_3)$ result from the distributive property. Referring to the sample space S_C, one can write:

- $A_1 \cap A_2 \cap A_3 = \{x \in \mathbf{R}_+ | 50 \text{m}^3/\text{s} \leq x \leq 60 \text{ m}^3/\text{s}\}$
- $A_1 \cup A_2 \cup A_3 = \{x \in \mathbf{R}_+ | x \geq 0\} \equiv S_C$
- $(A_1 \cup A_2) \cup A_3 = A_1 \cup (A_2 \cup A_3) = S_C$
- $(A_1 \cap A_2) \cap A_3 = A_1 \cap (A_2 \cap A_3) = \{x \in \mathbf{R}_+ | 50 \text{m}^3/\text{s} \leq x \leq 60 \text{m}^3/\text{s}\}$
- $(A_1 \cup A_2) \cap A_3 = (A_1 \cap A_3) \cup (A_2 \cap A_3) = \{x \in \mathbf{R}_+ | 50 \text{m}^3/\text{s} \leq x \leq 60 \text{m}^3/\text{s}\}$
- $(A_1 \cap A_2) \cup A_3 = (A_1 \cup A_3) \cap (A_2 \cup A_3) = \{x \in \mathbf{R}_+ | x \geq 30 \text{m}^3/\text{s}\}$

Operations between simple and compound events, contained in a sample space, can be more easily visualized through Venn diagrams, as illustrated in Fig. 3.1. These diagrams, however, are not quite adequate to measure or interpret probability relations among events.

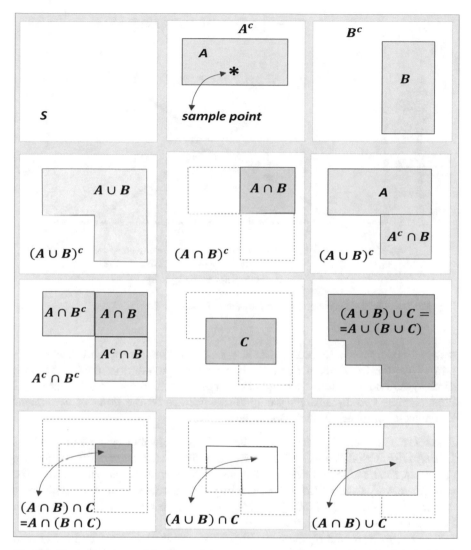

Fig. 3.1 Venn diagrams and basic operations among events in a sample space (adapted from Kottegoda and Rosso 1997)

As derived from the basic operations, the sample space can also be expressed as the union of an exhaustive set of mutually and collectively exclusive events. Alluding to Fig. 3.1, as the mutually and collectively exclusive events $(A \cap B^c)$, $(A \cap B)$, $(A^c \cap B)$, and $(A^c \cap B^c)$ form such an exhaustive set, in the sense it encompasses all possible outcomes, it is simple to conclude their union results in the sample space S.

When the random experiment involves simultaneous observations of several variables, the preceding notions have to be extended to a multidimensional sample space. In hydrology, there are many examples of associations between concurrent

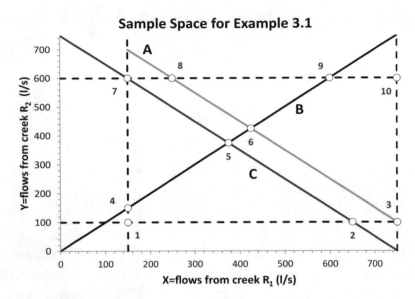

Fig. 3.2 Two-dimensional sample space for events of Example 3.1

observations from two (or more) variables, such as the number of rainy days and the rainfall depths for a given duration, or the annual number of floods, the peak discharges, and the flood hydrograph volumes, to name a few. Example 3.1 illustrates the two-dimensional sample space formed by the flows of two creeks immediately upstream of their confluence.

Example 3.1 Stream R_3 is formed by the confluence of tributary creeks R_1 and R_2. During a dry season, the flows X from R_1, immediately upstream of the confluence, vary from 150 l/s to 750 l/s, whereas the flows Y from creek R_2, also upstream of the confluence, vary from 100 to 600 l/s. The two-dimensional sample space is given by the set $S = \{(x,y) \in \mathbf{R}_+ | 150 \leq x \leq 750, 100 \leq y \leq 600\}$, which is depicted in Fig. 3.2 (adapted from Shahin et al. 1993).

The events A, B, and C, shown in Fig. 3.2, are defined as $A = \{R_3$ flows exceed 850 l/s$\}$, $B = \{R_1$ flows exceed R_2 flows$\}$, and $C = \{R_3$ flows are less than 750 l/s$\}$. The intersection between A and B corresponds to the event $A \cap B = \{(x,y) \in S | x + y > 850$ and $x > y\}$ and can be distinguished in Fig. 3.2 by the polygon formed by points 3, 6, 9, and 10. The union $A \cup B = \{(x,y) \in S | x + y > 850$ and/or $x > y\}$ corresponds to the polygon enclosed by points 1, 4, 9, 10, and 3, whereas $A \cap C = \varnothing$. Take the opportunity given by Example 3.1 to graphically identify the events $(A \cup C)^c$, $(A \cup C)^c \cap B$ and $A^c \cap C^c$.

3.2   Notion and Measure of Probability

Having defined the sample space and random events, the next step is to associate a probability with each of these events, as a relative measure of its respective likelihood to occur, between the extremes of 0 (or impossibility) and 1 (or certainty). In spite of being a rather intuitive notion, the mathematical concept of probability had a slow historic evolution, incorporating gradual revisions that were made necessary to reconcile its different views and interpretations.

The early definition of probability, named classic or a priori, had its origin in the works of French mathematicians of the seventeenth century, like Blaise Pascal (1623–1662) and Pierre de Fermat (1601–1665), within the context of games of chance. According to this definition, if a finite sample space S contains n_S equally likely and mutually exclusive outcomes from a random experiment, from which n_A are related to an event A, the probability of A is given by

$$P(A) = \frac{n_A}{n_S} \qquad (3.1)$$

This is an a priori definition because it assumes, before the facts, that the outcomes are equally likely and mutually exclusive. For instance, in a coin tossing experiment, in which it is known beforehand that the coin is fair, the probability of heads or tails is clearly 0.5.

There are situations where the classical definition of probability is adequate, while in others, two intrinsic limitations may arise. The first one refers to its impossibility of accommodating the scenario of outcomes that are not equally likely, whereas the second one concerns the absent notion of infinite sample spaces. These limitations have led to a broader definition of probability, known as frequentist (or objective or statistical), which is attributed to the Austrian mathematician Richard von Mises (1883–1953), who early in the twentieth century used it as a foundation for his theory of probability and statistics (Papoulis 1991). According to this, if a random experiment is repeated a large number of times n, under rigorously identical conditions, and an event A, contained into the sample space S, has occurred n_A times, then, the frequentist probability of A is given by

$$P(A) = \lim_{n \to \infty} \frac{n_A}{n} \qquad (3.2)$$

The notion implied by this definition is illustrated in Fig. 3.3, as referring to the probability of heads, in an actual coin tossing experiment, in which no assumption regarding the coin fairness or unbiasedness is made in advance. In this figure, Eq. (3.2) has been used to calculate the probability of heads, as the number of coin tosses increases from 1 to 100.

The frequentist definition, although broader and more generic than the classical, still has limitations. One is related to the natural question of how large the value of n must be in order to converge to some constant value of $P(A)$. For the example

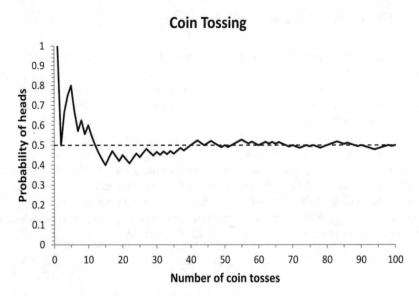

Fig. 3.3 Illustration of frequentist probability from a coin tossing experiment

illustrated in Fig. 3.3, this limitation is made evident by the impossibility of finding an indisputable value for the probability of heads at the end of the coin tossing experiment. Furthermore, even if $P(A)$ converges to some value, will we get the same limiting value, if the entire experiment is repeated a second time? Defendants of the frequentist concept usually respond to this objection by stating the convergence of $P(A)$ should be interpreted as an axiom. However, presuming beforehand that the convergence will necessarily occur is a complex assumption to make (Ross 1988). Another limitation refers to the physical impossibility of repeating an experiment over a large number of times under rigorously identical conditions. Finally, an objection to both classical and frequentist definitions relates to their common difficulty of accommodating the idea of subjective probability, as in the case of a renowned expert in a particular field of knowledge expressing his/her *degree of belief* in some possible occurrence.

These drawbacks have led the Russian mathematician Andrey Kolmogorov (1903–1987) to propose that the theory of probability should be developed from simple and self-evident axioms, very much like algebra and geometry. In 1933, he published his book setting out the axiomatic foundations of probability, which are used nowadays by mathematicians and statisticians (Kolmogorov 1933). This modern axiomatic approach to probability is based on three simple axioms and states that, given a sample space S, the probability function $P(.)$ is a non-negative number that must satisfy the following conditions:

1. $0 \leq P(.) \leq 1$
2. $P(S) = 1$

3. For any sequence of mutually exclusive events E_1, E_2, ..., the probability of their union is equal to the sum of their respective individual probabilities, or,

$$P\left(\bigcup_{i=1}^{\infty} E_i\right) = \sum_{i=1}^{\infty} P(E_i).$$

The three conditions listed above are, in fact, axioms from which all mathematical properties concerning the probability function $P(.)$ must be deduced. Kolmogorov's axiomatic approach forms the logical essence of the modern theory of probability and has the advantage of accommodating all previous definitions and the notion of subjective probability as well. The reader interested in more details on the many interpretations of probability is referred to Rowbottom (2015).

The following are immediate corollaries of the axioms:

- $P(A^c) = 1 - P(A)$.
- $P(\varnothing) = 0$.
- If A and B are two events contained in sample space S and $A \subset B$, then $P(A) \leq P(B)$.
- If A_1, A_2, ..., A_k are events in sample space S, then $P\left(\bigcup_{i=1}^{k} A_i\right) \leq \sum_{i=1}^{k} P(A_i)$.

  This corollary is referred to as Boole's inequality.
- If A and B are events in sample space S, then $P(A \cup B) = P(A) + P(B) - P(A \cap B)$. This is termed the addition rule of probabilities.

Example 3.2 Two events can potentially induce the failure of a dam, located in an earthquake-prone area: the occurrence of a flood larger than the spillway flow capacity, which is referred to as event A, and/or the structural collapse resulting from a destructive earthquake, event B. Suppose that, based on regional past data, the annual probabilities of these events have been estimated as $P(A) = 0.02$ and $P(B) = 0.01$. Knowing only these estimates, assess the dam failure probability in any given year (adapted from Kottegoda and Rosso 1997).

Solution The failure of the dam can possibly result from the isolated action of floods or earthquakes, or from the combined action of both. In other terms, a dam failure is a compound event resulting from the union of A and B, whose probability is given by $P(A \cup B) = P(A) + P(B) - P(A \cap B)$. Although $P(A \cap B)$ is unknown, it can be anticipated that the probability of simultaneous occurrence of both inducing dam-failure events has a very small value. Based on this and on Boole's inequality, a conservative estimate of the annual dam failure probability is given by $P(A \cup B) \cong P(A) + P(B) = 0.02 + 0.01 = 0.03$.

3.3 Conditional Probability and Statistical Independence

The probability of an event A occurring can be modified by the previous or concurrent occurrence of another event B. For instance, the probability that the mean hourly flow of a given catchment will exceed 50 m^3/s, in the next 6 h, is certainly affected by the fact that it has already surpassed 20 m^3/s. This is a simple example of conditional probability, denoted by $P(A|B)$, that A will occur given that

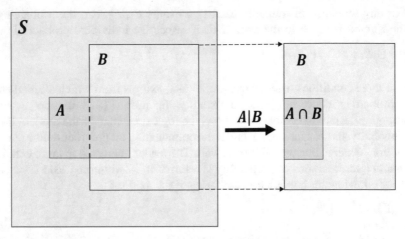

Fig. 3.4 Venn diagram depicting sample space reduction by conditioning

B has already happened or is on the verge of happening. Provided $P(B)$ exists and is not zero, the probability $P(A|B)$ is defined as

$$P(A|B) = \frac{P(A \cap B)}{P(B)} \tag{3.3}$$

The Venn diagram, depicted in Fig. 3.4, if interpreted in terms of areas representing probabilities, depicts the concept behind Eq. (3.3). In fact, given that event B has occurred or is certain to occur, the sample space must be reduced from the original S to B. Thereafter, the probability of A, given B, needs to be updated by means of Eq. (3.3).

The following corollaries can be derived from Kolmogorov's axioms and from the definition of conditional probability:

- If $P(B) \neq 0$, then, for any given event A, $0 \leq P(A|B) \leq 1$.
- If two events A_1 and A_2 are disjoint and if $P(B) \neq 0$, then $P(A_1 \cup A_2|B) = P(A_1|B) + P(A_2|B)$.
- As a particular case of previous proposition, it can be written that $P(A|B) + P(A|B^c) = 1$.
- If $P(B) \neq 0$, then $P(A_1 \cup A_2|B) = P(A_1|B) + P(A_2|B) - P(A_1 \cap A_2|B)$.

Equation (3.3) can be rewritten in the form of $P(A \cap B) = P(B) P(A|B)$ and, since $P(A \cap B) = P(B \cap A)$, it follows that $P(B \cap A) = P(A) P(B|A)$. This is the multiplication rule of probabilities, which can be generalized to more than two events. For instance, for exactly three events, the multiplication rule of probabilities is given by

$$P(A \cap B \cap C) = P(A)\, P(B|A)\, P(C|A \cap B) \tag{3.4}$$

If the probability of A is not affected by the occurrence of B and vice-versa, or if $P(A|B) = P(A)$ and $P(B|A) = P(B)$, then these two events are considered statistically independent. For independent A and B, the rule of multiplication of probabilities results in

$$P(A \cap B) = P(B \cap A) = P(B)\, P(A) = P(A)P(B) \tag{3.5}$$

By generalizing Eq. (3.5), one can state that for k mutually and collectively independent events in a sample space, denoted by A_1, A_2, ..., A_k, the probability of their simultaneous or concurrent occurrence is $P(A_1 \cap A_2 \cap \ldots \cap A_k) = P(A_1)\, P(A_2) \ldots P(A_k)$.

Example 3.3 Suppose a town is located downstream of the confluence of rivers R_1 and R_2 and can possibly be flooded by high waters from R_1 (event A), or from R_2 (event B), or from both. If $P(A)$ is the triple of $P(B)$, if $P(A|B) = 0.6$, and if the probability of the town being flooded is 0.01, calculate (a) the probability of a flood coming from the river R_2 and (b) the probability of a flood coming only from river R_1, given the town has been flooded.

Solution (a) The probability of the town being flooded is given by the probability of the union $P(A \cup B) = P(A) + P(B) - P(A \cap B)$ of events A and B. Substituting the given data, then $P(A \cup B) = 3P(B) + P(B) - P(B)P(A|B)$ and $0.01 = 3\,P(B) + P(B) - 0.6P(B)$. Solving for $P(B)$ and $P(A)$, $P(B) = 0.003$ and $P(A) = 0.009$. (b) The probability of a flood coming only from river R_1, given the town has been flooded, can be written as $P[(A \cap B^c)|(A \cup B)] = \frac{P[(A \cap B^c) \cap (A \cup B)]}{P(A \cup B)}$. A simple Venn diagram for the compound event in the numerator of the right-hand side of this equation shows it is equivalent to $P[(A \cap B^c)]$, which, in turn, is equal to $P(A)[1 - P(B|A)]$. In the resulting equation, only $P(B|A)$ is unknown. However, this unknown probability can be derived from $P(A)\, P(B|A) = P(B)\, P(A|B) \Rightarrow 3P(B)\, P(B|A) = P(B)\, P(A|B) \Rightarrow P(B|A) = P(A|B)/3 = .0.2$. By substituting the remaining values in the original equation, it results in $P[(A \cap B^c)|(A \cup B)] = \frac{0.009(1-0.2)}{0.01} = 0.72$.

3.4 Law of Total Probability and Bayes' Formula

Suppose the sample space S of some random experiment is the result of the union of k mutually exclusive and exhaustive events B_1, B_2, ..., B_k, whose respective probabilities of occurrence are all different from zero. Also, consider some event A, such as the one illustrated in Fig. 3.5, whose probability is given by $P(A) = P(B_1 \cap A) + P(B_2 \cap A) + \ldots + P(B_k \cap A)$. By employing the definition

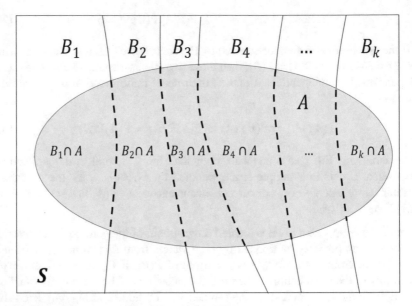

Fig. 3.5 Venn diagram for the law of total probability

of conditional probability for each term of the right-hand side of this equation, it follows that

$$P(A) = P(B_1)\,P(A|B_1) + P(B_2)\,P(A|B_2) + \ldots + P(B_k)\,P(A|B_k)$$

$$= \sum_{i=1}^{k} P(B_i)\,P(A|B_i) \tag{3.6}$$

Equation (3.6) is the formal expression of the law of total probability.

Example 3.4 The water supply system of a city uses two distinct and complementary reservoirs: number one, with the storage capacity of 150,000 l, and a probability of coming into operation of 0.7, and number two, with 187,500 l of storage, with a probability of 0.3. The city's daily water consumption is a random variable whose probabilities of equaling or exceeding the volumes of 150,000 and 187,500 l are respectively 0.3 and 0.1. Knowing that when one of the reservoirs is active, the other is inactive, answer the following questions: (a) what is the probability of failure of the water supply system on any given day? and (b) supposing the daily water consumptions in consecutive days are statistically independent among themselves, what is the probability of failure of the system on any given week?

Solution (a) Consider the failure to fulfill the city daily water consumption is represented by event A, while events B and B^c respectively denote the active operation of reservoirs one and two. Application of Eq. (3.6), with $k=2$, results in $\quad P(A) = P(A|B)\,P(B) + \quad P(A|B^c)\,P(B^c) = 0.3 \times 0.7 + 0.1 \times 0.3 = 0.24$.

(b) The probability of not fulfilling the city's water supply demand in a given week is equivalent to the probability of having, at least, one failure in 7 days, which is equal to the complement, with respect to 1, of the probability of no daily failure in a week. Therefore, the answer is given by $[1-(1-0.24)^7] = [1-(0.76)^7] = 0.8535$.

Bayes' formula, named after the English philosopher and Presbyterian minister Thomas Bayes (1702–1761), results from a combination of the multiplication rule of probabilities and the law of total probability. In effect, considering again the situation depicted in Fig. 3.5, one can express the probability of any one of the mutually exclusive and exhaustive events, say, for example, B_j, conditioned on the occurrence of A, as

$$P(B_j|A) = \frac{P(B_j \cap A)}{P(A)} \tag{3.7}$$

From the rule of multiplication of probabilities, the numerator of the right-hand side of Eq. (3.7) can be expressed as $P(A|B_j) P(B_j)$, whereas the denominator can be put in terms of the law of total probability. The resulting equation is Bayes' formula, or

$$P(B_j|A) = \frac{P(A|B_j) P(B_j)}{\sum_{i=1}^{k} P(A|B_i) P(B_i)} \tag{3.8}$$

Bayes' formula creates an important logical framework to review or update prior probabilities, as new information is added to the existing ones. Suppose, for instance, the event B_j be a possible hypothesis about some subject matter and let $P(B_j)$ represent the degree of belief in B_j before the occurrence of experiment A. The probability (or degree of belief) in a particular hypothesis B_j is assessed a priori, by an expert opinion or by some other means, and is unconditional. After the occurrence of experiment A, new evidence is collected and will influence the prior probability $P(B_j)$ by conditioning. The result of conditioning B_j to A is the posterior probability $P(B_j|A)$, whose evaluation is enabled by Bayes' formula. Today, Bayesian Statistics represents an important and independent branch of Mathematical Statistics, with a plethora of applications in many fields of knowledge. Bayes' formula and its applications in Statistical Hydrology are the core topics of Chap. 11.

Example 3.5 A meteorological satellite sends a set of binary codes ("0" or "1") to describe the development of a storm. However, electrical interferences on the sent signals can possibly lead to transmission errors. Suppose a sent message containing 80 % of digits "0," has been transmitted and also that there is a probability of 85 % of a given "0" or "1" has been correctly received by the earth station. If a "1" has been received, what is the probability of a "0" had been transmitted? (adapted from Larsen and Marx 1986).

Solution Let T_0 or T_1 respectively represent the events that digit "0" or "1" has been transmitted. Analogously, let R_0 or R_1 denote the reception of a "0" or of a "1," respectively. According to the given data, $P(T_0) = 0.8$, $P(T_1) = 0.2$, $P(R_0|T_0) = 0.85$, $P(R_1|T_1) = 0.85$, $P(R_0|T_1) = 0.15$ and $P(R_1|T_0) = 0.15$. The sought probability is $P(T_0|R_1)$, which can be easily calculated by means of Bayes' formula. Equation (3.8), as particularized for the problem on focus, results in $P(T_0|R_1) = [P(R_1|T_0) P(T_0)]/[P(R_1|T_0) P(T_0) + P(R_1|T_1) P(T_1)]$. With the given data, $P(T_0|R_1) = (0.15 \times 0.8)/(0.15 \times 0.8 + 0.85 \times 0.2) = 0.4138$.

3.5 Random Variables

A random variable is a function X associating a numerical value with each outcome of a random experiment. Although different outcomes can possibly share the same value of X, there is only one single value of the random variable associated with each outcome. In order to facilitate understanding of the concept of a random variable, consider the experiment of simultaneously flipping two coins, distinguishable one from another. The sample space for this random experiment is $S = \{hh, tt, ht, th\}$, where h designates "heads" and t "tails." The mutually exclusive and exhaustive events $A = \{hh\}$, $B = \{tt\}$, $C = \{ht\}$, and $D = \{th\}$ are assumed equally likely, and, hence, each one occurs with probability 0.25. Let X be defined as the random variable "number of heads." Mapping the sample space S for X allows assigning to the variable X its possible numerical values: $x = 2$, $x = 1$, or $x = 0$. The extreme values of X, 0 and 2, are respectively associated with the occurrences of events A and B, whereas $x = 1$ corresponds to the union of events C and D.

Beyond simply associating the possible outcomes to values of X, it is necessary to assign a probability to each of its numerical values. Hence, $P(X = 2) = P(A) = 0.25$, $P(X = 0) = P(B) = 0.25$, and $P(X = 1) = P(C \cup D) = P(C) + P(D) = 0.50$. These probabilities are generically denoted by $p_X(x)$, which is equivalent to $P(X = x)$, and are illustrated in the charts of Fig. 3.6.

For the example illustrated in Fig. 3.6, the random variable X is viewed as discrete since it can take on only integer values and also because it is associated with a finite and countable sample space. The chart on the left of Fig. 3.6 refers to $p_X(x)$, which is the Probability Mass Function (PMF) and gives the probability that the random variable X takes on the argument x. The chart on the right of Fig. 3.6 represents $P_X(x)$, which denotes the Cumulative Distribution Function (CDF) and indicates the probability that the random variable X be equal to or less than the argument x, or in formal terms, $P_X(x) = P(X \leq x) = \sum_{\text{all } x_i \leq x} p_X(x_i)$. For a discrete random variable X, the probability mass function $p_X(x)$ exhibits the following properties:

1. $p_X(x) \geq 0$ for any value of x
2. $\sum_{\text{all } x} p_X(x) = 1$

Inversely, if a function $p_X(x)$ possesses properties (1) and (2), then it is a PMF.

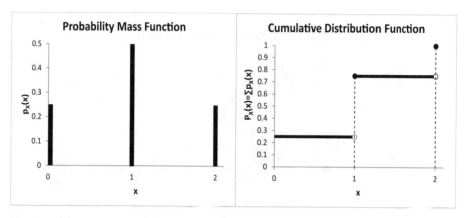

Fig. 3.6 Probability distribution functions of random variable X

On the other hand, if the random variable X can take on any real number, then it belongs to the continuous type and as such, the analogous function to the discrete-case PMF, is termed the Probability Density Function (PDF). This non-negative function, henceforth denoted by $f_X(x)$, is depicted in Fig. 3.7 and represents the limiting case of a relative frequency histogram as the sample size goes to infinity and the bin width tends to zero. The value taken by the PDF at the argument x_0, or $f_X(x_0)$, does not give the probability of X at x_0. Actually, it gives the rate at which the probability that X does not exceed x_0 changes in the vicinity of that argument. As shown in Fig. 3.7, the area enclosed by the points a and b, located on the horizontal axis of the domain of X, and their images, $f_X(a)$ and $f_X(b)$, read on the vertical axis of the counter-domain, is the probability of X being comprised between a and b. Thus, for a PDF $f_X(x)$, it is valid to write

$$P(a < X \le b) = \int_a^b f_X(x)\,dx \qquad (3.9)$$

Further, if the integration's lower bound of Eq. (3.9) continually approaches b and ultimately coincides with it, the result would be equivalent to the "area of a line" on the real plane, which, by definition, is zero. Therefore, for any continuous random variable X, $P(X = x) = 0$.

Similarly to the discrete case, the Cumulative Distribution Function (CDF) of a continuous random variable X, denoted by $F_X(x)$, gives the probability that X does not exceed the argument x, or $P(X \le x)$, or $P(X < x)$. For the general domain $-\infty < x < \infty$,

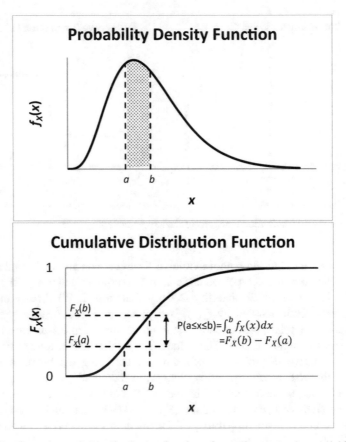

Fig. 3.7 Density and cumulative distribution function of a continuous random variable

$$F_X(x) = \int\limits_{-\infty}^{x} f_X(x)\, dx \qquad (3.10)$$

Inversely, the corresponding PDF can be obtained by differentiation of $F_X(x)$, or

$$f_X(x) = \frac{dF_X(x)}{dx} \qquad (3.11)$$

The CDF of a continuous random variable is a non-decreasing function, for which $F_X(-\infty) = 0$ and $F_X(+\infty) = 1$.

The PMF and PDF functions, and their corresponding cumulative distribution functions, describe completely the probabilistic behavior of discrete and continuous random variables, respectively. In particular, density functions of continuous random variables X can possibly have a great variety of shapes; Fig. 3.8 depicts some of them. As a general requisite, in order to be a PDF, a given function must be non-negative and its integration over the whole domain of X must be equal to one.

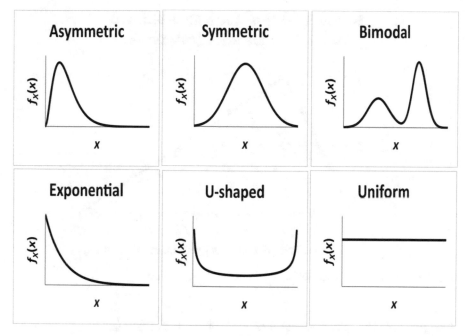

Fig. 3.8 Some of the possible shapes of a probability density function

Example 3.6 In Spring, the weekly mean dissolved-oxygen concentrations of a river reach, located in the American state of Wisconsin, is supposed to be distributed according to the triangular PDF, as shown in Fig. 3.9. Dissolved-Oxygen (DO) concentration is a bounded random variable. Answer the following questions: (a) What is the probability of having a weekly DO level of less than 7 mg/l? (b) Certain species of freshwater fish, such as carp, require a minimum DO level of 8.5 mg/l. What is the probability this river reach will provide such an ideal DO requirement for carp, during the spring months?

Solution (a) As with any density function, $f_X(x)$ must integrate to one, over the domain of X, which, in this case, spans from 6 to 10 mg/l. This allows the calculation of the unknown ordinate y, indicated in Fig. 3.9, giving $[y(10-6)]/2 = 1 \Rightarrow y = 1/2$. The sought probability $P(X < 7 \text{ mg/l})$ is, then, given by the area of the triangle formed by the points $(6,0)$, $(7,0)$, and $(7,y)$, which results in $P(X < 7) = (1 \times 0.5)/2 = 0.25$. (b) What is asked is the probability $P(X > 8.5 \text{ mg/l})$. One of the possibilities of calculating it is to find the area of the triangle formed by the points $(8.5,0)$, $(8.5,z)$, and $(10,0)$. However, ordinate z is unknown, but can be calculated using the property of similar triangles, which in this case gives $y/z = 3/1.5$ and results in $z = 0.25$. Finally, $P(X > 8.5) = (1.5 \times 0.25)/2 = 0.1875$.

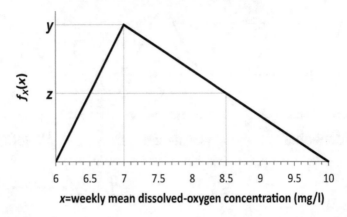

Example 3.6 - PDF of Weekly Mean Dissolved-Oxygen Concentration in Spring Months

x=weekly mean dissolved-oxygen concentration (mg/l)

Fig. 3.9 PDF of weekly mean DO concentration in Spring months

Example 3.7 The mathematical function defined by $f_X(x) = \frac{1}{\theta}\exp\left(-\frac{x}{\theta}\right)$, for $x \geq 0$ and $\theta > 0$, is the parametric form of the exponential probability density function. The numerical value of parameter θ specifies a particular PDF from the family of functions with the exponential parametric form. (a) Prove that, regardless of the numerical value of θ, the given function is indeed a PDF; (b) express the CDF $F_X(x)$; (c) calculate $P(X > 3)$, for the particular case where $\theta = 2$; and (d) plot the graphs of $f_X(x)$ and $F_X(x)$, versus x, for $\theta = 2$.

Solution

(a) Because $x \geq 0$ and $\theta \geq 0$, the function is non-negative, which is the first requirement for a PDF. The second necessary and sufficient condition is $\int\limits_0^\infty \frac{1}{\theta}\exp\left(-\frac{x}{\theta}\right)dx = 1$. Solving the integral equation $\int\limits_0^\infty \frac{1}{\theta}\exp\left(-\frac{x}{\theta}\right)dx = -\exp\left(-\frac{x}{\theta}\right)\big]_0^\infty = 1$, thus proving the exponential function is, in fact, a PDF.

(b) $F_X(x) = \int\limits_0^x \frac{1}{\theta}\exp\left(-\frac{x}{\theta}\right)dx = -\exp\left(-\frac{x}{\theta}\right)\big]_0^x = 1 - \exp\left(-\frac{x}{\theta}\right)$.

(c) $P(X > 3) = 1 - P(X < 3) = 1 - F_X(3) = 1 - \left[1 - \exp\left(-\frac{3}{2}\right)\right] = 0.2231$.

(d) Graphs of $f_X(x)$ and $F_X(x)$, versus x, for $\theta = 2$: Fig. 3.10.

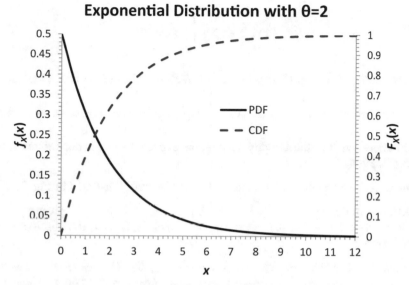

Fig. 3.10 PDF and CDF for the exponential distribution with parameter $\theta = 2$

3.6 Descriptive Measures of Random Variables

The population of a random variable X is entirely known, from a statistical point of view, by the complete specification of its PMF $p_X(x)$, in the discrete case, or of its PDF $f_X(x)$, in the continuous case. Analogously to the descriptive statistics of a sample drawn from the population, covered in Chap. 2, the shape characteristics of $p_X(x)$ or $f_X(x)$ can also be summarized by the population descriptive measures. These are usually obtained by the mean values, as weighted by $p_X(x)$ or $f_X(x)$, of functions of the random variables and include the expected value, the variance, and the coefficients of skewness and kurtosis, among others.

3.6.1 Expected Value

The expected value of X is the result of weighting, by $p_X(x)$ or $f_X(x)$, all possible values of the random variable. The expected value is denoted by $E[X]$ and is equivalent to the population mean value μ_X, thus indicating the abscissa of the centroid of the functions $p_X(x)$ or $f_X(x)$. Formally, $E[X]$ is defined by

$$E[X] = \mu_X = \sum_{\text{all } x_i} x_i \, p_X(x_i) \tag{3.12}$$

for discrete X; and

$$E[X] = \mu_X = \int_{-\infty}^{+\infty} x f_X(x)\, dx \tag{3.13}$$

for continuous X. In Eq. (3.13), in order to $E[X]$ to exist and be finite, the integral must be absolutely convergent, i.e., $\int_{-\infty}^{+\infty} |x| f_X(x)\, dx < \infty$. For some exceptional distributions, such as the Cauchy density, the expected value is not defined (Johnson et al. 1994).

Example 3.8 Calculate the expected value for the PMF illustrated in Fig. 3.6.

Solution Application of Eq. (3.12) gives $E[X] = \mu_X = 0 \times 0.25 + 1 \times 0.5 + 2 \times 0.25 = 1$ which is, in fact, the abscissa of the PMF centroid.

Example 3.9 Consider an exponential random variable X, whose PDF is given by $f_X(x) = \frac{1}{\theta} \exp\left(-\frac{x}{\theta}\right)$, for $x \geq 0$ and $\theta \geq 0$, as in Example 3.7. (a) Calculate the expected value of X; and (b) show the PDF is positively asymmetric distribution, based only on the population measures of central tendency, namely, the mean, the mode, and the median.

Solution (a) For the exponential distribution $E[X] = \mu_X = \int_0^{+\infty} x f_X(x)\, dx = \int_0^{+\infty} \frac{x}{\theta} \exp\left(-\frac{x}{\theta}\right) dx$. This integral equation can be solved by parts. In fact, by making $dv = \frac{1}{\theta} \exp\left(-\frac{x}{\theta}\right) dx \Rightarrow v = -\exp\left(-\frac{x}{\theta}\right)$ and $u = x \Rightarrow du = dx$. Substituting these into the equation of the expected value, it follows that $\int_0^{\infty} u\, dv = uv]_0^{\infty} - \int_0^{\infty} v\, du = -x\exp\left(-\frac{x}{\theta}\right)]_0^{\infty} - \theta\exp\left(-\frac{x}{\theta}\right)]_0^{\infty} = \theta$. Thus, for the exponential parametric form, the population mean is given by the parameter θ. For other parametric forms, μ_X is, in general, a function of one or more parameters that fully specify a particular distribution. For the specific case of $\theta = 2$ (see graphs of Example 3.7), the abscissa of the PDF's centroid is $x = 2$. (b) The mean μ_X of an exponential random variable is θ and, thus, a positive real number. The mode m_X is the x value that corresponds to the largest PDF ordinate; in this case, $m_X = 0$. The median u_X corresponds to the abscissa for which $F_X(x) = 0.5$. Since in this case, $F_X(x) = 1 - \exp\left(-\frac{x}{\theta}\right)$ (see Example 3.7), the inverse of $F_X(x)$, also known as the quantile curve, can be easily derived as $x = -\theta \ln(1 - F)$. For $F_X(x) = 0.5$, $u_X = -\theta \ln(1 - 0.5) = 0.6932\theta$. Therefore, one concludes that $m_X < u_X < \mu_X$, which is a key feature of positively asymmetric distributions. In fact, as it will be proved later on in this section, the skewness coefficient of an exponential distribution is $\gamma = +2$.

The concept of expected value can be generalized to compute the expectation of any real-valued function of a random variable. Given a real-valued function $g(X)$, of the random variable X, the mathematical expectation $E[g(X)]$ is formally defined as

$$E[g(X)] = \sum_{\text{all } x_i} g(x_i) p_X(x_i) \qquad (3.14)$$

for discrete X. In case of a continuous variable X, $E[g(X)]$ is defined as

$$E[g(X)] = \int_{-\infty}^{+\infty} g(x) f_X(x)\, dx \qquad (3.15)$$

Again, in Eq. (3.15), in order for $E[g(X)]$ to exist, the integral must be absolutely convergent. The mathematical expectation is an operator and has the following properties:
1. $E[c] = c$, for a constant c.
2. $E[cg(X)] = cE[g(X)]$, for a constant c.
3. $E[c_1 g_1(X) \pm c_2\, g_2(X)] = c_1 E[g_1(X)] \pm c_2 E[g_2(X)]$, for constant c_1 and c_2, and real-valued functions $g_1(X)$ and $g_2(X)$.
4. $E[g_1(X)] \geq E[g_2(X)]$, if $g_1(X) \geq g_2(X)$.

Example 3.10 The mathematical expectation $E[X - \mu_X]$ is named 1st-order central moment and corresponds to the mean value of the deviations of x from the mean μ_X, weighted by the PDF (or PMF) of X. Use the expected value properties to show that the 1st-order central moment is null.

Solution $E[X - \mu_X] = E[X] - E[\mu_X]$. Since μ_X is constant, from property (1), it is simple to conclude that $E[X - \mu_X] = \mu_X - \mu_X = 0$.

The application of the expectation operator to the deviations of x from a reference abscissa location a, as raised to the kth power, i.e., $E\left[(X - a)^k\right]$, is generically referred to as the moment of order k. Two special cases are of most interest: (a) if the reference location a is equal to zero, the moments are said to be about the origin and are denoted by μ_X, if $k = 1$ and μ'_k, for $k \geq 2$; and (b) if $a = \mu_X$, the moments are named central and are represented by μ_k. The moments about the origin are formally defined as

$$\mu_X = E[X] \text{ e } \mu'_k = \sum_{\text{all} x_i} x^k p_X(x_i) \qquad (3.16)$$

for discrete X. For a continuous random variable X,

$$\mu_X = E[X] \text{ e } \mu'_k = \int\limits_{-\infty}^{+\infty} x^k f_X(x) \, dx \qquad (3.17)$$

In a similar way, the central moments are given by

$$\mu_1 = 0 \text{ e } \mu_k = \sum_{\text{all } x_i} (x - \mu_X)^k p_X(x_i), \text{ if } k \geq 2 \qquad (3.18)$$

for discrete X. For a continuous random variable X,

$$\mu_1 = 0 \text{ e } \mu_k = \int\limits_{-\infty}^{+\infty} (x - \mu_X)^k f_X(x) \, dx, \text{ if } k \geq 2 \qquad (3.19)$$

These quantities are termed moments by analogy to moments from classical mechanics. In particular, μ_X corresponds to the abscissa of the PDF centroid (or PMF centroid), by analogy to the center of mass of a solid body, whereas moment μ_2 is mathematically equivalent to the moment of inertia with respect to a vertical axis through the centroid.

3.6.2 Variance

The population variance of a random variable X, denoted by Var[X] or σ_X^2, is defined as the central moment of order 2, or μ_2. The variance is the descriptive measure most used to characterize the dispersion of functions $p_X(x)$ and $f_X(x)$ around their respective means. Formally, Var[X], or σ_X^2, is given by

$$\text{Var}[X] = \sigma_X^2 = \mu_2 = E\left[(X - \mu_X)^2\right] = E\left[(X - E[X])^2\right] \qquad (3.20)$$

By expanding the squared term on the right-hand side of Eq. (3.20) and using the properties of mathematical expectation, one can rewrite it as

$$\text{Var}[X] = \sigma_X^2 = \mu_2 = E\left[X^2\right] - (E[X])^2 \qquad (3.21)$$

Thus, the population variance Var[X] is equal to the expected value of the square of X minus the square of the expected value of X. It has the same units of X^2 and the following properties:

1. Var[c] $= 0$, for a constant c.
2. Var[cX] $= c^2$Var[X], for a constant c.
3. Var[$cX + d$] $= c^2$Var[X], for constant c and d.

Similarly to the sample descriptors, the standard-deviation σ_X is defined as the positive square root of the variance and possesses the same units of X. As a relative dimensionless measure of dispersion of $p_X(x)$ or $f_X(x)$, the coefficient of variation CV_X is given by the expression

$$CV_X = \frac{\sigma_X}{\mu_X} \tag{3.22}$$

Example 3.11 Calculate the variance, the standard-deviation, and the coefficient of variation for the PMF shown in Fig. 3.6.

Solution Application of Eq. (3.21) requires previous knowledge of $E[X^2]$. This is calculated as $E[X^2] = 0^2 \times 0.25 + 1^2 \times 0.5 + 2^2 \times 0.25 = 1.5$. To return to Eq. (3.21), $Var[X] = \sigma^2{}_X = 1.5 - 1^2 = 0.5$. The standard-deviation is $\sigma_X = 0.71$ and the coefficient of variation is $CV_X = 0.71/1 = 0.71$.

Example 3.12 Consider the exponential random variable X, as in Example 3.9. Calculate the variance, the standard-deviation, and the coefficient of variation of X.

Solution The expected value of an exponential variable is θ (see the solution to Example 3.9). Again, application of Eq. (3.21) requires the previous calculation of $E[X^2]$. By definition, $E[X^2] = \int\limits_0^{+\infty} x^2 f_X(x)\,dx = \int\limits_0^{+\infty} \frac{x^2}{\theta} \exp\left(-\frac{x}{\theta}\right) dx$, which can be solved by integration by parts. Thus, by making $dv = \frac{x}{\theta}\exp\left(-\frac{x}{\theta}\right) dx \Rightarrow v = -x\exp\left(-\frac{x}{\theta}\right) - \theta\exp\left(-\frac{x}{\theta}\right)$, as in Example 3.9, and $u = x \Rightarrow du = dx$, the resulting equation is $\int\limits_0^\infty u\,dv = uv\Big|_0^\infty - \int\limits_0^\infty v\,du$. Solving it results in $\int\limits_0^{+\infty} \frac{x^2}{\theta}\exp\left(-\frac{x}{\theta}\right) dx = 0 - \int\limits_0^\infty \left[-x\exp\left(-\frac{x}{\theta}\right) - \theta\exp\left(-\frac{x}{\theta}\right)\right] dx = \theta E[X] + \theta^2 = 2\theta^2$. Now, to return to Eq. (3.21), $Var[X] = 2\theta^2 - \theta^2 = \theta^2$. Finally, $\sigma = \theta$ and $CV_X = 1$.

3.6.3 Coefficients of Skewness and Kurtosis

The coefficient of skewness γ of a random variable X is a dimensionless measure of asymmetry, defined as

$$\gamma = \frac{\mu^3}{(\sigma_X)^3} = \frac{E\left[(X - \mu_X)^3\right]}{(\sigma_X)^3} \tag{3.23}$$

The numerator of the right-hand side of Eq. (3.23) is the central moment of order 3, which shall reflect the equivalence between the average summations of

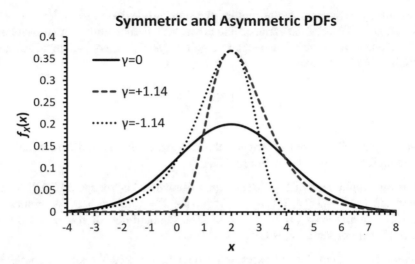

Fig. 3.11 Examples of symmetric and asymmetric density functions

positive and negative cubic deviations of X from its mean μ_X, in case of a symmetric distribution, or, otherwise, the numerical predominance of one over the other. The average summation is then scaled by the standard-deviation raised to power 3, in order to make γ a relative dimensionless index. In the case of equivalence, the numerator and the coefficient of skewness will both be zero, thus indicating a symmetric distribution. Contrarily, if the upper tail of the X density (or mass) function is more elongated than its lower tail or, in other words, if there is more dispersion among the values of X that are larger than the mode m_X, as compared to the ones that are smaller than it, then the positive cubic deviations will prevail over the negative ones. This will result in a positive coefficient of skewness, whose numerical value gives a relative dimensionless index of how right-skewed the distribution is. A similar reasoning applies to negative coefficients of skewness and left-skewed distributions. Figure 3.11 illustrates three distinct densities: one with a null coefficient of skewness, one right-skewed with $\gamma = +1.14$, and one left-skewed with $\gamma = -1.14$.

According to a recent interpretation (Westfall 2014), the coefficient of kurtosis of a random variable X, usually denoted by k, reflects the *tail extremity* of a density (or mass) function, in the sense of how prone it is to produce outliers. The classical notion refers to the coefficient of kurtosis k as measuring both *peakedness* (or flatness) and tail weight of the distribution. This notion can still be applied as a scaled dimensionless index for comparing unimodal symmetric distributions. The coefficient k is formally defined by the following equation:

$$\kappa = \frac{\mu_4}{(\sigma_X)^4} = \frac{E\left[(X - \mu_X)^4\right]}{(\sigma_X)^4} \tag{3.24}$$

For symmetric distributions, the coefficient of excess kurtosis, given by $(k-3)$, can be used as a relative index with respect to a perfectly symmetric distribution of reference, with coefficient of kurtosis $k = 3$.

Example 3.13 Consider the exponential random variable X, as in Example 3.9. Calculate the distribution's coefficients of skewness and kurtosis.

Solution Proceeding from the integrations by parts done for the calculations of $E[X]$ and $E[X^2]$, as in Examples 3.9 and 3.12, it is possible to generalize that, for any integer k, the mathematical statement $E[X^k] = \int_0^\infty \frac{x^k}{\theta} \exp\left(-\frac{x}{\theta}\right) dx = \theta^k \Gamma(k+1)$ is valid, where $\Gamma(.)$ denotes the Gamma function (see Appendix 1 for a brief review on the Gamma function). If the argument of the Gamma function is an integer, the result $\Gamma(k+1) = k!$ holds. Applying it to the moments about the origin of orders 3 and 4, it follows that $E[X^3] = 6\theta^3$ and $E[X^4] = 24\theta^4$. For the calculation of the coefficient of skewness, it is necessary first to expand the cube in the numerator of Eq. (3.23) and then proceed by using the expectation properties, to obtain $\gamma = \frac{E[X^3] - 3E[X^2]E[X] + 2(E[X])^3}{(\sigma_X)^3}$. Substituting the moments already calculated, the resulting coefficient of skewness is $\gamma = 2$. In an analogous way, the coefficient of kurtosis of the exponential distribution can be expressed as $\kappa = \frac{E[X^4] - 4E[X^3]E[X] + 6E[X^2](E[X])^2 - 3(E[X])^4}{(\sigma_X)^4}$. Finally, with the moments already calculated, $k = 9$.

3.6.4   Moment Generating Function

The probabilistic behavior of a random variable is completely specified by its density (or mass) function. This, in turn, can be completely determined by all its moments. A possible way to successfully find the moments of a density (or mass) function is through the moment generating function (MGF). The MGF of a random variable X is a function, usually designated by $\varphi(t)$, of argument t, defined in the interval $(-\varepsilon, \varepsilon)$ around $t = 0$, that allows the successive calculation of the moments about the origin of X, for any order $k \geq 1$. The function $\varphi(t)$ is formally defined as

$$\varphi(t) = E[e^{tX}] = \begin{cases} \displaystyle\sum_{\text{all } x} e^{tx} p_X(x), & \text{for discrete } X \\[2ex] \displaystyle\int_{-\infty}^{\infty} e^{tx} f_X(x)\, dx, & \text{for continuous } X \end{cases} \qquad (3.25)$$

The function $\varphi(t)$ is named moment generating because its kth derivative, with respect to t and evaluated at $t=0$, yields the moment μ'_k of the density (or mass) on focus. Supposing, for instance, that $k=1$, it follows from differentiation that

$$\varphi'(t) = \frac{d}{dt} E[e^{tX}] = E\left[\frac{d\,e^{tX}}{dt}\right] = E[Xe^{tX}] \quad \Rightarrow \quad \varphi'(t=0) = E[X] = \mu_X \quad (3.26)$$

In an analogous way, one can derive that $\varphi''(0) = E[X^2] = \mu'_2$, $\varphi'''(0) = E[X^3] = \mu'_3$ and so forth, up to $\varphi^k(0) = E[X^k] = \mu'_k$. In fact, as a summarizing statement, expansion of the MGF $\varphi(t)$, of a random variable X, into a Maclaurin series of integer powers of t (see Appendix 1 for a brief review), yields

$$\varphi(t) = E[e^{tX}] = E\left[1 + Xt + \frac{1}{2!}(Xt)^2 + \dots\right] = 1 + \mu'_1 t + \frac{1}{2!}\mu'_2 t + \dots \quad (3.27)$$

Example 3.14 The distribution with PMF $p_X(x) = e^{-\nu}\frac{\nu^x}{x!}$, $x = 0, 1, \dots$, is known as the Poisson distribution, with parameter $\nu > 0$. Use the MGF to calculate the mean and the variance of the Poisson discrete random variable.

Solution Equation (3.25), as applied to the Poisson mass function, gives $\varphi(t) = E[e^{tX}] = \sum_{x=0}^{\infty} \frac{e^{tx} e^{-\nu} \nu^x}{x!} = e^{-\nu} \sum_{x=0}^{\infty} \frac{(\nu e^t)^x}{x!}$. Using the identity $\sum_{k=0}^{\infty} \frac{a^k}{k!} = e^a$, one can write $\varphi(t) = e^{-\nu} e^{\nu \exp(t)} = \exp[\nu(e^t - 1)]$, whose derivative with respect to t is $\varphi'(t) = \nu e^t \exp[\nu(e^t - 1)]$ and $\varphi''(t) = (\nu e^t)^2 \exp[\nu(e^t - 1)] + \nu e^t \exp[\nu(e^t - 1)]$. For $t=0$, $E[X] = \varphi'(0) = \nu$ and $E[X^2] = \varphi''(0) = \nu^2 + \nu$. Recalling that $\text{Var}(X) = E[X^2] - (E[X])^2$, one concludes that $\mu_X = \text{Var}(X) = \nu$.

Example 3.15 The Normal is the best-known probability distribution and is at the origin of important theoretical results from statistics. The Normal PDF is given by $f_X(x) = \frac{1}{\sqrt{2\pi}\theta_2} \exp\left[-\frac{1}{2}\left(\frac{x-\theta_1}{\theta_2}\right)^2\right]$, where θ_1 and θ_2 are parameters that respectively define the central location and the scale of variation of X. The X domain spans from $-\infty$ to $+\infty$. Following substitution and algebraic manipulation, the MGF for the Normal distribution is expressed as $\varphi(t) = E[e^{tX}] = \frac{1}{\sqrt{2\pi}\theta_2} \int_{-\infty}^{\infty} \exp\left[-\frac{x^2 - 2\theta_1 x + \theta_1^2 - 2\theta_2^2 tx}{2\theta_2^2}\right] dx$. Use this expression of the MGF to calculate μ_X and $\text{Var}(X)$ for a Normal random variable.

Solution In the integrand of $\varphi(t)$, the term $x^2 - 2\theta_1 x + \theta_1^2 - 2\theta_2^2 tx$ can be rewritten as $x^2 - 2(\theta_1 + \theta_2^2 t)x + \theta_1^2$. This term will not be altered by introducing the algebraic artifice $\left[x - (\theta_1 + \theta_2^2 t)\right]^2 - (\theta_1 + \theta_2^2 t)^2 + \theta_1^2 = \left[x - (\theta_1 + \theta_2^2 t)\right]^2 - \theta_2^4 t^2 - 2\theta_1 \theta_2^2 t$. Replacing it back in the MGF equation and rewriting it with the constant terms outside the integrand, $\varphi(t) = \exp\left[\frac{\theta_2^4 t^2 + 2\theta_1\theta_2^2 t}{2\theta_2^2}\right]\frac{1}{\sqrt{2\pi}\theta_2}\int\limits_{-\infty}^{\infty}\exp\left[-\frac{[x-(\theta_1+\theta_2^2 t)]^2}{2\theta_2^2}\right]dx.$

Now, let us define a new variable, given by $Y = \frac{x-(\theta_1+\theta_2^2 t)^2}{\theta_2^2}$, which is also normally distributed, but with parameters $\theta_1 + \theta_2^2 t$ and θ_2. As with any PDF, the integral over the entire domain of the random variable must be equal to one, or

$\frac{1}{\sqrt{2\pi}\theta_2}\int\limits_{-\infty}^{\infty}\exp\left[-\frac{[x-(\theta_1+\theta_2^2 t)]^2}{2\theta_2^2}\right]dx = 1,$ which makes the MGF $\varphi(t) =$

$\exp\left[\frac{\theta_2^4 t^2 + 2\theta_1\theta_2^2 t}{2\theta_2^2}\right].$ The derivatives of $\varphi(t)$ are $\varphi'(t) = (\theta_1 + t\theta_2^2)\exp\left[\frac{\theta_2^2 t^2}{2} + \theta_1 t\right]$ and $\varphi''(t) = (\theta_1 + t\theta_2^2)^2\exp\left[\frac{\theta_2^2 t^2}{2} + \theta_1 t\right] + \theta_2^2\exp\left[\frac{\theta_2^2 t^2}{2} + \theta_1 t\right].$ At $t = 0$, $\varphi'(0) = \theta_1 \Rightarrow E[X] = \theta_1$ and $\varphi''(0) = \theta_1^2 + \theta_2^2 \Rightarrow E[X^2] = \theta_1^2 + \theta_2^2.$ Recalling that $\text{Var}(X) = E[X^2] - (E[X])^2$, one finally concludes that $\mu_X = \theta_1$ and $\text{Var}(X) = \sigma_X^2 = \theta_2$. As a result, the Normal PDF is most often written as $f_X(x) = \frac{1}{\sqrt{2\pi}\sigma_X}\exp\left[-\frac{1}{2}\left(\frac{x-\mu_X}{\sigma_X}\right)^2\right].$

3.7 Joint Probability Distributions of Random Variables

So far, the focus of this chapter has been kept on the main features of probability distributions of a single random variable. There are occasions, though, when one is interested in the joint probabilistic behavior of two or more random variables. In this section, the definitions and arguments developed for one single random variable are extended to the case of two variables. Denoting these by X and Y, their joint cumulative probability distribution function is defined as

$$\left.\begin{array}{c}F_{X,Y}(x,y)\\P_{X,Y}(x,y)\end{array}\right\} = P(X \leq x, Y \leq y) \qquad (3.28)$$

The probability distribution that describes the behavior of only variable X can be derived from $F_{X,Y}(x,y)$ or from $P_{X,Y}(x,y)$. In effect, for the continuous case, the CDF of X can be put in terms of the joint CDF as

$$F_X(x) = P(X \leq x) = P(X \leq x, Y \leq \infty) = F_{X,Y}(x, \infty) \qquad (3.29)$$

Likewise for Y, one can write

$$F_Y(x) = P(Y \leq y) = P(X \leq \infty, Y \leq y) = F_{X,Y}(\infty, x) \qquad (3.30)$$

$F_X(x)$ and $F_Y(y)$ are named marginal distributions of X and Y, respectively.

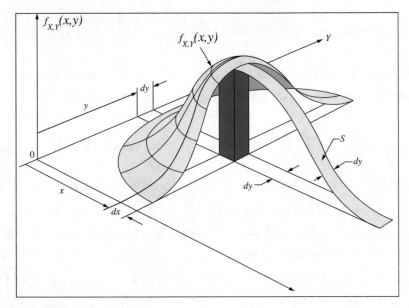

Fig. 3.12 3D example chart of a joint PDF of two continuous random variables (adapted from Beckmann 1968)

If X and Y are continuous, their joint probability density function is defined as

$$f_{X,Y}(x,y) = \frac{\partial^2}{\partial x \partial y} F_{X,Y}(x,y) \tag{3.31}$$

Figure 3.12 depicts a 3-dimensional example chart of a joint probability density distribution of two continuous random variables X and Y.

As for any density, the joint PDF $f_{X,Y}(x,y)$ must be a non-negative function. In complete analogy to the univariate functions, the volume bounded by the surface given by $f_{X,Y}(x,y)$ and the plane XY must be equal to one, or

$$\int_{-\infty}^{\infty} \int_{-\infty}^{\infty} f_{X,Y}(x,y)\, dx dy = 1 \tag{3.32}$$

The marginal density of X can be graphically visualized by projecting the joint density $f_{X,Y}(x,y)$ into the plane formed by the vertical and the X axes. In mathematical terms,

$$f_X(x) = \int_{-\infty}^{\infty} f_{X,Y}(x,y)\, dy \tag{3.33}$$

Likewise, the marginal density of Y, describing only the probabilistic behavior of Y, regardless of how X varies, can be derived from the joint PDF as

$$f_Y(y) = \int\limits_{-\infty}^{\infty} f_{X,Y}(x,y)\,dx \tag{3.34}$$

Thus, one can write the following mathematical statements:

$$F_X(\infty) = \int\limits_{-\infty}^{\infty} f_X(x)\,dx = 1 \text{ and } F_Y(\infty) = \int\limits_{-\infty}^{\infty} f_Y(y)\,dy = 1 \tag{3.35}$$

and

$$F_X(x) = \int\limits_{-\infty}^{x} f_X(x)\,dx = P(X \le x) \text{ and } F_Y(y) = \int\limits_{-\infty}^{y} f_Y(y)\,dy = P(Y \le y) \tag{3.36}$$

This logical framework can be extended to joint and marginal probability mass functions of two discrete random variables X and Y. For these, the following are valid relations:

$$P_{X,Y}(x,y) = P(X \le x, Y \le y) = \sum_{x_i \le x} \sum_{y_j \le y} P_{X,Y}\left(x_i, y_j\right) \tag{3.37}$$

$$p_X(x_i) = P(X = x_i) = \sum_j P_{X,Y}\left(x_i, y_j\right) \tag{3.38}$$

$$p_Y\left(y_j\right) = P\left(Y = y_j\right) = \sum_i P_{X,Y}\left(x_i, y_j\right) \tag{3.39}$$

$$P_X(x) = P(X \le x_i) = \sum_{x_i \le x} p_X(x_i) = \sum_{x_i \le x} \sum_j P_{X,Y}\left(x_i, y_j\right) \tag{3.40}$$

$$P_Y(y) = P\left(Y \le y_j\right) = \sum_{y_j \le y} p_Y\left(y_j\right) = \sum_{y_j \le y} \sum_i P_{X,Y}\left(x_i, y_j\right) \tag{3.41}$$

Example 3.16 Suppose that $f_{X,Y}(x,y) = 2x\exp(-x^2 - y)$ for $x \ge 0$ and $y \ge 0$. (a) Check whether $f_{X,Y}(x,y)$ is indeed a joint PDF. (b) Calculate $P(X > 0.5, Y > 1)$.

Solution (a) As the joint PDF $f_{X,Y}(x,y)$ is a non-negative function, it suffices to check whether the second necessary condition, given by Eq. (3.32), holds. Thus,

$$\int\limits_{-\infty}^{\infty} \int\limits_{-\infty}^{\infty} f_{X,Y}(x,y)\,dx\,dy = 2\int\limits_{0}^{\infty} x\exp(-x^2)\,dx \int\limits_{0}^{\infty} \exp(-y)\,dy = -\exp(-x^2)]_0^\infty(-e^{-y})]_0^\infty = 1.$$

Therefore, one concludes $f_{X,Y}(x,y)$ is actually a density. (b) $P(X > 0.5, Y > 1)$

$$= \int_{0,5}^{\infty} 2x \exp(-x^2)\, dx \int_{1}^{\infty} \exp(-y)\, dy = \exp(-1.25) = 0.2865.$$

The probability distribution of one of the two variables, under constraints imposed on the other, is termed conditional distribution. For the simpler case of discrete variables, the joint PMF of X, as conditioned on the occurrence $Y = y_0$, is a direct extension of Eq. (3.3), or

$$p_{X|Y=y_0} = \frac{p_{X,Y}(x, y_0)}{p_Y(y_0)} \tag{3.42}$$

For continuous variables, however, the concept of conditional distribution requires more attention. In order to better explain it, let us consider the events $x < X < x + dx$, denoted by A, and $y < Y < y + dy$, represented by B. The conditional probability density function $f_{X|Y}(x|y)$, as multiplied by dx, is equivalent to the conditional probability $P(A|B)$, or

$$f_{X|Y}(x|y)\, dx = P(x < X < x + dx | y < Y < y + dy) = P(A|B) \tag{3.43}$$

Note in this equation that X only is a random variable, since Y was kept fixed and inside the interval $(y, y + dy)$, thus showing that $f_{X|Y}(x|y)$ is indeed univariate. Now, by virtue of applying Eq. (3.3), the probability of the joint occurrence of events A and B is written as $P(A \cap B) = P(A|B)\, P(B) = f_{X,Y}(x, y)\, dx\, dy$ and if $P(B) = P(y < Y < y + dy) = f_Y(y)\, dy$, then, one can define the conditional density $f_{X|Y}(x|y)$ as

$$f_{X|Y}(x|y) = \frac{f_{X,Y}(x, y)}{f_Y(y)} \tag{3.44}$$

having the same properties that any probability density function should have. Employing the same line of reasoning as before and the law of total probability, it is simple to show that Bayes' formula, as applied to two continuous random variables, can be expressed as

$$f_{X|Y}(x|y) = \frac{f_{Y|X}(y|x) f_X(x)}{f_Y(y)} \text{ or } f_{X|Y}(x|y) = \frac{f_{Y|X}(y|x) f_X(x)}{\int_{-\infty}^{\infty} f_{Y|X}(y|x) f_X(x)\, dx} \tag{3.45}$$

Making reference to Fig. 3.12 and in the light of the new definitions, one can interpret Eq. (3.44) as the ratio between the volume of the prism $f_{X,Y}(x,y) dx dy$, hatched in the figure, and the volume of the band S, enclosed by the surface $f_{X,Y}(x,y)$ and the interval $(y, y + dy)$. Yet, there is the special case where X and Y are continuous random variables and one wishes to determine the PDF of X, conditioned on $Y = y_0$. In such a case, because Y is a fixed value, the band S becomes a flat slice and must be referred to as an area instead of a volume. Under these conditions, Eq. (3.44), for $Y = y_0$, is rewritten as

$$f_{X|Y}(x|Y = y_0) = \frac{f_{X,Y}(x, y_0)}{f_Y(y_0)} \tag{3.46}$$

As a consequence of Eq. (3.5), the random variables X and Y are considered statistically independent if the probability of any occurrence related to one of them is not affected by the other. This is summarized by

$$P(X \leq x_0, Y \leq y_0) = P(X \leq x_0)P(Y \leq y_0) \tag{3.47}$$

In terms of the joint CDF, the variables X and Y are independent if

$$P_{X,Y}(x_0, y_0) = P_X(x_0)P_Y(y_0) \text{ or } F_{X,Y}(x_0, y_0) = F_X(x_0)F_Y(y_0) \tag{3.48}$$

In the case of discrete variables, the independence condition is reduced to

$$P_{X,Y}(x, y) = p_X(x)p_Y(y) \tag{3.49}$$

whereas for continuous variables,

$$f_{X,Y}(x, y) = f_X(x)f_Y(y) \tag{3.50}$$

Therefore, the necessary and sufficient condition for two random variables to be considered statistically independent is that their joint density (or mass) function be the product of their marginal density (or mass) functions.

Example 3.17 Consider the following non-negative functions of X and Y: $f(x, y) = 4xy$, for $(0 \leq x \leq 1, 0 \leq y \leq 1)$ and $g(x, y) = 8xy$, for $(0 \leq x \leq 1, 0 \leq y \leq 1)$. (a) For the first function, check whether or not it is a density, and check whether X and Y are statistically independent. (b) Do the same for $g(x,y)$.

Solution (a) In order for $f(x, y) = 4xy$ be a density, the necessary and sufficient condition is $\int_0^1 \int_0^1 4xy\, dx\, dy = 1$. Solving the integral, $\int_0^1 \int_0^1 4xy\, dx\, dy = 4 \int_0^1 x\, dx \int_0^1 y\, dy = 1$. This proves that $f(x, y) = 4xy$ is a joint density. In order to check whether X and Y are independent, the necessary and sufficient condition is given by Eq. (3.50), requiring, for its verification, the calculation of the marginal densities. Marginal density of X: $f_X(x) = \int\limits_0^1 f_{X,Y}(x,y)\,dy = 4\int\limits_0^1 xy\,dy = 2x$. Marginal

of Y: $f_Y(y) = 4\int\limits_0^1 xy\,dx = 2y$. Therefore, because the joint PDF is the product of the

marginal densities, X and Y are statistically independent. (b) Proceeding in a similar way for the function $g(x,y) = 8xy$, it is verified that it is actually a joint density function. The marginal densities are $g_X(x) = 4x$ and $g_Y(y) = 4y^3$. In this case, because $g_{X,Y}(x,y) \neq g_X(x)g_Y(y)$, the variables are not independent.

The definition and properties of the mathematical expectation can be extended to joint probability distribution functions. In fact, Eqs. (3.14) and (3.15), which define mathematical expectations in broad terms, can be applied to a generic real-valued function $g(X, Y)$ of two random variables X and Y, by means of

$$E[g(X,Y)] = \begin{cases} \displaystyle\sum_x \sum_y g(x,y)p_{X,Y}(x,y) \text{ for discrete } X \text{ and } Y \\ \displaystyle\int\limits_{-\infty}^{\infty} \int\limits_{-\infty}^{\infty} g(x,y)f_{X,Y}(x,y)\,dx\,dy \text{ for continuous } X \text{ and } Y \end{cases} \tag{3.51}$$

By imposing $g(X,Y) = X^r Y^s$ in Eq. (3.51), it is possible to expand, for the bivariate case, the definition of moments $\mu'_{r,s}$, of orders r and s, about the origin. Likewise, by substituting $g(X,Y) = (X - \mu_X)^r (Y - \mu_Y)^s$ in Eq. (3.51), the central moments $\mu_{r,s}$ of orders r and s are defined as well. The following particular cases are easily recognized: (a) $\mu'_{1,0} = \mu_X$; (b) $\mu'_{0,1} = \mu_Y$; (c) $\mu_{2,0} = \text{Var}[X] = \sigma_X^2$ and (d) $\mu_{0,2} = \text{Var}[Y] = \sigma_Y^2$.

The central moment $\mu_{r=1,s=1}$ receives the special name of covariance of X and Y, and gives a measure of how strong the linear association between these two variables is. Formally, the covariance of X and Y is defined as

$$\text{Cov}[X,Y] = \sigma_{X,Y} = E[(X - \mu_X)(Y - \mu_Y)] = E[XY] - E[X]E[Y] \tag{3.52}$$

Note that when X and Y are statistically independent, it is clear that $E[XY] = E[X] \times E[Y]$ and, thus, application of Eq. (3.52) results in a covariance of zero. Conversely, if $\text{Cov}[X,Y] = 0$, the variables X and Y are not necessarily independent. In such a case, because $\text{Cov}[X,Y] = 0$, X and Y are linearly independent. However, they might exhibit some form of nonlinear dependence.

As the covariance possesses the same units as the ones resulting from the product of X and Y units, it is more practical to scale it by dividing it by $\sigma_X \sigma_Y$. This scaled measure of covariance is termed coefficient of correlation, is denoted by $\rho_{X,Y}$ and formally defined as

$$\rho_{X,Y} = \frac{\text{Cov}[X, Y]}{\sigma_X \sigma_Y} = \frac{\sigma_{X,Y}}{\sigma_X \sigma_Y} \tag{3.53}$$

Just as with its sample estimate $r_{X,Y}$, from Sect. 2.4.1 of Chap. 2, the coefficient of correlation is bounded by the extreme values of -1 and $+1$. Again, if variables X and Y are independent, then $\rho_{X,Y} = 0$. However, the converse is not necessarily true since X and Y might be associated by some functional relation other than the linear.

The following statements ensue from the plain application of the expectation properties to two or more random variables:

1. $E[aX + bY] = aE[X] + bE[Y]$, where a and b are constants.
2. $\text{Var}[aX + bY] = a^2\text{Var}[X] + b^2\text{Var}[Y] + 2ab\text{Cov}[X, Y]$, for linearly dependent X and Y.
3. $\text{Var}[aX + bY] = a^2\text{Var}[X] + b^2\text{Var}[Y]$, for linearly independent X and Y.
4. In the case of k random variables $X_1, X_2, \ldots, X_k, E[a_1X_1 + a_2X_2 + \ldots + a_kX_k] = a_1E[X_1] + a_2E[X_2] + \ldots + a_kE[X_k]$, where $a_1, a_2, \ldots, a_k$ are constants.
5. In the case of k random variables X_1, X_2, ..., X_k,

$$\text{Var}[a_1X_1 + a_2X_2 + \ldots + a_kX_k] = \sum_{i=1}^{k} a_i^2\text{Var}[X_i] + 2\sum_{i<j} a_ia_j\text{Cov}[X_i, X_j].$$

6. For k independent variables, $\text{Var}[a_1X_1 + a_2X_2 + \ldots + a_kX_k] = \sum_{i=1}^{k} a_i^2\text{Var}[X_i]$

Example 3.18 Consider a simple random sample (SRS) of N points, drawn from a population with mean μ and variance σ^2. Define Y as the arithmetic mean value of the N sample points. Calculate the mean and the variance of Y. Recall from Sect. 1.5 that the combination of the attributes of equally likely and statistically independent sample points defines an SRS.

Solution The arithmetic mean is given by $Y = \frac{X_1}{N} + \frac{X_2}{N} + \ldots + \frac{X_N}{N}$, where X_1, X_2, ..., X_N denote the sample elements (or sample points). As it is a simple random sample, its elements can be viewed as realizations of N independent random variables, all drawn from the same population of mean μ and variance σ^2. Using properties (4) and (6), previously listed in this section, with $a_1 = a_2, \ldots, a_N = (1/N)$, $E[X_1] = E[X_2] = \ldots = E[X_N] = \mu$, and $\text{Var}[X_1] = \text{Var}[X_2] = \ldots = \text{Var}[X_N] = \sigma^2$, it follows that $E[Y] = \frac{N\mu}{N} = \mu$ and $\text{Var}[Y] = \frac{N\sigma^2}{N^2} = \frac{\sigma^2}{N}$ or $\sigma_Y = \frac{\sigma}{\sqrt{N}}$.

Example 3.19 Show that the joint moment generating function of two statistically independent random variables X and Y is equal to the product of their individual moment generating functions.

Solution The joint MGF of the X and Y variables is given by $\varphi_{X,Y}(t_1, t_2) = E[\exp(t_1X + t_2Y)]$. The moments about the origin, of orders r and s, can be calculated from the joint MGF through the rth derivative with respect to t_1 and the sth derivative with respect to t_2, both evaluated at $t_1 = t_2 = 0$. However, if the variables are statistically independent, one can write $\varphi_{X,Y}(t_1, t_2) = E[\exp(t_1 X + t_2 Y)] = E[\exp(t_1 X)]E[\exp(t_2 Y)] = \varphi_X(t_1)\varphi_Y(t_2)$, which renders the calculation much simpler. Therefore, if two random variables are statistically independent, the joint MGF is the product of their individual MGFs. Conversely, if the joint MGF is equal to the product of the individual MGFs, then the variables are statistically independent.

The definition of expected value can also be extended to the random variable X conditioned on Y and vice-versa. In fact, if two discrete random variables X and Y, with joint PMF $p_{X,Y}(x, y)$ and marginal PMFs $p_X(x)$ and $p_Y(y)$, then the following conditional means are defined:

$$E[X|Y = y_0] = \sum_{\text{all } x_i} x_i \frac{p_{X,Y}(x_i, y_0)}{p_Y(y_0)} = \sum_{\text{all } x_i} x_i p_{X|Y}(x_i|y_0) \qquad (3.54)$$

$$E[Y|X = x_0] = \sum_{\text{all } y_j} y_j \frac{p_{X,Y}(x_0, y_j)}{p_X(x_0)} = \sum_{\text{all } y_j} y_j p_{Y|X}(y_j|x_0) \qquad (3.55)$$

If X and Y are continuous with joint PDF $f_{X,Y}(x, y)$ and marginal densities $f_X(x)$ and $f_Y(y)$, in like manner, the conditional means are given by

$$E[X|Y = y_0] = \int_{-\infty}^{\infty} x \frac{f_{X,Y}(x, y_0)}{f_Y(y_0)} = \int_{-\infty}^{\infty} x f_{X|Y}(x|Y = y_0) \qquad (3.56)$$

$$E[Y|X = x_0] = \int_{-\infty}^{\infty} y \frac{f_{X,Y}(x_0, y)}{f_X(x_0)} = \int_{-\infty}^{\infty} y f_{Y|X}(y|x = x_0) \qquad (3.57)$$

3.8 Probability Distributions of Functions of Random Variables

Suppose a given variable Y is linked to a random variable X, through a functional monotonically increasing or decreasing relation $Y = g(X)$, as in $Y = \ln(X)$ or $Y = \exp(-X)$, respectively, for $X > 0$. As well as being a function of a random variable, Y also is a random variable. If the probability distribution of X and the function $Y = g(X)$ are known, the distribution of Y can be derived.

If X is a discrete random variable, with mass function given by $p_X(x)$, the goal, in this case, is to derive the PMF $p_Y(y)$ of Y. For an increasing or decreasing monotonic function $Y = g(X)$, there exists a one-to-one correspondence (or a bijection) between Y and X, being valid to state that to each $g(x) = y$ there corresponds a unique $x = g^{-1}(y)$ and, thus, $P(Y = y) = P[X = g^{-1}(y)]$, or, generically,

$$p_Y(y) = p_X\left[g^{-1}(y)\right] \qquad (3.58)$$

If X is a continuous random variable, with density $f_X(x)$ and cumulative distribution $F_X(x)$, further discussion is needed. As before, the aim is to calculate $P(Y \le y)$ or $P[g(X) \le y]$. If the function $Y = g(X)$ is monotonically increasing, there exists a one-to-one correspondence between Y and X, and it is right to assert that to each $g(x) \le y$ corresponds a unique $x \le g^{-1}(y)$ and, thus,

$$P(Y \le y) = P\left[X \le g^{-1}(y)\right] \text{ ou } F_Y(y) = F_X\left[g^{-1}(y)\right] \qquad (3.59)$$

Contrarily, if the function $Y = g(X)$ decreases monotonically, to each $g(x) \le y$ there is only one $x \ge g^{-1}(y)$ and, thus,

$$P(Y \le y) = 1 - P\left[X \le g^{-1}(y)\right] \text{ ou } F_Y(y) = 1 - F_X\left[g^{-1}(y)\right] \qquad (3.60)$$

In both cases, the density of Y can be derived through differentiation of the CDF with respect to y. However, because densities are always non-negative and must integrate to one over the entire domain of X, it is necessary to take the absolute value of the derivative of $g^{-1}(y)$, with respect to y. In formal terms,

$$f_Y(y) = \frac{d}{dy}F_Y(y) = F'_X\left[g^{-1}(y)\right]\left|\frac{d\left[g^{-1}(y)\right]}{dy}\right| = f_X\left[g^{-1}(y)\right]\left|\frac{d\left[g^{-1}(y)\right]}{dy}\right| \qquad (3.61)$$

Example 3.20 A geometric discrete random variable X has a mass function $p_X(x) = p(1-p)^{x-1}$, for $x = 1, 2, 3, \ldots$ and $0 \le p \le 1$. Suppose X represents the occurrence in year x, and not before x, of a flood larger than the design flood of a cofferdam, built to protect a dam's construction site. In any given year, the probability of this extraordinary flood occurring, as related to the cofferdam failure, is p. Suppose further that the cofferdam has been recently heightened and that the time to failure, in years, has increased to $Y = 3X$. Calculate the probability of the time to failure, under the new scenario of the heightened cofferdam (adapted from Kottegoda and Rosso 1997).

Solution With $Y = 3X \Rightarrow g^{-1}(Y) = Y/3$ in Eq. (3.58), the resulting expression is $p_Y(y) = p(1-p)^{[(y/3)-1]}$, for $y = 3, 6, 9, \ldots$ e $0 \le p \le 1$. Hence, the conclusion is that the probabilities of having a failure after $1,2,3 \ldots$ years, before the cofferdam heightening, are equivalent to probabilities of failure after $3, 6, 9, \ldots$ years, under the new scenario.

Example 3.21 Suppose X is a Normal random variable with parameters μ and σ. Let Y define a new variable through $Y = \exp(X)$. Determine the PDF of Y.

Solution The Normal distribution (see Example 3.15) refers to an unbounded random variable. As x varies from $-\infty$ to $+\infty$, y varies from 0 to $+\infty$. Therefore, the density of Y must refer only to $y \ge 0$. With reference to Eq. (3.61), the inverse function is given by $x = g^{-1}(y) = \ln(y)$ and, thus, $|d[g^{-1}(y)]/dy| = 1/y$. Substituting these into Eq. (3.61), $f_Y(y) = \frac{1}{y\sigma\sqrt{2\pi}}\exp\left[-\frac{(\ln y - \mu)^2}{2\sigma^2}\right]$, for $y \geq 0$. This is known as the Lognormal distribution. It describes how $Y = \exp(X)$ is distributed when X is a Normal random variable.

The transformation given by Eq. (3.61) can be extended to the case of bivariate density functions. For this, consider a transformation of $f_{X,Y}(x,y)$ into $f_{U,V}(u,v)$, where $U = u(X,Y)$ and $V = v(X,Y)$ represent continuously differentiable bijective functions. In this case, one can write

$$f_{U,V}(u,v) = f_{X,Y}[x(u,v), y(u,v)]\,|J| \tag{3.62}$$

where J denotes the Jacobian, as calculated by the following determinant:

$$J = \begin{vmatrix} \dfrac{\partial x}{\partial u} & \dfrac{\partial x}{\partial v} \\[2mm] \dfrac{\partial y}{\partial u} & \dfrac{\partial y}{\partial v} \end{vmatrix} \tag{3.63}$$

The bounds of U and V depend on their relations with X and Y, and must be carefully determined for each particular case.

An important application of Eq. (3.62) refers to determining the distribution of the sum of two random variables $U = X + Y$, given the joint density $f_{X,Y}(x,y)$. To make it simpler, an auxiliary variable $V = X$ is created, so as to obtain the following inverse functions: $x(u,v) = v$ and $y(u,v) = u - v$. For these, the Jacobian becomes

$$J = \begin{vmatrix} 0 & 1 \\ 1 & -1 \end{vmatrix} = -1 \tag{3.64}$$

Substituting these quantities into Eq. (3.62), it follows that:

$$f_{U,V}(u,v) = f_{X,Y}[v, u - v] \tag{3.65}$$

Note, however, that what is actually sought is the marginal distribution of U. This can be determined by integrating the joint density, as in Eq. (3.65), over the domain $[A,B]$ of variable V. Thus,

$$f_U(u) = \int_A^B f_{X,Y}(v, u - v)\,dv = \int_A^B f_{X,Y}(x, u - x)\,dx \tag{3.66}$$

For the particular situation in which X and Y are independent, $f_{X,Y}(x,y) = f_X(x)f_Y(y)$ and Eq. (3.66) becomes

$$f_U(u) = \int_A^B f_X(x)f_Y(u - x)\,dx \tag{3.67}$$

The operation contained in the right-hand side of Eq. (3.67) is known as the convolution of functions $f_X(x)$ and $f_Y(y)$.

Example 3.22 The distribution of a random variable X is named uniform if its density is given by $f_X(x) = 1/a$, for $0 \leq x \leq a$. Suppose two independent uniform random variables X and Y are both defined in the interval $[0,a]$. Determine the density of $U = X + Y$.

Solution Application of Eq. (3.67) to this specific case is simple, except for the definition of the integration bounds A and B. The following conditions need to be abided by: $0 \leq u - x \leq a$ and $0 \leq x \leq a$. These inequalities can be algebraically manipulated and transformed into $u - a \leq x \leq u$ and $0 \leq x \leq a$. Thus, the integration bounds become $A = \text{Max}(u-a,0)$ and $B = \text{Min}(u,a)$, which imply two possibilities: $u < a$ and $u > a$. For $u < a$, $A = 0$ and $B = u$, and Eq. (3.67) turns itself into $f_U(u) =$

$$\frac{1}{a^2} \int_0^u dx = \frac{u^2}{a^2}, \text{ for } 0 \leq u \leq a. \text{ For } u > a, A = u - a \text{ and } B = a, \text{ and Eq. (3.67)}$$

becomes $f_U(u) = \dfrac{1}{a^2} \displaystyle\int_{u-a}^a dx = \dfrac{2a - u}{a^2}$, for $a \leq u \leq 2a$. Take the opportunity of this example to plot the density $f_U(u)$ and see that the PDF of the sum of two independent uniform random variables is given by an isosceles triangle.

3.9 Mixed Distributions

Consider a random variable X whose probabilistic behavior is described by the composition of m distributions, denoted by $f_i(x)$ and respectively weighted by parameters λ_i, for $i = 1, \ldots, m$, such that $\sum_{i=1}^m \lambda_i = 1$. Within this context, the density function of X is of the mixed type and given by

$$f_X(x) = \sum_{i=1}^m \lambda_i f_i(x) \tag{3.68}$$

The corresponding CDF is

$$F_X(x) = \int_{-\infty}^{x} \sum_{i=1}^{m} \lambda_i f_i(x)\, dx \qquad (3.69)$$

In hydrology, the mixed distributions approach meets its application field in the study of random variables whose outcomes may result from different causal mechanisms. For instance, short-duration rain episodes can possibly be caused by the passing of cold fronts over a region or from local convection, thus affecting the lifting of humid air into the atmosphere and the storm characteristics, such as its duration and intensity. Suppose X represents the annual maximum rainfall intensities, for some fixed sub-daily duration, at some site. If rains are related to frontal systems, their intensities ought to be described by some density function $f_1(x)$, whereas, if convection is their prevalent causal mechanism, they ought to be described by $f_2(x)$. If the proportion of rains caused by frontal systems is given by λ_1 and the proportion of convective storms is $\lambda_2 = (1-\lambda_1)$, then the overall probabilistic behavior of annual maxima of sub-daily rainfall intensities will be described by the combination of densities $f_1(x)$ and $f_2(x)$, as respectively weighted by λ_1 and λ_2, through Eq. (3.68). This same idea may be applied to other hydrologic phenomena, such as floods produced by different mechanisms, like rainfall and snowmelt.

Exercises

1. The possible values of the water heights H, relative to the mean water level, at each of two rivers A and B, are: $H = -3, -2, -1, 0, 1, 2, 3, 6$ m.

   (a) Consider the following events for river A: $A_1 = \{H_A > 0\}$, $A_2 = \{H_A = 0\}$ and $A_3 = \{H_A \leq 0\}$. List all possible pairs of mutually exclusive events among A_1, A_2 and A_3.

   (b) At each river, define the following events: normal water level $N = \{-1 \leq H \leq 1\}$, drought water level $D = \{H < 1\}$ and flood water level $F = \{H > 1\}$. Use the ordered pair (h_A, h_B) to identify the sample points that define the joint water levels in rivers A and B, respectively; e.g.: $(3, -1)$ defines the concurrent condition $h_A = 3$ and $h_B = -1$. Determine the sample points for the events $N_A \cap N_B$ and $(F_A \cup D_A) \cap N_B$ (adapted from Ang and Tang 1975).

2. Consider the cross-section of a gravity dam, as shown in Fig. 3.13. The effective storage volume V of the reservoir created by the dam varies from zero to full capacity c $(0 \leq V \leq c)$, depending on the time-varying inflows and outflows. The effective storage volume V has been discretized into volumes stored at levels between w_1 and w_2, w_2 and w_3, w_3 and w_4, w_4 and c, and then respectively identified as events A_1, A_2, A_3, and A_4, so that the annual frequencies of average daily water levels can be accordingly counted and

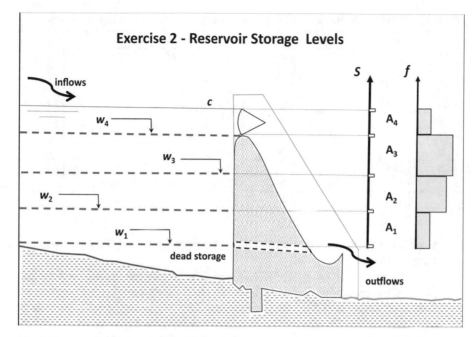

Fig. 3.13 Reservoir storage levels for Exercise 2

grouped into one of the four events. Identify the water levels w for each of the following events: (a) $(A_4)^c \cap (A_1)^c$; (b) $(A_3 \cup A_2)^c \cap (A_1)^c$; (c) $[A_4 \cup (A_1 \cup A_2)^c]^c$; and $(A_1 \cap A_2)^c$ (adapted from Kottegoda and Rosso 1997).

3. If the occurrence of a rainy day has probability of 0.25 and is statistically independent of it raining in the previous and in the following days, calculate

 (a) the probability of 4 rainy days in a week;
 (b) the probability that the next 4 days will be rainy; and
 (c) the probability of 4 rainy days in a row and 3 dry days in a week.

4. The river R is located close to the city C and, in any given year, reaches or exceeds the flood stage with a probability of 0.2. Parts of the city are flooded every year with a probability of 0.1. Past observations show that when river R is at or above flood stage, the probability of C being inundated increases to 0.2. Given that,

 (a) calculate the annual probability of a flood in river R or inundations in city C; and
 (b) calculate the probability of a flood in river R and inundations in city C.

5. A gravity retaining wall can possibly fail either by sliding along its contact surface with the foundations (event S) or by overturning (event O) or by both. If $P(S) = 3P(O)$, $P(O|S) = 0.7$, and the probability of failure of the wall is 10^{-3}, (a) determine the probability that sliding will occur; and (b) if the wall fails,

Table 3.1 DO and BOD measurements at 38 sites along the Blackwater River, in England

DO	BOD	DO	BOD	DO	BOD	DO	BOD
8.15	2.27	6.74	3.83	7.28	3.22	8.46	2.82
5.45	4.41	6.9	3.74	7.44	3.17	8.54	2.79
6.05	4.03	7.05	3.66	7.59	3.13	8.62	2.76
6.49	3.75	7.19	3.58	7.73	3.08	8.69	2.73
6.11	3.37	7.55	3.16	7.85	3.04	8.76	2.7
6.46	3.23	6.92	3.43	7.97	3	9.26	2.51
6.22	3.18	7.11	3.36	8.09	2.96	9.31	2.49
6.05	4.08	7.28	3.3	8.19	2.93	9.35	2.46
6.3	4	7.44	3.24	8.29	2.89	Average	Average
6.53	3.92	7.6	3.19	8.38	2.86	7.5	3.2

what is the probability that only sliding has occurred? (adapted from Ang and Tang 1975).

6. The Blackwater River, in central England, is regularly monitored for pollution at 38 sites along its course. Table 3.1 lists concurrent measurements of Dissolved Oxygen (DO) and Biochemical Oxygen Demand (BOD), both in mg/l, taken at the 38 sites of the Blackwater River. Owing to regional similarities in water uses, one can assume data refer to the same population (adapted from Kottegoda and Rosso 1997).

Knowing that the sample average values of DO and BOD measurements are respectively 7.5 and 3.2 mg/l, one can define the following events: $B_1 = \{DO \leq 7.5$ and $BOD > 3.2\}$; $B_2 = \{DO > 7.5$ and $BOD > 3.2\}$; $B_3 = \{DO > 7.5$ and $BOD \leq 3,2\}$; and $B_4 = \{DO \leq 7,5$ and $BOD \leq 3.2\}$. Based on the DO and BOD data, consider the reference event defined by the variation of both variables within the interval [average value -1 standard-deviation, average value $+1$ standard-deviation]. The standard-deviations of DO and BOD are respectively equal to 1.0 and 0.5 mg/l, which specifies the reference event as $A = \{6.5 < DO < 8.5$ and $2.7 < BOD < 3.7\}$. Under this setup,

(a) Make a scatterplot of DO versus BOD, and identify events B_1, B_2, B_3, B_4, and A on your chart;
(b) Estimate the probabilities of events B_i, $i = 1,...,4$, by their respective relative frequencies;
(c) Employ the law of total probability to calculate the probability that DO and BOD lie outside the boundaries of the reference event A; and
(d) Use Bayes' formula to calculate the probability that DO and BOD lie inside the boundaries defined by events B_1 to B_4, knowing that both are inside the reference event A.

7. A river splits into branches A and B to form a river island. The river bifurcation occurs downstream of the effluent discharge from a sewage treatment plant, whose efficiency is under scrutiny by the river regulation agency. Dissolved

oxygen (DO) concentrations in branches A and B are indicative of eventual pollution caused by the effluent discharge. Experts estimate the probabilities that river branches A and B have DO levels below regulatory standards are 2/5 and 3/4, respectively. They further estimate the probability of at least one of the two river branches being polluted is 4/5.

(a) Determine the probability of branch A being polluted, given that branch B is polluted;
(b) Determine the probability of branch B being polluted, knowing that branch A is polluted.

8. The probabilities of monthly rainfall depths larger than 60 mm in January, February, ..., December are respectively 0.24; 0.31; 0.30; 0.45; 0.20; 0.10; 0.05; 0.05; 0.04; 0.06; 0.10; and 0.20. Suppose a record of monthly rainfall depth, higher than 60 mm, is chosen at random. Calculate the probability this record refers to the month of July.

9. If the PDF of a random variable X is $f_X(x) = c(1 - x^2)$, $-1 \leq x \leq 1$, for constant c,

(a) Calculate c;
(b) Determine the CDF of X; and
(c) Calculate $P(X \leq 0.75)$.

10. In a small catchment, the probability of it raining on a given day is 0.60. If it rains, precipitation depth is an exponential random variable with $\theta = 10$ mm. Depending on the antecedent soil moisture condition in the catchment, a rainfall depth of less than 20 mm can possibly cause the creek to overflow. The probability of such an event to occur is 0.10. If it rains more than 20 mm, the probability that the creek overflows is 0.90. Knowing the creek has overflowed, what is the probability a rainfall depth of more than 20 mm has occurred?

11. Determine the mean and variance of a geometric random variable with mass function given by $p_X(x) = p(1 - p)^{x-1}$, for $x = 1, 2, 3, \ldots$ and $0 \leq p \leq 1$.

12. Under which conditions is the statement $P(X \leq E[X]) = 0.50$ valid?

13. Show that $E[X^2] \geq (E[X])^2$.

14. If X and Z are random variables, show that

(a) $E\left(\frac{X - \mu_X}{\sigma_X}\right) = 0$;

(b) $\text{Var}\left(\frac{X - \mu_X}{\sigma_X}\right) = 1$; and

(c) $\rho_{X,Z} = \text{Cov}\left(\frac{X - \mu_X}{\sigma_X}, \frac{Z - \mu_Z}{\sigma_Z}\right)$

15. A simple random sample of 36 points has been drawn from a population of a Normal variable X, with $\mu_X = 4$ and $\sigma_X = 3$. Determine the expected value and the variance of the sample arithmetic mean.

16. The probability mass function of the binomial distribution is given by $p_X(x)$
$= \binom{N}{x} p^x (1-p)^{N-x}$, $x = 0, 1, 2, \ldots, N$. Employ the moment generating
function to calculate the mean and variance of the binomial variable with
parameters N and p. Remember that, according to Newton's binomial theorem,
$$(a+b)^N = \sum_{k=0}^{N} \binom{N}{k} a^k b^{N-k}.$$

17. X and Y are two independent random variables with densities respectively given
by $\lambda_1 \exp(-x\lambda_1)$ and $\lambda_2 \exp(-y\lambda_2)$, for $x \geq 0$ and $y \geq 0$. For these,

   (a) determine the MGF of $Z = X + Y$; and
   (b) determine the mean and variance of Z from the MGF.

18. Suppose the joint PDF of X and Y is $f_{X,Y}(x,y) = \frac{\exp(-x/y)\,\exp(-y)}{y}$; $0 < x < \infty$, $0 < y < \infty$.

   (a) Calculate $P(X < 2 | Y = 3)$;
   (b) calculate $P(Y > 3)$; and
   (c) determine $E[X | Y = 4]$.

19. Suppose that the rainfall's duration and intensity are respectively denoted by
X and Y and that their joint probability density function is given by
$f_{X,Y}(x,y) = [(a+cy)(b+cx) - c] \exp(-ax - by - cxy)$, for x, $y \geq 0$ and
parameters a, $b \geq 0$ and $0 \leq c \leq 1$. Suppose that $a = 0.07$ h^{-1}, $b = 1.1$ h/mm
and $c = 0.08$ mm^{-1} for a specific site. What is the probability that the intensity
of rainfall that lasts for 6 h will exceed 3 mm/h?

20. Suppose, in Exercise 19, that $c = 0$. For this specific case, show X and Y are
statistically independent.

21. Consider the PDF $f_X(x) = 0,35$, $0 \leq X \leq a$. (a) Find the PDF of $Y = \ln(X)$
and define the domain of Y. (b) Plot a graph of $f_Y(y)$ versus y.

22. An earth dam must have a freeboard above the maximum pool level so that
waves, due to wind action, cannot wash over the crest of the dam and start
eroding the embankment. According to Linsley et al. (1992), *wind setup is the
tilting of the reservoir water surface caused by the movement of the surface
toward the leeward shore under the action of wind*. Wind setup may be
estimated by $Z = FV^2/(1500\ d)$, where Z = rise above the still-water level
in cm; V = wind speed in km/h; F = fetch or length of water surface over which
wind blows in m; and d = average depth of the reservoir along the fetch in
m. (a) If wind speed V is an exponential random variable with mean v_0, for
$v \geq 0$, determine the PDF of Z. (b) If $v_0 = 30$ km/h, $F = 300$ m, and $d = 10$ m,
calculate $P(Z > 30$ cm).

23. The PDF of a Gamma distribution, with parameters α and λ, is given by
$$f_X(x) = \frac{\lambda^\alpha x^{\alpha-1} \exp(-\lambda x)}{\Gamma(\alpha)} \ , \quad x, \alpha, \lambda > 0, \quad \text{where} \quad \Gamma(\alpha) = \int_0^\infty t^{\alpha-1} \exp(-t)\,dt$$

denotes the Gamma function [see Appendix 1 for a brief review on the properties of $\Gamma(.)$]. Suppose X and Y are independent Gamma random variables with parameters (α_1, λ_1) and (α_2, λ_2), respectively. Determine the joint PDF and the marginal densities of $U = X + Y$ and $V = X/(X + Y)$.

24. Suppose that from all 2-h duration rainfalls over a region, 55 % of them are produced by local convection, whereas 45 % by the passing of frontal systems. Let X denote the rainfall intensity for both types of rain-producing mechanisms. Assume intensities of both rainfall types are exponentially distributed with parameters $\theta = 15$ mm/h, for convective storms, and $\theta = 8$ mm/h, for frontal rains. (a) Determine and plot the PDF of X. (b) Calculate $P(X > 25$ mm/h$)$.

References

Ang H-SA, Tang WH (1975) Probability concepts in engineering planning and design, volume I: basic principles. Wiley, New York

Beckmann P (1968) Elements of applied probability theory. Harcourt, Brace and World, Inc., New York

Johnson NL, Kotz S, Balakrishnan N (1994) Continuous univariate distribution, vol 1. Wiley, New York

Kolmogorov AN (1933) Grundbegriffe der wahrscheinlichkeitrechnung. Ergbnisse der Mathematik. English translation: Kolmogorov AN (1950) Foundations of the theory of probability. Chelsea Publ. Co., New York

Kottegoda NT, Rosso R (1997) Statistics, probability, and reliability for civil and environmental engineers. McGraw-Hill, New York

Larsen RJ, Marx ML (1986) An introduction to mathematical statistics and its applications. Prentice-Hall, Englewood Cliffs (NJ)

Linsley RK, Franzini JB, Freyberg DL, Tchobanoglous G (1992) Water resources engineering, 4th edn. McGraw-Hill, New York

Papoulis A (1991) Probability, random variables, and stochastic processes, 3rd edn. McGraw-Hill, New York

Ross S (1988) A first course in probability, 3rd edn. Macmillan, New York

Rowbottom D (2015) Probability. Wiley, New York

Shahin M, van Oorschot HJL, de Lange SJ (1993) Statistical analysis in water resources engineering. Balkema, Rotterdam

Westfall PH (2014) Kurtosis as peakedness, 1905-2014. Am Stat 68(3):191–195

Chapter 4
Discrete Random Variables: Probability Distributions and Their Applications in Hydrology

Mauro Naghettini

4.1 Bernoulli Processes

Consider a random experiment with only two possible and dichotomous outcomes: "success," designated by the symbol S, and "failure," denoted by F. The related sample space is given by the set {S,F}. Such a random experiment is known as a Bernoulli trial. Suppose a random variable X is associated with a Bernoulli trial, so that $X = 1$, if the outcome is S, or $X = 0$, in case it is F. Suppose further that the probability of a success occurring is $P(X = 1) = p$, which inevitably implies that $P(X - 0) = 1 - p$. Under these assumptions, X defines a Bernoulli random variable, whose probability mass function (PMF) is given by

$$p_X(x) = p^x(1 - p)^{1-x}, \text{ for } x = 0, 1 \text{ and } 0 \le p \le 1 \qquad (4.1)$$

with expected value and variance respectively given by $E[X] = p$ and $\text{Var}[X] = p(1-p)$.

Now, in a more general context, suppose the time scale in which a hypothetical stochastic process evolves has been discretized into fixed-width time intervals, as, for instance, into years, indexed by $i = 1, 2, \ldots$ Also, suppose that within each time interval, either one "success" occurs, with probability p, or one "failure" occurs with probability $(1-p)$ and, in addition, that these probabilities are not affected by previous occurrences and do not vary with time. Such a discrete-time stochastic process, made of a sequence of independent Bernoulli trials, is referred to as a Bernoulli process, named after the Swiss mathematician Jakob Bernoulli (1654–1705), whose book *Ars Conjectandi* had great influence on the early developments of the calculus of probabilities and combinatorics.

M. Naghettini (✉)
Universidade Federal de Minas Gerais, Belo Horizonte, Minas Gerais, Brazil
e-mail: mauronag@superig.com.br

© Springer International Publishing Switzerland 2017
M. Naghettini (ed.), *Fundamentals of Statistical Hydrology*,
DOI 10.1007/978-3-319-43561-9_4

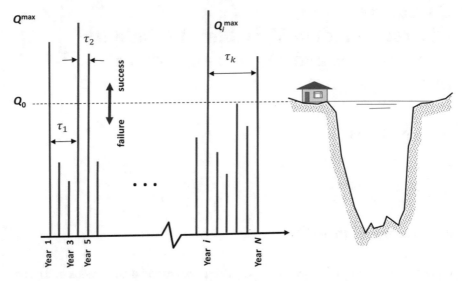

Fig. 4.1 Annual peak discharges as an illustration of a Bernoulli process

To better illustrate the application of Bernoulli processes in hydrology, consider the river cross-section depicted in Fig. 4.1, where discharge Q_0 corresponds to the flood stage or to the riverbanks stage, above which the rising waters start to flow onto the floodplains. For a given year, indexed by i, only the peak discharge Q_i^{max}, the largest among all instantaneous flows recorded in that particular time span, is selected to make one of the N sample points from the reduced hydrologic time series of annual peak discharges Q^{max}, shown in Fig. 4.1. For any given year i, $1 \leq i \leq N$, one can define as a "success" the event $\{S : Q_i^{max} > Q_0\}$ and as a "failure" its complement event $\{F : Q_i^{max} \leq Q_0\}$. The term "success," used in the context of a flood, is clearly a misnomer because it is certainly an undesirable event. Nonetheless, as it is conventionally and extensively employed in statistical literature, for the sake of clarity, hereinafter, we will continue referring to the occurrence of a flood as a success, being one of the possible outcomes of a Bernoulli trial.

As regarding the premise of statistical independence among time-sequential events, from the very nature of flood-producing mechanisms, it is fairly reasonable to admit as true the hypothesis that the probability of a success or a failure occurring in any given year is not affected by what has occurred in preceding years. Furthermore, to fulfill the requirements for such an independent sequence to be considered a Bernoulli process, it suffices to admit the annual probability for the event $\{S : Q_i^{max} > Q_0\}$ to occur is time-invariant and equal to p.

As associated with Bernoulli processes, one distinguishes the following discrete random variables, generically designated by Y:

1. The variable is said to be *binomial* if Y refers to the number of "successes" in N independent trials;

2. The variable is *geometric* if Y refers to the number of independent trials necessary for one single success to occur; and
3. The variable is negative binomial if Y refers to the number of independent trials necessary for r successes to occur.

The probability distributions associated with these three discrete random variables are detailed in the sections that follow.

4.1.1 Binomial Distribution

Consider a random experiment consisting of a sequence of N independent Bernoulli trials. For each trial, the probability of a success S occurring is constant and equal to p, so that the probability of a failure F is $(1-p)$. The sample space of such an experiment consists of 2^N points, each corresponding to one of all possible combinations formed by grouping the S's and F's outcomes from the N trials (see Appendix 1 for a brief review on elementary combinatorics). For a single trial, the Bernoulli variable, denoted by X, will assume the value $X=1$, if the outcome is a success, or $X=0$, if it is a failure. If N trials take place, a randomly selected realization could possibly be made of the sequence {S, F, S, S, ... , F, F}, for instance, which would result in $X_1=1$, $X_2=0$, $X_3=1$, $X_4=1$, ... , $X_{N-1}=0$, $X_N=0$. This setup fully characterizes the Bernoulli process.

Based on the described Bernoulli process, consider the discrete random variable Y representing the number of successes that have occurred among the N trials. It is evident that Y can take on any values from 0, 1, ... , N and also that $Y = \sum_{i=1}^{N} X_i$. As resulting from the independence assumption among the Bernoulli trials, each point with y successes and $(N-y)$ failures, in the sample space, may occur with probability $p^y(1-p)^{N-y}$. However, the y successes and the $(N-y)$ failures can possibly be combined from $N!/[y!\,(N-y)!]$ different ways, each with probability $p^y(1-p)^{N-y}$. Therefore, the PMF of Y can be written as

$$p_Y(y) = \frac{N!}{y!\,(N-y)!}\, p^y\,(1-p)^{N-y} = \binom{N}{y}\, p^y\,(1-p)^{N-y},$$

$$y = 0,\,1,\ \ldots,N \text{ and } 0 < p < 1 \tag{4.2}$$

which is known as the binomial distribution, with parameters N and p. Note that the Bernoulli distribution is a particular case of the binomial, with parameters $N=1$ and p. The PMFs for the binomial distribution with parameters $N=8$, $p=0.3$, $p=0.5$, and $p=0.7$ are depicted in the charts of Fig. 4.2. It is worth noting in this figure that the central location and shape of the binomial PMF experience significant changes as parameter p is altered while N is kept constant.

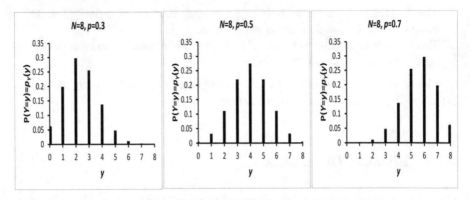

Fig. 4.2 Examples of mass functions for the binomial distribution

The binomial cumulative distribution function (CDF) gives the probability that Y is equal to or less than the argument y. It is defined as

$$F_Y(y) = \sum_{i=0}^{y} \binom{N}{i} p^i (1-p)^{N-i}, \quad y = 0, 1, 2, \ldots, N \tag{4.3}$$

The expected value, variance, and coefficient of skewness for a binomial variable Y (see Exercise 16 of Chap. 3) are as follows:

$$E[Y] = Np \tag{4.4}$$

$$\mathrm{Var}[Y] = Np(1-p) \tag{4.5}$$

$$\gamma = \frac{1-2p}{\sqrt{Np(1-p)}} \tag{4.6}$$

The binomial PMF is symmetrical for $p = 0.5$, right-skewed for $p < 0.5$, and left-skewed- for $p > 0.5$, as illustrated in Fig. 4.2.

Example 4.1 Counts of *Escherichia Coli* for 10 water samples collected from a lake, as expressed in hundreds of organisms per 100 ml of water ($10^2/100$ ml), are 17, 21, 25, 23, 17, 26, 24, 19, 21, and 17. The arithmetic mean value and the variance calculated for the 10 samples are respectively equal to 21 and 10.6. Suppose N represents the number of all different organisms that are present in a sample (in analogy to $N =$ the number of Bernoulli trials) and let p denote the fraction of N that corresponds to *E. Coli* (in analogy to $p =$ the probability of success). If X is the number of *E. Coli*, in ($10^2/100$ ml), estimate $P(X = 20)$ (adapted from Kottegoda and Rosso 1997).

Solution In this case, the true population values for the mean and the variance are not known, but can be estimated by their corresponding sample values, or

$\hat{\mu}_Y = \bar{y}$ and $\hat{\sigma}_Y^2 = S_y^2$, where the symbol "$^\wedge$" indicates "estimate." By making $(1-p)$ explicit in Eq. (4.5), it follows that $1 - p = \frac{\text{Var}[Y]}{Np} = \frac{\text{Var}[Y]}{E[Y]} \Rightarrow 1 - \hat{p}\,\frac{S_y^2}{\bar{y}}$ $= \frac{10.6}{21} = 0.505 \Rightarrow \hat{p} = 0.495$. As $E[Y] = Np$, N can be estimated by $(21/0.495) =$ 43. Finally $P(y = 20) = p_Y(20) = \binom{43}{20} 0.495^{20} 0.505^{23} = 0.1123$. This example shows that Bernoulli processes and the binomial distribution are not restricted to a discretized time scale, but are also applicable to a space scale or to generic trials that can possibly yield only one of two possible dichotomous outcomes.

Example 4.2 In the situation depicted in Fig. 4.1, suppose that $N = 10$ years and that the probability of Q_0 being exceeded by the annual peak flow, in any given year, is $p = 0.25$. Answer the following questions: (a) what is the probability that Q_0 will be exceeded in exactly 2 of the next 10 years? and (b) what is the probability that Q_0 will be exceeded at least in 2 of the next 10 years?

Solution It is easy to see that the situation illustrated in Fig. 4.1 conforms perfectly to a discrete-time Bernoulli process and also that the variable $Y = number\ of$ *"successes" in N years* is a binomial random variable. (a) The probability that Q_0 will be exceeded in exactly 2 of the next 10 years can be calculated directly from Eq. (4.2), or $p_Y(2) = \frac{10!}{2!8!} 0.25^2 (1 - 0.25)^8 = 0.2816$. (b) The probability that Q_0 will be exceeded at least in 2 of the next 10 years can be calculated by adding up the respective probabilities that Q_0 will be exceeded in exactly 2, 3, 4, $\ldots$, 10 of the next 10 years. However, this is equivalent to the complement, with respect to 1, of the summation of the respective probabilities of exactly 1 success and no success in 10 years. Therefore $P(Y \geq 2) = 1 - P(Y < 2) = 1 - p_Y(0) - p_Y(1) = 0.7560$.

The binomial distribution exhibits the additive property, which means that if Y_1 and Y_2 are binomial variables, with parameters respectively equal to (N_1, p) and (N_2, p), then the variable $(Y_1 + Y_2)$ will be also binomial, with parameters $(N_1 + N_2, p)$. The additive property can be extended to three or more binomial variates. The term *variate* applies to the case in which the probability distribution of a random variable is known or specified, and is extensively used henceforth. Haan (1977) points out that another important characteristic of Bernoulli processes, in general, and of binomial variates, in particular, is that the probability of any combination of successes and failures, for a sequence of N trials, does not depend on the time scale origin, from which the outcomes are being counted. This is derived from the assumption of independence among distinct trials and also from the premise of a time-constant probability of success p.

4.1.2 Geometric Distribution

For a Bernoulli process, the geometric random variable Y is defined as the number of trials necessary for one single success to occur. Hence, if the variable takes on the value $Y = y$, that means that $(y-1)$ failures took place before the occurrence of a success exactly in the yth trial. The mass and cumulative probability functions for the geometric distribution are respectively defined as in the following equations:

$$p_Y(y) = p(1-p)^{y-1}, \quad y = 1, 2, 3, \ldots \text{ and } 0 < p < 1 \tag{4.7}$$

$$P_Y(y) = \sum_{i=1}^{y} p(1-p)^{i-1}, \quad y = 1, 2, 3, \ldots \tag{4.8}$$

where the probability of success p denotes its single parameter.

The expected value of a geometric variate can be derived as follows:

$$E[Y] = \sum_{y=1}^{\infty} y p(1-p)^{y-1} = p \sum_{y=1}^{\infty} y(1-p)^{y-1}$$

$$= p \sum_{y=1}^{\infty} \frac{d}{d(1-p)}(1-p)^y = p \frac{d}{d(1-p)} \sum_{y=1}^{\infty}(1-p)^y \tag{4.9}$$

As for the previous equation, it can be shown that the sum of the infinite geometric series $\sum_{y=1}^{\infty}(1-p)^y$, for $0 < p < 1$, with both first term and multiplier equal to $(1-p)$, converges to $(1-p/p)$. Thus, substituting this term back into Eq. (4.9) and taking the derivative, with respect to $(1-p)$, it follows that:

$$E[Y] = \frac{1}{p} \tag{4.10}$$

Therefore, the expected value of a geometric variate is the reciprocal of the probability of success p of a Bernoulli trial. The variance of a geometric variate can be derived by similar mathematical artifice and is given by

$$\text{Var}[Y] = \frac{1-p}{p^2} \tag{4.11}$$

The coefficient of skewness of a geometric distribution is

$$\gamma = \frac{2-p}{\sqrt{1-p}} \tag{4.12}$$

The geometric PMFs with parameters $p = 0.3$, $p = 0.5$, and $p = 0.7$ are illustrated in Fig. 4.3.

Taking advantage of the notional scenario depicted in Fig. 4.1, we shall now introduce a concept of great importance in hydrology, which is the *return period*. In Fig. 4.1, consider the number of years between consecutive successes, denoted by

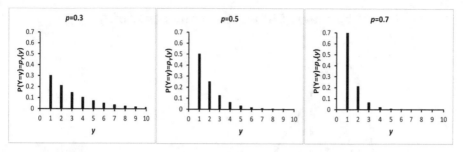

Fig. 4.3 Examples of mass functions for the geometric distribution

the variable τ and named herein as the *recurrence time interval*. Thus, with reference to Fig. 4.1 and placing the time scale origin at the year with the first success, $\tau_1 = 3$ years are needed for the event $\left\{ S : Q_{i=4}^{max} > Q_0 \right\}$ to recur. Thereafter, from the year of the second success, $\tau_2 = 2$ years are counted until the next success, and so forth up to $\tau_k = 5$ years of recurrence time interval. If we add the supposition that, for instance, $N = 50$ years and that 5 successes have occurred during this time span, the mean value for the recurrence time interval would be $\bar{\tau} = 10$ years, implying that, on average, the discharge Q_0 is exceeded once at every 10-year period.

It is evident that the variable *recurrence time interval* perfectly fits the definition of a geometric random variable and, as such, Eqs. (4.10) and (4.11) should apply. In particular, for Eq. (4.10), the return period, denoted by T and given in years, is defined as the expected value of the geometric variate τ, the recurrence time interval. Formally,

$$T = E[\tau] = \frac{1}{p} \tag{4.13}$$

Thus, the return period T does not refer to a *physical time*. In fact, T is a measure of central tendency of the physical times, which were termed in here as the recurrence time intervals. In other words, the return period T, associated with a specific reference event defining a success in a yearly indexed Bernoulli process, corresponds to the mean time interval, in years, necessary for the event to occur, which might take place in any given year, and is equal to the reciprocal of the annual probability of success.

In hydrology, the return period concept is frequently employed in the probabilistic modeling of annual maxima, such as the annual maximum daily rainfalls, and annual means, such as the annual mean flows. These are continuous random variables described by probability density functions (PDF), such as the one depicted in Fig. 4.4. If, as referring to the X variable in Fig. 4.4, a reference quantile x_T is defined so that the "success" represents the exceedance of X over x_T, then, the return period T is the average number of years necessary for the event $\{X > x_T\}$ to occur once, in any given year. From Eq. (4.13), the return period is the reciprocal of $P(X > x_T)$, the hatched area in Fig. 4.4.

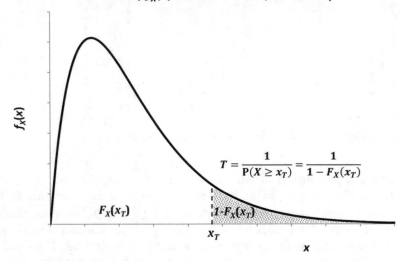

The density $f_X(x)$ and the return period of x_T

$$T = \frac{1}{P(X \geq x_T)} = \frac{1}{1 - F_X(x_T)}$$

$F_X(x_T)$ $1 - F_X(x_T)$

x_T

x

Fig. 4.4 Graphical representation of the return period for annual maxima

Example 4.3 The annual maximum daily rainfalls, denoted by X, are exponentially distributed with parameter $\theta = 20$ mm (see Example 3.7). Determine (a) the return period of $x_T = 60$ mm; and (b) the maximum daily rainfall of return period $T = 50$ years.

Solution (a) The CDF of an exponential variate X is $F_X(x) = 1 - \exp\left(-\frac{x}{\theta}\right)$, as derived in Example 3.7. For $\theta = 20$ mm, at $X = x_T = 60$ mm, the corresponding CDF value is $F_X(x_T) = 1 - \exp\left(-\frac{60}{20}\right) = 0.9502$. $F_X(x)$ and T are related by $T = \frac{1}{1 - F_X(x_T)}$ and thus, the return period of $x_T = 60$ mm is $T = 20$ years. (b) For $T = 50$ years, the corresponding CDF value is 0.98. The quantile function $x(F)$ is the inverse of $F_X(x)$, or $x(F) = F_X^{-1}(x) = -\theta\ln(1 - F)$. Thus, for $F = 0.98$ and $\theta = 20$ mm, the sought maximum daily rainfall is $x_{T=50} = 78.24$ mm.

An important extension of the return period concept refers to the definition of hydrologic risk of failure, as employed in the design of hydraulic structures for flood mitigation. On the basis of a reference quantile x_T, of return period T, the hydrologic risk of failure is defined as the probability that x_T be equaled or exceeded at least once in an interval of N years. In general, the reference quantile x_T corresponds to the design flood of a given hydraulic structure, whereas the interval of N years relates to the structure's expected service life. One of the possible ways to derive the expression for the hydrologic risk of failure, here denoted by R, makes use of the binomial distribution. In effect, the probability that the event $\{X \geq x_T\}$ will occur at least once in a period of N years and ultimately cause the structure to fail, is equivalent to the complement, with respect to 1, of the probability that it will not occur during this time interval. Therefore, using the notation Y for the number of times the event $\{X \geq x_T\}$ will occur in N years, from the Eq. (4.2), one can write

$$R = P(Y \geq 1) = 1 - P(Y = 0) = 1 - \binom{N}{0} p^0 (1-p)^{N-0} \qquad (4.14)$$

For the reference quantile x_T, of return period T, the probability p that the event $\{X \geq x_T\}$ will occur in any given year is, by definition, $p = 1/T$. Substituting this result into Eq. (4.14), then, it follows the formal definition for the hydrologic risk of failure as

$$R = 1 - \left(1 - \frac{1}{T}\right)^N \qquad (4.15)$$

If the hydrologic risk of failure is previously fixed, as a function of the importance and dimensions of the hydraulic structure, as well as of the expected consequences its eventual collapse would have for the populations, communities, and properties located in the downstream valley, one can make use of Eq. (4.15) to determine for which return period the design flood quantile should be estimated. For instance, in the case of a dam, whose service life is expected to be N years and that entails a hydrologic risk of failure R, the return period T of the design flood of the dam spillway should be derived from the application of Eq. (4.15). The graph of Fig. 4.5 facilitates and illustrates such a possible application of Eq. (4.15).

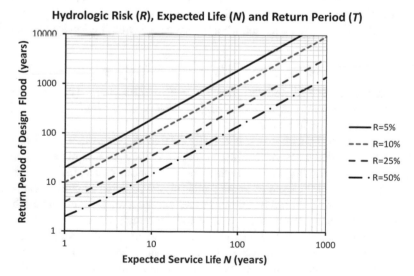

Fig. 4.5 Return period of the design flood as a function of the hydrologic risk of failure and of the expected service life of the hydraulic structure

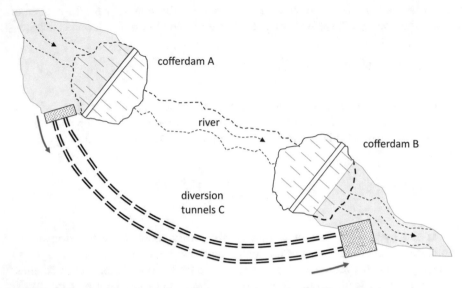

Fig. 4.6 River diversion scheme for a dam construction, as used in Example 4.4

Example 4.4 Figure 4.6 shows a sketch of the diversion scheme of a river during the projected construction of a dam. Two cofferdams, marked as A and B in this figure, should protect the dam construction site against flooding, during the projected construction schedule, as the river is diverted from its natural course into tunnels C, carved through rocks along the right riverbank. Suppose the civil engineering works will last 5 years and that the construction company has fixed as 10 % the risk that the dam site will be flooded at least once during this time schedule. Based on these elements, what should be the return period for the tunnels' design flood?

Solution Inversion of Eq. (4.15), for T, results in $T = \frac{1}{1-(1-R)^{1/N}}$. With $R = 0.10$ and $N = 5$ years, the inverted equation yields $T = 47.95$ years. Therefore, in this case, the tunnels C must have a cross section capable of conveying a design-flood discharge of return period equal to 50 years.

The return period concept is traditionally related to annual maxima, but it can certainly be extended to the probabilistic modeling of annual minima and of annual mean values as well. For the latter case, no substantial conceptual changes are needed. However, in the case of annual minima, the success, as conceptualized in the Bernoulli process, should be adapted to reflect the annual flows that are below some low threshold z_T. Thus, the return period for annual minima must be understood as the average time interval, in years, necessary for the event $\{Z < z_T\}$, a drought even more severe than z_T, to recur, in any given year. Supposing that Z represents a continuous random variable, characteristic of the annual drought flow, such as the annual minimum 7-day mean discharge (Q_7), it is verified that the return period T, associated with the reference quantile z_T, must correspond to the

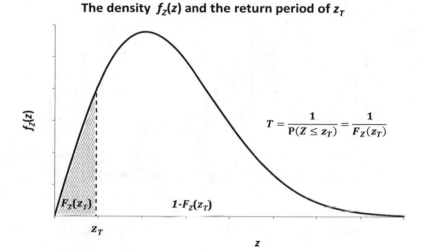

The density $f_Z(z)$ and the return period of z_T

$$T = \frac{1}{P(Z \leq z_T)} = \frac{1}{F_Z(z_T)}$$

$F_Z(z_T)$ $1-F_Z(z_T)$

z_T

z

$f_Z(z)$

Fig. 4.7 Extension of the return period concept to annual minima

reciprocal value of $P(Z < z_T)$ or the reciprocal of $F_Z(z_T)$. Figure 4.7 illustrates the extension of the return period concept to the case of annual minima, by means of a hypothetical density function $f_Z(z)$.

The geometric distribution has the special property of *memorylessness*. That is to say that, if m failures have already occurred after m or more trials, the distribution of the total number of trials $(m + n)$ before the occurrence of the first success will not be changed. In fact, by equating the conditional probability $P(Y \geq m + n | Y \geq m)$ and using the convergence properties of infinite geometric series, with both first term and multiplier equal to $(1-p)$, it immediately follows that $P(Y \geq m + n | Y \geq m) = P(Y \geq n)$.

4.1.3 Negative Binomial Distribution

Still referring to a Bernoulli process, the random variable Y is defined as a negative binomial (or a Pascal) if it counts the number of trials needed for the occurrence of exactly r successes. Its mass function can be derived by thinking of the intersection of two independent events: A, referring to the occurrence of the rth success at the yth trial, with $y \geq r$, and B referring to $(r-1)$ successes that have occurred in the previous $(y-1)$ trials. By definition of a Bernoulli process, event A may occur with probability p, in any trial. As regards the event B, its probability is given by the binomial distribution applied to $(r-1)$ successes in $(y-1)$ trials, or $P(B) = \binom{y-1}{r-1} p^{r-1}(1-p)^{y-r}$. Therefore, by calculating $P(A \cap B) = P(A)$ $P(B)$, it follows that

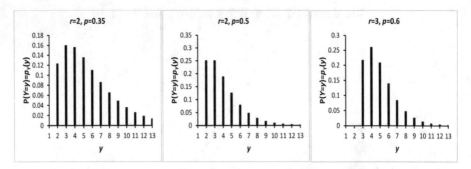

Fig. 4.8 Examples of mass functions for the negative binomial distribution

$$p_Y(y) = \binom{y-1}{r-1} p^r (1-p)^{y-r}, \text{ with } y = r, r+1, \ldots \quad (4.16)$$

Equation (4.16) gives the mass function of the negative binomial distribution, with parameters r and p. Some examples of PMFs for a negative binomial variate are shown in Fig. 4.8. As the negative binomial actually derives from the sum of r independent geometric variables, it is straightforward to show, using the properties of mathematical expectation, that its expected value and variance are respectively given by

$$E[Y] = \frac{r}{p} \quad (4.17)$$

and

$$\text{Var}(Y) = \frac{r(1-p)}{p^2} \quad (4.18)$$

Example 4.5 To return to the river diversion scheme in Example 4.4, suppose tunnels C have been designed for a flood of return period 10 years. Answer the following questions: (a) what is the probability that the second flood onto the dam site will occur in the 4th year after construction works have begun? (b) what is the hydrologic risk of failure for this new situation?

Solution (a) The probability the dam site will be flooded for a second time in the fourth year of construction can be calculated directly from Eq. (4.16), with $r = 2$, $y = 4$, and $p = 1/T = 0,10$, or $p_Y(4) = \binom{4-1}{2-1} 0.1^2 0.9^{4-2} = 0.0243$. (b) The hydrologic risk of failure for the new situation, with $N = 5$ and $T = 10$, is $R = 1 - \left(1 - \frac{1}{T}\right)^N = 1 - 0.90^5 = 0.41$ which is, therefore, unacceptably high.

4.2   Poisson Processes

The *Poisson processes* are among the most important stochastic processes. They are addressed in this section as a limiting case of a Bernoulli process that evolves over time, although the arguments can be extended to a length, an area or a volume. Following Shahin et al. (1993), consider a time span t, which is subdivided into N nonoverlapping subintervals, each of length t/N. Suppose that each subinterval is sufficiently short that a given event (or success) can occur at most once in the time interval t/N and that the probability of more than one success occurring within it is negligible. Consider further that the probability a success occurring in t/N is p, which is then supposed constant for each subinterval and is not affected by the occurrences in other nonoverlapping subintervals. Finally, suppose that the mean number of successes that have occurred in a time span t, are proportional to the length of the time span, the proportionality constant being equal to λ. Under these conditions, one can write $p = \lambda t/N$.

The number of occurrences (or successes) Y in the time span t is equal to the number of subintervals within which successes have occurred. If these subintervals are seen as a sequence of N independent Bernoulli trials, then

$$p_Y(y) = \binom{N}{y} \left(\frac{\lambda t}{N}\right)^y \left(1 - \frac{\lambda t}{N}\right)^{N-y} \tag{4.19}$$

In this equation, if $p = \lambda t/N$ is sufficiently small and N sufficiently large, so that $Np = \lambda t$, then it can be shown that

$$\lim_{N \to \infty} \binom{N}{y} \left(\frac{\lambda t}{N}\right)^y \left(1 - \frac{\lambda t}{N}\right)^{N-y} = \frac{(\lambda t)^y}{y!} e^{-\lambda t}, \text{ for } y = 0, 1, \ldots \text{ and } \lambda t > 0 \tag{4.20}$$

Making $\nu = \lambda t$ in Eq. (4.20), one finally gets the Poisson PMF as given by

$$p_Y(y) = \frac{\nu^y}{y!} e^{-\nu}, \text{ for } y = 0, 1, \ldots \text{ and } \nu > 0 \tag{4.21}$$

where parameter ν denotes the mean number of occurrences in the time span t. The Poisson distribution is named after the French mathematician and physicist Siméon Denis Poisson (1781–1840). The Poisson occurrences are often referred to as *arrivals*.

The Poisson CDF is written as

$$P_Y(y) = \sum_{i=0}^{y} \frac{\nu^i}{i!} e^{-\nu}, \text{ for } y = 0, 1, \ldots \tag{4.22}$$

As shown in Example 3.14 of Chap. 3, the mean and variance of a Poisson variate are respectively given by

$$E[Y] = \nu \text{ or } E[Y] = \lambda t \tag{4.23}$$

$$\text{Var}[Y] = \nu \text{ or } \text{Var}[Y] = \lambda t \tag{4.24}$$

Similarly to the mathematical derivation of $E[Y]$ and $\text{Var}[Y]$, it can be shown that the coefficient of skewness of a Poisson distribution is

$$\gamma = \frac{1}{\sqrt{\nu}} \text{ or } \gamma = \frac{1}{\sqrt{\lambda t}} \tag{4.25}$$

Figure 4.9 gives some examples of Poisson mass functions.

The parameter ν represents both the mean number and the variance of Poisson arrivals in the time span t. The proportionality constant λ is usually referred to as the intensity of the Poisson process and represents the mean arrival rate per unit time. Although described as a limiting case of a discrete-time Bernoulli process, the Poisson process is more general and evolves over a continuous-time scale. If its parameters ν and λ are constant in time, the Poisson process is said to be homogeneous or stationary. Otherwise, for the nonhomogeneous Poisson processes, $\lambda(t)$ is a

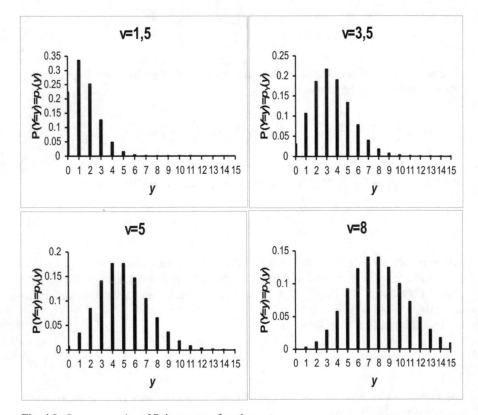

Fig. 4.9 Some examples of Poisson mass functions

function of time and the mean number of arrivals ν, in the time interval $[t_1, t_2]$, will be given by the integral of $\lambda(t)$ from t_1 to t_2.

It follows from the previous derivation of the Poisson distribution that it can be employed as an approximation to the binomial, provided N is sufficiently large and p sufficiently small. In practice, it is possible to approximate the binomial by the Poisson distribution, with parameter $\nu = Np$, from values as large as 20 for N and as small as 0.1 for p. In fact, provided the probability of success p is small enough ($p < 0.1$), it suffices to prescribe an average number of occurrences in a given time span. In analogy to the binomial, the Poisson distribution also has the additive property, meaning that if variables Y_1 and Y_2 are Poisson-distributed, with respective arrival rates λ_1 and λ_2, then $(Y_1 + Y_2)$ is also a Poisson variate, with parameter $(\lambda_1 + \lambda_2)$. Poisson processes are much more general and complex than the brief description given here, which was limited to introducing the Poisson discrete random variable. Readers interested in a broader description of Poisson processes may consult Ross (1989).

Example 4.6 Water transportation in rivers and canals make use of dam-and-lock systems for raising and lowering boats, ships, and barges between stretches of water that are not leveled. A lock is a concrete chamber, equipped with gates and valves for filling or emptying it with water, inside which a number of vessels are allowed to enter to complete the raising/lowering operation in an organized and timely manner. This operation is called locking through. Suppose that barges arrive at a lock at an average rate of 4 per hour. If the arrival of barges is a Poisson process, (a) calculate the probability that 6 barges will arrive in the next 2 h; and (b) if the lock master has just locked through all of the barges at the lock, calculate the probability she/he can take a 15-min break without another barge arriving (adapted from Haan 1977).

Solution (a) At the average rate (or intensity) of $\lambda = 4\,\mathrm{h}^{-1}$ and for $t = 2$ h $\Rightarrow \lambda t = \nu = 8$. Substituting values in Eq. (4.21), $P(Y = 6) = p_Y(6) = (8)^6 \frac{e^{-8}}{6!} = 0.1221$. (b) The lock master can take her/his 15-min break if no barge arrives in this time interval. The probability of no arrival in 0.25 h, at the rate of $\lambda = 4\,\mathrm{h}^{-1}$, for $t = 0.25$ h, so that $\lambda t = \nu = 1$, is given by $P(Y = 0) = p_Y(0) = (1)^0 \frac{e^{-1}}{0!} = 0.3679$.

4.3   Other Distributions of Discrete Random Variables

Other distributions of discrete random variables, that are useful for solving some hydrologic problems, are not directly related to Bernoulli and Poisson processes. These include the hypergeometric and the multinomial distributions, which are described in the subsections that follow.

4.3.1 Hypergeometric Distribution

Suppose a set of N items, from which A possess a given attribute a (for instance, of red color, or of positive sign, or of high quality, etc.) and $(N-A)$ possess the attribute b (for instance, of blue color, or of negative sign, or of low quality, etc.). Consider a sample of n items is drawn, without replacement, from the set of N items. Finally, consider that the discrete random variable Y refers to the number of items possessing attribute a, contained in the drawn sample of n items. The probability that Y will be equal to y items of the a type, is given by the hypergeometric distribution, whose mass function, with parameters N, A, and n, is expressed as

$$p_Y(y) = \frac{\binom{A}{y}\binom{N-A}{n-y}}{\binom{N}{n}}, \text{ with } 0 \le y \le A;\ y \le n;\ y \ge A - N + n \qquad (4.26)$$

The CDF for the hypergeometric distribution is

$$P_Y(y) = \sum_{i=0}^{y} \frac{\binom{A}{i}\binom{N-A}{n-i}}{\binom{N}{n}} \qquad (4.27)$$

The denominator of Eq. (4.26) gives the total number of possibilities of drawing a sample of size n, from the set of N items. The numerator, in turn, gives the number of possibilities of drawing samples of y items, of the a type, forcing the remaining $(n-y)$ items to be of the b type. It can be shown that the expected value and variance of a hypergeometric variate are respectively given by

$$E[Y] = \frac{nA}{N} \qquad (4.28)$$

and

$$\text{Var}[Y] = \frac{nA(N-A)(N-n)}{N^2(N-1)} \qquad (4.29)$$

If $n < 0.1\,N$, the hypergeometric distribution can successfully approximate a binomial distribution with parameters n and $p = A/N$.

Example 4.7 In February, 1935, 18 rainy days were counted among the daily records of a rainfall gauging station. Suppose the occurrence of a rainy day does not change the probability of it raining on the next day. If a sample of 10 days is selected at random, from the records of February 1935, (a) what is the probability that 7 out of the 10-day sample were rainy? (b) what is the probability that at least 6 were rainy days?

Solution

(a) Using Eq. (4.28), for the mass function of the hypergeometric distribution, with $N = 28$, $A = 18$, and $n = 10$ gives

$$p_Y(7) = \frac{\binom{18}{7}\binom{28-18}{10-7}}{\binom{28}{10}} = 0.2910$$

(b) The probability that at least 6 from the 10-day sample were rainy can be written as $P(Y \geq 6) = 1 - P(Y < 6) = 1 - P_Y(5)$, or $P(Y \geq 6) = 1 - p_Y(0) + p_Y(1) - p_Y(2) - p_Y(3) - p_Y(4) - p_Y(5) = 0.7785$.

4.3.2   Multinomial Distribution

The multinomial distribution is a generalization of the binomial, for the case where a random experiment can yield r distinct mutually exclusive and collectively exhaustive events $a_1, a_2, \ldots, a_r$, with respective probabilities of occurrence given by $p_1, p_2, \ldots, p_r$, such that $\sum p_i = 1$. The multinomial discrete variables are denoted by $Y_1, Y_2, \ldots, Y_r$, where Y_i represents the number of occurrences of the outcome related to a_i, in a sequence of N independent trials. The joint mass function of the multinomial distribution is given by

$$P(Y_1 = y_1, Y_2 = y_2, \ldots, Y_r = y_r) = p_{Y_1, Y_2, \ldots, Y_r}(y_1, y_2, \ldots, y_r)$$

$$= \frac{N!}{y_1! y_2! \ldots y_r!} p_1^{y_1} p_2^{y_2} \cdots p_r^{y_r} \qquad (4.30)$$

where $\sum y_i = N$ and $N, p_1, p_2, \ldots, p_r$ are parameters. The marginal mass function of each variable Y_i is a binomial with parameters N and p_i.

Example 4.8 At a given location, years are considered below normal (a_1) if their respective annual total rainfall depths are lower than 300 mm and normal (a_2) if the annual total rainfall depths lie between 300 and 1000 mm. Frequency analysis of annual rainfall records shows that the probabilities of outcomes a_1 and a_2 are, respectively, 0.4 and 0.5. Considering a randomly selected period of 15 years, calculate the probability that 3 below normal and 9 normal years will occur.

Solution In order to complete the sample space, it is necessary to define the third event, denoted by a_3, as corresponding to the above normal years, with annual total rainfall depths larger than 1000 mm. As events are collectively exhaustive, the probability of a_3 is $1-0.4-0.5=0.1$. Out of the 15 years, 3 should correspond to a_1 and 9 to a_2, thus only 3 remaining to event a_3. The sought probability is then given by

$$P(Y_1 = 3; Y_2 = 9; Y_3 = 3) = p_{Y_1,Y_2,Y_3}(3,9,3) = \frac{15!}{3!9!3!}0.4^3 0.5^9 0.1^3 = 0.0125.$$

4.4 Summary for Probability Distributions of Discrete Random Variables

What follows is a summary of the main characteristics of the six probability distributions of discrete random variables introduced in this chapter. Not all characteristics listed in the summary have been formally derived in the previous sections of this chapter, as one can use the mathematical principles that are common to all distribution to make the desired proofs. This summary is intended to serve as a brief reference item for the main probability distributions of discrete random variables.

4.4.1 Binomial Distribution

Notation: $Y \sim B(N,p)$
Parameters: N (positive integer) and p $(0 < p < 1)$
PMF: $p_Y(y) = \binom{N}{y} p^y (1-p)^{N-y}$, $y = 0, 1, \ldots, N$
Mean: $E[Y] = Np$
Variance: $\text{Var}[Y] = Np(1-p)$
Coefficient of skewness: $\gamma = \frac{1-2p}{\sqrt{Np(1-p)}}$
Coefficient of Kurtosis: $\kappa = 3 + \frac{1-6p(1-p)}{Np(1-p)}$
Moment Generating Function: $\phi(t) = (pe^t + 1 - p)^N$

4.4.2 Geometric Distribution

Notation: $Y \sim Ge(p)$
Parameters: p $(0 < p < 1)$
PMF: $p_Y(y) = p(1-p)^{y-1}$, $y = 1, 2, 3, \ldots$
Mean: $E[Y] = \frac{1}{p}$
Variance: $\text{Var}[Y] = \frac{1-p}{p^2}$

Coefficient of Skewness: $\gamma = \frac{2-p}{\sqrt{1-p}}$

Coefficient of Kurtosis: $\kappa = 3 + \frac{p^2-6p+6}{1-p}$

Moment Generating Function: $\phi(t) = \frac{p e^t}{1-(1-p)e^t}$

4.4.3   Negative Binomial Distribution

Notation: $Y \sim \text{NB}(r, p)$

Parameters: r (positive integer) and p $(0 < p < 1)$

PMF: $p_Y(y) = \binom{y-1}{r-1} p^r (1-p)^{y-r}, \quad y = r, r+1, \ldots$

Mean: $E[Y] = \frac{r}{p}$

Variance: $\text{Var}(Y) = \frac{r(1-p)}{p^2}$

Coefficient of Skewness: $\gamma = \frac{2-p}{\sqrt{r(1-p)}}$

Coefficient of Kurtosis: $\kappa = 3 + \frac{p^2-6p+6}{r(1-p)}$

Moment Generating Function: $\phi(t) = \left[\frac{p e^t}{1-(1-p)e^t} \right]^r$

4.4.4   Poisson Distribution

Notation: $Y \sim \text{P}(\nu)$

Parameters: ν $(\nu > 0)$

PMF: $p_Y(y) = \frac{\nu^y}{y!} e^{-\nu}, \quad y = 0, 1, \ldots$

Mean: $E[X] = \nu$

Variance: $\text{Var}[X] = \nu$

Coefficient of Skewness: $\gamma = \sqrt{\frac{1}{\nu}}$

Coefficient of Kurtosis: $\kappa = 3 + \frac{1}{\nu}$

Moment Generating Function: $\phi(t) = \exp[\nu(e^t - 1)]$

4.4.5   Hypergeometric Distribution

Notation: $Y \sim \text{H}(N, A, n)$

Parameters: N, A, and n (all positive integers)

PMF: $p_Y(y) = \frac{\binom{A}{y}\binom{N-A}{n-y}}{\binom{N}{n}}$, with $0 \le y \le A$; $y \le n$; $y \ge A - N + n$

Mean: $E[Y] = \frac{nA}{N}$

Variance: $\text{Var}[Y] = \frac{nA\,(N-A)\,(N-n)}{N^2\,(N-1)}$

Coefficient of Skewness: $\gamma = \frac{(N-2A)\,(N-2n)\,\sqrt{N-1}}{(N-2)\,\sqrt{nA(N-A)\,(N-n)}}$

Kurtosis: $\kappa = \left[\frac{N^2(N-1)}{n\,(N-2)\,(N-3)\,(N-n)}\right]\left[\frac{N\,(N+1)-6N(N-n)}{A\,(N-A)} + \frac{3n\,(N-n)\,(N+6)}{N^2} - 6\right]$

Moment Generating Function: no analytic form

4.4.6 Multinomial Distribution

Notation: $Y_1, Y_2, \ldots, Y_r \sim M\,(N, p_1, p_2, \ldots, p_r)$

Parameters: $N, y_1, y_2, \ldots, y_r$ (all positive integers) and $p_1, p_2, \ldots, p_r$ ($p_i > 0$ and $\sum p_i = 1$)

PMF: $p_{Y_1, Y_2, \ldots, Y_r}(y_1, y_2, \ldots, y_r) = \frac{N!}{y_1! y_2! \ldots y_r!} p_1^{y_1} p_2^{y_2} \cdots p_r^{y_r}$

Mean (of marginal PMF): $E[Y_i] = N p_i$

Variance (of marginal PMF): $\text{Var}[Y_i] = N p_i\,(1 - p_i)$

Coefficient of Skewness (of marginal PMF): $\gamma\,(Y_i) = \frac{1-2p_i}{\sqrt{N p_i\,(1-p_i)}}$

Coefficient of Kurtosis (of marginal PMF): $\kappa\,(Y_i) = 3 + \frac{1-6p_i\,^i(1-p_i)}{N p_i(1-p_i)}$

Moment Generating Function: $\varphi(t) = \left[\sum_{i=1}^{r} p_i\,e^{t_i}\right]^N$

Exercises

1. Consider a binomial distribution, with $N = 20$ and $p = 0.1$, and its approximation by a Poisson distribution with $\nu = 2$. Plot the two mass functions and comment on the differences between them.

2. Solve Exercise 1 with (a) $N = 20$ and $p = 0.6$; and (b) with $N = 8$ and $p = 0.1$.

3. Suppose the daily concentrations of a pollutant, in a river reach, are statistically independent. If 0.15 is the probability that the concentration exceeds 6 mg/m^3 on any given day, calculate (a) the probability the concentration will exceed 6 mg/m^3 in exactly two of the next 3 days; and (b) the probability the concentration will exceed 6 mg/m^3 for a maximum of two of the next three days.

4. If a marginal embankment has been designed to withstand the 20-year return period flood, calculate (a) the probability that the area protected by the embankment will be flooded at least once in the next 10 years; (b) the probability the protected area will be flooded at least three times in the next 10 years;

Exercise 6 - Levees near the confluence of rivers A and B

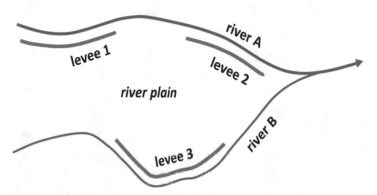

Fig. 4.10 Sketch of levees to protect against floods in between-rivers plains

and (c) the probability the protected area will be flooded no more than three times in the next 10 years.

5. Suppose the expected service life of a detention pond is 25 years. (a) What should be the return period for the design flood such that there is a 0.9 probability it will not be exceeded during the detention-pond expected service life? (b) What should be the return period for the design flood such that there is a 0.75 probability it will be exceeded at most once during the detention-pond expected service life?

6. The locations of three levees built along the banks of rivers A and B, to control floods in the plains between rivers, are shown in the sketch of Fig. 4.10. The levees have been designed according to the following: the design flood for levee 1 has a return period of 10 years; for levee 2, 20 years, and for levee 3, 25 years. Supposing that flood events in the two rivers and the occurrence of failures in levees 1 and 2 are statistically independent, (a) calculate the annual probability the plains between rivers will be flooded, due exclusively to floods from river A; (b) calculate the annual probability the plains between rivers will be flooded; (c) calculate the annual probability the plains between rivers will not be flooded in 5 consecutive years; and (d) considering a period of 5 consecutive years, calculate the probability that the third inundation of the plains will occur in the 5th year (adapted from Ang and Tang 1975).

7. Consider that a water treatment plant takes raw water directly from a river through a simply screened intake installed at a low water level. Suppose the discrete random variable X refers to the annual number of days the river water level is below the intake's level. Table 4.1 shows the empirical frequency distribution of X, based on 20 years of observations. (a) Assuming the expected value of X can be well approximated by the sample arithmetic average, estimate the parameter ν of a Poisson distribution for X. (b) Plot on the same chart the empirical and the Poisson mass functions for X and comment on the differences

Table 4.1 Empirical frequencies for the annual number of days the intake remains dry

$x \rightarrow$	0	1	2	3	4	5	6	7	8	>8
$f(X=x)$	0.0	0.06	0.18	0.2	0.26	0.12	0.09	0.06	0.03	0.0

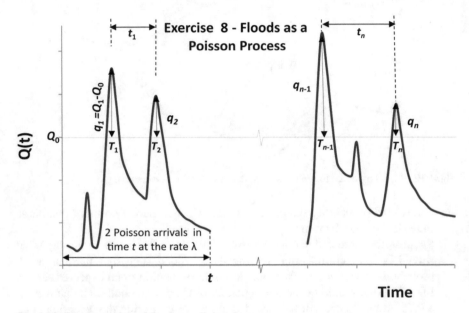

Fig. 4.11 Representation of floods as a Poisson process

between them. (c) Employ the Poisson distribution to calculate $P(3 \leq X \leq 6)$. (d) Employ the Poisson distribution to calculate $P(X \geq 8)$.

8. Flood hydrographs are characterized by a quick rising of discharges up to the flood peak, followed by a relatively slower flow recession, until the occurrence of a new flood, and so forth, as depicted in the graph of Fig. 4.11. In this figure, suppose a high threshold Q_0 is defined and that the differences $q_i = Q_i\text{-}Q_0$, between the flood hydrograph peaks Q_i and Q_0, are referred to as exceedances over the threshold. Under some conditions, it is possible to show that the exceedances over a high threshold follow a Poisson process (Todorovic and Zelenhasic 1970). This is actually the most frequent representation for constructing flood models for partial duration series, or peaks-over-threshold (POT) models, to be detailed in Chap. 8. In this representation, the number of exceedances over Q_0, during a time Δt, is a discrete random variable and Poisson-distributed with constant arrival rate λ. However, the variable t, denoting the time elapsed between two consecutive Poisson arrivals, as exemplified by the realization t_1 in Fig. 4.11, is continuous and non-negative. Show that the probability distribution of t is exponential, with density function $f_T(t) = \lambda \exp(-\lambda t)$.

9. With reference to Fig. 4.11 and the solution to Exercise 8, it is also possible to derive the probability distribution for the continuous time t necessary to the arrival of the nth Poisson occurrence. This can be done by noting that $t = t_1 + t_2 + \ldots + t_n$ is the sum of an integer number n of exponential variates t_i, $i = 1$, 2, $\ldots$, n. Show that the distribution of t has as its density the function $f_T(t) = \lambda^n t^{n-1} e^{-\lambda t} / (n-1)!$, which is the Gamma density, for integer values of n. Hint: use the result of Exercise 8 and the methods described in Sect. 3.7 of Chap. 3 to derive the time required for two Poisson arrivals. From the previous result, extend your derivation to three Poisson arrivals. Proceed up until a point where a repetition pattern can be recognized and induction can be used to draw the desired conclusion.

10. A manufacturing company has bid to deliver compact standardized water treatment units for rural water supply. Based on previous experiences, it is estimated that 10 % of units are defective. If the bid consists of delivering 5 units, determine the minimum number of units to be manufactured so that there is a 95 % certainty that no defective units are delivered. It is assumed that the delivery of a unit is an independent trial and that the existence of possible defects in a unit is not affected by eventual defects in other units (adapted from Kottegoda and Rosso 1997).

11. A regional study of low flows for 25 catchments is being planned. The hydrologist in charge of the study does not know that 12 of the 25 catchments have inconsistent data. Suppose that in a first phase of the regional study, 10 catchments have been selected. Calculate (a) the probability that 3 out of the 10 selected catchments have inconsistent data; (b) the probability that at least 3 out of the 10 selected catchments have inconsistent data; and (c) the probability that all 10 selected catchments have inconsistent data.

12. At a given location, the probability that any of days in the first fortnight of January will be rainy is 0.20. Assuming rainy or dry days as independent trials, calculate (a) the probability that, in January of the next year, only the 2nd and the 3rd days will be rainy; (b) the probability of a sequence of at least two consecutive rainy days occurring, only in the period from the 4th to the 7th of January of next year; (c) denoting by Z the number of rainy days within the 4-day period of item (b), calculate the probability mass function for the variable Z; (d) calculate $P(Z > 2)$ and $P(Z \geq 2)$; and (e) the first three central moments of Z (adapted from Shahin et al. 1993).

References

H-SA A, Tang WH (1975) Probability concepts in engineering planning and design, volume I: basic principles. Wiley, New York

Haan CT (1977) Statistical methods in hydrology. Iowa University Press, Ames, IA

Kottegoda NT, Rosso R (1997) Statistics, probability, and reliability for civil and environmental engineers. McGraw-Hill, New York

Ross S (1989) Introduction to probability models, 4th edn. Academic Press, Boston
Shahin M, van Oorschot HJL, de Lange SJ (1993) Statistical analysis in water resources engineering. Balkema, Rotterdam
Todorovic PE, Zelenhasic E (1970) A stochastic model for flood analysis. Water Resour Res 6 (6):411–424

Chapter 5
Continuous Random Variables: Probability Distributions and Their Applications in Hydrology

Mauro Naghettini and Artur Tiago Silva

5.1 Uniform Distribution

A continuous random variable X, defined in the subdomain $\{x \in \Re | a \leq x < b\}$, is uniformly distributed if the probability of it being comprised in some interval $[m, n]$, contained in $[a, b]$, is directly proportional to the length $(m-n)$. Denoting the proportionality constant by ρ, then,

$$P(m \leq X \leq n) = \rho\,(m - n) \quad \text{if } a \leq m \leq n \leq b \tag{5.1}$$

Since $P(a \leq X \leq b) = 1$, it is clear that $\rho = 1/(b - a)$. Therefore, for any interval $a \leq x \leq b$, the uniform cumulative distribution function (CDF) is given by

$$F_X(x) = \frac{x - a}{b - a} \tag{5.2}$$

If $x < a, F_X(x) = 0$ and, if $x > b, F_X(x) = 1$. The probability density function (PDF) for the uniform variate is obtained by differentiating Eq. (5.2), with respect to x. Thus,

$$f_X(x) = \frac{1}{b - a} \quad \text{if } a \leq x \leq b \tag{5.3}$$

M. Naghettini (✉)
Universidade Federal de Minas Gerais, Belo Horizonte, Minas Gerais, Brazil
e-mail: mauronag@superig.com.br

A.T. Silva
CERIS, Instituto Superior Técnico, Universidade de Lisboa, Lisbon, Portugal
e-mail: artur.tiago.silva@tecnico.ulisboa.pt

© Springer International Publishing Switzerland 2017
M. Naghettini (ed.), *Fundamentals of Statistical Hydrology*,
DOI 10.1007/978-3-319-43561-9_5

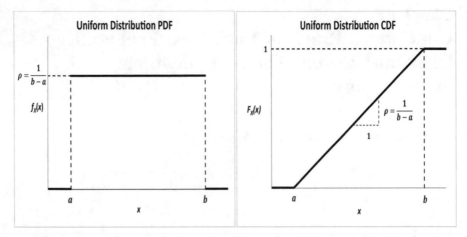

Fig. 5.1 PDF and CDF of a uniform distribution

The uniform distribution is also known as rectangular. Figure 5.1 depicts both the PDF and the CDF for a uniform variate.

The mean and the variance of a uniform distribution are respectively given by

$$E[X] = \frac{a+b}{2} \tag{5.4}$$

and

$$\mathrm{Var}[X] = \frac{(b-a)^2}{12} \tag{5.5}$$

When the subdomain of the random variable X is defined as $[0,1]$, the resulting uniform distribution, designated as the unit uniform distribution or unit rectangular distribution, encounters its main application, which is that of representing the distribution of $X = F_Y(y)$, where $F_Y(y)$ denotes any cumulative distribution function for a continuous random variable Y. In effect, since $0 \leq F_Y(y) = P(Y \leq y) \leq 1$ for any probability distribution, then the unit uniform distribution can be used for generating uniform random numbers x, which, in turn, may be interpreted as the non-exceedance probability $F_Y(y)$ and thus employed to obtain the quantities $y = F_{Y=X}^{-1}(y)$, distributed according to $F_Y(y)$, provided that $F_Y(y)$ be explicitly invertible and can be expressed in analytical form.

Random number generation is essential to simulating a large number of realizations of a given random variable, distributed according to a specified probability distribution function, with the purpose of analyzing numerous outcomes that are statistically similar to those observed. In Sect. 5.13, which contains a summary of the probability distributions described in this chapter, the most common approaches to generate random numbers for some distributions are included. Usually, the techniques used to generate a large number of realizations of a random variable (or of a stochastic process) are grouped under the general designation of *Monte Carlo simulation methods*. The reader interested in details on random number generation and on Monte Carlo methods, as applied to the solution of engineering problems, should consult Ang and Tang (1990) and Kottegoda and Rosso (1997).

Example 5.1 Denote by X the minimum daily temperature at a given location and suppose that X is uniformly distributed in the range from 16 to 22 °C. (a) Calculate the mean and variance of X. (b) Calculate the daily probability that X will exceed 18 °C. (c) Given that the minimum temperature on a sunny day has been persistently higher than 18 °C, calculate the probability that X will exceed 20 °C during the rest of the day?

Solution (a) The mean and the variance are obtained from the direct application of Eqs. (5.4) and (5.5), with $a = 16$ and $b = 22$ °C. Thus, $E[X] = 19$ °C and $\text{Var}[X] = 3$ (°C)^2. (b) $P(X > 18\ °C) = 1 - P(X < 18\ °C) = 1 - F_X(18) = 2/3$. (c) The density of X is $f_X(x) = 1/6$ for the subdomain $16 \leq X \leq 22$. However, according to the wording of the question, the minimum temperature on that particular day has remained persistently above 18 °C. As the sample space has been reduced from any real number in the range 16–22 °C to any real number within [18, 22], the probability density function must reflect such particular conditions and needs to be altered to $f_X^R(x) = 1/(22 - 18) = 1/4$, valid for the subdomain $18 \leq X \leq 22$. Thus, for the latter conditions, $P(X > 20 | X > 18) = 1 - F_X^R(20) = 1 - (20 - 18)/(22 - 18) = 1/2$.

5.2 Normal Distribution

The normal or Gaussian distribution was formally derived by the German mathematician Carl Friedrich Gauss (1777–1855), following earlier developments by astronomers and other mathematicians in their search for an *error curve* for the systematic treatment of discrepant astronomical measurements of the same phenomenon (Stahl 2006). The normal distribution is utilized to describe the behavior of a continuous random variable that fluctuates in a symmetrical manner around a central value. Some of its mathematical properties, which are discussed throughout this section, make the normal distribution appropriate for modeling the sum of a large number of independent random variables. Furthermore, the normal distribution is at the foundations of the construction of confidence intervals, statistical hypotheses testing, and correlation and regression analysis, which are topics covered in later chapters.

The normal distribution is a two-parameter model, whose PDF and CDF are respectively given by

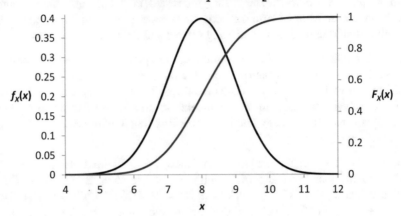

Fig. 5.2 PDF and CDF for the normal distribution, with $\theta_1 = 8$ and $\theta_2 = 1$

$$f_X(x) = \frac{1}{\sqrt{2\pi\theta_2^2}} \exp\left[-\frac{1}{2}\left(\frac{x - \theta_1}{\theta_2}\right)^2\right] \quad \text{for } -\infty < x < \infty \tag{5.6}$$

and

$$F_X(x) = \int_{-\infty}^{x} \frac{1}{\sqrt{2\pi\theta_2^2}} \exp\left[-\frac{1}{2}\left(\frac{x - \theta_1}{\theta_2}\right)^2\right] dx \tag{5.7}$$

where θ_1 and θ_2 are, respectively, location and scale parameters. Figure 5.2 illustrates the PDF and CDF for the normal distribution, with parameters $\theta_1 = 8$ and $\theta_2 = 1$.

The expected value, variance, and coefficient of skewness for the normal distribution of X (see solution to Example 3.15 of Chap. 3), with parameters θ_1 and θ_2, are respectively given by

$$E[X] = \mu = \theta_1 \tag{5.8}$$
$$\text{Var}[X] = \sigma^2 = \theta_2^2 \tag{5.9}$$

and

$$\gamma = 0 \tag{5.10}$$

As a result of Eqs. (5.8) and (5.9), the normal density function is usually expressed as

$$f_X(x) = \frac{1}{\sqrt{2\pi}\sigma} \exp\left[-\frac{1}{2}\left(\frac{x-\mu}{\sigma}\right)^2\right] \quad \text{for} \ -\infty < x < \infty \qquad (5.11)$$

and X is said to be normally distributed with mean μ and standard deviation σ, or, synthetically, that $X \sim N(\mu,\sigma)$. Therefore, the mean μ of a normal variate X is equal to its location parameter, around which the values of X are symmetrically scattered. The degree of scatter around μ is given by the scale parameter, which is equal to the standard deviation σ of X. Figure 5.3 exemplifies the effects that marginal variations of the location and scale parameters have on the normal distribution.

By employing the methods described in Sect. 3.7 of Chap. 3, it can be shown that, if $X \sim N(\mu_X, \sigma_X)$, then the random variable $Y = aX + b$, resulting from a linear combination of X, is also normally distributed with mean $\mu_Y = a\mu_X + b$ and standard deviation $\sigma_Y = a\sigma_X$, or, $Y \sim N(\mu_Y = a\mu_X + b, \quad \sigma_Y = a\sigma_X)$. This is termed the reproductive property of the normal distribution and can be extended to any linear combination of N independent and normally distributed random variables X_i, $i = 1, 2, \ldots, N$, each with its own respective parameters μ_i and σ_i.

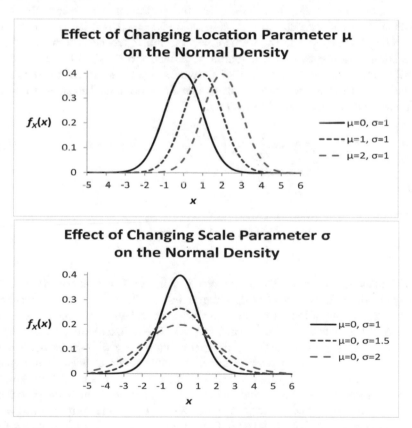

Fig. 5.3 Effects of marginal variations of the location and scale parameters on the normal density, for $X \sim N(\mu,\sigma)$

In fact, by extending the result given in the solution to Example 3.19 of Chap. 3, it can be shown that $Y = \sum_{i=1}^{N} a_i X_i + b$ follows a normal distribution with parameters

$\mu_Y = \sum_{i=1}^{N} a_i \mu_i + b$ and $\sigma_Y = \sqrt{\sum_{i=1}^{N} a_i^2 \sigma_i^2}$. As a particular case, (see the solution to Example 3.18 of Chap. 3), if Y denotes the arithmetic mean value of N normal variates X_i, all with common mean μ_X and standard deviation σ_X, then $Y \sim N(\mu_X, \sigma_X/\sqrt{N})$.

The CDF of the normal distribution, given by the integral Eq. (5.7), does not have an analytical solution. In effect, in order to calculate the function $F_X(x)$ for a particular pair of parameter values $\theta_1 = \mu$ and $\theta_2 = \sigma$, it is necessary to numerically integrate the function $F_X(x)$ over the desired subdomain of X; for a different pair of parameter values, another numerical integration would be required. Such an inconvenience can be overcome by linearly transforming the normal variable X, with parameters μ and σ, into $Z = (X - \mu)/\sigma$. In fact, using the reproductive property of the normal distribution, for the particular case where the linear combination coefficients are $a = 1/\sigma$ and $b = -\mu/\sigma$, it is clear that $Z \sim N(\mu_Z = 0, \sigma_Z = 1)$. The transformed variable Z receives the name of standard normal variate and its distribution is termed the standard (or unit) normal distribution. Note that as deviations of X from its mean μ are scaled by its standard deviation σ, the standard variate Z is always dimensionless. The standard normal PDF and CDF are respectively given by

$$f_Z(z) = \frac{1}{\sqrt{2\pi}} \exp\left(-\frac{z^2}{2}\right), \quad -\infty < z < \infty \tag{5.12}$$

and

$$F_Z(z) = \Phi(z) = \int_{-\infty}^{z} \frac{1}{\sqrt{2\pi}} \exp\left(-\frac{z^2}{2}\right) dz \tag{5.13}$$

The standard normal CDF $F_Z(z)$ receives the special notation $\Phi(z)$ and is calculated through numerical integration of Eq. (5.13). In general, the results of the numerical integration for different nonnegative values of argument z are organized in tables, such as that shown in Table 5.1. This a double-entry table in which the value of z_0 is found by crossing the first-column entry with the first-row entry, whereas the corresponding reading inside the table refers to $\Phi(z_0)$. For negative values of z_0, the areas $\Phi(z_0)$ are derived from the readings of Table 5.1 by using the symmetry properties of the normal distribution. In order to evaluate the probability $P(X \le x_0)$, for a nonstandard normal variate $X \sim N(\mu_X, \sigma_X)$, the corresponding value of $z_0 = (x_0 - \mu_X)/\sigma_X$ is first calculated. Then, with the value of $\Phi(z_0)$, obtained from Table 5.1, it suffices to make $P(X \le x_0) = \Phi(z_0)$. This operation is facilitated in

Table 5.1 Values of $F_Z(z_0) = \Phi(z_0)$ for the standard normal distribution

Standard Normal Distribution $\mu_Z=0,\ \sigma_Z=1$

$$F_Z(z_0) = \Phi(z_0) = \int_{-\infty}^{z_0} \frac{1}{\sqrt{2\pi}} \exp\left(-\frac{z^2}{2}\right) dz$$

z_0	0.00	0.01	0.02	0.03	0.04	0.05	0.06	0.07	0.08	0.09
0.0	0.5000	0.5040	0.5080	0.5120	0.5160	0.5199	0.5239	0.5279	0.5319	0.5359
0.1	0.5398	0.5438	0.5478	0.5517	0.5557	0.5596	0.5606	0.5675	0.5714	0.5753
0.2	0.5793	0.5832	0.5871	0.5910	0.5948	0.5987	0.6026	0.6064	0.6103	0.6141
0.3	0.6179	0.6217	0.6255	0.5293	0.6331	0.6368	0.6406	0.6443	0.6480	0.6517
0.4	0.6554	0.6591	0.6628	0.5664	0.6700	0.6736	0.6772	0.6808	0.6844	0.6879
0.5	0.6915	0.6950	0.6985	0.7019	0.7054	0.7088	0.7123	0.7157	0.7190	0.7224
0.6	0.7257	0.7291	0.7324	0.7357	0.7389	0.7422	0.7454	0.7486	0.7517	0.7549
0.7	0.7580	0.7611	0.7642	0.7673	0.7704	0.7734	0.7764	0.7794	0.7823	0.7852
0.8	0.7881	0.7910	0.7939	0.7967	0.7995	0.8023	0.8051	0.8078	0.8106	0.8133
0.9	0.8159	0.8186	0.8212	0.8238	0.8264	0.8289	0.8315	0.8340	0.8365	0.8389
1.0	0.8413	0.8438	0.8461	0.8585	0.8508	0.8531	0.8554	0.8577	0.8599	0.8621
1.1	0.8643	0.8665	0.8686	0.8708	0.8729	0.8749	0.8770	0.8790	0.8810	0.8830

(continued)

Table 5.1 (continued)

z_0	0.00	0.01	0.02	0.03	0.04	0.05	0.06	0.07	0.08	0.09
1.2	0.8849	0.8869	0.8888	0.8907	0.8925	0.8944	0.8962	0.8980	0.8997	0.9015
1.3	0.9032	0.9049	0.9066	0.9082	0.9099	0.9115	0.9137	0.9147	0.9162	0.9177
1.4	0.9192	0.9207	0.9222	0.9236	0.9251	0.9265	0.9279	0.9292	0.9306	0.9319
1.5	0.9332	0.9345	0.9357	0.9370	0.9382	0.9394	0.9406	0.9418	0.9429	0.9441
1.6	0.9452	0.9463	0.9474	0.9484	0.9495	0.9505	0.9515	0.9525	0.9535	0.9545
1.7	0.9554	0.9564	0.9573	0.9582	0.9591	0.9599	0.9608	0.9616	0.9625	0.9633
1.8	0.9641	0.9649	0.9656	0.9664	0.9671	0.9678	0.9686	0.9693	0.9699	0.9706
1.9	0.9713	0.9719	0.9726	0.9732	0.9738	0.9744	0.9750	0.9756	0.9761	0.9767
2.0	0.9772	0.9778	0.9783	0.9788	0.9793	0.9798	0.9803	0.9808	0.9812	0.9817
2.1	0.9821	0.9826	0.9830	0.9834	0.9838	0.9842	0.9846	0.9850	0.9854	0.9857
2.2	0.9861	0.9864	0.9868	0.9871	0.9875	0.9878	0.9881	0.9884	0.9887	0.9890
2.3	0.9893	0.9896	0.9898	0.9901	0.9904	0.9906	0.9909	0.9911	0.9913	0.9916
2.4	0.9918	0.9920	0.9922	0.9925	0.9927	0.9929	0.9931	0.9932	0.9934	0.9936
2.5	0.9938	0.9940	0.9941	0.9943	0.9945	0.9946	0.9948	0.9949	0.9951	0.9952
2.6	0.9953	0.9955	0.9956	0.9957	0.9959	0.9960	0.9961	0.9962	0.9963	0.9964
2.7	0.9965	0.9966	0.9967	0.9968	0.9969	0.9970	0.9971	0.9972	0.9973	0.9974
2.8	0.9974	0.9975	0.9976	0.9977	0.9977	0.9978	0.9979	0.9979	0.9980	0.9981
2.9	0.9981	0.9982	0.9982	0.9983	0.9984	0.9984	0.9985	0.9985	0.9986	0.9986
3.0	0.9987	0.9987	0.9987	0.9988	0.9988	0.9989	0.9989	0.9989	0.9990	0.9990
3.1	0.9990	0.9991	0.9991	0.9991	0.9992	0.9992	0.9992	0.9992	0.9993	0.9993
3.2	0.9993	0.9993	0.9994	0.9994	0.9994	0.9994	0.9994	0.9995	0.9995	0.9995
3.3	0.9995	0.9995	0.9995	0.9996	0.9996	0.9996	0.9996	0.9996	0.9996	0.9997
3.4	0.9997	0.9997	0.9997	0.9997	0.9997	0.9997	0.9997	0.9997	0.9997	0.9998

MS Excel through the function NORM.DIST(.). In the statistical software R, the equivalent function is pnorm(.). Inversely, if the goal is to calculate the quantile x_0, for the non-exceedance probability P, one should first read (or interpolate), in Table 5.1, the value of z_0 that matches $\Phi(z) = P$ and, then, find the quantile $x = \mu_X + z\sigma_X$. This operation is performed in MS Excel by the function NORM. INV(.), and in R by the function qnorm(.).

Example 5.2 Suppose the annual mean flows Q of a main tributary of the Amazon river are normally distributed with mean value 10,000 m^3/s and standard deviation 5000 m^3/s. Calculate (a) the probability that next year's mean flow will be less than 5000 m^3/s; and (b) the annual mean flow of return period $T = 50$ years.

Solution (a) The sought probability $P(Q < 5000)$ is equal to $P\{z < [(5000-10,000)/5000]\}$, or $\Phi(-1)$. Readings in Table 5.1 give $\Phi(z)$ only for nonnegative entries z. Since the normal distribution is symmetric, one can write $\Phi(-1) = 1 - \Phi(+1) = 1 - 0.8413 = 0.1587$; (b) The definition of return period applies to annual mean values in the same way described for annual maxima, or, in other terms, $T = 1/P(Q > q)$. With $T = 50$ years, $P(Q > q) = 1/50 = 0.02$ and, thus, $\Phi(z) = 1 - 0.02 = 0.98$. This reading in Table 5.1 corresponds to the entry value $z = 2.054$. Finally, the annual mean flow q, of return period $T = 50$ years, corresponds to the quantile $q = 10,000 + 2.054 \times 5000 = 20,269$ m^3/s. Now, solve this example using the MS Excel functions.

The $\Phi(z_0 = 1)$ reading from Table 5.1, of 0.8413, shows that 68.26 % of the whole area below the normal density is comprised between one standard deviation below and above the mean. Proceeding in the same way, one can notice that 95.44 % of the area below the density lies between two standard deviations from each side of the mean, whereas 99.74 % of the area is contained between the bounds μ-3σ and μ+3σ. Although the normal variate can vary from $-\infty$ to $+\infty$, the tiny probability of 0.0013, of having a value below (μ-3σ), discloses the range of the distribution applicability to nonnegative hydrologic variables. In fact, as long as $\mu_X > 3\sigma_X$, the likelihood of a negative value of X is negligible. Both $\Phi(z)$ and its inverse can be approximated by functions of easy implementation in computer codes. According to Abramowitz and Stegun (1972), an accurate approximation to $\Phi(z)$, for $z \geq 0$, is given by

$$\Phi(z) \cong 1 - \left(b_1 t + b_2 t^2 + b_3 t^3 + b_4 t^4 + b_5 t^5\right) f_Z(z) \tag{5.14}$$

where $f_Z(z)$ denotes the normal density function and t is an auxiliary variable given by

$$t = \frac{1}{1 + rz} \tag{5.15}$$

in which $r = 0.2316419$. The coefficients b_i are

$$b_1 = 0.31938153$$
$$b_2 = -0.356563782$$
$$b_3 = 1.781477937 \qquad (5.16)$$
$$b_4 = -1.821255978$$
$$b_5 = 1.330274429$$

According to Abramowitz and Stegun (1972), the inverse of $\Phi(z)$, here denoted by $z(\Phi)$, for $\Phi \geq 0,5$, can be approximated by

$$z(\Phi) \cong m - \frac{c_0 + c_1 m + c_2 m^2}{1 + d_1 m + d_2 m^2 + d_3 m^3} \qquad (5.17)$$

where m is an auxiliary variable given by:

$$m = \sqrt{\ln\left[\frac{1}{(1-\Phi)^2}\right]} \qquad (5.18)$$

and the coefficients c_i and d_i are the following:

$$c_0 = 2.515517$$
$$c_1 = 0.802853$$
$$c_2 = 0.010328$$
$$d_1 = 1.432788 \qquad (5.19)$$
$$d_2 = 0.189269$$
$$d_3 = 0.001308$$

An important application of the normal distribution stems from the *central limit theorem* (CLT). According to the *classical or strict variant* of this theorem, if S_N denotes the sum of N *independent and identically distributed* random variables $X_1, X_2, \ldots, X_N$, all with the same mean μ and the same standard deviation σ, then, the variable

$$Z_N = \frac{S_N - N\mu}{\sigma\sqrt{N}} \qquad (5.20)$$

tends asymptotically to be distributed according to a standard normal distribution, i. e., for sufficiently large values of N, $Z_N \sim N(0,1)$. For practical purposes, if $X_1, X_2, \ldots, X_N$ are independent, with identical and symmetrical (or moderately skewed) distributions, values of N close to 30, or even lesser, are sufficient to allow convergence of Z_N to a standard normal variate.

As stated earlier, a particular result stemming from the reproductive property of the normal distribution is that, if Y represents the arithmetic mean of N normally distributed variables X_i, all with mean μ_X and standard deviation σ_X, then $Y \sim N(\mu_X, \sigma_X/\sqrt{N})$. Application of Eq. (5.20) to the variable Y (see Example 5.3) shows that the same result could be achieved with the use of the central limit theorem, although, in this case, the variables X_i are not required to be normally distributed. The only condition to be observed is that the number N of summands X_i must be sufficiently large to allow convergence to a normal distribution. Kottegoda and Rosso (1997) suggest that if the common distribution of the X_i summands departs moderately from the normal, the convergence is relatively fast. However, if the departure from a bell-shaped density is pronounced, then values of N larger than 30 may be required to guarantee convergence.

The CLT in its classical or strict version has little applicability in hydrology. In fact, the very idea that a given hydrologic variable be the result of the sum of a large number of independent and identically distributed random variables, in most cases, contradicts the reality of hydrologic phenomena. Take as an example the annual total rainfall depth, obtained from the summation of daily rainfalls over the year. To assume that the daily rainfalls are independent and identically distributed, with the same mean and the same standard deviation for all days of the year is clearly not realistic, from a hydrological point of view. This fact hinders the application of the strict version of the CLT to annual total rainfall. In contrast, the *generalized variant* of the CLT is general enough to be used with some hydrologic variables. According to this variant of the CLT, if X_i ($i = 1, 2, ..., N$) denote independent variables, each one with its respective mean and variance, given by μ_i and σ_i^2, then, the variable

$$Z_N = \frac{S_N - \sum_{i=1}^{N} \mu_i}{\sqrt{\sum_{i=1}^{N} \sigma_i^2}} \tag{5.21}$$

tends to be a standard normal variate, as N increases to infinity, under the condition that none of the summands X_i has a dominant effect on the sum S_N. The rigorous mathematical proof of the generalized variant of the CLT was undertaken by the Russian mathematician Aleksandr Lyapunov (1857–1918).

The premise of independence is still required in the generalized version of the CLT. However, Benjamin and Cornell (1970) point out that, when $N \to \infty$, Z_N tends to be normally distributed even if the summands X_i are not strictly and collectively independent. The only condition for this to hold is that the summands be jointly distributed in such a manner that the correlation coefficient between any given summand and the vast majority of the others is null. The practical importance of this generalized variant of the CLT lies in the fact that, once general conditions are set out, the convergence of the sum of a large number of random summands, or, by extension, of the arithmetic mean, to a normal distribution can be established without the exact knowledge of the marginal distributions of X_i or of their joint distribution.

The extended generalized version of the CLT, with a few additional considerations of a practical nature, is applicable to some hydrologic variables. To return to the example of the annual total rainfall depth, it is quite plausible to assume that, within a climatic region where the rain episodes are not too clustered in a few days (or weeks) of the year, there should not be a dominant effect of one or more daily rainfalls, over the annual total rainfall. Furthermore, if one disregards the persistent multiday rainfall episodes that may occur as large-scale atmospheric systems move over a region, then the hypothesis of independence among most of X_i variables may well be true. Therefore, under these particular conditions, and supposing that $N = 365$ (or 366) be large enough to allow convergence, which will largely depend on the shapes of the marginal distributions of X_i, one can assume that, in many cases, annual total rainfalls can be suitably described by the normal distribution. Such a situation is depicted in Fig. 5.4, which shows the normal density superimposed over the histogram of the annual total rainfalls recorded from 1767 to 2014, at the Radcliffe Meteorological Station, in Oxford, England. On the other hand, employing similar arguments to justify fitting the normal distribution to annual mean flows seems more complicated, probably due to the stronger statistical dependence among consecutive summands X_i and to the effects of other hydrologic phenomena on the annual mean discharges.

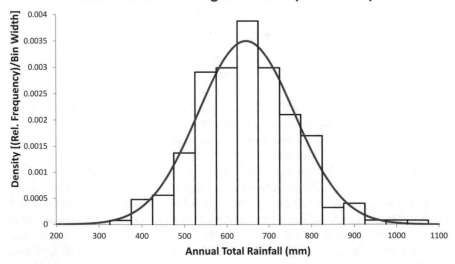

Fig. 5.4 Histogram and normal density for the annual total rainfall depths (mm), recorded at Oxford (England). The normal distribution was calculated using the sample mean and standard deviation. Note that relative frequencies have been divided by the bin width to be expressed in density units

Example 5.3 A plan for monitoring dissolved oxygen (DO) concentrations is being prepared for a river reach downstream of a large reservoir. The monitoring plan will consist of a systematic program of regular weekly measurements of DO concentration at a specific river section. The random variable DO concentration, here denoted by X, is physically bounded from below by zero and from above by the maximum DO concentration at saturation, which depends mainly on the water temperature. Suppose that eight preliminary weekly measurements of DO concentration have yielded $\bar{x} = 4\,mg/l$ and $s_X = 2\,mg/l$. Based only on the available information, how many weekly measurements will be necessary so that the difference between the sample mean and the true population mean of X be at most 0.5 mg/l, with a confidence of 95 %?

Solution As opposed to the normal random variable, X, in this case, is double-bounded and, as a result of its strong dependence on the streamflows, its PDF will probably be asymmetric. Suppose that X_i denotes the DO concentration measured at the ith week of the planned N-week monitoring program. Given that the monitoring cross-section is located in a river reach with highly regulated flows and that the time interval between measurements is the week, it is plausible to assume that X_i and X_j, for $i \neq j$ and $i,j \leq N$, are statistically independent and identically distributed, with common mean μ and standard deviation σ, even if their marginal and joint distributions are not known. Thus, it is also plausible to admit that the arithmetic average of N terms X_i, deemed as independent and identically distributed (IID) random variables, will tend to be normally distributed, as N grows sufficiently to allow convergence. In other terms, it is possible to apply the strict version of the CLT. Accordingly, by making the sum over N IID variables as $S_N = N\bar{x}$, where $\bar{x}$ denotes the arithmetic average of X_i, and substituting it into Eq. (5.20), one can write $Z_N = \frac{N\bar{x} - N\mu}{\sigma\sqrt{N}} = \frac{\bar{x}-\mu}{\sigma/\sqrt{N}} \sim \mathbf{N}(0,1)$. In order to guarantee the 95 % confidence level, the following probability statement is needed: $P\left(z_{2.5\%} \leq \frac{\bar{x}-\mu}{\sigma/\sqrt{N}} \leq z_{97.5\%}\right) = 0.95$. Readings from Table 5.1 give $z_{0.975} = 1.96$ and, by symmetry, $z_{0.025} = -1.96$. Substituting these into the equation for $P(.)$ and algebraically manipulating the inequality so that the term of the difference between the mean values be explicit, then $P\left(|\bar{x} - \mu| \leq 1.96\sigma/\sqrt{N}\right) = 0.95$. Assuming that σ can be estimated by $s_X = 2$ mg/l and recalling that $|\bar{x} - \mu| = 0.5\,mg/l$, the resulting inequality is $(1.96 \times 2)/\sqrt{N} \geq 0.5$ or $N \geq 61.47$. Therefore, a period of at least 62 weeks of DO monitoring program is necessary to keep the difference, between the sample mean and the true population mean of X, below 0.5 mg/l, with a 95 % confidence level.

In Chap. 4, the binomial discrete random variable X, with parameter p, is introduced as the sum of N independent Bernoulli variables. As a result of the CLT, if N is large enough, one can approximate the binomial by the normal distribution. Recalling that the mean and the variance of a binomial variate are Np and $Np(1-p)$, respectively, then the variable defined by

$$Z = \frac{X - Np}{\sqrt{Np(1-p)}} \qquad (5.22)$$

tends to be distributed according to a standard normal $N(0,1)$, as N increases. The convergence is faster for p values around 0.5. For p close to 0 or 1, larger values of N are required.

In an analogous way, one can approximate the Poisson variate X, with mean and variance equal to ν, by the standard normal distribution, through the variable

$$Z = \frac{X - \nu}{\sqrt{\nu}} \qquad (5.23)$$

when $\nu > 5$. Note, however, that, in both cases, since the probability mass function is approximated by the density of a continuous random variable, the so-called *continuity correction* must be performed. In fact, for the discrete case, when $X = x$, the ordinate of the mass function is a line or a point, which, in the continuous case, must be approximated by the area below the density function between the abscissae $(x-0.5)$ and $(x+0.5)$, for integer values of x.

5.3   Lognormal Distribution

Consider a random variable X that results from the multiplicative action of a great number of independent random components X_i $(i = 1,2,\ldots,N)$, or $X = X_1.X_2 \ldots X_N$. In such a case, from the CLT, the variable $Y = \ln(X)$, such that $Y = \ln(X_1) + \ln(X_2) + \ldots + \ln(X_N)$, will tend to be normally distributed, with parameters μ_Y and σ_Y, as N becomes large enough to allow convergence. Under these conditions, the variable X is said to follow a lognormal distribution, with parameters $\mu_{\ln(X)}$ and $\sigma_{\ln(X)}$, denoted by $X \sim \mathbf{LN}\left(\mu_{\ln(X)}, \sigma_{\ln(X)}\right)$ or $X \sim \mathbf{LNO2}\left(\mu_{\ln(X)}, \sigma_{\ln(X)}\right)$. By applying Eq. (3.61) to $Y = \ln(X)$, it is easy to determine that the probability density function of the lognormal variate X is given by

$$f_X(x) = \frac{1}{x\sigma_{\ln(X)}\sqrt{2\pi}} \exp\left\{ -\frac{1}{2}\left[\frac{\ln(X) - \mu_{\ln(X)}}{\sigma_{\ln(X)}}\right]^2 \right\} \text{ para } x \geq 0 \qquad (5.24)$$

The calculations of probabilities and quantiles for the lognormal distribution, considering $Y = \ln(X)$ for the function argument and, inversely, $X = \exp(Y)$ for the corresponding quantile, are similar to those described for the normal distribution. Note that, as $\log_{10}(X) = 0.4343\ln(X)$, common or decimal logarithms can also be used to derive the lognormal density. In such a case, Eq. (5.24) must be multiplied

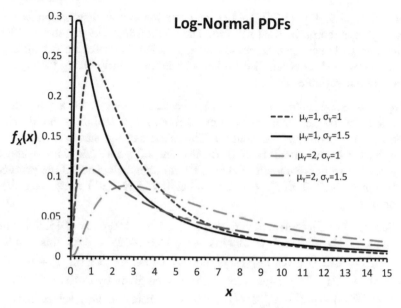

Fig. 5.5 Examples of lognormal densities

by 0.4343 and the quantiles are calculated with $x = 10^y$ instead of $x = \exp(y)$. The lognormal distribution is sometimes referred to as Galton's law of probabilities after British scientist Francis Galton (1822–1911). Figure 5.5 shows examples of lognormal densities, for some specific values of parameters.

The expected value and the variance of a lognormal variate are respectively given by

$$E[X] = \mu_X = \exp\left[\mu_{\ln(X)} + \frac{\sigma_{\ln(X)}^2}{2}\right] \tag{5.25}$$

and

$$\text{Var}[X] = \sigma_X^2 = \mu_X^2\left[\exp\left(\sigma_{\ln(X)}^2\right) - 1\right] \tag{5.26}$$

By dividing both sides of Eq. (5.26) by μ_X^2 and extracting the square root of the resulting terms, one gets the expression for the coefficient of variation of a lognormal variate

$$\text{CV}_X = \sqrt{\exp\left[\sigma_{\ln(X)}^2\right] - 1} \tag{5.27}$$

The coefficient of skewness of the lognormal distribution is

$$\gamma = 3\,\text{CV}_X + (\text{CV}_X)^3 \tag{5.28}$$

Since $\mathrm{CV}_X > 0$, Eq. (5.28) implies that the lognormal distribution is always positively asymmetric or right skewed. The relationship between the coefficients of variation and skewness, expressed by Eq. (5.28), is useful for calculating the parameters of the lognormal distribution without previously taking the logarithms of the original variable X.

Example 5.4 Suppose that, from the long record of rainfall data at a given site, it is reasonable to assume that the wettest 3-month total rainfalls be distributed according to a lognormal distribution. The mean and the standard deviation for the wettest 3-month total rainfall depths are 600 and 150 mm. Supposing X denotes the wettest 3-month total rainfall variable, calculate (a) the probability $P(400 \text{ mm} < X < 700 \text{ mm})$; (b) the probability $P(X > 300 \text{ mm})$; and (c) the median of X.

Solution (a) The coefficient of variation of X is $\mathrm{CV} = 150/600 = 0.25$. Substituting this value in Eq. (5.27), one obtains $\sigma_{\ln(X)} = 0.246221$. With this result and $\mu_X = 600$, Eq. (5.25) gives $\mu_{\ln(X)} = 6.366617$. Thus, $X \sim \mathbf{LN}(\mu_{\ln(X)} = 6.366617, \sigma_{\ln(X)} = 0.246221)$. The sought probability is then given by $P(400 < X < 700) = \Phi\left(\frac{\ln 700 - 6.366617}{0.246221}\right) - \Phi\left(\frac{\ln 400 - 6.366617}{0.246221}\right) = 0.7093$, where the $\Phi(.)$ values have been linearly interpolated through the readings from Table 5.1. (b) The sought probability is $P(X > 300) = 1 - P(X < 300) = 1 - \Phi\left(\frac{\ln 300 - 6.366617}{0.246221}\right) = 0.9965$. (c) The transformed variable $Y = \ln(X)$ is distributed according to a normal distribution, which is symmetric, with all its central tendency measures coinciding at a single abscissa. As a result, the median of Y is equal to its mean value, or $y_{md} = 6.366617$. It should be noted, however, that, as the median corresponds to the central point that partitions the sample into 50 % of values above and below it, then, the logarithmic transformation, as a strictly increasing function, will not change the relative position (or the rank) of the median, with respect to the other points. Hence, the median of $\ln(X)$ is equal to the natural logarithm of the median of X, or $y_{md} = \ln(x_{md})$, and, inversely, $x_{md} = \exp(y_{md})$. It is worth noting, however, that this is not valid for the mean and other mathematical expectations. Therefore, for the given data, the median for the 3-month total rainfall depths is $x_{md} = \exp(y_{md}) = \exp(6.366617) = 582.086$ mm.

The three-parameter lognormal distribution (**LN3** or **LNO3**) is similar to the two-parameter model, except by a lower-bound, denoted by a, which is subtracted from X. In other words, the random variable $Y = \ln(X - a)$ is distributed according to a normal distribution with mean μ_Y and standard-deviation σ_Y. The corresponding PDF is

$$f_X(x) = \frac{1}{(x-a)\sigma_Y\sqrt{2\pi}}\exp\left\{-\frac{1}{2}\left[\frac{\ln(x-a) - \mu_Y}{\sigma_Y}\right]^2\right\} \tag{5.29}$$

The mean and standard deviation of an LN3 variate are respectively given by

$$E[X] = a + \exp\left(\mu_Y + \frac{\sigma_Y^2}{2}\right) \tag{5.30}$$

and

$$\text{Var}[X] = \sigma_X^2 = \left[\exp(\sigma_Y^2) - 1\right] \exp\left(2\mu_Y + \sigma_Y^2\right) \tag{5.31}$$

According to Kite (1988), the coefficient of variation of the auxiliary random variable $(X-a)$ can be written as

$$\text{CV}_{(X-a)} = \frac{1 - \sqrt[3]{w^2}}{\sqrt[3]{w}} \tag{5.32}$$

where w is defined by the following function of the coefficient of skewness, γ_X, of the original variable X:

$$w = \frac{-\gamma_X + \sqrt{\gamma_X^2 + 4}}{2} \tag{5.33}$$

Kite (1988) also showed that the lower bound a can be obtained from $\text{CV}_{(X-a)}$, $E[X]$, and σ_X, by means of the following equation:

$$a = E[X] - \frac{\sigma_X}{\text{CV}_{(X-a)}} \tag{5.34}$$

The procedure suggested by Kite (1988) for the calculation of the LN3 param-
eters follows the sequence: (1) with $E[X]$, σ_X, and γ_X, calculate w and $\text{CV}_{(X-a)}$,
using Eqs. (5.33) and (5.32), respectively; (2) calculate a with Eq. (5.34); and
(3) the two remaining parameters, μ_Y and σ_Y, are the solutions to the system formed
by Eqs. (5.30) and (5.31).

The proposition of the lognormal distribution as a probabilistic model relies on
the extension of the CLT to a variable that results from the multiplicative action of
independent random components. There is evidence that hydraulic conductivity in
porous media (Freeze 1975), raindrop sizes in a storm (Ajayi and Olsen 1985), and
other geophysical variables (see Benjamin and Cornell 1970, Kottegoda and Rosso
1997, and Yevjevich 1972) may result from this kind of process. However, relying
solely on this argument to endorse the favored use of the lognormal distribution for
modeling the most common hydrologic variables, such as those related to floods
and droughts, is rather controversial. The controversy arises from the difficulty of
understanding and clearly pointing out such a multiplicative action of multiple
components. Besides, the requirements of independence among multiplicands and
convergence to the normal distribution, inherent to the CLT, are often difficult to be
verified and met. Nonetheless, these are not arguments to rule out the lognormal
distribution from Statistical Hydrology. On the contrary, since its variate is always

positive and its coefficient of skewness a non-fixed positive quantity, the lognormal distribution is potentially a reasonable candidate for modeling annual maximum (or mean) flows, annual maximum daily rainfalls, and annual, monthly or 3-month total rainfall depths, among other hydrologic variables.

5.4 Exponential Distribution

The solution to Exercise 8 of Chap. 4 shows that the continuous time, between two consecutive Poisson arrivals, is an exponentially distributed random variable. In addition to this mathematical fact, the exponential distribution, also known as the negative exponential distribution, has many possible applications in distinct fields of knowledge, and, particularly, in hydrology. The probability density function of the one-parameter exponential distribution is given by

$$f_X(x) = \frac{1}{\theta} \exp\left(-\frac{x}{\theta}\right) \text{ or } f_X(x) = \lambda \exp(-\lambda x), \text{ for } x \geq 0 \qquad (5.35)$$

where θ (or $\lambda = 1/\theta$) denotes its single parameter. If $X \sim \mathbf{E}(\theta)$ or $X \sim \mathbf{E}(\lambda)$, the corresponding CDF is

$$F_X(x) = 1 - \exp\left(-\frac{x}{\theta}\right) \text{ or } F_X(x) = 1 - \exp(-\lambda x) \qquad (5.36)$$

The expected value, variance, and coefficient of skewness of an exponential variate (see Examples 3.12 and 3.13 of Chap. 3) are respectively given by

$$E[X] = \theta \text{ or } E[X] = \frac{1}{\lambda} \qquad (5.37)$$

$$\text{Var}[X] = \theta^2 \text{ or } \text{Var}[X] = \frac{1}{\lambda^2} \qquad (5.38)$$

and

$$\gamma = 2 \qquad (5.39)$$

Note that the coefficient of skewness for the exponential distribution is a positive constant. Figure 5.6 depicts examples of exponential PDFs and CDFs, for $\theta = 2$ and $\theta = 4$.

Since the exponential shape arises from the distribution of the continuous time interval until some specific event happens, it has been used in practice as a model for lifetimes of machine components and the waiting time, starting from now, until a flood or an earthquake occurs, among others applications. The main justification for the use of the exponential distribution in this context is that it is the only model, among the distributions of continuous random variables, with the *memorylessness*

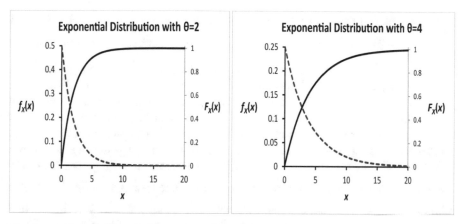

Fig. 5.6 PDFs and CDFs for the Exponential distribution, with $\theta = 2$ and $\theta = 4$

property. Such a property states that, if an exponential variate X is used for modeling the lifetime of some electronic component, for example, X is characterized by a *lack of memory* if the conditional probability that the component survives for at least $v + t$ hours, given it has survived t hours, is the same as the initial probability that it survives for at least v hours. Since, from Eq. (5.36), $\exp[-\lambda(v + t)] = \exp(-\lambda v)\exp(-\lambda t)$, it is apparent that the exponential distribution exhibits the property of *memorylessness*, as $P(X > v + t \mid X > t) = P(X > v)$.

Example 5.5 With reference to the representation of flood events as a Poisson process, as described in Exercise 8 of Chap. 4, consider that, on average, 2 floods over the threshold $Q_0 = 60$ m³/s, occur every year. Suppose that the exceedances $(Q - Q_0)$ are exponentially distributed with mean 50 m³/s. Calculate the annual maximum flow of return period $T = 100$ years.

Solution This is a Poisson process with constant $v = \int_0^1 \lambda(t)\, dt = 2$, where the limits of integration 0 and 1 denote, respectively, the beginning and the end of the water year, $\lambda(t)$ the Poisson arrival rate, and v the annual mean number of arrivals. When they occur, the exceedances $X = (Q - Q_0)$ are exponentially distributed, with CDF $G_X(x) = 1 - \exp(-x/\theta)$ and $\theta = 50$ m³/s. In order to calculate the annual maximum flows associated with a given return period, it is necessary, first, to derive the CDF of the annual maximum exceedances, as denoted by $F_{X\max}(x)$, since $T = 1/(1 - F_{X\max})$. Then, if the goal is to determine the distribution of the annual maximum exceedances x, one has to consider that each one of the 1, 2, 3, ... ∞ independent exceedances, which may occur in a year, must be less than or equal to x, as x represents the annual maximum value. Thus, $F_{X\max}(x)$ can be determined by weighting the probability of having N independent exceedances within a year, given by $[G_X(x)]^N$, by the mass function of the annual number of exceedances N, which has been supposed to be Poisson-distributed with parameter v. Therefore, $F_{X\max}(x)$

$$= \sum_{N=0}^{\infty} [G_X(x)]^N \frac{v^N e^{-v}}{N!} = \sum_{N=0}^{\infty} [v G_X(x)]^N \frac{e^{-v}}{N!}.$$ By multiplying and dividing the right-

hand side of this equation by $e^{-\nu G(x)}$, one gets $F_{X\max}(x) = \exp\{-\nu[1 - G_X(x)]\}$ $\sum_{N=0}^{\infty} \frac{[\nu G_X(x)]^N \exp[-\nu G_X(x)]}{N!}$. The summation, on the right-hand side of this equation, is equal to 1, as it refers to the sum over the entire domain of a Poisson mass function, with parameter $\nu G_X(x)$. After algebraic manipulation, one gets $F_{X\max}(x) = \exp\{-\nu[1 - G_X(x)]\}$, which is the fundamental equation for calculating annual probabilities for partial duration series, with Poisson arrivals (see Sect. 1.4 of Chap. 1). For the specific problem focused on this example, the CDF for the exceedances is supposed to be exponential, or $G_X(x) = 1 - \exp(-x/\theta)$, whose substitution in the fundamental equation results in the *Poisson-Exponential model* for partial duration series. Formally, $F_{Q\max}(q) = \exp\left\{-\nu \exp\left(-\frac{q - Q_0}{\theta}\right)\right\}$, where $Q\max = Q_0 + X$ represents the annual maximum flow. Recalling that $a = b e^c \Leftrightarrow \ln(a) = \ln(b) + c \Leftrightarrow a = \exp[\ln(b) + c]$, one finally gets $F_{Q\max}(q) = \exp\{-\exp[-\frac{1}{\theta}(q - Q_0 - \theta\ln\nu)]\}$, which is the expression of the CDF for the Gumbel distribution, with parameters θ and $[Q_0 + \theta \ln(\nu)]$, to be further detailed in Sect. 5.7 of this chapter. In summary, the modeling of partial duration series, with Poisson arrivals and exponentially distributed exceedances over a given threshold, leads to the Gumbel distribution for the annual maxima. The quantile function or the inverse of the CDF for the Gumbel distribution is $q(F) = Q_0 + \theta\ln(\nu) - \theta\ln[-\ln(F)]$. For the quantities previously given, $T = 100$ $\Rightarrow F_{Q\max} = 1 - 1/100 = 0.99$; $\theta = 50$; $\nu = 2$, and $Q_0 = 60 \text{ m}^3/\text{s}$, one finally obtains $q(F = 0.99) = 289.8 \text{ m}^3/\text{s}$, which is the 100-year flood discharge.

5.5 Gamma Distribution

The solution to Exercise 9 of Chap. 4 has shown that the probability distribution of the time t for the Nth Poisson occurrence is given by the density $f_T(t) = \lambda^N t^{N-1} e^{-\lambda t}/(N - 1)!$, which is the gamma PDF for integer values of the parameter N. Under these conditions, the gamma distribution results from the sum of N independent exponential variates, with common parameter λ or $\theta = 1/\lambda$.

The gamma distribution, for integer N, is also known as Erlang's distribution, after the Danish mathematician Agner Krarup Erlang (1878–1929). In more general terms, parameter N does not need to be an integer and, without this restriction, the two-parameter gamma PDF is given by

$$f_X(x) = \frac{1}{\theta \Gamma(\eta)} \left(\frac{x}{\theta}\right)^{\eta-1} \exp\left(-\frac{x}{\theta}\right) \text{ for } x \geq 0; \theta \text{ and } \eta > 0 \qquad (5.40)$$

where the real numbers θ and η denote, respectively, the scale and shape parameters. Synthetically, for a gamma variate, $X \sim \textbf{Ga}(\theta,\eta)$. In Eq. (5.40), $\Gamma(\eta)$ represents the normalizing factor that ensures the density integrate to one as

$x \to \infty$. This normalizing factor is given by the complete gamma function $\Gamma(.)$, for the argument η, or

$$\Gamma(\eta) = \int_0^\infty x^{\eta-1} e^{-x} dx \tag{5.41}$$

If η is an integer number, the complete gamma function $\Gamma(\eta)$ is equivalent to $(\eta-1)!$. The reader is referred to Appendix 1 for a brief review of the properties of the gamma function. Appendix 2 contains tabulated values for the gamma function, for $1 \le \eta \le 2$. The recurrence relation $\Gamma(\eta+1) = \eta\Gamma(\eta)$ allows extending the calculation for other values of η. The gamma CDF is expressed as

$$F_X(x) = \int_0^x \frac{1}{\theta\Gamma(\eta)} \left(\frac{x}{\theta}\right)^{\eta-1} \exp\left(-\frac{x}{\theta}\right) dx \tag{5.42}$$

Similarly to the normal CDF, the integral in Eq. (5.42) cannot be calculated analytically. As a result, calculating probabilities for the gamma distribution requires numerical approximations. A simple approximation, generally efficient for values of η higher than 5, makes use of the gamma variate scaled by θ. In effect, due to the scaling property of the gamma distribution, if $X \sim \mathbf{Ga}(\theta,\eta)$, it can be shown that $\xi = x/\theta$ is also gamma distributed, with scale parameter $\theta_\xi = 1$ and shape parameter η. As such, the CDF of X can be written as the ratio

$$F_X(x) = \frac{\int_0^\xi \xi^{\eta-1} e^{-\xi} d\xi}{\int_0^\infty \xi^{\eta-1} e^{-\xi} d\xi} = \frac{\Gamma_i(\xi,\eta)}{\Gamma(\eta)} \tag{5.43}$$

between the incomplete gamma function, denoted by $\Gamma_i(\xi,\eta)$, and the complete gamma function $\Gamma(\eta)$. Kendall and Stuart (1963) show that, for $\eta \ge 5$, this ratio is well approximated by the standard normal CDF, $\Phi(u)$, calculated at the point u, which is defined by

$$u = 3\sqrt{\eta} \left(\sqrt[3]{\frac{\xi}{\eta}} - 1 + \frac{1}{9\eta} \right) \tag{5.44}$$

Example 5.6 applies this approximation procedure for calculating $F_X(x)$.

The expected value, variance and the coefficient of skewness for the gamma distribution are respectively given by

$$E[X] = \eta\theta \tag{5.45}$$

$$\text{Var}[X] = \eta\theta^2 \tag{5.46}$$

and

$$\gamma = \frac{2}{\sqrt{\eta}} \tag{5.47}$$

Figure 5.7 depicts some examples of gamma densities, for selected values of θ and η. Note in this figure that the effect of changing parameter θ, which possesses the same dimension as that of the gamma variate, is that of compressing or expanding the density, through the scaling of X. In turn, the great diversity of shapes, as shown by the gamma densities, is granted by the dimensionless parameter η. As illustrated in Fig. 5.7, for decreasing values of η, the gamma density becomes more skewed to the right, as one would expect from Eq. (5.47). For $\eta = 1$, the density intersects the vertical axis at ordinate $1/\theta$ and configures the particular case in which the gamma becomes the exponential distribution with parameter θ. As the shape parameter η grows, the gamma density becomes less skewed and its mode is increasingly shifted to the right. For very large values of η, as a direct result of Eq. (5.47), the gamma distribution tends to be symmetric. In fact, when $\eta \to \infty$, the gamma variate tends to be normally distributed. This is an expected result since, as $\eta \to \infty$, the gamma distribution would be the sum of an infinite number of independent exponential variables, which, by virtue of the CLT, tends to follow a normal distribution.

Given that it is defined only for nonnegative real-valued numbers, has non-fixed positive coefficients of skewness, and may exhibit a great diversity of shapes, the

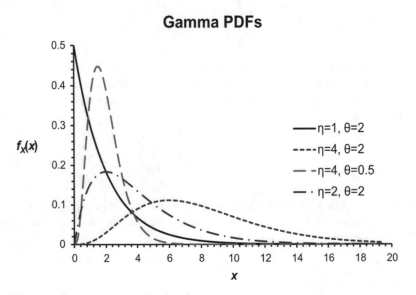

Fig. 5.7 Examples of gamma density functions

gamma distribution is a potentially strong candidate to model hydrological and hydrometeorological variables. In particular, Haan (1977) lists a number of successful applications of the gamma distribution to rainfall-related quantities, considering daily, monthly, and annual durations. Haan (1977) also gives an example of modeling annual mean flows with the use of the gamma distribution. On the other hand, Vlcek and Huth (2009), based on fittings of the gamma distribution to 90 samples of annual maximum daily rainfall depths, recorded at several locations across Europe, are skeptical and contend its use as a favored probabilistic model for extreme daily precipitations.

The exponential and gamma probability distributions are related to the discrete Poisson processes evolving in continuous time. The exponential distribution models the continuous time interval between two successive Poisson arrivals, whereas the gamma distribution refers to the continuous waiting time for the nth Poisson happening. Remember that in Chap. 4 the geometric and negative binomial discrete random variables have definitions that are conceptually similar to those attributed to the exponential and the gamma variates, respectively, with the caveat that their evolution takes place at discrete times. As such, the geometric and negative binomial models can be thought as the respective discrete analogues to the exponential and gamma distributions.

Example 5.6 Recalculate the probabilities sought in items (a) and (b) of Example 5.4, using the gamma distribution.

Solution First, the numerical values for parameters η and θ must be calculated. By combining Eqs. (5.45) and (5.46), the scale parameter θ can be directly calculated using $\mathrm{Var}[X] = E[X]\theta \Rightarrow \theta = \mathrm{Var}[X]/E[X] = (150)^2/600 = 37.5\,\mathrm{mm}$. Substituting this value into one of the two equations, it follows that $\eta = 16$. (a) $P(400 < X < 700) = F_X(700) - F_X(400)$. To calculate probabilities for the gamma distribution, one needs first to scale the variable by the parameter θ, or, for $x = 700$, $\xi = x/\theta = 700/37.5 = 18.67$. This quantity in Eq. (5.44), with $\eta = 16$, results in $u = 0.7168$. Table 5.1 gives $\Phi(0.7168) = 0.7633$, and thus $P(X < 700) = 0.7633$. Proceeding in the same way for $x = 400$, $P(X < 400) = 0.0758$. Therefore, $P(400 < X < 700) = 0.7633 - 0.0758 = 0.6875$. (b) $P(X > 300) = 1 - P(X < 300) = 1 - F_X(300)$. For $x = 300$, $\xi = x/\theta = 300/37.5 = 8$. Equation (5.44), with $\eta = 16$, results in $u = -2.3926$ and, finally, $\Phi(-2.3926) = 0.008365$. Hence, $P(X > 300) = 1 - 0.008365 = 0.9916$. Note that these results are not very different from those obtained in the solution to Example 5.4. The software MS Excel has the built-in function GAMMA.DIST(.) which returns the non-exceedance probability for a quantile, given the parameters. In R, the appropriate function is pgamma(.). Repeat the solution to this example, using the MS Excel resources.

5.6 Beta Distribution

The beta distribution models the probabilities of a double-bounded continuous random variable X. For the standard beta distribution, X is defined in the domain interval [0,1]. As such, the beta density is expressed as

$$f_X(x) = \frac{1}{B(\alpha,\beta)} x^{\alpha-1}(1-x)^{\beta-1} \text{ for } 0 \leq x \leq 1, \alpha > 0, \beta > 0 \qquad (5.48)$$

where α and β are parameters, and $B(\alpha,\beta)$ denotes the complete beta function given by

$$B(\alpha,\beta) = \int_0^1 t^{\alpha-1}(1-t)^{\beta-1}dt = \frac{\Gamma(\alpha)\Gamma(\beta)}{\Gamma(\alpha+\beta)} \qquad (5.49)$$

The short notation is $X \sim \mathbf{Be}(\alpha,\beta)$. The beta CDF is written as

$$F_X(x) = \frac{1}{B(\alpha,\beta)} \int_0^x x^{\alpha-1}(1-x)^{\beta-1}dx = \frac{B_i(x,\alpha,\beta)}{B(\alpha,\beta)} \qquad (5.50)$$

where $B_i(x,\alpha,\beta)$ denotes the incomplete beta function. When $\alpha = 1$, Eq. (5.50) can be solved analytically. However, for $\alpha \neq 1$, calculating probabilities for the beta distribution requires numerical approximations for the function $B_i(x,\alpha,\beta)$, such as the ones described in Press et al. (1986). Figure 5.8 depicts some possible shapes for the beta density.

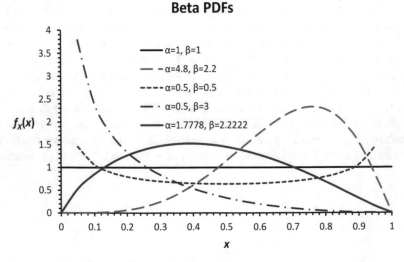

Fig. 5.8 Examples of densities for the beta distribution

The mean and variance of a beta variate are respectively given by

$$E[X] = \frac{\alpha}{\alpha + \beta} \tag{5.51}$$

and

$$\mathrm{Var}[X] = \frac{\alpha\beta}{(\alpha + \beta)^2(\alpha + \beta + 1)} \tag{5.52}$$

In Fig. 5.8, one can notice that the uniform distribution is a particular case of the beta distribution, for $\alpha = 1$ and $\beta = 1$. Parameter α controls the beta density values as it approaches the variate lower bound: if $\alpha < 1$, $f_X(x) \rightarrow \infty$ as $x \rightarrow 0$; if $\alpha = 1$, $f_X(0) = 1/B(1,\beta)$; and if $\alpha > 1$, $f_X(0) = 0$. Analogously, parameter β controls the beta density values as it approaches the variate upper bound. For equal values of both parameters, the beta density is symmetric. If both parameters are larger than 1, the beta density is unimodal. The great variety of shapes of the beta distribution makes it useful for modeling double-bounded continuous random variables. Example 5.7 refers to the modeling of dissolved oxygen concentrations with the beta distribution.

Example 5.7 Dissolved oxygen (DO) concentrations, measured in weekly intervals at a river cross section and denoted by X, are lower-bounded by 0 and upper-bounded by the concentration at saturation, which depends on many factors, especially, on the water temperature. Suppose the upper bound is 9 mg/l and that the DO concentration mean and variance are respectively equal to 4 mg/l and 4 $(\mathrm{mg/l})^2$. If the DO concentrations are scaled by the upper bound, or $Y = X/9$, one can model the transformed variable Y using the standard beta distribution. Use such a model to calculate the probability that the DO concentration be less than or equal to 2 mg/l.

Solution Recalling the properties of mathematical expectation, from Sects. 3.6.1 and 3.6.2 of Chap. 3, the mean and variance of the transformed variable Y are respectively equal to 4/9 and 4/81. Solving the system formed by Eqs. (5.51) and (5.52), one obtains $\alpha = 1.7778$ and $\beta = 2.2222$. Note that the beta density, for these specific values of α and β, is also plotted in Fig. 5.8. The probability that X is lower than 2 mg/l is equal to the probability that Y is lower than 2/9. To calculate $P[Y \leq (2/9)]$, by means of Eq. (5.50), it is necessary to perform a numerical approximation for the incomplete beta function, with the specified arguments $B_i[(2/9), \alpha = 1.7778, \beta = 2.2222]$. Besides the algorithm given in Press et al. (1986), the software MS Excel has the built-in function BETA.DIST(.), which implements the calculation as in Eq. (5.50). The equivalent function in R is pbeta (). Using one of these functions to calculate the sought probability, one obtains $P[Y \leq (2/9)] = 0.1870$. Thus, the probability that X is lower than or equal to 2 mg/l is 0.1870.

5.7   Extreme Value Distributions

A special category of probability distributions arise from the classical theory of extreme values, whose early developments are due to the pioneering works of the French mathematician Maurice Fréchet (1878–1973) and the British statisticians Ronald Fisher (1890–1962) and Leonard Tippet (1902–1985), followed by contributions made by the Russian mathematician Boris Gnedenko (1912–1995) and further consolidation by the German mathematician Emil Gumbel (1891–1966). The extreme-value theory is currently a very important and active branch of Mathematical Statistics, with developments of great relevance and applicability in the fields of actuarial sciences, economics, and engineering. The goal of this section is to introduce the principles of the extreme-value theory and present its main applications to hydrologic random variables. The reader interested in the mathematical foundations of classical extreme-value theory should consult Gumbel (1958). For an updated introduction to the topic, Coles (2001) is recommended reading. For general applications of the extreme-value theory to engineering, Castillo (1988) and Ang and Tang (1990) are very useful references.

5.7.1   Exact Distributions of Extreme Values

The maximum and minimum values of a sample of size N, from the random variable X, which is distributed according to a fully specified distribution $F_X(x)$, are also random variables and have their own probability distributions. These are generally related to the distribution $F_X(x)$, designated parent distribution, of the initial variate X. For the simple random sample $\{x_1, x_2, \ldots, x_N\}$, x_i denotes the ith record from the N available records for the variable X. Since it is not possible to predict the value of x_i before its occurrence, one can assume that x_i represents the realization of the random variable X_i, as corresponding to the ith random drawn from the population X. By generalizing this notion, one can interpret the sample $\{x_1, x_2, \ldots, x_N\}$ as the joint realization of N independent and identically distributed random variables $\{X_1, X_2, \ldots, X_N\}$. Based on this rationale, the theory of extreme values has the main goal of determining the probability distributions of the maximum $Y = \max\{X_1, X_2, \ldots, X_N\}$ and of the minimum $Z = \min\{X_1, X_2, \ldots, X_N\}$ of X.

If the parent distribution $F_X(x)$ is completely known, the distribution of Y can be derived from the fact that, if $Y = \max\{X_1, X_2, \ldots, X_N\}$ is equal to or less than y, then all random variables X_i must also be equal to or less than y. As all variables X_i are assumed independent and identically distributed, according to the function $F_X(x)$ of the initial variate X, the cumulative probability distribution of Y can be derived by equating

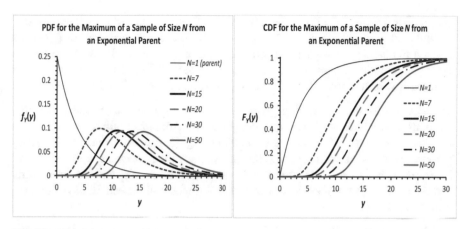

Fig. 5.9 Exact distributions for the maximum of a sample from an exponential parent

$$F_Y(y) = P(Y \le y) = P\left[(X_1 \le y) \cap (X_2 \le y) \cap \ldots \cap (X_N \le y)\right]$$
$$= [F_X(y)]^N \tag{5.53}$$

Thus, the density function of Y is

$$f_Y(y) = \frac{dF_Y(y)}{dy} = N[F_X(y)]^{N-1}f_X(y) \tag{5.54}$$

Equation (5.53) indicates that, since $F_X(y) < 1$ for any given y, $F_Y(y)$ decreases with N and, thus, both the cumulative and the density functions of Y are shifted to the right, along the X axis, as N increases. This is illustrated in Fig. 5.9 for the particular case where the parent probability density function is $f_X(x) = 0.25$ $\exp(-0.25x)$. Notice in this figure that the mode of Y shifts to the right as N increases, and that, even for moderate values of N, the probability that the maximum Y comes from the upper-tail of the parent density function $f_X(x)$ is high.

By employing similar arguments, one can derive the cumulative probability and density functions for the sample minimum $Z = \min\{X_1, X_2, \ldots, X_N\}$. Accordingly, the CDF of Z is given by

$$F_Z(z) = 1 - [1 - F_X(z)]^N \tag{5.55}$$

whereas its density is

$$f_z(z) = N[1 - F_X(z)]^{N-1}f_X(z) \tag{5.56}$$

Contrary to the sample maximum distributions, the functions $F_Z(z)$ and $f_Z(z)$ are shifted to the left along the X axis as N increases.

Equations (5.53)–(5.56) provide the exact distributions for the extreme values of a sample of size N, drawn from the population of the initial variate X, for which the parent distribution $F_X(x)$ is completely known. These equations show that the exact extreme-value distributions depend not only on the parent distribution $F_X(x)$ but also on the sample size N. In general, with an exception made for some simple parent distributions, such as the exponential used in Fig. 5.9, it is not always straightforward to derive the analytical expressions for $F_Y(y)$ and $F_Z(z)$.

Example 5.8 Assume that, in a given region, during the spring months, the time intervals between rain episodes are independent and exponentially distributed with a mean of 4 days. With the purpose of managing irrigation during the spring months, farmers need to know the maximum time between rain episodes. If 16 rain episodes are expected for the spring months, calculate the probability that the maximum time between successive episodes will exceed 10 days (adapted from Haan 1977).

Solution The prediction of 16 rain episodes for the spring months implies that 15 time intervals, separating successive rains, are expected, making $N = 15$ in Eq. (5.53). Denoting the maximum time interval between rains by $T_{\max}$, the sought probability is $P(T_{\max} > 10) = 1 - P(T_{\max} < 10) = 1 - F_{T_{\max}}(10)$. For the exponential parent distribution, with $\theta = 4$, Eq. (5.53) gives $F_{T_{\max}}(10) = [F_T(10)]^{15} = [1 - \exp(-10/4)]^{15} = 0.277$. Thus, $P(T_{\max} > 10) = 1 - 0.277 = 0.723$. The probability density function for $T_{\max}$ is derived by direct application of Eq. (5.54), or,

$$f_{T_{\max}}(t_{\max}) = 15\left[1 - \exp\left(-\tfrac{t_{\max}}{4}\right)\right]^{15-1}\left[\tfrac{1}{4}\exp\left(-\tfrac{t_{\max}}{4}\right)\right].$$ The cumulative and density functions, for $N = 15$, are highlighted in the plots of Fig. 5.9.

5.7.2 Asymptotic Distributions of Extreme Values

The practical usefulness of the statistical analysis of extremes is greatly enhanced by the asymptotic extreme-value theory, which focuses on determining the limiting forms of $F_Y(y)$ and $F_Z(z)$ as N tends to infinity, without previous knowledge of the exact shape of the parent distribution $F_X(x)$. In many practical situations, $F_X(x)$ is not completely known or cannot be determined from the available information, which prevents the application of Eqs. (5.53) and (5.56) and, thus, the derivation of the exact distributions of extreme values. The major contribution of the asymptotic extreme-value theory was to demonstrate that the limits $\lim_{N\to\infty} F_Y(y)$ and $\lim_{N\to\infty} F_Z(z)$ converge to some functional forms, despite the incomplete knowledge of the parent distribution $F_X(x)$. The convergence to the limiting forms actually depends on the tail characteristics of the parent distribution $F_X(x)$ in the direction of the desired extreme, i.e., it depends on the upper tail of $F_X(x)$, if the interest is on the maximum Y, or on the lower tail of $F_X(x)$, if the interest is on the minimum Z. The central portion of $F_X(x)$ does not have a significant influence on the convergence of $\lim_{N\to\infty} F_Y(y)$ and $\lim_{N\to\infty} F_Z(z)$.

Assume that $\{X_1, X_2, \ldots, X_N\}$ represents a collection of N independent random variables, with common parent distribution $F_X(x)$. For $Y = \max\{X_1, X_2, \ldots, X_N\}$ and $Z = \min\{X_1, X_2, \ldots, X_N\}$, let the linearly transformed variables Y_N and Z_N be defined as $Y_N = (Y - b_N)/a_N$ and $Z_N = (Z - b_N)/a_N$, where $a_N > 0$ and real-valued b_N are N-dependent sequences of normalizing constants intended to avoid degeneration of the limit distributions $\lim_{N\to\infty} F_Y(y)$ and $\lim_{N\to\infty} F_Z(z)$, or, in other words, to prevent the limits to take only values 0 and 1, as $N \to \infty$. The Fisher-Tippett theorem (Fisher and Tippett 1928), following the pioneering work by Fréchet (1927), and later complemented by Gnedenko (1943), establishes that, for a wide class of distributions $F_X(x)$, the limits $\lim_{N\to\infty} F_{Y_N}(y)$ and $\lim_{N\to\infty} F_{Z_N}(z)$ converge to only three types of functional forms, depending on the tail shape of the parent distribution $F_X(x)$ in the direction of the desired extreme. The three limiting forms are generally referred to and formally expressed as

- *Extreme-Value type I or EV1 or the double exponential form*

   (a) for maxima: $\exp[-e^{-y}]$, for $-\infty < y < \infty$; or
   (b) for minima: $1 - \exp(-e^z)$, for $-\infty < y < \infty$,
   if the initial variate X is unbounded in the direction of the desired extreme and if its probability density function decays as an exponential tail in the direction of the desired extreme;

- *Extreme-Value type II or EV2 or the simple exponential form*

   (a) for maxima: $\exp(-y^{-\gamma})$, for $y > 0$, and 0, for $y \leq 0$; or
   (b) for minima: $1 - \exp[-(-z)^{-\gamma}]$, for $z < 0$, and 1, for $z \geq 0$,
   if the initial variate X is unbounded in the direction of the desired extreme and if its probability density function decays as a polynomial (or Cauchy-Pareto or heavy) tail in the direction of the desired extreme; and

- *Extreme-Value type III or EV3 or the exponential form, with an upper bound for maxima or a lower bound for minima*

   (a) for maxima: $\exp[-(-y)^\gamma]$, for $y < 0$, and 1, for $y \geq 0$; or
   (b) for minima: $1 - \exp(-z^\gamma)$, for $z > 0$, and 0, for $z \leq 0$,
   when X is bounded in the direction of the desired extreme.

In the reduced expressions given for the three asymptotic forms, the exponent γ denotes a positive constant.

Taking only the case for maxima as an example, the parent distribution of the initial variate X has an exponential upper tail if it has no upper bound and if, for large values of x, the ordinates of $f_X(x)$ and of $1 - F_X(x)$ are small and the derivative $f'_X(x)$ is also small and negative, in such a way that the relation $f_X(x)/[1 - F_X(x)] = -f'_X(x)/f_X(x)$ holds (Ang and Tang 1990). In other words, the parent distribution is exponentially tailed if $F_X(x)$, in addition to being unbounded, approaches 1 at least as fast as the exponential distribution does as $x \to \infty$. In turn, $F_X(x)$ has a

Fig. 5.10 Examples of
upper-tail types for density
functions

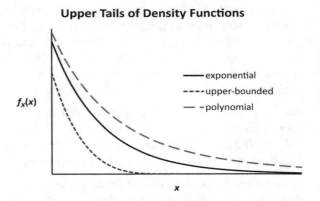

Upper Tails of Density Functions

$f_X(x)$

——— exponential
- - - - upper-bounded
— - polynomial

x

polynomial (or Cauchy-Pareto or heavy) upper tail if it is unbounded to the right
and if $\lim\limits_{x \to \infty} x^k [1 - F_X(x)] = a$, where a and k are positive numbers. In other words,
the parent distribution has a polynomial (Cauchy-Pareto, subexponential or heavy)
upper tail if $F_X(x)$, in addition to being unbounded from above, approaches 1 less
fast than the exponential distribution does, as $x \to \infty$. Finally, if X is bounded from
above by the value w, such that $F_X(w) = 1$, the limiting distribution of its maximum
converges to the Extreme-Value type III. Figure 5.10 illustrates the three types of
upper tails, by contrasting three parent density functions. Analogous explanations,
concerning the lower tails of the parent distribution, can be made for the minima
(Ang and Tang 1990).

The tail shape of the parent distribution, in the direction of the desired extreme,
determines to which of the three asymptotic forms the distribution of the maximum
(or the minimum) will converge. For maxima, the convergence to the EV type I
takes place if $F_X(x)$ is, for instance, exponential, gamma, normal, lognormal or the
EV1 itself; to the EV type II if $F_X(x)$ is log-gamma (the gamma distribution for the
logarithms of X), Student's t distribution (a sampling distribution to be described
later in this chapter) or the EV2 itself; and to the EV type III if $F_X(x)$ is uniform,
beta or the EV3 itself. For minima, the convergence to the EV type I takes place if
$F_X(x)$ is, for instance, normal or the EV1 itself; to the EV type II if $F_X(x)$ is Student's
t distribution or the EV2 itself; and to the EV type III if $F_X(x)$ is uniform,
exponential, beta, lognormal, gamma or the EV3 itself.

The three asymptotic forms from the classical extreme-value theory encounter
numerous applications in hydrology, despite its fundamental assumption, that of
independent and identically distributed (IID) initial variables, not fully conforming
to the hydrological reality. In order to provide a practical context, let Y and
Z respectively refer to the annual maximum and the annual minimum of the mean
daily discharges represented in the set $\{X_1, X_2, \ldots, X_{365}\}$, which, for rigorous
observance of the assumption of IID variables, must be independent among themselves and have a common and identical probability distribution. Independence
among daily flows is an unreasonable assumption, since the correlation between

successive daily discharges is usually very high. In turn, admitting, for instance, that, in a region of pronounced seasonality, the daily flows on January 16th have the same distribution, with the same mean and variance, as have the August 19th flows is clearly unrealistic. In addition to these departures from the basic assumption of IID variables, it is uncertain whether or not $N = 365$ (or 366) is large enough to allow convergence to the maximum or to the minimum.

Analogously to relaxing the basic premises required to apply the central limit theorem for the sum of a great number of random variables, further developments from the asymptotic theory of extreme values have attempted to accommodate eventual departures from the original assumption of IID variables. Juncosa (1949) proved that more general classes of limiting distribution exist if one drops the restriction that original random variables are identically distributed. Later on, Leadbetter (1974, 1983) demonstrated that for dependent initial variables, with some restrictions imposed on the long-range dependence structure of stationary sequences of exceedances over a high threshold, the asymptotic forms from the extreme-value theory remain valid. Thus, under somewhat weak regularity conditions, the assumption of independence among the original variables may be dropped (Leadbetter et al. 1983; Coles 2001).

In most applications of extremal distributions in hydrology, the determination of the appropriate asymptotic type is generally made on the basis of small samples of annual extreme data, without any information concerning the parent distribution of the original variables. However, according to Papalexiou et al. (2013), in such cases the actual behavior of the upper tail of the parent distribution might not be captured by the available data. To demonstrate this, the authors studied the tail behavior of non-zero daily rainfall depths, recorded at 15,029 rainfall gauging stations, spread over distinct geographic and climatic regions of the world, with samples of large sizes, ranging from 50 to 172 years of records. They concluded that, despite the widespread use of light-tailed distributions for modeling daily rainfall, 72.6 % of gauging stations showed upper tails of the polynomial or subexponential type, whereas only 27.4 % of them exhibited exponential or upper-bounded (or hyperexponential) upper tails. On the basis of these arguments and by relaxing the fundamental assumptions of the classical extreme-value theory, Papalexiou and Koutsoyiannis (2013) investigated the distributions of annual maxima of daily rainfall records from 15,135 gauging stations across the world. The authors' conclusion is categorically favorable to the use of the limiting distribution Extreme-Value type II, also known as the Fréchet distribution.

Although these recent studies present strong arguments in favor of using the Extreme-Value type II asymptotic distribution as a model for the annual maximum daily rainfalls, it seems that further research is needed in the pursuit of general conclusions on the extreme behavior of other hydrologic variables. As far as rainfall is concerned, the extension of the conclusions to extremes of sub-daily durations, generally more skewed than daily rainfall extremes, or to less-skewed extremes of m-day durations ($m > 1$), is certainly not warranted without previous verification. As far as discharges are concerned, additional difficulties for applying asymptotic extreme-value results may arise since the usual stronger correlation among successive daily flows can possibly affect the regularity conditions, under which the independence assumption can be disregarded (Perichi and Rodríguez-Iturbe 1985). There are other issues of concern in respect of the application of extremal asymptotic distributions in hydrology, such as the slow convergence to the maximum (or minimum), as noted by Gumbel (1958) and, more recently, by Papalexiou and Koutsoyiannis (2013), and the fact that the three asymptotic forms are not exhaustive (see Juncosa 1949, Benjamin and Cornell 1970, and Kottegoda and Rosso 1997). Nevertheless, the extremal distributions are the only group of probability distributions that offers theoretical arguments that justify their use in the modeling of hydrologic maxima and minima, although such arguments might be somewhat debatable for a wide variety of applications.

The Extreme-Value type I for maxima, also known as the $Gumbel_{max}$ distribution for maxima, has been, and still is a popular model for the hydrologic frequency analysis of annual maxima. The Extreme-Value type II for maxima, also known as the $Fréchet_{max}$ distribution for maxima, had a considerably limited use in hydrologic frequency analysis, as compared to the first asymptotic type, until recent years. However, it is expected that its use as a model for hydrologic annual maxima will increase in the future, following the results published by Papalexiou and Koutsoyiannis (2013). The Extreme-Value type III for maxima, also known as the $Weibull_{max}$ distribution for maxima, is often disregarded as a model for the hydrologic frequency analysis of annual maxima, since one of its parameters is, in fact, the variate's upper-bound, which is always very difficult to estimate from small samples of extreme data. Thus, as for the models for maxima to be described in the subsections that follow, focus will initially be on the $Gumbel_{max}$ and $Fréchet_{max}$ distributions, and posteriorly on the Generalized Extreme-Value (GEV) distribution, which condenses in a single parametric form all three extremal types. Regarding the minima, the most frequently used models in hydrology practice, namely, the $EV1_{min}$ or the Gumbel distribution for minima, and the $EV3_{min}$ or the Weibull distribution for minima, are highlighted.

5.7.2.1   Gumbel Distribution for Maxima

The Extreme-Value type I for maxima, also referred to as $Gumbel_{max}$, Fisher-Tippet I or double exponential for maxima, arises from the classical extreme-value theory, as the asymptotic form for the maximum of a set of IID initial variates, with an exponential upper tail. Over the years, it has probably been the most used probability distribution in frequency analysis of hydrologic variables, with particular emphasis on its application for estimating IDF (Intensity-Duration-Frequency) relations for storm rainfalls at a given location.

The CDF of the $Gumbel_{max}$ is given by

$$F_Y(y) = \exp\left[-\exp\left(-\frac{y-\beta}{\alpha}\right)\right] \text{ for } -\infty < y < \infty, \ -\infty < \beta < \infty, \alpha > 0$$

(5.57)

where α denotes the scale parameter and β the location parameter. This latter parameter actually corresponds to the mode of Y. The distribution's density function is

$$f_Y(y) = \frac{1}{\alpha} \exp\left[-\frac{y-\beta}{\alpha} - \exp\left(-\frac{y-\beta}{\alpha}\right)\right]$$

(5.58)

The expected value, variance, and coefficient of skewness are respectively given by

$$E[Y] = \beta + 0.5772\alpha$$

(5.59)

$$\text{Var}[Y] = \sigma_Y^2 = \frac{\pi^2\alpha^2}{6}$$

(5.60)

and

$$\gamma = 1.1396$$

(5.61)

It is worth noting that the Gumbel$_{\max}$ distribution has a constant and positive coefficient of skewness. Figure 5.11 depicts some examples of Gumbel$_{\max}$ densities, for selected values of parameters α and β.

The inverse function, or quantile function, for the Gumbel$_{\max}$ distribution is

$$y(F) = \beta - \alpha\ln[-\ln(F)] \text{ or } y(T) = \beta - \alpha\ln\left[-\ln\left(1 - \frac{1}{T}\right)\right]$$

(5.62)

where T denotes the return period, in years, and F represents the annual non-exceedance probability. By replacing y for the expected value $E[Y]$ in Eq. (5.62), it follows that the Gumbel$_{\max}$ mean has a return period of $T = 2.33$ years. In early developments of regional flood frequency analysis, the quantile y $(T = 2.33)$ was termed *mean annual flood* (Dalrymple 1960).

Example 5.9 Denote by X the random variable *annual maximum daily discharge*. Assume that, at a given location, $E[X] = 500$ m^3/s and $E[X^2] = 297025$ (m^3/s)^2. Employ the Gumbel$_{\max}$ model to calculate: (a) the annual maximum daily discharge of return period $T = 100$ years; and (b) given that the annual maximum daily discharge is larger than 600 m^3/s, the probability that X exceeds 800 m^3/s in any given year.

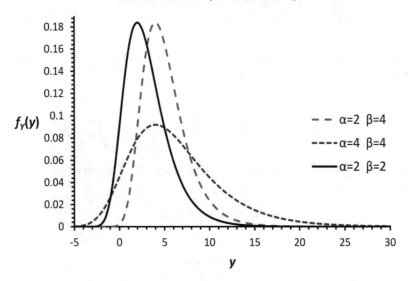

Fig. 5.11 Examples of Gumbel$_{max}$ density functions

Solution (a) Recalling that $Var[X] = E[X^2] - (E[X])^2$, it follows that $Var[X] = 47025 \ (m^3/s)^2$. Solving the system formed by Eqs. (5.59) and (5.60), the parameter values are $\alpha = 169.08 \ m^3/s$ and $\beta = 402.41 \ m^3/s$. Substituting these values in Eq. (5.62), the annual maximum daily discharge of return period $T = 100$ years, according to the Gumbel$_{max}$ model, is $x(100) = 1180 \ m^3/s$. (b) Let the events A and B represent the outcomes $\{X > 600 \ m^3/s\}$ and $\{X > 800 \ m^3/s\}$, respectively. Thus, the sought probability is $P(B|A)$, which is the same as $P(B|A) = P(B \cap A)/P(A)$. Since $P(B \cap A) = P(B)$, the numerator of the previous equation corresponds to $P(B) = 1 - F_X(800) = 0.091$, whereas the denominator, in turn, to $P(A) = 1 - F_X(600) = 0.267$. Thus, $P(B|A) = 0.34$.

5.7.2.2 Fréchet Distribution for Maxima

The Fréchet$_{max}$ distribution arises, from the classical extreme-value theory, as the asymptotic form for the maximum of a set of IID initial variates, with polynomial (or heavy) tail. It is also known as log-Gumbel$_{max}$ due to the fact that, if $Z \sim$ Gumbel$_{mx}(\alpha, \beta)$, then $Y = \ln(Z) \sim$ Fréchet$_{max}[1/\alpha, \exp(\beta)]$. The distribution is named after the French mathematician Maurice Fréchet (1878–1973), one of the pioneers of classical extreme-value theory. In recent years, heavy-tailed distributions, such as the Fréchet model, are being increasingly recommended as extremal distributions for hydrologic maxima (Papalexiou and Koutsoyiannis 2013).

The standard two-parameter Fréchet$_{max}$ CDF is expressed as

$$F_Y(y) = \exp\left[-\left(\frac{\tau}{y}\right)^\lambda\right] \text{ for } y > 0, \ \tau, \lambda > 0 \qquad (5.63)$$

where τ represents the scale parameter and λ is the shape parameter. The Fréchet$_{max}$ density function is given by

$$f_Y(y) = \frac{\theta}{\tau}\left(\frac{\tau}{y}\right)^{\lambda+1}\exp\left[-\left(\frac{\tau}{y}\right)^\lambda\right] \qquad (5.64)$$

The expected value, variance, and coefficient of variation of Y are respectively given by

$$E[Y] = \tau\Gamma\left(1 - \frac{1}{\lambda}\right) \text{ for } \lambda > 1 \qquad (5.65)$$

$$\text{Var}[Y] = \sigma_Y^2 = \tau^2\left[\Gamma\left(1 - \frac{2}{\lambda}\right) - \Gamma^2\left(1 - \frac{1}{\lambda}\right)\right] \text{ for } \lambda > 2 \qquad (5.66)$$

and

$$\text{CV}_Y = \sqrt{\frac{\Gamma(1 - 2/\lambda)}{\Gamma^2(1 - 1/\lambda)} - 1} \text{ for } \lambda > 2 \qquad (5.67)$$

It is worth noting that the shape parameter λ depends only on the coefficient of variation of Y, a fact that makes the calculation of Fréchet$_{max}$ parameters a lot easier. In effect, if CV_Y is known, Eq. (5.67) can be solved for λ, by using numerical iterations. Eq. (5.65) is then solved for τ. Figure 5.12 shows examples of the Fréchet$_{max}$ density function, for some specific values of τ and λ.

The Fréchet$_{max}$ quantile function, for the annual non-exceedance probability F, is

$$y(F) = \tau[-\ln(F)]^{1/\lambda} \qquad (5.68)$$

or, in terms of the return period T,

$$y(T) = \tau\left[\ln\left(\frac{T}{T-1}\right)\right]^{-1/\lambda} \qquad (5.69)$$

As mentioned earlier, the Gumbel$_{max}$ and Fréchet$_{max}$ distributions are related by the logarithmic transformation of variables. In effect, if Y is a Fréchet$_{max}$ variate, with parameters τ and λ, the transformed variable $\ln(Y)$ follows a Gumbel$_{max}$ distribution, with parameters $\alpha = 1/\lambda$ and $\beta = \ln(\tau)$. This mathematical fact implies that, for a given return period, the corresponding quantile, as calculated with

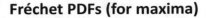

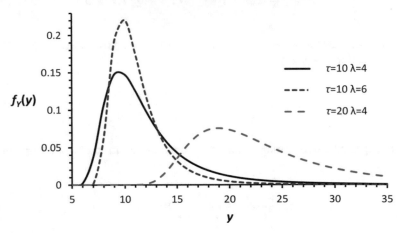

Fig. 5.12 Examples of Fréchet$_{max}$ density functions

Fréchet$_{max}$, is much larger than that calculated with Gumbel$_{max}$. The three-parameter Fréchet distribution, with an added location parameter, is a particular case of the GEV model, which is described in the next subsection.

5.7.2.3 Generalized Extreme-Value (GEV) Distribution for Maxima

The Generalized Extreme-Value (GEV) distribution for maxima was introduced by Jenkinson (1955) as the condensed parametric form for the three limiting distributions for maxima. The CDF of the GEV distribution is given by

$$F_Y(y) = \exp\left\{ -\left[1 - \kappa\left(\frac{y-\beta}{\alpha}\right) \right]^{1/\kappa} \right\} \tag{5.70}$$

where κ, α, and β denote, respectively, the parameters of shape, scale, and location. The value and sign of κ determine the asymptotic extremal type: if $\kappa < 0$, the GEV becomes the Extreme-Value type II, with the domain in $y > \beta + \alpha/\kappa$, whereas if $\kappa > 0$, the GEV corresponds to the Extreme-Value type III, with the domain in $y < \beta + \alpha/\kappa$. If $\kappa = 0$, The GEV becomes the Gumbel$_{max}$, with scale parameter α and shape parameter β. The GEV probability density function is expressed as

$$f_Y(y) = \frac{1}{\alpha}\left[1 - \kappa\left(\frac{y-\beta}{\alpha}\right) \right]^{1/\kappa - 1} \exp\left\{ -\left[1 - \kappa\left(\frac{y-\beta}{\alpha}\right) \right]^{1/\kappa} \right\} \tag{5.71}$$

Figure 5.13 illustrates the three possible shapes for the GEV distribution, as functions of the value and sign of κ.

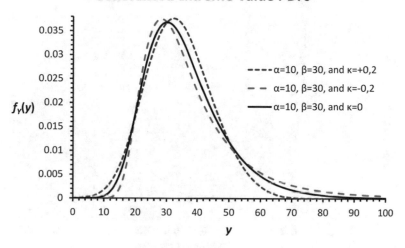

Fig. 5.13 Examples of GEV density functions

The moments of order r for the GEV distribution only exist for $\kappa > -1/r$. As a result, the mean of a GEV variate is not defined for $\kappa < -1$, the variance does not exist for $\kappa < -1/2$, and the coefficient of skewness exists only for $\kappa > -1/3$. Under these restrictions, the mean, variance, and coefficient of skewness for a GEV variate are respectively defined as

$$E[Y] = \beta + \frac{\alpha}{\kappa}[1 - \Gamma(1 + \kappa)] \tag{5.72}$$

$$\text{Var}[Y] = \left(\frac{\alpha}{\kappa}\right)^2 \left[\Gamma(1 + 2\kappa) - \Gamma^2(1 + \kappa)\right] \tag{5.73}$$

and

$$\gamma = \frac{\kappa}{|\kappa|}\left\{\frac{-\Gamma(1 + 3\kappa) + 3\Gamma(1 + \kappa)\Gamma(1 + 2\kappa) - 2\Gamma^3(1 + \kappa)}{\left[\Gamma(1 + 2\kappa) - \Gamma^2(1 + \kappa)\right]^{3/2}}\right\} \tag{5.74}$$

Equation (5.74) shows that the GEV shape parameter κ depends only on the coefficient of skewness γ of Y. This is a one-to-one dependence relation, which is depicted in the graph of Fig. 5.14, for $\kappa > -1/3$. It is worth noting in this figure that the point marked with the cross corresponds to the $\text{Gumbel}_{\max}$ distribution, for which $\kappa = 0$ and $\gamma = 1.1396$.

The calculation of the GEV parameters should start from Eq. (5.74), which needs to be solved for κ, given that the value of the coefficient of skewness γ is known. The solution is performed through numerical iterations, starting from suitable initial

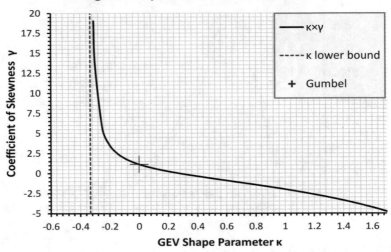

Fig. 5.14 Relation between the shape parameter κ and the coefficient of skewness γ of a GEV variate, valid for $\kappa > -1/3$

values, which may be obtained from the graph in Fig. 5.14. Calculations then proceed for α, by making it explicit in Eq. (5.73), or

$$\alpha = \sqrt{\frac{\kappa^2 \text{Var}[Y]}{\Gamma(1+2\kappa) - \Gamma^2(1+\kappa)}} \qquad (5.75)$$

and, finally, by rearranging Eq. (5.72), β is calculated with

$$\beta = E[Y] - \frac{\alpha}{\kappa}[1 - \Gamma(1+\kappa)] \qquad (5.76)$$

With the numerical values for the GEVparameters, the quantiles are given by

$$x(F) = \beta + \frac{\alpha}{\kappa}[1 - (-\ln F)^\kappa] \qquad (5.77)$$

or, in terms of the return period T,

$$x(T) = \beta + \frac{\alpha}{\kappa}\left\{1 - \left[-\ln\left(1 - \frac{1}{T}\right)\right]^\kappa\right\} \qquad (5.78)$$

Example 5.10 Employ the GEV model to solve Example 5.9, supposing the coefficient of skewness of X is $\gamma = 1.40$.

Solution (a) With $\gamma = 1.40$ in Fig. 5.14, one can notice that the value of κ that satisfies Eq. (5.74) lies somewhere between -0.10 and 0. The solution to this example is achieved by using the MS Excel software. First, choose an Excel cell, which is termed *cell M*, to contain the value of the shape parameter κ and assign to it an initial value between -0.10 and 0, say -0.08. MS Excel has the built-in function GAMMALN(w) which returns the natural logarithm of the gamma function for the argument w, such that EXP((GAMMALN(w)) be equal to $\Gamma(w)$. Choose another Excel cell, here called *cell N*, and assign to it the resulting subtraction of both sides of Eq. (5.74), making use of the GAMMALN(.) function. Note that, if the result of this subtraction is zero, one has the value of κ that satisfies Eq. (5.74). Then, indicate to the *Solver* Excel component that the goal is to make the result, assigned to *cell N*, which contains the rearranged Eq. (5.74), as close as possible to zero, by changing the content of *cell M*, which is assigned to κ. Proceeding in this way, one obtains $\kappa = -0.04$. With this result and knowing that Var$[X] = 47025$ (m³/s)² in Eq. (5.75), the solution for the scale parameter is $\alpha = 159.97$. Finally, for the location parameter, Eq. (5.76) gives $\beta = 401.09$. Then, by making use of Eq. (5.78), one obtains $x(100) = 1209$ m³/s. (b) Using the same representations for events A and B, as in the solution to Example 5.9, $P(B|A) = 0.345$.

Example 5.11 Solve Example 5.5 for the case where the exceedances $(Q-Q_0)$ have the mean and standard deviation respectively equal to 50 and 60 m³/s and are distributed according to a Generalized Pareto (GPA) distribution.

Solution This is a Poisson process with $\nu = 2$ as the annual mean number of arrivals. When they occur, the exceedances $X = (Q-Q_0)$ follow a Generalized Pareto (GPA) distribution, with CDF $G_X(x) = 1 - \left[1 - \kappa\left(\frac{x}{a}\right)\right]^{1/\kappa}$, where κ and α denote, respectively, the shape and scale parameters. For $\kappa > 0$, the variate is upper-bounded by α/κ and, for $\kappa < 0$ it is unbounded to the right, with a polynomial upper tail. If $\kappa = 0$, the GPA becomes the exponential distribution, in the form of $G_X(x) = 1 - \exp(-x/\alpha)$, for which case the solution is given in Example 5.5. The GPA distribution is named after Vilfredo Pareto (1848–1923), an Italian civil engineer and economist who first used the GPA, in the field of economics, as the income distribution. An important theorem related to the GPA parametric form, introduced by Pickands (1975), states that if we consider only the values of a generic random variable W that exceed a sufficiently high threshold w_0, then, the conditional exceedances distribution $F(W - w_0|W \geq w_0)$ converges to a GPA as w_0 increases. This result has been used to characterize the upper tails of probability distributions, in connection with the extreme-value limiting forms. The density upper tails depicted in Fig. 5.10 are, in fact, GPA densities for positive, negative, and null values of the shape parameter κ. For a GPA variate X, the following equations must hold: $\alpha = \frac{E[X]}{2}\left[\frac{(E[X])^2}{\text{Var}[X]} + 1\right]$ and $\kappa = \frac{1}{2}\left[\frac{(E[X])^2}{\text{Var}[X]} - 1\right]$. Solving these equations, with $E[X] = 50$ and Var$[X] = 3600$, one obtains $\alpha = 42.36$ and $\kappa = -0.153$. Thus, for this case, the conditional exceedance distribution is unbounded to the right and has a polynomial upper tail. Similarly to Example 5.5, in order to calculate flood flow quantiles associated with a specific return period, it is necessary to derive the CDF of the annual maximum exceedances, denoted by $F_{X\max}(x)$, which, in terms of the partial results from Example 5.5, is given by $F_{X_{\max}}(x) = \exp\{-\nu[1 - G_X(x)]\}$. If $G_X(x)$ is a GPA, then the *Poisson-Pareto model* for partial duration series, formally expressed as $F_{Q_{\max}}(q) = \exp\left\{-\nu\left[1 - \kappa\left(\frac{q-Q_0}{\alpha}\right)\right]^{1/\kappa}\right\}$, where $Q_{\max} = Q_0 + X$ represents the annual maximum discharge. After algebraic simplifications, similar to those from Example 5.5, one obtains $F_{Q_{\max}}(q) = \exp\{-[1 - \kappa(\frac{q-\beta}{\alpha^*})]\}$, which is the GEV cumulative distribution function, with scale parameter $\alpha^* = \alpha(\nu)^{-\kappa}$ and location parameter $\beta = Q_0 + (\alpha - \alpha^*)/\kappa$. The GEV shape parameter κ is identical to that of the GPA distribution. In conclusion, modeling partial duration series with the number of flood occurrences following a Poisson law and exceedances distributed as a GPA leads to a GEV as the probability distribution of annual maxima. The specific quantities for this example are: $F_{Q_{\max}} = 1 - 1/100 = 0.99$; $\nu = 2$, $Q_0 = 60$ m^3/s, $\alpha = 42.36$, $\kappa = -0.153$, $\alpha^* = 47.1$, and $\beta = 90.96$. Inverting the GEV cumulative function, it follows that $q(F) = \beta + \frac{\alpha^*}{\kappa}\{1 - [-\ln(F)]^{\kappa}\}$. Substituting the numerical values in the equation, one obtains the 100-year flood as $q(F = 0.99) = 342.4$ m^3/s.

5.7.2.4 Gumbel Distribution for Minima

The Gumbel$_{\min}$ arises from the classical extreme-value theory as the asymptotic form for the minimum of a set of IID initial variates, with an exponential lower tail. The distribution has been used, although not frequently, as an extremal distribution for modeling annual minima of drought-related variables, such as the Q_7, the lowest average discharge of 7 consecutive days in a given year.

The cumulative distribution function for the Gumbel$_{\min}$ is

$$F_Z(z) = 1 - \exp\left[-\exp\left(\frac{z - \beta}{\alpha}\right)\right] \text{ for } -\infty < z < \infty, \ -\infty < \beta < \infty, \ \alpha > 0 \tag{5.79}$$

where α represents the scale parameter and β the location parameter. Analogously to the Gumbel$_{\max}$, β is, in fact, the mode of Z. The probability density function of the Gumbel$_{\min}$ distribution is given by

$$f_Z(z) = \frac{1}{\alpha}\exp\left[\frac{z - \beta}{\alpha} - \exp\left(\frac{z - \beta}{\alpha}\right)\right] \tag{5.80}$$

The mean, variance, and coefficient of skewness of a Gumbel$_{\min}$ variate are respectively expressed as

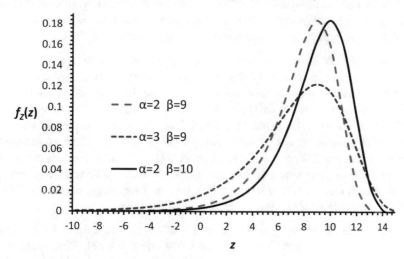

Fig. 5.15 Examples of Gumbel$_{min}$ density functions

$$E[Z] = \beta - 0.5772\alpha \qquad (5.81)$$

$$\text{Var}[Z] = \sigma_Z^2 = \frac{\pi^2 \alpha^2}{6} \qquad (5.82)$$

and

$$\gamma = -1.1396 \qquad (5.83)$$

It is worth noting that the Gumbel$_{min}$ distribution is skewed to the left with a fixed coefficient of $\gamma = -1.1396$. The Gumbel$_{min}$ and Gumbel$_{max}$ probability density functions, both with identical parameters, are symmetrical with respect to a vertical line crossing the abscissa axis at the common mode β. Figure 5.15 shows examples of the Gumbel$_{min}$ distribution, for some specific values of parameters α and β.

The Gumbel$_{min}$ quantile function is written as

$$z(F) = \beta + \alpha \ln[-\ln(1 - F)] \quad \text{or} \quad y(T) = \beta + \alpha \ln\left[-\ln\left(1 - \frac{1}{T}\right)\right] \qquad (5.84)$$

where T denotes the return period, in years, and F represents the annual non-exceedance probability. Remember that, for annual minima, the return period is the reciprocal of F, or $T = 1/P(Z \leq z) = 1/F_Z(z)$. It is worth noting that, depending on the numerical values of the distribution parameters and on the target return period, calculation of Gumbel$_{min}$ quantiles can possibly yield negative numbers, as in the examples depicted in Fig. 5.15. This is one major disadvantage of the Gumbel$_{min}$ distribution, which has prevented its spread use as a model for low-flow frequency analysis. The other is related to the fixed negative coefficient of skewness. This fact seems to be in disagreement with low-flow data samples, which may exhibit, in many cases, histograms moderately skewed to the right.

Example 5.12 The variable *annual minimum 7-day mean flow*, denoted by Q_7, is the lowest average discharge of 7 consecutive days in a given year and is commonly used for the statistical analysis of low flows. The yearly values of Q_7 from 1920 to 2014, as reduced from the daily flows for the Dore River at Saint-Gervais-sous-Meymont, in France, are listed in Table 2.7 of Chap. 2. The sample mean and standard deviation are $\bar{z} = 1.37$ m^3/s and $s_Z = 0.787$ m^3/s, and these are assumed as good estimates for $E[Z]$ and $\sigma[Z]$, respectively. $Q_{7,10}$ denotes the minimum 7-day mean flow of 10-year return period and has been used as a drought characteristic flow in water resources engineering. Employ the Gumbel$_{min}$ model to estimate $Q_{7,10}$ for the Dore River at Saint-Gervais-sous-Meymont.

Solution With $E[Z] = 1.37$ m^3/s and $\sigma[Z] = 0.787$ m^3/s, solutions to the system formed by Eqs. (5.81) and (5.82) give $\alpha = 0.614$ and $\beta = 1.721$. With these results and by making $T = 10$ years in Eq. (5.84), the estimate of $Q_{7,10}$ obtained with the Gumbel$_{min}$ model is $z(T = 10) = 0.339$ m^3/s.

5.7.2.5 Weibull Distribution for Minima

The Extreme-Value type III distribution arises from the classical extreme-value theory as the asymptotic form for the minimum of a set of IID initial variates with a lower-bounded tail. The Extreme-Value type III distribution for minima is also known as Weibull$_{min}$, after being first applied for the analysis of the strength of materials to fatigue, by the Swedish engineer Waloddi Weibull (1887–1979). Since low flows are inevitably bounded by zero in the most severe cases, the Weibull$_{min}$ distribution is a natural candidate to model hydrologic minima. If low-flows are lower-bounded by zero, the EV3 distribution is referred to as the two-parameter Weibull$_{min}$. On the other hand, if low-flows are lower-bounded by some value ξ, the EV3 distribution is referred to as the three-parameter Weibull$_{min}$.

The cumulative distribution function for the two-parameter Weibull$_{min}$ is given by

$$F_Z(z) = 1 - \exp\left[-\left(\frac{z}{\beta}\right)^\alpha\right] \text{ para } z \geq 0, \beta \geq 0 \text{ e } \alpha > 0 \qquad (5.85)$$

where β and α are, respectively, scale and shape parameters. If $\alpha = 1$, the Weibull$_{min}$ becomes the one-parameter exponential distribution with scale parameter β. The probability density function of the two-parameter Weibull$_{min}$ distribution is expressed as

$$f_Z(z) = \frac{\alpha}{\beta}\left(\frac{z}{\beta}\right)^{\alpha-1} \exp\left[-\left(\frac{z}{\beta}\right)^{\alpha}\right] \tag{5.86}$$

The mean and variance of a two-parameter Weibull$_{min}$ variate are, respectively, given by

$$E[Z] = \beta\Gamma\left(1 + \frac{1}{\alpha}\right) \tag{5.87}$$

and

$$\text{Var}[Z] = \beta^2\left[\Gamma\left(1 + \frac{2}{\alpha}\right) - \Gamma^2\left(1 + \frac{1}{\alpha}\right)\right] \tag{5.88}$$

The coefficients of variation and skewness of a two-parameter Weibull$_{min}$ variate are, respectively,

$$\text{CV}_Z = \frac{\sqrt{\Gamma\left(1 + \frac{2}{\alpha}\right) - \Gamma^2\left(1 + \frac{1}{\alpha}\right)}}{\Gamma\left(1 + \frac{1}{\alpha}\right)} = \frac{\sqrt{B(\alpha) - A^2(\alpha)}}{A(\alpha)} \tag{5.89}$$

and

$$\gamma = \frac{\Gamma\left(1 + \frac{3}{\alpha}\right) - 3\Gamma\left(1 + \frac{2}{\alpha}\right)\Gamma\left(1 + \frac{1}{\alpha}\right) + 2\Gamma^3\left(1 + \frac{1}{\alpha}\right)}{\sqrt{\left[\Gamma\left(1 + \frac{2}{\alpha}\right) - \Gamma^2\left(1 + \frac{1}{\alpha}\right)\right]^3}} \tag{5.90}$$

Figure 5.16 illustrates examples of two-parameter Weibull$_{min}$ density functions.

Calculation of parameters and probabilities for the two-parameter Weibull$_{min}$ distribution is performed by first solving Eq. (5.89) for α, either through a numerical iterations procedure, similar to the one used to calculate the GEV shape parameter (see solution to Example 5.10), or by tabulating (or regressing) possible values of α and the auxiliary function $A(\alpha) = \Gamma(1 + 1/\alpha)$ against CV_Z. Analysis of the dependence of α and $A(\alpha)$ on the coefficient of variation CV_Z leads to the following correlative relations:

$$\alpha = 1.0079(\text{CV})^{-1.084}, \quad \text{for } 0.08 \leq \text{CV}_Z \leq 2 \tag{5.91}$$

and

$$A(\alpha) = -0.0607(\text{CV}_Z)^3 + 0.5502(\text{CV}_Z)^2 - 0.4937(\text{CV}_Z) \\ + 1.003, \text{ for } 0.08 \leq \text{CV}_Z \leq 2 \tag{5.92}$$

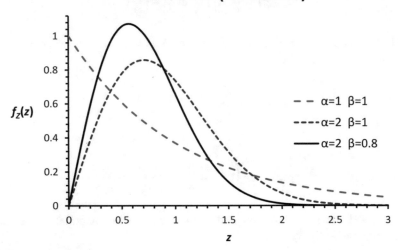

Fig. 5.16 Examples of density functions for the two-parameter Weibull$_{min}$ distribution

which provide good approximations to the numerical solution of Eq. (5.89). Once α and $A(\alpha)$ have been determined, parameter β can be calculated from Eq. (5.87), or

$$\beta = \frac{E[Z]}{A(\alpha)} \tag{5.93}$$

With both parameters known, the two-parameter Weibull$_{min}$ quantiles are determined by

$$z(F) = \beta \ [-\ln(1 - F)]^{\frac{1}{\alpha}} \text{ or } z(T) = \beta \left[-\ln\left(1 - \frac{1}{T}\right)\right]^{\frac{1}{\alpha}} \tag{5.94}$$

Example 5.13 Employ the Weibull$_{min}$ model to estimate $Q_{7,10}$ for the Dore River at Saint-Gervais-sous-Meymont, with $E[Z] = 1.37$ m^3/s and $\sigma[Z] = 0.787$ m^3/s.

Solution For $E[Z] = 1.37$ m^3/s and $\sigma[Z] = 0.787$ m^3/s, it follows that $CV_Z = 0.5753$. Eqs. (5.91) and (5.92) yield, respectively, $\alpha = 1.8352$ and $A(\alpha) = 0.9126$. Equation (5.93) gives $\beta = 1.5012$. Finally, with $\alpha = 1.8352$, $\beta = 1.5012$, and by making $T = 10$ years in Eq. (5.94), the estimate of $Q_{7,10}$ obtained with the Weibull$_{min}$ model is $z(T = 10) = 0.44$ m^3/s.

For the three-parameter Weibull$_{min}$, the density and cumulative distribution functions become

$$f_Z(z) = \alpha \left(\frac{z-\xi}{\beta-\xi}\right)^{\alpha-1} \exp\left[-\left(\frac{z-\xi}{\beta-\xi}\right)^{\alpha}\right] \text{ for } z > \xi, \beta \geq 0 \text{ e } \alpha > 0 \qquad (5.95)$$

and

$$F_Z(z) = 1 - \exp\left[-\left(\frac{z-\xi}{\beta-\xi}\right)^{\alpha}\right] \qquad (5.96)$$

The mean and variance of a three-parameter Weibull$_{\text{min}}$ variate are, respectively,

$$E[Z] = \xi + (\beta - \xi)\Gamma\left(1 + \frac{1}{\alpha}\right) \qquad (5.97)$$

and

$$\text{Var}[Z] = (\beta - \xi)^2 \left[\Gamma\left(1 + \frac{2}{\alpha}\right) - \Gamma^2\left(1 + \frac{1}{\alpha}\right)\right] \qquad (5.98)$$

According to Haan (1977), the following relations hold for a three-parameter Weibull$_{\text{min}}$ distribution:

$$\beta = E[Z] + \sigma_Z C(\alpha) \qquad (5.99)$$

and

$$\xi = \beta - \sigma_Z D(\alpha) \qquad (5.100)$$

where

$$C(\alpha) = D(\alpha)\left[1 - \Gamma\left(1 + \frac{1}{\alpha}\right)\right] \qquad (5.101)$$

and

$$D(\alpha) = \frac{1}{\sqrt{\Gamma\left(1 + \frac{2}{\alpha}\right) - \Gamma^2\left(1 + \frac{1}{\alpha}\right)}} \qquad (5.102)$$

The expression for the coefficient of skewness, given by Eq. (5.90) for the two-parameter Weibull$_{\text{min}}$, still holds for the three-parameter variant of this distribution. It is worth noting that the coefficient of skewness is a function of α only, and this fact greatly facilitates the procedures for calculating the parameters of the distribution. These are: (1) first, with the coefficient of skewness γ, α is determined by solving Eq. (5.90), through numerical iterations; (2) then, $C(\alpha)$ and $D(\alpha)$ are calculated with Eqs. (5.101) and (5.102), respectively; and, finally, (3) β and ξ are determined from Eqs. (5.99) and (5.100).

5.8   Pearson Distributions

The influential English mathematician Karl Pearson (1857–1936) devised a system of continuous probability distributions, whose generic density functions can be written in the form

$$f_X(x) = \exp\left[-\int_{-\infty}^{x} \frac{a+x}{c_0 + c_1 x + c_2 x^2} \, dx \right] \tag{5.103}$$

where specific values of coefficients c_0, c_1, and c_2, for the quadratic function in the integrand's denominator, define eight distinct types of probability distributions, with different levels of skewness and kurtosis, including the normal, gamma, beta, and Student's t distributions. The distribution types are commonly referred to as Pearson's type 0, type I and so forth, up to type VII (Pollard 1977). From this system of distributions, the one belonging to the gamma family, referred to as Pearson type III, is among the most used in hydrology, particularly for the frequency analysis of annual maximum rainfall and runoff. In this section, the Pearson type III and its related log-Pearson type III distributions are described along with comments on their applications for hydrologic variables.

5.8.1   Pearson Type III Distribution

A random variable X is distributed according to a Pearson type III (P-III or sometimes P3) if the deviations $(X-\xi)$ follows a gamma distribution with scale parameter α and shape parameter β. If the P-III location parameter ξ is null, it reduces to a two-parameter gamma distribution, as described in Sect. 5.5. As a result, the P-III distribution is also referred to as the three-parameter gamma. The P-III density function is given by

$$f_X(x) = \frac{1}{\alpha \Gamma(\beta)} \left(\frac{x-\xi}{\alpha} \right)^{\beta-1} \exp\left(-\frac{x-\xi}{\alpha} \right) \tag{5.104}$$

The variable X is defined in the range $\xi < x < \infty$. In general, the scale parameter α can take negative or positive values. However, if $\alpha < 0$, the P-III variate is upper-bounded. The cumulative distribution function for the P-III distribution is written as

$$F_X(x) = \frac{1}{\alpha \Gamma(\beta)} \int_{\xi}^{x} \left(\frac{x-\xi}{\alpha} \right)^{\beta-1} \exp\left(-\frac{x-\xi}{\alpha} \right) dx \tag{5.105}$$

The P-III cumulative distribution function can be evaluated by the same procedure used for the two-parameter gamma distribution of X, described in Sect. 5.5,

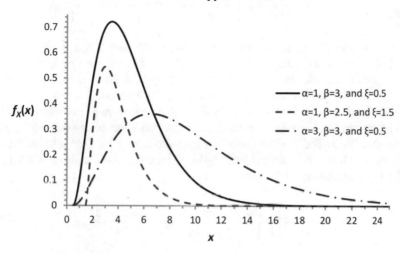

Fig. 5.17 Examples of Pearson type III density functions

except that, for the P-III, the variable to consider must refer to the deviations $(X-\xi)$. Figure 5.17 illustrates different shapes the P-III distribution can exhibit, for three distinct sets of parameters. One can notice in Fig. 5.17 that even slight variations of the shape parameter β can cause substantial changes in the skewness of the distribution. Furthermore, it is possible to conclude that increasing the scale parameter α increases the scatter of X, whereas changing the location parameter ξ causes the origin of X to be shifted.

The mean, variance and coefficient of skewness of a P-III variate are respectively written as

$$E[X] = \alpha\beta + \xi \tag{5.106}$$

$$\mathrm{Var}[X] = \alpha^2\beta \tag{5.107}$$

and

$$\gamma = \frac{2}{\sqrt{\beta}} \tag{5.108}$$

5.8.2 Log-Pearson Type III Distribution

If the variable $\ln(X)$ follows a Pearson type III model, then X is distributed according to the log-Pearson type III (LP-III). The probability density function of an LP-III distribution is

$$f_X(x) = \frac{1}{\alpha x \Gamma(\beta)} \left[\frac{\ln(x) - \xi}{\alpha}\right]^{\beta-1} \exp\left[-\frac{\ln(x) - \xi}{\alpha}\right] \qquad (5.109)$$

where ξ, α, and β are, respectively, location, scale, and shape parameters. The LP-III distribution can display a great variety of shapes. According to Rao and Hamed (2000), in order to be used in the frequency analysis of hydrologic maxima, only the LP-III distributions with $\beta > 1$ and $\alpha > 0$ are of interest. In effect, a negative coefficient of skewness implies that $\alpha < 0$ and, therefore, that an LP-III-distributed X is upper-bounded. Such a condition is considered by many hydrologists as inadequate for modeling hydrologic maxima (see Rao and Hamed 2000, Papalexiou and Koutsoyiannis 2013). The cumulative distribution function for the log-Pearson type III is given by

$$F_X(x) = \frac{1}{\alpha \Gamma(\beta)} \int_0^x \frac{1}{x} \left[\frac{\ln(x) - \xi}{\alpha}\right]^{\beta-1} \exp\left[-\frac{\ln(x) - \xi}{\alpha}\right] dx \qquad (5.110)$$

By making $y = [\ln(x) - \xi]/\alpha$ in Eq. (5.110), the CDF for the LP-III becomes

$$F_Y(y) = \frac{1}{\Gamma(\beta)} \int_0^y y^{\beta-1} \exp(-y)\, dy \qquad (5.111)$$

which can be solved by using Eq. (5.42), with $\theta = \alpha = 1$ and $\eta = \beta$, along with the method described in Sect. 5.5. The mean of a log-Pearson type III variate is

$$E[X] = \frac{e^\xi}{(1 - \alpha)^\beta} \qquad (5.112)$$

The higher-order moments for the LP-III are complex. Bobée and Ashkar (1991) derived the following expression for the LP-III moments about the origin

$$\mu'_r = \frac{e^{r\xi}}{(1 - r\alpha)^\beta} \qquad (5.113)$$

where r denotes the moment order. It should be noted, however, that moments of order r do not exist for $\alpha > 1/r$. Calculations of the parameters of the LP-III distribution can be performed using the indirect or direct methods. The former is considerably easier and consists of calculating the parameters of the Pearson type III distribution, as applied to the logarithms of X, or, in other terms, applying Eqs. (5.106)–(5.108) to the transformed variable $Z = \ln(X)$. The direct method, which does not involve the logarithmic transformation of X, is more complex and is not covered here. The reader interested in such a specific topic should consult the references Bobée and Ashkar (1991), Kite (1988), and Rao and Hamed (2000).

The LP-III distribution has an interesting peculiarity, as derived from the long history of its use as a recommended model for flood frequency analysis in the United States of America. Early applications of the LP-III distribution date back to 1960s, when Beard (1962) employed it, with the logarithmic transformation of flood discharges. Later, the United States Water Resources Committee performed a comprehensive comparison of different probability distributions and recommended the LP-III distribution as a model to be systematically employed by the US federal agencies for flood frequency analysis (WRC 1967, Benson 1968). Since the publication of the first comparison results by WRC (1967), four comprehensive revisions on the procedures for using the LP-III model were carried out by the United States Water Resources Committee and published as *bulletins* (WRC 1975, 1976, 1977, and 1981). The latest complete revision was published in 1981 under the title *Guidelines for Determining Flood Flow Frequency—Bulletin 17B*, which can be downloaded from the URL http://water.usgs.gov/osw/bulletin17b/dl_flow.pdf [accessed: 8th January 2016]. The Hydrologic Engineering Center of the United States Army Corps of Engineers developed the HEC-SSP software to implement hydrologic frequency analysis using the *Bulletin 17B* guidelines. Both the SSP software and user's manual are available from http://www.hec.usace.army.mil/software/hec-ssp/ [accessed: 3rd March 2016]. At the time this book is being written, the *Hydrologic Frequency Analysis Work Group* (http://acwi.gov/hydrology/Frequency/), of the Subcommittee on Hydrology of the United States Advisory Committee on Water Information, is preparing significant changes and improvements to Bulletin 17B that may lead to the publication of a new Bulletin 17C soon.

Since the recommendation of the LP-III by WRC, the distribution has been an object of great interest and a source of some controversy among statistical hydrologists and the core theme of many research projects. These generally have covered a wide range of specific subjects, from comparative studies of methods for estimating parameters, quantiles, and confidence intervals, to the reliable estimation of the coefficient of skewness, which is a main issue related to the LP-III distribution. A full discussion of these LP-III specificities, given the diversity and amount of past research, is clearly beyond the scope of this introductory text. The reader interested in details on the LP-III distribution and its current status as a model for flood frequency analysis in the United States should start by reading updated critical reviews on the topic as given in Stedinger and Griffis (2008, 2011) and England (2011).

5.9 Special Probability Distributions Used in Statistical Hydrology

The list of probability distributions used in hydrology is long and diverse. The descriptions given in the preceding sections have covered the main probability distributions, with one to three parameters, that hydrologists have traditionally employed in frequency analysis of hydrologic random variables. As seen in

Chaps. 8 and 10, frequency analysis of hydrologic variables can be performed either at a single site or at many sites within a hydrologically homogenous region. In the latter case, a common regional probability distribution, in most cases with three parameters, is fitted to scaled data from multiple sites. The choice of the best-fitting regional three-parameter model is generally made on the basis of differences between statistical descriptors of fitted and theoretical models, taking into account the dispersion and bias introduced by multisite sampling fluctuations and cross-correlation. To account for bias and dispersion, a large number of homogenous regions are simulated, with data randomly generated from a hypothetical population, whose probability distribution is assumed to be a generic four-parameter function, from which any candidate three-parameter model is a particular case. The four-parameter Kappa distribution encompasses some widely used two-parameter and three-parameter distributions as special cases and serves the purpose of deciding on the best-fitting regional model. In this same context of regional frequency analysis, the five-parameter Wakeby distribution is also very useful, as it is considered a robust model with respect to misspecification of the regional probability distribution. Both distributions are briefly described in the subsections that follow.

Another type of special probability distributions used in Statistical Hydrology refers to those arising from mixed-populations of the hydrologic variable of interest. Examples of mixed-populations in hydrology may include (1) floods caused by different mechanisms such as tropical cyclones or thunderstorms, as reported by Murphy (2001), or generated by different processes, such as rainfall or snowmelt, as in Waylen and Woo (1984); and (2) heavy storm rainfalls as associated with convective cells or frontal systems. Modeling random variables, which are supposedly drawn from mixed-populations, generally results in compound or mixed-distributions. One compound distribution that has received much attention from researchers over the years is the TCEV (two-component extreme value) distribution. The TCEV model is also briefly described as an example of compound distributions used in Statistical Hydrology.

5.9.1 Kappa Distribution

The four-parameter Kappa distribution was introduced by Hosking (1988). Its CDF, PDF, and quantile function are respectively given by

$$F_X(x) = \left\{ 1 - h \left[1 - k \frac{(x - \xi)}{\alpha} \right]^{\frac{1}{k}} \right\}^{\frac{1}{h}} \qquad (5.114)$$

$$f_X(x) = \frac{1}{\alpha} \left[1 - k \frac{(x - \xi)}{\alpha} \right]^{\frac{1}{k} - 1} [F_X(x)]^{1 - h} \qquad (5.115)$$

and

$$x(F) = \xi + \frac{\alpha}{k}\left[1 - \left(\frac{1 - F^h}{h}\right)^k\right]$$ (5.116)

where ξ and α denote location and scale parameters, respectively, and k and h are shape parameters. The Kappa variate is upper-bounded at $\xi + \alpha/k$ if $k > 0$ or unbounded from above if $k \leq 0$. Furthermore, it is lower-bounded at $\xi + \alpha(1 - h^{-k})/k$ if $h > 0$, or at $\xi + \alpha/k$ if $h \leq 0$ and $k < 0$, or unbounded from below if $h \leq 0$ and $k \geq 0$. Figure 5.18 depicts some examples of Kappa density functions.

As shown in Fig. 5.18, the four-parameter Kappa distribution can exhibit a variety of shapes and includes, as special cases, the exponential distribution, when $h = 1$ and $k = 0$, the Gumbel$_{max}$ distribution, when $h = 0$ and $k = 0$, the uniform distribution, when $h = 1$ and $k = 1$, the Generalized Pareto distribution, when $h = 1$, the GEV distribution, when $h = 1$, and, the three-parameter

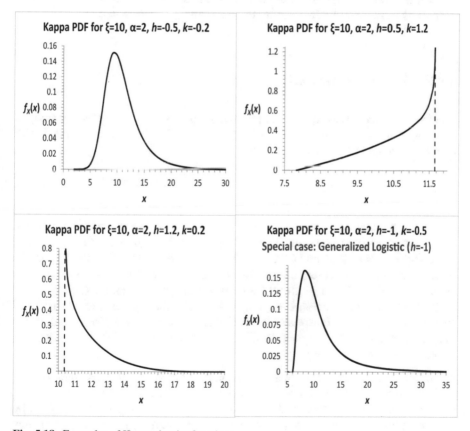

Fig. 5.18 Examples of Kappa density functions

Generalized Logistic distribution, which is one of the graphs depicted in Fig. 5.18, when $h = -1$.

According to Hosking (1988), for $h \neq 0$ and $k \neq 0$, moments of order r from the four-parameter Kappa distribution can be calculated from the following equation:

$$E\left[\left(1 - k\frac{x - \xi}{\alpha}\right)^r\right] = \begin{cases} h^{-(1+rk)}\frac{\Gamma(1+rk)\Gamma(1/h)}{\Gamma(1+rk+1/h)} & \text{for } h > 0 \\ (-h)^{-(1+rk)}\frac{\Gamma(1+rk)\Gamma(-rk-1/h)}{\Gamma(1-1/h)} & \text{for } h < 0 \end{cases} \quad (5.117)$$

where $\Gamma(.)$ denotes the gamma function. The first four moments of the Kappa distribution are not always sufficient to calculate its four parameters, as some combinations of the coefficients of skewness and kurtosis may correspond to distinct pairs of parameters h and k. Because the four-parameter Kappa distribution encompasses many two-parameter and three-parameter distributions as special cases, it is very useful as a general model to generate artificial data in order to compare the fit of less-complex distributions to actual data. In Chap. 10, on regional frequency analysis, the four-parameter Kappa distribution is revisited and its usefulness to regional hydrologic frequency analysis is demonstrated.

5.9.2   Wakeby Distribution

The five-parameter Wakeby distribution was proposed for flood frequency analysis by Houghton (1978). According to reparametrization by Hosking and Wallis (1997), its quantile function is given by

$$x(F) = \xi + \frac{\alpha}{\beta}\left[1 - (1 - F)^\beta\right] - \frac{\gamma}{\delta}\left[1 - (1 - F)^{-\delta}\right] \quad (5.118)$$

where ξ is the location parameter, and α, β, γ, and δ denote the other parameters. The Wakeby distribution is analytically defined only by its quantile function, given in Eq. (5.118), as its PDF and CDF cannot be expressed in explicit form. For details on moments and parameters of the Wakeby distribution, the reader is referred to Houghton (1978), Hosking and Wallis (1997), and Rao and Hamed (2000).

By virtue of its five parameters, the Wakeby distribution has a great variety of shapes and properties that make it particularly suitable for regional hydrologic frequency analysis. Hosking and Wallis (1997) points out the following attributes of the Wakeby distribution:

- For particular sets of parameters, it can emulate the shapes of many right-skewed distributions, such as $Gumbel_{max}$, lognormal, and Pearson type III;
- Analogously to the Kappa distribution, the diversity of shapes the Wakeby distribution can attain makes it particularly useful in regional frequency

analysis, as a robust model, with respect to misspecification of the regional probability distribution;

- For $\delta > 0$, the Wakeby distribution is heavy tailed, thus agreeing with recent findings in the frequency analysis of maximum daily rainfall;
- The distribution has a finite lower bound which is physically reasonable for most hydrologic variables; and
- The explicit form of the quantile function facilitates its use for generating Wakeby-distributed random samples.

The use of the Wakeby distribution as a general model, from which other less-complex are particular cases of, is addressed in Chap. 10 in the context of assessing the accuracy of regional frequency estimates.

5.9.3   TCEV (Two-Component Extreme Value) Distribution

The solution to Example 5.5 showed that the annual probabilities of flood discharges, under the Poisson-Exponential representation for flood exceedances over a high threshold Q_0, is given by $F_{Q\max}(q) = \exp\{-\exp[-\frac{1}{\theta}(q - Q_0 - \theta\ln\nu)]\}$, which is the expression of the CDF for the Gumbel distribution, with parameters θ and $[Q_0 + \theta\ln(\nu)]$. Assume now that a generic random variable X results from two independent Poisson processes, with parameters given by ν_1 and ν_2, as corresponding to exponentially distributed exceedances over the thresholds x_1 and x_2, respectively, with mean values θ_1 and θ_2. As the Poisson-Exponential processes are assumed independent, the annual probabilities of X are given by the product of two Gumbel cumulative distribution functions, each with its own parameters. The resulting compound CDF is

$$F_X(x) = \exp\left[-e^{-\frac{x-x_1-\theta_1\ln(\nu_1)}{\theta_1}} - e^{-\frac{x-x_2-\theta_2\ln(\nu_2)}{\theta_2}}\right] = \exp\left[-\nu_1 e^{-\frac{x-x_1}{\theta_1}} - \nu_2 e^{-\frac{x-x_2}{\theta_2}}\right] \quad (5.119)$$

The TCEV (two-component extreme value) distribution was introduced for flood frequency analysis by Rossi et al. (1984) to model the compound process resulting from two Poisson-Exponential processes: one for the more frequent floods, referred to as *ordinary* floods, and the other for the rare or *extraordinary* floods. The TCEV cumulative distribution function arises from the extension of Eq. (5.119) to lower values of X, by imposing $x_1 = x_2 = 0$ (Rossi et al. 1984), thus resulting in

$$F_X(x) = \exp\left[-\nu_1 e^{-\frac{x}{\theta_1}} - \nu_2 e^{-\frac{x}{\theta_2}}\right] \quad (5.120)$$

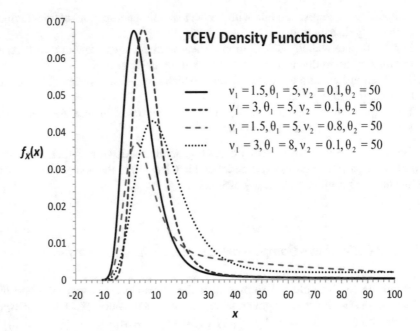

Fig. 5.19 Examples of TCEV density functions

where the annual mean number of ordinary and extraordinary floods are respectively given by ν_1 and ν_2, with $\nu_1 > \nu_2$, and corresponding mean exceedances equal to θ_1 and θ_2, with $\theta_1 < \theta_2$. The density function for the TCEV distribution is

$$f_X(x) = \left(\frac{\nu_1}{\theta_1}e^{-\frac{x}{\theta_1}} - \frac{\nu_2}{\theta_2}e^{-\frac{x}{\theta_2}}\right)\exp\left[-\nu_1 e^{-\frac{x}{\theta_1}} - \nu_2 e^{-\frac{x}{\theta_2}}\right] \qquad (5.121)$$

Figure 5.19 depicts some examples of the TCEV density function, for some sets of parameter values.

The moments for the TCEV distribution are complex and given by sums of the gamma function and its derivatives. Beran et al. (1986) derived expressions for the mean and higher-order noncentral moments of a TCEV variate. The most interesting application of the TCEV distribution relates to its use as a regional model for flood frequency analysis. The regional approach for flood frequency analysis with the TCEV model has been widely used in Europe, as reported by Fiorentino et al. (1987) and Cannarozzo et al. (1995).

Example 5.14 At a given location, sea wave heights depend on the prevalent direction of coming storms. For eastern storms, the annual maximum wave heights are Gumbel-distributed with scale and location parameters equal to 0.5 and 3 m, respectively. For northern storms, the annual maximum wave heights are also distributed as a Gumbel variate with scale parameter 0.30 m and location parameter

2.70 m. Determine the annual maximum sea wave height that will occur in any given year with 10 % exceedance-probability independently of storm direction (adapted from Kottegoda and Rosso 1997).

Solution Let X denote the annual maximum sea wave height, independently of storm direction. For the given data, Eq. (5.119) yields $F_X(x) = \exp\left[-e^{-\frac{x-3}{0.5}} - e^{-\frac{x-2.70}{0.30}}\right]$, for $x \geq 3$. For a 10 % exceedance-probability, $F_X(x_0) = 0.90$. However, as the compound CDF does not have an explicit inverse function, an approximate iterative solution for x_0 is required. Using the MS Excel *Solver* component, as in Example 5.10, the approximate solution is found at $x_0 = 4.163$ m. Take the opportunity to compare the probability distributions by plotting on the same chart the Gumbel CDFs for eastern and northern storms, and the compound CDF.

5.10 Sampling Distributions

Up to this point, most of the probability distributions described here, by virtue of their attributes of shape and/or theoretical justification, serve the purpose of modeling hydrologic random variables. Other statistical problems, such as, for instance, the construction of hypotheses tests and confidence intervals for population descriptors, require other specific probability distributions. These are generally termed sampling distributions as they refer to the distribution of a given statistic, deemed as a random variable, when it is derived from a finite-sample. In this context, a sampling distribution is thought of as the distribution of a particular statistic for all possible samples of size N drawn from the population. In general, it depends on the underlying distribution of the population being sampled and on the sample size. In this section, the following important sampling distributions are described: chi-square χ^2, Student's t, and Snedecor's F. They are related to normal populations.

5.10.1 Chi-Square (χ^2) Distribution

Suppose that, for $X_i \sim N(\mu,\sigma)$, $Z_i = \frac{X_i - \mu}{\sigma}$, $i = 1, 2, \ldots, N$, denotes a set of N independent random variables, distributed as a standard normal $N(0,1)$. Under these conditions, it can be shown that the random variable Y, as defined by

$$Y = \sum_{i=1}^{N} Z_i^2 \qquad (5.122)$$

follows a χ^2 distribution, whose density function is given by

$$f_{\chi^2}(y) = \frac{1}{2\Gamma(\nu/2)} \left(\frac{y}{2}\right)^{\frac{k}{2}-1} \exp\left(-\frac{y}{2}\right) \text{ for } y \text{ and } \nu > 0 \qquad (5.123)$$

where ν denotes a parameter.

The parameter ν is known as the *number of degrees of freedom* as an allusion to the same concept from mechanics, referring to the possible number of independent parameters that define the position and orientation of a rigid body in space. By comparing Eqs. (5.40) and (5.123), one can determine that the χ^2 distribution is a special case of the gamma distribution, with $\eta = \nu/2$ and $\theta = 2$. As a result, the χ^2 cumulative distribution function can be written in terms of the gamma CDF [Eq. (5.43)] as

$$F_{\chi^2}(y) = \frac{\Gamma_i(u = y/2, \eta = \nu/2)}{\Gamma(\eta = \nu/2)} \qquad (5.124)$$

and also be calculated as the ratio between the incomplete and complete gamma functions, as described in Sect. 5.5. Appendix 3 presents a table of the χ^2 cumulative distribution function, for different numbers of degrees of freedom.

The mean, variance and coefficient of skewness of the χ^2 distribution are respectively given by

$$E[\chi^2] = \nu \qquad (5.125)$$

$$\text{Var}[\chi^2] = 2\nu \qquad (5.126)$$

and

$$\gamma = \frac{2}{\sqrt{\nu/2}} \qquad (5.127)$$

Figure 5.20 illustrates some possible shapes of the χ^2 density function, for selected values of parameter ν.

If, differently from the previous formulation, the variables Z_i are defined as $Z_i = \frac{X_i - \bar{x}}{\sigma}$, $i = 1, 2, \ldots, N$, where X_i denote the elements from a simple random sample from a normal population with sample mean $\bar{x}$, then, it can be shown that the variable $Y = \sum_{i=1}^{N} Z_i^2$ is distributed according to a χ^2 distribution, with $\nu = (N-1)$ degrees of freedom. In such a case, one degree of freedom is said to be lost from having the population mean μ been previously estimated by the sample mean $\bar{x}$. Furthermore, recalling that the sample variance is given by $s_X^2 = \sum_{i=1}^{N} (X_i - \bar{x})/(N - 1)$ and also that $Y = \sum_{i=1}^{N} (X_i - \bar{x})^2/\sigma^2$, it is clear that

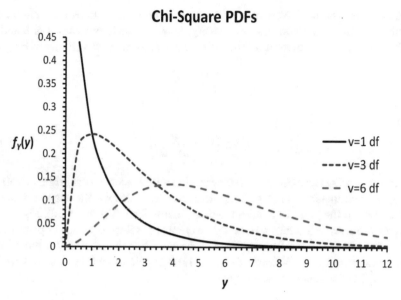

Fig. 5.20 Examples of χ^2 density functions

$$Y = (N - 1)\frac{s_X^2}{\sigma_X^2} \qquad (5.128)$$

follows a χ^2 distribution with $\nu = (N-1)$ degrees of freedom. This result is extensively used in Chap. 7, in formulating hypotheses tests and constructing confidence intervals for the variance of normal populations.

5.10.2 Student's t Distribution

If $U \sim N(0,1)$ and $V \sim \chi^2(\nu)$ are two independent random variables, then, it can be shown that the variable T, defined as $T = U\sqrt{\nu}/\sqrt{V}$, is distributed according to the density function given by

$$f_T(t) = \frac{\Gamma[(\nu + 1)/2]\,(1 + t^2/\nu)^{-(\nu+1)/2}}{\sqrt{\pi\nu}\,\Gamma(\nu/2)} \qquad \text{for } -\infty < t < \infty \text{ and } \nu > 0 \quad (5.129)$$

which is known as Student's t distribution, due to the English chemist and statistician William Gosset (1876–1937), who used to sign his papers under the pen name of Student. The distribution parameter is denoted by ν, which is also referred to as the number of degrees of freedom. The cumulative distribution function, given by the integral of the density, from $-\infty$ to t, can only be evaluated through numerical integration techniques. Appendix 4 presents a table for Student's t cumulative distribution function, under the form of $F_T(t) = \alpha$, for different values of ν and α.

The mean and variance of a Student's t variate are respectively given by

$$E[T] = 0 \tag{5.130}$$

and

$$\text{Var}[T] = \frac{\nu}{\nu - 2} \tag{5.131}$$

The Student's t is a symmetrical distribution with respect to the origin of variable t and, as parameter ν grows, it approaches very closely the standard normal distribution, to the point of being indistinguishable from it (for $\nu > 30$). Figure 5.21 depicts examples of Student's t density functions for selected values of parameter ν.

The Student's t is usually employed as the sampling distribution for the mean of a random sample of size N drawn from a normal population with unknown variance. In fact, if the T variable is expressed as

$$T = \frac{\bar{x} - \mu_X}{\sqrt{s_X^2/N}} \tag{5.132}$$

being, in the sequence, multiplied and divided by σ_X, one obtains

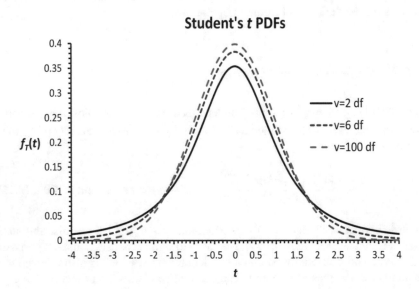

Fig. 5.21 Examples of Student's t density functions

$$T = \frac{\frac{\bar{x}-\mu_X}{\sigma_X/\sqrt{N}}}{\sqrt{s_X^2/\sigma_X^2}} = \frac{U\sqrt{N-1}}{\sqrt{V}} \tag{5.133}$$

which corresponds to the given definition of T. Remember that $U = (\bar{x} - \mu_X)/(\sigma\sqrt{N})$ is distributed as a standard normal (see Example 5.3) and also that $V = (N-1)s_X^2/\sigma_X^2$ follows a χ^2 distribution with $(N-1)$ degrees of freedom, as formally expressed in Eq. (5.128). By comparing Eq. (5.133) with the definition of a Student's t variate, one thus determines that the sampling distribution of the mean of a random sample of size N, drawn from a normal population with unknown variance, is indeed the Student's t distribution with $(N-1)$ degrees of freedom. In this case, one degree of freedom is said lost from having the population variance σ_X^2 been previously estimated by s_X^2.

Example 5.15 To return to the solution to Example 5.3, after determining that, in fact, the population variance of the variable dissolved oxygen concentration was estimated by the sample variance, calculate the probability that the 8-week monitoring program yields a sample mean that will differ from the population mean by at least 0.5 mg/l.

Solution The arguments used to solve Example 5.3 are still valid, except for the fact that, now, the variable $T = \frac{\bar{x}-\mu_X}{s_X/\sqrt{N}} \sim$ Student's t, with $N-1=7$ degrees of freedom. The sought probability corresponds to the inequality $P(|\bar{x} - \mu_X| > 0.5)$. Dividing both sides of this inequality by $s_X/\sqrt{N}$, one gets $P\left(|T| > \frac{0.5}{s_X/\sqrt{N}}\right)$ or $P(|T| > 0.707)$, or still $1 - P(|T| < 0.707)$. To calculate probabilities or quantiles for the Student's t distribution, one can make use of the table given in Appendix 4 or respectively employ the MS Excel built-in functions T.DIST.2T(.) or T.INV.2T(.), for two-tailed t, or T.DIST(.) or T.INV(.), for left-tailed t. In R, the appropriate functions are pt(.) and qt(.). In particular, for $\nu = 7$ and $t = 0.707$, the function T.DIST.2 T(0.707;7) returns 0.502. Hence, the probability that the 8-week monitoring program yields a sample mean that will differ from the population mean by at least 0.5 mg/l is $(1-0.502) = 0.498$, which is a rather high probability, thus suggesting the need of a longer monitoring program.

5.10.3 Snedecor's F Distribution

If $U \sim \chi^2$, with m degrees of freedom, and $V \sim \chi^2$, with n degrees of freedom, are two independent random variables, then, it can be shown that the variable defined as

$$Y = \frac{U/m}{V/n} \tag{5.134}$$

follows an F distribution, with parameters $\gamma_1 = m$ and $\gamma_2 = n$, and density function given by

$$f_F(f) = \frac{\Gamma[(\gamma_1 + \gamma_2)/2]}{\Gamma(\gamma_1/2)\,\Gamma(\gamma_2/2)}\gamma_1^{\gamma_1/2}\gamma_2^{\gamma_2/2}f^{(\gamma_1-2)/2}(\gamma_2 + \gamma_1 f)^{-(\gamma_1+\gamma_2)/2} \text{ for } \gamma_1,\gamma_2,f > 0$$

(5.135)

The cumulative distribution function, given by the integral of Eq. (5.135), from 0 to f, can be evaluated only by numerical methods. Appendix 5 presents a table for the Snedecor's F cumulative distribution function, for different values of γ_1 and γ_2, which are also termed degrees of freedom of the numerator and denominator, respectively. The mean and variance of a Snedecor's F variate are respectively given by

$$E[F] = \frac{\gamma_1}{\gamma_2 - 2}$$

(5.136)

and

$$\text{Var}[X] = \frac{\gamma_2^2(\gamma_1 + 2)}{\gamma_1(\gamma_2 - 2)(\gamma_2 - 4)}$$

(5.137)

Figure 5.22 shows some examples of Snedecor's F densities, for particular parametric sets.

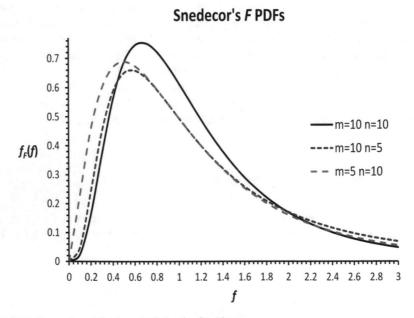

Fig. 5.22 Examples of Snedecor's F density functions

This distribution was introduced by the American statistician George Snedecor (1881–1974) as the sampling distribution for the ratio between variances from two distinct normal populations. The term F, used for the Snedecor's variate, is to honor the famous British statistician Ronald Fisher (1890–1962). The Snedecor's F distribution has been used in conventional statistics for testing hypotheses concerning sample variances from normal populations, particularly for the analysis of variance (ANOVA) and of residuals from regression. In the MS Excel software, the built-in functions that correspond to Snedecor's F distribution are F.DIST(.) and F.INV(.), for left-tailed F, and F.DIST.RT(.) and F.INV.RT(.), for right-tailed F. In R, the appropriate functions are pf(.) and qf(.).

5.11   Bivariate Normal Distribution

The joint distribution of two normal random variables is known as the bivariate normal. Formally, if X and Y are normally distributed, with respective parameters μ_X, σ_X, μ_Y, and σ_Y, and the correlation coefficient between the two is denoted as ρ, then, the bivariate normal joint density function is given by

$$
f_{X,Y}(x,y) = \frac{1}{2\pi\sigma_X\sigma_Y\sqrt{1-\rho^2}}
$$
$$
\times \exp\left\{-\frac{1}{2(1-\rho^2)}\left[\left(\frac{x-\mu_X}{\sigma_X}\right)^2 - 2\rho\frac{(x-\mu_X)(y-\mu_Y)}{\sigma_X\sigma_Y} + \left(\frac{y-\mu_Y}{\sigma_Y}\right)^2\right]\right\}
$$

$$(5.138)$$

for $\infty < x < \infty$ and $\infty < y < \infty$. The joint probability $P(X < x, Y < y)$ is given by the double integration of the bivariate normal density, for the appropriate integration limits, and its calculation requires numerical methods. Some computer programs designed to implement calculations for the bivariate normal are available on the Internet. The URL http://stat-athens.aueb.gr/~karlis/morematerial.html [accessed: 13th January 2016] offers a list of topics related to the bivariate normal distribution. In addition, this URL makes available the software *bivar1b.exe*, which calculates the joint CDF for variables X and Y.

The panels of Fig. 5.23 illustrate different shapes that the bivariate normal joint density function can assume, for four sets of parameters. Note that, when X and Y are independent, the volume of the joint density is symmetrically distributed over the plane defined by the variables' domains. As the linear dependence between the variables grows, the pairs (x,y) and their respective non-exceedance probabilities, as given by the volumes below the bivariate joint density surface, concentrate along the projections of the straight lines on the plane xy. These straight lines set out the dependence of X on Y, and, inversely, of Y on X. In R, the mvtnorm package (Genz et al. 2009), which implements the multivariate normal distribution, has a specific function for the bivariate normal distribution.

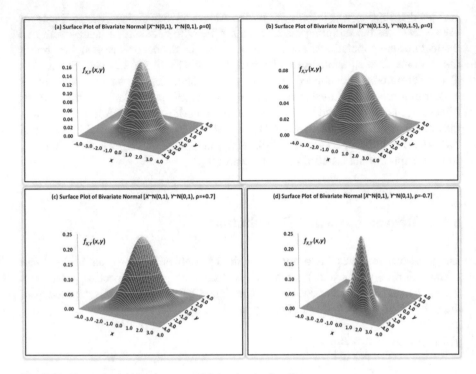

Fig. 5.23 Examples of bivariate normal joint density functions

By applying Eqs. (3.33) and (3.34) to the bivariate normal density one can determine that the marginal distributions of X and Y are indeed their respective univariate normal densities. Conditional probabilities for the bivariate normal distribution can be calculated from Eq. (3.44).

5.12 Bivariate Distributions Using Copulas

What follows is a brief presentation of the main aspects of bivariate analysis using copulas. A thorough introduction to copulas is beyond the scope of this chapter, since there are entire books devoted to that subject such as Nelsen (2006) and Salvadori et al. (2007). The interested reader is referred to those texts.

In engineering practice it is often necessary to conduct statistical analyses of hydrological events which are characterized by more than one variable. For example, river floods may be characterized by the joint distribution of peak flows and flood volumes, and extreme rainfall events are often characterized by the combined effects of the event's duration and mean rainfall intensity. The bivariate normal distribution, presented in Sect. 5.11 is rarely adequate for hydrological applications, particularly when dealing with extreme phenomena, since the underlying random variables can hardly be considered symmetrically distributed. While there are other multivariate probability distributions available in the technical literature, which are usually straightforward extensions of well-known univariate distributions such as the exponential or the gamma distributions, they usually suffer from several formal limitations, such as the need of prescribing the marginal distributions and of describing the dependence structure between variables (e.g., Nelsen 2006). A convenient way to overcome these difficulties is through the use of copulas. Accordingly, the use of copulas in hydrological applications has increased very rapidly in recent years (Salvadori and De Michele 2010).

A bivariate copula, also termed 2-copula, is a bivariate distribution function $C(u, v) = P(U \leq u, V \leq v)$ with support on the unit square space $[0, 1]^2$ and with standard uniform marginals, such that $C(u, 0) = 0$, $C(u, 1) = u$, $C(0, v) = 0$, $C(1, v) = v$, and $C(u_2, v_2) - C(u_2, v_1) - C(u_1, v_2) + C(u_1, v_1) \geq 0$ for all $0 \leq u_1 \leq u_2 \leq 1$ and $0 \leq v_1 \leq v_2 \leq 1$.

Consider the random variables X and Y with respective continuous CDFs $F_X(x)$ and $F_Y(y)$, for x and y real, and joint CDF denoted as $F_{X,Y}(x, y) = P(X \leq x, Y \leq y)$. Since $F_X(x)$ and $F_Y(y)$ are standard uniform distributions (see Scct. 5.1), it follows that $U = F_X(x)$ and $V = F_Y(y)$. Sklar's theorem (Sklar 1959) states that there must exist a bivariate copula, C such that

$$F_{X,Y}(x, y) = C(F_X(x), F_Y(y)) \tag{5.139}$$

and

$$C(u, v) = F_{X,Y}\left(F_X^{-1}(x), F_Y^{-1}(y)\right) \tag{5.140}$$

An important and practical result from Sklar's theorem is that marginal distributions and copula can be considered separately. Therefore, by using a copula, one can obtain the joint distribution of two variables whose marginal distributions are from different families, which, in hydrologic practice, can occur very often.

Copula notation is particularly useful for defining and calculating several types of joint and conditional probabilities of X and Y. Under the conditions stated earlier, one is able to write (Serinaldi 2015)

$$P(U > u \cap V > v) = 1 - u - v + C(u, v) \tag{5.141}$$

$$P(U > u \cup V > v) = 1 - C(u, v) \tag{5.142}$$

$$P(U > u | V > v) = (1 - u)(1 - u - v + C(u, v)) \tag{5.143}$$

$$P(U > u | V \leq v) = 1 - \frac{C(u, v)}{u} \tag{5.144}$$

Table 5.2 Summary characteristics of three one-parameter Archimedean copulas

Family	Parameter space	$\varphi(t)$	$\varphi^{-1}(s)$
Clayton	$\alpha \in [-1, 0[\cup]0, +\infty]$	$\dfrac{1}{\alpha}(t^{-\alpha} - 1)$	$(1 + \alpha s)^{-1/\alpha}$
Frank	$\alpha \in [-\infty, 0[\cup]0, +\infty]$	$-\ln\left(\dfrac{e^{-\alpha t} - 1}{e^{-\alpha} - 1}\right)$	$-\dfrac{1}{\alpha}\ln(1 + e^{-s}(e^{-\alpha} - 1))$
Gumbel–Hougaard	$\alpha \in [1, +\infty[$	$(-\ln(t))^{\alpha}$	$\exp(-s^{1/\alpha})$

$$P(U > u | V = v) = 1 - \frac{\partial C(u, v)}{\partial u} \tag{5.145}$$

Further definitions of conditional probabilities using copula notation are given by Salvadori et al. (2007).

There are several types of copula functions in the technical literature. Archimedean copulas are widely used in hydrological applications due to their convenient properties. An Archimedean copula is constructed using a generator φ such that

$$C(u, v) = \varphi^{-1}[\varphi(u) + \varphi(v)] \tag{5.146}$$

Table 5.2 shows the generator and generator inverse for some common one-parameter Archimedean copulas $C_\alpha(u, v)$. For details on how to fit copulas to data, the interested reader is referred to the works by Nelsen (2006) and Salvadori et al. (2007).

Example 5.16 In order to exemplify the construction of a bivariate distribution using a copula, we revisit the case study of the Lehigh River at Stoddartsville (see Sect. 1.4). Figure 5.24 shows the scatterplot of the annual maximum peak discharges (variable Y) and the corresponding 5-day flood volumes (variable X). It should be noted that, as systematic streamflow records only started in the 1944 water year, the 5-day flood volume value, that would correspond to the flood peak 445 m³/s on May 22nd, 1942, is not available and thus is not plotted on the chart of Fig. 5.24. Obtain the joint PDF and CDF of (X, Y) using a copula.

Solution Consider that $X \sim \mathbf{GEV}(\alpha = 10.5777, \quad \beta = 4.8896, \quad \kappa - 0.1799)$ and Y follows a lognormal distribution, that is $Y \sim \mathbf{LN}(\mu_Y = 4.2703, \sigma_Y = 0.8553)$. By calculating the non-exceedance probabilities of the observations on the scatterplot of Fig. 5.24, they can be transformed into the $[0, 1]^2$ space, as shown in Fig. 5.25.

The copula is fitted to the (U, V) pairs of points. In this example we used the Gumbel–Hougaard copula (Table 5.2). Estimation of the copula parameter α used the method of maximum likelihood, which is formally described in Sect. 6.4 of the next chapter, and was carried out in R using the function fitCopula of the copula package (Yan 2007). Figure 5.26 shows the copula density and the distribution functions ($\hat{\alpha} = 3.2251$). Finally, using Sklar's theorem (Eq. 5.139), the joint distribution of (X, Y) is obtained and plotted in Fig. 5.27.

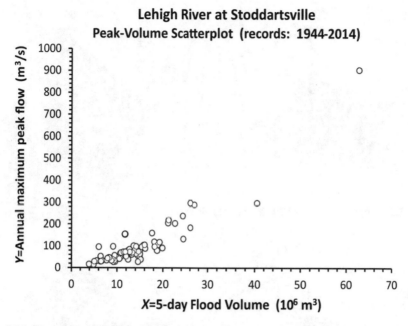

Fig. 5.24 Scatterplot of annual peak discharges and corresponding 5-day flood volumes of the Lehigh River at Stoddartsville from 1943/44 to 2013/14

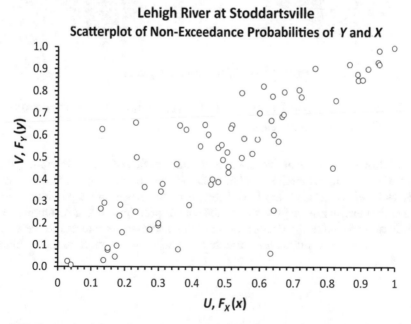

Fig. 5.25 Scatterplot of non-exceedance probabilities of annual peak discharges Y and corresponding flood volumes X, of the Lehigh River at Stoddartsville (1943/44-2013/14)

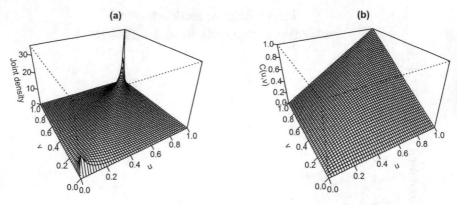

Fig. 5.26 Copula density and distribution functions

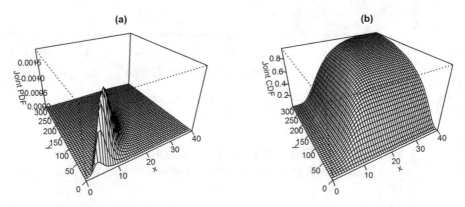

Fig. 5.27 Joint PDF and CDF of the variables shown in Fig. 5.24

5.13 Summary for Probability Distributions of Continuous Random Variables

What follows is a summary of the main characteristics of some probability distributions of continuous random variables introduced in this chapter. Not all characteristics listed in the summary have been formally derived in the previous sections of this chapter, as one can use the mathematical principles that are common to all distributions to make the desired proofs. This summary is intended to serve as a brief reference for the main probability distributions of continuous random variables.

5.13.1 Uniform Distribution

Notation: $X \sim \mathbf{U}(a, b)$
Parameters: a and b
PDF: $f_X(x) = \frac{1}{b-a}$ for $a \leq x \leq b$
Mean: $E[X] = \frac{a+b}{2}$

Variance: $\mathrm{Var}[X] = \frac{(b-a)^2}{12}$
Coefficient of Skewness: $\gamma = 0$
Coefficient of Kurtosis: $k = 1.8$
Moment Generating Function: $\varphi(t) = \frac{e^{bt} - e^{at}}{t(b-a)}$

Random Number Generation: algorithms to generate independent unit uniform random numbers $U \sim \mathbf{U}(0, 1)$ are standard in statistical software (e.g., MS Excel built-in function RAND).

5.13.2 Normal Distribution

Notation: $X \sim \mathbf{N}(\mu, \sigma)$
Parameters: μ and σ
PDF: $f_X(x) = \frac{1}{\sqrt{2\pi}\sigma} \exp\left[-\frac{1}{2}\left(\frac{x-\mu}{\sigma}\right)^2\right]$ for $-\infty < x < \infty$
Mean: $E[X] = \mu$
Variance: $\mathrm{Var}[X] = \sigma^2$
Coefficient of Skewness: $\gamma = 0$
Coefficient of Kurtosis: $k = 3$

Moment Generating Function: $\varphi(t) = \exp\left(\mu t + \frac{\sigma^2 t^2}{2}\right)$

Random Number Generation: If $U_1 \sim \mathbf{U}(0, 1)$ and $U_2 \sim \mathbf{U}(0, 1)$ are independent, then $\sin(2\pi U_2)\sqrt{-2\ln(U_1)}$ and $\cos(2\pi U_2)\sqrt{-2\ln(U_1)}$ are also independent and follow $\mathbf{N}(0,1)$.

5.13.3 Lognormal Distribution (2 Parameters)

Notation: $X \sim \mathbf{LN}(\mu_Y, \sigma_Y)$ or $X \sim \mathbf{LN2}(\mu_Y, \sigma_Y)$ or $X \sim \mathbf{LNO2}(\mu_Y, \sigma_Y)$
Parameters: μ_Y and σ_Y, with $Y = \ln(X)$
PDF: $f_X(x) = \frac{1}{x \sigma_{\ln(X)}\sqrt{2\pi}} \exp\left\{-\frac{1}{2}\left[\frac{\ln(X) - \mu_{\ln(X)}}{\sigma_{\ln(X)}}\right]^2\right\}$ para $x > 0$

Mean: $E[X] = \mu_X = \exp\left[\mu_{\ln(X)} + \frac{\sigma_{\ln(X)}^2}{2}\right]$

Variance: $\text{Var}[X] = \sigma_X^2 = \mu_X^2\left[\exp\left(\sigma_{\ln(X)}^2\right) - 1\right]$

Coefficient of Variation: $\text{CV}_X = \sqrt{\exp\left[\sigma_{\ln(X)}^2\right] - 1}$

Coefficient of Skewness: $\gamma = 3\,\text{CV}_X + (\text{CV}_X)^3$

Coefficient of Kurtosis: $\kappa = 3 + \left(e^{\sigma_{\ln(X)}^2} - 1\right)\left(e^{3\sigma_{\ln(X)}^2} + 3e^{2\sigma_{\ln(X)}^2} + 6e^{\sigma_{\ln(X)}^2} + 6\right)$

Random Number Generation: $\exp\{\mu + \sigma[N(0,1)]\} \sim \text{LN}(\mu, \sigma)$

5.13.4 Exponential Distribution (1 Parameter)

Notation: $X \sim \mathbf{E}(\theta)$

Parameter: θ

PDF: $f_X(x) = \frac{1}{\theta}\exp\left(-\frac{x}{\theta}\right)$, $x \geq 0$

Quantile Function: $x(F) = -\theta\,\ln(1 - F)$

Mean: $E[X] = \theta$

Variance: $\text{Var}[X] = \theta^2$

Coefficient of Skewness: $\gamma = 2$

Coefficient of Kurtosis: $\kappa = 9$

Moment Generating Function: $\varphi(t) = \frac{1}{1-\theta t}$ para $t < \frac{1}{\theta}$

Random Number Generation: $-\theta\,\ln[\mathbf{U}(0,1)] \sim \mathbf{E}(\theta)$

5.13.5 Gamma Distribution (2 Parameters)

Notation: $X \sim \mathbf{Ga}(\theta, \eta)$

Parameters: θ and η

PDF: $f_X(x) = \frac{1}{\theta\Gamma(\eta)}\left(\frac{x}{\theta}\right)^{\eta-1}\exp\left(-\frac{x}{\theta}\right)$ para x, θ e $\eta > 0$

Mean: $E[X] = \eta\theta$

Variance: $\text{Var}[X] = \eta\theta^2$

Coefficient of Skewness: $\gamma = \frac{2}{\sqrt{\eta}}$

Coefficient of Kurtosis: $\kappa = 3 + \frac{6}{\eta}$

Moment Generating Function: $\varphi(t) = \left(\frac{1}{1-\theta t}\right)^{\eta}$ para $t < \frac{1}{\theta}$

Random Number Generation: for integer η: $-\theta\,\ln\left[\prod_{i=1}^{\eta} U_i\right] \sim \mathbf{Ga}(\theta, \eta)$ where

$U_i \sim \mathbf{U}(0,1)$

5.13.6 Beta Distribution

Notation: $X \sim \mathbf{Be}(\alpha,\beta)$
Parameters: α and β
PDF: $f_X(x) = \frac{1}{B(\alpha,\beta)}x^{\alpha-1}(1-x)^{\beta-1}$ for $0 \le x \le 1$, $\alpha > 0$, $\beta > 0$ and $B(\alpha,\beta) = \int\limits_0^1 t^{\alpha-1}(1-t)^{\beta-1}dt$

Mean: $E[X] = \frac{\alpha}{\alpha+\beta}$
Variance: $\mathrm{Var}[X] = \frac{\alpha\beta}{(\alpha+\beta)^2(\alpha+\beta+1)}$
Coefficient of Skewness: $\gamma = \frac{2(\beta-\alpha)\sqrt{\alpha+\beta+1}}{\sqrt{\alpha\beta}(\alpha+\beta+2)}$
Coefficient of Kurtosis: $\kappa = \frac{3(\alpha+\beta+1)\left[2(\alpha+\beta)^2+\alpha\beta(\alpha+\beta-6)\right]}{\alpha\beta(\alpha+\beta+2)(\alpha+\beta+3)}$
Random Number Generation: if $g_1 = -\ln\left[\prod\limits_{i=1}^{\nu} U_i\right] \sim \mathbf{Ga}(1,\nu)$, $g_2 \sim \mathbf{Ga}(1,\omega)$ and $U_i \sim \mathbf{U}(0,1)$, then $w = \frac{g_1}{g_1+g_2} \sim \mathbf{Be}(\nu,\omega)$

5.13.7 Gumbel$_{max}$ Distribution

Notation: $Y \sim \mathbf{Gu}_{\max}(\alpha,\beta)$
Parameters: α and β
PDF: $f_Y(y) = \frac{1}{\alpha}\exp\left[-\frac{y-\beta}{\alpha} - \exp\left(-\frac{y-\beta}{\alpha}\right)\right]$
Quantile Function: $y(F) = \beta - \alpha \ln[-\ln(F)]$
Mean: $E[Y] = \beta + 0.5772\alpha$
Variance: $\mathrm{Var}[Y] = \sigma_Y^2 = \frac{\pi^2\alpha^2}{6}$
Coefficient of Skewness: $\gamma = 1.1396$
Coefficient of Kurtosis: $\kappa = 5.4$
Random Number Generation: $\beta - \alpha\{\ln[-\ln U(0,1)]\} \sim \mathbf{Gu}_{\max}(\alpha,\beta)$

5.13.8 GEV Distribution (Maxima)

Notation: $Y \sim \mathbf{GEV}(\alpha,\beta,\kappa)$
Parameters: α, β, and κ
PDF: $f_Y(y) = \frac{1}{\alpha}\left[1 - \kappa\left(\frac{y-\beta}{\alpha}\right)\right]^{1/\kappa-1}\exp\left\{-\left[1 - \kappa\left(\frac{y-\beta}{\alpha}\right)\right]^{1/\kappa}\right\}$ if $\kappa \ne 0$ or
$f_Y(y) = \frac{1}{\alpha}\exp\left[-\frac{y-\beta}{\alpha} - \exp\left(-\frac{y-\beta}{\alpha}\right)\right]$ if $\kappa = 0$
Quantile Function: $x(F) = \beta + \frac{\alpha}{\kappa}\left[1 - (-\ln F)^{\kappa}\right]$

Mean: $E[Y] = \beta + \frac{\alpha}{\kappa}[1 - \Gamma(1 + \kappa)]$

Variance: $\text{Var}[Y] = \left(\frac{\alpha}{\kappa}\right)^2 [\Gamma(1 + 2\kappa) - \Gamma^2(1 + \kappa)]$

Coefficient of Skewness: $\gamma = \frac{\kappa}{|\kappa|} \left\{ \frac{-\Gamma(1+3\kappa)+3\Gamma(1+\kappa)\,\Gamma(1+2\kappa)-2\Gamma^3(1+\kappa)}{[\Gamma(1+2\kappa)-\Gamma^2(1+\kappa)]^{3/2}} \right\}$

Random Number Generation: use quantile functions with F coming from $\mathbf{U}(0,1)$.

5.13.9   Gumbel$_{min}$ Distribution

Notation: $Z \sim \mathbf{Gu}_{\min}(\alpha,\beta)$

Parameters: α and β

PDF: $f_Z(z) = \frac{1}{\alpha}\exp\left[\frac{z-\beta}{\alpha} - \exp\left(\frac{z-\beta}{\alpha}\right)\right]$

Quantile Function: $z(F) = \beta + \alpha\,\ln[-\ln(1 - F)]$

Mean: $E[Z] = \beta - 0.5772\alpha$

Variance: $\text{Var}[Z] = \sigma_Z^2 = \frac{\pi^2\alpha^2}{6}$

Coefficient of Skewness: $\gamma = -1.1396$

Coefficient of Kurtosis: $\kappa = 5.4$

Random Number Generation: use quantile functions with F coming from $\mathbf{U}(0,1)$.

5.13.10   Weibull$_{min}$ Distribution (2 Parameters)

Notation: $Z \sim \mathbf{W}_{\min}(\alpha,\beta)$

Parameters: α and β

PDF: $f_Z(z) = \frac{\alpha}{\beta}\left(\frac{z}{\beta}\right)^{\alpha-1}\exp\left[-\left(\frac{z}{\beta}\right)^\alpha\right]$

Quantile Function: $z(F) = \beta\,[-\ln(1 - F)]^{\frac{1}{\alpha}}$

Mean: $E[Z] = \beta\Gamma\left(1 + \frac{1}{\alpha}\right)$

Variance: $\text{Var}[Z] = \beta^2\left[\Gamma\left(1 + \frac{2}{\alpha}\right) - \Gamma^2\left(1 + \frac{1}{\alpha}\right)\right]$

Coefficient of Skewness: $\gamma = \frac{\Gamma\left(1+\frac{3}{\alpha}\right)-3\Gamma\left(1+\frac{2}{\alpha}\right)\Gamma\left(1+\frac{1}{\alpha}\right)+2\Gamma^3\left(1+\frac{1}{\alpha}\right)}{\sqrt{\left[\Gamma\left(1+\frac{2}{\alpha}\right)-\Gamma^2\left(1+\frac{1}{\alpha}\right)\right]^3}}$

Random Number Generation: use quantile functions with F coming from $\mathbf{U}(0,1)$.

5.13.11   Pearson Type III Distribution

Notation: $X \sim \mathbf{PIII}(\alpha,\beta,\xi)$ or $X \sim \mathbf{P\text{-}III}(\alpha,\beta,\xi)$ or $X \sim \mathbf{P3}(\alpha,\beta,\xi)$

Parameters: α, β, and ξ

PDF: $f_X(x) = \frac{1}{\alpha\Gamma(\beta)}\left(\frac{x-\xi}{\alpha}\right)^{\beta-1}\exp\left(-\frac{x-\xi}{\alpha}\right)$

Mean: $E[X] = \alpha\beta + \xi$
Variance: $\mathrm{Var}[X] = \alpha^2\beta$
Coefficient of Skewness: $\gamma = \frac{2}{\sqrt{\beta}}$
Coefficient of Kurtosis: $\kappa = 3 + \frac{6}{\sqrt{\beta}}$

5.13.12 Chi-Square (χ^2) Distribution

Notation: $Y \sim \chi^2(\nu)$
Parameter: ν
PDF: $f_{\chi^2}(y) = \frac{1}{2\Gamma(\nu/2)} \left(\frac{y}{2}\right)^{\frac{k}{2}-1} \exp\left(-\frac{y}{2}\right)$ para y e $\nu > 0$
Mean: $E[\chi^2] = \nu$
Variance: $\mathrm{Var}[\chi^2] = 2\nu$
Coefficient of Skewness: $\gamma = \frac{2}{\sqrt{\nu/2}}$

5.13.13 Student's t Distribution

Notation: $T \sim t(\nu)$
Parameter: ν
PDF: $f_T(t) = \frac{\Gamma[(\nu+1)/2]\left(1+t^2/\nu\right)^{-(\nu+1)/2}}{\sqrt{\pi\nu}\,\Gamma(\nu/2)}$ for $-\infty < t < \infty$ and $\nu > 0$
Mean: $E[T] = 0$
Variance: $\mathrm{Var}[T] = \frac{\nu}{\nu-2}$
Coefficient of Skewness: $\gamma = 0$

5.13.14 Snedecor's F Distribution

Notation: $F \sim F(\gamma_1, \gamma_2)$
Parameters: γ_1 and γ_2
PDF: $f_F(f) = \frac{\Gamma[(\gamma_1+\gamma_2)/2]}{\Gamma(\gamma_1/2)\,\Gamma(\gamma_2/2)} \gamma_1^{\gamma_1/2}\gamma_2^{\gamma_2/2} f^{(\gamma_1-2)/2}\left(\gamma_2 + \gamma_1 f\right)^{-(\gamma_1+\gamma_2)/2}$ for $\gamma_1, \gamma_2, f > 0$
Mean: $E[F] = \frac{\gamma_1}{\gamma_2-2}$
Variance: $\mathrm{Var}[X] = \frac{\gamma_2^2(\gamma_1+2)}{\gamma_1(\gamma_2-2)(\gamma_2-4)}$

Exercises

1. Suppose the daily mean concentration of iron in a river reach, denoted by X, varies uniformly between 2 and 4 mg/l. Questions: (a) calculate the mean and variance of X; (b) calculate the daily probability that X exceeds 3.5 mg/l; and (c) given that the iron concentration has exceeded 3 mg/l on a given day, calculate $P(X \geq 3.5 \text{ mg/l})$.

2. In addition to the approximation equations given in Sect. 5.2, the numerical integration of the standard normal density function can be done through any numerical integration method, such as for instance, the trapezoidal and Simpson's rules. However, the numerical calculation of improper integrals requires variable transformation, such that the limits of integration are finite. To do this, one can employ the following identity, under the condition that the function to be integrated decreases to zero as fast as $1/x^2$ does, as x tends to $(\pm)$ infinity:

$$\int_a^b f(x)\,dx = \int_{1/b}^{1/a} \frac{1}{t^2} f\left(\frac{1}{t}\right) dt, \quad \text{for } ab > 0 \qquad (5.147)$$

For the case in which definite integration is done from $-\infty$ up to a positive real number, two steps are needed. Consider, for instance, the following integration:

$$\int_{-\infty}^b f(x)\,dx = \int_{-\infty}^{-A} f(x)\,dx + \int_{-A}^b f(x)\,dx \qquad (5.148)$$

where $-A$ denotes a negative real number large enough to fulfill the condition required for the function decreasing to zero. The first integral on the right-hand side of Eq. (5.148) can be calculated through the mathematical artifice given in Eq. (5.147); for the second integral, Simpson's rule, for instance, can be used. What follows is a computer code, written in FORTRAN, to do the numerical integration of the standard normal density function, according to Eqs. (5.147) and (5.148). Compile this program (in FORTRAN or in any other computer programming language you are used to) to implement the numerical integration of the standard normal density for any valid argument.

```
C NUMERICAL INTEGRATION OF THE STANDARD NORMAL DENSITY FUNCTION
C
C THIS PROGRAM CALCULATES P(X<X), GIVEN X, WHERE X IS A STANDARD
C NORMAL VARIATE [X~N(0,1)]. THE CALCULATION IS DONE THROUGH THE
C NUMERICAL EVALUATION OF TWO INTEGRALS: I1, FROM -∞ TO -4, USING
C VARIABLE TRANSFORMATION AND I2, FROM -4 TO X, USING SIMPSON'S
C RULE, WITH A FIXED NUMBER OF 500 SEGMENTS. THE FINAL RESULT IS
```

```
C GIVEN BY THE SUM (I1+I2), MULTIPLIED BY THE SQUARE ROOT OF 1/2Π.
C
 PROGRAM NORMAL
 EXTERNAL FUNC,TRANSF
 WRITE(*,*)
 WRITE(*,*) 'INPUT ARGUMENT X OF THE STANDARD NORMAL VARIATE'
 WRITE(*,*)
 XL=-1./4.
 B=-4.
C DEFINING THE LOWER LIMIT OF -1/4 FOR I1 AND -4 TO I2
 XH=0.0
C DEFINING THE UPPER LIMIT OF 0 FOR INTEGRAL I1
        WRITE(*,*) 'STANDARD NORMAL CUMULATIVE DISTRIBUTION FUNC-
TION'
        WRITE(*,*) '-----------------------------------------'
        WRITE(*,*)
        WRITE(*,*) ' RESULTS FOR THE NUMERICAL INTEGRATION'
        WRITE(*,*)
        WRITE(*,*) ' X P(X<X)'
        WRITE(*,*) '------- -------------------'
        WRITE(*,*)
 CALL LEFTI(TRANSF,XL,XH,RES1)
 CALL RIGHTI(FUNC,B,C,RES2)
 RES=(RES1+RES2)/SQRT(2.*3.14592654)
 WRITE(*,'(2X,F8.3,11X,F7.3)') C,RES
 WRITE(*,*)
 WRITE(*,*)
 WRITE(*,*) 'DO YOU WISH TO RUN THE PROGRAM FOR ANOTHER X? YES-1,
NO=0'
 READ(*,*) IQ
 IF(IQ.EQ.1) GOTO 99
 END
C SUBROUTINE TO CALCULATE THE LEFT TAIL I1
 SUBROUTINE LEFTI(TRANSF,XL,XH,RES1)
        NN=49
        XHL=(XH-XL)/(FLOAT(NN)+1)
        SUM=TRANSF(XL+XHL/2.)
 DO 12 I=1,NN
        SUM=SUM+TRANSF(XL+XHL/2.+FLOAT(I)*XHL)
 12 CONTINUE
 RES1=SUM*XHL
        RETURN
        END
C SUBROUTINE TO CALCULATE THE RIGHT TAIL I2
 SUBROUTINE RIGHTI(FUNC,B,C,RES2)
```

```
      N=500
      XHR=ABS(C-B)/FLOAT(N)
      SUME=0.
      SUMO=0.
      DO 14 J=1,N-1,2
      SUMO=SUMO+FUNC(B+FLOAT(J)*XHR)
   14 CONTINUE
   DO 16 K=2,N-2,2
         SUME=SUME+FUNC(B+FLOAT(K)*XHR)
   16 CONTINUE
   RES2=(C-B)*(FUNC(B)+4.*SUMO+2.*SUME+FUNC(C))/(3*FLOAT(N))
      RETURN
      END
C NORMAL DENSITY FUNCTION
 FUNCTION FUNC(X)
      FUNC=EXP(-X*X/2.)
      RETURN
      END
C TRANSFORMED DENSITY FUNCTION
 FUNCTION TRANSF(X)
         TRANSF=EXP(-1./(2.*X*X))/(X*X)
      RETURN
      END
```

3. With reference to the computer code described in Exercise 2,

   (a) test it, by calculating $\Phi(-3.5)$, $\Phi(-1)$, $\Phi(0)$, $\Phi(1)$, and $\Phi(3.5)$, and comparing your results to the values given in Table 5.1;
   (b) if $X \sim N(300,180)$, use the program to calculate $P(220 \leq X \leq 390)$;
   (c) if $X \sim N(300,180)$, use the program to calculate $P(X < 450 | X > 390)$; and
   (d) repeat the solutions to items (a), (b), and (c), using Eq. (5.14).

4. Solve Exercise 7 of Chap. 4, employing the normal approximation to the Poisson distribution.

5. Download the annual mean flow data of the Lehigh River at Stoddartsville (USGS 01447500) from http://waterdata.usgs.gov/pa/nwis/inventory/?site_no=01447500, for the entire period of record. Fit a normal distribution to the flow data, by estimating its parameters from the sample mean and standard deviation. Plot on a single chart the histogram and the normal density, similarly to what has been done in Fig. 5.4. Comment on the possible application of the central limit theorem to annual mean flows.

6. Solve items (a) and (b) of Example 5.4, using the normal distribution. Plot the resulting density function. Calculate the 100-year return-period quantile.

7. Solve Exercise 5 for the two-parameter lognormal distribution.

8. In Example 5.4, assume the coefficient of skewness is equal to 1.5. Solve items (a) and (b), for the three-parameter lognormal distribution. Plot the resulting density function. Calculate the 100-year return-period quantile.

9. Solve items (a) and (b) of Example 5.4, using the Exponential distribution. Plot the resulting density function. Calculate the 100-year return-period quantile.

10. Solve items (a) and (b) of Example 5.4, using the gamma distribution. Plot the resulting density function. Calculate the 100-year return-period quantile.

11. The wind direction, at a given location, is a random variable X, measured in angles from the north direction, with mean and standard deviation respectively equal to 200° and 100°. Discuss the convenience of using the beta distribution as a model for X. Calculate the parameters for the beta distribution and the probability that X be comprised between 90° and 150°. Plot the resulting density function.

12. Solve Example 5.8 assuming the time between rains is normally distributed with a mean of 4 days and standard deviation of 2 days. Plot on the same chart the parent density and the maximum time density.

13. Solve items (a) and (b) of Example 5.4, using the Gumbel$_{max}$ distribution. Plot the resulting density function. Calculate the 100-year return-period quantile.

14. The annual maximum flows at a given river section are well described by the Gumbel$_{max}$ distribution with location and scale parameters respectively given by $\beta = 173$ m^3/s and $\alpha = 47$ m^3/s. At this river cross-section, the riverbank stage corresponds to discharge $Q_t = 250$ m^3/s. Knowing that river plains have been flooded, what is the probability the exceedance over the discharge Q_t will be less than or equal to 100 m^3/s?

15. The Alva River at Ponte de Mucela, in Portugal, has on average 3 exceedances over the reference discharge 65 m^3/s, per year. Statistical tests support the plausibility of the hypotheses (1) that the number of exceedances follows a Poisson distribution; (2) that the maximum exceedances are independent; and (3) that the exceedances are exponentially distributed. If the mean and standard deviation of exceedances are respectively equal to 72.9 and 76.7 m^3/s, calculate the 100-year return-period flood discharge.

16. Solve Example 5.9, for the Fréchet distribution of maxima. Plot the resulting density function. Compare your results with those given in the solution to Example 5.9.

17. Solve Exercise 15, assuming that hypothesis (3) is rejected.

18. Plot on the same chart the cumulative distribution function for the GEV, for the sets of parameters shown in Fig. 5.13. Discuss the GEV modeling of annual maximum discharges, when $\kappa > 0$ and $\kappa \leq 0$.

19. The mean, variance, and coefficient of skewness for the annual minimum daily flows for a large tributary of the Amazon River, in Brazil, are 694.6 m^3/s, 26186.62 (m^3/s)^2, and 1.1, respectively. Use the Gumbel$_{min}$ distribution to find the 25-year return-period minimum daily flow.

20. Solve Exercise 19 using the two-parameter Weibull$_{min}$ distribution.

21. Organize the relations given by Eqs. (5.101) and (5.102) in the form of tables of α, $C(\alpha)$, and $D(\alpha)$, and set up a practical scheme to calculate the parameters for the three-parameter Weibull$_{min}$ distribution.

22. Solve Exercise 19 using the three-parameter Weibull$_{min}$ distribution.

23. Solve items (a) and (b) of Example 5.4, using the Pearson type III distribution. Plot the resulting density function. Calculate the 100-year return-period quantile.

24. Solve items (a) and (b) of Example 5.4, using the Log-Pearson type III distribution. Plot the resulting density function. Calculate the 100-year return-period quantile.

25. Consider a chi-square distribution with $\nu = 4$. Calculate $P(\chi^2 > 5)$.

26. The daily concentrations of dissolved oxygen have been measured for 30 consecutive days. The sample yielded a mean of 4.52 mg/l and the standard deviation of 2.05 mg/l. Assuming the DO concentrations are normally distributed, determine the absolute value of the maximum error of estimate of the population mean μ, with 95 % probability. In other terms, determine d such that $P(|\overline{X} - \mu| \leq d) = 0.95$.

27. Consider a Snedecor's F distribution, with $\gamma_1 = 10$ and $\gamma_2 = 5$. Calculate $P(F > 2)$.

28. Consider the bivariate normal distribution with parameters $\mu_X = 2$, $\sigma_X = 2$, $\mu_Y = 1$, $\sigma_Y = 0.5$, and $\rho = 0.7$. Determine the conditional density $f_{Y|X}(y|x = 3)$.. Calculate the probability $P(Y < 3|X = 3)$.

29. *Buffon's needle problem* (adapted from Rozanov 1969). Suppose a needle is randomly tossed onto a plane, which has been ruled with parallel lines separated by a fixed distance L. The needle is a line segment of length $l \leq L$. The problem posed by the French naturalist and mathematician Georges-Louis Leclerc, Compte de Buffon (1707–1788), was to calculate the probability that the needle intersects one of the lines. In order to solve it, let X_1 denote the angle between the needle and the direction of the parallel lines and X_2 represent the distance between the bottom extremity of the needle and the nearest line above it, as shown in Fig. 5.28. The conditions of the needle tossing experiment are such that the random variable X_1 is uniformly distributed in the interval $[0,\pi]$ and X_2 is also a uniform variate in $[0,L]$. Assuming these two variables are independent, their joint density function is given by $f_{X_1,X_2}(x_1, x_2) = 1/\pi L$, $0 \leq x_1 \leq \pi$, $0 \leq x_2 \leq L$. Assume that event A refers to the needle intersecting one of the parallel lines, which will occur only if $X_2 \leq l\sin(X_1)$, or if the point (x_1, x_2) falls in region B, the shaded area in

Fig. 5.28. Thus, $P\{(X_1, X_2) \in B\} = \iint\limits_{B} \dfrac{dx_1\,dx_2}{\pi L} = \dfrac{2l}{\pi L}$, where $l\displaystyle\int\limits_{0}^{\pi} \sin x_1$

$dx_1 = 2l$ is the area of region B. The assumption of independence between the two variables can be tested experimentally. In fact, if the needle is repeatedly tossed n times onto the plane and if event A has occurred n_A times, then $\frac{n_A}{n} \approx \frac{2l}{\pi L}$, for a large number of needle tosses n. In this case, the quantity $\frac{2l}{L}\frac{n}{n_A}$ must be a

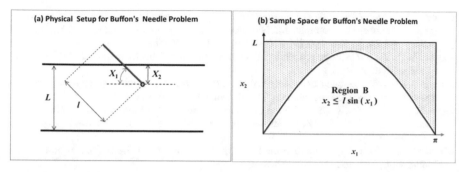

Fig. 5.28 Buffon's needle tossing experiment

good approximation to the number $\pi = 3.415$. . . It is possible to simulate the Buffon's needle tossing experiment by mean of the software *Buffon* available from the URL http://www.efg2.com/Lab/Mathematics/Buffon.htm [accessed: 13th January 2016]. Download and run the *Buffon* software, for increasing values of n, and determine the respective approximations to π. Plot your results on a chart with n on the horizontal axis and the π estimates on the vertical axis.

References

Abramowitz M, Stegun IA (1972) Handbook of mathematical functions. Dover, New York

Ajayi GO, Olsen RL (1985) Modelling of a raindrop size distribution for microwave and millimetre wave applications. Radio Sci 20(2):193–202

Ang H-SA, Tang WH (1990) Probability concepts in engineering planning and design, volume II: decision, risk, and reliability. Copyright Ang & Tang

Beard LR (1962) Statistical methods in hydrology (Civil Works Investigations Project CW-151). United States Army Engineer District. Corps of Engineers, Sacramento, CA

Benjamin JR, Cornell CA (1970) Probability, statistics, and decision for civil engineers. McGraw-Hill, New York

Benson MA (1968) Uniform flood-frequency estimating methods for federal agencies. Water Resour Res 4(5):891–908

Beran M, Hosking JRM, Arnell NW (1986) Comment on "TCEV distribution for flood frequency analysis". Water Resouces Res 22(2):263–266

Bobée B, Ashkar F (1991) The gamma family and derived distributions applied in hydrology. Water Resources Publications, Littleton, CO

Cannarozzo M, D'Asaro F, Ferro V (1995) Regional rainfall and flood frequency analysis for Sicily using the two component extreme value distribution. Hydrol Sci J 40(1):19–42. doi:10. 1080/02626669509491388

Castillo E (1988) Extreme value theory in engineering. Academic, Boston

Coles S (2001) An introduction to statistical modeling of extreme values. Springer, London

Dalrymple T (1960) Flood-frequency analyses, Manual of Hydrology: Part.3. Flood-flow Techniques, Geological Survey Water Supply Paper 1543-A. Government Printing Office, Washington

Fiorentino M, Gabriele S, Rossi F, Versace P (1987) Hierarchical approach for regional flood frequency analysis. In: Singh VP (ed) Regional flood frequency analysis. Reidel, Dordrecht

Fisher RA, Tippett LHC (1928) Limiting forms of the frequency distribution of the largest or smallest member of a sample. Mathematical proceedings of the Cambridge Philosophical Society, vol 24, pp 180–190. DOI:10.1017/S0305004100015681

Fréchet M (1927) Sur la loi de probabilité de l'écart maximum. Annales de la Societé Polonaise de Mathématique 6(92):116

Freeze RA (1975) A stochastic conceptual analysis of one-dimensional ground-water flow in non-uniform homogeneous media. Water Resour Res 11:725–741

Genz A, Bretz F, Miwa T, Mi X, Leisch F, Scheipl F, Hothorn T (2009). mvtnorm: Multivariate normal and t Distributions. R package version 0.9-8. http://CRAN.R-project.org/package = mvtnorm

Gnedenko B (1943) Sur la distribution limite du terme maximum d'une série aléatoire. Ann Mathematics Second Series 44(3):423–453

Gumbel EJ (1958) Statistics of Extremes. Columbia University Press, New York

Haan CT (1977) Statistical methods in hydrology. The Iowa University Press, Ames, IA

Hosking JRM (1988) The 4-parameter Kappa distribution. IBM Research Report RC 13412. IBM Research, Yorktown Heights, NY

Hosking JRM, Wallis JR (1997) Regional frequency analysis: an approach based on L-moments. Cambridge Cambridge University Press, Cambridge

Houghton J (1978) Birth of a parent: the Wakeby distribution for modeling flood flows. Water Resour Res 15(6):1361–1372

Jenkinson AF (1955) The frequency distribution of the annual maximum (or minimum) of meteorological elements. Q J Roy Meteorol Soc 81:158–171

Juncosa ML (1949) The asymptotic behavior of the minimum in a sequence of random variables. Duke Math J 16(4):609–618

Kendall MG, Stuart A (1963) The advanced theory of statistics, vol I, Distribution theory. Griffin, London

Kite GW (1988) Frequencyd and risk analysis in hydrology. Water Resources Publications, Fort Collins, CO

Kottegoda NT, Rosso R (1997) Statistics, probability, and reliability for civil and environmental engineers. McGraw-Hill, New York

Leadbetter MR (1974) On extreme values in stationary sequences. Probab Theory Relat Fields 28 (4):289–303

Leadbetter MR (1983) Extremes and local dependence in stationary sequences. Zeitschrift für Wahrscheinlichkeitstheorie und verwandte Gebiete 65:291–306

Leadbetter MR, Lindgren G, Rootzén H (1983) Extremes and related properties of random sequences and processes. Springer, New York

Murphy PJ (2001) Evaluation of mixed-population flood-frequency analysis. J Hydrol Eng 6:62–70

Nelsen RB (2006) An introduction to copulas, ser. Lecture notes in statistics. Springer, New York

Papalexiou SM, Koutsoyiannis D (2013) The battle of extreme distributions: a global survey on the extreme daily rainfall. Water Resour Res 49(1):187–201

Papalexiou SM, Koutsoyiannis D, Makropoulos C (2013) How extreme is extreme? An assessment of daily rainfall distribution tails. Hydrol Earth Syst Sci 17:851–862

Perichi LR, Rodríguez-Iturbe I (1985) On the statistical analysis of floods. In: Atkinson AC, Feinberg SE (eds) A celebration of statistics. Springer, New York, pp 511–541

Pickands J (1975) Statistical inference using extreme order statistics. Ann Stat 3(1):119–131

Pollard JH (1977) A handbook of numerical and statistical techniques. Cambridge University Press, Cambridge

Press W, Teukolsky SA, Vetterling WT, Flannery BP (1986) Numerical recipes in Fortran 77—the art of scientific computing. Cambridge University Press, Cambridge

Rao AR, Hamed KH (2000) Flood frequency analysis. CRC Press, Boca Raton, FL

Rossi FM, Fiorentino M, Versace P (1984) Two component extreme value distribution for flood frequency analysis. Water Resour Res 20(7):847–856

Rozanov YA (1969) Probability theory: a concise course. Dover, New York

Salvadori G, De Michele C (2010) Multivariate multiparameter extreme value models and return periods: a copula approach. Water Resour Res 46(10):1

Salvadori G, De Michele C, Kottegoda NT, Rosso R (2007) Extremes in nature: an approach using copulas, vol 56. Springer Science & Business Media

Serinaldi F (2015) Dismissing return periods! Stochastic Environ Res Risk Assessment 29 (4):1179–1189

Sklar M (1959) Fonctions de répartition à n dimensions et leurs marges. Publication de l'Institut de Statistique de l'Université de Paris 8:229–231

Stahl S (2006) The evolution of the normal distribution. Math Mag 79(2):96–113

Stedinger JR, Griffis VW (2008) Flood frequency analysis in the United States: time to update. J Hydrol Eng 13(4):199–204

Stedinger JR, Griffis VW (2011) Getting from here to where? Flood frequency analysis and climate. J Am Water Resour Assoc 47(3):506–513

Vlcek O, Huth R (2009) Is daily precipitation gamma-distributed? Adverse effects of an incorrect use of the Kolmogorov-Smirnov test. Atmos Res 93:759–766

Waylen PR, Woo MK (1984) Regionalization and prediction of floods in the Fraser river catchment. Water Resour Bull 20(6):941–949

WRC (1967, 1975, 1976, 1977, 1981) Guidelines for determining flood flow frequency, Bulletin 15 (1975), Bulletin 17 (1976), Bulletin 17A (1977), Bulletin 17B (1981). United States Water Resources Council-Hydrology Committee, Washington

Yan J (2007) Enjoy the joy of copulas: with a package copula. J Stat Software 21(4):1–21

Yevjevich VM (1972) Probability and statistics in hydrology. Water Resources Publications, Fort Collins, CO

Chapter 6
Parameter and Quantile Estimation

Mauro Naghettini

6.1 Introduction

In contrast to descriptive statistics, which is related only to a data sample, statistical inference seeks the underlying properties of the population from which data have been drawn. Statistical inference includes the validation of an assumed model for the population underlying distribution, the estimation of its parameters, the construction of confidence intervals, and the testing of hypotheses concerning population descriptors. The methods of statistical inference make the association between the physical reality of a sample of observed data and the abstract concept of a probabilistic model of a given random variable. In fact, population is a notional term as it would consist of an infinite (or a large finite) number of possibly observable outcomes of a random experiment, which actually have not yet occurred and, thus, do not exist, in the physical sense. In reality, the sample consists of a much smaller number of actually observed data, of size N, and denoted by $\{x_1, x_2, \ldots, x_N\}$, which are supposed to have been drawn from the population. The sample $\{x_1, x_2, \ldots, x_N\}$ does indeed represent the real facts from which are inferred the estimates of population descriptors, such as the mean, variance, and coefficient of skewness, the characteristics of the underlying probability distribution, and the estimates of its respective parameters. Figure 6.1 depicts a scheme illustrating the reasoning behind the methods of statistical inference. In this figure, the population, as associated with the sample space of a random experiment, is mapped by a continuous random variable X, whose density function $f_X(x)$ is defined by its parameters $\theta_1, \theta_2, \ldots, \theta_k$.

Neither $f_X(x)$ nor its set of parameters $\theta_1, \theta_2, \ldots, \theta_k$ is actually known and must be assumed, for the former, or estimated, for the latter. Assume that $f_X(x)$ could be

M. Naghettini (✉)
Universidade Federal de Minas Gerais, Belo Horizonte, Minas Gerais, Brazil
e-mail: mauronag@superig.com.br

© Springer International Publishing Switzerland 2017
M. Naghettini (ed.), *Fundamentals of Statistical Hydrology*,
DOI 10.1007/978-3-319-43561-9_6

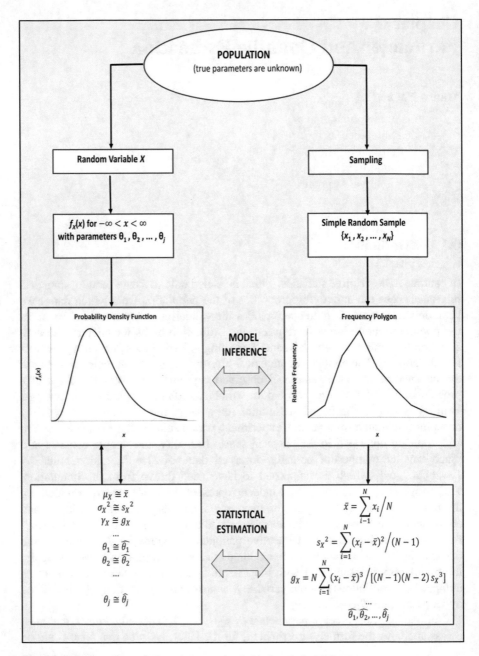

Fig. 6.1 Sample and population: the reasoning behind statistical inference

correctly prescribed from the application of some deductive law, such as the central limit theorem, or from the characteristics of the physical phenomenon being modeled, or even from the data sample descriptors, and, thus, no doubt concerning the analytical form of the population underlying distribution remains. However, even for this idealized situation, the estimates $\hat{\theta}_1, \hat{\theta}_2, \ldots, \hat{\theta}_k$, of the parameters $\theta_1, \theta_2, \ldots, \theta_k$, of $f_X(x)$, must be obtained from the only source of information that is available, which is the data sample.

The problem, as previously described, is termed parameter estimation and is freely used here to signify the act of obtaining estimates of the parameters and descriptors of the population underlying distribution. Among the classical methods of statistical inference, two possible pathways are envisaged for estimating parameters: point estimation and interval estimation. Point estimation refers to assigning one single numeric value to a given population parameter, from the analysis of a data sample. Interval estimation, in turn, utilizes the information contained in a data sample to make a broader statement concerning the probability, or the degree of confidence, with which an interval of numeric values will contain the true value of the population parameter or descriptor. In the sections that follow, the principles related to both pathways are described, with an enlarged focus on the parameter point estimation, as resulting from its more extended and frequent use in Statistical Hydrology.

6.2 Preliminaries on Parameter Point Estimation

As already mentioned, the starting point of parameter estimation is a data sample of size N, given by the elements $\{x_1, x_2, \ldots, x_N\}$. These denote the realizations of the random variables $\{X_1, X_2, \ldots, X_N\}$. In order to consider the sample as simple and random, or an SRS (Simple Random Sample), it is necessary that the variables $\{X_1, X_2, \ldots, X_N\}$ be independent and identically distributed, or IID variables, for short. More formally, if the common density of the independent variables $\{X_1, X_2, \ldots, X_N\}$ is $f_X(x)$, then the joint density function of the SRS is given by $f_{X_1, X_2, \ldots, X_N}(x_1, x_2, \ldots, x_N) = f_X(x_1) f_X(x_2) \ldots f_X(x_N)$. Thus, once the form of the density $f_X(x)$ is assumed or specified, the parameters $\theta_1, \theta_2, \ldots, \theta_k$ that fully describe it need to be estimated from the SRS $\{x_1, x_2, \ldots, x_N\}$, whose likelihood is given by the joint density $f_{X_1, X_2, \ldots, X_N}(x_1, x_2, \ldots, x_N)$.

Assume, for simplicity, that there is only one parameter θ to be estimated from the SRS $\{x_1, x_2, \ldots, x_N\}$. If all the available information is contained in the SRS, the estimate of θ must necessarily be a function $g(x_1, x_2, \ldots, x_N)$ of the observed data. As the sample elements $\{x_1, x_2, \ldots, x_N\}$ are the realizations of the random variables $\{X_1, X_2, \ldots, X_N\}$, one can interpret the function $g(x_1, x_2, \ldots, x_N)$ as a particular realization of the random variable $g(X_1, X_2, \ldots, X_N)$. If this function is utilized to estimate the parameter θ of $f_X(x)$, then, it is inevitable to distinguish the θ *estimator*, as denoted by Θ or $\hat{\theta}$, from the θ *estimate*, represented by $\hat{\theta}$. In fact, an estimate $\hat{\theta} = g(x_1, x_2, \ldots, x_N)$ is just a number or, in other terms, the realization of the estimator $\Theta = \hat{\theta} = g(X_1, X_2, \ldots, X_N)$. For instance, the estimates $\bar{x}$ and s_X^2, of the mean and variance, respectively calculated for a sample $\{x_1, x_2, \ldots, x_N\}$, are the realizations of the estimators $\overline{X}$ and S_X^2 of the IID variables $\{X_1, X_2, \ldots, X_N\}$. The generic estimator $\Theta = \hat{\underline{\theta}} = g(X_1, X_2, \ldots, X_N)$ is, in fact, a random variable, whose properties should be studied under the light of the probability theory. In this context, it is clearly inappropriate to raise the question whether a given θ estimate is better or worse than another estimate of θ, as, in such a case, one would be comparing two different numbers. However, it is absolutely legitimate and relevant to ask how the estimator $\Theta_1 = \hat{\underline{\theta}}_1 = g_1(X_1, X_2, \ldots, X_N)$ compares with its competitor estimator $\Theta_2 = \hat{\underline{\theta}}_2 = g_2(X_1, X_2, \ldots, X_N)$. The answer to this question is related to the properties of estimators.

Firstly, it is undesirable that an estimation procedure, as materialized by a given estimator from a sample of size N, yield estimates which, on their ensemble, are systematically larger or smaller than the true value of the parameter being estimated. In effect, what is desirable is that the mean of the estimates be equal to the parameter true value. Formally, a point estimator $\hat{\underline{\theta}}$ is said to be an unbiased estimator of the population parameter θ if

$$E\left[\hat{\underline{\theta}}\right] = \theta \tag{6.1}$$

From Eq. (6.1), it is clear that the property of unbiasedness does not depend on the sample size N. If the estimator is biased, then the bias is given by the difference $E\left[\hat{\underline{\theta}}\right] - \theta$. Example 6.1 illustrates the unbiasedness property for the sample arithmetic mean and the sample variance.

Example 6.1 Show that the sample mean and variance are unbiased estimators for the population μ and σ^2, respectively.

Solution Consider a simple random sample $\{x_1, x_2, \ldots, x_N\}$ of size N. The point estimator for the population mean, μ, is $\hat{\underline{\theta}} = \overline{X} = \frac{1}{N}(X_1 + X_2 + \ldots + X_N)$. In this case, Eq. (6.1) yields $E[\overline{X}] = \frac{1}{N}\{E[X_1] + E[X_2] + \ldots + E[X_N]\}$, or $E[\overline{X}] = \frac{1}{N}N\mu = \mu$. For the population variance, the estimator is $\hat{\underline{\theta}} = S^2 = \frac{1}{(N-1)}\sum_{i=1}^{N}(X_i - \overline{X})^2$. Application of Eq. (6.1), in this case, gives $E[S^2] = \frac{1}{N-1}E\left[\sum_{i=1}^{N}(X_i - \overline{X})^2\right] = \frac{1}{N-1}E\left[\sum_{i=1}^{N}(X_i - \mu)^2 - N(\overline{X} - \mu)^2\right]$. Recalling that the expected value of a sum is the sum of the expected values of the summands, then $E[S^2] = \frac{1}{N-1}\left\{\sum_{i=1}^{N}E\left[(X_i - \mu)^2\right] - NE\left[(\overline{X} - \mu)^2\right]\right\}$. In this equation, the expected value in the first term, between braces, corresponds to the variance of X, or σ^2, whereas the second term is equal to the variance of $\overline{X}$, or σ^2/N. Thus, $E[S^2] = \frac{1}{N-1}\left(N\sigma^2 - N\frac{\sigma^2}{N}\right) = \sigma^2$. Therefore, the sample arithmetic mean and sample variance are indeed unbiased estimators for μ and σ^2, for any value of N.

Note, however, that the sample standard deviation $S = \sqrt{S^2}$ is biased and its bias for normal populations is approximately equal to $\sigma/4N$ (Yevjevich 1972).

The second desirable property of an estimator is consistency. An estimator $\hat{\theta}$ is consistent if, as $N \to \infty$, it converges in probability to θ. Formally, an estimator $\hat{\theta}$ is consistent if, for any positive number ε,

$$\lim_{N \to \infty} P\left[\left|\hat{\theta} - \theta\right| \leq \varepsilon\right] = 1$$

or   (6.2)

$$\lim_{N \to \infty} P\left[\left|\hat{\theta} - \theta\right| > \varepsilon\right] = 0$$

In simple terms, the estimator $\hat{\theta}$ is consistent if it converges to θ as the sample size becomes very large. As opposed to unbiasedness, consistency depends on the sample size and is an asymptotic property of the sampling distribution of $\hat{\theta}$. Loosely speaking, if an estimator $\hat{\theta}$ is consistent, it is asymptotically unbiased and its variance $\mathrm{Var}\left[\hat{\theta}\right]$ converges to zero, as N approaches infinity.

To establish consistency for an estimator, in formal terms, the general result provided by Chebyshev's inequality, named after the Russian mathematician Pafnuty Chebyshev (1821–1894), is usually required. According to this, the probability that any randomly selected value of X, from any probability distribution with mean μ and variance σ^2, will deviate more than ε from μ, must obey the following inequality:

$$P(|X - \mu| \geq \varepsilon) \leq \frac{\sigma^2}{\varepsilon^2}$$

or   (6.3)

$$P(|X - \mu| < \varepsilon) > 1 - \frac{\sigma^2}{\varepsilon^2}$$

Assuming $\hat{\theta}$ is an unbiased estimator for θ, application of Chebyshev's inequality to $\hat{\theta}$ yields

$$P\left[\left|\hat{\theta} - \theta\right| \geq \varepsilon\right] \leq \frac{E\left[\left(\hat{\theta} - \theta\right)^2\right]}{\varepsilon^2}$$ (6.4)

If $E\left[\left(\hat{\theta} - \theta\right)^2\right]$ is assumed to converge to zero, as N approaches infinity, the right-hand side of Eq. (6.4) will tend to zero and the condition $\lim_{N \to \infty} P\left[\left|\hat{\theta} - \theta\right| > \varepsilon\right] = 0$ for consistency is satisfied.

If $\hat{\theta}$ is a biased estimator for θ, it can be shown that $E\left[\left(\hat{\theta} - \theta\right)^2\right] = \mathrm{Var}(\hat{\theta}) + \left[B(\hat{\theta})\right]^2$ (see Exercise 4 in this chapter), where $B(\hat{\theta})$ denotes the bias of $\hat{\theta}$.

In general, $E\left[\left(\hat{\theta} - \theta\right)^2\right]$ is referred to as the Mean Square Error (MSE), being equivalent to the variance, when there is no bias, or to the sum of the variance and the squared bias, when there is bias. With this change, Eq. (6.4) still holds and a general procedure for testing for consistency of a biased or an unbiased estimator can be outlined. First, Eq. (6.1) must be employed to check whether $\hat{\theta}$ is unbiased or not. If unbiased, calculate $\mathrm{Var}\left(\hat{\theta}\right)$ and check $\lim_{N\to\infty}\mathrm{Var}\left(\hat{\theta}\right)$: if the limit is zero, $\hat{\theta}$ is consistent; otherwise, it is not. If $\hat{\theta}$ is biased, calculate bias $B\left(\hat{\theta}\right)$ and check the limits of $\mathrm{Var}\left(\hat{\theta}\right)$ and $B\left(\hat{\theta}\right)$, as $N \to \infty$. If both converge to zero, the biased $\hat{\theta}$ is consistent. This general procedure is illustrated in the worked out examples that follow.

Example 6.2 Show that the sample arithmetic mean is a consistent estimator for the population mean μ.

Solution In Example 6.1, it was shown that the sample arithmetic mean $\hat{\theta} = \overline{X}$ is an unbiased estimator for μ. According to the general procedure for testing for consistency, previously outlined, as $\hat{\theta} = \overline{X}$ is unbiased, it suffices to calculate Var $\left(\overline{X}\right)$ and check its limit as $N \to \infty$. In Example 3.18, the variance of $\overline{X}$ was proved to be σ^2/N. Assuming σ is finite, as $N \to \infty$, the limit of σ^2/N tends to zero. As a conclusion $\hat{\theta} = \overline{X}$ is a consistent estimator for μ.

Example 6.3 Consider $\hat{\theta}_1 = \frac{1}{N}\sum_{i=1}^{N}\left(X_i - \overline{X}\right)^2$ and $\hat{\theta}_2 = \frac{1}{(N-1)}\sum_{i=1}^{N}\left(X_i - \overline{X}\right)^2$ as estimators for the population variance σ^2. In Example 6.1 it was shown that $\hat{\theta}_2$ is an unbiased estimator for σ^2. Likewise, it can be shown that $E\left[\hat{\theta}_1\right] = \frac{N-1}{N}\sigma^2 \neq \sigma^2$ and, thus, that $\hat{\theta}_1$ is a biased estimator for σ^2. Are $\hat{\theta}_1$ and $\hat{\theta}_2$ consistent?

Solution As $\hat{\theta}_2$ is unbiased, to be consistent, it suffices to check whether $\lim_{N\to\infty}\mathrm{Var}\left(\hat{\theta}_2\right) \to 0$. It can be shown that $\mathrm{Var}\left(\hat{\theta}_1\right) = \mu_4(N-1)^2/N^3 - \sigma^4(N-1)$ $(N-3)/N^3$, where μ_4 denotes the population central moment of order 4 (Weisstein 2016). As $\hat{\theta}_2 = [N/(N-1)]\hat{\theta}_1$, then $\mathrm{Var}\left(\hat{\theta}_2\right) = [N/(N-1)]^2\mathrm{Var}\left(\hat{\theta}_1\right) = \mu_4/$ $N - \sigma^4(N-3)/[N(N-1)]$. Assuming that μ_4 and σ are finite, it immediately follows that $\lim_{N\to\infty}\mathrm{Var}\left(\hat{\theta}_2\right) \to 0$ and that $\hat{\theta}_2$ is indeed a consistent estimator for σ^2. The same line of reasoning can be applied for estimator $\hat{\theta}_1$ but the bias $B\left(\hat{\theta}_1\right)$ must be taken into account. First, for $\mathrm{Var}\left(\hat{\theta}_1\right)$, as in the previously given expression, repeated applications of l'Hôpital's rule for limits lead convergence to zero of $\mathrm{Var}\left(\hat{\theta}_1\right)$ as $N \to \infty$. The bias $B\left(\hat{\theta}_1\right)$ is equal to the difference $B\left(\hat{\theta}_1\right) = E\left[\hat{\theta}_1\right] - \sigma^2$, which gives $B\left(\hat{\theta}_1\right) = -\sigma^2/N$, whose limit as $N \to \infty$ is also zero, thus showing that $\hat{\theta}_1$, although biased for finite N, becomes a consistent estimator for σ^2, as $N \to \infty$. For this particular case, as far as finite samples are concerned, the usual practice is to opt for the unbiased and consistent estimator $\hat{\underline{\theta}}_2$, in detriment of the biased yet consistent estimator $\hat{\underline{\theta}}_1$. A simpler alternative way to show the consistency property of estimators is to start directly from Eq. (6.2), with the aid of the Law of Large Numbers (LLN). The weak law of large numbers is a theorem from probability theory stating that for a sequence of IID random variables $\{X_1, X_2, \ldots, X_N\}$, with a common mean μ and finite variance σ^2, the arithmetic average $(X_1 + X_2 + \ldots + X_N)/N$ converges in probability to μ, as $N \to \infty$. The weak law of large numbers is usually proved on the basis of Chebyshev's inequality. The strong law of large numbers is, in fact, a rigorous generalization of the weak law, as it is based on asymptotic analysis and is stricter than convergence in probability (Mood et al. 1974). The consistency property of the sample mean, for example, follows immediately from the weak law of large numbers, as both methods lead to convergence in probability to μ. As regards the estimators $\hat{\underline{\theta}}_1$ and $\hat{\underline{\theta}}_2$, applying the LLN to test for consistency requires further arguments. The unbiased estimator $\hat{\theta}_2$ can be rewritten as $\hat{\theta}_2 = \frac{N}{(N-1)} \sum_{i=1}^{N} \frac{1}{N} \left(X_i^2 - \overline{X}^2 \right)$. From the LLN, since variables X_i are supposed IID, the first term in the summation $1/N \sum X_i^2 \to E[X_i^2]$, as $N \to \infty$, whereas the second term $\overline{X}^2 = [(X_1 + X_2 + \ldots + X_N)/N]^2 \to E^2[X]$. Therefore, the consistency of $\hat{\theta}_2$ is warranted by $\lim_{N \to \infty} \hat{\theta}_2 = N/(N-1)\{E[X^2] - E^2[X]\} = \sigma^2$. Making the same algebraic manipulations for the biased estimator $\hat{\theta}_1$, it follows that the limit $\lim_{N \to \infty} \hat{\theta}_1 = \{E[X^2] - E^2[X]\} = \sigma^2$, and as such, Eq. (6.2) is satisfied, thus proving that the biased $\hat{\theta}_1$ is a consistent estimator for σ^2.

The solution to Example 6.3 shows that one can possibly have estimators that are unbiased and consistent, biased and consistent, in addition to (un) biased and inconsistent. In fact, these two desirable properties of estimators are not always fulfilled by the same estimator and, on occasion, one has to decide which attribute is preferable in detriment of the other. In hydrology, as the sample sizes are usually finite and small, unbiasedness is certainly more important than consistency. Nevertheless, unbiased estimators are not always obtainable or easily obtainable, which leaves no choice but the use of biased estimators.

The third desirable property of an estimator is efficiency. Since there may be more than one unbiased estimator for a parameter θ, the one with the least variance is more desirable than its competitor(s). The efficiency of $\hat{\underline{\theta}}_1$ relative to $\hat{\underline{\theta}}_2$, where $\hat{\underline{\theta}}_1$ and $\hat{\underline{\theta}}_2$ are two unbiased estimators, is given by

$$\text{Eff}\left(\hat{\theta}_1, \hat{\theta}_2\right) = \frac{\text{Var}\left(\hat{\theta}_2\right)}{\text{Var}\left(\hat{\theta}_1\right)} \tag{6.5}$$

If $\text{Eff}\left(\hat{\theta}_1, \hat{\theta}_2\right) > 1$, then $\hat{\theta}_1$ is more efficient than $\hat{\theta}_2$ or, otherwise, the opposite. The most efficient unbiased estimator is the one with the least variance, among all unbiased estimators for the parameter θ. The notion of relative efficiency and

Eq. (6.5) can be extended to biased estimators by replacing the variances, in the right-hand side of the equation, for the corresponding mean square errors $E\left[\left(\hat{\theta}_i - \theta\right)^2\right]$. Example 6.4 illustrates such a possibility.

Example 6.4 Considering the estimators $\hat{\underline{\theta}}_1 = \frac{1}{N}\sum_{i=1}^{N}\left(X_i - \overline{X}\right)^2$ and $\hat{\underline{\theta}}_2 = \frac{1}{(N-1)}\sum_{i=1}^{N}\left(X_i - \overline{X}\right)^2$ for the variance σ^2 of a normal population with a true mean μ, find the relative efficiency $Eff\left(\hat{\underline{\theta}}_1, \hat{\underline{\theta}}_2\right)$.

Solution As $\hat{\underline{\theta}}_2$ is unbiased and $\hat{\underline{\theta}}_1$ is biased (see solutions to Examples 6.1 and 6.3), Eq. (6.5) needs to be rewritten in terms of the MSEs: $\text{Eff}\left(\hat{\underline{\theta}}_1, \hat{\underline{\theta}}_2\right) = \text{MSE}\left(\hat{\underline{\theta}}_2\right)/\text{MSE}\left(\hat{\underline{\theta}}_1\right)$. Because $\hat{\underline{\theta}}_2$ is an unbiased estimator, $\text{MSE}\left(\hat{\underline{\theta}}_2\right) = \text{Var}\left(\hat{\underline{\theta}}_2\right)$. However, the expression given in the solution to Example 6.3 for $\text{Var}\left(\hat{\underline{\theta}}_2\right)$ is valid for any distribution with a generic fourth central moment μ_4. In this present case, for which a normal distribution is assumed, $\mu_4 = 3\sigma^4$. Replacing this particular value of μ_4 in the equation, the result is $\text{Var}\left(\hat{\underline{\theta}}_2\right) = \text{MSE}\left(\hat{\underline{\theta}}_2\right) = 2\sigma^4/(N-1)$. The solution to Example 6.3 has also shown that $B\left(\hat{\underline{\theta}}_1\right) = -\sigma^2/N$. For $\mu_4 = 3\sigma^4$, $\text{Var}\left(\hat{\underline{\theta}}_1\right) = 2(N-1)\sigma^4/N^2$ and the $\text{MSE}\left(\hat{\underline{\theta}}_1\right) = (2N-1)\sigma^4/N^2$. Therefore, the relative efficiency is $\text{Eff}\left(\hat{\underline{\theta}}_1, \hat{\underline{\theta}}_2\right) = 2N^2/[(N-1)(2N-1)]$. For $N \geq 2$, it yields $\text{Eff}\left(\hat{\underline{\theta}}_1, \hat{\underline{\theta}}_2\right) > 1$ and shows that $\hat{\underline{\theta}}_1$ is relatively more efficient than $\hat{\underline{\theta}}_2$.

The properties of unbiasedness, consistency and efficiency should guide the selection of the adequate estimators. However, the property of sufficiency is also a desirable property that an estimator should have and deserves a brief explanation. In other words, an estimator $\hat{\underline{\theta}}$ is considered a sufficient estimator for θ, if it extracts as much information as possible about θ from the sample $\{x_1, x_2, \ldots, x_N\}$, so that no additional information can be conveyed by any other estimator (Kottegoda and Rosso 1997). In formal terms, an estimator $\hat{\underline{\theta}} = g(X_1, X_2, \ldots, X_N)$ is sufficient for θ, if, for all θ and for any sample point, the joint density function of $(X_1, X_2, \ldots, X_N)$, conditioned on $\hat{\underline{\theta}}$, does not depend on θ (see Exercise 7 in this chapter). A practical example of a sufficient estimator, for the measure of central tendency of a density function, is the sample mean, whereas other measures, such as the median, for example, are not: if one single sample point changes, the mean changes accordingly, while the median does not. For the reader interested in a rigorous treatment on the properties of statistical estimators, the following are excellent references on the subject: Cramér (1946), Rao (1973), and Casella and Berger (1990).

In applications of statistics to engineering and hydrologic problems, population moments of order higher than 3 are generally not used, as unbiased estimators for them are hard to find and also because their variances are too large since the samples sizes are typically small. Table 6.1 summarizes the common estimators

Table 6.1 Estimators for the population mean, variance, standard-deviation, and coefficient of variation, irrespectively of of the parent distribution of the original variable X

Population	Estimator	Bias	Variance of estimator
μ	$\overline{X} = \dfrac{\sum\limits_{i=1}^{N} X_i}{N}$	0	$\dfrac{\sigma^2}{N}$
σ^2	$S_X^2 = \dfrac{1}{(N-1)} \sum\limits_{i=1}^{N} (X_i - \overline{X})^2$	0	$\mu_4/N - \sigma^4(N-3)/[N(N-1)]$, where $\mu_4 = $ 4th order central moment
σ	$S_X = \sqrt{S_X^2}$	$O\left(\dfrac{1}{N}\right)$ for $N < 20$ 0 for $N \geq 20^a$	$\dfrac{\mu_4 - \sigma^4}{4\sigma^2 N} + O\left(\dfrac{1}{N^2}\right)$ for $N < 20$ 0 for $N \geq 20^a$
C_{V_X}	$\hat{C}_{V_X} = \left(1 + \dfrac{1}{4N}\right)\dfrac{S_X}{\overline{X}}$	0	–

[a] $O(1/N)$ and $O(1/N^2)$ are small quantities proportional to $1/N$ and $1/N^2$ (Yevjevich 1972)

Table 6.2 Estimators for the population mean, variance, standard-deviation, and coefficient of variation of a random variable X from a normal population

Population	Estimator	Bias	Variance of estimator
μ	$\overline{X} = \dfrac{\sum\limits_{i=1}^{N} X_i}{N}$	0	$\dfrac{\sigma^2}{N}$
σ^2	$S_X^2 = \dfrac{1}{(N-1)} \sum\limits_{i=1}^{N} (X_i - \overline{X})^2$	0	$\dfrac{2\sigma^4}{N-1}$
σ	$S_X = \sqrt{S_X^2}$	$\approx \dfrac{\sigma}{4N}$	$\approx \dfrac{\sigma^2}{2(N-1)}$
C_{V_X}	$\hat{C}_{V_X} = \left(1 + \dfrac{1}{4N}\right)\dfrac{S_X}{\overline{X}}$	0	$\approx \dfrac{C_{V_X}^2}{2N}$

for the mean, variance, standard deviation, and coefficient of variation of X from a generic population, along with their respective biases and variances. Table 6.2 does the same for a normal population.

As already mentioned, once the probability distribution is assumed to describe the data sample, the estimates of its parameters are found by some statistical method and, then, used to calculate the desired probabilities and quantiles. Among the many statistical methods for parameter estimation, the following are important: (1) the method of moments; (2) the method of maximum likelihood; (3) the method of L-moments; (4) the graphical method; (5) the least-squares estimation method; and (6) the method of maximum entropy. The former three are the most frequently used in Statistical Hydrology and are formally described in this chapter's sections that follow.

The graphical estimation method is an old-fashioned one, but is still useful and helpful in Statistical Hydrology; it is described later, in Chap. 8, as hydrologic frequency analysis with probability charts is introduced. The least-squares estimation method aims to estimate the parameters of any theoretical function by minimizing the sum of its squared differences, with respect to a given empirical curve or sample data. In this book, the least-squares method is not used for estimating parameters of a probability distribution function, in spite of estimation being totally feasible in practice. In effect, the least-squares method is more suitable for regression analysis and is formally described in Chap. 9.

The concept of entropy as a measure of information was introduced in the late 1940s by the distinguished American mathematician and electrical engineer Claude Shannon (1916–2001). Since then, the entropy theory has encountered many applications in science and engineering and, more recently, as a result of its particular ability to enable the estimation of probability distributions from limited data, it has found a fertile ground in Statistical Hydrology. However, as entropy-based applications in Statistical Hydrology are a sort of specialized subject and would require specific chapters to be suitably explained, they are not going to be covered in this introductory textbook. The reader will find in Singh (1997) an excellent review on the entropy concept and its applications in hydrology and water resources engineering.

The Maximum Likelihood Estimation (MLE) method is the most efficient among the methods currently used in Statistical Hydrology, as it generally yields parameters and quantiles with the smallest sampling variances. However, in many cases, the MLE's highest efficiency is only asymptotic and estimation from a small sample may result in estimators of relatively inferior quality (Rao and Hamed 2000). MLE estimators show the desirable properties of consistency, sufficiency, and are asymptotically unbiased. For finite samples, however, MLE estimators may be biased and, for small sample sizes, may be difficult to find. MLE estimation generally requires computer numerical solutions to systems of equations which are, in most cases, implicit and nonlinear.

The method of moments (MOM) is the simplest and perhaps the most intuitive. MOM estimators are consistent, but are often biased and less efficient than MLE estimators, especially for distributions with more than two parameters, which require estimation of higher order moments. As hydrologic samples are, in many cases, of small sizes, estimators of higher order moments are expected to be significantly biased. In this regard, Yevjevich (1972) comments that when MOM estimation is used for symmetrical distributions, the efficiencies of its estimators are comparable to those of MLE. However, for skewed distributions, as it is often the case for hydrologic variables, the efficiency of MOM estimators usually declines and they should be considered only as first approximations.

The L-moments method (L-MOM) yields parameter estimators of quality comparable to that of MLE estimators, with the advantage of requiring less computational effort to solve its systems of equations, which are usually much simpler than those involved in MLE. For small samples, L-MOM estimators are sometimes more accurate than MLE estimators (Rao and Hamed 2000). In the subsections that follow, the principles behind each of the three estimation methods are described, along with examples of their respective applications in hydrology.

6.3   Method of Moments (MOM)

The method of moments was introduced by the British statistician Karl Pearson (1857–1936). It consists of making the population moments equal to the sample moments. The result from this operation yields the MOM estimators for the distribution parameters. Formally, let $y_1, y_2, y_3, \ldots, y_N$ be the observed data from an SRS drawn from the population of a random variable Y, distributed according to the density $f_Y(y; \theta_1, \theta_2, \ldots, \theta_k)$ of k parameters. If μ_j and m_j respectively denote the population and sample moments, then the fundamental system of equations of the MOM estimation method is given by

$$\mu_j(\theta_1, \theta_2, \ldots, \theta_k) = m_j \quad \text{for} \quad j = 1, 2, \ldots, k \qquad (6.6)$$

The solutions $\hat{\theta}_1, \hat{\theta}_2, \ldots, \hat{\theta}_k$ to this system of k equations and k unknowns are the MOM estimators for parameters θ_j. The next four worked out examples illustrate applications of the MOM estimation method.

Example 6.5 Let $y_1, y_2, y_3, \ldots, y_n$ be an SRS sampled from the population of the random variable Y, with density given by $f_Y(y; \theta) = (\theta + 1)y^\theta$ for $0 \le y \le 1$, described by a single parameter θ. (a) Determine the MOM estimator for θ. (b) Assuming that an SRS from Y consists of the elements $\{0.2; 0.9; 0.05; 0.47; 0.56; 0.8; 0.35\}$, determine the MOM estimate for θ and the probability that Y is larger than 0.8.

Solution

(a) MOM fundamental equation: $\mu_1 = m_1$, just a single equation because there is only one parameter to estimate. Population moment: $\mu_1 = E(Y) =$
$\int_0^1 y(\theta + 1)y^\theta \, dy = \frac{\theta+1}{\theta+2}$. Sample moment: $m_1 = \frac{1}{n}\sum_{i=1}^{n} Y_i = \bar{Y}$. Thus, $\frac{\hat{\theta}+1}{\hat{\theta}+2}$
$= \bar{Y} \Rightarrow \hat{\theta} = \frac{2\bar{Y}-1}{1-\bar{Y}}$. This is the MOM estimator for θ.

(b) The SRS $\{0.2; 0.9; 0.05; 0.47; 0.56; 0.8; 0.35\}$ yields $\bar{y} = 0.4757$. Using the MOM estimator equation, it follows that $\hat{\theta} = \frac{2\times0.4757-1}{1-0.4757} = -0.0926$, which is
the estimate for θ. The CDF is $F_Y(y) = \int_0^y (\theta + 1)y^\theta dy = y^{\theta+1}$. With the estimate $\hat{\theta} = -0.0926$, $P(Y > 0.8) = 1 - F_Y(0.8) = 1 - 0.8167 = 0.1833$.

Example 6.6 Use the MOM estimation method to fit a binomial distribution, with $N = 4$, to the data given in Table 6.3. Also, calculate $P(X \ge 1)$. Recall that for a binomial discrete variable $E(X) = Np$, where $p =$ probability of success and $N =$ number of independent Bernoulli trials.

Table 6.3 Data for Example 6.6

X (number of successes)	0	1	2	3	4
Observed data for the specified X value	10	40	60	50	16

Solution The binomial distribution is defined by the parameters N and p. In this case, the parameter N was specified as 4, thus leaving p to be estimated. MOM equation gives $\mu_1 = m_1$, which, in this case, yields $N\hat{p} = \overline{X}$ or $\hat{p} = \overline{X}/4$. This is the MOM estimator for p. The MOM estimate for p demands the calculation of the sample mean $\bar{x}$, which for the given SRS, yields $\bar{x} = (0 \times 10 + 1 \times 40 + 2 \times 60 + 3 \times 50 + 4 \times 16)/176 = 2.12$ and, thus, $\hat{p} = 0.5313$. Finally, $P(X \geq 1) = 1 - P(X = 0) = 1 - \binom{4}{0} \times 0.5313^0 \times (1 - 0.5313)^4 = 0.9517$.

Example 6.7 Table 1.1 of Chap. 1 lists the annual maximum rainfall depths recorded at the rainfall gauging station of Ponte Nova do Paraopeba, in Brazil, for the water years 1940/41 to 1999/2000, with some missing data. For this sample, the following descriptive statistics have been calculated: $\bar{x} = 82.267$ mm, $s_X = 22.759$ mm, $s_X^2 = 517.988$ mm^2, and the sample coefficient of skewness $g_X = g = 0.7623$. (a) Determine the MOM estimators for the parameters of the Gumbel$_{max}$ distribution. (b) Calculate the MOM estimates for the parameters of the Gumbel$_{max}$ distribution. (c) Calculate the probability that the annual maximum rainfall at this location exceeds 150 mm, in any given year. (d) Calculate the annual maximum rainfall of return period $T = 100$ years.

Solution
(a) Assume $X \sim \mathbf{Gu_{max}}(\alpha,\beta)$. In this case, there are two parameters to be estimated and, thus, the first two central moments are needed: the mean and the variance of X, which are $E[X] = \beta + 0.5772\alpha$ and $\text{Var}[X] = \sigma_X^2 = \frac{\pi^2 \alpha^2}{6}$. Replacing the population moments for the sample moments and solving for α and β, one gets the MOM estimators for the Gumbel$_{max}$ distribution: $\hat{\alpha} = S_X/1.283$ and $\hat{\beta} = \overline{X} - 0.45 S_X$.

(b) The MOM estimates for α and β result from the substitution of $\overline{X}$ and S_X by their respective sample estimates $\bar{x} = 82.267$ and $s_X = 22.759$. Results: $\hat{\alpha} = 17.739$ and $\hat{\beta} = 72.025$.

(c) The sought probability is $1 - F_X(150) = 1 - \exp\left[-\exp\left(\frac{-150-\hat{\beta}}{\hat{\alpha}}\right)\right] = 0.0123$.

(d) The estimate of the 100-year quantile is given by $\hat{x}(T = 100) = \hat{\beta} - \hat{\alpha} \ln\left[-\ln\left(1 - \frac{1}{100}\right)\right] = 153.63$ mm.

Example 6.8 Solve Example 6.7 for the GEV distribution.

Solution
(a) Assume $X \sim \mathbf{GEV}(\alpha,\beta,\kappa)$. Now, there are three parameters to be estimated and, thus, the three first central moments are needed: the mean, the variance, and the

coefficient of skewness of X, respectively given by Eqs. (5.72), (5.73), and (5.74) of Chap. 5. As mentioned in Chap. 5, calculation of GEV parameters should start from Eq. (5.74), which needs to be solved for κ, by means of numerical iterations from an estimated value of the coefficient of skewness. An alternative way to solve Eq. (5.74) for κ is through regression equations of $\kappa \times \gamma$, as in the following examples, suggested by Rao and Hamed (2000): for $1.1396 < \gamma < 10$ (Extreme-Value Type II or Fréchet):

$$\kappa = 0.2858221 - 0.357983\gamma + 0.116659\gamma^2$$
$$- 0.022725\gamma^3 + 0.002604\gamma^4 - 0.000161\gamma^5 + 0000004\gamma^6$$

for $-2 < \gamma < 1.1396$ (Extreme-Value Type III or Weibull):

$$\kappa = 0.277648 - 0.322016\gamma + 0.060278\gamma^2$$
$$+0.016759\gamma^3 - 0.005873\gamma^4 - 0.00244\gamma^5 - 0.00005\gamma^6$$

and
for $-10 < \gamma < 0$ (Extreme-Value Type III or Weibull):

$$\kappa = -0.50405 - 0.00861\gamma + 0.015497\gamma^2$$
$$+0.005613\gamma^3 + 0.00087\gamma^4 + 0.000065\gamma^5.$$

Since $\hat{\gamma} = g = 0.7623$, the second equation is suitable for the present case and produces the first piece of MOM estimators, the shape parameter estimator $\hat{\kappa}$. Following it, the other two MOM estimators are: $\hat{\alpha} = \sqrt{\dfrac{\hat{\kappa}^2 s_X^2}{\Gamma\left(1+2\hat{\kappa}\right)-\Gamma^2\left(1+\hat{\kappa}\right)}}$ and $\hat{\beta} = \overline{X} - \hat{\alpha}_{\hat{\kappa}}\left[1 - \Gamma\left(1+\hat{\kappa}\right)\right]$.

(b) The MOM estimates for α, β, and κ follow from the substitution of $\overline{X}$, S_X, and $\hat{\gamma}$ by their respective sample estimates $\overline{x} = 82.267$, $s_X = 22.759$, and $g = 0.7623$, in the sequence outlined in (a). Results: $\hat{\kappa} = 0.072$, $\hat{\alpha} = 19.323$, and $\hat{\beta} = 72.405$.

(c) $1 - F_X(150) = 1 - \exp\left\{-\left[1 - \hat{\kappa}\left(\frac{150-\hat{\beta}}{\hat{\alpha}}\right)\right]^{1/\hat{\kappa}}\right\} = 0.0087.$

(d) 100-year quantile: $\dfrac{x(T)=\hat{\beta}+\hat{\alpha}}{\hat{\kappa}\left\{1-\left[-\ln\left(1-\frac{1}{T}\right)\right]^{\hat{\kappa}}\right\}=148.07\,\text{mm}}.$

6.4 Maximum Likelihood Estimation (MLE) Method

The maximum likelihood estimation (MLE) method was introduced by the eminent British statistician Ronald Fisher (1890–1962). It basically involves maximizing a function of the distribution parameters, known as the likelihood function. By equating this function to the condition it reaches its maximum value, it results in a system of an identical number of equations and unknowns, whose solutions yield the MLE estimators for the distribution parameters.

Let y_1, y_2, y_3,..., y_N represent the observed data from an SRS drawn from a population of the random variable Y, with the density function $f_Y(y; \theta_1, \theta_2, \ldots, \theta_k)$ of k parameters. The joint density function of the SRS, which is supposedly made of the IID variables Y_1, Y_2, Y_3, $\ldots$, Y_N, is given by $f_{Y_1, Y_2, \ldots, Y_N}(y_1, y_2, \ldots, y_N) = f_Y(y_1)f_Y(y_2) \ldots f_Y(y_N)$. In fact, this joint density is proportional to the probability that a particular SRS had been drawn from the population of density $f_Y(y; \theta_1, \theta_2, \ldots, \theta_k)$ and it is the likelihood function itself. For discrete Y, the likelihood function is the probability of the joint occurrence of y_1, y_2, y_3,..., y_N. Formally, the likelihood function is written as

$$L(\theta_1, \theta_2, \ldots, \theta_k) = \prod_{i=1}^{N} f_Y(y_i; \theta_1, \theta_2, \ldots, \theta_k) \qquad (6.7)$$

This is a function of the parameters θ_j only, as the arguments y_i represent the data from the SRS. The values of θ_j that maximize the likelihood function can be interpreted as the ones that also maximize the probability that the specific SRS in question, as a particular realization of Y_1, Y_2, Y_3, $\ldots$, Y_N, had been drawn from the population of Y, with density $f_Y(y; \theta_1, \theta_2, \ldots, \theta_k)$.

The search for the condition of maximum value for $L(\theta_1, \theta_2, \ldots, \theta_k)$ implies the following system of k equations and k unknowns:

$$\frac{\partial L(\theta_1, \theta_2, \ldots, \theta_k)}{\partial \theta_j} = 0 ; \qquad j = 1, 2, \ldots, k \qquad (6.8)$$

The solutions to this system of equations yield the MLE estimators $\hat{\underline{\theta}}_j$. Usually, maximizing the log-likelihood function $\ln[L(\theta)]$, instead of $L(\theta_1, \theta_2, \ldots, \theta_k)$, facilitates finding the solutions to the system given by Eq. (6.8). This is justified by the fact that the logarithm is a continuous and monotonic increasing function and, as such, maximizing the logarithm of a function is the same as maximizing the function itself. The next two worked out examples illustrate applications of the MLE method.

Example 6.9 Let y_1, y_2, y_3, $\ldots$, y_N be an SRS drawn from the population of a Poisson-distributed random variable Y, with parameter ν. Determine the MLE estimator for ν.

Solution The Poisson mass function is $p_Y(y) = \frac{\nu^y}{y!}e^{-\nu}$, for $y = 0, 1, \ldots$ and $\nu > 0$ and its respective likelihood functions is $L(\nu; Y_1, Y_2, \ldots, Y_N) = \prod_{i=1}^{N} \frac{\nu^{Y_i} \exp(-\nu)}{Y_i!} = \frac{\nu^{\sum_{i=1}^{N} Y_i} \exp(-N\nu)}{\prod_{i=1}^{N} Y_i!}$. The search for the value of ν that maximizes

$L\ (\nu\ )$ is greatly facilitated by the log-likelihood function, as in $\ln[L\ (\nu;\ Y_1, Y_2,\ \ldots\ , Y_N\ )] = -N\nu + \ln(\nu)\sum_{i=1}^{N} Y_i - \ln\left(\prod_{i=1}^{N} Y_i!\right)$. Taking the derivative of this function, with respect to ν, yields $\frac{d\ \ln[L(\nu;\ Y_1,\ Y_2,\ \ldots,\ Y_N)]}{d\nu} = -N + \frac{1}{\nu}\sum_{i=1}^{N} Y_i$. Equating this derivative to zero, it results in the MLE estimator for ν: $\hat{\underline{\nu}} = \frac{1}{N}\sum_{i=1}^{n} Y_i$ or $\hat{\underline{\nu}} = \overline{Y}$.

Example 6.10 Solve Example 6.7 with the MLE method.

Solution

(a) The likelihood function of an SRS of size N, drawn from a Gumbel$_{\max}$ population is given by $L(\alpha,\beta) = \frac{1}{\alpha^N}\exp\left[-\sum_{i=1}^{N}\left(\frac{Y_i-\beta}{\alpha}\right) - \sum_{i=1}^{N}\exp\left(-\frac{Y_i-\beta}{\alpha}\right)\right]$. Analogously to the solution of Example 6.9, finding the MLE estimators is greatly expedited by employing the log-likelihood function as in $\ln[L(\alpha,\beta)] = -N\ln(\alpha) - \frac{1}{\alpha}\sum_{i=1}^{N}(Y_i - \beta) - \sum_{i=1}^{N}\exp\left(-\frac{Y_i-\beta}{\alpha}\right)$. Taking the derivatives of this function, with respect to α and β, and equating both to zero, one gets the following system of equations:

$$\frac{\partial}{\partial\alpha}\ln[L(\alpha,\beta)] = -\frac{N}{\alpha} + \frac{1}{\alpha^2}\sum_{i=1}^{N}(Y_i - \beta) - \frac{1}{\alpha^2}\sum_{i=1}^{N}(Y_i - \beta)\exp\left(-\frac{Y_i-\beta}{\alpha}\right) = 0 \quad \text{(I)}$$

$$\frac{\partial}{\partial\beta}\ln[L(\alpha,\beta)] = \frac{N}{\alpha} - \frac{1}{\alpha}\sum_{i=1}^{N}\exp\left(-\frac{Y_i-\beta}{\alpha}\right) = 0 \quad \text{(II)}$$

Rao and Hamed (2000) suggest solving this system of equations as follows. First, by working with equation II, it follows that $\exp\left(\frac{\beta}{\alpha}\right) = \dfrac{N}{\sum_{i=1}^{N}\exp(-Y_i/\alpha)}$, which is then substituted into equation I. After simplification, $F(\alpha) = \sum_{i=1}^{N} Y_i\exp\left(-\frac{Y_i}{\alpha}\right) - \left(\frac{1}{N}\sum_{i=1}^{N} Y_i - \alpha\right)\sum_{i=1}^{N}\exp\left(-\frac{Y_i}{\alpha}\right) = 0$. This is a function of α only, but still cannot be solved analytically. To solve it, one has to employ Newton's method, by starting iterations from an initial value for α, so that the value for the next iteration be given by $\alpha_{j+1} = \alpha_j - F(\alpha_j)/F'(\alpha_j)$. In this equation, F' represents the derivative of F, with respect to α, or, in formal terms, $F'(\alpha) = \frac{1}{\alpha^2}\sum_{i=1}^{N} Y_i^2\exp\left(-\frac{Y_i}{\alpha}\right) + \sum_{i=1}^{N}\exp\left(-\frac{Y_i}{\alpha}\right) + \frac{1}{\alpha}\sum_{i=1}^{N} Y_i\exp\left(-\frac{Y_i}{\alpha}\right)$. The iterations end when $F(\alpha)$ is sufficiently close to zero, thus yielding the MLE estimator

$\hat{\underline{\alpha}}$. Then, the MLE estimator $\hat{\underline{\beta}}$ results from $\hat{\underline{\beta}} = \hat{\underline{\alpha}} \ln \left[\dfrac{N}{\sum\limits_{i=1}^{N} \exp(-Y_i/\alpha)} \right]$. These are

the MLE estimators for the parameters of the Gumbel$_{\max}$ distribution.

(b) The MLE estimates for α and β result from the substitution of the summations involved in the estimators equations by their respective values calculated from the sample data, but, as seen in (a), it requires iterations by Newton's method. The free software ALEA, developed by the Department of Hydraulic and Water Resources Engineering, of the Brazilian Federal University of Minas Gerais, includes a routine that implements not only the described procedure for finding MLE estimates for Gumbel$_{\max}$ parameters, but also many other routines for calculating parameters for the probability distributions most currently used in hydrology, using the MOM, MLE, and L-MOM methods. The ALEA software can be downloaded from the URL http://www.ehr.ufmg.br/downloads. For the data given in Table 1.1, the annual maximum rainfalls at the gauging station of Ponte Nova do Paraopeba, the MLE estimates, as calculated by the ALEA software, are $\hat{\alpha} = 19.4$ and $\hat{\beta} = 71.7$.

(c) The sought probability is $1 - F_X(150) = 1 - \exp\left[-\exp\left(\frac{-150 - \hat{\beta}}{\hat{\alpha}} \right) \right] = 0.0175$.

(d) 100-year quantile: $\hat{x}(T = 100) = \hat{\beta} - \hat{\alpha} \ln\left[-\ln\left(1 - \frac{1}{100} \right) \right] = 160.94\,\text{mm}$.

6.5 Method of L-Moments (L-MOM)

Greenwood et al. (1979) introduced the probability weighted moments (PWM) as defined by the following general expression:

$$M_{p,r,s} = E\left[X^p [F_X(x)]^r [1 - F_X(x)]^s \right] = \int_0^1 [x(F)]^p F^r (1 - F)^s dF \qquad (6.9)$$

where $x(F)$ denotes the quantile function and p, r, and s are real numbers. When r and s are null and p is a non-negative number, the PWMs $M_{p,0,0}$ are equivalent to the conventional moments about the origin μ'_p of order p. In particular, the PWMs $M_{1,0,s}$ and $M_{1,r,0}$ are the most useful for characterizing probability distributions. They are defined as

$$M_{1,0,s} = \alpha_s = \int_0^1 x(F)(1 - F)^s dF \qquad (6.10)$$

$$M_{1,r,0} = \beta_r = \int_0^1 x(F) F^r dF \tag{6.11}$$

Hosking (1986) showed that α_s and β_r, when expressed as linear functions of x, are sufficiently general to serve the purpose of estimating parameters of probability distributions, in addition to being less subject to sampling fluctuations and, thus, being more robust than the corresponding conventional moments. For an ordered set of IID random variables $X_1 \leq X_2 \leq \ldots \leq X_N$, unbiased estimators of α_s and β_r can be obtained through the following expressions:

$$\underline{a}_s = \hat{\underline{\alpha}}_s = \frac{1}{N} \sum_{i=1}^{N} \frac{\binom{N-i}{s}}{\binom{N-1}{s}} X_i \tag{6.12}$$

and

$$\underline{b}_r = \hat{\underline{\beta}}_r = \frac{1}{N} \sum_{i=1}^{N} \frac{\binom{i-1}{r}}{\binom{N-1}{r}} X_i \tag{6.13}$$

The PWMs α_s and β_r, as well as their corresponding sample estimators a_s and b_r, are related by

$$\alpha_s = \sum_{i=1}^{s} \binom{s}{i} (-1)^i \beta_i \text{ or } \beta_r - \sum_{i=1}^{r} \binom{r}{i} (-1)^i \alpha_i \tag{6.14}$$

Example 6.11 Given the annual mean discharges (m^3/s) for the Paraopeba River at Ponte Nova do Paraopeba (Brazil), listed in Table 6.4, for the calendar years 1990–1999, calculate the estimates of α_s and β_r , for $r,s \leq 3$.

Solution Table 6.4 also shows some partial calculations necessary to apply Eq. (6.12), for $s = 0,1,2$, and 3. The value of a_0 is obtained by dividing the sum of the 10 items in column 5 by $N \binom{N-1}{0} = 10$, thus resulting in $a_0 = 85.29$; one can notice that a_0 is equivalent to the sample arithmetic mean. Similar calculations for the values in columns 6–8, lead to the results $a_1 = 35.923$, $a_2 = 21.655$, and $a_3 = 15.211$. The values for b_r can be calculated either by Eq. (6.13) or derived from a_s, using Eq. (6.14). For the latter and for $r,s \leq 3$, it is clear that

Table 6.4 Annual mean flows (m³/s) for the Paraopeba River at Ponte Nova do Paraopeba

1	2	3	4	5	6	7	8
Calendar year	Annual mean flow Q (m³s)	Rank i	Ranked Q_i	$\binom{N-i}{0} Q_i$	$\binom{N-i}{1} Q_i$	$\binom{N-i}{2} Q_i$	$\binom{N-i}{3} Q_i$
1990	53.1	1	53.1	53.1	477.9	1911.6	4460.4
1991	112.1	2	57.3	57.3	458.4	1604.4	3208.8
1992	110.8	3	63.6	63.6	445.2	1335.6	2226
1993	82.2	4	80.9	80.9	485.4	1213.5	1618
1994	88.1	5	82.2	82.2	411	822	822
1995	80.9	6	88.1	88.1	352.4	528.6	352.4
1996	89.8	7	89.8	89.8	269.4	269.4	89.8
1997	114.9	8	110.8	110.8	221.6	110.8	–
1998	63.6	9	112.2	112.2	112.2	–	–
1999	57.3	10	114.9	114.9	–	–	–

$$\alpha_0 = \beta_0 \text{ or } \beta_0 = \alpha_0$$

$$\alpha_1 = \beta_0 - \beta_1 \text{ or } \beta_1 = \alpha_0 - \alpha_1$$

$$\alpha_2 = \beta_0 - 2\beta_1 + \beta_2 \text{ or } \beta_2 = \alpha_0 - 2\alpha_1 + \alpha_2$$

$$\alpha_3 = \beta_0 - 3\beta_1 + 3\beta_2 - \beta_3 \text{ or } \beta_3 = \alpha_0 - 3\alpha_1 + 3\alpha_2 - \alpha_3$$

In the above equations, by replacing the PWMs for their estimated quantities, one obtains $b_0 = 85.29$, $b_1 = 49.362$, $b_2 = 35.090$, and $b_3 = 27.261$.

The PWMs α_s and β_r, although likely to being used in parameter estimation, are not easy to interpret as shape descriptors of probability distributions. Given that, Hosking (1990) introduced the concept of L-moments, which are quantities that can be directly interpreted as scale and shape descriptors of probability distributions. The L-moments of order r, denoted by λ_r, are linear combinations of the PWMs α_s and β_r and formally defined as

$$\lambda_r = (-1)^{r-1} \sum_{k=0}^{r-1} p_{r-1, k} \alpha_k = \sum_{k=0}^{r-1} p_{r-1, k} \beta_k \tag{6.15}$$

where $p_{r-1, k} = (-1)^{r-k-1} \binom{r-1}{k} \binom{r+k-1}{k}$. Application of Eq. (6.15) for the L-moments of order less than 5 results in

$$\lambda_1 = \alpha_0 = \beta_0 \tag{6.16}$$

$$\lambda_2 = \alpha_0 - 2\alpha_1 = 2\beta_1 - \beta_0 \tag{6.17}$$

$$\lambda_3 = \alpha_0 - 6\alpha_1 + 6\alpha_2 = 6\beta_2 - 6\beta_1 + \beta_0 \tag{6.18}$$

$$\lambda_4 = \alpha_0 - 12\alpha_1 + 30\alpha_2 - 20\alpha_3 = 20\beta_3 - 30\beta_2 + 12\beta_1 - \beta_0 \tag{6.19}$$

The sample L-moments are denoted by l_r and are calculated by replacing α_s and β_r, as in Eqs. (6.16)–(6.19), for their respective estimates a_s and b_r.

The L-moment λ_1 is equivalent to the mean μ and, thus, is a population location measure. For orders higher than 1, the L-moment ratios are particularly useful in describing the scale and shape of probability distributions. As an analogue to the conventional coefficient of variation, one defines the coefficient τ as

$$\tau = \frac{\lambda_2}{\lambda_1} \tag{6.20}$$

which is interpreted as a population measure of dispersion or scale. Also as analogues to the conventional coefficients of skewness and kurtosis, one defines the coefficients τ_3 and τ_4 as

and
$$\tau_3 = \frac{\lambda_3}{\lambda_2} \tag{6.21}$$

$$\tau_4 = \frac{\lambda_4}{\lambda_2} \tag{6.22}$$

The sample L-moment ratios, denoted by t, t_3, and t_4, are calculated by replacing λ_r, as in Eqs. (6.20)–(6.22), for their estimates l_r. As compared to conventional moments, L-moments feature a number of advantages, among which the most important is the existence of variation bounds for τ, τ_3, and τ_4. In fact, if X is a non-negative continuous random variable, it can be shown that $0 < \tau < 1$. As for τ_3 and τ_4, it is a mathematical fact that these coefficients are bounded by $[-1,+1]$, as opposed to their corresponding conventional homologous, which can assume arbitrarily higher values. Other advantages of L-moments, as compared to conventional moments, are discussed by Hosking and Wallis (1997) and Vogel and Fennessey (1993).

The L-moments method (L-MOM) for estimating the parameters of probability distributions is similar to the conventional MOM method. In fact, as exemplified in Table 6.5, the L-moments and the L-moment ratios, namely λ_1, λ_2, τ, τ_3, and τ_4, can be related to the parameters of probability distributions, and vice-versa. The L-MOM method for parameter estimation consists of setting equal the population L-moments to the sample L-moments estimators. The results from this operation yield the estimators for the parameters of the probability distribution. Formally, let y_1, y_2, y_3,..., y_N be the sample data from an SRS drawn from the population of a random variable Y with density $f_Y(y; \theta_1, \theta_2, \dots, \theta_k)$, of k parameters. If $[\lambda_1, \lambda_2, \tau_j]$ and $[l_1, l_2, t_j]$ respectively denote the population L-moments (and their L-moment ratios) and homologous estimators, then the fundamental system of equations for the L-MOM estimation method is given by

$$\begin{aligned} \lambda_i(\theta_1, \theta_2, \dots, \theta_k) &= \underline{l_i} \text{ with } i = 1, 2 \\ \tau_j(\theta_1, \theta_2, \dots, \theta_k) &= \underline{t_j} \text{ with } j = 3, k - 2 \end{aligned} \tag{6.23}$$

The solutions $\hat{\theta}_1$, $\hat{\theta}_2$, $\dots$, $\hat{\theta}_k$ to this system, of k equations and k unknowns, are the L-MOM estimators for parameters θ_j.

Example 6.12 Find the L-MOM estimates for the Gumbel$_{max}$ distribution parameters, using the data given in Example 6.11.

Solution The solution to Example 6.11 showed that the PWM β_r estimates are $b_0 = 85.290$, $b_1 = 49.362$, $b_2 = 35.090$, and $b_3 = 27.261$. Here, there are two parameters to estimate and, thus, the first two L-moments, namely λ_1 and λ_2, are needed. These are given by Eqs. (6.16) and (6.17), and their estimates are $l_1 = b_0 = 85.29$ and $l_2 = 2b_1 - b_0 = 2 \times 49.362 - 85.29 = 13.434$. From the relations of Table 6.5, for the Gumbel$_{max}$ distribution, it follows that $\underline{\hat{\alpha}} = \underline{l_2}/\ln(2) \Rightarrow \hat{\alpha} = 19.381$ and $\hat{\beta} = \underline{l_1} - 0.5772\underline{\hat{\alpha}} \Rightarrow \hat{\beta} = 74.103$.

Table 6.5 L-moments and L-moment ratios for some probability distributions

Distribution	Parameters	λ_1	λ_2	τ_3	τ_4
Uniform	a,b	$\dfrac{a+b}{2}$	$\dfrac{b-a}{6}$	0	0
Exponential	θ	θ	$\dfrac{\theta}{2}$	$\dfrac{1}{3}$	$\dfrac{1}{6}$
Normal	μ,σ	μ	$\dfrac{\sigma}{\sqrt{\pi}}$	0	0.1226
Gumbel$_{max}$	$\alpha\beta$	$\beta + 0.5772\alpha$	$\alpha \ln(2)$	0.1699	0.1504

6.6 Interval Estimation

A point estimate of a parameter of a probability distribution, as shown in the preceding sections, is a number that will differ from the true unknown population value by a variable quantity, depending on the sample size and on the estimation method. The point estimation process, however, does not provide any measure of the estimation error. The issue of the estimate's reliability is addressed by the interval estimator, whose purpose is to assess the degree of confidence with which it will contain the true parameter value. In fact, a point estimator for a parameter θ is a function $\hat{\theta}$, which, as dependent on the random variable X, is also a random variable and, as such, should be described by its own probability density $f_{\hat{\theta}}(\hat{\theta})$. It is true that, if $\hat{\theta}$ is a continuous random variable, then $P(\hat{\theta} = \theta) = 0$, which would make such an equation worthless, as expressed in terms of an equality sign. However, by constructing the random variables L, corresponding to a lower bound, and U, as corresponding to an upper bound, and both as functions of the random variable $\hat{\theta}$, the point estimator for θ, it is possible to write the following probability statement:

$$P(L \leq \theta \leq U) = 1 - \alpha \tag{6.24}$$

where θ denotes the true population value and $(1-\alpha)$ represents the degree of confidence.

Since θ is the true parameter and not a random variable, one should exercise care to interpret Eq. (6.24). It would be misleading to interpret it as if $(1-\alpha)$ were the probability that parameter θ is contained between the limits of the interval. Precisely because θ is not a random variable, Eq. (6.24) must be correctly interpreted as being $(1-\alpha)$ the probability that the interval $[L,U]$ will contain the true population value of a parameter θ.

To make clear the probability statement given by Eq. (6.24), suppose one wants to estimate the mean μ of a population of X, with known standard deviation σ, and the arithmetic mean $\overline{X}$, of a sample of size N, is going to be used to this end. From the solution to Example 5.3 and, in general, from the central limit theorem applied to large samples, it is known that $\left(\frac{\overline{X}-\mu}{\sigma/\sqrt{N}}\right) \sim N(0,1)$. Thus, for the scenario

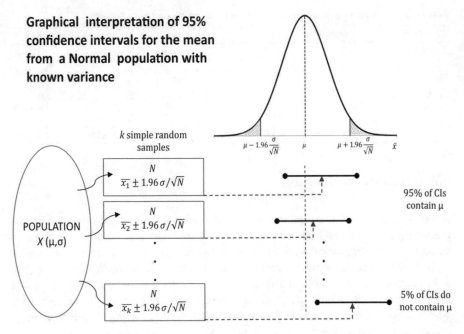

Fig. 6.2 Illustration for interpreting confidence intervals $(1-\alpha)=0.95$ for μ, for a population with known σ

described in Example 5.3, one can write $P\left(-1.96 < \frac{\bar{X}-\mu}{\sigma/\sqrt{N}} < +1.96\right) = 0.95$. In order to put such inequality in terms similar to those of Eq. (6.24), it is necessary to enclose parameter μ apart, in the center of the inequality, as in $P\left(\bar{X} - 1.96\frac{\sigma}{\sqrt{N}} < \mu < \bar{X} + 1.96\frac{\sigma}{\sqrt{N}}\right) = 0.95$. This expression can be interpreted as follows: if samples of the same size N are repeatedly drawn from a population and a confidence interval, such as $\left[\bar{X} - 1.96\sigma/\sqrt{N}, \bar{X} + 1.96\sigma/\sqrt{N}\right]$, is constructed for each sample, then 95 % of these intervals would contain the true parameter μ and 5 % would not. Figure 6.2 illustrates this interpretation, which is essential to interval estimation. Note in Fig. 6.2 that all k intervals, constructed from the k samples of size N, have the same width, but are centered at different points, with respect to parameter μ. If a specific sample yields the bounds $[l,u]$, these would be realizations of the random variables L and U, and, from this interpretation, would have a 95 % chance of containing μ.

The line of reasoning described in the preceding paragraphs can be generalized for constructing confidence intervals for a generic parameter θ, of a probability distribution, as estimated from a random sample $y_1, y_2, y_3, \ldots, y_N$, drawn from the population of Y. This general procedure, usually referred to as the pivotal method, can be outlined in the following steps:

- Select a pivot function $V = v(\theta, Y_1, Y_2, \ldots, Y_N)$, of the parameter θ and of IID variables $Y_1, Y_2, \ldots, Y_N$, whose density function $g_V(v)$ has θ as the only unknown parameter;
- Determine the constants v_1 and v_2, such that $P(v_1 < V < v_2) = 1 - \alpha$ or that $P(V < v_1) = \alpha/2$ and $P(V > v_2) = \alpha/2$;
- Using algebra, rewrite $v_1 < V < v_2$, so that the parameter θ be enclosed apart, into its center, by the inequality signs, and rewrite it as $P(L < \theta < U) = 1 - \alpha$;
- Considering the sample itself, replace the random variables $Y_1, Y_2, \ldots, Y_N$ for the observed data $y_1, y_2, y_3, \ldots, y_N$, and calculate the realizations l and u of variables L and U; and
- The $100(1-\alpha)\%$ confidence for the parameter θ is given by $[l, u]$.

The greatest difficulty in applying the pivotal method relates to the selection of a suitable pivot function, which is not always possible. Nevertheless, in some important practical cases, the pivot function and its respective density can be obtained. Some of these practical cases are listed in Table 6.6.

Example 6.13 Assume the daily water consumption of a community to be a normal variate X and that a sample of size 30 yielded $\bar{x} = 50$ m^3 and $s_X^2 = 256$ m^6. (a) Construct a $100(1-\alpha) = 95\%$ CI for the population mean μ. (b) Construct a $100(1-\alpha) = 95\%$ CI for the population variance σ^2.

Solution

(a) From Table 6.6, the pivot function for this case is $V = \frac{\bar{X}-\mu}{S/\sqrt{N}}$, which follows the Student's t distribution, with $\nu = 30 - 1 = 29$ degrees of freedom. In order to set out the probability statement $P(v_1 < V < v_2) = 0.95$, one determines from Student's t table of Appendix 4 that $-v_1 = v_2 = |t_{0.025, 29}| = 2.045$. As Student's t distribution is symmetrical, the quantiles corresponding to $\alpha/2 = 0.025$ and $1 - \alpha/2 = 0.975$ are identical in absolute value and thus one can

Table 6.6 Some pivot functions used to construct confidence intervals (CI) from samples of size N

Population of Y	CI for parameter	Attribute for the second parameter	Pivot function V	Distribution of V
Normal	μ	σ^2 Known	$\dfrac{\bar{Y} - \mu}{\sigma/\sqrt{N}}$	$N(0,1)$
Normal	μ	σ^2 Unknown	$\dfrac{\bar{Y} - \mu}{S/\sqrt{N}}$	Student's t_{n-1}
Normal	σ^2	μ Known	$\sum_{i=1}^{N} \left(\dfrac{Y_i - \mu}{\sigma} \right)^2$	χ_N^2
Normal	σ^2	μ Unknown	$(N-1)\dfrac{S^2}{\sigma^2}$	χ_{N-1}^2
Exponential	θ	–	$\dfrac{2N\bar{Y}}{\theta}$	χ_{2N}^2

write $P\left(-2.045 < \frac{\overline{X}-\mu}{S/\sqrt{30}} < 2.045\right) = 0.95$. Rearranging this so that the population mean μ stands alone in the center of the inequality, $P\left(\overline{X}-\right.$
$2.045\frac{S}{\sqrt{30}} < \mu < \overline{X} + 2.045\frac{S}{\sqrt{30}}) = 0.95$. Replacing $\overline{X}$ and S for their respective
realizations $\overline{x} = 50\ \text{m}^3$ and $s = \sqrt{256} = 16\ \text{m}^6$, the 95 % CI for μ is [44.03,
55.97].

(b) From Table 6.6, the pivot function for this case is $(N-1)\frac{S^2}{\sigma^2}$, whose distribution
is $\chi^2_{N-1=29}$. To establish $P(v_1 < V < v_2) = 0.95$, one determines from the
Chi-Square table of Appendix 3 that $v_1 = 16.047$, for $\alpha/2 = 0.025$ and
29 degrees of freedom, and that $v_2 = 45.722$, for $1 - \alpha/2 = 0.975$ and
29 degrees of freedom. Note that for the χ^2 distribution, the quantiles are not
symmetrical, with respect to its center. Thus, one can write $P(16.047 <$
$(30-1)\frac{S^2}{\sigma^2} < 45.722) = 0.95$. Rearranging this so that the population variance
σ^2 stands alone in the center of the inequality, $P\left(\frac{29S^2}{45.722} < \sigma^2 < \frac{29S^2}{16.047}\right) = 0.95$.
Replacing S^2 by its realization $s^2 = 256$, it follows that the 95 % CI for σ^2 is
[162.37,462.64]. If $100(1-\alpha)$ had changed to 90 %, the CI for σ^2 would be
[174.45; 419.24] and, thus, narrower but with a lesser degree of confidence.

The construction of confidence intervals for the mean and variance of a normal
population is greatly facilitated by the possibility of deriving their respective exact
sampling distribution functions, as are the cases of Student's t and χ^2 distributions.
In general, an exact sampling distribution function can be determined in explicit
form when the parent distribution of X exhibits the additive property. Examples of
distributions of this kind include the normal, gamma, binomial, and Poisson. For
other random variables, it is almost invariably impossible to determine, in explicit
form, the exact sampling distributions of many of their moment functions, such as
their coefficients of skewness and kurtosis, or their parameter point estimators $\hat{\theta}$.
For these cases, which unfortunately are manifold in hydrology, two alternatives
are possible: the methods that involve Monte Carlo simulation and the asymptotic
methods. For both, the results are only approximate but are in fact the only available
options for this kind of statistical inference problems.

The asymptotic methods, usually of more frequent use, yield results that are
valid as the sample size tends to infinity. These methods arise as attempts to apply
the central limit theorem to operations concerning a large number of sample-
derived quantities. Obviously, in practice, the sample is finite, but is natural to
raise the question of how large it must be such that the approximations are
reasonable. Although no concise and totally satisfying answer to this question
exists in statistical inference books, it can be frequently found suggested that
samples of sufficiently large sizes, of the order of $N > 50$, are acceptable for this
purpose.

Cramér (1946) showed that, under general conditions and for large samples, the
sampling distributions of moment functions and moment-based characteristics
asymptotically converge to a normal distribution with mean equal to the population

quantity being estimated and variance that can be written as c/N, where c depends on what is being estimated and on the estimation method. Mood, Graybill and Boes (1974) argue that any sequences of estimators of θ in a density $f_X(x;\theta)$ are approximately normally distributed with mean θ and variance that is a function of θ itself and of $1/N$. Hosking (1986) extended the results alike to both PWM and L-moment estimators, provided the parent variate has finite variance. Thus, once the normal distribution mean and variance of $\hat{\theta}$ have been estimated, one can determine approximate confidence intervals for the population parameter θ, with the same interpretation given to the ones calculated from the exact sampling distributions.

The results referred to in the preceding paragraph show that as approximate confidence intervals for a generic estimator $\hat{\theta}$ are constructed, the variance of the asymptotic normal distribution will depend on the reciprocal of the sample size, on θ itself, and also on the estimation method. For instance, if $\hat{\theta}$ is an MLE estimator of a single-parameter density $f_X(x;\theta)$, Mood et al. (1974) show that the variance of the asymptotic normal distribution is given by $\left\{ NE\left[\{\partial \ln[f_X(x;\ \theta)]/\partial\theta\}^2 \right] \right\}^{-1}$. However, if the distribution has more than one parameter, determining the variance of the asymptotic normal distribution becomes more complex as it is necessary to include the statistical dependence among the parameters. The reader interested in details on these issues should consult the references Cramér (1946), Rao (1973), and Casella and Berger (1990), for theoretical considerations, and Kaczmarek (1977), Kite (1988), and Rao and Hamed (2000), for examples and applications in hydrology and hydrometeorology. The next section, on constructing confidence intervals for quantiles, shows results that are also related to the reliability of point estimators $\hat{\theta}$.

6.7   Confidence Intervals for Quantiles

After having estimated the parameters of a probability distribution $F_X(x)$, another important objective of Statistical Hydrology is to estimate the quantile X_F, corresponding to the non-exceedance probability F, or, equivalently, the quantile X_T, corresponding to the return period T. The quantile X_F can be estimated by the inverse function $F^{-1}(x)$, which is denoted here by $\varphi(F)$, with the formal meaning of $x_F = \varphi(F)$ or $x_T = \varphi(T)$. It is clear that a point estimator, such as $\hat{X}_T$, contains errors that are inherent to the uncertainties resulting from the estimation of population characteristics and parameters from finite samples of size N. A measure frequently used to quantify the uncertainties of $\hat{X}_T$ and the reliability of quantile estimators is the standard error of estimate, denoted by S_T and formally defined as

$$S_T = \sqrt{E\left[\left\{ \hat{X}_T - E\left[\hat{X}_T\right] \right\}^2 \right]} \tag{6.25}$$

At this point, it is worth noting that the standard error of estimate takes into account only the errors that result from the estimation process from finite samples and do not include the errors that may originate from an incorrect choice of the probability distribution model. Hence, assuming that the probability distribution $F_X(x)$ has been correctly prescribed, the standard error of estimate should encompass the errors made during the estimation of $F_X(x)$ parameters. As a consequence, the three different estimation methods described here, namely, MOM, MLE, and L-MOM, will yield different standard errors. The most efficient estimation method, from a statistical point of view, is the one with the smallest value of S_T.

The asymptotic methods for sampling distributions show that, for large samples, the distribution of $\hat{X}_T$ is approximately normal, with mean value $\hat{X}_T$ and standard deviation S_T. As a result, the $100(1-\alpha)\%$ confidence intervals for the true population quantile X_T can be estimated by

$$\hat{X}_T \pm |z_{\alpha/2}| \hat{S}_T \tag{6.26}$$

where $z_{\alpha/2}$ denotes the standard normal variate, for the non-exceedance probability of $\alpha/2$. Applying the properties of mathematical expectation to Eq. (6.25), it can be shown that, for any probability distribution $F_X(x;\alpha,\beta)$, of two parameters, generically denoted by α and β, the square of the standard error of estimate can be expressed as

$$S_T^2 = \left(\frac{\partial x}{\partial \alpha}\right)^2 \text{Var}(\hat{\underline{\alpha}}) + \left(\frac{\partial x}{\partial \beta}\right)^2 \text{Var}(\hat{\underline{\beta}}) + 2\left(\frac{\partial x}{\partial \alpha}\right)\left(\frac{\partial x}{\partial \beta}\right) \text{Cov}(\hat{\underline{\alpha}},\hat{\underline{\beta}}) \tag{6.27}$$

Likewise, for a probability distribution $F_X(x;\alpha,\beta,\gamma)$, defined by three parameters, generically denoted by α, β, and γ, it can be shown that

$$S_T^2 = \left(\frac{\partial x}{\partial \alpha}\right)^2 \text{Var}(\hat{\underline{\alpha}}) + \left(\frac{\partial x}{\partial \beta}\right)^2 \text{Var}(\hat{\underline{\beta}}) + \left(\frac{\partial x}{\partial \gamma}\right)^2 \text{Var}(\hat{\underline{\gamma}})$$
$$+ 2\left(\frac{\partial x}{\partial \alpha}\right)\left(\frac{\partial x}{\partial \beta}\right) \text{Cov}(\hat{\underline{\alpha}},\hat{\underline{\beta}}) + 2\left(\frac{\partial x}{\partial \alpha}\right)\left(\frac{\partial x}{\partial \gamma}\right) \text{Cov}(\hat{\underline{\alpha}},\hat{\underline{\gamma}})$$
$$+ 2\left(\frac{\partial x}{\partial \beta}\right)\left(\frac{\partial x}{\partial \gamma}\right) \text{Cov}(\hat{\underline{\beta}},\hat{\underline{\gamma}}) \tag{6.28}$$

In Eqs. (6.27) and (6.28), the partial derivatives are calculated from the relation $x_T = \varphi(T)$ and, thus, depend on the analytical expression of the inverse function for the probability distribution $F_X(x)$. On the other hand, the variances and covariances of the parameter estimators will depend on which of the three estimation methods, among MOM, MLE, and L-MOM, has been used for parameter estimation. In the subsections that follow, the more general case, for a three-parameter distribution, is fully described for each of the three estimation methods.

6.7.1   Confidence Intervals for Quantiles (Estimation Method: MOM)

If the method of moments has been used to estimate parameters α, β, and γ, of $F_X(x; \alpha,\beta,\gamma)$, their corresponding variances and covariances must be calculated from the relations between the parameters and the population moments μ_1' (or μ_X), μ_2 (or σ_X^2), and μ_3 (or $\gamma_X \sigma_X^3$), which should be estimated by the sample moments m_1' (or $\overline{X}$), m_2 (or S_X^2), and m_3 (or $g_X S_X^3$), with γ_X and g_X respectively denoting the population and sample coefficients of skewness of X. Thus, by the method of moments, the quantile estimator $\hat{X}_T$ is a function of the sample moments m_1', m_2, and m_3, or $\hat{X}_T = f(m_1', m_2$ and $m_3)$, for a given return period T. Due to this peculiarity of the method of moments, Kite (1988) rewrites Eq. (6.28) as

$$
S_T^2 = \left(\frac{\partial \hat{X}_T}{\partial m_1'}\right)^2 \mathrm{Var}\left(m_1'\right) + \left(\frac{\partial \hat{X}_T}{\partial m_2}\right)^2 \mathrm{Var}(m_2) + \left(\frac{\partial \hat{X}_T}{\partial m_3}\right)^2 \mathrm{Var}(m_3) +
$$
$$
+ 2\left(\frac{\partial \hat{X}_T}{\partial m_1'}\right)\left(\frac{\partial \hat{X}_T}{\partial m_2}\right)\mathrm{Cov}\left(m_1', m_2\right) + 2\left(\frac{\partial \hat{X}_T}{\partial m_1'}\right)\left(\frac{\partial \hat{X}_T}{\partial m_3}\right)\mathrm{Cov}\left(m_1', m_3\right) +
$$
$$
+ 2\left(\frac{\partial \hat{X}_T}{\partial m_3}\right)\left(\frac{\partial \hat{X}_T}{\partial m_2}\right)\mathrm{Cov}(m_3, m_2)
$$

$$(6.29)$$

where the partial derivatives can be obtained by the analytical relations linking $\hat{X}_T$ and m_1', m_2, and m_3. Kite (1988) points out that the variances and covariances of m_1', m_2, and m_3 are given by expressions that depend on the population moments μ_2 to μ_5. These are:

$$
\mathrm{Var}\left(m_1'\right) = \frac{\mu_2}{N} \tag{6.30}
$$

$$
\mathrm{Var}(m_2) = \frac{\mu_4 - \mu_2^2}{N} \tag{6.31}
$$

$$
\mathrm{Var}(m_3) = \frac{\mu_6 - \mu_3^2 - 6\mu_4\mu_2 + 9\mu_2^3}{N} \tag{6.32}
$$

$$
\mathrm{Cov}\left(m_1', m_2\right) = \frac{\mu_3}{N} \tag{6.33}
$$

$$
\mathrm{Cov}\left(m_1', m_3\right) = \frac{\mu_4 - 3\mu_2^2}{N} \tag{6.34}
$$

$$\text{Cov}(m_2, m_3) = \frac{\mu_5 - 4\mu_3\mu_2}{N} \tag{6.35}$$

Kite (1988) suggests the solution to Eq. (6.29) be facilitated by expressing X_T as a function of the first two population moments and of the frequency factor, denoted by K_T, which is dependent upon the return period T and the $F_X(x)$ parameters. Formally, the frequency factor is defined as

$$K_T = \frac{X_T - \mu_1'}{\sqrt{\mu_2}} \tag{6.36}$$

By rearranging Eqs. (6.29)–(6.36), Kite (1988) finally proposes the following equation to calculate S_T^2 for MOM quantile estimators:

$$S_T^2 = \frac{\mu_2}{N}\left\{1 + K_T\gamma_1 + \frac{K_T^2}{4}(\gamma_2 - 1) + \frac{\partial K_T}{\partial \gamma_1}\left[2\gamma_2 - 3\gamma_1^2 - 6 + K_T\left(\gamma_3 - 6\gamma_1\frac{\gamma_2}{4} - 10\frac{\gamma_1}{4}\right)\right]\right\}$$
$$+ \frac{\mu_2}{N}\left[\left(\frac{\partial K_T}{\partial \gamma_1}\right)^2\left(\gamma_4 - 3\gamma_3\gamma_1 - 6\gamma_2 - 9\gamma_1^2\frac{\gamma_2}{4} + 35\frac{\gamma_1^2}{4} + 9\right)\right] \tag{6.37}$$

where,

$$\gamma_1 = \gamma_X = \frac{\mu_3}{\mu_2^{3/2}} \quad \text{(population's coefficient of skewness)} \tag{6.38}$$

$$\gamma_2 = \kappa = \frac{\mu_4}{\mu_2^2} \quad \text{(population's coefficient of kurtosis)} \tag{6.39}$$

$$\gamma_3 = \frac{\mu_5}{\mu_2^{5/2}} \tag{6.40}$$

and

$$\gamma_4 = \frac{\mu_6}{\mu_2^3} \tag{6.41}$$

Note that for a two-parameter distribution, the frequency factor no longer depends on the moment of order 3 and, thus, the partial derivatives in Eq. (6.37) are null, and it then reduces to

$$S_T^2 = \frac{\mu_2}{N}\left\{1 + K_T\gamma_1 + \frac{K_T^2}{4}(\gamma_2 - 1)\right\} \tag{6.42}$$

Finally, the estimation of the confidence intervals for the quantile X_T, with parameters estimated by the method of moments from a sample of size N, is performed by initially replacing $\gamma_1, \gamma_2, \gamma_3, \gamma_4, K_T$, and $\partial K_T / \partial \gamma_1$, as in Equation 6.37, for the population values (or expressions) that are valid for the probability distribution being studied, and, then, μ_2, for its respective sample estimate. Following that, one extracts the square root of S_T^2 and, then, applies Eq. (6.26), for a previously specified confidence level of $100(1-\alpha)\%$. Example 6.14 illustrates the calculations for the two-parameter Gumbel$_{max}$ distribution. Other examples and applications can be found in Kite (1988) and Rao and Hamed (2000). The software ALEA (http://www.ehr.ufmg.br/downloads) implements the calculations of confidence intervals for quantiles, estimated by the method of moments, for the most used probability distributions in current Statistical Hydrology.

Example 6.14 With the results and MOM estimates found in the solution to Example 6.7, estimate the 95 % confidence interval for the 100-year quantile.

Solution From the solution to Example 6.7, $X \sim \text{Gu}_{max}(\hat{\alpha} = 17.739, \hat{\beta} = 72.025)$ and $N = 55$. The Gumbel$_{max}$ distribution, with population coefficients of skewness and kurtosis fixed and respectively equal to $\gamma_1 = 1.1396$ and $\gamma_2 = 5.4$, is a two-parameter distribution, for which is valid Eq. (6.42). Replacing the expressions valid for this distribution, first, those of moments $\mu_1' = \beta + 0.5772\alpha$ and $\mu_2 = \frac{\pi^2 \alpha^2}{6}$, and, then, that of quantiles $X_T = \beta - \alpha \ln\left[-\ln\left(1 - \frac{1}{T}\right)\right]$, into Eq. (6.36), it is easy to see that $K_T = -0.45 - 0.7797 \ln\left[-\ln\left(1 - 1/T\right)\right]$ and that, for $T = 100$ years, $K_T = 3.1367$. Returning to Eq. (6.42), substituting the values for K_T, $\gamma_1 = 1.1396$, $\gamma_2 = 5.4$, and $\hat{\mu}_2 = \frac{\pi^2 \hat{\alpha}^2}{6} = 517.6173$, the result is $\hat{S}_{T=100}^2 = 144.908$ and, thus, $\hat{S}_{T=100} = 12.038$. With this result, the quantile estimate $x_{T=100} = 153.160$, and $z_{0.025} = -1.96$ in Eq. (6.26), the 95 % confidence interval is [130.036, 177.224]. The correct interpretation of this CI is that the probability that these bounds, estimated with the method of moments, will contain the true population 100-year quantile is 0.95.

6.7.2 Confidence Intervals for Quantiles (Estimation Method: MLE)

If the parameters α, β, and γ, of $F_X(x;\alpha,\beta,\gamma)$, have been estimated by the maximum likelihood method, the partial derivatives, as in Eqs. (6.27) and (6.28), should be calculated from the relation $x_T = \varphi(T)$ and, thus, will depend on the analytical expression for the inverse of $F_X(x)$. In turn, according to Kite (1988) and Rao and Hamed (2000), the variances and covariances of the parameter estimators are the elements of the following symmetric matrix, known as the covariance matrix:

$$
I = \begin{bmatrix} \mathrm{Var}(\hat{\underline{\alpha}}) & \mathrm{Cov}\left(\hat{\underline{\alpha}},\hat{\underline{\beta}}\right) & \mathrm{Cov}\left(\hat{\underline{\alpha}},\hat{\underline{\gamma}}\right) \\[2mm] & \mathrm{Var}\left(\hat{\underline{\beta}}\right) & \mathrm{Cov}\left(\hat{\underline{\beta}},\hat{\underline{\gamma}}\right) \\[2mm] & & \mathrm{Var}\left(\hat{\underline{\gamma}}\right) \end{bmatrix} \tag{6.43}
$$

which is calculated by the inverse of the following square matrix, the Hessian matrix,

$$
M = \begin{bmatrix} -\dfrac{\partial^2 \ln L}{\partial \alpha^2} & -\dfrac{\partial^2 \ln L}{\partial \alpha \partial \beta} & -\dfrac{\partial^2 \ln L}{\partial \alpha \partial \gamma} \\[3mm] & -\dfrac{\partial^2 \ln L}{\partial \beta^2} & -\dfrac{\partial^2 \ln L}{\partial \beta \partial \gamma} \\[3mm] & & -\dfrac{\partial^2 \ln L}{\partial \gamma^2} \end{bmatrix} \tag{6.44}
$$

where L denotes the likelihood function. Letting D represent the determinant of matrix $\mathbf{M}$, then, the variance of $\hat{\underline{\alpha}}$, for example, will be given by the determinant of the matrix that will remain after having eliminated the first line and the first column of $\mathbf{M}$, divided by D. In other terms, the variance of $\hat{\underline{\alpha}}$ is calculated as

$$
\mathrm{Var}(\hat{\underline{\alpha}}) = \frac{\dfrac{\partial^2 \ln L}{\partial \beta^2} \times \dfrac{\partial^2 \ln L}{\partial \gamma^2} - \left(\dfrac{\partial^2 \ln L}{\partial \beta \partial \gamma}\right)^2}{D} \tag{6.45}
$$

After all elements of the covariance matrix $\mathbf{I}$ are calculated, one goes back to Eq. (6.28) and estimates S_T^2. After that, the square root of S_T^2 is extracted and Eq. (6.26) should be applied to a previously specified confidence level $100(1-\alpha)\%$. Example 6.15, next, shows the calculation procedure for the Gumbel$_{\max}$ distribution. Other examples and applications can be found in Kite (1988) and Rao and Hamed (2000). The software ALEA (http://www.ehr.ufmg.br/downloads) implements the calculations of confidence intervals for quantiles, estimated by the method of maximum likelihood, for the most used probability distributions in current Statistical Hydrology.

Example 6.15 With the results and MLE estimates found in the solution to Example 6.10, estimate the 95 % confidence interval for the 100-year quantile.

Solution The log-likelihood function $\ln(L)$ for the Gumbel$_{\max}$ probability distribution is written as $\ln[L(\alpha,\beta)] = -N\ln(\alpha) - \frac{1}{\alpha}\sum_{i=1}^{N}(Y_i - \beta) - \sum_{i=1}^{N}\exp\left(-\frac{Y_i-\beta}{\alpha}\right)$. Kimball (1949), cited by Kite (1988), developed the following approximate expressions for the second-order partial derivatives: $\frac{\partial^2 \ln L}{\partial \alpha^2} = -\frac{1.8237N}{\alpha^2}$; $\frac{\partial^2 \ln L}{\partial \beta^2} = -\frac{N}{\alpha^2}$

and $\frac{\partial^2 \ln L}{\partial \alpha \partial \beta} = \frac{0.4228N}{\alpha^2}$, which are the elements of matrix $\mathbf{M}$, having, in this case, 2×2 dimensions. By inverting matrix $\mathbf{M}$, as described before, one gets the following elements of the covariance matrix: $\mathrm{Var}(\hat{\underline{\alpha}}) = 0.6079 \frac{\alpha^2}{N}$, $\mathrm{Var}(\hat{\underline{\beta}}) = 1.1087 \frac{\alpha^2}{N}$, and $\mathrm{Cov}(\hat{\underline{\alpha}}, \hat{\underline{\beta}}) = 0.2570 \frac{\alpha^2}{N}$. As the quantile function for the Gumbel$_{\mathrm{max}}$ is $Y_T = \beta - \alpha \ln\left[-\ln\left(1 - \frac{1}{T}\right)\right]$, the partial derivatives in Eq. (6.27) are $\frac{\partial Y_T}{\partial \alpha} = -\ln\left[-\ln\left(1 - \frac{1}{T}\right)\right] = W$ and $\frac{\partial Y_T}{\partial \beta} = 1$. Returning to Eq. (6.27), with the calculated variances, covariances, and partial derivatives, one obtains the variance for the MLE quantiles for Gumbel$_{\mathrm{max}}$ as $S_T^2 = \frac{\alpha^2}{N}\left(1.1087 + 0.5140W + 0.6079W^2\right)$. With the results $\hat{\alpha} = 19.4$ and $\hat{\beta} = 71.7$, from the solution to Example 6.10, and $W = 4.60$, for $T = 100$ years, one obtains $S_T^2 - 130.787$ and, $S_I = 11.436$. By comparing these with the results from the solution to Example 6.14, it is clear that the MLE estimators have smaller variance and, thus, are deemed more reliable than MOM estimators. With the calculated value for S_T, the quantile estimate $x_{T=100} = 160.940$, and $z_{0.025} = -1.96$ in Eq. (6.26), the 95 % confidence interval for the 100-year MLE quantile is [138.530, 183.350]. The correct interpretation of this CI is that the probability that these bounds, estimated with the method of maximum likelihood, will contain the true population 100-year quantile is 0.95.

6.7.3 Confidence Intervals for Quantiles (Estimation Method: L-MOM)

Similarly to the previous case, if the parameters α, β, and γ, of $F_X(x;\alpha,\beta,\gamma)$, have been estimated by the method of L-moments, the partial derivatives, as in Eqs. (6.27) and (6.28), should be calculated from the relation $x_T = \varphi(T)$ and, thus, depend on the analytical expression for the inverse of $F_X(x)$. In turn, the variances and covariances are the elements of the covariance matrix, exactly as in Eq. (6.43). Its elements, however, must be calculated from the covariance matrix for the PWMs α_r and β_r, for $r = 1$, 2, and 3. Hosking (1986) showed that the vector $\mathbf{b} = (b_1, b_2, b_3)^T$ is asymptotically distributed as a multivariate normal, with means $\boldsymbol{\beta} = (\beta_1, \beta_2, \beta_3)^T$ and covariance matrix given by $\mathbf{V}/N$. The expressions for evaluating matrix $\mathbf{V}$ and, then, the standard error S_T, are quite complex and can be found in Hosking (1986) and Rao and Hamed (2000), for some probability distributions. The software ALEA (http://www.ehr.ufmg.br/downloads) implements the calculations of confidence intervals for quantiles, estimated by the method of L-moments, for some probability distributions.

Example 6.16 With the results and L-MOM estimates found in the solution to Example 6.12, estimate the 95 % confidence interval for the 100-year quantile.

Solution Hosking (1986) presents the following expressions for the variances and covariances of the L-MOM estimators for parameters α and β of the Gumbel$_{max}$ distribution: $\text{Var}(\hat{\alpha}) = 0.8046 \frac{\alpha^2}{N}$, $\text{Var}(\hat{\beta}) = 1.1128 \frac{\alpha^2}{N}$, and $\text{Cov}(\hat{\alpha}, \hat{\beta}) = 0.2287 \frac{\alpha^2}{N}$. The partial derivatives in Eq. (6.27) are $\frac{\partial Y_T}{\partial \alpha} = -\ln\left[-\ln\left(1 - \frac{1}{T}\right)\right] = W$ and $\frac{\partial Y_T}{\partial \beta} = 1$. With the calculated variances, covariances, and partial derivatives in Eq. (6.27), one gets $S_T^2 = \frac{\alpha^2}{N}\left(1.1128 + 0.4574W + 0.8046W^2\right)$. With the results $\hat{\alpha} = 19.381$ and $\hat{\beta} = 74.103$, from the solution to Example 6.12, and $W = 4.60$, for $T = 100$ years, one obtains $S_T^2 = 760.39$ and $S_T = 27.58$. Note that, in this case, the sample size is only 10 and because of that S_T^2 is much larger than its homologous in Examples 6.14 and 6.15. The 100-year L-MOM quantile is $\hat{y}(T = 100) = \hat{\beta} - \hat{\alpha} \ln\left[-\ln\left(1 - \frac{1}{100}\right)\right] = 163.26$. With the calculated value for S_T, the quantile estimate $\hat{y}(T = 100) = 163.26$ and $z_{0.025} = -1.96$ in Eq. (6.26), the 95 % confidence interval for the 100-year MLE quantile is [109,21; 217,31]. The correct interpretation of this CI is that the probability that these bounds, estimated with the method of L-moments, will contain the true population 100-year quantile is 0.95.

The confidence intervals as calculated by the normal distribution are approximate because they are derived from asymptotic methods. Meylan et al. (2008) present some arguments in favor of using the normal approximation for the sampling distribution of X_T. They are summarized as follows: (1) as resulting from the central limit theorem, the normal distribution is the asymptotic form of a large number of sampling distributions; (2) an error of second order is made in case the true sampling distribution departs significantly from the normal distribution; and (3) the eventual differences between the true sampling distribution and the normal distribution will be significant only for high values of the confidence level $100(1-\alpha)\%$.

6.8   Confidence Intervals for Quantiles by Monte Carlo Simulation

For the two-parameter distributions and for the most used estimation methods, the determination of S_T, in spite of requiring a number of calculation steps, is not too demanding. However, for three-parameter distributions, the determination of S_T, in addition to being burdensome and complex, requires the calculation of covariances among the mean, the variance and the coefficient of skewness. The software ALEA (www.ehr.ufmg.br/downloads) implements the calculations of confidence intervals for quantiles from the probability distributions most commonly used in Statistical Hydrology, as estimated by MOM and MLE, under the hypothesis of asymptotic approximation by the normal distribution. For quantiles estimated by L-MOM, the software ALEA calculates confidence intervals only for some probability distributions.

Hall et al. (2004) warn, however, that for three-parameter distributions, the nonlinear dependence of the quantiles and the third-order central moment can make the quantile sampling distribution depart from the normal and, thus, yield underestimated or overestimated quantile confidence intervals, especially for large return periods. In addition to this difficulty, the fact that there are no simple expressions for the terms that are necessary to calculate S_T should be considered, taking into account the whole set of probability distributions and estimation methods used in the frequency analysis of hydrologic random variables.

An alternative to the normal approximation is given by the Monte Carlo simulation method. Especially for three-parameter distributions, it is less arduous than the normal approximation procedures, but it requires computer intensive methods. In fact, it makes use of Monte Carlo simulation of a large number of synthetic samples, of the same size N as the original sample. For each synthetic sample, the target quantile is estimated by the desired estimation method and, then, the empirical probability distribution of all quantile estimates serves the purpose of yielding $100(1-\alpha)\%$ confidence intervals. This simulation procedure is detailed in the next paragraphs.

Assume that a generic probability distribution $F_X(x|\theta_1, \theta_2, \ldots, \theta_k)$ has been fitted to the sample $\{x_1, x_2, \ldots, x_N\}$, such that its parameters $\theta_1, \theta_2, \ldots, \theta_k$ were estimated by some estimation method designated as EM (EM may be MOM, MLE, L-MOM, or any other estimation method not described in here). The application of the Monte Carlo simulation method for constructing confidence intervals about the estimate of the quantile $\hat{X}_F$, as corresponding to the non-exceedance probability F or the return period $T = 1/(1-F)$, requires the following sequential steps:

1. Generate a unit uniform number from $\mathbf{U}(0,1)$, and denote it by u_i;
2. Estimate the quantile $\hat{X}_F$, corresponding to u_i, by means of the inverse function $\hat{X}_{Fi} = F^{-1}(u_i) = \varphi(u_i)$, using the estimates for parameters $\theta_1, \theta_2, \ldots, \theta_k$ as yielded by the method EM for the original sample $\{x_1, x_2, \ldots, x_N\}$;
3. Repeat steps (1) and (2) up to $i = N$, thus making up one of the W synthetic samples of variable X;
4. Every time a new synthetic sample with size N is produced, apply the EM method to it in order to estimate parameters $\theta_1, \theta_2, \ldots, \theta_k$, and use the estimates to calculate the quantile $\hat{X}_F = F^{-1}(x) = \varphi(F)$, corresponding to a previously specified non-exceedance probability (e.g., $F = 0.99$ or $T = 100$ years);
5. Repeat steps (1) to (4) for the second, third, and so forth up to the Wth synthetic sample, where W should be a large number, for instance, $W = 5000$;
6. For the specified non-exceedance probability F, after the W distinct quantile estimates $\hat{X}_{F,j}, j = 1, \ldots W$ are gathered, rank them in ascending order;
7. The interval bounded by the ranked quantiles the closest to rank positions $W(\alpha/2)$ and $W(1-\alpha/2)$ gives the $100(1-\alpha)\%$ confidence interval for quantile X_F; and
8. Repeat steps (4) to (7) for other values of the non-exceedance probability F, as desired.

Besides being a simple and easy-to-implement alternative, given the vast computing resources available nowadays, the construction of confidence intervals
through Monte Carlo simulation does not assume the sampling distribution of $\hat{X}_F$
must necessarily be normal. In fact, the computer intensive technique allows the
empirical distribution of quantiles to be shaped by a large number of synthetic series,
all with statistical descriptors similar to the ones observed from the original. For
example, a three-parameter lognormal (LNO3) distribution has been fitted to the
sample of 73 annual peak discharges of the Lehigh River at Stoddartsville,
listed in Table 7.1 of Chap. 7, using the L-MOM estimation method (see
Sect. 6.9.11). The L-MOM parameters estimates for the LNO3 are $\hat{a} = 21.5190$,
$\hat{\mu}_Y = 3.7689$, and $\hat{\sigma}_Y = 1.1204$. Assuming the LNO3 distribution, with the
L-MOM-estimated parameters, represents the true parent distribution, 1000 samples
of size 73 were generated from such a hypothesized population. For each sample, the
L-MOM method was employed to estimate the LNO3 parameters, producing a total
of 1000 sets of estimates, which were then used to calculate 1000 different quantiles
for each return period of interest. The panels (a), (b), (c), and (d) of Fig. 6.3 depict the
histograms of the 1000 quantiles estimates (m^3/s) for the return periods 10, 20,
50, and 100 years, respectively. For each return period, these quantile estimates
should be ranked and then used to calculate the respective confidence intervals,
according to step (7) previously referred to in this section. If more precision is
required, a larger number of synthetic samples are needed.

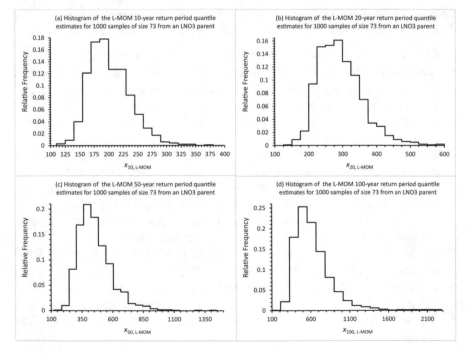

Fig. 6.3 Histograms of quantiles for the return periods 10, 20, 50, and 100 years, as estimated with
the L-MOM methods, 1000 samples drawn from an assumed LNO3($\hat{a}$, $\hat{\mu}_Y$, $\hat{\sigma}_Y$) parent for the
annual peak flows of the Lehigh River at Stoddartsville

6.9   Summary of Parameter Point Estimators

The following is a summary of equations for estimating parameters, by the MOM and MLE methods, for some probability distributions, of discrete and continuous random variables, organized in alphabetical order. In all cases, equations are based on a generic simple random sample $\{X_1, X_2, \dots, X_N\}$ of size N. For a few cases, equations for estimating parameters by the L-MOM method are also provided. For distributions and estimation methods not listed in this summary and for equations for the standard errors of estimate (for calculating confidence intervals for quantiles from three-parameter distributions), the following references are suggested: Kite (1988), Rao and Hamed (2000), and Hosking and Wallis (1997). Good sources for algorithms and computer codes for parameter estimation are Hosking (1991) and the R archive at https://cran.r-project.org/web/packages/.

6.9.1   Bernoulli Distribution

MOM: $\hat{\underline{p}} = \overline{X}$

MLE: $\hat{\underline{p}} = \overline{X}$

L-MOM: $\hat{\underline{p}} = \underline{l_1}$

6.9.2   Beta Distribution

MOM:

$\hat{\underline{\alpha}}$ and $\hat{\underline{\beta}}$ are the solutions to the system:

$$\overline{X} = \frac{\alpha}{\alpha + \beta} \text{ and }$$

$$S_X^2 = \frac{\alpha\beta}{(\alpha + \beta)^2(\alpha + \beta + 1)}$$

MLE:

$\hat{\underline{\alpha}}$ and $\hat{\underline{\beta}}$ are the solutions to the system:

$$\frac{\partial}{\partial \alpha}[\ln \Gamma(\alpha) - \ln \Gamma(\alpha + \beta)] = \frac{1}{N}\sum_{i=1}^{N} \ln(X_i)$$

$$\frac{\partial}{\partial \beta}[\ln \Gamma(\beta) - \ln \Gamma(\alpha + \beta)] = \frac{1}{N}\sum_{i=1}^{N} \ln(1 - X_i)$$

6.9.3 Binomial Distribution

For a known number m of independent Bernoulli trials:
MOM: $\hat{\underline{p}} = \overline{X}/m$
MLE: $\hat{\underline{p}} = \overline{X}/m$
L-MOM: $\hat{\underline{p}} = l_1/m$

6.9.4 Exponential Distribution

MOM: $\hat{\underline{\theta}} = \overline{X}$
MLE: $\hat{\underline{\theta}} = \overline{X}$
L-MOM: $\hat{\underline{\theta}} = \underline{l_1}$

6.9.5 Gamma Distribution

MOM:

$$\hat{\underline{\theta}} = \frac{S_X^2}{\overline{X}}$$

$$\hat{\underline{\eta}} = \frac{\overline{X}^2}{S_X^2}$$

MLE:

$\hat{\underline{\eta}}$ is the solution to equation

$$\ln \eta - \frac{\partial}{\partial \eta} \ln \Gamma(\eta) = \ln \overline{X} - \frac{1}{N} \sum_{i=1}^{N} \ln X_i \qquad (6.46)$$

After solving (6.46), $\hat{\underline{\theta}} = \overline{X}/\hat{\underline{\eta}}$.
Solution to Eq. (6.46) can be approximated by (Rao and Hamed 2000):

$$\hat{\underline{\eta}} = \frac{0.5000876 + 0.1648852y - 0.054427y^2}{y} \text{ if } 0 \leq y \leq 0.5772, \text{ or}$$

$$\hat{\underline{\eta}} = \frac{8.898919 + 9.059950y + 0.9775373y^2}{y(17.7928 + 11.968477y + y^2)} \text{ if } 0.5772 < y \leq 17$$

where $y = \ln \overline{X} - \frac{1}{N} \sum_{i=1}^{N} \ln X_i$

L-MOM:

$\hat{\underline{\eta}}$ is the solution (Newton' method) to equation

$$t = \frac{l_2}{l_1} = \frac{\Gamma(\eta + 0,5)}{\sqrt{\pi}\Gamma(\eta + 1)} \tag{6.47}$$

After solving (6.47), $\hat{\underline{\theta}} = l_1/\hat{\underline{\eta}}$.

Hosking (1990) propose the solution to η, as in Eqn. (6.47), be obtained as follows.

$$\hat{\underline{\eta}} = \frac{1 - 0.3080z}{z - 0.05812z^2 + 0.01765z^3} \text{ if } 0 < t < 0.5 \text{ and } z = \pi t^2, \text{ or}$$

$$\hat{\underline{\eta}} = \frac{0.7213z - 0.5947z^2}{1 - 2.1817z + 1.2113z^2} \text{ if } 0.5 \leq t < 1 \text{ and } z = 1 - t$$

6.9.6 Geometric Distribution

MOM: $\hat{\underline{p}} = 1/\overline{X}$
MLE: $\hat{\underline{p}} = 1/\overline{X}$
L-MOM: $\hat{\underline{p}} = 1/l_1$

6.9.7 GEV Distribution

MOM:

Alternative 1: solve Eq. (5.74) for κ, by replacing γ for the sample coefficient of skewness g_X. The solution is iterative, by Newton's method, or as suggested in the solution to Example 5.10.

Alternative 2 (Rao and Hamed 2000):
For the sample coefficient of skewness in the range $1.1396 < g_X < 10$ ($g = g_X$):

$$\hat{\underline{\kappa}} = 0.2858221 - 0.357983g + 0.116659g^2 - 0.022725g^3$$
$$+ 0.002604g^4 - 0.000161g^5 + 0.000004g^6$$

For the sample coefficient of skewness in the range $-2 < g_X < 1.1396$ ($g = g_X$):

$$\hat{\underline{\kappa}} = 0.277648 - 0.322016g + 0.060278g^2 + 0.016759g^3$$
$$- 0.005873g^4 - 0.00244g^5 - 0.00005g^6$$

For the sample coefficient of skewness in the range $-10 < g_X < 0$ $(g = g_X)$:

$$\hat{\kappa} = -0.50405 - 0.00861g + 0.015497g^2 + 0.005613g^3$$
$$+ 0.00087g^4 + 0.000065g^5$$

Then, $\underline{\hat{\alpha}} = \dfrac{S_X \hat{\kappa}}{\sqrt{\Gamma(1 + 2\hat{\kappa}) - \Gamma^2(1 + \hat{\kappa})}}$ and $\underline{\hat{\beta}} = \overline{X} - \frac{\hat{\alpha}}{\hat{\kappa}}[1 - \Gamma(1 + \hat{\kappa})]$

MLE:

$\underline{\hat{\alpha}}, \underline{\hat{\beta}}$, and $\underline{\hat{\kappa}}$ are the solutions, by Newton's method, to the system:

$$\frac{1}{\alpha}\left[\sum_{i=1}^{N}\exp(-Y_i - \kappa Y_i) - (1 - \kappa)\sum_{i=1}^{N}\exp(\kappa Y_i)\right] = 0 \qquad (6.48)$$

$$\frac{1}{\kappa\alpha}\left[\sum_{i=1}^{N}\exp(-Y_i - \kappa Y_i) - (1 - \kappa)\sum_{i=1}^{N}\exp(\kappa Y_i) + N - \sum_{i=1}^{N}\exp(-Y_i)\right] = 0$$
$$(6.49)$$

$$\frac{1}{\kappa^2}\left[\sum_{i=1}^{N}\exp(-Y_i - \kappa Y_i) - (1 - \kappa)\sum_{i=1}^{N}\exp(\kappa Y_i) + N - \sum_{i=1}^{N}\exp(-Y_i)\right]$$
$$+ \frac{1}{\kappa}\left[-\sum_{i=1}^{N}Y_i + \sum_{i=1}^{N}Y_i\exp(Y_i) + N\right] = 0 \qquad (6.50)$$

where $Y_i = \frac{1}{\kappa}\ln\left[1 - \kappa\left(\frac{X_i - \beta}{\alpha}\right)\right]$. The solution to this system of equations is complicated. The reader should consult the references Prescott and Walden (1980) and Hosking (1985), respectively, for the algorithm of the solution and the corresponding FORTRAN programming code.

L-MOM (Hosking et al. 1985):

$$\underline{\hat{\kappa}} = 7.8590C + 2.9554C^2, \text{ where } C = 2/\left(3 + t_3\right) - \ln 2/\ln 3$$

$$\underline{\hat{\alpha}} = \frac{l_2 \hat{\kappa}}{\Gamma(1 + \hat{\kappa})\left(1 - 2^{-\hat{\kappa}}\right)}$$

$$\underline{\hat{\beta}} = l_1 - \frac{\hat{\alpha}}{\hat{\kappa}\left[1 - \Gamma(1 + \hat{\kappa})\right]}$$

6.9.8 Gumbel$_{max}$ Distribution

MOM:

$$\hat{\alpha} = 0.7797\,S_X$$

$$\underline{\hat{\beta}} = \overline{X} - 0,45 S_X$$

MLE (Rao and Hamed 2000):
$\hat{\alpha}$ and $\hat{\beta}$ are the solutions to the system of equations:

$$\frac{\partial}{\partial \alpha} \ln\left[L\left(\alpha, \beta\right)\right] = -\frac{N}{\alpha} + \frac{1}{\alpha^2}\sum_{i=1}^{N}(X_i - \beta) - \frac{1}{\alpha^2}\sum_{i=1}^{N}(X_i - \beta)\exp\left(-\frac{X_i - \beta}{\alpha}\right) = 0$$

$$(6.51)$$

$$\left[L\left(\alpha, \beta\right)\right] = \frac{N}{\alpha} - \frac{1}{\alpha}\sum_{i=1}^{N}\exp\left(-\frac{X_i - \beta}{\alpha}\right) = 0 \qquad (6.52)$$

Combining both equations, it follows that

$$F(\alpha) = \sum_{i=1}^{N} X_i \exp\left(-\frac{X_i}{\alpha}\right) - \left(\frac{1}{N}\sum_{i=1}^{N} X_i - \alpha\right)\sum_{i=1}^{N}\exp\left(-\frac{X_i}{\alpha}\right) = 0 \qquad (6.53)$$

Solution to (6.53), by Newton's method, yields $\hat{\alpha}$.

Then, $\hat{\beta} = \hat{\alpha} \ln\left[\dfrac{N}{\sum_{i=1}^{N}\exp(-X_i/\hat{\alpha})}\right]$.

L-MOM:

$$\underline{\hat{\alpha}} = \frac{l_2}{\ln 2}$$

$$\underline{\hat{\beta}} = l_1 - 0.5772\,\hat{\alpha}$$

6.9.9 Gumbel$_{min}$ distribution

MOM:

$$\underline{\hat{\alpha}} = 0.7797\,S_X$$

$$\underline{\hat{\beta}} = \overline{X} + 0.45 S_X$$

L-MOM:

$$\underline{\hat{\alpha}} = \frac{l_2}{\ln 2}$$

$$\underline{\hat{\beta}} = \underline{l_1} + 0.5772\underline{\hat{\alpha}}$$

6.9.10   Lognormal Distribution (2 parameters, with $Y = lnX$)

MOM:

$$\underline{\hat{\sigma}_Y} = \sqrt{\ln\left(CV_X^2 + 1\right)}$$

$$\underline{\hat{\mu}_Y} = \ln\overline{X} - \frac{\hat{\sigma}_Y^2}{2}$$

MLE:

$$\underline{\hat{\mu}_Y} = \overline{Y}$$

$$\underline{\hat{\sigma}_Y} = \sqrt{\frac{N-1}{N}} S_Y$$

L-MOM (Rao and Hamed 2000):

$$\underline{\hat{\sigma}_Y} = 2\,erf^{-1}(t)$$

$$\underline{\hat{\mu}_Y} = \ln \underline{l_1} - \frac{\hat{\sigma}_Y^2}{2}$$

where $erf(w) = \frac{2}{\sqrt{\pi}} \int\limits_0^w e^{-u^2}\,du$. The inverse $erf^{-1}(t)$ is equal to $u/\sqrt{2}$, being u the standard normal variate corresponding to $\Phi[(t+1)/2]$.

6.9.11   Lognormal distribution [3 parameters, with $Y = ln(X - a)$]

MOM:

$$\underline{\hat{\sigma}_Y} = \sqrt{\ln\left(CV_{X-a}^2 + 1\right)} \text{ where } CV_{X-a} = \frac{1 - w^{2/3}}{\sqrt[3]{w}} \text{ and } w = \frac{-\gamma_X + \sqrt{\gamma_X^2 + 4}}{2}$$

$$\hat{\mu}_Y = \ln\left(\frac{S_X}{CV_{X-a}}\right) - \frac{1}{2}\ln\left(CV_{X-a}^2 + 1\right)$$

MLE: see Rao and Hamed (2000)

L-MOM (Hosking 1990):

$$\hat{\sigma}_Y = 0.999281z - 0.006118z^3 + 0.000127z^5 \text{ where } z = \sqrt{\frac{8}{3}} \, \Phi^{-1}\left(\frac{1 + t_3}{2}\right)$$

$$\hat{\mu}_Y = \ln\left(\frac{l_2}{erf(\hat{\sigma}_Y/2)}\right) - \frac{\hat{\sigma}_Y^2}{2} \text{ where } erf(w) = \frac{2}{\sqrt{\pi}}\int_0^w e^{-u^2}\,du$$

$$\hat{a} = l_1 - \exp\left(\hat{\mu}_Y + \hat{\sigma}_Y^2/2\right)$$

6.9.12   Log-Pearson Type III Distribution

MOM (Kite 1988, Rao and Hamed 2000):

If $\mu'_r = \frac{\exp(\xi r)}{(1-r\alpha)^\beta}$ are estimated by $m'_r = \frac{\sum_{i=1}^N X_i^r}{N}$, $\hat{\alpha}$, $\hat{\beta}$, and $\hat{\xi}$ are the solutions to:

$$\ln m'_1 = \xi - \beta \ln(1 - \alpha)$$
$$\ln m'_2 = 2\xi - \beta \ln(1 - 2\alpha)$$
$$\ln m'_3 = 3\xi - \beta \ln(1 - 3\alpha)$$

To find the solutions, Kite (1988) suggests:

- Define $B = \dfrac{\ln m'_3 - 3\ln m'_1}{\ln m'_2 - 2\ln m'_1}$, $A = \frac{1}{\alpha} - 3$ and $C = \dfrac{1}{B - 3}$
- For $3.5 < B < 6$, $A = -0.23019 + 1.65262C + 0.20911C^2 - 0.04557C^3$
- For $3.0 < B \leq 3.5$, $A = -0.47157 + 1.99955C$
- $\hat{\alpha} = \dfrac{1}{A + 3}$
- $\hat{\beta} = \dfrac{\ln m'_2 - 2\ln m'_1}{\ln(1 - \hat{\alpha})^2 - \ln(1 - 2\hat{\alpha})}$
- $\hat{\xi} = \ln m'_1 + \hat{\beta} \ln(1 - \hat{\alpha})$

MLE:

$\hat{\alpha}$, $\hat{\beta}$ and $\hat{\xi}$ are the solutions, by Newton's method, to the system of equations:

$$\sum_{i=1}^N (\ln X_i - \xi) = N\alpha\beta$$

$$N \Psi(\beta) = \sum_{i=1}^{N} \ln \left[\left(\ln X_i - \xi \right) / \alpha \right]$$

$$N = \alpha \left(\beta - 1 \right) \sum_{i=1}^{N} \frac{1}{\ln X_i - \xi}$$

where $\Psi(\beta) = \dfrac{\Gamma'(\beta)}{\Gamma(\beta)}$, which, according to Abramowitz and Stegun (1972), can be

approximated by $\Psi(\beta) \cong \ln \beta - \dfrac{1}{2\beta} - \dfrac{1}{12\beta^2} + \dfrac{1}{120\beta^4} - \dfrac{1}{252\beta^6} + \dfrac{1}{240\beta^8} - \dfrac{1}{132\beta^{10}}.$

L-MOM:

Estimates for the L-MOM can be obtained by using the same estimation procedure described for the Pearson Type III, with the transformation $Z_i = \ln(X_i)$.

6.9.13 Normal Distribution

MOM:

$$\hat{\mu}_X = \overline{X}$$

$$\hat{\sigma}_X = S_X$$

MLE:

$$\hat{\mu}_X = \overline{X}$$

$$\hat{\sigma}_X = \sqrt{\frac{N-1}{N}} S_X$$

L-MOM:

$$\hat{\mu}_X = l_1$$

$$\hat{\sigma}_X = \sqrt{\pi} l_2$$

6.9.14 Pearson Type III Distribution

MOM:

$$\hat{\beta} = \left(\frac{2}{g_X} \right)^2$$

$$\hat{\alpha} = \sqrt{\frac{S_X^2}{\hat{\beta}}}$$

$$\hat{\xi} = \overline{X} - \sqrt{S_X^2 \hat{\beta}}$$

MLE:

$\hat{\alpha}$, $\hat{\beta}$, and $\hat{\xi}$ are the solutions, by Newton's method, to the system of equations:

$$\sum_{i=1}^{N} (X_i - \xi) = N\alpha\beta$$

$$N\Psi(\beta) = \sum_{i=1}^{N} \ln\left[(X_i - \xi) /\alpha \right]$$

$$N = \alpha(\beta - 1) \sum_{i=1}^{N} \frac{1}{X_i - \xi}$$

where $\Psi(\beta) = \dfrac{\Gamma'(\beta)}{\Gamma(\beta)}$ (see log-Pearson Type III in Sect. 6.9.12).

L-MOM:

For $t_3 \geq 1/3$ and $t_m = 1 - t_3$, $\hat{\beta} = \dfrac{0.36067 t_m - 0.5967 t_m^2 + 0.25361 t_m^3}{1 - 2.78861 t_m + 2.56096 t_m^2 - 0.77045 t_m^3}$.

Para $t_3 < 1/3$ and $t_m = 3\pi t_3^2$, $\hat{\beta} = \dfrac{1 + 0.2906 t_m}{t_m + 0.1882 t_m^2 + 0.0442 t_m^3}$.

$$\hat{\alpha} = \sqrt{\pi} l_2 \frac{\Gamma(\hat{\beta})}{\Gamma(\hat{\beta} + 0.5)}$$

$$\hat{\xi} = l_1 - \hat{\alpha}\hat{\beta}$$

6.9.15   Poisson Distribution

MOM:

$$\hat{\nu} = \overline{X}$$

MLE:

$$\hat{\nu} = \overline{X}$$

6.9.16 Uniform Distribution

MOM:

$$\hat{\underline{a}} = \overline{X} - \sqrt{3}S_x$$

$$\hat{\underline{b}} = \overline{X} + \sqrt{3}S_x$$

MLE:

$$\hat{\underline{a}} = \text{Min}\,(X_i)$$

$$\hat{\underline{b}} = \text{Max}\,(X_i)$$

L-MOM:

$\hat{\underline{a}}$ and $\hat{\underline{b}}$ are the solutions to $\underline{l}_1 = (a+b)/2$ and $\underline{l}_2 = (b-a)/6$.

6.9.17 Weibull$_{min}$Distribution

MOM:

$\hat{\underline{\alpha}}$ and $\hat{\underline{\beta}}$ are the solutions to the system of equations:

$$\overline{X} = \beta\Gamma\left(1+\frac{1}{\alpha}\right)$$

$$S_X^2 = \beta^2\left[\Gamma\left(1+\frac{2}{\alpha}\right) - \Gamma^2\left(1+\frac{1}{\alpha}\right)\right]$$

(See Sect. 5.7.2.5 of Chap. 5).

MLE:

$\hat{\underline{\alpha}}$ and $\hat{\underline{\beta}}$ are the solutions, by Newton's method, to the system of equations:

$$\beta^{-\alpha} = \frac{N}{\displaystyle\sum_{i=1}^{N}X_i^{\alpha}}$$

$$\alpha = \frac{N}{\beta^{-\alpha}\displaystyle\sum_{i=1}^{N}X_i^{\alpha}\ln\,(X_i) - \displaystyle\sum_{i=1}^{N}\ln\,(X_i)}$$

Exercises

1. Given the density $f_X(x) = x^{\theta-1} \exp(-x)/\Gamma(\theta)$, $x > 0$, $\theta > 0$, determine the value of c, such that cX is an unbiased estimator for θ. Remember the following property of the Gamma function: $\Gamma(\theta + 1) = \theta\Gamma(\theta)$.

2. Assume $\{Y_1, Y_2, \ldots, Y_N\}$ is an SRS from the variable Y with mean μ. Under which conditions $W = \sum_{i=1}^{N} a_i Y_i$ is an unbiased estimator for μ? (adapted from Larsen and Marx 1986).

3. Assume X_1 and X_2 make up an SRS of size 2 from an exponential distribution with density $f_X(x) = (1/\theta)\exp(-x/\theta)$, $x \geq 0$. If $Y = \sqrt{X_1 X_2}$ is the geometric mean of X_1 and X_2, show that $W = 4Y/\pi$ is an unbiased estimator for θ (adapted from Larsen and Marx 1986).

4. Show that the mean square error is given by $E\left[\left(\hat{\theta} - \theta\right)^2\right] = \mathrm{Var}(\hat{\theta}) + \left[B(\hat{\theta})\right]^2$.

5. Suppose W_1 and W_2 are two unbiased estimators for a parameter θ, with respective variances $\mathrm{Var}(W_1)$ and $\mathrm{Var}(W_2)$. Assume X_1, X_2, and X_3 represent an SRS of size 3, from an exponential distribution with parameter θ. Calculate the relative efficiency of $W_1 = (X_1 + 2X_2 + X_3)/4$ with respect to $W_2 = \overline{X}$ (adapted from Larsen and Marx 1986).

6. Suppose $\{X_1, X_2, \ldots, X_N\}$ is an SRS from $f_X(x; \theta) = 1/\theta$, for $0 < x < \theta$, and that $W_N = X_{\max}$. Show that W_N is a biased but consistent estimator for θ. To solve this exercise remember that the exact distribution for the maximum of a SRS can be obtained using the methods described in Sect. 5.7.1 of Chap. 5 (adapted from Larsen and Marx 1986).

7. Recall that an estimator $W = h(X_1, X_2, \ldots, X_N)$ is deemed sufficient for θ, if, for all θ and for any sample values, the joint density of $(X_1, X_2, \ldots, X_N)$, conditioned to w, does not depend on θ. More precisely, W is sufficient if $f_{X_1}(x_1)f_{X_2}(x_2)\ldots f_{X_N}(x_N)/f_W(w)$ does not depend on θ. Consider the estimator W_N, described in Exercise 6, and show it is sufficient (adapted from Larsen and Marx 1986).

8. The two-parameter exponential distribution has density function given by $f_X(x) = (1/\theta)\exp[-(x - \xi)/\theta]$, $x \geq \xi$, where ξ denotes a location parameter. Determine the MOM and MLE estimators for ξ and θ.

9. Table 1.3 of Chap. 1 lists the annual maximum mean daily discharges (m^3/s) of the Shokotsu River recorded at Utsutsu Bridge (ID # 301131281108030-www1.river.go/jp), in Hokkaido, Japan. Employ the methods described in this chapter to calculate (a) the estimates for the parameters of the gamma distribution, by the MOM, MLE, and L-MOM methods; (b) the probability that the annual maximum daily flow will exceed 1500 m^3/s, in any given year, using the MOM, MLE, and L-MOM parameter estimates; (c) the flow quantile for the return period of 100 years, using the MOM, MLE, and L-MOM parameter

Table 6.7 Experimental
Manning coefficients for
plastic tubes (from Haan
1965)

0.0092	0.0085	0.0083	0.0091
0.0078	0.0084	0.0091	0.0088
0.0086	0.0090	0.0089	0.0093
0.0081	0.0092	0.0085	0.0090
0.0085	0.0088	0.0088	0.0093

estimates; and (d) make a comparative analysis of the results obtained in (b), (c), and (d).

10. Solve Exercise 9 for the one-parameter exponential distribution.
11. Solve Exercise 9 for the two-parameter lognormal distribution.
12. Solve Exercise 9 for the $Gumbel_{max}$ distribution.
13. Solve Exercise 9 for the GEV distribution.
14. Solve Exercise 9 for the Pearson type III distribution.
15. Solve Exercise 9 for the log-Pearson type III distribution.
16. Data given in Table 6.7 refer to the Manning coefficients for plastic tubes, determined experimentally by Haan (1965). Assume this sample has been drawn from a normal population with parameters μ and σ. (a) Construct a 95 % confidence interval for the mean μ. (b) Construct a 95 % confidence interval for the variance σ^2.
17. Solve Exercise 16 for the 90 % confidence level. Interpret the differences between results for 95 and 90 % confidence levels.
18. Assume the true variance in Exercise 16 is known and equal to the value estimated from the sample. Under this assumption, solve item (a) of Exercise 16 and interpret the new results.
19. Assume the true mean in Exercise 16 is known and equal to the value estimated from the sample. Under this assumption, solve item (b) of Exercise 16 and interpret the new results.
20. Table 1.3 of Chap. 1 lists the annual maximum mean daily discharges (m^3/s) of the Shokotsu River recorded at Utsutsu Bridge, in Japan. Construct the 95 % confidence intervals for the $Gumbel_{max}$ quantiles of return periods 2, 50, 100, and 500 years, estimated by the MOM, MLE, and L-MOM. Decide on which estimation method is more efficient. Interpret the results from the point of view of varying return periods.
21. The reliability of the MOM, MLE, and L-MOM estimators for parameters and quantiles has been the object of many studies. These generally take into account the main properties of estimators and allow comparative studies among them. The references Rao and Hamed (2000), Kite (1988), and Hosking (1986) make syntheses of the many studies on this subject. The reader is asked to read these syntheses and make her/his own on the main characteristics of MOM, MLE, and L-MOM estimators for parameters and quantiles, for the distributions exponential, $Gumbel_{max}$, GEV, Gamma, Pearson type III, log-Pearson type III, and lognormal, as applied to the frequency analysis of annual maxima of hydrologic events.

22. Table 2.7 of Chap. 2 lists the Q_7 flows, in m^3/s, for the Dore River at Saint-Gervais-sous-Meymont, in France, from 1920 to 2014. Fit a Weibull$_{min}$ distribution for these data, employing the MOM and MLE methods, and use their respective estimates to calculate the reference flow $Q_{7,10}$. Use MS Excel and the method described in Sect. 6.8 to construct a 95 % confidence interval for $Q_{7,10}$, for both MOM and MLE, on the basis of 100 synthetic samples.

23. Write and compile a computer program to calculate the 95 % confidence intervals for quantiles, through Monte Carlo simulation, as described in Sect. 6.8, for the Gumbel$_{max}$ distribution, considering the MOM, MLE, and L-MOM estimation methods. Run the program for the data of Table 1.3 of Chap. 1, for return periods $T = 2$, 10, 50, 100, 500, and 1000 years. Plot your results, with the return periods in abscissa, in log-scale, and the estimated quantiles and respective confidence intervals, in ordinates.

24. Solve Exercise 23 for the GEV distribution.

References

Abramowitz M, Stegun IA (1972) Handbook of mathematical functions. Dover, New York

Casella G, Berger R (1990) Statistical inference. Duxbury Press, Belmont, CA

Cramér H (1946) Mathematical methods of statistics. Princeton University Press, Princeton

Greenwood JA, Landwehr JM, Matalas NC, Wallis JR (1979) Probability weighted moments: definition and relation to parameters expressible in inverse form. Water Resour Res 15 (5):1049–1054

Haan CT (1965) Point of impending sediment deposition for open channel flow in a circular conduit. MSc Thesis, Purdue University

Hall MJ, Van Den Boogard HFP, Fernando RC, Mynet AE (2004) The construction of confidence intervals for frequency analysis using resampling techniques. Hydrol Earth Syst Sci 8 (2):235–246

Hosking JRM (1985) Algorithm AS 215: maximum-likelihood estimation of the parameters of the generalized extreme-value distribution. J Roy Stat Soc C 34(3):301–310

Hosking JRM (1986) The theory of probability weighted moments. Research Report RC 12210. IBM Research, Yorktown Heights, NY

Hosking JRM (1990) L-moments: analysis and estimation of distributions using linear combinations of order statistics. J R Stat Soc B 52:105–124

Hosking JRM (1991) Fortran routines for use with the method of L-moments—Version 2. Research Report 17097. IBM Research, Yorktown Heights, NY

Hosking JRM, Wallis JR (1997) Regional frequency analysis: an approach based on L-moments. Cambridge University Press, Cambridge

Kaczmarek Z (1977) Statistical methods in hydrology and meteorology. Report TT 76-54040. National Technical Information Service, Springfield, VA

Kimball BF (1949) An approximation to the sampling variation of an estimated maximum value of a given frequency based on fit of double exponential distribution of maximum values. Ann Math Stat 20(1):110–113

Kite GW (1988) Frequency and risk analysis in hydrology. Water Resources Publications, Fort Collins, CO

Kottegoda NT, Rosso R (1997) Statistics, probability, and reliability for civil and environmental engineers. McGraw-Hill, New York

Larsen RJ, Marx ML (1986) An introduction to mathematical statistics and its applications. Prentice-Hall, Englewood Cliffs, NJ

Meylan P, Favre AC, Musy A (2008) Hydrologie fréquentielle—une science prédictive. Presses Polytechniques et Universitaires Romandes, Lausanne

Mood AM, Graybill FA, Boes DC (1974) Introduction to the theory of statistics, International, 3rd edn. McGraw-Hill, Singapore

Prescott P, Walden AT (1980) Maximum likelihood estimation of the parameters of the generalized extreme-value distribution. Biometrika 67:723–724

Rao AR, Hamed KH (2000) Flood frequency analysis. CRC Press, Boca Raton, FL

Rao CR (1973) Linear statistical inference and its applications. Wiley, New York

Singh VP (1997) The use of entropy in hydrology and water resources. Hydrol Process 11:587–626

Vogel RM, Fennessey NM (1993) L-moment diagrams should replace product-moment diagrams. Water Resour Res 29(6):1745–1752

Weisstein EW (2016) Sample variance distribution. From MathWorld—A Wolfram Web Resource. http://mathworld.wolfram.com/SampleVarianceDistribution.html. Accessed 20 Jan 2016

Yevjevich VM (1972) Probability and statistics in hydrology. Water Resources Publications, Fort Collins, CO

Chapter 7
Statistical Hypothesis Testing

Mauro Naghettini

7.1 Introduction

Together with parameter estimation and confidence interval construction, the hypotheses testing techniques are among the most relevant and useful methods of inferential statistics, for making decisions concerning the value of some population parameter or the shape of the probability distribution, from a data sample. In general, these tests start by setting out a hypothesis, in the form of a conjectural statement on the statistical properties of the random variable population. This hypothesis can be established, for instance, as a prior premise concerning the value of some population parameter, such as the population mean or variance. The decision of rejecting or not rejecting the hypothesis will depend on confronting the conjectural statement with the physical reality imposed by the data sample. Rejecting the hypothesis implies the need for revising the initial conjecture, as resulting from its discordance with the reality. Contrarily, not rejecting the hypothesis means that the sample data do not reveal sufficient evidence to discard the plausibility of the conjectural statement. It is worth noting that not rejecting does not mean accepting as true the hypothesis being tested. Once again, the truth would only be known if the entire population could be suitably sampled.

For being an inference concerning a random variable, the decision of rejecting or not rejecting a hypothesis is made on the probabilistic terms of a *significance level* α. For instance, by collating the appropriate data, one can possibly reject the hypothesis that the mean flow over the last 30 years, observed at a given gauging station, has decreased. By rejecting it, one is not stating that flows have remained stable or have increased, but that the variation of flows, over the considered period, stems merely from the natural fluctuations of data, without important effects on the

M. Naghettini (✉)
Universidade Federal de Minas Gerais Belo Horizonte, Minas Gerais, Brazil
e-mail: mauronag@superig.com.br

© Springer International Publishing Switzerland 2017
M. Naghettini (ed.), *Fundamentals of Statistical Hydrology*,
DOI 10.1007/978-3-319-43561-9_7

value of the population mean. In this case, such a variation of flows is said nonsignificant. However, by examining the same data, another person, perhaps more concerned with the consequences of his/her decision, might reach to a different conclusion, that the differences between the data and the conjectural statement, as implied by the hypothesis, are indeed significant. Hypothesis tests are sometimes termed significance tests.

In order to remove the subjectivity that may be embedded in decision making in hypotheses testing, in relation to how significant the differences are, the significance level α is usually specified beforehand, so that the uncertainties inherent to hypotheses testing can be taken into account, in an equal manner, by two different analysts. Accordingly, based on the same significance level α, both analysts would have made the same decision for the same test, with the same data. The significance level α, of a hypothesis test, is complementary to the probability $(1-\alpha)$ that a confidence interval $[L,U]$ contains the true value of the population parameter θ. Actually, the confidence interval $[L,U]$ establishes the bounds for the so-called test statistic, within which the hypothesis on θ cannot be rejected. Contrarily, if the calculated values for the test statistic fall outside the bounds imposed by $[L,U]$, then the hypothesis on θ must be rejected, at the significance level α. Thus, according to this interpretation, the construction of a $(1-\alpha)$ confidence interval represents the inverse operation of testing a hypothesis on the parameter θ, at the significance level α.

In essence, testing a hypothesis is to collect and interpret empirical evidence that justify the decision of rejecting or not rejecting some conjecture (1) on the true value of a population parameter or (2) on the shape of the underlying probability distribution, taking into account the probabilities that wrong decisions can possibly be made, as a result of the uncertainties that are inherent in the random variable under analysis. Hypothesis tests can be categorized either as parametric or nonparametric. They are said to be parametric if the sample data are assumed to have been drawn from a normal population or from another population, whose parent probability distribution is known or specified. On the other hand, nonparametric tests do not assume a prior probability distribution function for describing the population, from which the data have been drawn. In fact, nonparametric tests are not formulated on the basis of the sample data themselves, but on the basis of selected attributes or characteristics associated with them, such as, for instance, their ranking orders or the counts of positive and negative differences between each sample element and the sample median. In relation to the nature of the hypothesis being tested, significance tests on a true value of a population parameter are useful in many areas of applied statistics, whereas tests on the shape of the underlying probability distribution are often required in Statistical Hydrology. The latter are commonly referred to as goodness-of-fit tests.

7.2 The Elements of a Hypothesis Test

In general, the sequential steps to test a hypothesis are:

- Develop the hypothesis to be tested, denoting it as H_0 and designating it as the *null hypothesis*. This can possibly be, for example, a conjecture stating that the mean annual total rainfall μ_0, over the last 30 years, did not deviate significantly from the mean annual total rainfall μ_1, over the previous period of 30 years. If the null hypothesis is not false, any observed difference between the mean annual rainfall depths is due to fluctuations of the data sampled from the same population. For this example, the null hypothesis can be stated as $H_0:\{\mu_0-\mu_1=0\}$.
- Develop the *alternative hypothesis* and denote it as H_1. For the example given in the previous step, the alternative hypothesis, which is opposed to H_0, is expressed as $H_1: \mu_0-\mu_1 \neq 0$.
- Specify a *test statistic T*, suitable for the null and alternative hypotheses formulated in previous steps. For the example given in the first step, the test statistic should be based on the difference $T = \overline{X}_0 - \overline{X}_1$, between the sample means for the corresponding time periods of the population means being tested.
- Specify the *probability distribution* of the test statistic, which is an action that must take into consideration not only the null hypothesis but also the underlying probability distribution of the population from which the sample has been drawn. For the example given in the first step, annual total rainfall depths, as stemming from the Central Limit Theorem, can possibly be assumed as normally distributed. As seen in Chap. 5, for normal populations, sampling distributions for means and variances are known and explicit, and, thus, it is possible to infer the probability distribution for the test statistic T.
- Specify the *region of rejection R*, or *critical region R*, for the test statistic. Specifying R depends on the previous definition of the significance level α, which, as mentioned earlier, plays the role of removing the degree of subjectivity associated with decision making under uncertainty. For the example being discussed, if the significance level is arbitrarily fixed as $100\alpha = 5\%$, this would define the bounds $[T_{0.025}, T_{0.975}]$, below and above which, respectively, begins the region of rejection R.
- Check if the test statistic $\hat{T}$, estimated from the data sample, falls inside or outside the region of rejection R. For the example being discussed, if $\hat{T} < T_{0.025}$ or $\hat{T} > T_{0.975}$, the null hypothesis H_0 must be rejected in favor of the alternative hypothesis H_1. In such a case, one can interpret that the difference $\mu_1-\mu_0$ is significant at the $100\alpha = 5\%$ level. On the contrary, if $\hat{T}$ is contained in the interval bounded by $[T_{0.025}, T_{0.975}]$, the correct decision would be not rejecting the null hypothesis H_0, thus implying that no empirical evidence of significant differences between the two means, μ_1 and μ_0, was found.

For the general sequential steps, listed before, the example makes reference to the differences between μ_0 and μ_1, which can be positive or negative, which implies that the critical region R extends through both tails of the probability distribution of the test statistic. In such a case, the test is considered two-sided or bilateral or two-tailed. Had both hypotheses been formulated in a different manner, such as H_0:$\{\mu_0 = 760\}$ against H_1:$\{\mu_1 = 800\}$, the test would be termed one-sided or uni-lateral or one-tailed, as the critical region would extend through only one tail of the probability distribution of the test statistic, in this case, the upper tail. Had the hypotheses been H_0:$\{\mu_0 = 760\}$ against H_1:$\{\mu_1 = 720\}$, the test would be also one-sided, but the critical region would extend through the lower tail of the probability distribution of the test statistic. Thus, besides being more specific than bilateral tests, the one-sided tests are further categorized as lower-tail or upper-tail tests.

From the general steps for setting up a hypothesis test, one can deduce that, in fact, there is a close relationship between the actions of testing a hypothesis and of constructing a confidence interval. To make it clear, let the null hypothesis be H_0: $\{\mu = \mu_0\}$, on the mean of a normal population with known variance σ^2. Under these conditions, it is known that, for a sample of size N, the test statistic should be $T = (\overline{X} - \mu_0)/\sigma/\sqrt{N}$, which follows a standard normal distribution. In such a case, if the significance level is fixed as $\alpha = 0.05$, the two sided-test would be defined for the critical region R, extending through values of T smaller than $T_{\alpha/2=0.025} = $ spi2;$z_{0.025} = -1.96$ and larger than $T_{1-\alpha/2=0.975} = z_{0.975} = +1.96$. If, at this significance level, H_0 is not rejected, such a decision would be justified by the fact that either $\hat{T} > T_{0.025}$ or $\hat{T} < T_{0.975}$, or equivalently by the respective circum-stance that either $\overline{X} > \mu_0 - 1.96\sigma/\sqrt{N}$ or $\overline{X} < \mu_0 + 1.96\sigma/\sqrt{N}$. Rearranging these inequalities, one can write them under the form $\overline{X} - 1.96\sigma/\sqrt{N} < \mu_0 < \overline{X} + 1.96\sigma/\sqrt{N}$, which is the expression of the 100 $(1-\alpha) = 95\%$ confidence interval for the mean μ_0. By means of this example, one can see that, in mathematical terms, testing a hypothesis and constructing a confidence interval are closely related procedures. In spite of this mathematical relationship, they are intended for distinct purposes: while the confidence interval sets out how accurate the current knowledge on μ is, the test of hypothesis indicates whether or not is plausible to assume the value μ_0 for μ.

According to what has been outlined so far, the rejection of the null hypothesis happens when the estimate of the test statistic falls within the critical region. The decision of rejecting the null hypothesis is equivalent to stating that the test statistic is significant. In other terms, in the context of testing H_0:$\{\mu = \mu_0\}$ at $\alpha = 0.05$, if the difference between the hypothesized and empirical means is large and occurs randomly with probability of less than 5% (or in less than 5 out of 100 identical tests that could be devised), then this result would be considered significant and the hypothesis must be rejected. Nonetheless, the lack of evidence to reject the null hypothesis does not imply acceptance of H_0 as being true, as the actual truth may lie elsewhere. In fact, the decision to not reject H_0, in some cases, may indicate the eventual need for its reformulation, to make it stricter or narrower, followed by additional testing procedures.

Consider now that the null hypothesis is actually true and, as such, its probability of being rejected is given by

$$P(T \in R | H_0 \text{ is true}) = P(T \in R | H_0) = \alpha \tag{7.1}$$

It is clear that if one has rejected a true hypothesis, an incorrect decision would have been made. The error resulting from such a decision is termed *Type I Error*. From Eq. (7.1), the probability that a type I error occurs is expressed as

$$P(\text{Type I Error}) = P(T \in R | H_0) = \alpha \tag{7.2}$$

In the absence of this type of error, or, if a true hypothesis H_0 is not rejected, the probability that a correct decision has been made is complementary to the type I error. Formally,

$$P(T \notin R | H_0) = 1 - \alpha \tag{7.3}$$

As opposed to that, the action of not rejecting the null hypothesis when it is actually false, since it is known that H_1 is true, is another possible incorrect decision. The error resulting from such a decision is termed *Type II Error*. The probability that a type II error occurs is expressed as

$$P(\text{Type II Error}) = P(T \notin R | H_1) = \beta \tag{7.4}$$

In the absence of this type of error, or, if a false hypothesis H_0 is rejected, the probability that such a correct decision has been made is complementary to the type I error. Formally,

$$P(T \in R | H_1) = 1 - \beta \tag{7.5}$$

The probability, complementary to β, as given by Eq. (7.5), is designated the *Power of the Test* and, as seen later on in this section, is an important criterion to compare different hypothesis tests.

The type I and type II errors are strongly related. In order to demonstrate this, consider the graph of Fig. 7.1, which depicts a one-sided test of the null hypothesis $H_0:\{\mu = \mu_0\}$ against the alternative hypothesis $H_1:\{\mu = \mu_1\}$, where μ denotes the mean of a normal population and $\mu_1 > \mu_0$.

If the estimated test statistic $\hat{T}$ is larger than T_{critical}, the null hypothesis must be rejected, at the significance level α. In such a case, assuming that H_0 is actually true, the decision to reject it is incorrect and the probability of committing such an error is α. On the other hand, if the estimated test statistic $\hat{T}$ is smaller than T_{critical}, the null hypothesis must not be rejected, at the significance level α. Now, assuming that H_1 is actually true, the decision to not reject H_0 is also incorrect and the probability of committing such an error is β. In the graph of Fig. 7.1, it is clear that decreasing α makes the value of T_{critical} shift to the right of its initial location, thus causing β to

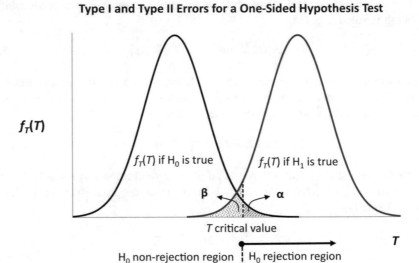

Type I and Type II Errors for a One-Sided Hypothesis Test

Fig. 7.1 Illustration of type I and type II errors for a one-sided test of a hypothesis

increase. Thus, it is clear that decreasing the probability of committing a type I error has the adverse effect of increasing the probability of committing a type II error. The opposite situation is equally true.

It is certain that, by setting up a hypothesis test, no one is willing to make a wrong decision of any kind. However, as uncertainties are present and wrong decisions might be made, the desirable and logical solution is that of minimizing the probabilities of committing both types of errors. Such an attempt unfolds the additional difficulties brought in by the strong dependence between α and β, and by the distinct characteristics of type I and type II errors, thus forcing a compromise solution in the planning of the decision-making process related to a hypothesis test. In general, such a compromise solution starts with the prior prescription of a sufficiently low significance level α, such that β falls in acceptable values. Such a strategy is justified by the fact that it is possible to previously specify the significance level α, whereas, for the probability β, such a possibility does not exist. In fact, the alternative hypothesis is more generic and broader than the null hypothesis. For example, the alternative hypothesis $H_1:\{\mu_0-\mu_1 \neq 0\}$ is ill-defined as it encompasses the union of many hypotheses (e.g.: $H_1:\mu_0-\mu_1 < 0$, $H_1:\mu_0-\mu_1 < 2$, $H_1:\mu_0-\mu_1 > 2$, or $H_1:\mu_0-\mu_1 > 0$, among others), whereas the null hypothesis $H_0:\{\mu-\mu_1 = 0\}$ is completely defined. In other terms, while α will depend only on the null hypothesis, β will depend on which of the alternative hypotheses is actually true, which is obviously not known a priori. In general, it is a frequent practice to prescribe a prior significance level α of 5 %, which seems reasonable and acceptable as it is equivalent to state that at most 5 wrong decisions, out of 100 possible decisions, are made. If the consequences of a type I error are too severe, one can choose an even smaller significance level, such as $\alpha = 0.01$, to the detriment of an increased unknown value of β.

Although the probability β depends on which alternative hypothesis, among those encompassed by H_1, is actually true and, thus, cannot be anticipated, it is instructive to investigate the behavior of β, under different true hypotheses. Such an investigation is performed by means of the quantity $(1 - \beta)$, which, as previously mentioned, is termed the power of the test. In Fig. 7.1, the power of the test, for the specific alternative hypothesis $H_1:\{\mu = \mu_1\}$, can be visualized by the area below the density function of the test statistic, under H_1, to the right of abscissa T_{critical}. For another alternative hypothesis, for example, $H_1:\{\mu = \mu_2\}$, it is clear that the power of the test would have a different value. The relationships between β, or $(1-\beta)$, and a sequence of alternative hypotheses, define, respectively, the *operating characteristic curve*, and the *power curve*, which serve the purposes of distinguishing and comparing different tests.

In order to exemplify the construction of an operating characteristic curve and a power curve, consider the two-sided test for the mean of a sample of varying size N, taken from a normal population of parameters μ and σ, or, let $H_0:\{\mu = \mu_0\}$ be tested against a set of alternative hypotheses $H_1:\{\mu \neq \mu_0\}$. Once more, the test statistic is $T = (\overline{X} - \mu_0)/\sigma/\sqrt{N}$, which follows the normal distribution $N(0,1)$. The numerator of the test statistic can be altered to express deviations $\mu_0 + k$ from μ_0, where k denotes a real number. As such, with $T = k\sqrt{N}/\sigma$, the test actually refers to $H_0:\{\mu = \mu_0\}$, when $k = 0$, against a number of standardized shifts $k\sqrt{N}/\sigma$, with respect to zero, or equivalently, against a set of deviations $\mu_0 \pm k$, with respect to μ_0. For this example, while the probability distribution associated with the null hypothesis is $N(\mu_0, \sigma/\sqrt{N})$, such that the variable $(\overline{X} - \mu_0)\sqrt{N}/\sigma$ is distributed as $N(0,1)$, the distributions associated with the alternative hypotheses H_1 are given by $N(\mu_0 + k, \sigma/\sqrt{N})$, such that $[\overline{X} - (\mu_0 + k)]\sqrt{N}/\sigma$ follow $N(k\sqrt{N}/\sigma, 1)$. These distributions are depicted in Fig. 7.2, with deviations of ± 3 standardized units, with respect to the standard normal distribution $N(0,1)$. In Fig. 7.2, one sees the type I error ($\alpha = 0.05$), as shaded areas for the two-sided test against $H_1:\{\mu \neq 0\}$, and the type II errors, if the true hypothesis were $H_1:\{\mu = +3\}$ or $H_1:\{\mu = -3\}$.

The type II error corresponds to the nonrejection of H_0, when H_1 is true, which will happen when the test statistic satisfies the condition $z_{\alpha/2} \leq T \leq +z_{1-\alpha/2}$, where $z_{\alpha/2}$ and $z_{1-\alpha/2}$ represent the bounds of the critical region. The probability β of committing a type II error should be calculated on the basis of the test statistic distribution when H_1 is true, or, in formal terms, as $\beta = \Phi(z_{1-\alpha/2} - k\sqrt{N}/\sigma) - \Phi(z_{\alpha/2} - k\sqrt{N}/\sigma)$, where $\Phi(.)$ denotes the standard normal CDF and $k\sqrt{N}/\sigma$ represents the mean, under H_1. Thus, one can notice that β depends on α and N, and on the different alternative hypotheses as given by k/σ. Such a dependence on multiple factors can be depicted graphically by means of the operating characteristic curve, illustrated in Fig. 7.3, for $\alpha = 0.05$ ($z_{0.025} = -1.96$), sample sizes N varying from 1 to 50, and $k/\sigma = 0.25, 0.50, 0.75$, and 1.

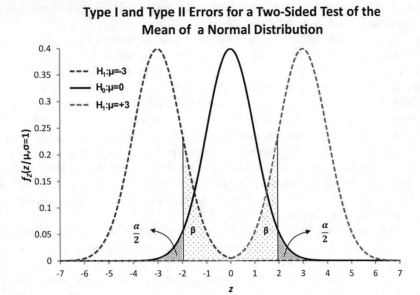

Fig. 7.2 Probabilities α and β for a two-sided test for the sample mean from a normal population

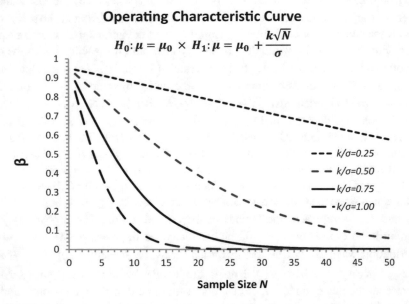

Fig. 7.3 Examples of operating characteristic curves for hypotheses tests

By looking at the operating characteristic curve of Fig. 7.3, one can notice that for samples of a fixed size N, the probability of committing a type II error decreases as k/σ increases. This is equivalent to saying that small deviations from the hypothesized mean are hard to detect, which leads to higher probabilities of making

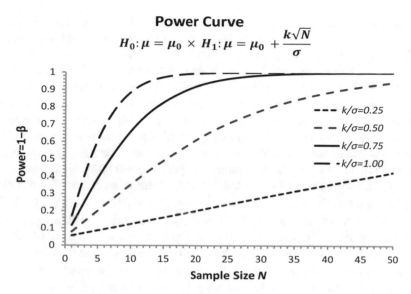

Fig. 7.4 Examples of power curves for hypotheses tests

the incorrect decision of not rejecting a false null hypothesis. One can also note that β decreases with increasing N, thus showing the relatively lower probabilities of committing a type II error, when tests are based on samples of larger sizes.

The power curve (or power function) is given by the complement of the probability β, with respect to 1, and is depicted in Fig. 7.4, for the example being discussed. The power of the test, as previously defined, represents the probability of making the correct decision of rejecting a false null hypothesis, in favor of the true alternative hypothesis. Figure 7.4 shows that, for samples of the same size, the probability of not committing a type II error increases, as k/σ increases. Likewise, the power of the test increases, with increasing sample sizes.

Figures 7.3 and 7.4 show that if, for example, the respective probabilities of committing type I and type II errors were both kept fixed at $100\alpha = 5\%$ and $100\beta = 10\%$, and if the null hypothesis $H_0:\{\mu = \mu_0\}$ were being tested against the alternative hypothesis $H_1:\{\mu = \mu_0 + 0.5\sigma\}$, then a sample of size 42 would be required. If a sample with at least 42 elements were not currently available or if gathering additional data were too onerous, then one could search for a trade-off solution between the reliability of the test, imposed by α and β, and the cost and willingness to wait for additional sampling. In hydrology practice, the decisions concerning the population characteristics of a random variable are usually made on the basis of samples of fixed sizes and, usually, *post hoc* power analysis for the hypothesis test being used is not carried out, an exception being made for assessing and comparing the power of different tests or in highly sensitive cases. Thus, the remaning sections of this chapter will keep the focus on hypotheses tests based on samples of fixed size, with the previous specification of low significance levels, such as $100\alpha = 5\%$ or 10%, implicitly accepting the resulting probabilities β. The reader interested in power analysis and more elaborate tests and methods, should consult Hoel et al. (1971), Mood et al. (1974), and Ramachandran and Tsokos (2009), for books of intermediate level of complexity, or Bickel and Doksum (1977) and Casella and Berger (1990) for more in-depth texts.

Instead of previously choosing the value of the significance level α, an alternative manner to perform a statistical hypothesis test is to compute the so-called p-value, based on the estimated test statistic $\hat{T}$, and make the decision of rejecting or not rejecting H_0 by comparing the p-value with current-practice significance levels. The p-value is defined as the lowest level of significance at which the null hypothesis would have been rejected and is very useful for making decisions on H_0, without resorting to reading tables of specific probability distributions. For example, if, for a one-sided lower tail test, the p-value of an estimated test statistic $\hat{T}$ is calculated as the *non-exceedance probability* $p = P(T < \hat{T} | H_0) = 0.02$, then H_0 would be rejected at $\alpha = 0.05$, because $p < \alpha$ and the estimated test statistic would lie in the critical region, but would not be rejected at $\alpha = 0.01$, for the opposite reason. For a one-sided upper tail test, the p-value is given by the *exceedance probability* $p = P(T > \hat{T} | H_0)$ and should be compared to an assumed value of α; if $p < \alpha$, H_0 would be rejected at the significance level α and, otherwise, not rejected. For a two-sided test, if the T distribution is symmetric, $p = 2P(T > |\hat{T}| \, | H_0)$ should be compared to α and, if $p < \alpha$, H_0 would be rejected at the significance level α and, otherwise, not rejected. If the T distribution is not symmetric, the calculation of p should be performed for both tails, by making $p = 2s$, where $s = \min[P(T < \hat{T} | H_0), P(T > \hat{T} | H_0)]$, and then compared to α; if $p < \alpha$, H_0 would be rejected at the level α and, otherwise, not rejected. Ramachandran and Tsokos (2009) note that the p-value can be interpreted as a measure of support for H_0: the lower its value, the lower the support. In a typical decision making, if the p-value drops below the significance level of the test, the support for the null hypothesis is not sufficient. The p-value approach to decision making in statistical hypothesis testing is employed in the vast majority of statistical software.

7.3 Some Parametric Tests for Normal Populations

Most of the statistical methods that concern parametric hypotheses tests refer to normal populations. This assessment can be justified, first, by the possibility of deducing the exact sampling distributions for normally distributed variables, and, second, by the power and extension of the central limit theorem. In the subsections that follow, the descriptions of the main parametric tests for the testing of hypotheses for normal populations, along with their assumptions and test statistics, are provided. For these tests to yield rigorous results, their underlying assumptions must hold true. In some special cases, as resulting from the application of the central limit theorem to large samples, one may consider extending parametric tests for nonnormal populations. It must be pointed out, however, that the results, from such extensions, will only be approximate and the degree of approximation will be given by the difference between the true significance level, as evaluated by the rate of rejections of a true hypothesis from Monte Carlo simulations, and the nominal specified value for α.

7.3.1 Parametric Tests for the Mean of a Single Normal Population

The underlying assumption for the hypotheses tests described in this subsection is that the simple random sample $\{x_1, x_2, \ldots, x_N\}$, has been drawn from a normal population with unknown mean μ. The normal variance σ^2, as the distribution second parameter, plays an important role in testing hypotheses on the mean μ. In fact, whether or not the population variance σ^2 is known determines the test statistic to be used in the test.

- H_0: $\mu = \mu_1$ against H_1: $\mu = \mu_2$. Population variance σ^2: known.

  Test statistic: $Z = \frac{\bar{x} - \mu_1}{\sigma/\sqrt{N}}$

  Probability distribution of the test statistic: standard normal $N(0,1)$

  Test type: one-sided at significance level α

  Decision:

  If $\mu_1 > \mu_2$, reject H_0 if $\frac{\bar{x} - \mu_1}{\sigma/\sqrt{N}} < -z_{1-\alpha}$

  If $\mu_1 < \mu_2$, reject H_0 if $\frac{\bar{x} - \mu_1}{\sigma/\sqrt{N}} > +z_{1-\alpha}$

- H_0: $\mu = \mu_1$ against H_1: $\mu = \mu_2$. Population variance σ^2: unknown and estimated by s_X^2.

  Test statistic: $T = \frac{\bar{x} - \mu_1}{s_X/\sqrt{N}}$

  Probability distribution of the test statistic: Student's t with $\nu = N-1$

  Test type: one-sided at significance level α

  Decision:

  If $\mu_1 > \mu_2$, reject H_0 if $\frac{\bar{x} - \mu_1}{s_X/\sqrt{N}} < -t_{1-\alpha, \nu=N-1}$

  If $\mu_1 < \mu_2$, reject H_0 if $\frac{\bar{x} - \mu_1}{s_X/\sqrt{N}} > +t_{1-\alpha, \nu=N-1}$

- H_0: $\mu = \mu_0$ against H_1: $\mu \neq \mu_0$. Population variance σ^2: known.

  Test statistic: $Z = \frac{\bar{x} - \mu_0}{\sigma/\sqrt{N}}$

  Probability distribution of the test statistic: standard normal $N(0,1)$

  Test type: two-sided at significance level α

  Decision:

  Reject H_0 if $\left| \frac{\bar{x} - \mu_0}{\sigma/\sqrt{N}} \right| > z_{1-\alpha/2}$

- H_0: $\mu = \mu_0$ against H_1: $\mu \neq \mu_0$. Population variance σ^2: unknown and estimated by s_X^2.

Test statistic: $T = \frac{\bar{x} - \mu_0}{s_X/\sqrt{N}}$

Probability distribution of the test statistic: Student's t with $\nu = N - 1$

Test type: two-sided at significance level α

Decision:

Reject H_0 if $\left| \frac{\bar{X} - \mu_0}{s_X/\sqrt{N}} \right| > t_{1-\alpha/2, \nu = N-1}$

Example 7.1 Consider the time series of annual total rainfalls recorded since 1767, at the Radcliffe Meteorological Station, in Oxford (England), retrievable from the URL http://www.geog.ox.ac.uk/research/climate/rms/rain.html. Figure 5.4 shows that the normal distribution fitted to the sample data closely matches the empirical histogram, thus it being plausible to assume data have been drawn from a normal population. For this example, assume that the available sample begins in the year 1950 and ends in 2014. Test the hypothesis that the population mean is 646 mm, at $100\alpha = 5\%$.

Solution The sample of size $N = 65$ yields $\bar{x} = 667.21$ mm and $s_X^2 = 13108.22$ mm^2, with no other information regarding the population variance, which is, then, deemed unknown and, as such, must be estimated by the sample variance. The null hypothesis $H_0:\{\mu = 646\}$ should be tested against the alternative hypothesis $H_1:\{\mu \neq 646\}$, thus setting up for a two-sided hypothesis test. The test statistic is $T = (\bar{X} - 646)/s_X/\sqrt{N}$ and follows a Student's t distribution with $\nu = 65 - 1 = 64$ degrees of freedom. Substituting the sample mean and variance in the given equation, the estimate of the test statistic is $\hat{T} = 1.4936$. Either the table of Student's t quantiles of Appendix 4 or the MS Excel built-in function T.INV (0.975;64) returns the critical value $T_{critical} = 1.9977$. Since $\hat{T} < T_{critical}$, the estimate of the test statistic does not fall in the critical region and, thus, the decision is for the nonrejection of the null hypothesis. Alternatively, using the p-value approach, since the Student's t distribution is symmetric, one needs first to calculate the probability $p = 2P(T > |\hat{T}| \, |H_0)$ and then compare it to $\alpha = 0.05$. For $P(T > 1.4936|H_0)$, the MS Excel built-in function T.DIST.RT(1.4936;64), for the right tail of Student's t distribution, returns the value 0.07. Then, $p = 2 \times 0.07 = 0.14$, which is larger than $\alpha = 0.05$, thus confirming the decision of not rejecting the null hypothesis. This decision should be interpreted as follows: based on the available sample, there is no empirical evidence that the population mean differs from 646 mm and, if 100 such tests had been performed in identical conditions, no more than 5 would have led to a different conclusion.

Example 7.2 Solve the Example 7.1, assuming the population variance σ^2 is known and equal to 13045.89 mm^2.

Solution The added information concerning the population variance has the important effect of changing the test statistic and, therefore, its probability distribution and the decision rules for the test. For this example, it is still a two-sided test at the significance level of $100\alpha = 5\%$, but the test statistic changes to $Z = \frac{\bar{X} - 646}{\sigma/\sqrt{N}}$, which follows a standard normal distribution. Substituting the sample mean and the population variance in the given equation, the estimate of the test statistic is $\hat{Z} = 1.4971$. Table 5.1 of Chap. 5 gives $Z_{critical} = z_{0.975} = 1.96$. As $\hat{Z} < Z_{critical}$, the null hypothesis H_0 should not be rejected in favor of H_1. Because in both cases N is a relatively large number, the decision and the values for the test statistics do not differ much for Examples 7.1 and 7.2. However, for sample sizes of less than 30, the differences might begin to increase considerably due to the distinct probability distributions of the test statistic, namely the Student's t and the standard normal.

7.3.2   Parametric Tests for the Means of Two Normal Populations

The underlying assumption for the hypotheses tests described in this subsection is that the simple random samples $\{x_1, x_2, \ldots, x_N\}$ and $\{y_1, y_2, \ldots, y_M\}$, of sizes N and M, have been drawn from normal populations with unknown means μ_X and μ_Y, respectively. The distributions' other parameters, namely, the variances σ_X^2 and σ_Y^2, play an important role in testing hypotheses on the means. Whether or not the populations' variances are known and/or equal determine the test statistic to be used in the test. Tests described in this subsection are two-sided, but they can be easily modified to one-sided, by altering the alternative hypothesis H_1 and the significance level α.

- H_0: $\mu_X - \mu_Y = \delta$ against H_1: $\mu_X - \mu_Y \neq \delta$.
  Populations' variances σ_X^2 and σ_Y^2: known

  Test statistic: $Z = \dfrac{(\bar{X}-\bar{Y})-\delta}{\sqrt{\frac{\sigma_X^2}{N}+\frac{\sigma_Y^2}{M}}}$

  Probability distribution of the test statistic: standard normal $N(0,1)$
  Test type: two-sided at significance level α
  Decision:

  Reject H_0 if $\left| \dfrac{(\bar{X}-\bar{Y})-\delta}{\sqrt{\frac{\sigma_X^2}{N}+\frac{\sigma_Y^2}{M}}} \right| > z_{1-\frac{\alpha}{2}}$

- H_0: $\mu_X - \mu_Y = \delta$ against H_1: $\mu_X - \mu_Y \neq \delta$.
  Populations' variances σ_X^2 and σ_Y^2: equal, unknown, and estimated by s_X^2 and s_Y^2.

  Test statistic: $T = \dfrac{(\bar{X}-\bar{Y})-\delta}{\sqrt{(N-1)s_X^2+(M-1)s_Y^2}} \sqrt{\dfrac{NM(N+M-2)}{N+M}}$

  Probability distribution of the test statistic: Student's t with $\nu = N+M-2$
  Test type: two-sided at significance level α
  Decision:

  Reject H_0 if $\left| \dfrac{(\bar{X}-\bar{Y})-\delta}{\sqrt{(N-1)s_X^2+(M-1)s_Y^2}} \sqrt{\dfrac{NM(N+M-2)}{N+M}} \right| > t_{1-\frac{\alpha}{2}, \nu=N+M-2}$

- H_0: $\mu_X - \mu_Y = \delta$ against H_1: $\mu_X - \mu_Y \neq \delta$.

Populations' variances σ_X^2 and σ_Y^2: unequal, unknown, and estimated by s_X^2 and s_Y^2.

Test statistic: $T = \dfrac{(\overline{X}-\overline{Y})-\delta}{\sqrt{(s_X^2/N)+(s_Y^2/M)}}$

Probability distribution of the test statistic: according to Casella and Berger (1990) the test statistic distribution is approximated by the Student's t, with ν given by

$$\nu = \frac{\left[(s_X^2/N)+(s_Y^2/M)\right]^2}{\left[\dfrac{(s_X^2/N)^2}{N-1}+\dfrac{(s_Y^2/M)^2}{M-1}\right]} \text{ degrees of freedom.}$$

Test type: two-sided at significance level α

Decision:

Reject H_0 if $\left| \dfrac{(\overline{X}-\overline{Y})-\delta}{\sqrt{(s_X^2/N)+(s_Y^2/M)}} \right| > t_{1-\frac{\alpha}{2},\nu}$

Example 7.3 Consider the annual total rainfalls recorded since 1767, at the Radcliffe Meteorological Station, in Oxford (England), retrievable from the URL http://www.geog.ox.ac.uk/research/climate/rms/rain.html. Split the sample into two subsamples of equal sizes: one, denoted by X, for the period 1767 to 1890, and the other, denoted by Y, from 1891 to 2014. Test the hypothesis, at $100\alpha = 5\%$, that the mean annual rainfall depths, for the time periods 1767–1892 and 1893–2014, do not differ significantly.

Solution Assume that, for the time periods 1767–1890 and 1891–2014, annual total rainfalls are normally distributed, with respective means μ_X and μ_Y, and unequal and unknown variances σ_X^2 and σ_Y^2. The subsamples X and Y, each with 124 elements, yield respectively $\overline{x} = 637.17$ mm and $s_X^2 = 13336.46$ mm^2, and $\overline{y} = 654.75$ mm and $s_Y^2 = 12705.63$ mm^2. For this example, the null hypothesis is H_0:$\{\mu_X - \mu_Y = \delta = 0\}$, which is to be tested against the alternative hypothesis H_1: $\mu_X - \mu_Y = \delta \neq 0$. As the variances are assumed unequal and unknown, the test statistic should be $T = \dfrac{(\overline{x}-\overline{y})}{\sqrt{((s_X^2/124))+(s_Y^2/124)}}$, whose probability distribution can be approximated by the Student's t, with $\nu = \dfrac{123\left[(s_X^2/124)+(s_Y^2/124)\right]^2}{(s_X^2/124)^2+(s_Y^2/124)^2} = 246$ degrees of freedom. Substituting the sample values into the test statistic equation, the result is $|T| = 1.213$. The MS Excel function T.INV(0.975; 246) returns $t_{0.975,\nu=246} = 1.97$. As $1.213 < 1.97$, the null hypothesis H_0 cannot be rejected in favor of H_1. Now, solve this example, by assuming (1) the populations' variances are known; (2) the population's variances are equal yet unknown; (3) the sample has been split into subsamples V, from 1767 to 1899, and U, from 1900 to 2014; and (4) only the last 80 years of records are available. Comment on the differences found in testing these hypotheses.

7.3.3   Parametric Tests for the Variance of a Single Normal Population

The underlying assumption for the hypotheses tests described in this subsection is that the simple random sample $\{x_1, x_2, \ldots, x_N\}$, has been drawn from a normal population with unknown variance σ^2. Whether or not the population's mean μ is known determines the test statistic to be used. Tests described in this subsection are two-sided, but can be easily modified to one-sided, by altering the alternative hypothesis H_1 and the significance level α.

- $H_0: \sigma^2 = \sigma_0^2$ against $H_1: \sigma^2 \neq \sigma_0^2$. Population mean μ: known.

  Test statistic: $Q = \dfrac{\sum_{i=1}^{N}(X_i - \mu)^2}{\sigma_0^2} = N \dfrac{s_{\bar{x}}^2}{\sigma_0^2}$

  Probability distribution of the test statistic: χ^2 with $\nu = N$
  Test type: two-sided at significance level α
  Decision:
  Reject H_0 if $N \dfrac{s_{\bar{x}}^2}{\sigma_0^2} < \chi^2_{\frac{\alpha}{2}, N}$ or if $N \dfrac{s_{\bar{x}}^2}{\sigma_0^2} > \chi^2_{1-\frac{\alpha}{2}, N}$

- $H_0: \sigma^2 = \sigma_0^2$ against $H_1: \sigma^2 \neq \sigma_0^2$. Population mean μ: unknown. Estimated by $\overline{X}$.

  Test statistic: $K = \dfrac{\sum_{i=1}^{N}(X_i - \overline{X})^2}{\sigma_0^2} = (N-1) \dfrac{s_{\bar{x}}^2}{\sigma_0^2}$

  Probability distribution of the test statistic: χ^2 with $\nu = N - 1$
  Test type: two-sided at significance level α
  Decision:
  Reject H_0 if $(N-1) \dfrac{s_{\bar{x}}^2}{\sigma_0^2} < \chi^2_{\frac{\alpha}{2}, N-1}$ or if $(N-1) \dfrac{s_{\bar{x}}^2}{\sigma_0^2} > \chi^2_{1-\frac{\alpha}{2}, N-1}$

Example 7.4 Consider again the annual rainfall data observed at the Radcliffe Meteorological Station, in Oxford (England). Assume that the available sample begins in the year 1950 and ends in 2014. Test the null hypothesis that the population variance σ_0^2 is 13,000 mm^2 against the alternative $H_1 : \{\sigma_1^2 < 13,000\,\text{mm}^2\}$, at $100\alpha = 5\%$.

Solution Again, the basic assumption is that the annual total rainfall depths are normally distributed. The sample of size $N = 65$ yields $\bar{x} = 667.21$ mm and $s_X^2 = 13108.22$ mm^2, with no other information regarding the population mean, which must then be estimated by the sample mean. For this example, the null hypothesis $H_0: \{\sigma_0^2 = 13000\}$ should be tested against $H_1 : \{\sigma_1^2 < 13,000\}$, which is a one-sided lower tail test, with the test statistic $K = (N-1)s_x^2/\sigma_0^2$ following a chi-square distribution with 64 degrees of freedom. Substituting the sample estimates into the test statistic equation, it follows that $K = 64.53$. The MS Excel built-in function CHISQ.INV(0.05;64) returns $\chi^2_{0.05, 64} = 44.59$. As $64.53 > 44.59$, the null hypothesis H_0 should not be rejected in favor of H_1, at $100\alpha = 5\%$. Now, solve this example, by assuming (1) the population mean is known and equal to 646 mm;

(2) the alternative hypothesis has changed to $\sigma_1^2 > 13,000 \, \text{mm}^2$; and (3) the alternative hypothesis has changed to $\sigma_1^2 \neq 13,000 \, \text{mm}^2$. Use the p-value approach to make your decisions.

7.3.4 Parametric Tests for the Variances of Two Normal Populations

The underlying assumption for the hypotheses tests described in this subsection is that the simple random samples $\{x_1, x_2, \ldots, x_N\}$ and $\{y_1, y_2, \ldots, y_M\}$, of sizes N and M, have been drawn from normal populations with unknown variances σ_X^2 and σ_Y^2, respectively. Whether or not the populations' means μ_X and μ_Y are known determines the number of degrees of freedom for the test statistic to be used in the test. Tests described in this subsection are two-sided, but can be easily modified to one-sided, by altering the alternative hypothesis H_1 and the significance level α.

- $H_0: \frac{\sigma_X^2}{\sigma_Y^2} = 1$ against $H_1: \frac{\sigma_X^2}{\sigma_Y^2} \neq 1$

  Populations' means μ_X and μ_Y: known

  Test statistic: $\varphi = \frac{s_X^2/\sigma_X^2}{s_Y^2/\sigma_Y^2}$

  Probability distribution of the test statistic: Snedecor's F, with $\nu_1 = N$ and $\nu_2 = M$

  Test type: two-sided at significance level α

  Decision:

  Reject H_0 if $\varphi < F_{N,M,\alpha/2}$ or if $\varphi > F_{N,M,1-\alpha/2}$

- $H_0: \frac{\sigma_X^2}{\sigma_Y^2} = 1$ against $H_1: \frac{\sigma_X^2}{\sigma_Y^2} \neq 1$

  Populations' means μ_X and μ_Y: unknown and estimated by $\bar{x}$ and $\bar{y}$

  Test statistic: $f = \frac{s_X^2/\sigma_X^2}{s_Y^2/\sigma_Y^2}$

  Probability distribution of the test statistic: Snedecor's F, with $\nu_1 = N - 1$ and $\nu_2 = M - 1$

  Test type: two-sided at significance level α

  Decision:

  Reject H_0 if $f < F_{N-1,M-1,\alpha/2}$ or if $f > F_{N-1,M-1,1-\alpha/2}$

Example 7.5 A constituent dissolved in the outflow from a sewage system has been analyzed 7 and 9 times through procedures X and Y, respectively. Test results for procedures X and Y yielded standard deviations $s_X = 1.9$ and $s_Y = 0.8 \, \text{mg/l}$, respectively. Test the null hypothesis that procedure Y is more precise than procedure X, at $100\alpha = 5\%$ (adapted from Kottegoda and Rosso 1997)

Solution Assuming data have been drawn from two normal populations, the null hypothesis to be tested is $H_0: \frac{\sigma_X^2}{\sigma_Y^2} = 1$ against the alternative $H_1: \begin{cases} \frac{\sigma_X^2}{\sigma_Y^2} > \end{cases}$

1 or $\sigma_X^2 > \sigma_Y^2$}. Therefore, this is a one-sided upper tail test at $\alpha = 0.05$. The test statistic is $f = \frac{s_X^2/\sigma_X^2}{s_Y^2/\sigma_Y^2}$, following the Snedecor's F distribution with $\gamma_1 = 7 - 1 = 6$ and $\gamma_2 = 9 - 1 = 8$ degrees of freedom for the numerator and the denominator, respectively. Substituting the sample values into the test statistic equation, it results in $f = 5.64$. The table of F quantiles, in Appendix 5, reads $F_{6,8,0.95} = 3.58$. As $5.64 > 3.58$, the decision is for rejecting the null hypothesis in favor of the alternative hypothesis, at $\alpha = 005$. In other words, the empirical evidence has shown that procedure Y is more precise, as the variance associated with its results is smaller than that of the results from the competing procedure X.

7.4   Some Nonparametric Tests Useful in Hydrology

The previously described parametric tests require that the underlying probability distributions of the populations of variables X and Y be normal. In fact, as long as a random variable is normally distributed, it is possible to derive its exact sampling distributions and, therefore, the probability distributions for the most common test statistics. However, as one attempts to apply parametric tests for nonnormal populations, the main consequence will fall upon the significance level, with the violation of its nominal specified value α. For instance, if T denotes the test statistic $T = (\overline{X} - \mu_0)/s_X/\sqrt{N}$ for a random variable X, whose probabilistic behavior departs from normality, the true probability of committing a type I error will not be necessarily equal to the nominal value α. If such a situation arises, one could write

$$\int_{-\infty}^{-t_{\alpha/2}} f_T(t|H_0)dt + \int_{t_{\alpha/2}}^{\infty} f_T(t|H_0)dt \neq \alpha \qquad (7.6)$$

where $f_T(t)$ represents the true yet unknown probability distribution of the test statistic $T = (\overline{X} - \mu_0)/s_X/\sqrt{N}$, for non-Gaussian X.

The science of Mathematical Statistics presents two possible solutions for the problem posed by Eq. (7.6). The first refers to demonstrating, through Monte Carlo simulations, that even if a random variable X departs from normality, the true density $f_T(t|H_0)$, of the test statistic, might behave similarly to the assumed density under normality. For instance, Larsen and Marx (1986) give some examples showing that if the actual probability distribution of X is not too skewed or if the sample size is not too small, the Student's t distribution can reasonably approximate the true density $f_T(t|H_0)$, for testing hypotheses on the population mean of X. In such cases, one is able to state that Student's t is *robust*, with respect to moderate departures from normality. Given the usually highly skewed probability distributions associated with hydrologic variables and the usual small size samples, such a possible solution to the problem posed by Eq. (7.6) seems to be of limited application in Statistical Hydrology.

The second possible solution to the problem posed by Eq. (7.6), is given by the possibility of substituting the conventional test statistics by others, whose probability distributions remain invariant, under H_0, to the diverse shapes the underlying population distribution can take. The inferential statistical procedures that exhibit such characteristics, with particular emphasis on hypotheses tests, are termed nonparametric. The general elements of setting up a hypothesis test, described in Sects. 7.1 and 7.2, remain the same for the nonparametric tests. The specific feature that differentiates nonparametric from parametric tests lies in conceiving the test statistic as invariant to the probability distribution of the original random variable. In fact, the test statistics for most nonparametric tests are based on characteristics that can be derived from the original data but do not directly include the data values in the calculations. These characteristics can be, for instance, the number of positive and negative differences between the hypothesized median and the sample values, or the correlation coefficient between the ranking orders of data from two samples, or the number of turning points along a sequence of time indices, among many others.

The number and variety of nonparametric tests have increased substantially since their introduction in the early 1940s. This section does not have the objective of advancing through the mathematical foundations of nonparametric statistics and nonparametric tests. The reader interested in those details should consult specialized references, such as Siegel (1956), Gibbons (1971), and Hollander and Wolfe (1973). The subsections that follow contain descriptions, followed by worked out examples, of the main nonparametric tests employed in Statistical Hydrology. These tests are intended to check on whether or not the fundamental assumptions that are necessary for hydrologic frequency analysis hold for a given sample. Recall that the basic premise that allows the application of statistical methods to a collection of data, reduced from a hydrologic time series, is that it is a simple random sample, drawn from a single population, whose probability distribution is not known. This basic premise reveals the implicit assumptions of randomness, independence, homogeneity, and stationarity that a hydrologic sample must hold. For the usually short samples of skewed hydrologic data, these implicit assumptions can be tested through the use of nonparametric tests.

7.4.1 Testing the Randomness Hypothesis

In the context of Statistical Hydrology, the essential assumption is that a sample is drawn at random from the population, with an equal chance of independently drawing each of its elements. The sample values are thought of as realizations of random variables, representing the 1st draw, the 2nd draw, and so forth up to the Nth draw. A sample such as this is being referred to throughout this book as simple random and the related stochastic process as purely random. Nonrandomness in a sample of a hydrologic variable can arise from statistical dependence among its elements, nonhomogeneities, and nonstationarities. Causes can be either natural or man-made. Natural causes are mostly related to climate fluctuations, earthquakes, and other disasters, whereas anthropogenic causes are associated with land-use changes, construction of large-reservoir dams upstream, and human-induced climate change. There are a number of tests specifically designed for detecting serial dependence, nonhomogeneities and nonstationarities. In this subsection, a general test for randomness is described.

The general assumption of randomness for a sample of a hydrologic variable cannot be unequivocally demonstrated, but it can be checked through a nonparametric hypothesis test. NERC (1975) suggests that the hypothesis that data have been sampled at random can be tested by counting the number of turning points they make throughout time. A turning point is either a peak or a trough in a time plot of the concerned hydrologic variable. The heuristics of the test is that too many or too few turning points are indicative of possible nonrandomness.

For a purely random stochastic process, as realized by a time series of N observed values, it can be shown that the expected number of turning points, denoted by p, is given by

$$E[p] = \frac{2(N-2)}{3} \tag{7.7}$$

with variance approximated as

$$\mathrm{Var}[p] = \frac{16N - 29}{90} \tag{7.8}$$

For large samples, Yule and Kendall (1950) proved that the distribution of p can be approximated by a normal distribution. Thus, for a large sample and under the null hypothesis H_0:{the sample data are random}, the standardized test statistic

$$T = \frac{p - E[p]}{\sqrt{\mathrm{Var}[p]}} \tag{7.9}$$

follows a standard normal distribution. For a relatively large sample of a hydrologic variable, say $N > 30$, the number of peaks or troughs p is counted on a graph of the sample values against their respective occurrence times or time indices. The sample estimate for the standardized test statistic is designated by $\hat{T}$. As a two-sided test at the significance level α, the decision would be to reject the null hypothesis, if $|\hat{T}| > z_{1-\alpha/2}$. See Example 7.6 for an application of this test.

7.4.2 Testing the Independence Hypothesis

The term independence essentially means that no observed value in a sample can affect the occurrence or the non-occurrence of any other sample element. In the context of hydrologic variables, the natural water storages in the catchment, for example, can possibly determine the occurrence of high flows, following a sequence of high flows, or, contrarily, persistent low flows, following a sequence of low flows. The statistical dependence between time-contiguous flows in a sample, for the given example, will highly depend on the time interval that separates the consecutive data: strong dependence is expected for daily intervals, whereas weak or no dependence is expected for seasonal or annual intervals. As frequently mentioned throughout this book, Statistical Hydrology deals mostly with yearly values, abstracted as annual means, totals, maxima, or minima, from the historical series of hydrologic data. Samples abstracted for monthly, seasonal and other time intervals, as well as non-annual extreme data samples, such as in partial duration series, can also be employed in Statistical Hydrology. In any case, however, the assumption of statistical independence among the sample data must be previously checked. A simple nonparametric test that is frequently used for such a purpose is the one proposed by Wald and Wolfowitz (1943), to be described next in this subsection.

Let $\{x_1, x_2, \ldots, x_N\}$ represent a sample of size N from X, and $\{x'_1, x'_2, \ldots, x'_N\}$ denote the sequence of differences between the ith sample value x_i and the sample mean $\bar{x}$. The test statistic for the Wald–Wolfowitz nonparametric test is calculated as

$$R = \sum_{i=1}^{N-1} x'_i x'_{i+1} + x'_1 x'_N \qquad (7.10)$$

Notice that the statistic R is a quantity that is proportional to the serial correlation coefficient between the successive elements of the sample.

For large N and under the null hypothesis H_0, that sample data are independent, Wald and Wolfowitz (1943) proved that R follows a normal distribution with mean

$$E[R] = -\frac{s_2}{N-1} \qquad (7.11)$$

and variance given by

$$\mathrm{Var}[R] = \frac{s_2^2 - s_4}{N-1} + \frac{s_2^2 - 2s_4}{(N-1)(N-2)} - \frac{s_2^2}{(N-1)^2} \qquad (7.12)$$

where the sample estimate of s_r is $\hat{s}_r = \sum_{i=1}^{N} (x'_i)^r$.

Hence, under the null hypothesis H_0:{sample data are independent], the standardized test statistic for the Wald–Wolfowitz test is

$$T = \frac{R - E[R]}{\sqrt{\text{Var}[R]}} \tag{7.13}$$

which follows a standard normal distribution. The sample estimate for the standardized test statistic is designated by $\hat{T}$. As a two-sided test at the significance level α, the decision would be to reject the null hypothesis, if $|\hat{T}| > z_{1-\alpha/2}$. As a test based on the serial correlation coefficient, the Wald–Wolfowitz test can also be used to check for non-stationarity, as introduced by linear trends throughout time (see Bobée and Ashkar 1991, WMO 2009). See Example 7.6 for an application of this test.

7.4.3   Testing the Homogeneity Hypothesis

In essence, the term homogeneity refers to the attribute that all elements of a sample come from a single population. A flood data sample, for instance, may eventually consist of floods produced by ordinary rainfalls, of moderate intensities and volumes, and by floods produced by extraordinary rainfalls, of high intensities and volumes, associated with particularly extreme hydrometeorological conditions, such as the passage of hurricanes or typhoons over the catchment. For such a case, there are certainly two different populations, as distinguished by the flood producing mechanism, and one would end up with a nonhomogeneous (heterogeneous) flood data sample. Bobée and Ashkar (1991) point out, however, that the variability usually present in extreme data samples, such as flood and maximum rainfall data, is, in some cases, so high that can make the task of deciding on its homogeneity difficult. In general, it is much easier to detect heterogeneities in samples of annual totals or annual mean values than in extreme data samples.

Heterogeneities may also be associated with flow data, observed at some gauging station, before and after the construction of a large-reservoir dam upstream, since highly regulated flows, as compared to natural flows, will necessarily have a larger mean with a smaller variance. The decision on whether or not a data sample can be considered homogenous is often made with the help of a nonparametric test proposed by Mann and Whitney (1947), to be described next in this subsection.

Given a data sample $\{x_1, x_2, \ldots, x_N\}$, of size N, one first needs to split it into two subsamples: $\{x_1, x_2, \ldots, x_{N_1}\}$ of size N_1 and $\{x_{N_1+1}, x_{N_1+2}, \ldots, x_N\}$ of size N_2, such that $N_1 + N_2 = N$ and that N_1 and N_2 are approximately equal, with $N_1 \leq N_2$. The next step is to rank the complete sample, of size N, in ascending order, noting each sample element's ranking order m and whether it comes from the first or the second subsample. The intuitive idea behind the Mann–Whitney test is that, if the sample is not homogeneous, the ranking orders of the elements coming from the first sub-sample will be consistently higher (or lower) than those of the second subsample.

The Mann–Whitney test statistic is given by the lowest value V between the quantities

$$V_1 = N_1 N_2 + \frac{N_1(N_1 + 1)}{2} - R_1 \qquad (7.14)$$

and

$$V_2 = N_1 N_2 - V_1 \qquad (7.15)$$

where R_1 denotes the sum of the ranking orders of all elements from the first subsample.

For N_1 and N_2 both larger than 20 and under the null hypothesis H_0:{sample is homogeneous}, Mann and Whitney (1947) proved that V follows a normal distribution with mean

$$E[V] = \frac{N_1 N_2}{2} \qquad (7.16)$$

and variance given by

$$\mathrm{Var}[V] = \frac{N_1 N_2 (N_1 + N_2 + 1)}{12} \qquad (7.17)$$

Hence, under the null hypothesis H_0:{sample is homogeneous}, the standardized test statistic for the Mann–Whitney test is

$$T = \frac{V - E[V]}{\sqrt{\mathrm{Var}[V]}} \qquad (7.18)$$

which follows a standard normal distribution. The sample estimate for the standardized test statistic is designated by $\hat{T}$. As a two-sided test at the significance level α, the decision must be to reject the null hypothesis, if $\left|\hat{T}\right| > z_{1-\alpha/2}$. See Example 7.6 for an application of this test.

A possible shortcoming of applying the Mann–Whitney test in identifying heterogeneities in a hydrologic sample relates to choosing the point at which the complete sample is split into two subsamples, determining not only the hypothetical breakpoint but also the subsamples sizes. These and the breakpoint may not coincide with the actual duration and onset of heterogeneities that may exist in the sample. An interesting modification of the Mann–Whitney test, introduced as an algorithm by Mauget (2011), samples data rankings over running time windows, of different widths, thus allowing the identification of heterogeneities of arbitrary onset and duration. The approach was successfully applied to annual area-averaged temperature data for the continental USA and seems to be a promising alternative for identifying heterogeneities in large samples of hydrologic variables.

7.4.4 Testing the Stationarity Hypothesis

The term stationarity refers to the notion that the essential statistical properties of the sample data, including their probability distribution and related parameters, are invariant with respect to time. Nonstationarities include monotonic and nonmonotonic trends in time or an abrupt change (or a jump) at some point in time. The flow regulation by a large human-made reservoir can drastically change the statistical properties of natural flows and, thus, introduce an abrupt change into the time series of the wet-season mean flows, in particular, at the time the reservoir initiated its operation. An example of such a jump in flow series is given in Chap. 12, along with the description of the hypothesis test proposed by Pettitt (1979), specifically designed to detect and identify abrupt changes in time series.

Nonmonotonic trends in hydrologic time series are generally related to climate fluctuations operating at interannual, decadal, or multidecadal scales. Climate fluctuations can occur in cycles, such as the solar activity cycle of approximately 11 years, with changes in total solar irradiance and sunspots. Other climate oscillations occur with a quasi-periodic frequency, such as ENSO, the El Niño Southern Oscillation, which recurs every 2–7 years, causing substantial changes in heat fluxes over the continents, oceans, and atmosphere. Detecting and modeling the influence of quasi-periodic climate oscillations on hydrologic variables are complex endeavors and require appropriate specific methods. Chapter 12 presents some methods for modeling nonstationary hydrologic variables, under the influence of quasi-periodic climate oscillations.

Monotonic trends in hydrologic time series are associated with gradual changes taking place in the catchment. A monotonic trend can possibly appear in the flow time series of a small catchment that has experienced, over the years, a slow time-evolving urbanization process. Gradual changes in temperature or precipitation over an area, natural or human-induced, can also unfold a monotonic trend in hydrologic time series. As mentioned in Chap. 1, the notion of heterogeneity, as applied particularly to time series, encompass that of nonstationarity: a nonstationary series is nonhomogeneous with respect to time, although a nonhomogeneous (or heterogeneous) series is not necessarily nonstationary. The decision on whether or not a data sample can be considered stationary, with respect to monotonic linear or nonlinear trends, can be made with the help of a nonparametric test based on Spearman's rank order correlation coefficient.

The Spearman's rank order correlation coefficient, denoted by ρ and named after the British psychologist Charles Spearman (1863–1945), is a nonparametric measure of the statistical dependence between two random variables. In the context of Spearman's ρ, statistical dependence is not restricted to linear dependence, as in conventional Pearson's linear correlation coefficient. The essential idea of the test based on Spearman's ρ is that a monotonic trend, linear or nonlinear, hidden in a hydrologic time series X_t, evolving in time t, can be detected by measuring the degree of correlation between the rank orders m_t, for the sequence X_t, and the corresponding time indices T_t, for $T_t = 1, 2, \ldots, N$. The Spearman's rank order correlation coefficient is formally given by

$$r_S = 1 - \frac{6 \sum_{t=1}^{N} (m_t - T_t)^2}{N^3 - N} \tag{7.19}$$

The test statistic for monotonic trend, based on the Spearman's ρ, is calculated as

$$T = r_S \sqrt{\frac{N-2}{1-r_S^2}} \tag{7.20}$$

According to Siegel (1956), for $N > 10$ and under the null hypothesis of no correlation between m_t and T_t, the probability distribution of T can be approximated by the Student's t distribution, with $(N-2)$ degrees of freedom.

Before computing r_S for a given sample, it is necessary to check whether or not it contains *ties*, or two or more equal values of X_t, for different rank orders. A simple correction that can be made is to assign the average rank to each of tied values, prior to the calculation of r_S. For example, if, for a sample of size 20, the 19th and the 20th ranked values of X_t are equal, both must be assigned to their average rank order of 19.5. The sample estimate for the test statistic is designated by $\hat{T}$. As a two-sided test at the significance level α, the decision would be to reject the null hypothesis, if $|\hat{T}| > t_{1-\alpha/2, N-2}$. See Example 7.6 for an application of this test.

In recent years, a growing interest in climate change has led to a profusion of papers on the topic of detecting nonstationarities in hydrologic time series, including a number of different tests and computer programs to perform the calculations. One nice example of these is the software TREND, available from http://www.toolkit.net.au/tools/TREND [Accessed: 10th February 2016], and developed under the eWater initiative of the University of Canberra, in Australia.

Example 7.6 Consider the flood peak data sample of the Lehigh River at Stoddartsville (USGS gauging station 01447500) for the water-years 1941/42 to 2013/2014, listed in Table 7.1. For this sample, test the following hypotheses: (a) randomness; (b) independence; (c) homogeneity; and (d) stationarity (for the absence of a linear or a nonlinear monotonic trend), at the significance level $\alpha = 0.05$.

Solution

(a) Test of the randomness hypothesis. The plot of annual peak discharges versus time is depicted in Fig. 7.5.

By looking at the graph of Fig. 7.5, one can count the number of turning points as $p = 53$, with 27 troughs and 26 peaks. For $N = 73$, Eqs. (7.7) and (7.8) give the estimates of $E[p]$ and $Var[p]$ respectively equal to 47.33 and 12.66. With these values in Eq. (7.9), the estimate of the standardized test statistic is $\hat{T} = 1.593$. At the significance level $\alpha = 0.05$, the critical value of the test statistic is $z_{0.975} = 1.96$. Therefore, as $|\hat{T}| < z_{0.975}$, the decision is not to reject

Table 7.1 Annual peak discharges of the Lehigh River at Stoddartsville (m^3/s) and auxiliary quantities for performing the nonparametric tests of Wald–Wolfowitz, Mann–Whitney, and Spearman's ρ

Water year	T_t	X_t	m_t	Subsample	$x_i' = X_t - \bar{x}$	Ranked X_t
1941/42	1	445	72	1	341.9	14.0
1942/43	2	70.0 (estimated)	36	1	−32.7	15.9
1943/44	3	79.3	43.5	1	−23.4	27.2
1944/45	4	87.5	47	1	−15.2	27.6
1945/46	5	74.8	42	1	−27.9	28.1
1946/47	6	159	63	1	55.8	28.3
1947/48	7	55.2	26	1	−47.5	29.5
1948/49	8	70.5	37	1	−32.2	29.5
1949/50	9	53.0	25	1	−49.7	31.7
1950/51	10	205	66	1	102.6	32.3
1951/52	11	65.4	31	1	−37.3	33.1
1952/53	12	103	56	1	0.38	33.4
1953/54	13	34.0	13	1	−68.7	34.0
1954/55	14	903	73	1	800.6	38.2
1955/56	15	132	61	1	28.9	38.2
1956/57	16	38.5	16	1	−64.2	38.5
1957/58	17	101	55	1	−2.17	41.1
1958/59	18	52.4	24	1	−50.3	42.8
1959/60	19	97.4	53.5	1	−5.28	45.6
1960/61	20	46.4	21	1	−56.3	45.6
1961/62	21	31.7	9	1	−71.0	46.4
1962/63	22	62.9	29	1	−39.8	47.0
1963/64	23	64.3	30	1	−38.4	51.8
1964/65	24	14.0	1	1	−88.7	52.4
1965/66	25	15.9	2	1	−86.8	53.0
1966/67	26	28.3	6	1	−74.4	55.2
1967/68	27	27.2	3	1	−75.5	56.4
1968/69	28	47.0	22	1	−55.7	60.9
1969/70	29	51.8	23	1	−50.9	62.9
1970/71	30	33.4	12	1	−69.3	64.3
1971/72	31	90.9	49	1	−11.8	65.4
1972/73	32	218	67	1	115.4	66.0
1973/74	33	80.4	45	1	−22.3	68.5
1974/75	34	60.9	28	1	−41.8	68.5
1975/76	35	68.5	33.5	1	−34.2	69.9
1976/77	36	56.4	27 Sum = 1247.5	1	−46.3	70.0
1977/78	37	84.7	46	2	−18.0	70.5
1978/79	38	69.9	35	2	−32.8	72.5
1979/80	39	66.0	32	2	−36.7	73.1

(continued)

Table 7.1 (continued)

Water year	T_t	X_t	m_t	Subsample	$x_i' = X_t - \bar{x}$	Ranked X_t
1980/81	40	38.2	14.5	2	−64.5	73.1
1981/82	41	32.3	10	2	−70.4	74.2
1982/83	42	119	60	2	16.0	74.8
1983/84	43	105	58	2	2.08	79.3
1984/85	44	237	68	2	134.6	79.3
1985/86	45	117	59	2	14.0	80.4
1986/87	46	95.7	51	2	−6.98	84.7
1987/88	47	41.1	17	2	−61.0	87.5
1988/89	48	88.6	48	2	−14.1	88.6
1989/90	49	29.5	7.5	2	−73.2	90.9
1990/91	50	45.6	19.5	2	−57.1	95.4
1991/92	51	28.1	5	2	−74.6	95.7
1992/93	52	104	57	2	1.52	96.6
1993/94	53	68.5	33.5	2	−34.2	97.4
1994/95	54	29.5	7.5	2	−73.2	97.4
1995/96	55	204	65	2	100.9	101
1996/97	56	97.4	53.5	2	−5.28	103
1997/98	57	38.2	14.5	2	−64.5	104
1998/99	58	72.5	38	2	−30.2	105
1999/2000	59	33.1	11	2	−69.6	117
2000/01	60	45.6	19.5	2	−57.1	119
2001/02	61	154	62	2	51.4	132
2002/03	62	79.3	43.5	2	−23.4	154
2003/04	63	289	69	2	186.1	159
2004/05	64	184	64	2	81.4	184
2005/06	65	297	70.5	2	194.6	204
2006/07	66	73.1	39.5	2	−29.6	205
2007/08	67	96.6	52	2	−6.13	218
2008/09	68	73.1	39.5	2	−29.6	237
2009/10	69	74.2	41	2	−28.5	289
2010/11	70	297	70.5	2	194.6	297
2011/12	71	27.6	4	2	−75.1	297
2012/13	72	42.8	18	2	−60.0	445
2013/14	73	95.4	50 Sum = 1453.5	2	−7.26	903

the null hypothesis H_0 that observed data have been sampled at random from the population.

(b) Test of the independence hypothesis. The sixth column of Table 7.1 lists the differences between each flood peak X_t and the complete-sample mean $\bar{x} = 102.69 \ \mathrm{m}^3/\mathrm{s}$. These are the main values to calculate the Wald–Wolfowitz test statistic by means of Eq. (7.10). The result is $R = 14645.00$. The differences

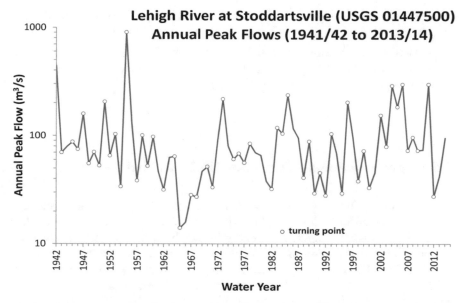

Fig. 7.5 Graph of annual peak discharges versus water-year for the Lehigh River at Stoddartsville (USGS gauging station 01447500)

$x_i' = X_t - \bar{x}$ in Table 7.1 also give $s_2 = 1188516.11$ and $s_4 = 443633284532.78$, which, when substituted into Eqs. (7.11) and (7.12), yield the estimates of $E[R]$ and $\mathrm{Var}[R]$ equal to -16507.17 and 13287734392, respectively. With these inserted in Eq. (7.13), the estimated standardized test statistic is $\hat{T} = 0.270$. At the significance level $\alpha = 0.05$, the critical value of the test statistic is $z_{0.975} = 1.96$. Therefore, as $|\hat{T}| < z_{0.975}$, the decision is not to reject the null hypothesis H_0 that observed sample data are independent.

(c) Test of the homogeneity hypothesis. The fourth column of Table 7.1 lists the ranking orders m_t, which are the basic values for calculating the Mann–Whitney test statistics through Eqs. (7.14) and (7.15), alongside the sum of the ranking orders for the 36 points from subsample 1, which is $R_1 = 1247.5$. The test statistic is the smallest value between V_1 and V_2, which, in this case, is $V = V_2 = 608.5$. Substitution of R_1 and V into Eqs. (7.16) and (7.17) give estimates of $E[V]$ and $\mathrm{Var}[V]$ respectively equal to 666 and 8214. With these in Eq. (7.18), the estimated standardized test statistic is $\hat{T} = -0.634$. At the significance level $\alpha = 0.05$, the critical value of the test statistic is $z_{0.975} = 1.96$. Therefore, as $|\hat{T}| < z_{0.975}$, the decision is not to reject the null hypothesis H_0 that observed sample data are homogeneous.

(d) Test of the stationarity hypothesis, for the absence of a monotonic trend. The fourth column of Table 7.1 lists the ranking orders m_t and the second column, the time indices T_t. Both are necessary to calculate the test statistics for the Spearman's ρ test, through Eqs. (7.19) and (7.20). The estimated Spearman's correlation coefficient is $r_S = 0.0367$. With this in Eq. (7.20), the estimated test statistics is $\hat{T} = 0.310$. At the significance level $\alpha = 0.05$, the critical value of the test statistic is $t_{0.975,71} = 1.994$. Therefore, as $\left|\hat{T}\right| < t_{0.975,71}$, the decision is not to reject the null hypothesis H_0 that observed sample data are stationary, with respect to a monotonic trend in time.

7.5 Some Goodness-of-Fit Tests Useful in Hydrology

The previous sections have described procedures to test hypotheses on population parameters and on attributes a simple random sample must have. Another important class of hypothesis tests refers to checking the suitability of a conjectured shape for the probability distribution of the population against the reality imposed by the sample points. This class of hypotheses test is generally referred to as Goodness-of-Fit (GoF) tests. As with any hypothesis test, a GoF test begins by setting up a conjectural statement under the null hypothesis, such as $H_0: \{X$ follows a Poisson distribution with parameter $\nu\}$, which is then tested against the opposite alternative hypothesis H_1. GoF test statistics are generally based on differences between empirical and hypothesized (or expected) frequencies or between empirical and hypothesized (or expected) quantiles.

In the context of Statistical Hydrology, the ubiquitous situation is that the probability distribution of the population is not known from prior knowledge and that inference should be supported mostly by the information contained in the data sample, which is usually of small size. As opposed to other fields of application of statistical inference, in hydrology, the available samples are of fixed sizes and increasing them, by intensifying monitoring programs, might be very expensive, time-demanding, and, sometimes, ineffective. In order to make inferences on the population probability distribution, one can also have some additional help from observing the physical characteristics of the phenomenon being modeled, such as lower and upper bounds, and from deductive arguments, stemming from the central limit theorem and the asymptotic extreme-value theory, as seen in Chap. 5. Discussion of these specific topics is left to the next chapter, which deals with hydrologic frequency analysis. In this section, the focus is on describing the most useful GoF tests.

As seen previously in this chapter, the decision to not reject the null hypothesis does not mean accepting it as true. In fact, such a decision signifies that no sufficient evidence was found in the data sample to disprove the null hypothesis and the probabilities of making incorrect decisions are α and β. In GoF tests, by not rejecting the null hypothesis, one cannot claim the true probability distribution of the population is that under H_0. Furthermore, by applying the same GoF test to different probability distributions under H_0, one cannot discriminate or categorize the candidate distributions by comparing the values of their respective test statistics or p-values, since such a comparison cannot be carried out on an equal basis. The main reasons for that are: (1) the candidate distributions have different number of parameters, possibly estimated by different methods; (2) critical values for the test statistic usually depend on the hypothesized distribution under H_0; (3) the number and extension of partitions of the sample space, as required by some GoF tests, might differ one test from another; (4) type I and type II errors for GoF tests under distinct null hypotheses are of a different nature and not directly comparable to each other; and (5) finally, and perhaps the most important reason, GoF tests have not been conceived for such a comparative analysis.

GoF tests are techniques from inferential statistics that can prove very helpful in deciding on the fit of a single hypothesized probability distribution model to a sample and should be used as such. The main GoF tests that are currently employed in Statistical Hydrology are the chi-square (or chi-squared), Kolmogorov–Smirnov, Anderson–Darling, and the Probability Plot Correlation Coefficient (PPCC). Their descriptions and respective worked out examples are the subject of the subsections that follow.

7.5.1   The Chi-Square (χ^2) GoF Test

Assume $A_1, A_2, \ldots, A_r$ represents a collection of mutually and collectively exhaustive disjoint events, such that their union defines the entire sample space. Assume also the null hypothesis $H_0:\{P(A_i)=p_i,$ for $i=1, 2, \ldots, r\}$, such that $\sum_{i=1}^{r} p_i = 1$. Under these conditions, suppose that, from a number N of random experiments, the absolute frequencies pertaining to events $A_1, A_2, \ldots, A_r$ be given by the quantities $\rho_1, \rho_2, \ldots, \rho_r$, respectively. If the null hypothesis is true, then the joint probability distribution of the variables $\rho_1, \rho_2, \ldots, \rho_r$ is the multinomial (see Sect. 4.3.2, of Chap. 4), with mass function given by

$$P(\rho_1 = O_1, \rho_2 = O_2, \ldots, \rho_r = O_r | H_0) = \frac{N!}{O_1! O_2! \ldots O_r!} p_1^{O_1} p_2^{O_2} \ldots p_r^{O_r} \quad (7.21)$$

where $\sum_{i=1}^{r} O_i = N$.

Consider, then, the following statistic:

$$\chi^2 = \sum_{i=1}^{r} \frac{(O_i - Np_i)^2}{E_i} = \sum_{i=1}^{r} \frac{(O_i - E_i)^2}{E_i} \quad (7.22)$$

defined by the realizations O_i, of variables ρ_i, and by their respective expected values $E_i = E[\rho_i]$, which, under the null hypothesis, are equal to Np_i. Hence, the statistic χ^2 expresses the sum of the squared differences between the realizations of the random variables ρ_i and their respective expected values.

In Sect. 5.10.1, of Chap. 5, it is posited that the sum of the squared differences between N independent and normally distributed variables, and their common mean value μ, follows the χ^2 with $\nu = N$ degrees of freedom. Although the similarity between the chi-square variate and the statistic given in Eq. (7.22) may appear evident, the latter contains the sum of r variables that are not necessarily independent and normally distributed. However, the asymptotic theory of statistics and probability comes to our assistance, once again, showing that, as N tends to infinity, the statistic defined by Eq. (7.22) tends to follow the chi-square distribution, with $\nu = (r-1)$ degrees of freedom. In formal terms,

$$\lim_{N \to \infty} P\left(\chi^2 < x | H_0\right) = \int_0^x \frac{x^{(r-3)/2} e^{-x/2}}{2^{(r-1)/2} \Gamma[(r-1)/2]} dx \qquad (7.23)$$

Thus, for large N, one can employ such a result to test the null hypothesis H_0, under which the expected relative frequencies of variables ρ_i be given by Np_i, with p_i calculated by the hypothesized probability distribution. A high value of the χ^2 statistic would reveal large differences between observed and expected frequencies, and a poor fit by the distribution hypothesized under H_0. Otherwise, a low value of χ^2 would be indicative of a good fit.

It is instructive to note that the limiting distribution, as given in Eq. (7.23), does not depend on p_i. In fact, it depends only on the number of partitions r of the sample space. This is a positive feature of the chi-square test as the generic probability distribution of the test statistic remains unchanged for distinct null hypotheses, provided the number of partitions r had been correctly specified. In practice, the chi-square GoF test usually provides good results for $N > 50$ and for $Np_i \geq 5$, with $i = 1, 2, \ldots, r$. If the probabilities p_i are calculated from a distribution with k estimated parameters, it is said that k degrees of freedom have been lost. In other terms, in such a case, parameter ν, of the probability distribution of the test statistic χ^2, would be $\nu = (r-k-1)$. The chi-square GoF test is a one-sided upper tail test and the decision would be for rejecting $H_0:\{X$ follows a hypothesized probability distribution$\}$ if $\hat{\chi}^2 > \chi^2_{1-\alpha, \nu}$. Examples 7.7 and 7.8 illustrate applications of the chi-square GoF test for discrete and continuous random variables, respectively.

Example 7.7 A water treatment plant withdraws raw water directly from a river through a simply screened intake installed at a low water level. Assume the discrete random variable X refers to the annual number of days the river water level is below the intake's level. Table 7.2 shows the empirical frequency distribution of X, based

Table 7.2 Empirical frequencies for the annual number of days the intake remains dry

$x \to$	0	1	2	3	4	5	6	7	8	>8
$f(x_i)$	0.0	0.06	0.18	0.2	0.26	0.12	0.09	0.06	0.03	0.0

Table 7.3 Empirical and expected absolute frequencies for the chi-square GoF test

x_i	$O_i = 50f(x_i)$	$E_i = 50p(x_i)$	$O_i - E_i$	$(O_i - E_i)^2/E_i$
0	0	1.0534	−1.0534	1.0534
1	3	4.0661	−1.0661	0.2795
2	9	7.8476	1.1524	0.1692
3	10	10.097	−0.0973	0.0009
4	13	9.7439	3.2561	1.0880
5	6	7.5223	−1.5223	0.3080
6	4.5	4.8393	−0.3393	0.0238
7	3	2.6685	0.3315	0.0412
8	1.5	1.2876	0.2124	0.0350
>8	0	0.8740	−0.8740	0.8740
Total	50	50	–	3.8733

on 50 years of observations. Employ the method of moments for parameter estimation to fit a Poisson distribution to the sample, then calculate the expected frequencies according to this model, and test its goodness-of-fit, at $\alpha = 0.05$.

Solution The Poisson mass function is $p_X(x) = e^{-\lambda}\lambda^x/x!$, for $x = 0, 1, \ldots$ and $\lambda > 0$, with expected value $E[X] = \lambda$. The sample mean can be calculated by weighting the x values by their respective empirical frequencies $f(x_i)$, yielding $\bar{x} = 3.86$. Thus, the estimate of parameter λ, by the method of moments, is $\hat{\lambda} = 3.86$. The expected absolute frequencies E_i, for x_i, shown in Table 7.3, were calculated by multiplying the Poisson probabilities $p_X(x_i)$ by the sample size $N = 50$. Similar calculations were made for the empirical absolute frequencies O_i.

Table 7.3 presents additional results that are required to calculate the test statistic χ^2. These are the simple differences and the scaled squared differences between empirical and expected absolute frequencies. The total sum of the last column of Table 7.3 gives the estimated value of the test statistic as $\hat{\chi}^2 = 3.8733$. For this example, the total number of partitions of the sample space is taken as $r = 10$ ($x_i = 0, 1, \ldots, >8$). With only one parameter estimated from the sample, $k = 1$, which defines the number of degrees of freedom for the probability distribution of the test statistic as $\nu = (r - k - 1) = 8$. The chi-square GoF test is a one-sided upper tail test, for which the critical region, at $\alpha = 0.05$ and $\nu = 8$, begins at $\chi^2_{0.95,8} = 15.51$ [Appendix 3 or MS Excel function CHISQ.INV(0.95;8)]. As $\hat{\chi}^2 < \chi^2_{0.95, \nu=8}$, the decision is not to reject the null hypothesis H_0 that the random variable X follows a Poisson probability distribution. For this example, even with $N = 50$, some empirical and expected absolute frequencies, for some partitions, are lower than the recommended minimum of 5. This is a possible shortcoming of the chi-square GoF test as very low frequencies, particularly in the distribution tails, can affect the overall results and even the decision making. In some cases, including this example, such an undesired situation can be corrected by merging some partitions. For instance, expected frequencies for $x = 0$ and $x = 1$ can be aggregated into the total frequency of 5.1195, if both partitions are merged into a new one defined by $x \leq 1$. Likewise, the upper-tail partitions can also be merged into a new one defined by $x \geq 6$. Of course, changing partitions will lead to new values of r and ν, and a new estimate for the test statistic. It is suggested that the reader solve this example, with the recommended partition changes.

Example 7.8 Refer to Fig. 5.4 in which the normal density fitted to the 248 annual total rainfalls, observed at the Radcliffe Meteorological Station, in Oxford (England) since 1767, has been superimposed over the empirical histogram. Use the chi-square GoF procedure to test the null hypothesis that data are normally distributed, at the significance level $\alpha = 0.05$.

Solution In the case of a continuous random variable, the mutually and collectively exhaustive disjoint partitions of the sample space are generally defined through bin intervals, inside which the empirical and expected absolute frequencies are counted and calculated. For this particular sample, as shown in Fig. 5.4, the number of bins had already been defined as $r = 15$, most of which with width 50 mm. Table 7.4 summarizes the calculations for the chi-square GoF test.

With $r = 15$, empirical frequencies O_i vary around acceptable values, except for the first and four of the last five bins, which count less than 5 occurrences each. However, for the sake of clarity, partitions are kept as such in the solution to this example. In order to calculate the frequencies, as expected from the normal distribution, its parameters μ and σ need to be estimated. The sample gives $\bar{x} = 645.956$ and $s_X^2 = 13045.89$, which, by the method of moments, yield the estimates $\hat{\mu} = 645.956$ and $\hat{\sigma} = 114.219$. Therefore, the expected relative frequency for the ith bin can be calculated as $p_i = \Phi[(\text{UB} - \hat{\mu})/\hat{\sigma}] - \Phi[(\text{LB} - \hat{\mu})/\hat{\sigma}]$, where UB and

Table 7.4 Empirical and expected absolute frequencies for the chi-square GoF test of normally distributed annual total rainfalls, as measured at the Radcliffe Station (England)

Bin interval (mm)	O_i	E_i	$O_i - E_i$	$(O_i - E_i)^2 / E_i$
<375	1	2.1923	−1.1923	0.6485
(375,425)	6	4.3862	1.6138	0.5938
(425,475)	7	10.0947	−3.0947	0.9487
(475,525)	17	19.2381	−2.2381	0.2604
(525,575)	36	30.3605	5.5695	1.0476
(575,625)	37	39.6772	−2.6772	0.1806
(625,675)	48	42.9407	5.0593	0.5961
(675,725)	37	38.4851	−1.4851	0.0573
(725,775)	26	28.5634	−2.5634	0.2301
(775,825)	21	17.5555	3.4445	0.6758
(825,875)	4	8.9349	−4.9349	2.7256
(875,925)	5	3.7655	1.2345	0.4047
(925,975)	1	1.3140	−0.3140	0.0750
(975,1025)	1	0.3796	0.6204	1.0137
>1075	1	0.1122	0.8878	7.0253
Total	248	248	–	16.4833

LB represent the bin's upper and lower bounds, respectively, and $\Phi(.)$ denotes the CDF of the standard normal distribution. The absolute frequency E_i, for bin i, is given by the product of p_i by the sample size $N = 248$. Then, the simple and scaled squared differences between empirical and expected frequencies are calculated. The summation through the last column of Table 7.4 yields the estimated value for the test statistics as $\hat{\chi}^2 = 16.48$. As the total number of partitions is $r = 15$ and two parameters have been estimated from the sample $(k = 2)$, then $\nu = (r-k-1) = 12$ degrees of freedom for the distribution of the test statistic. The chi-square GoF test is a one-sided upper tail test, for which, the critical region, at $\alpha = 0.05$ and $\nu = 12$, begins at $\chi^2_{0.95, 12} = 21.03$ [from Appendix 3 or MS Excel function CHISQ.INV $(0.95;12)$]. As $\hat{\chi}^2 < \chi^2_{0.95, \nu=12}$, the decision is not to reject the null hypothesis H_0 that the annual total rainfalls at the Radcliffe Station follow a normal probability distribution. Once again, very low frequencies, in both tails of the distribution, have affected the estimation of the test statistic, this time severely. Note that the fraction of the test statistic that corresponds to the last bin is almost half of the total sum, which, if a bit larger, would lead to the decision of rejecting the null hypothesis. As in the previous example, this shortcoming of the chi-square GoF test can be overcome by merging some partitions. For instance, the first bin can be merged into the second, whereas the last five bins can be rearranged into two new bins, of different widths so that at least 5 occurrences are counted within each. It is suggested that the reader solve this example, with the recommended partition changes.

7.5.2   The Kolmogorov–Smirnov (KS) GoF Test

The Kolmogorov–Smirnov (KS) nonparametric GoF test is based on the maximum difference between the values of the empirical and expected cumulative distributions, as calculated for all points in a sample of a continuous random variable. As originally proposed by Kolmogorov (1933), the test is not applicable to discrete random variables.

Assume X represent a random variable, from whose population the sample data $\{X_1, X_2, \ldots, X_N\}$ have been drawn. The null hypothesis to be tested is $H_0:\{P(X < x) = F_X(x)\}$, where $F_X(x)$ denotes a probability distribution function with known parameters. In order to implement the KS test, one needs first to rank the sample data in ascending order to obtain the sequence $\{x_{(1)}, x_{(2)}, \ldots, x_{(m)}, \ldots x_{(N)}\}$, where $1 \leq m \leq N$ designates the ranking orders. For each sample point $x_{(m)}$, the empirical CDF is calculated as the proportion of data that does not exceed $x_{(m)}$. Its calculation is made in two different ways: first, as

$$F_N^1(x_m) = \frac{m}{N} \tag{7.24}$$

and second as

$$F_N^2(x_m) = \frac{m-1}{N} \tag{7.25}$$

The next step in the KS GoF test is to calculate the expected CDF $F_X(x)$, as hypothesized under H_0, for every $x_{(m)}$, $1 \leq m \leq N$. The KS test statistic is given by

$$D_N = \max\{D_N^+, D_N^-\} \tag{7.26}$$

where

$$D_N^+ = \max\left|F_N^1[x_{(m)}] - F_X[x_{(m)}]\right|; \quad m = 1, 2, \ldots, N \tag{7.27}$$

and

$$D_N^- = \max\left|F_X[x_{(m)}] - F_N^2[x_{(m)}]\right|; \quad m = 1, 2, \ldots, N \tag{7.28}$$

and corresponds to the largest absolute difference between the empirical and expected cumulative probabilities.

If H_0 is true, as $N \to \infty$, the statistic D_N would tend to zero. On the other hand, if N is finite, the statistic D_N would be of the order of $1/\sqrt{N}$ and, thus, the quantity $\sqrt{N}D_N$ would not tend to zero, even for high values of N. Smirnov (1948) determined the limiting distribution of the random variable $\sqrt{N}D_N$, which, under H_0, is expressed as

$$\lim_{N \to \infty} \mathrm{P}\left(\sqrt{N}D_N \leq z\right) = \frac{\sqrt{2\pi}}{z} \sum_{k=1}^{\infty} \exp\left[-\frac{(2k-1)^2\pi^2}{8z^2}\right] \tag{7.29}$$

Therefore, for samples of sizes larger than 40, the critical values for the test statistic D_N are $1.3581/\sqrt{N}$, at the significance level $\alpha = 0.05$, and $1.6276/\sqrt{N}$, at $\alpha = 0.01$. These results are from the sum of the first five terms in the summation of Eq. (7.29) and remain unaltered from the sixth term on. For samples of sizes smaller than 40, the critical values of D_N should be taken from Table 7.5. Critical values for the KS test do not change with the hypothesized distribution $F_X(x)$ under H_0, provided the $F_X(x)$ parameters are known, i.e., not estimated from the sample.

As parameters estimates are obtained from the sample, Monte Carlo simulations show that critical values for the KS GoF test statistic are too conservative, with respect to the magnitude of type I error, and may lead to incorrect nonrejections of the null hypothesis (Lilliefors 1967, Vlcek and Huth 2009). Lilliefors (1967) published a new table of critical values for the KS test statistic, which is valid to test specifically the null hypothesis of normal data, under H_0, with parameters estimated from the sample. The most frequently used values are reproduced in Table 7.6. Taking $\alpha = 0.05$ and $N = 30$ as an example, the original table (Table 7.5) would give the critical value of 0.242 for the test statistic, whereas the corrected

Table 7.5 Critical values of $D_{N,\alpha}$ for the KS GoF test, as in Smirnov (1948)

N	$D_{N,0.10}$	$D_{N,0.05}$	$D_{N,0.02}$	$D_{N,0.01}$	N	$D_{N,0.10}$	$D_{N,0.05}$	$D_{N,0.02}$	$D_{N,0.01}$
10	0.369	0.409	0.457	0.489	26	0.233	0.259	0.290	0.311
11	0.352	0.391	0.437	0.468	27	0.229	0.254	0.284	0.305
12	0.338	0.375	0.419	0.449	28	0.225	0.250	0.279	0.300
13	0.325	0.361	0.404	0.432	29	0.221	0.246	0.275	0.295
14	0.314	0.349	0.390	0.418	30	0.218	0.242	0.270	0.290
15	0.304	0.338	0.377	0.404	31	0.214	0.238	0.266	0.285
16	0.295	0.327	0.366	0.392	32	0.211	0.234	0.262	0.281
17	0.286	0.318	0.355	0.381	33	0.208	0.231	0.258	0.277
18	0.279	0.309	0.346	0.371	34	0.205	0.227	0.254	0.273
19	0.271	0.301	0.337	0.351	35	0.202	0.224	0.251	0.269
20	0.265	0.294	0.329	0.352	36	0.199	0.221	0.247	0.265
21	0.259	0.287	0.321	0.344	37	0.196	0.218	0.244	0.262
22	0.253	0.281	0.314	0.337	38	0.194	0.215	0.241	0.258
23	0.247	0.275	0.307	0.330	39	0.191	0.213	0.238	0.255
24	0.242	0.269	0.301	0.323	40	0.189	0.210	0.235	0.252
25	0.238	0.264	0.295	0.317	>40	$1.22/\sqrt{N}$	$1.36/\sqrt{N}$	$1.52/\sqrt{N}$	$1.63/\sqrt{N}$

Table 7.6 Critical values for the KS GoF test statistic $D_{N,\alpha}$, specific for the normal distribution under H_0, with parameters estimated from the sample (Lilliefors 1967)

N	$D_{N, 0.10}$	$D_{N, 0.05}$	$D_{N, 0.01}$	N	$D_{N, 0.10}$	$D_{N, 0.05}$	$D_{N, 0.01}$
4	0.352	0.381	0.417	14	0.207	0.227	0.261
5	0.315	0.337	0.405	15	0.201	0.220	0.257
6	0.294	0.319	0.364	16	0.195	0.213	0.250
7	0.276	0.300	0.348	17	0.189	0.206	0.245
8	0.261	0.285	0.331	18	0.184	0.200	0.239
9	0.249	0.271	0.311	19	0.179	0.195	0.235
10	0.239	0.258	0.294	20	0.174	0.190	0.231
11	0.230	0.249	0.284	25	0.165	0.180	0.203
12	0.223	0.242	0.275	30	0.144	0.161	0.187
13	0.214	0.234	0.268	>30	$0.805/\sqrt{N}$	$0.886/\sqrt{N}$	$1.031/\sqrt{N}$

value from Lilliefors' table (Table 7.6) would be 0.161, which, depending on the sample estimate of D_N, say 0.18, for instance, would lead to the wrong decision of not rejecting H_0. Example 7.9 illustrates an application of the KS GoF test for the null hypothesis of normally distributed annual mean flows of the Lehigh River at Stoddartsville.

Example 7.9 The first two columns of Table 7.11 list the annual mean flows of the Lehigh River at Stoddartsville (USGS gauging station 01447500) for the water-years 1943/44 to 2014/2015. Test the null hypothesis that these data have been sampled from a normal population, at the significance level $\alpha = 0.05$.

Solution The third and fourth columns of Table 7.11 list the ranking orders and the flows sorted in ascending order, respectively. The empirical frequencies are obtained by direct application of Eqs. (7.24) and (7.25). For example, for the tenth-ranked flow, $F_N^1(3.54) = 10/72 = 0.1389$ and $F_N^2(3.54) = (10 - 1)/72 = 0.1250$. The sample of size $N = 72$ yields $\bar{x} = 5.4333 \, \mathrm{m}^3/\mathrm{s}$ and $s_X = 1.3509 \, \mathrm{m}^3/\mathrm{s}$, which are the MOM estimates for parameters μ and σ, respectively. The sixth column of Table 7.11 gives the expected frequencies, under the null hypothesis H_0:{data have been sampled from a normal population}, calculated as $F_X(x_m) = \Phi[(x_m - \hat{\mu})/\hat{\sigma}]$. Figure 7.6 depicts the empirical and expected frequencies plotted against the ranked annual mean flows. In Fig. 7.6, it is also highlighted the absolute maximum value of the differences between the expected and empirical frequencies, these calculated with Eqs. (7.24) and (7.25). The absolute maximum difference $\hat{D}_N = 0.0863$ is the estimated KS test statistic. At $\alpha = 0.05$ and for $N = 72$, Table 7.6 yields the critical value $D_{N,0.05} = 0.886/\sqrt{N} = 0.886/\sqrt{72} = 0.1044$. The KS GoF test is a one-sided upper tail test, for which, the critical region, at $\alpha = 0.05$, begins at $D_{N,0.05} = 0.1044$. As $\hat{D}_N < D_{N,0.05}$, the decision is not to reject the null hypothesis H_0, that data have been sampled from a normal population.

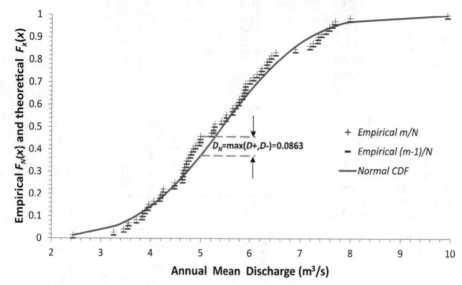

Fig. 7.6 Empirical and normal frequencies for the KS GoF test for annual mean flows

In order to test H_0:{data have been drawn from a normal distribution, with parameters estimated from the sample} without the use of tables, Stephens (1974) proposes the following equation for the critical values of $D_{N,\alpha}$:

$$D_{N,a} = \frac{k(\alpha)}{\sqrt{N} - 0.01 + 0.85/\sqrt{N}} \tag{7.30}$$

where $k(0.10) = 0.819$, $k(0.05) - 0.895$, $k(0.025) = 0.955$, and $k(0.01) = 1.035$.

With the exponential distribution hypothesized under H_0, Stephens (1974) proposes

$$D_{N,a} = \frac{k(\alpha)}{\sqrt{N} + 0.26 + 0.50/\sqrt{N}} + \frac{0.2}{N} \tag{7.31}$$

where $k(0.10) = 0.990$, $k(0.05) = 1.094$, $k(0.025) = 1.190$, and $k(0.01) = 1.308$.

As for H_0:{data have been drawn from a Gumbel$_{max}$ distribution, with parameters estimated from the sample}, Chandra et al. (1981) provide Table 7.7 for the critical values of $\sqrt{N}D_{N,\alpha}$. Notice that the estimated test statistic $\hat{D}_N$ must be multiplied by $\sqrt{N}$ before being compared to the values given in Table 7.7. If it exceeds the tabulated value, H_0 must be rejected, at the level α.

Table 7.7 Critical values for the test statistic $\sqrt{N}D_{N,\alpha}$, specific for the Gumbel$_{max}$ hypothesized distribution under H_0, with parameters estimated from the sample (Chandra, Singpurwalla and Stephens, 1981)

N	Upper significance level α			
	0.10	0.05	0.025	0.01
10	0.760	0.819	0.880	0.994
20	0.779	0.843	0.907	0.973
50	0.790	0.856	0.922	0.988
∞	0.803	0.0874	0.939	1.007

Table 7.8 Critical values for the test statistic $D_{N,\alpha}$, specifically for the two-parameter gamma hypothesized distribution under H_0, with parameters estimated from a sample of size 25, or 30, or asymptotic

Gamma shape η	Upper significance level α			
	0.10	0.05	0.01	N
=2	0.176	0.190	0.222	25
	0.161	0.175	0.203	30
	0.910	0.970	1.160	N
=3	0.166	0.180	0.208	25
	0.151	0.165	0.191	30
	0.860	0.940	1.080	N
=4	0.164	0.178	0.209	25
	0.148	0.163	0.191	30
	0.830	0.910	1.060	N
≥8	0.159	0.173	0.203	25
	0.146	0.161	0.187	30
	0.810	0.890	1.040	N

Asymptotic values are to be multiplied by $1/\sqrt{N}$ (from Crutcher 1975)

If Z is a Weibull$_{min}$ variate, with scale and shape parameters ω and ψ, respectively, such that $F_Z(z) = 1 - \exp[-(z/\omega)^{\psi}]$, the critical values of Table 7.7 can also be used for the purpose of testing if the sample data come from a Weibull$_{min}$ population. However, before doing so, one needs to transform the Weibull$_{min}$ variate into $Y = -\ln(Z)$ and take into account the mathematical fact that $Y \sim$ Gumbel$_{max}$ with location $-\ln(\omega)$ and scale $1/\omega$. Then, the KS GoF test is performed for the transformed variable Y.

Critical values for the GEV distribution under H_0 are hard to obtain. Wang (1998) comments on this and provides critical values for the case where the GEV parameters are estimated through LH-moments, a generalization of the L-moments described in Sect. 6.5 of Chap. 6.

Crutcher (1975) presents tables of critical values of the KS test statistic $D_{N,\alpha}$ for sample sizes $N = 25$, 30, and ∞, for exponential, normal, Gumbel$_{max}$, or two-parameter gamma, under H_0. The tabulated critical values for the gamma distribution under H_0, with estimated parameters θ, of scale, and η, of shape, are reproduced in Table 7.8.

7.5.3 The Anderson–Darling (AD) GoF Test

The capabilities of the chi-square and KS GoF tests to discern false hypotheses are particularly diminished in the distribution tails, as a result of both the small number of observations and the relatively larger estimation errors that are usually found in these partitions of the sample space. Alternatively, the Anderson–Darling (AD) GoF nonparametric test, introduced by Anderson and Darling (1954), is designed to give more weight to the distribution tails, where the largest and smallest data points can have a strong impact on the quality of curve fitting. Analogously to the KS procedure, the AD GoF test is based on the differences between the empirical $[F_N(x)]$ and expected $[F_X(x)]$ cumulative probability distributions of a continuous random variable. In contrast to the KS procedure, the AD GoF gives more weight to the tails by dividing the squared differences between $F_N(x)$ and $F_X(x)$ by $\sqrt{F_X(x)[1 - F_X(x)]}$. As such, the AD test statistic is written as

$$A^2 = \int\limits_{-\infty}^{\infty} \frac{[F_N(x) - F_X(x)]^2}{F_X(x)[1 - F_X(x)]} f_X(x)\, dx \tag{7.32}$$

where $f_X(x)$ denotes the density function under the null hypothesis. Anderson and Darling (1954) demonstrated that A^2 can be estimated as

$$A_N^2 = -N - \sum_{m=1}^{N} \frac{(2m - 1)\left\{\ln F_X(x_m) + \ln[1 - F_X(x_{N-m+1})]\right\}}{N} \tag{7.33}$$

where $\{x_1, x_2, \ldots, x_m, \ldots x_N\}$ represents the set of data ranked in ascending order.

The larger the A_N^2 statistic, the more dissimilar the empirical $[F_N(x)]$ and expected $[F_X(x)]$ distributions, and the higher the support to reject the null hypothesis. The probability distribution of the AD test statistic depends on the distribution $F_X(x)$ that is hypothesized under H_0. If $F_X(x)$ is the normal distribution, the critical values of A^2 are given in Table 7.9.

According to D'Agostino and Stephens (1986), in the context of using the critical values of Table 7.9, the test statistic, as calculated with Eq. (7.33), must be multiplied by the correction factor $(1 + 0.75/N + 2.25/N^2)$ and is valid for $N > 8$.

If the probability distribution, hypothesized under H_0, is Gumbel$_{\max}$, the critical values of A^2 are those listed in Table 7.10.

In this case, D'Agostino and Stephens (1986) point out that the test statistic, as calculated with Eq. (7.33), must be multiplied by the correction factor

Table 7.9 Critical values of AD test statistic $A^2{}_\alpha$ if the probability distribution, hypothesized under H_0, is normal or lognormal (from D'Agostino and Stephens 1986)

α	0.1	0.05	0.025	0.01
$A^2_{\mathrm{crit},\alpha}$	0.631	0.752	0.873	1.035

Table 7.10 Critical values of AD test statistic A^2_α if the probability distribution, hypothesized under H_0, is Gumbel$_{max}$ (from D'Agostino and Stephens 1986)

α	0.1	0.05	0.025	0.01
$A^2_{crit,\alpha}$	0.637	0.757	0.877	1.038

$(1 + 0.2/\sqrt{N})$. Table 7.10 can also be employed for testing the exponential and the two-parameter Weibull$_{min}$ under H_0. For the latter, the Weibull$_{min}$ variate transformation into Gumbel$_{max}$, as described in the previous subsection, is needed.

Example 7.10 Solve Example 7.9 with the AD GoF test.

Solution Table 7.11 shows the partial calculations necessary to estimate the A^2 statistic.

In Table 7.11, the hypothesized $F_X(x)$ under H_0 is the normal distribution and the expected frequencies are calculated as $\Phi[(x - 5.4333)/1.3509]$. The uncorrected test statistic A^2_N can be calculated as $A^2_N = -N - \sum_{i=1}^{N} S_i/N = -72 - (-5220.23)/72 = 0.5032$. For this example, the correction factor is $(1 + 0.75/N + 2.25/N^2) = 1.0109$. Therefore, the corrected test statistic is $A^2_N = 0.5087$. The Table 7.9 reading, for $\alpha = 0.05$, is $A^2_{crit,0.05} = 0.752$, which defines the lower bound of the critical region for the AD one-sided upper tail test. As $A^2_N < A^2_{crit,0.05}$, the decision is not to reject the null hypothesis H_0, that data have been sampled from a normal population.

Özmen (1993) provides polynomial functions for calculating $A^2_{crit,\alpha}$ for the Pearson III distribution under H_0, assuming that the shape parameter β is known or specified, and the location and scale parameters are estimated by the method of maximum likelihood. The general polynomial equation is of the form

$$A^2_{crit,\alpha} = A + BN + C\alpha + D\alpha N + EN^2 + F\alpha^2 \qquad (7.34)$$

where the polynomial coefficients are given in Table 7.12, for different assumed values of the shape parameter β, $0.01 \le \alpha \le 0.20$, and $5 \le N \le 40$.

Ahmad et al. (1998) modified the AD GoF test statistic to give more weight to the differences between $F_N(x)$ and $F_X(x)$ at the upper tail, compared to those at the lower tail. The authors justify their proposed modification in the context of the greater interest hydrologists usually have in estimating quantiles of high return periods. The Modified Anderson–Darling (MAD) test statistic is given by

$$AU^2_N = \frac{N}{2} - 2\sum_{m=1}^{N} [F_X(x_m)] - \sum_{m=1}^{N} \left(2 - \frac{2m-1}{N}\right) \ln[1 - F_X(x_m)] \qquad (7.35)$$

for the upper tail, whereas

$$AL^2_N = -\frac{3N}{2} - 2\sum_{m=1}^{N} [F_X(x_m)] - \sum_{m=1}^{N} \left(\frac{2m-1}{N}\right) \ln[F_X(x_m)] \qquad (7.36)$$

Table 7.11 Calculations of the AD GoF test statistic for the annual mean flows (X_t), in m³/s, of the Lehigh River at Stoddartsville, for water years 1943/44 to 2014/2015

Water year	X_t	m	x_m	x_{N-m+1}	$v = F_X[x_m]$	$\ln(v)$	$t = 1 - F_X[x_{N-m+1}]$	$\ln(t)$	$S_i = (2m-1)[\ln v + \ln t]$
1943/44	4.24	1	2.44	9.94	0.0134	−4.3140	0.0004	−7.7538	−12.0678
/45	6.07	2	3.26	7.99	0.0540	−2.9185	0.0293	−5.5302	−19.3463
/46	5.86	3	3.46	7.69	0.0724	−2.6259	0.0476	−3.0458	−28.3586
/47	6.33	4	3.54	7.63	0.0808	−2.5156	0.0516	−2.9636	−38.3542
/48	5.29	5	3.57	7.57	0.0834	−2.4847	0.0564	−2.8742	−48.2295
/49	4.65	6	3.72	7.50	0.1021	−2.2818	0.0627	−2.7700	−55.5692
/50	5.26	7	3.84	7.42	0.1187	−2.1314	0.0711	−2.6440	−62.0792
1950/51	6.38	8	3.85	7.35	0.1208	−2.1139	0.0776	−2.5567	−70.0591
/52	7.25	9	3.89	7.34	0.1272	−2.0620	0.0784	−2.5450	−78.3179
/53	5.79	10	3.95	7.25	0.1366	−1.9907	0.0888	−2.4216	−83.8334
/54	4.18	11	3.97	7.19	0.1394	−1.9706	0.0968	−2.3353	−90.4234
/55	6.42	12	4.07	6.90	0.1563	−1.8559	0.1387	−1.9757	−88.1268
/56	6.33	13	4.18	6.51	0.1772	−1.7302	0.2133	−1.5450	−81.8798
/57	4.26	14	4.20	6.42	0.1805	−1.7118	0.2321	−1.4607	−85.6603
/58	5.66	15	4.24	6.38	0.1884	−1.6694	0.2424	−1.4171	−89.5096
/59	4.07	16	4.26	6.33	0.1918	−1.6514	0.2530	−1.3743	−93.7987
/60	6.51	17	4.48	6.33	0.2401	−1.4265	0.2537	−1.3716	−92.3381
1960/61	4.64	18	4.48	6.23	0.2408	−1.4238	0.2770	−1.2836	−94.7577
/62	3.26	19	4.64	6.22	0.2788	−1.2772	0.2813	−1.2684	−94.1880
/63	3.84	20	4.64	6.14	0.2795	−1.2747	0.3014	−1.1993	−96.4852
/64	3.85	21	4.65	6.07	0.2816	−1.2671	0.3192	−1.1421	−98.7772
/65	2.45	22	4.67	6.00	0.2852	−1.2546	0.3381	−1.0844	−100.5779
/66	3.54	23	4.68	5.92	0.2895	−1.2397	0.3598	−1.0222	−101.7839
/67	4.76	24	4.72	5.92	0.2981	−1.2102	0.3606	−1.0200	−104.8196

(continued)

Table 7.11 (continued)

Water year	X_t	m	x_m	x_{N-m+1}	$v = F_X[x_m]$	$\ln(v)$	$t = 1 - F_X[x_{N-m+1}]$	$\ln(t)$	$S_i = (2m-1)[\ln v + \ln t]$
/68	4.67	25	4.72	5.90	0.2996	−1.2053	0.3637	−1.0103	−108.6162
/69	4.48	26	4.76	5.88	0.3091	−1.1740	0.3700	−0.9941	−110.5739
/70	4.99	27	4.80	5.86	0.3188	−1.1432	0.3772	−0.9775	−112.2648
1970/71	5.25	28	4.82	5.79	0.3241	−1.1268	0.3980	−0.9293	−113.0880
/72	7.35	29	4.85	5.78	0.3339	−1.0969	0.3981	−0.9212	−115.0280
/73	7.57	30	4.89	5.72	0.3447	−1.0652	0.4151	−0.8791	−114.7175
/74	6.14	31	4.97	5.66	0.3657	−1.0059	0.4332	−0.8366	−112.3889
/75	6.22	32	4.99	5.66	0.3713	−0.9909	0.4340	−0.8347	−115.0080
/76	5.66	33	4.99	5.53	0.3720	−0.9887	0.4705	−0.7539	−113.2696
/77	5.92	34	5.21	5.52	0.4353	−0.8318	0.4739	−0.7468	−105.7642
/78	7.35	35	5.25	5.43	0.4469	−0.8055	0.5015	−0.6902	−103.2071
/79	5.92	36	5.26	5.29	0.4485	−0.8018	0.5432	−0.6103	−100.2607
/80	4.64	37	5.29	5.26	0.4568	−0.7835	0.5515	−0.5951	−100.6385
1980/81	3.89	38	5.43	5.25	0.4985	−0.6961	0.5531	−0.5921	−96.6151
/82	4.68	39	5.52	5.21	0.5261	−0.6422	0.5647	−0.5715	−93.4508
/83	5.52	40	5.53	4.99	0.5295	−0.6359	0.6280	−0.4653	−86.9938
/84	7.19	41	5.66	4.99	0.5560	−0.5692	0.6287	−0.4640	−83.6917
/85	4.82	42	5.66	4.97	0.5668	−0.5677	0.6343	−0.4553	−84.9102
/86	6.90	43	5.72	4.89	0.5849	−0.5364	0.6553	−0.4226	−81.5122
/87	5.72	44	5.78	4.85	0.6019	−0.5076	0.6661	−0.4063	−79.5118
/88	4.72	45	5.79	4.82	0.6052	−0.5022	0.6759	−0.3916	−79.5560
/89	4.85	46	5.86	4.80	0.6228	−0.4735	0.6812	−0.3839	−78.0244
/90	5.43	47	5.88	4.76	0.6300	−0.4621	0.6909	−0.3698	−77.3680
1990/91	4.89	48	5.90	4.72	0.6363	−0.4521	0.7004	−0.3561	−76.7822
/92	4.48	49	5.92	4.72	0.6394	−0.4472	0.7019	−0.3540	−77.7194
/93	5.78	50	5.92	4.68	0.6402	−0.4460	0.7105	−0.3418	−77.9861

/94	7.42	51	6.00	4.67	0.6619	-0.4126	0.7148	-0.3357	-75.5858
/95	3.97	52	6.07	4.65	0.6808	-0.3844	0.7184	-0.3307	-73.6662
/96	7.50	53	6.14	4.64	0.6986	-0.3587	0.7205	-0.3279	-72.0848
/97	6.00	54	6.22	4.64	0.7187	-0.3302	0.7212	-0.3269	-70.3139
/98	5.21	55	6.23	4.48	0.7230	-0.3244	0.7592	-0.2755	-65.3890
/99	3.46	56	6.33	4.48	0.7463	-0.2926	0.7599	-0.2746	-62.9650
99/2000	5.53	57	6.33	4.26	0.7470	-0.2917	0.8082	-0.2129	-57.0223
/01	3.57	58	6.38	4.24	0.7576	-0.2776	0.8116	-0.2087	-55.9257
/02	3.95	59	6.42	4.20	0.7680	-0.2640	0.8195	-0.1991	-54.1879
/03	7.99	60	6.51	4.18	0.7867	-0.2399	0.8228	-0.1951	-51.7684
/04	7.69	61	6.90	4.07	0.8613	-0.1493	0.8437	-0.1700	-38.6284
/05	6.23	62	7.19	3.97	0.9032	-0.1018	0.8606	-0.1501	-30.9824
/06	7.63	63	7.25	3.95	0.9112	-0.0930	0.8634	-0.1469	-29.9815
/07	5.88	64	7.35	3.89	0.9215	-0.0817	0.8728	-0.1361	-27.6579
/08	5.90	65	7.35	3.85	0.9224	-0.0807	0.8792	-0.1287	-27.0176
/09	4.72	66	7.42	3.84	0.9289	-0.0737	0.8813	-0.1263	-26.2080
/10	4.80	67	7.50	3.72	0.9373	-0.0647	0.8979	-0.1077	-22.9309
2010/11	9.94	68	7.57	3.57	0.9435	-0.0581	0.9166	-0.0870	-19.5957
/12	4.97	69	7.63	3.54	0.9484	-0.0530	0.9192	-0.0843	-18.8076
/13	4.99	70	7.69	3.46	0.9524	-0.0487	0.9276	-0.0751	-17.2157
/14	4.20	71	7.99	3.26	0.9707	-0.0297	0.9460	-0.0555	-12.0219
/15	3.72	72	9.94	2.44	0.9996	-0.0004	0.9866	-0.0135	-1.9877
Total	-	-	-	-	-	-	-	-	-5220.2326

Table 7.12 Polynomial coefficients of Eq. (7.34) (from Özmen 1993)

Shape β	A	B	C	D	E	F
0.5	0.88971	0.16412	−9.86652	−0.20142	0	30.0303
1.0	1.01518	0.07580	−9.16860	−0.13433	−0.0004447	32.0377
1.5	1.00672	0.02154	−7.93372	−0.08029	0	27.9847
2.0	0.98379	0.01225	−6.27139	−0.01974	−0.0001560	19.0327
2.5	0.96959	0.01292	−6.14333	−0.01919	−0.0001650	18.4079
3.0	0.96296	0.01366	−6.28657	−0.03032	−0.0001237	19.6629
3.5	0	0.08490	0	−0.11941	−0.0013400	0
4.0	1.05828	0.01003	−6.95325	−0.03406	0	21.8651

is valid for the lower tail, such that $A_N^2 = AU_N^2 + AL_N^2$, where A_N^2 is the conventional AD statistic, as given by Eq. (7.33).

Based on a large number of Monte Carlo simulations, Heo et al. (2013) derived regression equations for the critical values of the MAD upper-tail test statistic, assuming GEV, GLO (Generalized Logistic, as described in Sect. 5.9.1, as a special case of Kappa distribution), and GPA (Generalized Pareto) as hypothesized distributions under H_0. The shape parameters of the three distributions were specified within the range of possible values, whereas the other parameters were estimated from synthetically generated samples of different sizes. The parameter estimation method used in the experiment was not mentioned in the paper. For GEV and GLO under H_0, the regression equation has the general form of

$$AU_{\text{crit},\alpha}^2 = a + \frac{b}{N} + \frac{c}{N^2} + d\kappa + e\kappa^2 \tag{7.37}$$

whereas for GPA, it is given by

$$AU_{\text{crit},\alpha}^2 = \frac{1}{a + b/N + c/N^2 + d\kappa + e\kappa^2} \tag{7.38}$$

where κ denotes the specified (assumed) value for the shape parameter, N is the sample size ($10 \leq N \leq 100$), and a, b, c, d, and e are regression coefficients given in Table 7.13, for the significance levels 0.05 and 0.10.

7.5.4 The Probability Plot Correlation Coefficient (PPCC) GoF Test

The PPCC GoF test was introduced by Filliben (1975), as a testing procedure for normally distributed data under the null hypothesis. Later on, Filliben's test has been adapted to accommodate other theoretical distributions under H_0. Modified as such, the test is generally referred to as the Probability Plot Correlation Coefficient

Table 7.13 Regression coefficients for Eqs. (7.37) and (7.38) (from Heo et al. 2013)

Distribution	α	Regression coefficients				
		a	b	c	d	e
GEV	0.05	0.2776	-0.2441	1.8927	-0.0267	0.1586
	0.10	0.2325	-0.1810	1.8160	-0.0212	0.1315
GLO	0.05	0.3052	-0.6428	1.1345	0.0662	0.2745
	0.10	0.2545	-0.5823	1.6324	0.0415	0.2096
GPA	0.05	4.4863	24.6003	-142.5599	-2.6223	0.3976
	0.10	5.3332	28.5881	-157.7515	-3.1744	0.4888

(PPCC) GoF test. Given the data sample $\{x_1, x_2, \ldots, x_n\}$ of the random variable X and assuming the null hypothesis that such a sample has been drawn from a population whose probability distribution is $F_X(x)$, the PPCC GoF test statistic is formulated on the basis of the linear correlation coefficient r, between the data ranked in ascending order, as denoted by the sequence $\{x_1, x_2, \ldots, x_m, \ldots x_N\}$, and the theoretical quantiles $\{w_1, w_2, \ldots, w_m, \ldots w_N\}$, which are calculated as $w_m = F_X^{-1}(1 - q_m)$, where q_m denotes the empirical probability corresponding to the ranking order m. Formally, the PPCC test statistic is calculated as

$$r = \frac{\sum_{m=1}^{N} (x_m - \bar{x})(w_m - \bar{w})}{\sqrt{\sum_{m=1}^{N} (x_m - \bar{x})^2 \sum_{m=1}^{N} (w_m - \bar{w})^2}} \tag{7.39}$$

where $\bar{x} = \frac{1}{N}\sum_{l=1}^{N} x_i$ and $\bar{w} = \frac{1}{N}\sum_{i=1}^{N} w_i$. According to Filliben (1975), GoF tests based on the correlation coefficient are invariant to the method used to estimate the location and scale parameters.

The intuitive idea behind the PPCC GoF test is that an eventual strong linear association between x_m and w_m is supportive to the assumption that $F_X(x)$ is a plausible model for the X population. The null hypothesis $H_0{:}\{r = 1$, as implicit by $X \sim F_X(x)\}$, should be tested against the alternative $H_1{:}\{r < 1, \neq F_X(x)\}$, thus making the PPCC GoF test a one-sided lower tail test. The critical region for H_0, at the significance level α, begins at $r_{\text{crit},\alpha}$, below which, if $r < r_{\text{crit},\alpha}$, the null hypothesis must be rejected in favor of H_1.

In the PPCC test statistic formulation, as in Eq. (7.39), the specification of $F_X(x)$, in the form of $w_m = F_X^{-1}(1 - q_m)$ is implicit. The empirical probabilities q_m, as corresponding to the ranking orders m, are usually termed *plotting positions* and vary according to the specification of $F_X(x)$. In general, the different formulae for calculating plotting positions aim to obtain unbiased quantiles or unbiased probabilities, with respect to a target distribution $F_X(x)$. These formulae can be written in the following general form:

Table 7.14 Formulae for plotting positions q_m

Name	Plotting position formula	a	Statistical justification
Weibull	$q_m = \dfrac{m}{N+1}$	0	Unbiased exceedance probabilities for all distributions
Blom	$q_m = \dfrac{m-0.375}{N+0.25}$	0.375	Unbiased normal quantiles
Cunnane	$q_m = \dfrac{m-0.40}{N+0.2}$	0.40	Approximately quantile-unbiased
Gringorten	$q_m = \dfrac{m-0.44}{N+0.12}$	0.44	Optimized for Gumbel
Median	$q_m = \dfrac{m-0.3175}{N+0.365}$	0.3175	Median exceedance probabilities for all distributions
Hazen	$q_m = \dfrac{m-0.50}{N}$	0.5	None

Source: adapted from an original table published in Stedinger et al. (1993)

Table 7.15 Critical values $r_{\text{crit},\alpha}$ for the normal distribution, with $a = 0.375$ in Eq. (7.40)

N	$\alpha = 0.10$	$\alpha = 0.05$	$\alpha = 0.01$
10	0.9347	0.9180	0.8804
15	0.9506	0.9383	0.9110
20	0.9600	0.9503	0.9290
30	0.9707	0.9639	0.9490
40	0.9767	0.9715	0.9597
50	0.9807	0.9764	0.9664
60	0.9835	0.9799	0.9710
75	0.9865	0.9835	0.9757
100	0.9893	0.9870	0.9812

Source: adapted from an original table published in Stedinger et al. (1993)

$$q_m = \frac{m-a}{N+1-2a} \tag{7.40}$$

where a varies with the specification of $F_X(x)$. Table 7.14 summarizes the different formulae for calculating plotting positions, with the indication of the particular value of a, in Eq. (7.40), to be used with the specified target distribution $F_X(x)$.

As quantiles w_m vary with $F_X(x)$, it is clear that the probability distribution of the test statistic will also vary with the specification of $F_X(x)$, under H_0. Table 7.15 lists the critical values $r_{\text{crit},\alpha}$, for the case where $F_X(x)$ is specified as the normal distribution, with plotting positions q_m calculated with Blom's formula. For a lognormal variate, the critical values in Table 7.15 remain valid for the logarithms of the original variable. Example 7.11 illustrates the application of the PPCC GoF test for the normal distribution under H_0, or the original Filliben's test.

Example 7.11 Solve Example 7.9 with Filliben's GoF test.

Solution Table 7.11 lists the empirical quantiles x_m, as ranked in ascending order. The theoretical quantiles w_m are calculated by the inverse function $\Phi^{-1}(q_m)$ of the

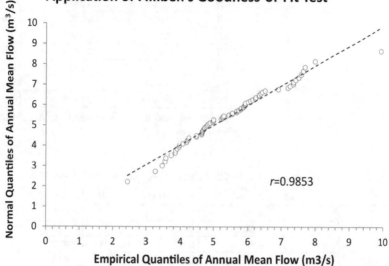

Fig. 7.7 Linear association between empirical and theoretical quantiles for Filliben's test

normal distribution, with estimated mean 5.4333 and standard deviation 1.3509, and argument q_m given by Blom's formula for plotting positions. In order to exemplify such a calculation, consider the ranking order $m = 10$, which in Blom's formula yields $q_{m=10} = 0.1332$, with $a = 0.375$ and $N - 72$. The inverse function $\Phi^{-1}(0.1332)$ can be easily calculated through the MS Excel built-in function NORM.INV(0.1332; 5.4333; 1.3509; TRUE) which returns $w_{10} = 3.9319$. Such calculations must proceed for all ranking orders up to $m = 72$. The linear correlation coefficient between empirical (x_m) and theoretical (w_m) quantiles can be calculated with Eq. (7.39) or through MS Excel function CORREL(.). Figure 7.7 depicts the plot of theoretical versus empirical quantiles, the linear regression, and the corresponding value of the correlation coefficient $r = 0.9853$, which is the estimated value of Filliben's test statistic. From Table 7.15, with $\alpha = 0.05$, and using linear interpolation for critical values between $N = 60$ and $N = 75$, the result is $r_{crit,0.05} = 0.9828$. As r is slightly higher than $r_{crit,0.05}$, the decision is not to reject the null hypothesis H_0, that data have been sampled from a normal population, at the significance level of 5%.

Table 7.16 lists the critical values $r_{crit,\alpha}$, for the case where the parent $F_X(x)$ is specified as the Gumbel$_{max}$ distribution, with plotting positions q_m calculated with Gringorten's formula. Table 7.16 can also be employed for testing the exponential and the two-parameter Weibull$_{min}$ under H_0. For the latter, the Weibull$_{min}$ variate transformation into Gumbel$_{max}$, as described in the Sect. 7.5.2, is needed.

Table 7.16 Critical values $r_{crit,\alpha}$ for the Gumbel$_{max}$ distribution, with $a = 0.44$ in Eq. (7.40)

N	$\alpha = 0.10$	$\alpha = 0.05$	$\alpha = 0.01$
10	0.9260	0.9084	0.8630
20	0.9517	0.9390	0.9060
30	0.9622	0.9526	0.9191
40	0.9689	0.9594	0.9286
50	0.9729	0.9646	0.9389
60	0.9760	0.9685	0.9467
70	0.9787	0.9720	0.9506
80	0.9804	0.9747	0.9525
100	0.9831	0.9779	0.9596

Source: adapted from an original table published in Stedinger et al. (1993)

Table 7.17 Critical values $r_{crit,\alpha}$ for the GEV distribution, with $a = 0.40$ in Eq. (7.40)

α	N	$\kappa = -0.30$	$\kappa = -0.20$	$\kappa = -0.10$	$\kappa = 0$	$\kappa = 0.10$	$\kappa = 0.20$
0.01	5	0.777	0.791	0.805	0.817	0.823	0.825
0.01	10	0.836	0.845	0.856	0.866	0.876	0.882
0.01	20	0.839	0.855	0.878	0.903	0.923	0.932
0.01	30	0.834	0.858	0.890	0.92	0.942	0.953
0.01	50	0.825	0.859	0.902	0.939	0.961	0.970
0.01	100	0.815	0.866	0.92	0.959	0.978	0.985
0.05	5	0.853	0.863	0.869	0.874	0.877	0.880
0.05	10	0.881	0.890	0.900	0.909	0.916	0.920
0.05	20	0.898	0.912	0.926	0.938	0.948	0.953
0.05	30	0.903	0.920	0.937	0.952	0.961	0.967
0.05	50	0.908	0.929	0.950	0.965	0.974	0.979
0.05	100	0.914	0.940	0.963	0.978	0.985	0.989
0.10	5	0.888	0.892	0.896	0.899	0.901	0.903
0.10	10	0.904	0.912	0.920	0.927	0.932	0.936
0.10	20	0.920	0.932	0.943	0.952	0.958	0.962
0.10	30	0.928	0.941	0.953	0.962	0.969	0.973
0.10	50	0.935	0.950	0.963	0.973	0.979	0.982
0.10	100	0.944	0.961	0.974	0.983	0.988	0.991

Table 7.17 lists the critical values $r_{crit,\alpha}$, for the case where the parent $F_X(x)$ is specified as the GEV distribution, with plotting positions q_m calculated with Cunnane's formula. The critical values of Table 7.17 were obtained by Chowdhury et al. (1991) through a large number of Monte Carlo simulations, of samples of different sizes drawn from GEV populations, with specified (assumed) values for the shape parameter κ.

Heo et al. (2008) proposed regression equations to approximate the critical values of PPCC GoF tests with the normal, Gumbel$_{max}$, Pearson type III, GEV, and three-parameter Weibull$_{min}$ as hypothesized distributions under H_0.

The equations facilitate the application of PPCC GoF tests, without the use of extensive tables. Some of these regression equations are reproduced next.

- Distribution hypothesized under H_0: normal (Blom's formula for plotting position)

$$\ln\left(\frac{1}{1 - r_{\text{crit},\alpha}}\right) = 1.29 + 0.283 \ln \alpha$$

$$+ \left(0.887 - 0.751\alpha + 3.21\alpha^2\right)\ln N \text{ for } 0.005 \leq \alpha < 0.1$$

$$(7.41)$$

- Distribution hypothesized under H_0: Gumbel$_{\max}$ (Gringorten's formula for plotting position)

$$\ln\left(\frac{1}{1 - r_{\text{crit},\alpha}}\right) = 2.54\,\alpha^{0.146}N^{0.152-0.00993\,\ln\alpha} \text{ for } 0.005 \leq \alpha < 0.1 \qquad (7.42)$$

- Distribution hypothesized under H_0: Pearson type III (Blom's formula for plotting position)

$$\ln\left(\frac{1}{1 - r_{\text{crit},\alpha}}\right) = (a + b\gamma)N^{c+d\gamma+e\gamma^2} \text{ for } 0.5 \leq \gamma < 5.0 \qquad (7.43)$$

where γ denotes the specified coefficient of skewness. The regression coefficients a to e are given in Table 7.18, for $\alpha = 0.01$, 0.05, and 0.10.
- Distribution hypothesized under H_0: GEV (Cunnane's formula for plotting position)

$$\ln\left(\frac{1}{1 - r_{\text{crit},\alpha}}\right) = (1.527 - 0.7656\kappa + 2.2284\kappa^2 - 3.824\kappa^3)N^{0.1986+0.3858\kappa-0.5985\kappa^2}$$

for $\alpha = 0.05$, $-0.20 \leq \kappa < 0.25$ or $3.5351 \geq \gamma > 0.0872$

$$(7.44)$$

and

Table 7.18 Regression coefficients for Eq. (7.43)

Regression coefficient	Significance level α		
	0.01	0.05	0.10
a	1.37000	1.73000	1.92000
b	−0.05080	−0.08270	−0.08640
c	0.23700	0.20700	0.19400
d	−0.03400	−0.02000	−0.01580
e	0.00356	0.00223	0.00166

Table 7.19 Critical values $r_{crit,\alpha}$ for the 2-p Weibull$_{min}$, with $a = 0.44$ in Eq. (7.40)

N	$\alpha = 0.01$	$\alpha = 0.05$	$\alpha = 0.10$	N	$\alpha = 0.01$	$\alpha = 0.05$	$\alpha = 0.10$
10	0.8680	0.9091	0.9262	50	0.9399	0.9647	0.9728
15	0.8930	0.9274	0.9420	55	0.9445	0.9669	0.9745
20	0.9028	0.9384	0.9511	60	0.9468	0.9693	0.9762
25	0.9140	0.9460	0.9577	65	0.9489	0.9709	0.9775
30	0.9229	0.9525	0.9625	70	0.9518	0.9722	0.9786
35	0.9281	0.9559	0.9656	80	0.9552	0.9742	0.9803
40	0.9334	0.9599	0.9687	90	0.9579	0.9766	0.9821
45	0.9374	0.9630	0.9713	100	0.9606	0.9777	0.9831

Adapted from Vogel and Kroll (1989)

$$\ln\left(\frac{1}{1 - r_{crit,\alpha}}\right) = (1.695 - 0.5205\kappa + 1.229\kappa^2 - 2.809\kappa^3)N^{0.1912+0.2838\kappa-0.3765\kappa^2}$$

for $\alpha = 0.10$, $-0.25 \leq \kappa < 0.25$ or $5.6051 \geq \gamma > 0.0872$

$$(7.45)$$

where κ is the GEV assumed shape parameter and γ denotes the specified coefficient of skewness.

Vogel and Kroll (1989) developed critical values for the PPCC GoF test for the two-parameter Weibull$_{min}$ parent distribution, under H_0, and plotting positions calculated with Gringorten's formula. Part of these critical values are given in Table 7.19, for significance levels $\alpha = 0.01$, 0.05, and 0.10.

Kim et al. (2008) provide charts of the critical statistics versus sample sizes, at 1 and 5 % significance levels, for the PPCC testing of the GLO and GPA parent distributions.

7.5.5 Some Comments on GoF Tests

In general, goodness-of-fit tests are ineffective in discerning differences between empirical and theoretical probabilities (or quantiles) in the distribution tails. Such a drawback is critical in hydrologic frequency analysis, since the usual short samples might contain only a few extreme data points, if any, and further because its main interest is exactly to infer the characteristics of distribution tails. For instance, the chi-square GoF test, as applied to continuous random variables, needs the previous specification of the number of bins and the bin widths, which might severely affect the test statistic estimation, especially in the distribution tails, as seen in Example 7.8, and even change the decision making.

In addition, GoF tests have other shortcomings. In the case of the KS GoF test, the mere observation of its critical values, as in Tables 7.5 and 7.6, reveals that the differences allowed between the empirical and theoretical distributions are too great before a false null hypothesis is rejected. For instance, for a relatively large sample, say $N = 80$, at $\alpha = 0.05$, the corresponding critical value in Table 7.6 would read $0.886/\sqrt{N} \approx 0.10$, which means that a 10 % difference, between empirical and theoretical probabilities, would be admissible before rejecting H_0. A 10 % difference is certainly highly permissible when a decision concerning a costly and potentially impacting water resources project is at stake. Another potential disadvantage of the KS GoF test is not having tables of critical values for all parent probability distributions of current use in Statistical Hydrology.

The AD GoF test is an interesting alternative to previous tests, especially for giving more weight to the distribution tails and for having a competitive advantage in power analysis, as compared to conventional GoF tests, such as the chi-square and KS (Heo et al. 2013). It exhibits the potential disadvantage of not being invariant to the estimation method of location and scale parameters. For a three-parameter distribution, calculation of the AD GoF test, requires the specification of an assumed true value of the shape parameter. The Modified AD or MAD GoF test seems to be a promising alternative as far as frequency analysis of hydrologic maxima is concerned.

The PPCC GoF test, as the remaining alternative, has good qualities such as the intuitiveness and simple formulation of its test statistic and its favorable power analysis, when compared to other GoF tests, as reported in Chowdhury et al. (1991), Vogel and McMartin (1991), and Heo et al. (2008). Unlike the AD test, PPCC GoF tests are invariant to the estimation method of location and scale parameters, which is seen as another potential advantage. In spite of the favorable comparative analysis, the power of GoF tests, here including the PPCC, is considerably lowered for samples of reduced sizes, as usually encountered in at-site hydrologic frequency analysis (Heo et al. 2008). Tables of critical values for the PPCC test are currently available for many parent probability distributions of use in Hydrology. It is worth noting, however, that the tables or regression equations for the critical values of three-parameter distributions were based on an assumed or specified value for the shape parameter (or coefficient of skewness).

GoF tests, as any other hypothesis test, aim to check whether or not the differences between empirical and theoretical realizations are significant, assuming that they arise from the parent distribution hypothesized under H_0. Therefore, the eventual decision to not reject the null hypothesis, at a given significance level, does not imply that data have been indeed sampled from the hypothesized parent distribution. This, in principle, is unknown and might be any of the many distributions that are contained in the alternative hypothesis. In addition, test statistics for GoF tests have probability distributions that change with the hypothesized parent distribution under H_0, thus yielding critical and p-values that are not comparable with each other. As such, GoF tests cannot be used to choose the best-fit distribution for a given sample. Other alternatives, such as conventional moment diagrams and L-moment diagrams, to be discussed in the next chapter, seem more useful in choosing the best-fit model.

7.6 Test for Detecting and Identifying Outliers

For a given sample, a data point is considered an outlier if it departs significantly
from the overall tendency shown by the other data points. Such a departure might
have an origin in measurement or data processing errors but might also arise from
indeterminate causes. In any case, the presence of outliers in a sample may
drastically affect parameter and quantile estimation and the fitting of a candidate
probability distribution. In Sect. 2.1.5, an ad hoc procedure to identify outliers is
described, by means of the Interquartile Range (IQR). This, although practical and
useful, is merely of an exploratory nature and does not constitute itself a formal
hypothesis test, with a prescribed significance level.

Among the available formal procedures to detect and identify outliers, the
Grubbs–Beck (GB) test, introduced by Grubbs (1950, 1969) and complemented
by Grubbs and Beck (1972), is one of the most frequently utilized in Statistical
Hydrology. Since the original test was developed for outliers of samples drawn
from a normal population, in order to use the GB tables of critical values, it is an
accepted practice to assume that the natural logarithms of the variable being
studied, including annual maxima and minima, are normally distributed (WRC
1981). As such, the quantities x_U and x_L, given in Eqs. (7.46) and (7.47), respectively define the upper and lower bounds, above and below which a possible outlier
would lie in a ranked data sample. Formally,

$$x_U = \exp\left(\bar{y} + k_{N,\alpha} s_Y\right) \tag{7.46}$$

and

$$x_L = \exp\left(\bar{y} - k_{N,\alpha} s_Y\right) \tag{7.47}$$

where $\bar{y}$ and s_Y denote the mean and standard deviation of the natural logarithms of
the data points x_i from a sample of size N of the random variable X, and $k_{N,\alpha}$ denotes
the critical value of the GB test statistic.

The critical values of the GB test statistic, for $100\alpha = 5\,\%$ and $100\alpha = 10\,\%$, and
sample sizes within the interval $10 \leq N \leq 120$, are respectively approximated by the
following equations:

$$k_{N,\alpha=0.05} = -5.2269 + 8.768\,N^{1/4} - 3.8063\,N^{1/2} + 0.8011\,N^{3/4} - 0.0656\,N \tag{7.48}$$

and

$$k_{N,\alpha=0.10} = -3.8921 + 6.7287\,N^{1/4} - 2.7691\,N^{1/2} + 0.5639\,N^{3/4} - 0.0451N \tag{7.49}$$

According to the GB test, at $\alpha = 0.05$ or 0.10, and $k_{N,\alpha}$ approximated by Eqs. (7.48) or (7.49), the data points higher than x_U or lower than x_L deviate significantly from the overall tendency of the remaining data points in the sample. Following the criticism that the GB test rarely identifies more than one low outlier in a flood data sample, when more than one does indeed exist (Lamontagne et al. 2012), research is under way to develop and evaluate new tests to detect multiple low outliers in flood data samples (see Spencer and McCuen 1996, Lamontagne and Stedinger 2016).

Once an outlier is detected and identified, the decision of keeping it or removing it from the sample is a matter of further investigation. If the careful scrutiny of an outlier data point is conclusive and characterizes it as an incorrect measurement or observation, it certainly must be expunged from the sample. The action of removing low outliers from a sample is also justified in frequency analysis of hydrologic maxima when the focus is on estimating the upper-tail quantiles, as unusually small floods can possibly affect the estimation of large floods. However, if a high outlier is identified in a sample of hydrologic maxima, it certainly should not be removed, since it can possibly result from natural causes, such as the occurrence of extraordinary hydrometeorological conditions, as compared to the ordinary data points, and be decisive in defining the distribution upper tail. The same reasoning, in the opposite direction, can be applied to hydrologic minima.

Exercises

1. Consider the test of the null hypothesis H_0: $p = 0.5$, against H_1: $p > 0.5$, where p denotes the probability of success in 18 independent trials of a Bernoulli process. Assume the decision is to reject the null hypothesis, if the discrete random variable Y, denoting the number of successes in 18 trials, is equal to or larger than 13. Calculate the power of the test, given by $[1 - \beta(p)]$, for different values of $p > 0.5$, and make a plot of $[1 - \beta(p)]$ against p.
2. Solve Exercise 1 for H_1: $p \neq 0.5$.
3. Table 7.20 lists the annual total rainfall depths, in mm, measured at the gauging station of Tokyo, Japan, from 1876 to 2015 (in Japan, the water year coincides with the calendar year). Assume this sample has been drawn from a normal population, with known variance equal to $65{,}500$ mm^2. Test H_0:$\mu_0 = 1500$ mm against H_1:$\mu_1 = 1550$ mm, at $\alpha = 0.05$.
4. Solve Exercise 3, assuming the normal variance is not known.
5. Solve Exercise 3 for H_1:$\mu_1 \neq 1550$ mm.
6. Solve Exercise 5, assuming the normal variance is not known.
7. Considering the data sample given in Table 7.20, split it into two subsamples of equal sizes. Test the null hypothesis that the population means for both subsamples do not differ from each other by 15 mm, at $\alpha = 0.05$.
8. Considering the data sample given in Table 7.20, assume the normal mean is known and equal to 1540 mm. Test H_0: $\sigma_0 = 260$ mm against H_1: $\sigma_1 < 260$ mm, at $\alpha = 0.05$.

Table 7.20 Annual total rainfall (mm) observed at the gauging station of Tokyo by Japan Meteorological Agency (Courtesy: Dr. S Oya, Swing Corporation, for data retrieval)

Year	Rainfall	Year	Rainfall	Year	Rainfall	Year	Rainfall	Year	Rainfall
1876	1756	1904	1382	1932	1690	1960	1282	1988	1516
1877	1317	1905	1330	1933	1011	1961	1260	1989	1938
1878	1764	1906	1520	1934	1247	1962	1256	1990	1513
1879	1493	1907	1640	1935	1657	1963	1575	1991	2042
1880	1686	1908	1692	1936	1628	1964	1140	1992	1620
1881	1444	1909	1512	1937	1359	1965	1392	1993	1873
1882	1478	1910	1751	1938	2230	1966	1644	1994	1132
1883	1553	1911	1867	1939	1750	1967	1023	1995	1220
1884	1315	1912	1734	1940	1094	1968	1491	1996	1334
1885	1532	1913	1597	1941	2155	1969	1343	1997	1302
1886	1290	1914	1694	1942	1470	1970	1122	1998	1547
1887	1250	1915	1927	1943	1391	1971	1439	1999	1622
1888	1379	1916	1931	1944	1329	1972	1628	2000	1603
1889	1319	1917	1308	1945	1616	1973	1150	2001	1491
1890	1958	1918	1337	1946	1236	1974	1581	2002	1295
1891	1221	1919	1534	1947	1038	1975	1541	2003	1854
1892	1715	1920	2194	1948	1757	1976	1558	2004	1750
1893	1161	1921	2025	1949	1782	1977	1454	2005	1482
1894	1321	1922	1411	1950	1952	1978	1030	2006	1740
1895	1398	1923	1697	1951	1590	1979	1454	2007	1332
1896	1374	1924	1475	1952	1641	1980	1578	2008	1858
1897	1497	1925	1713	1953	1519	1981	1464	2009	1802
1898	1712	1926	1177	1954	1771	1982	1576	2010	1680
1899	1649	1927	1445	1955	1554	1983	1341	2011	1480
1900	1188	1928	1750	1956	1657	1984	880	2012	1570
1901	1589	1929	1909	1957	1500	1985	1517	2013	1614
1902	1754	1930	1476	1958	1805	1986	1458	2014	1808
1903	1912	1931	1565	1959	1626	1987	1089	2015	1782

9. Solve Exercise 8 assuming the normal mean is not known.
10. Considering the data sample given in Table 7.20, split it into two subsamples of equal sizes. Test the null hypothesis that the population variances for both subsamples do not differ from each other, at $\alpha = 0.05$.
11. Table 7.21 lists the annual maximum daily rainfall depths, in mm, measured at the gauging station of Kochi, in southern Japan, from 1886 to 2015. Kochi is located on Shikoku Island, one of four main islands that form the Japanese archipelago, and directly faces the Pacific Ocean, with frequent and intense rainfalls, some triggered by the passing of typhoons over the region. Test the null hypothesis that the Kochi observed rainfall are random, at $\alpha = 0.05$.
12. Test the null hypothesis that Kochi observed rainfall data, as listed in Table 7.21, are independent, at $\alpha = 0.05$.

Table 7.21 Annual maximum daily rainfall (mm) observed at the gauging station of Kochi, located on Shikoku Island, by Japan Meteorological Agency (courtesy: Dr. S Oya, Swing Corporation, for data retrieval)

Year	Max 1 day	Year	Max 1 day	Year	Max 1 day	Year	Max 1 day	Year	Max 1 day
1886	117	1912	246	1938	239	1964	76	1990	197
1887	212	1913	157	1939	254	1965	194	1991	115
1888	161	1914	239	1940	112	1966	260	1992	180
1889	160	1915	156	1941	149	1967	156	1993	194
1890	293	1916	231	1942	136	1968	175	1994	158
1891	172	1917	171	1943	205	1969	163	1995	210
1892	187	1918	195	1944	149	1970	177	1996	121
1893	180	1919	184	1945	173	1971	291	1997	214
1894	129	1920	364	1946	291	1972	337	1998	629
1895	128	1921	101	1947	218	1973	122	1999	217
1896	172	1922	260	1948	226	1974	155	2000	211
1897	240	1923	139	1949	237	1975	295	2001	230
1898	200	1924	114	1950	212	1976	525	2002	106
1899	255	1925	344	1951	166	1977	126	2003	154
1900	242	1926	166	1952	152	1978	250	2004	244
1901	160	1927	162	1953	218	1979	136	2005	173
1902	170	1928	200	1954	263	1980	221	2006	188
1903	184	1929	130	1955	195	1981	136	2007	315
1904	139	1930	126	1956	371	1982	192	2008	231
1905	115	1931	162	1957	174	1983	145	2009	142
1906	200	1932	210	1958	130	1984	216	2010	236
1907	179	1933	169	1959	162	1985	176	2011	160
1908	169	1934	124	1960	158	1986	111	2012	223
1909	127	1935	184	1961	191	1987	158	2013	196
1910	120	1936	159	1962	176	1988	265	2014	372
1911	232	1937	173	1963	203	1989	163	2015	211

13. Test the null hypothesis that Kochi observed rainfall data, as listed in Table 7.21, are homogeneous, at $\alpha = 0.05$.
14. Test the null hypothesis that Kochi observed rainfall data, as listed in Table 7.21, are stationary, at $\alpha = 0.05$.
15. Use the chi-square, KS, AD, and Filliben GoF tests, at $\alpha = 0.05$, to test the null hypothesis that Tokyo annual total rainfall depths (mm), as listed in Table 7.20, are normally distributed. Estimate parameters by the method of moments.
16. Counts of *Escherichia Coli* for 10 water samples collected from a lake, as expressed in hundreds of organisms per 100 ml of water ($10^2/100$ ml), are 17, 21, 25, 23, 17, 26, 24, 19, 21, and 17. The arithmetic mean value and the variance calculated for the 10 samples are respectively equal to 21 and 10.6. Assume N represents the number of all different organisms that are present in a

sample (in analogy to $N=$ the number of Bernoulli trials) and let p denote the fraction of N that corresponds to $E.$ $Coli$ (in analogy to $p=$ the probability of success). Fit a binomial distribution to $Y=$ hundreds of $E.$ $Coli$ organisms per 100 ml of water. Use the chi-square GoF test, at $\alpha=0.10$, to check the binomial fit to the data sample.

17. Use the chi-square GoF procedure, at $\alpha=0.05$, to test the annual maximum daily rainfall (mm), at the Kochi gauging station, given in Table 7.21, for the two-parameter lognormal, one-parameter exponential, Gumbel$_{max}$, GEV, Pearson type III, and log-Pearson type III models. Use the method of moments to estimate the distributions' parameters.

18. Use the KS GoF procedure, at $\alpha=0.05$, to test the annual maximum daily rainfall (mm), at the Kochi gauging station, given in Table 7.21, for the two-parameter lognormal, one-parameter exponential, Gumbel$_{max}$, Pearson type III, and log-Pearson type III models. Use the method of moments to estimate the distributions' parameters.

19. Use the AD or MAD procedure, as appropriate, at $\alpha=0.05$, to test the annual maximum daily rainfall (mm), at the Kochi gauging station, given in Table 7.21, for the two-parameter lognormal, one-parameter exponential, Gumbel$_{max}$, GEV, Pearson type III, and log-Pearson type III models. Use the method of moments to estimate the distributions' parameters.

20. Use the PPCC GoF procedure, at $\alpha=0.05$, to test the annual maximum daily rainfall (mm), at the Kochi gauging station, given in Table 7.21, for the two-parameter lognormal, one-parameter exponential, Gumbel$_{max}$, GEV, Pearson type III, and log-Pearson type III models. Use the method of moments to estimate the distributions' parameters.

21. Use the chi-square GoF procedure, at $\alpha=0.05$, to test the annual peak discharges of the Lehigh River at Stoddartsville (m^3/s), given in Table 7.1, for the two-parameter lognormal, one-parameter exponential, Gumbel$_{max}$, GEV, Pearson type III, and log-Pearson type III models. Use the method of moments to estimate the distributions' parameters.

22. Use the KS GoF procedure, at $\alpha=0.05$, to test the annual peak discharges of the Lehigh River at Stoddartsville (m^3/s), given in Table 7.1, for the two-parameter lognormal, one-parameter exponential, Gumbel$_{max}$, Pearson type III, and log-Pearson type III models. Use the method of moments to estimate the distributions' parameters.

23. Use the AD and MAD GoF procedure, as appropriate, at $\alpha=0.05$, to test the annual peak discharges of the Lehigh River at Stoddartsville (m^3/s), given in Table 7.1, for the two-parameter lognormal, one-parameter exponential, Gumbel$_{max}$, GEV, Pearson type III, and log-Pearson type III models. Use the method of moments to estimate the distributions' parameters.

24. Use the PPCC GoF procedure, at $\alpha=0.05$, to test the annual peak discharges of the Lehigh River at Stoddartsville (m^3/s), given in Table 7.1, for the two-parameter lognormal, one-parameter exponential, Gumbel$_{max}$, GEV, Pearson type III, and log-Pearson type III models. Use the method of moments to estimate the distributions' parameters.

25. Table 2.7 of Chap. 2 lists the Q_7 flows, in m^3/s, for the Dore River at Saint-Gervais-sous-Meymont, in France, from 1920 to 2014. Fit a Weibull$_{min}$ distribution to these data, employing the MOM estimation method. Use the PPCC GoF test, at $\alpha = 0.05$, to determine whether or not the Q_7 flow data can be modeled by a Weibull$_{min}$ distribution.

26. Table 2.6 of Chap. 2 lists the independent peak discharges of the Greenbrier River at Alderson (West Virginia, USA) that exceeded the threshold 17000 cfs. Fit a Generalized Pareto Distribution (GPA) to these data, employing the MOM estimation method (see Example 5.11). Use the MAD GoF test, at $\alpha = 0.05$, to determine whether or not the peaks-over-threshold flood data can be modeled by a GPA distribution.

27. For the data listed in Table 7.21, use both the Interquartile Range (IQR) exploratory procedure and the Grubbs–Beck test, at $\alpha = 0.05$ and $\alpha = 0.10$, to detect and identify possible outliers. Compare and comment on your results.

28. Use the GB test, at $\alpha = 0.05$, to detect possible outliers in the sample of annual peak discharges of the Lehigh River at Stoddartsville (m^3/s), given in Table 7.1. Read about the Spencer-McCuen test, as described in Spencer and McCuen (1996) and commented on Lamontagne and Stedinger (2016), and apply it to the data in Table 7.1. Compare with the GB test and comment on your results.

References

Ahmad MI, Sinclair CD, Spurr BD (1998) Assessment of flood frequency models using empirical distribution function statistics. Water Resour Res 24(8):1323–1328

Anderson TW, Darling DA (1954) A test of goodness of fit. J Am Stat Assoc 49:756–769

Bickel PJ, Doksum K (1977) Mathematical statistics: basic ideas and selected topics. Prentice Hall, Englewood Cliffs, NJ

Bobée B, Ashkar F (1991) The Gamma family and derived distributions applied in hydrology. Water Resources Publications, Littleton, CO

Casella G, Berger R (1990) Statistical inference. Duxbury Press, Belmont, CA

Chandra M, Singpurwalla MD, Stephens MA (1981) Kolmogorov statistics for tests of fit for the extreme value and Weibull distribution. J Am Stat Assoc 76(375):729–731

Chowdhury JU, Stedinger JR, Lu L-H (1991) Goodness of fit tests for regional generalized extreme value flood distributions. Water Resour Res 27(7):1765–1776

Crutcher HL (1975) A note on the possible misuse of the Kolmogorov-Smirnov test. J Appl Meteorol 14:1600–1603

D'Agostino RB, Stephens M (1986) Goodness-of-fit techniques. Marcel Dekker, New York

Heo J-H, Kho YW, Shin H, Kim and Kim T (2008) Regression equations of probability plot correlation coefficient test statistic from several probability distributions. J Hydrol 355:1–15

Heo J-H, Shin H, Nam W, Om J, Jeong C (2013) Approximation of modified Anderson-Darling test statistic for extreme value distributions with unknown shape parameter. J Hydrol 499:41–49

Hoel PG, Port SC, Stone CJ (1971) Introduction to statistical theory. Houghton Mifflin, Boston

Hollander M, Wolfe DA (1973) Nonparametric statistical methods. Wiley, New York

Gibbons JD (1971) Nonparametric statistical inference. McGraw-Hill, New York

Grubbs FE, Beck G (1972) Extension of sample sizes and percentage points for significance tests of outlying observations. Technometrics 14(4):847–854

Grubbs FE (1969) Procedures for detecting outlying observations in samples. Technometrics 11(1):1–21

Grubbs FE (1950) Sample criteria for testing outlying observations. Ann Math Stat 21(1):27–58

Filliben JJ (1975) The probability plot correlation coefficient test for normality. Technometrics 17(1):111–117

Kim S, Kho Y, Shin H, Heo J-H (2008) Derivation of probability plot correlation coefficient test statistics for the generalized logistic and the generalized Pareto distributions. Proceedings of the world environmental and water resources congress 2008 Ahupua'a, 1:10. American ociety of Civil Engineers

Kolmogorov A (1933) Sulla determinazione empirica di una legge di distribuzione. Giornale dell' Istituto Italiano degli Attuari 4:83–91

Kottegoda NT, Rosso R (1997) Statistics, probability, and reliability for civil and environmental engineers. McGraw-Hill, New York

Lamontagne JR, Stedinger JR (2016) Examination of the Spencer-McCuen outlier detection test for log-Pearson type III distributed data. J Hydrol Eng 21(3):1–7, 04015069

Lamontagne JR, Stedinger JR, Berenbrock C, Veilleux AG, Ferris JC, Knifong DL (2012) Development of regional skew for selected flood durations for the Central Valley Region, California, based on data through water year 2008. Report 2012-5130. United States Geological Survey, Reston, VA

Larsen RJ, Marx ML (1986) An introduction to mathematical statistics and its applications. Prentice-Hall, Englewood Cliffs, NJ

Lilliefors HW (1967) On the Kolmogorov-Smirnov test for normality with mean and variance unknown. J Am Stat Assoc 62(318):399–402

Mann HB, Whitney DR (1947) On the test of whether one of two random variables is stochastically larger than the other. Ann Math Stat 18:50–60

Mauget S (2011) Time series analysis based on running Mann-Whitney Z statistics. J Time Series Anal 32:47–53

Mood AM, Graybill FA, Boes DC (1974) Introduction to the theory of statistics, international, 3rd edn. McGraw-Hill, Singapore

NERC (1975) Flood studies report, vol 1. National Environmental Research Council, London

Özmen T (1993) A modified Anderson-Darling goodness-of-fit test for the gamma distribution with unknown scale and location parameter. Thesis of MSc in Operation Research, Air Force Institute of Technology, Wright-Patterson Air Force Base, OH

Pettitt AN (1979) A non-parametric approach to the change-point problem. Appl Stat 28(2):126

Ramachandran KM, Tsokos CP (2009) Mathematical statistics with applications. Elsevier Academic Press, Burlington, MA

Siegel S (1956) Nonparametric statistics for the behavioral sciences. McGraw-Hill, New York

Spencer CS, McCuen RH (1996) Detection of outliers in Pearson type III data. J Hydrol Eng 1(2):2–10

Stedinger JR, Vogel RM, Foufoula-Georgiou E (1993) Frequency analysis of extreme events. In: Maidment DR (ed) Chapter 18 in handbook of hydrology. McGraw-Hill, New York

Stephens MA (1974) EDF statistics for goodness of fit and some comparisons. J Am Stat Assoc 69(347):730–737

Smirnov N (1948) Table for estimating the goodness-of-fit of empirical distributions. Ann Math Stat 19:279–281

Vlcek O, Huth R (2009) Is daily precipitation Gamma-distributed? Adverse effects of an incorrect use of the Kolmogorov-Smirnov test. Atmos Res 93:759–766

Vogel RM, Kroll CN (1989) Low flow frequency analysis using probability-plot correlation coefficients. J Water Resour Plann Manag 115(3):338–357

Vogel RM, McMartin DE (1991) Probability plot goodness-of-fit and skewness estimation procedures for the Pearson type 3 distribution. Water Resourc Res 27(12):3149–3158

Wald A, Wolfowitz J (1943) An exact test for randomness in the non-parametric case based on serial correlation. Ann Math Stat 14:378–388

Wang QJ (1998) Approximate goodness-of-fit tests of fitted generalized extreme value distributions using LH moments. Water Resour Res 34(12):3497–3502

WMO (2009) Guide to hydrological practices. WMO No. 168, vol II, 6th edn. World Meteorological Organization, Geneva

WRC (1981) Guidelines for determining flood flow frequency—Bulletin 17B. United States Water Resources Council-Hydrology Committee, Washington

Yule GY, Kendall MG (1950) An introduction to the theory of statistics. Charles Griffin & Co., London

Chapter 8
At-Site Frequency Analysis of Hydrologic Variables

Mauro Naghettini and Eber José de Andrade Pinto

8.1 Introduction

Analysis and estimation of flood flows and related rainfall intensities and depths have always been ubiquitous problems in the domain of water resources engineering. Throughout history, the natural attraction that floodplains and river valleys have exerted on human societies, due to the favorable conditions to develop and maintain activities, such as agriculture, fishing, transportation, and convenient access to local water resources, has induced humans to occupy and use them. However, the economic and social benefits resulting from the occupation and use of river valleys and floodplains are frequently offset by the negative effects of flood-induced disasters, such as the loss of lives and the material damage to riverine communities and properties. In fact, it should not be a surprise that rivers occasionally reclaim their own dynamic constructions which are their valleys and plains. However, it is an unfortunate surprise to acknowledge that human societies occasionally disregard the fact that occupying the floodplains means to coexist with risk.

Flood-risk reduction and mitigation of flood-induced damages can be achieved by human actions on the fluvial system, such as the construction of reservoirs and levees, regulations controlling land use in flood hazard areas, and the implementation of alert systems and protection measures for riverine structures and properties. Estimation of flood flows is essential not only for planning these interventions but also for designing and operating the engineering structures to control and manage

M. Naghettini (✉)
Universidade Federal de Minas Gerais, Belo Horizonte, Minas Gerais, Brazil
e-mail: mauronag@superig.com.br

E.J.d.A. Pinto
CPRM Serviço Geológico do Brasil, Universidade Federal de Minas Gerais,
Belo Horizonte, Minas Gerais, Brazil
e-mail: eber.andrade@cprm.gov.br

© Springer International Publishing Switzerland 2017
M. Naghettini (ed.), *Fundamentals of Statistical Hydrology*,
DOI 10.1007/978-3-319-43561-9_8

water since their structural safety depends much on reliable estimates of flood characteristics. Engineers and hydrologists are often asked to estimate relevant characteristics of flood flows, such as the peak discharges, the volume and duration of flood hydrographs, the flooded areas, the associated precipitation depths and intensities, as well as their corresponding critical values for design and/or operation purposes (ASCE 1996), and further relate them to exceedance probabilities or return periods. Equally important is the estimation of low-flow quantiles and associated probabilities, as necessary to develop and manage water resources during prolonged droughts and periods of water scarcity.

Frequency analysis of hydrologic variables, broadly defined here as the quantification of the expected number of occurrences of an event of a given magnitude, is perhaps the earliest and most frequent application of probability and statistics in the field of water resources engineering. In brief, the methods of frequency analysis, as applied to maxima, aim to estimate the probability with which a random variable will be equal to or greater than a given quantile, from a sample of observed data (Kidson and Richards 2005). Frequency analysis of minima is similar, but non-exceedance probabilities are of concern. If only data recorded at a single streamflow (or rainfall) gauging station are available, an at-site frequency analysis is being carried out. Otherwise, if other observations of the variable, as recorded at distinct gauging stations within a specified region, are jointly employed for statistical inference, then the frequency analysis is said to be regional.

At-site frequency analysis of hydrologic random variables has received much attention from hydrologists over the years. New probability distribution models and improved techniques of statistical inference have been proposed, in the pursuit of more reliable estimates of rare quantiles. However, the relatively short samples of hydrologic maxima seem to impose a limit to the degree of statistical sophistication that can be employed in at-site frequency analysis. Given this, the regional frequency analysis is certainly an alternative that seeks to balance the limited temporal distribution of hydrologic data, as recorded at a single site, with a more detailed characterization of the variable with data from multiple locations. Potter (1987), Bobée and Rasmussen (1995), and Hosking and Wallis (1997) are among those who presented the many advantages and arguments in favor of regional methods, as compared to at-site frequency analysis. Despite these arguments, it is plausible to expect that a number of decisions on the occurrence of rare hydrologic events will be made still on the basis of a single sample of hydrologic data. This expectation may be justified either by (1) the scarcity or even the absence of adequate hydrologic data within a given region; or (2) the expedite way with which engineering solutions are occasionally proposed to problems involving hydrologic rare events. It appears then that at-site frequency analysis will stay as a current engineering method for some time and thus deserves a full description of its methods and limitations.

Hydrologic frequency analysis, either at a site or concurrently at multiple sites, is generally performed with the reduced extreme-value series, which are collections of maximum or minimum values that have occurred during a time period. If the time period is a year, extreme-value series are annual; otherwise, they are said to be non-annual, as extremes can be abstracted within a season or within a varying time interval (Chow 1964). As seen in previous chapters, the frequency analyses of annual total values (or annual mean values) are, in most cases, applications of fitting the normal distribution to sample data and are not the main focus of this chapter. Amongst the non-annual extreme-value series, the partial-duration series are those in which only the independent extreme values that are higher, in the case of maxima, or lower, in the case of minima, than a specified reference threshold are selected from the records. Sections 8.2 and 8.3 of this chapter deal mostly with the frequency analysis of annual extreme values, with a particular focus on annual maxima, whereas Sect. 8.4 is devoted to the partial duration series. The mathematical relations between the two types of frequency analyses are also given in Sect. 8.4.

The sample of extreme data to be considered for frequency analysis should be representative of the variation expected for the random quantity of interest, be previously screened for occasional or systematic measurement errors (observational, and/or processing), and be large enough to allow some reliable extrapolation. In addition, as seen in previous chapters, data must have been drawn at random from a single population, such that it constitutes a simple random sample. In this regard, the nonparametric tests of randomness, independence, homogeneity, and stationarity, as seen in Chap. 7, play the important role of checking the statistical attributes data must exhibit so that a frequency analysis can be performed afterwards.

Hydrologic frequency analyses can also be categorized as graphical or analytical. The former type consists of graphing the ranked data against a conveniently distorted scale of probabilities, and of extending the resulting linearized relation, limited by how reliable the analyst judges the data to be. This rather subjective approach used to be current practice in the early days of Statistical Hydrology and, of course, has been totally superseded by analytical frequency analysis, in which a theoretical probability model, as fitted to the data sample, completely defines the random variable behavior throughout its entire domain of variation. However, some elements of graphical frequency analysis, such as the construction of probability charts (or probability papers), are still useful if used together with analytical frequency analysis. That is the main justification for including Sect. 8.2, on graphical frequency analysis, in this chapter.

In carrying out a full analytical frequency analysis of a hydrologic variable, beyond the usual problems of inferring parameters and quantiles from short samples, lies the nontrivial task of identifying the most adequate and robust probability model for a case in particular. No consensual and unequivocal set of rules exists to respond to such a task. In fact, the true probability distribution from which a particular sample has been drawn is most likely a complicated mixture of many other probability distributions and will never be entirely known. As remarked by Stedinger et al. (1993), even if the functional representation of such a mixture of distributions were known, it would likely have too many parameters to be of practical use. The issue is then to choose the most parsimonious and robust distribution, among a set of candidate models, capable of yielding reasonably reliable quantile estimates for the engineering problem at hand. Some general guidelines for making such a choice, together with a few objective aiding tools are provided in Sect. 8.3.

In summary the steps necessary to complete an at-site frequency analysis of a hydrologic variable are as follows:

- Have an option to work with the annual extreme-value series or with the partial duration series;
- Screen data for measurement, observational, or processing errors;
- Test data for randomness, independence, homogeneity, and stationarity;
- Detect and identify possible outliers in the sample, and assess actions relating thereto;
- Define a set of candidate probability distribution functions, estimate their parameters, quantiles, and confidence intervals, and assess their respective goodness-of-fit indicators;
- Make use of the general guidelines and objective aiding tools to select the most parsimonious and robust model among the set of fitting candidate probability distribution functions;
- Extend the selected model up to a credible limit of extrapolation, compatible with the sample size and estimated uncertainties; and
- If a further extension is necessary, consider improving the at-site study by resorting to regional frequency analysis, described in Chap. 10.

Some of these steps are detailed and discussed in the sections that follow. Prior to that, however, it is instructive to describe previously used techniques pertaining to graphical frequency analysis that can be helpful in successfully completing some of the steps previously outlined. These techniques are described next, in Sect. 8.2.

8.2 Graphical Frequency Analysis

In brief, the graphical frequency analysis consists of plotting the ranked data, on the vertical axis, against their respective empirical probabilities, on the horizontal axis. Associating each data point to its respective exceedance probability, or equivalently to its return period, is uncertain as it depends on the sample size and also on the existence of possible outliers in the sample. For instance, if the true 500-year flood peak is found amidst the 80 data points that make a given sample, it would be impossible to assign the correct probability or return period to that particular observation, based only on the data recorded at that location. On the other hand, no one would have the prior knowledge about the true return period associated with that particular point, based only on the data observed at that site. Sampling is thus a definite source of uncertainty in probability plotting.

Early applications of graphical frequency analysis were likely motivated by the idea of extrapolating the overall tendency shown by the plot of data against empirical probabilities to rarer events. One issue of concern here is how far the extrapolation should go before unreliable results are obtained or before a new tendency, beyond the range of past observations, is produced. The uncertainties, arising either from extrapolation or from sampling issues, can be partially reduced by (a) the construction of the so-called probability papers on which linearized probability plots can be drawn; and (b) the development of improved plotting-position formulae to assign empirical probabilities to ranked data. These are the topics to be covered in the next subsections.

8.2.1   Probability Paper

Except for the uniform distribution, when a cumulative distribution function $F_X(x)$ is plotted against x on an arithmetic scale, a curved line will result, such as that highlighted in Fig. 8.1, valid for the standard normal distribution $\Phi(z)$ versus z. However, any probability distribution that is fully described by location and scale parameters would plot as a straight line if the arithmetic scale is conveniently transformed into a probability scale. In order to perform such a graphical lineari-zation, one should first draw a straight line between two points of the curved line, covering the desired range, as depicted in Fig. 8.1 (Haan 1977). Then, the

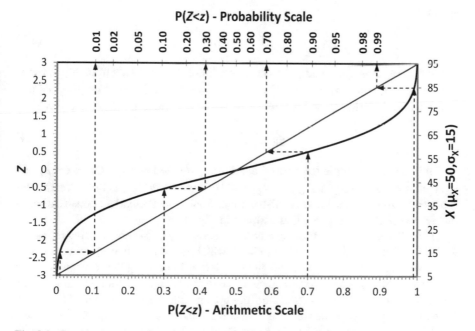

Fig. 8.1 Graphical construction of a normal probability paper (adapted from Haan 1977)

probability scale is constructed on the upper secondary horizontal axis, by shifting the probabilities, as read on the original arithmetic scale on the lower primary axis, to the left or to the right, as exemplified by the dashed arrowhead lines shown in Fig. 8.1 for the normal distribution. Also shown in this figure is the linearized normal plot that would have been obtained, should the generic normal variate X, on the right secondary vertical axis, have mean $\mu_X = 50$ and standard deviation $\sigma_X = 15$. If all labels are removed from Fig. 8.1, except the probability scale on the upper horizontal axis, one would have a normal probability paper.

The mathematical formalism for constructing probability papers can be set up by starting from the analytic equation of a straight line relating the variable X, on the vertical axis, to its cumulative distribution function $F_X(x)$, on the horizontal axis. Formally,

$$X = ag[F_X(x)] + b \tag{8.1}$$

where a is the angular coefficient, b denotes the value of X when $g[F_X(x)] = 0$, and $g[F_X(x)]$ represents the linearizing function for the CDF $F_X(x)$. Notice that for $F_X(x)$ being linear, it must be completely definable by the parameters β and α, of location and scale, respectively. For any CDF as such, the corresponding standard variate Z can be written as

$$Z = \frac{X - \beta}{\alpha} \tag{8.2}$$

such that

$$X = \alpha Z + \beta \tag{8.3}$$

By comparing Eqs. (8.1) and (8.3), one immediately sees they are of identical forms and that $a = \alpha$, $b = \beta$, and $g[F_X(x)] = Z$. As $Z = F_Z^{-1}(z)$, the function $g[F_X(x)]$ that linearizes the relation between X and $F_X(x)$ can be written in the general form

$$g[F_X(x)] = F_Z^{-1}(z) \tag{8.4}$$

Taking as an example the normal distribution defined by the parameters μ_X and σ_X, then $a = \sigma_X$, $b = \mu_X$, and $g[F_X(x)] = \Phi^{-1}(z)$, where $\Phi^{-1}(z)$ denotes the inverse function of the standard normal distribution. Such a mathematical construction of the normal probability paper is illustrated in Fig. 8.2.

In Fig. 8.2, one notices that the auxiliary secondary horizontal axis depicting the variation of the standard normal variate Z, which is the only horizontal axis that is actually on arithmetic scale, serves the purpose of linearizing the relation between X and $F_X(x)$, through the function $g[F_X(x)] = \Phi^{-1}(z)$. The other horizontal axes, the lower with the CDF $F_X(x)$ itself and the upper with the complement $[1-F_X(x)]$, have their respective scales distorted, in correspondence to the values of Z, so that the relation of X either with $F_X(x)$ or $[1-F_X(x)]$ plots as a straight line, with

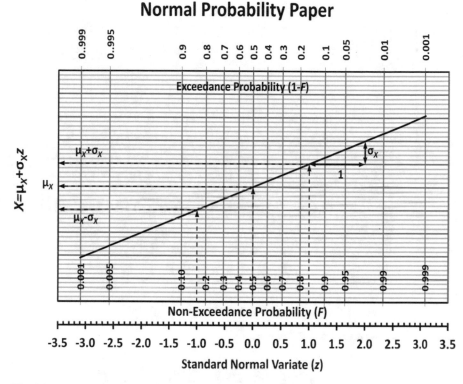

Fig. 8.2 Mathematical construction of a normal probability paper

coefficients $a = \mu_X$ and $b = \sigma_X$. Now, if the auxiliary horizontal axis of the standard variate Z and all labels are removed from Fig. 8.2, except the probability scale on the lower horizontal axis, the resulting chart would be a generic normal probability paper, as depicted in Fig. 8.3. On such a paper one could easily plot the N ranked observations of an assumed normally distributed variable X, according to a conveniently chosen scale on the vertical axis, against their corresponding empirical non-exceedance probabilities. These can be calculated through an appropriate plotting position formula selected among those listed in Table 7.14.

As previously mentioned, probability papers can be constructed for any probability distribution that can be completely defined by location and scale parameters. Besides the normal probability paper, the others that are most frequently used in Statistical Hydrology are the two-parameter lognormal, exponential, and Gumbel$_{max}$ papers. The lognormal is identical to the normal paper, except for the scale of variable X, on the vertical axis, which is transformed into a logarithmic scale. A generic lognormal probability paper is shown in Fig. 8.4, where the non-exceedance probabilities $F_X(x)$ appear on the lower horizontal axis and the return periods T, obtained as $T = 1/[1 - F_X(x)]$, on the auxiliary upper horizontal axis.

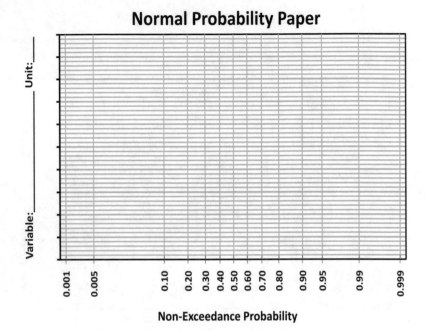

Fig. 8.3 A generic normal probability paper

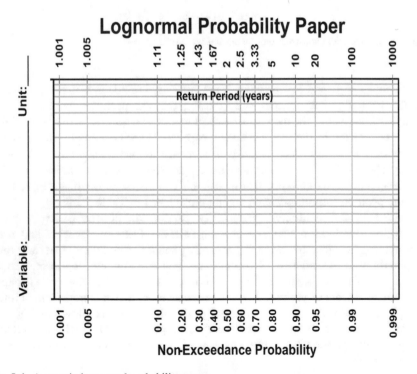

Fig. 8.4 A generic lognormal probability paper

Table 8.1 Elements necessary for constructing probability papers

Distribution	Variable	Location	Scale	a	b	$g[F_X(x)]$
Normal	X	μ_X	σ_X	σ_X	μ_X	$\Phi^{-1}(z)$
Lognormal	$Y = \ln X$	μ_Y	σ_Y	σ_Y	μ_Y	$\Phi^{-1}(z)$
Exponential	X	0	θ	θ	0	$-\ln(1 - F)$
Gumbel$_{max}$	X	β	α	α	β	$-\ln[-\ln(F)]$

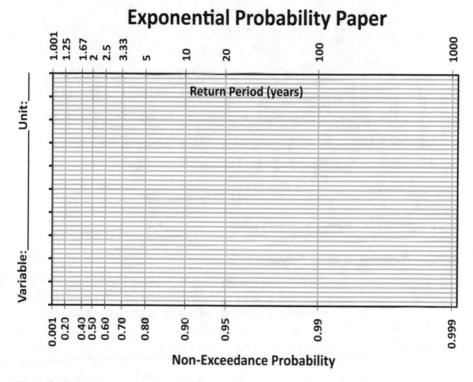

Fig. 8.5 A generic exponential probability paper

Table 8.1 summarizes the coefficients a and b, the transformation of X, if applicable, and the linearizing function $g[F_X(x)]$ that are necessary to construct the probability papers most frequently used in Statistical Hydrology. The steps that should be followed to construct these papers are similar to those described for the normal probability paper. Figures 8.5 and 8.6 show the Exponential and Gumbel$_{max}$ probability papers, respectively. All these papers have been drawn using a few resources of MS Excel, such as white fonts and white colors to *hide* labels and axes, manual labeling of markers to define probability and return period scales, and "Error Bars" (from "Layout" within the "Chart Tools" menu) to extend vertical lines starting from the markers on the horizontal axis.

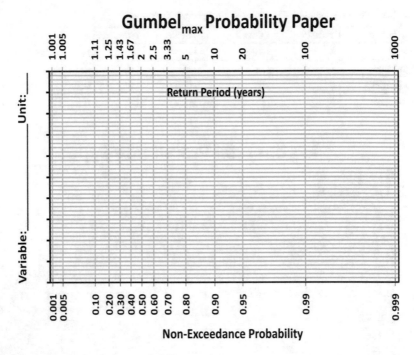

Fig. 8.6 A generic Gumbel$_{max}$ probability paper

8.2.2 *Empirical Distribution*

A probability plot is a graph that associates the magnitudes of ranked data, on the vertical axis, with their respective empirical probabilities, or plotting positions, on the horizontal axis. The resulting curve is referred to as the empirical distribution. If the population were known, the plotting position associated with a given data point would be a mere fraction between the number of data that have not exceeded it and the total number of data points. Thus, in such a case, the plotting position of the largest data point would be 1 and the plotting position of the lowest would be 0. However, for actual samples, as the smallest and largest data are always waiting to happen, plotting positions with extreme values, such as 0 and 1, should be avoided.

Gumbel (1958) postulated the following attributes a formula for calculating plotting positions should exhibit:

- The plotting position must be such that all observed data can be plotted.
- The plotting position should be bounded by $(m - 1)/N$ and m/N, where m denotes the ranking order of a given data point and N the total number of observations. It is customary to rank maximum (or total or mean) data in descending order, such that $m = 1$ for the largest observation and plotting positions refer to exceedance probabilities, and rank minimum data in ascending order, such that

$m = 1$ for the smallest observation and plotting positions refer to non-exceedance probabilities.

- In the case of annual extreme-value series, the return period associated with the largest observation, for maxima, or with the smallest observation, for minima, should converge to N.
- The observations should be equally spaced on the empirical probability scale.
- The plotting position formula should be intuitive, analytically simple, and easy to use.

As previously outlined in Sect. 7.5.4, within the context of PPCC GoF test, many plotting position formulae have been proposed for use in hydrologic frequency analysis. Some are summarized in Table 7.14, according to their respective statistical justifications, and are all particular cases of Cunnane's general formula given by Eq. (7.40). The different plotting position formulae yield similar values for the central portion of the empirical distribution but they can diverge by a great amount in the tails. As noted by Stedinger et al. (1993), all plotting position formulae give crude estimates of the unknown probabilities associated with the largest and, by symmetry, with the smallest observations. These authors suggest that the actual exceedance probability for the largest observed datum in a sample of size N lies between $0.29/N$ and $1.38/(N + 2)$ nearly 50 % of the time. To give an example, in such conditions, the true return period of the largest observation in an annual-maxima sample of size 50 would be somewhere between 38 and 172 years, in half of the time. The reader interested in plotting position formulae should consult Cunnane (1978), for a full review, and Stedinger et al. (1993), for an elucidative discussion.

The plotting of an empirical distribution can be made on one of the probability papers previously described. It consists of the following sequential steps:

(a) Rank data in descending order, if maxima (or total or mean values), or in ascending order, if minima, and denote each data point as x_m, $m = 1, 2, \ldots, N$, where m represents the ranking order.
(b) Calculate the respective plotting position for each ranked data point, according to an appropriately selected formula, among those of Table 7.14, and to the choice made for the probability paper. Notice that if data refer to maxima (or total or mean values), the resulting plotting position would result in an exceedance probability; otherwise, in a non-exceedance probability. For both cases, plotting positions are represented as q_m.
(c) Set an adequate scale on the vertical axis of the probability paper and plot all pairs $[q_m, x_m]$ for $m = 1, 2, \ldots, N$.

Example 8.1 Table 8.2 lists part of the 248 annual total rainfall depths (mm) observed at the Radcliffe Meteorological Station, in Oxford, England, from 1767 to 2014, retrievable from http://www.geog.ox.ac.uk/research/climate/rms/rain.html. Table 8.2 also lists part of the data, as ranked in descending order, and some partial calculations necessary to plot the empirical distribution. Among these are the plotting positions in the last column, as calculated by Blom's formula, which

Table 8.2 Elements for plotting the empirical distribution of the annual total rainfall depths observed at the Radcliffe Meteorological Station, in Oxford, England, 1767–2014

Year	Rainfall x (mm)	Ranked data (x_m)	Rank order (m)	Blom's $q_m P(X \geq x_m)$
1767	777.1	1034.7	1	0.0025
1768	861.3	978.9	2	0.0065
1769	595.8	962.8	3	0.0106
1770	636.1	913.1	4	0.0146
1771	411.8	895.9	5	0.0186
...	...	...	m	$q_m = \frac{m-0.375}{N+0.25}$
2010	624.7	417.7	244	0.9814
2011	476.6	411.8	245	0.9854
2012	978.9	408.7	246	0.9894
2013	642.4	379.5	247	0.9935
2014	797.1	353.4	248	0.9975

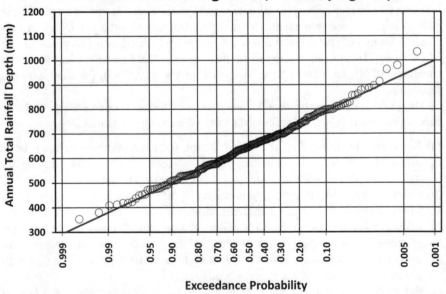

Fig. 8.7 Empirical and theoretical distributions of the annual total rainfalls at the Radcliffe Station, in Oxford, England, on normal probability paper

appears adequate since Example 7.8 has not discarded the null hypothesis of normally distributed data. Plot the corresponding empirical distribution.

Solution Figure 8.7 depicts the empirical distribution graph for the annual total rainfall depths, in mm, observed at the Radcliffe Meteorological Station, in Oxford,

England, from 1767 to 2014, resulting from plotting the pairs $[q_m, x_m]$, in black circles, on the normal probability paper. Superimposed over the empirical distribution, lies the theoretical normal distribution fitted to the sample, with parameters estimated by the method of moments. Notice that, as expected, the fitted normal distribution plots as a straight line on the normal probability paper. However, even for such a huge sample, relative to the usual sample sizes found in hydrology, prudence must be exercised in extending the normal straight line much beyond the range of observed data. On one side, the plotting positions calculated for the largest observations would shift to the right of their current positions, should these same data remain the largest for the next, say 50 years, thus reinforcing the hypothesis of normally distributed data; the opposite could also possibly be true. On the other side, based on the argument about plotting positions raised by Stedinger et al. (1993), the true exceedance probability associated with the largest annual rainfall over the 248 years of measurements at the Radcliffe Meteorological Station lies between 0.0012 (or $T = 855$ years) and 0.0055 (or $T = 181$ years) nearly 50 % of the time. This example illustrates the main uncertainties related to graphical frequency analysis, arising either from extrapolation or from sampling issues.

Sometimes it is helpful to plot different theoretical distributions, as fitted to the same sample, on a single probability paper, overlaid on the empirical distribution. Such a practical approach to visually compare how different distributions fit the data is an example of the joint use of graphical and analytical frequency analyses and is used as an aiding tool in later sections of this chapter. It is worth noting, however, that two-parameter distributions, other than the one used to construct the probability paper, and three-parameter distributions will plot as curves on the graph. Once more, in performing such a useful comparative analysis, one should not disregard the inescapable uncertainties arising from extrapolation or sampling issues. For plotting several distributions of annual maxima on a single chart, the exponential probability paper is a convenient choice since, as the linearizing function is $g(F) = -\ln(1 - F) = \ln(T)$, it suffices to log-transform the return period scale. Furthermore, with respect to the exponential straight line, as plotted on an exponential paper, distributions with heavy upper tails will slope upwards, whereas distributions with light (hyperexponential) upper tails will slope downwards.

8.2.3   Plotting of Historic Events

Measuring rainfall depths is usually less demanding than gauging river flows since it does not require the previous definition of rating curves and in-river measurement of flow velocities. As a result, rainfall gauging networks are usually denser and have longer periods of records than flow gauging networks, on a worldwide scale. Bayliss and Reed (2001) report that, up until 1999, the mean length of annual flood peak records, available for regional flood studies in the UK, was just 23.4 years, which is probably much less than the rainfall homologous statistic.

However, as large floods usually leave marks on river reaches, on river cross sections, and on human-made constructions, informal records on past events can be retrieved from several sources and then incorporated into flood frequency analysis.

Possible sources of informal (or non-systematic) records of past floods include peak levels marked on plaques, bridges, and buildings; church and local authorities archives; drawings and blueprints of past engineering works; and newspapers. These are termed historical floods and should be distinguished from the so-called paleofloods. Paleoflood reconstruction, by tracing and interpreting geological and botanical evidences that remained preserved long after the passage of floods, is another possible source of informal records of extreme floods that have occurred in the past. A new science, named Paleoflood Hydrology, has emerged from the development of new techniques and methods, over recent decades, to reconstruct ancient floods and incorporate them into frequency analysis. The reader is referred to House et al. (2002) and to Benito and Thorndycraft (2004) for a full account on the developments of Paleoflood Hydrology. In Chap. 11 an example of Bayesian flood frequency analysis, with the incorporation of paleofloods, illustrates such an interesting topic. This subsection focuses on how to assign empirical probabilities (or plotting positions) to historical floods.

Historic flood marks indicate the water levels that have risen above a fixed threshold during some historic period. For some, occurrences of floods greater than such a reference threshold are known but the magnitudes of flood peaks are usually unknown or poorly defined. Flood series that contain such a type of data are said to be censored and the threshold is the censoring value (NERC 1975, Bayliss and Reed 2001). Stedinger and Cohn (1986) refer to such a flood series as a binomially censored series since each flood should belong to one of the two possible states: either above or below the threshold. For some other historic floods, both their occurrences and magnitudes above the threshold are known or can be determined. In both cases, censored historic series should be analyzed alongside systematic series. Assigning empirical probabilities or plotting positions to such a combined series requires an attentive interpretation of flood records.

Following Hirsch (1987), assume that, in a period of N years, there are g known flood occurrences, of which, the k largest are singled out. Of these N years, s years correspond to the systematic period of records, such that $s < N$. Of the k largest floods in N years, e have occurred within the systematic period of records, with $e < k$ and $e < s$, such that $g = k + s - e$. The intrinsic assumption here is that a threshold Q_0 has been defined such that the k largest floods are larger than or equal to it and the remainder are smaller than it. Figure 8.8 illustrates such a scheme for considering both systematic and historic flood records.

Hirsch (1987) analyzed and compared plotting-position formulae for flood records of combined systematic and historic information. These are termed exceedance formulae and are generalizations of the plotting-position formula introduced by Hirsch and Stedinger (1986). The generalized exceedance formula can be written as

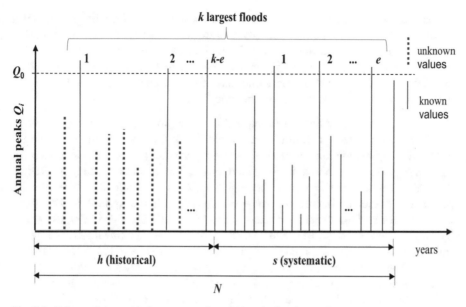

Fig. 8.8 Scheme for considering systematic and historic flood records

$$
\begin{cases}
q_m = \dfrac{m-a}{k+1-2a}\left(\dfrac{k}{N}\right) & \text{for} & m = 1,\cdots,k & \text{(a)} \\[3mm]
q_i = \dfrac{k}{N} + \dfrac{m-k-a}{s-e+1-2a}\left(\dfrac{N-k}{N}\right) & \text{for} & m = k+1,\cdots,k+s-e & \text{(b)}
\end{cases}
\qquad (8.5)
$$

where a denotes the constant in Cunnane's general plotting position formula, as in Eq. (7.40) and according to recommendations given in Table 7.14, N is the total number of years resulting from the union of systematic and historic flood records, such that $N = h + s$, and k represents the total number of floods that have exceeded the threshold Q_0.

The system of Eq. (8.5) allows the plotting of empirical probabilities associated with both systematic and historical floods. Equation (8.5a) should be applied to all floods above the threshold, from the systematic and historical records, whereas Eq. (8.5b) should be applied to the systematic floods below the threshold. As for the specific value of a that should be used in the system of Eq. (8.5), Hirsch (1987) points out that $a = 0$, as related to Weibull plotting-position formula, appears to be more robust as regards probability unbiasedness, whereas $a = 0.44$ and $a = 0.5$, as respectively related to Gringorten and Hazen formulae, appear to be preferable, as quantile unbiasedness is concerned. Example 8.2 illustrates the use of the system of Eq. (8.5).

Example 8.2 Bayliss and Reed (2001) compiled 15 floods, from 1822 to 1998, that have exceeded the reference threshold $Q_0 = 265$ m^3/s, for the River Avon at Evesham Worcestershire, in England. Their sources of information were newspapers, engineering and scientific journals, and archives from the Severn River Authority. The systematic flow records span from 1937 to 1938. They were ranked in descending order and are listed in Table 8.3, alongside the compiled historic floods. Use the system of Eq. (8.5), with $a = 0.44$ as related to Gringorten's plotting position formula, to plot on the same chart the systematic floods only, and the combined records of systematic and historic floods.

Solution The fourth column of Table 8.3 lists the plotting positions q_m as calculated with the Gringorten formula (see Table 7.14); the fifth column of Table 8.3 gives the corresponding empirical return periods, in years, calculated as $T_m = 1/q_m$. The combined historic and systematic records, already ranked in descending order, are listed in the eighth column of Table 8.3. The plotting positions, in the ninth column, were calculated with the system of Eq. (8.5): Eq. (8.5a) for all floods that have exceeded the threshold $Q_0 = 265$ m^3/s, as defined by Bayliss and Reed (2001), and Eq. (8.5b) for all systematic floods below the threshold. The parameters used in the equations are: $N = 177$ years (1998 − 1882 + 1); $h = 115$ years; $s = 62$ years; $k = 19$ floods above 265 m^3/s; $e = 4$ floods in the systematic period of records above 265 m^3/s; $k\text{-}e = 15$ historic floods above 265 m^3/s; and $a = 0.44$. Figure 8.9 depicts the empirical distribution for the systematic floods only, and the combined records of systematic and historic floods, on exponential probability paper.

8.3 Analytical Frequency Analysis

Conventional frequency analysis of sample data of a random variable, whose analytical form of its probability distribution is known or can reliably be assumed, consists of estimating the distribution parameters, using the estimation method that best combines the attributes of efficiency and accuracy, and of estimating the desired quantiles and their respective confidence intervals. In the case of hydrologic maxima (and minima and means, with respectively lesser emphasis), the analytical form of the parent distribution is not known or cannot be unequivocally assumed, thus making the data sample the only piece of objective information that is actually known, except for the cases where historical and paleoflood evidences are available. Such a complicating factor has led hydrologists to work with a range of candidate probability distributions, as described in Chap. 5, to model hydrologic random variables. These distributions vary in shape, skewness, kurtosis, number of parameters, and variable domain, and no reliable objective criterion has been established to select the "best" model among them. Some distributions are motivated by the central limit theorem, a few by the extreme-value theory, and others have no theoretical justifications. Since, in most cases, theoretical justifications are rather debatable and estimation uncertainties are high, no consensus has been built among hydrologists as far as recommended models are concerned. Therefore, on a typical hydrologic frequency analysis, in addition to screening data, estimating

Table 8.3   Systematic and historic floods of the River Avon at Evesham Worcestershire

				Combined systematic and historic flood records (1822–1998)			
Rank order	Year	Q (m³/s)	Gringorten	T (years)	Rank order	Year	Q (m³/s)	Eq. (8.5)	T (years)
1	1997	427	0.0090148	110.93	1	1997	427	0.003144	318.07
2	1967	362	0.0251127	39.82	2	1900[a]	410	0.008758	114.18
3	1946	356	0.0412106	24.27	3	1848[a]	392	0.014373	69.58
4	1939	316	0.0573084	17.45	4	1852[a]	370	0.019987	50.03
5	1981	264	0.0734063	13.62	5	1829[a]	370	0.025601	39.06
6	1959	246	0.0895042	11.17	6	1882[a]	364	0.031215	32.04
7	1958	244	0.1056021	9.47	7	1967	362	0.03683	27.15
8	1938	240	0.1216999	8.22	8	1946	356	0.042444	23.56
9	1979	231	0.1377978	7.26	9	1923[a]	350	0.048058	20.81
10	1980	216	0.1538957	6.50	10	1875[a]	345	0.053672	18.63
11	1960	215	0.1699936	5.88	11	1931[a]	340	0.059287	16.87
12	1978	214	0.1860914	5.37	12	1888[a]	336	0.064901	15.41
13	1992	213	0.2021893	4.95	13	1874[a]	325	0.070515	14.18
14	1942	201	0.2182872	4.58	14	1939	316	0.076129	13.14
15	1968	199	0.2343851	4.27	15	1935[a]	306	0.081744	12.23
16	1987	192	0.2504829	3.99	16	1932[a]	298	0.087358	11.45
17	1954	191	0.2665808	3.75	17	1878[a]	296	0.092972	10.76
18	1971	189	0.2826787	3.54	18	1885[a]	293	0.098586	10.14
19	1940	187	0.2987766	3.35	19	1895[a]	290	0.104201	9.60
20	1941	184	0.3148744	3.18	20	1981	264	0.115946	8.62
21	1950	182	0.3309723	3.02	21	1959	246	0.131304	7.62
22	1976	177	0.3470702	2.88	22	1958	244	0.146663	6.82
23	1984	175	0.3631681	2.75	23	1938	240	0.162022	6.17
24	1974	173	0.3792659	2.64	24	1979	231	0.177381	5.64
25	1989	163	0.3953638	2.53	25	1980	216	0.19274	5.19
26	1970	157	0.4114617	2.43	26	1960	215	0.208099	4.81
27	1982	155	0.4275596	2.34	27	1978	214	0.223457	4.48
28	1998	150	0.4436574	2.25	28	1992	213	0.238816	4.19
29	1949	149	0.4597553	2.18	29	1942	201	0.254175	3.93
30	1965	148	0.4758532	2.10	30	1968	199	0.269534	3.71
31	1985	145	0.4919511	2.03	31	1987	192	0.284893	3.51
32	1993	143	0.5080489	1.97	32	1954	191	0.300252	3.33
33	1991	139	0.5241468	1.91	33	1971	189	0.31561	3.17
34	1956	139	0.5402447	1.85	34	1940	187	0.330969	3.02
35	1957	138	0.5563426	1.80	35	1941	184	0.346328	2.89
36	1973	136	0.5724404	1.75	36	1950	182	0.361687	2.76
37	1990	134	0.5885383	1.70	37	1976	177	0.377046	2.65
38	1966	131	0.6046362	1.65	38	1984	175	0.392405	2.55
39	1952	130	0.6207341	1.61	39	1974	173	0.407763	2.45
40	1951	130	0.6368319	1.57	40	1989	163	0.423122	2.36

(continued)

Table 8.3 (continued)

Systematic flood records (1937–1998)				Combined systematic and historic flood records (1822–1998)				
Rank order	Year	Q (m³/s)	Gringorten	T (years)	Rank order	Year	Q (m³/s)	Eq. (8.5)	T (years)
41	1986	129	0.6529298	1.53	41	1970	157	0.438481	2.28
42	1994	124	0.6690277	1.49	42	1982	155	0.45384	2.20
43	1977	124	0.6851256	1.46	43	1998	150	0.469199	2.13
44	1963	117	0.7012234	1.43	44	1949	149	0.484558	2.06
45	1988	116	0.7173213	1.39	45	1965	148	0.499916	2.00
46	1995	114	0.7334192	1.36	46	1985	145	0.515275	1.94
47	1972	113	0.7495171	1.33	47	1993	143	0.530634	1.88
48	1944	103	0.7656149	1.31	48	1991	139	0.545993	1.83
49	1983	103	0.7817128	1.28	49	1956	139	0.561352	1.78
50	1969	94.9	0.7978107	1.25	50	1957	138	0.576711	1.73
51	1955	93.9	0.8139086	1.23	51	1973	136	0.592069	1.69
52	1961	92.3	0.8300064	1.20	52	1990	134	0.607428	1.65
53	1948	91.4	0.8461043	1.18	53	1966	131	0.622787	1.61
54	1953	86.3	0.8622022	1.16	54	1952	130	0.638146	1.57
55	1945	86.3	0.8783001	1.14	55	1951	130	0.653505	1.53
56	1962	67.9	0.8943979	1.12	56	1986	129	0.668864	1.50
57	1947	67.1	0.9104958	1.10	57	1994	124	0.684222	1.46
58	1937	47.0	0.9265937	1.08	58	1977	124	0.699581	1.43
59	1964	41.0	0.9426916	1.06	59	1963	117	0.71494	1.40
60	1975	35.9	0.9587894	1.04	60	1988	116	0.730299	1.37
61	1996	31.9	0.9748873	1.03	61	1995	114	0.745658	1.34
62	1943	7.57	0.9909852	1.01	62	1972	113	0.761017	1.31
					63	1944	103	0.776375	1.29
					64	1983	103	0.791734	1.26
					65	1969	94.9	0.807093	1.24
					66	1955	93.9	0.822452	1.22
					67	1961	92.3	0.837811	1.19
					68	1948	91.4	0.85317	1.17
					69	1953	86.3	0.868528	1.15
					70	1945	86.3	0.883887	1.13
					71	1962	67.9	0.899246	1.11
					72	1947	67.1	0.914605	1.09
					73	1937	47.0	0.929964	1.08
					74	1964	41.0	0.945323	1.06
					75	1975	35.9	0.960681	1.04
					76	1996	31.9	0.97604	1.02
					77	1943	7.57	0.991399	1.01

[a]Historic flood

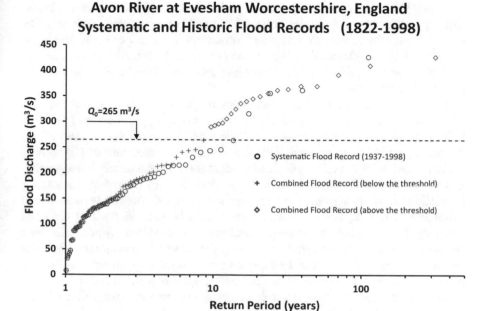

Fig. 8.9 Empirical distribution of systematic and historic flood records

parameters, quantiles and confidence intervals, and testing hypotheses, hydrologists must choose the probability distribution, amongst some candidates, that most adequately models the random variable being studied. This configures a type of ad hoc analysis. The guidelines for such an ad hoc analysis are outlined in the subsections that follow.

8.3.1 Data Screening

The results from frequency analysis strongly depend on the amount, type, and quality of data, which are used as a pivotal source of information in the statistical procedures that follow. In this context, no matter how good and sophisticated a stochastic model is, it will never improve the eventual poor-quality data used to estimate its parameters. The hydrologist should assess the quality of the available hydrologic data, before going to further steps of hydrologic frequency analysis.

In some countries, where hydrometric services have not reached full maturity and regularity, prior consistency analysis of hydrologic data is regarded as a basic requirement that must precede hydrologic frequency analysis. Under the risk of possibly undermining and biasing parameter and quantile estimation, statistical inference assumes that gross and/or systematic errors in collected observations are not admissible and, as such, incorrect data must be corrected prior to the analysis. Whenever applicable, hydrologists should look for inconsistencies in raw data, by comparing observations collected at neighboring gauging stations, or by checking the criteria used to define and extend rating curves, or by assessing the overall quality of collected data. Errors can occur also in data coding, storage, and retrieval. General guidelines for consistency analysis of hydrologic data can be found in WMO (1994).

It has been a fundamental assumption throughout this textbook that the sample of hydrologic data, to be considered for frequency analysis, must be one of an infinite number of possible samples of random and independent data drawn from the same population. In essence, sample data must be realizations of IID random variables and hold the assumptions of randomness, independence, homogeneity, and stationarity, as discussed in Sect. 7.4 of Chap. 7. Because hydrologic data are generally skewed, non-parametric methods should be used to test these fundamental assumptions. In fact, non-parametric tests should be used in the first place, as routine procedures before frequency analysis is carried out. Among the tests designed to check the plausibility of these fundamental assumptions, the ones described in Sect. 7.4 are widely accepted and utilized very often in hydrologic practice. However, if a sample does not pass a specific test, it should be further scrutinized in search of strong hydrologic evidences that may justify discarding it from subsequent analysis.

The reliability of parameter and quantile estimates is intrinsically related to the sample size and representativeness. As mentioned in Sect. 1.4, for the sample of annual peak flows of the Lehigh River at Stoddartsville, sample representativeness cannot be assessed through an objective measure or tested by a specific procedure, since one would have to know beforehand what to expect from the population variability. In some cases, by comparing data from a specific sample with data from other larger samples, collected at nearby gauging stations, one should be able to conclude whether or not the time span covered by the records possibly refers to a predominantly wet (or dry) period, as opposed to a period from which natural variability is expected. However, to return to the case study of the Lehigh River, had the regional flow records started only in 1956/57 instead of 1941/42, it would be hard for someone, in charge of analyzing in 2014 such a shortened sample, to think it plausible that a flood peak three times greater than the largest peak discharge observed in almost 60 years, from 1956/57 to 2013/14, could have happened on August, 19th 1955. As noted in Sect. 1.4, frequency analysis based on such an unrepresentative shortened-sample would severely lower the variance and bias the results.

Benson (1960) used a paradigm 1000-year record of annual maximum peak flows, whose probabilities were, by construction, exactly determinable, to demonstrate that in order to estimate the 10-year return-period flood, within 25 % of the true value, in 95 % of the time, a sample of at least 18 annual records would be necessary. For estimating the 50-year and 100-year return-period floods, under the same conditions, the minimum sample sizes would increase to 39 and 48 annual records, respectively.

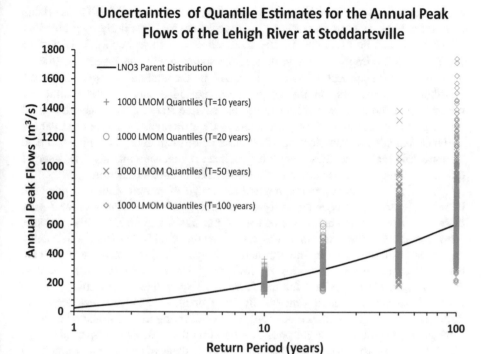

Fig. 8.10 Uncertainties of quantiles, as estimated for samples drawn from a hypothesized LNO3 population of annual peak discharges

Bearing in mind Benson's experiment, consider again the example of generating synthetic samples for the annual flood discharges of the Lehigh River at Stoddartsville, described in Sect. 6.8 of Chap. 6. Recall that 1000 samples of size 73 have been drawn from a population distributed according to a three-parameter lognormal distribution with parameter values $\hat{a} = 21.5190$, $\hat{\mu}_Y = 3.7689$, and $\hat{\sigma}_Y = 1.1204$. For each sample, the L-MOM method was employed to estimate the LNO3 parameters, thus resulting in 1000 sets of estimates, which were then used to calculate a group of 1000 different quantiles, for each of the return periods 10, 20, 50, and 100 years, as shown in the histograms of Fig. 6.3. A different look at this same issue is provided by the chart of Fig. 8.10, where a broader perspective of the uncertainties entailed by quantile estimation is put into evidence. The LNO3 parent is superimposed over the 1000 estimated T-year quantiles, which exhibit an increasing scatter as the return period augments, to the extent of tripling the assumedly true value of the 100-year return period quantile.

Benson's experiment and Fig. 8.10 illustrate that the reliability of quantile estimates strongly depends on the sample size and on the target return period. The total uncertainties associated with the T-year quantile arise either from a possibly incorrect selection of the parent probability distribution or from the effects of sampling errors on parameter estimation. Those arising from the former source are hard to objectively quantify and, in general, are not accounted for. Those arising from estimation, assuming a correct choice was made for the underlying probability distribution, can be evaluated by the methods described in Sect. 6.7. Equations (6.27) and (6.28) imply that the variance S_T^2 associated with the T-year quantile X_T depends on the reciprocal of the sample size N, on the method used to estimate the distribution parameters, on the target return period T, and on the number of parameters to be estimated. The smaller the sample size, or the larger the target return period or the number of parameters to be estimated, the larger the quantile variance. An adverse combination of all these factors can make a quantile estimate so unreliable that it becomes useless for engineering decision making. The issue of quantile reliability assessment is further discussed in later subsections.

Extending a frequency curve beyond the range of sample data is a potential benefit of hydrologic frequency analysis. However, since the variance of the T-year quantile varies with the reciprocal of the sample size N and dramatically increases with increasing T, the question that arises is how far should one extend a frequency curve, as resulting from at-site frequency analysis, before obtaining unreliable estimates of quantiles? The book *Hydrology of Floods in Canada—A Guide to Planning and Design*, edited for the Associate Committee on Hydrology of the National Research Council Canada, by Watt et al. (1988), provides empirical guidance for such a complex question, that is of relying only on data of a single flow gauging station for the purpose of estimating the T-year return-period design flood. According to this guidance, the frequency analysis of annual maximum flood discharges observed at a single station should be performed only for samples of sizes at least equal to 10 and be extrapolated up to a maximum return period of $T = 4N$, where N denotes the sample size. The referenced book also recognizes the difficulty of providing general guidance on such an issue, given the uncertainties arising from both parameter estimation and the choice of the parent probability distribution.

The presence of outliers in a data sample can seriously affect the estimation of parameters and quantiles of a theoretical probability distribution. The detection and identification of low and high outliers in a data sample can be performed by the method described in Sect. 7.6. If a low outlier is detected in records of maximum values, the usual recommendation is to delete it from the sample and recalculate the statistics, since the low outlier can possibly bias the estimation of the probability distribution upper tail, which is of major interest in a frequency analysis of maxima. In some cases, the distribution should be truncated below a certain low threshold level and be adjusted accordingly (WRC 1981). If a high outlier is detected in a record of maximum values, it should be compared to historic information at nearby sites and, if the high outlier is confirmed to be the maximum value over an extended period of time, it should be retained in the sample and treated by the methods described in Sect. 8.2.3. Otherwise, the high outlier should also be retained in the sample and two actions are possible. The first is to look for an appropriate probabilistic model capable of better fitting the empirical distribution upper tail. The second is to resort to the methods of regional frequency analysis, described in

Chap. 10. The same line of reasoning, in the opposite direction, can possibly be applied to minima.

8.3.2   Choosing the Probability Model

8.3.2.1   Theoretical Aspects

As seen in previous chapters, there exists a nonextensive set of candidate probability distributions that are usually employed to model maxima of hydrologic variables. In this set are the extremal distributions, as referring to those derived from the asymptotic extreme-value theory, here including the Generalized Extreme-Value and generalized Pareto models, and the non-extremal distributions encompassing the lognormal (with 2 or 3 parameters), exponential, Gamma, Pearson Type III, and LogPearson Type III models. Other models described in Chap. 5, such as those with more than three parameters, like the Kappa and Wakeby distributions, and compound models, such as the TCEV, are not usually employed in the context of at-site frequency analysis, given the uncertainties added by the increased number of parameters. No one distribution from the set of candidates can be seen as an analytical form that is universally accepted or consensually preferable, as far as the modeling of hydrologic maxima at a site is concerned. In general, the choice of one model rather than another follows some general, however debatable, theoretical criteria and a few objective analytical tools. The former are discussed in this subsection.

As regards the upper-bounded probability distributions, it is a physical fact that some random quantities have limits that are intrinsically and simply defined, such as the concentration of dissolved oxygen in a water body, whose variation is limited by the concentration at saturation, which depends on many factors, the most important of which is the water temperature. Other quantities also have an upper bound, which may not be known or knowable a priori, as resulting from insufficient information on the many factors that may influence the physical phenomenon of interest. The very existence of upper bounds for extreme rainfall and flood magnitudes, and the ability to determine them in a given region or catchment are sources of long-standing controversies in flood hydrology (Horton 1936, Yevjevich 1968, Klemeš 1987, Yevjevich and Harmancioglu 1987). These controversies have split flood analysis into two separate approaches: the one in which mostly unbounded distributions are fitted to data with the implicit assumption of a nonzero probability for any future larger flood, regardless of how large it might be, and the opposing one, which is best represented by PMP/PMF-based analysis. PMP stands for probable maximum precipitation, whereas PMF stands for probable maximum flood.

In short, the PMF is regarded as a potential upper bound for floods at a river cross section, resulting from a hypothetical storm of critical duration and depth named PMP (probable maximum precipitation), led by the most severe but plausible hydrologic conditions. WMO (1986) defines the PMP as the potential highest depth of rainfall, of a specified duration, that is meteorologically possible to occur over a given area, located in a certain geographic region, during a given season of the year. The prevailing method for determining the PMP over an area makes use of local meteorological maximization of historic events, with or without storm transposition from one geographic area to another (WMO 1986). The PMF's implicit assumption is the existence of physical limits to both the supply of precipitation and the basin hydrologic response. However, neither the very concept of PMP-PMF nor that their estimates are unequivocal is universally accepted. Actually, PMP-PMF estimates depend greatly on the quantity and the quality of hydrological and hydrometeorological observations, a fact that makes them highly susceptible to the uncertainties imposed by the available data and the modeling tools. In spite of these complicating issues, the PMP-PMF approach is extensively used worldwide for designing large hydraulic structures, such as spillways of large dams. As with any deterministic method, the major drawback of the PMP-PMF-based approach is that it does not provide probability estimates for risk assessment and risk-informed decisions.

Assuming that upper–bounds for maximum rainfalls and floods do exist, it appears widely accepted that their estimation is uncertain or at least hampered by the limited human knowledge of extreme-flood-producing mechanisms and by the difficulty of adequately quantifying the variation, in time and space, of influential variables. On the other hand, one can possibly hypothesize that a flood discharge of 100,000 m^3/s would never occur in a catchment of drainage area of a few square kilometers. Such a notion of physical implausibility has led some hydrologists, such as Boughton (1980) and Laursen (1983), to recommend only upper-bounded probability distributions as models for extreme rainfalls and related floods. Hosking and Wallis (1997) oppose such a recommendation and argue that if the goal of flood frequency analysis is to estimate the 100-year return-period quantile in a catchment of a few square kilometers, it would be irrelevant to consider as physically impossible the occurrence of a flood-peak flow of 100,000 m^3/s. They add that imposing an upper bound to the probability distribution of floods and maximum rainfalls can possibly compromise the estimation of the quantiles that are actually relevant for frequency analysis. Hosking and Wallis (1997) conclude that, by choosing an unbounded probability distribution for hydrologic maxima, the implicit assumptions are: (1) the upper bound of a competing upper-bound model is not known or cannot be accurately determined; and (2) within the range of return periods of practical interest, an unbounded model better approximates the true parent distribution than an upper-bounded model would do.

The discussion and controversies related to the existence of hydrologic upper bounds and the ability to estimate them, as well as their inclusion in extreme rainfall and flood frequency analyses are far from being exhausted. However, some recent works (see Botero and Francés 2010, Fernandes et al. 2010) seem to point towards a theoretical reconciliation between the two opposing views of PMP-PMF estimates and extreme flood frequency analysis. In particular, Fernandes et al. (2010) employed the Bayesian paradigm to account for the uncertainties on flood upper-bound estimates by eliciting a prior probability distribution, using PMF estimates

Table 8.4 Relative scale of upper-tail weights for some distributions

Upper tail	Form of $f_X(x)$ for large x	Distributions
Heavy	x^{-A}	Generalized extreme value, generalized Pareto, and generalized Logistic with shape parameter $\kappa < 0$
↑	$x^{-A \ln x}$	Lognormal
	$\exp(-x^A)$ $0 < A < 1$	Weibull with shape parameter $\alpha < 1$
	$x^A \exp(-Bx)$	Pearson Type III with positive skewness
	$\exp(-x)$	Exponential, Gumbel$_{max}$
↓	$\exp(-x^A), A > 1$	Weibull with shape parameter $\alpha \geq 1$
Light	Finite upper bound	Generalized extreme value, generalized Pareto, and generalized logistic with shape parameter $\kappa > 0$; Pearson type III with negative skewness

A and B are arbitrary positive constants (adapted from Hosking and Wallis 1997)

transposed to the site of interest. Such an approach is fully described as an example in Chap. 11, in the context of an introduction of Bayesian analysis and its applications to hydrologic variables.

Another source of controversy in frequency analysis of hydrologic maxima relates to the weight or heaviness of the distribution's upper tail, which controls the intensity with which probability increases as quantile augments, for large return periods. As seen in Sect. 5.7.2, the upper-tail weight reflects the rate at which the density function $f_X(x)$ decreases as x tends to infinity, relative to the rate with which the exponential density does. The upper-tail is said to be heavy if the density approaches 0 less fast than the exponential does as $x \to \infty$, and, it is said to be light, otherwise. In fact, in more general terms, the upper-tail weights of distinct densities can be graded by relativizing their respective densities' decreasing rates as $x \to \infty$. Table 8.4 shows such a relative scale of upper-tail weights for the distributions most frequently used in Statistical Hydrology.

For many applications of frequency analysis of hydrologic maxima, estimation of the distribution upper tail is of paramount importance, since curve extrapolation for large return periods is the primary motivation. However, as commented previously, the usual sizes of hydrologic samples are invariably too small to allow a reliable estimation of the distribution upper tail. Hosking and Wallis (1997) argue that, if no sufficient reasons, both from the theoretical or empirical sides, exist to recommend only one kind of tail weight, for frequency analysis of hydrologic maxima, it is advisable to use a set of candidate distributions that covers the full spectrum of tail weights. On the other hand, Papalexiou and Koutsoyiannis (2013) categorically recommend the use of heavy-tailed probability distributions in the frequency analysis of annual maximum daily rainfalls. In particular, Papalexiou and Koutsoyiannis (2013) favor the GEV distribution, with negative shape parameter, for modeling annual maximum daily rainfalls. Serinaldi and Kilsby (2014) attempted to reconcile some opposing views concerning the tail behavior of extreme daily rainfalls, under the framework of high exceedances over a high threshold and the generalized Pareto distribution. They concluded that the heaviness of extreme daily rainfalls distributions, for varying thresholds and record lengths, may be ascribed to a complex blend of extreme and nonextreme values, and fluctuations of the parent distributions. In such a context, it seems the previous advice by Hosking and Wallis (1997), of using a broad range of candidate distributions, should still prevail.

As far as probability models for hydrologic maxima are of concern, the issue of a possible lower bound is sometimes raised. However, as opposed to the upper bound, the lower bound of maxima is easier to be estimated and, in the most extreme case, it may be fixed at zero. Hosking and Wallis (1997) point out that if the quantiles of interest are close to zero, it may be worth requiring that the probability distribution be bounded from below by zero; when such a requirement is imposed, models, such as the generalized Pareto and Pearson Type III, retain a convenient form. For other cases, where a variable can assume values that are usually much larger than zero, more realistic results are obtained by fitting a distribution that has a lower bound greater than zero.

In some rare cases, in arid regions, a sample of annual mean (or even maximum) values may contain zero values. For these cases, Hosking and Wallis (1997) suggest that a possible alternative to model such a sample is to resort to a mixed distribution, as defined by

$$F_X(x) = \begin{cases} 0 & \text{if } x < 0 \\ p + (1 - p)G_X(x) & \text{if } x \geq 0 \end{cases} \tag{8.6}$$

where p denotes the proportion of zero values in the sample and $G_X(x)$ denotes the cumulative probability distribution function of the nonzero values. This same approach can possibly be adapted to model annual mean (or maximum) values below and above a greater-than-zero lower bound, as would be the situation of a sample with low outliers (see, for example, WRC 1981).

The Gumbel$_{max}$, Fréchet, and Weibull extremal distributions, or their condensed analytical form given by the GEV, arise from the classical asymptotic extreme-value theory and are the only ones, among the usual candidate models of hydrologic maxima, for which theoretical justifications can be provided (see Sect. 5.7.2). Recall, however, that the convergence to the limiting forms of asymptotic extreme-value theory requires a large collection of initial IID variables. The major objections to the strict application of classical asymptotic extreme-value theory to hydrologic maxima are: (1) the need for framing the initial variables as IID (see Sect. 5.7.2 and Perichi and Rodríguez-Iturbe 1985); (2) the slow convergence to the limiting forms (Papalexiou and Koutsoyiannis 2013); and (3) the assumed exhaustiveness of the three asymptotic forms (see Benjamin and Cornell 1970, Kottegoda and Rosso 1997, Juncosa 1949).

As detailed in Sect. 5.7.2, the works of Juncosa (1949) and Leadbetter (1974, 1983) allowed the relaxation, under some conditions, of the basic premise at the origin of classical extreme-value theory, namely the assumption of IID initial variables.

Although relaxing the basic assumption makes the results of asymptotic extreme-value theory conceivably applicable to some hydrologic variables, it is worth noting that the sequence of nonzero daily values of the initial variables will certainly be a large number, but not a guarantee of convergence to one of the three limiting forms (Papalexiou and Koutsoyiannis 2013), neither are the asymptotes exhaustive, as the extreme-value distributions of some initial variables do not necessarily converge to one of the three limiting forms. In spite of these difficulties, extremal distributions seem the only ones that find some theoretical grounds, though not definitive, that justify their application to modeling hydrologic maxima (and minima).

As seen in Sect. 5.2, the extended version of the central limit theorem (CLT), with a few additional considerations of a practical nature, is applicable to some hydrologic variables, such as the annual total rainfall depths and, in some cases, to annual mean flows. The extension of the CLT to the logarithm of a strictly positive random variable, which is conceptualized as resulting from the multiplicative action of a large number of variables, has led Chow (1954) to propose the lognormal distribution as a model for hydrologic extremes. Stedinger et al. (1993) comment that some processes, such as the dilution of a solute in a solution and other processes (see Sect. 5.3), result from the multiplicative action of intervening variables. However, in the case of floods and extreme rainfalls, such a multiplicative action is not evident. These objections, however, do not provide arguments to discard the lognormal distribution from the set of candidate models for hydrologic maxima. Since its variate is always positive, with a non-fixed positive coefficient of skewness, the lognormal distribution is potentially a good candidate for modeling annual maximum (or mean) flows, annual maximum daily rainfalls, and annual, monthly, or 3-month total rainfall depths.

As regards the number of parameters of a candidate probability distribution, the principle of parsimony should apply. This general principle states that, from two competing statistical models with equivalent capacity of explaining a given phenomenon, one should always prefer the one with fewer parameters, as additional parameters add estimation uncertainties. For instance, if both $Gumbel_{max}$ and GEV models are employed to estimate the 100-year flood at a given site and if their respective estimates are not too different from each other, the $Gumbel_{max}$ should be preferred. Adding a third parameter to a general probability distribution certainly grants shape flexibility to it and improves its ability to fit empirical data. However, estimation of the third parameter usually requires the estimation of the coefficient of skewness, which is very sensitive to sample fluctuations, since its calculation involves deviations from the sample mean raised to the third power. In this context, parameter estimation through robust methods, such as L-MOM, and regional frequency analysis gain importance and usefulness in Statistical Hydrology.

It is worth mentioning that, in some particular cases, there might be valid arguments for choosing a less parsimonious model. In the case of Gumbel vs GEV, for example, Coles et al. (2003) argue that the lack of data makes statistical modeling and the accounting of uncertainty particularly important. They warn against the risks of adopting a Gumbel model without continuing to take account of the uncertainties such a choice involved, including those related to model extrapolation. Finally, they advise that it is always a best choice to work with the GEV model in place of the Gumbel, unless there is additional information supporting the Gumbel selection.

The previous considerations reveal there is not a complete and consensual set of rules to select a single probability distribution, or even a family of distributions, from a list of candidates to model hydrologic maxima. For at-site frequency analysis, the lack of such a set of rules refers the analyst to a variety of criteria that basically aim to assess the adequacy of a theoretical distribution to the data sample. These criteria include: (1) goodness-of-fit (GoF) tests, as the ones described in Sect. 7.5, with special emphasis on PPCC tests; (2) comparison of theoretical and empirical moment-ratio diagrams, and of some information measures, to be discussed in the next subsections; and (3) graphical assessment of goodness-of-fit by plotting theoretical and empirical distributions on appropriate probability papers, as described in Sect. 8.2.

Similar arguments may also apply to the frequency analysis of hydrologic minima. However, the range of candidate distributions for modeling minima is narrower and may include no more than the Gumbel$_{min}$ and Weibull$_{min}$, and, for some, the lognormal and Log-Pearson Type III models. In addition to that, the extrapolation of probabilistic models of hydrologic minima is less demanding, since decision-making for drought-related water-resources projects requires return periods of the order of 10 years. Arguments from the asymptotic extreme value theory, as outlined in Sect. 5.7.2, seem to favor the Weibull$_{min}$ as a limiting form for left-bounded parent distributions. When compared to Gumbel$_{min}$, which has a constant negative coefficient of skewness of -1.1396 and can possibly yield negative quantiles, the two and three-parameter Weibull$_{min}$ appear as featured candidates for modeling hydrologic minima.

8.3.2.2 Moment-Ratio Diagrams

As seen in Chap. 3, the moments of a given distribution can be written as functions of its parameters. Moreover, higher order moments can be expressed as functions of lower order moments. For instance, the coefficient of skewness of the two-parameter lognormal distribution is written as $\gamma = 3\,CV_X + (CV_X)^3$ and is, thus, a unique function of the coefficient of variation CV, which in turn is a function of the first and second order moments. In general, for any distribution, the coefficients of skewness and kurtosis are respectively written as $\gamma = \mu_3/\sigma^{3/2}$ and $\kappa = \mu_4/\sigma^2$, and they represent characteristic quantities of a particular distribution. For distributions of fixed shape, such as most two-parameter distributions, the coefficients of skewness and kurtosis are constant and given in Table 8.5 for some well-known distributions.

For distributions of variable shape, such as three-parameter distributions and the lognormal, the coefficients of skewness and kurtosis can be related to each other through unique functions and a chart depicting these functions can be drawn. Such a

Table 8.5 Coefficients of skewness (γ) and kurtosis (κ) for two-parameter distributions

Distribution	Skewness γ	Kurtosis κ
Normal	0	3
Exponential	2	9
Gumbel$_{max}$	1.1396	5.4002

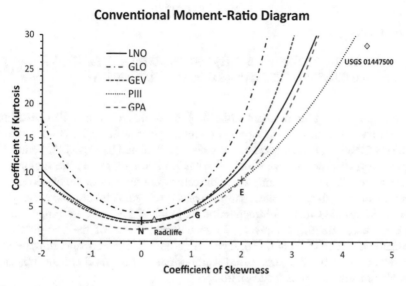

Fig. 8.11 Conventional moment-ratio diagram for some distributions

chart is termed a moment-ratio diagram and is illustrated in Fig. 8.11 for the distributions most-frequently used in frequency analysis of hydrologic maxima. In such a diagram, notice that variable-shape distributions such as the lognormal (LNO), generalized logistic (GLO), generalized extreme value (GEV), Pearson Type III (PIII), and generalized Pareto (GPA) plot as curves, whereas fixed-shape distributions such as the normal (N), exponential (E), and Gumbel$_{max}$ (G) plot as points.

The $\gamma \times \kappa$ relationships plotted in the diagram of Fig. 8.11 can be derived from the summary of distributions' properties, given in Sect. 5.13 of Chap. 5, or, alternatively, approximated by the following polynomial functions:

(a) for LNO

$$\kappa = 3 + 0.025653\gamma + 1.720551\gamma^2 + 0.041755\gamma^3 + \\ + 0.046052\gamma^4 - 0.00478\gamma^5 + 0.000196\gamma^6 \tag{8.7}$$

(b) for GLO

$$\kappa = 4.2 + 2.400505\gamma^2 + 0.244133\gamma^4 - 0.00933\gamma^6 + 0.002322\gamma^8 \tag{8.8}$$

(c) for GEV

$$\kappa = 2.695079 + 0.185768\gamma + 1.753401\gamma^2 + 0.110735\gamma^3 + \\ + 0.037691\gamma^4 + 0.0036\gamma^5 + 0.00219\gamma^6 + 0.000663\gamma^7 + 0.000056\gamma^8 \quad (8.9)$$

(d) for PIII

$$\kappa = 3 + 1.5\gamma^2 \quad (8.10)$$

(e) for GPA

$$\kappa = 1.8 + 0.292003\gamma + 1.34141\gamma^2 + 0.090727\gamma^3 + 0.022421\gamma^4 + \\ + 0.004\gamma^5 + 0.000681\gamma^6 + 0.000089\gamma^7 + 0.000005\gamma^8 \quad (8.11)$$

The idea of using the moment-ratio diagram to choose a probability distribution that best fits a data sample is graphical and consists of plotting the sample estimates of the coefficients of skewness and kurtosis, as calculated by Eqs. (2.13) and (2.14), respectively, and then locate which distribution is the closest to the sample point. However, results are misleading since the estimators for γ and κ are biased and their sampling errors, given the short samples of hydrologic data, are excessively high. This is exemplified in the moment-ratio diagram of Fig. 8.11 by pinpointing the $\gamma \times \kappa$ estimates for the sample of 73 annual peak flows of the Lehigh River at Stoddartsville (USGS 01447500), listed in Table 7.1. The corresponding point lies outside the range of commonly used distributions and no valid conclusions can be drawn from such a rather futile exercise.

On the other hand, as the sample size becomes larger to the extent of yielding reliable estimates of γ and κ, such as the 248 annual total rainfalls, recorded at the Radcliffe Meteorological Station, in Oxford, England, the moment-ratio diagram can be useful. Notice that by pinpointing the Radcliffe $\gamma \times \kappa$ estimates on the diagram of Fig. 8.11, one can easily notice that the normal, lognormal, and GEV are featured candidates to model such a sample. Unfortunately, the Radcliffe sample is one of a kind among hydrologic samples. An alternative to make the moment-ratio diagram useful to select a parent distribution is to employ it in the context of a hydrologically homogeneous region, by plotting (or averaging) the $\gamma \times \kappa$ pairs, estimated for a large number of sites located within the region, and checking if they cluster around one of the depicted theoretical curves (for an example of such a procedure, see Rao and Hamed 2000).

Analogous to the idea of conventional moment-ratio diagram, the L-moment-ratio diagram is constructed on the basis of the L-moment homologous measures of skewness and kurtosis. These are the L-moment-ratio τ_3 (sometimes termed L-Skewness), given by Eq. (6.21), and τ_4 (or L-Kurtosis), given by Eq. (6.22). Recall from Sect. 6.5 that, unlike conventional moments, the more robust estimators of the rth-order L-moments do not require deviations from the sample mean to be raised to the power r. Like conventional moments, the L-moment ratios τ_3 and τ_4 are constant for fixed-shape distributions and are given in Table 8.6 for the normal, exponential, and Gumbel$_{max}$ models. Figure 8.12 depicts the L-moment-ratio

Table 8.6 Coefficients of L-Skewness τ_3 and L-Kurtosis τ_4 for two-parameter distributions

Distribution	L-Skewness τ_3	L-Kurtosis τ_4
Normal	0	0.1226
Exponential	1/3	1/6
Gumbel$_{max}$	0.1699	0.1504

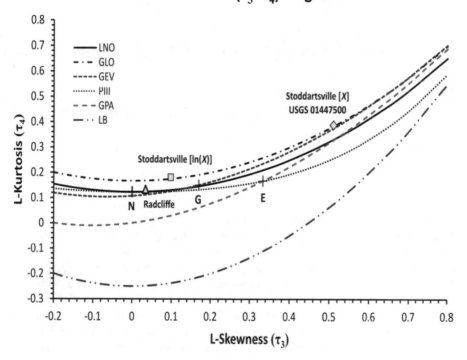

Fig. 8.12 L-moment-ratio diagram for some distributions

diagram with the theoretical curves for the variable-shape lognormal (LNO), Generalized Logistic (GLO), generalized extreme value (GEV), Pearson Type III (PIII), and generalized Pareto (GPA) distributions, and the points for the fixed-shape normal (N), exponential (E), and Gumbel$_{max}$ (G) distributions. Also plotted in Fig. 8.12 is the theoretical lower-bound (LB) for the $\tau_3 \times \tau_4$ relationship.

The $\tau_3 \times \tau_4$ relationships plotted on the diagram of Fig. 8.12 can be approximated by the following polynomial functions:

(a) for LNO (two and three parameters)

$$\tau_4 = 0.12282 + 0.77518\tau_3^2 + 0.12279\tau_3^4 - 0.13638\tau_3^6 + 0.11368\tau_3^8 \qquad (8.12)$$

(b) for GLO

$$\tau_4 = 0.16667 + 0.83333\tau_3^2 \tag{8.13}$$

(c) for GEV

$$\tau_4 = 0.10701 + 0.11090\tau_3 + 0.84838\tau_3^2 - 0.06669\tau_3^3 + \\ + 0.00567\tau_3^4 - 0.04208\tau_3^5 + 0.03763\tau_3^6 \tag{8.14}$$

(d) for PIII

$$\tau_4 = 0.1224 + 0.30115\tau_3^2 + 0.95812\tau_3^4 - 0.57488\tau_3^6 + 0.19383\tau_3^8 \tag{8.15}$$

(e) for GPA

$$\tau_4 = 0.20196\tau_3 + 0.95924\tau_3^2 - 0.20096\tau_3^3 + 0.04061\tau_3^4 \tag{8.16}$$

(f) for LB

$$\tau_4 = -0.25 + 1.25\tau_3^2 \tag{8.17}$$

In a comprehensive comparative study of the two types of moment-ratio diagrams, Vogel and Fennessey (1993) advocate replacing the conventional moment-ratio diagram with the L-moment-ratio diagram, for most applications of goodness of fit in hydrology. The authors highlight the following outperforming qualities of L-moment-ratio diagram: (1) estimators of τ_3 and τ_4 are nearly unbiased for all sample sizes and for all parent distributions; (2) sample estimates of τ_3 and τ_4, denoted by t_3 and t_4, and calculated as described in Sect. 6.5, are far less sensitive to the eventual presence of outliers in the samples; and (3) the L-moment-ratio diagram allows a better graphical discrimination among the distributions, making easier the identification of the parent distribution.

The exercise of pinpointing the $\tau_3 \times \tau_4$ estimates, for the sample of 73 annual peak flows of the Lehigh River at Stoddartsville (USGS 01447500) and the 248 annual total rainfalls, recorded at the Radcliffe Meteorological Station, on the L-moment-ratio diagram of Fig. 8.12 was also performed. For the Lehigh River flood peaks, the GLO, GEV, LNO, and GPA distributions appear as plausible parent distributions, whereas for the rainfalls at Radcliffe, the previous result, from the conventional moment-ratio diagram, seems to be confirmed, with a stronger tendency towards the LNO. As with its predecessor, the moment-ratio diagram is more useful for selecting a parent distribution if employed in the context of regional frequency analysis of a hydrologically homogeneous region, by plotting (or averaging) the $\tau_3 \times \tau_4$ pairs, estimated for a large number of sites located within the region. In this regard, Hosking and Wallis (1997) propose a unified method for regional frequency analysis with L-moments, including a formal goodness-of-fit measure for choosing the regional parent distribution. Such a unified method will be described in Chap. 10.

8.3.2.3 Information Measures

As mentioned earlier, increasing the number of parameters in order to achieve model flexibility and greater accuracy also increases the estimation uncertainty. Some information measures seek to summarize into a single metric these two opposing objectives of model selection, namely, greater accuracy and less estimation uncertainty. The first information measure is the Akaike Information Criterion (AIC), introduced by Akaike (1974) and given by the following expression:

$$AIC = -2\ln\left[\prod_{i=1}^{N} f_X(x_i; \Theta)\right] + 2p \tag{8.18}$$

where p is the number of parameters of the model and $\ln\left[\prod_{i=1}^{N} f_X(x_i; \Theta)\right]$ denotes the log-likelihood function of the density $f_X(x_i; \Theta)$, evaluated at the point of maximum likelihood, for a sample of size N, defined by the MLE estimates of the p parameters contained in vector Θ. The first term in the right-hand side of Eq. (8.18) measures the estimation accuracy, whereas the second term measures the estimation uncertainty due to the number of estimated parameters. The lower the AIC value the better the model given by $f_X(x_i; \Theta)$. As a result, the AIC score can be used to compare how different models fit the same data sample, if their respective parameters are estimated by the MLE method.

Calenda et al. (2009) remark that the AIC, as a measure based on the likelihood function, an asymptotically unbiased quantity, provides accurate results for samples of at least 30 records. When the sample size N, divided by the number of MLE-estimated parameters p, is smaller than 40, or $N/p < 40$, Calenda et al. (2009) suggest correcting Eq. (8.18) to

$$AIC_c = -2\ln\left[\prod_{i=1}^{N} f_X(x_i; \Theta)\right] + 2p\left(\frac{N}{N - p - 1}\right) \tag{8.19}$$

The second information measure is the Bayesian Information Criterion (BIC), which is very similar to AIC, but has been developed in the context of Bayesian statistical analysis, to be introduced in Chap. 11. It is formally defined as

$$\text{BIC} = -2\ln\left[\prod_{i=1}^{N} f_X(x_i; \Theta)\right] + p\ln(N) \qquad (8.20)$$

The second term of the right-hand side of Eq. (8.20) shows that BIC, as compared to AIC, puts more weight on the estimation uncertainty, especially for models with a high number of parameters and/or small sample sizes. As with the AIC, the lower the BIC score the better the model given by $f_X(x_i; \Theta)$. It is worth noting, however, that both scores can be applied only in the context of maximum-likelihood estimation. Calculations of AIC and BIC scores are illustrated in Example 8.3.

Laio et al. (2009) compared model selection techniques for the frequency analysis of hydrologic extremes. In addition to AIC and BIC scores, they used a third measure denoted by ADC, which is based on the Anderson-Darling test statistic, described in Sect. 7.5.3. Their main conclusions are: (a) no criterion performs consistently better than its competitors; (b) all criteria are effective at identifying from a sample a true two-parameter parent distribution and less effective when the parent is a three-parameter distribution; (c) BIC is more inclined than AIC and ADC towards selecting the more parsimonious models; and (d) AIC and BIC usually select the same model, whereas ADC, in many cases, selects a different model. As a final remark, they note that the obtained results do not ensure a definitive conclusion, as it remains unclear which criterion or combination of criteria should be adopted in practical applications.

8.3.3 Estimating Quantiles with Frequency Factors

Once the parent distribution has been chosen and its parameters estimated by an efficient and accurate estimation method, selected among MOM, MLE, and L-MOM, the quantile estimates x_T, for different return periods T, can be computed. The most efficient estimation method is the one that returns the narrowest confidence interval (or the smallest standard error of estimate S_T) for x_T, for a fixed confidence level $(1-\alpha)$. Estimation of confidence intervals for quantiles can be performed by the methods described in Sect. 6.7. As stemming from its mathematical properties, the method of maximum likelihood is, in most cases, the most efficient estimation method. Albeit the MLE estimates are asymptotically unbiased, for small samples, they can possibly be biased and, thus, yield inaccurate quantile estimates, especially in the distribution tails. Therefore, the most adequate estimation method is the one that combines efficiency and accuracy, since concession to one attribute in the detriment of the other would be equivalent to seeking precisely-estimated inaccurate quantiles.

According to Chow (1964), the quantile X_T of a random variable X, for the return period T, can be written as a deviation ΔX added to the mean μ. Formally,

$$X_T = \mu + \Delta X \tag{8.21}$$

The deviation ΔX depends on the analytical form and shape characteristics of the probability distribution $F_X(x)$, on its parameters, and on the return period T. Still according to Chow (1964), the deviation ΔX can be written as a function of the distribution's standard deviation σ, as $\Delta X = K_T^D \sigma$, where K_T^D denotes the frequency factor, whose variation with T is specific for a given distribution identified as D. As such, Eq. (8.21) can be rewritten as

$$X_T^D = \mu + K_T^D \sigma \tag{8.22}$$

Equation (8.22) is valid for the population's mean and standard deviation, which must be estimated from the data sample. However, instead of using the sample estimates $\bar{x}$ and s_X, which would apply only to MOM parameter estimates, the correct use of Eq. (8.22) requires one to estimate μ and σ from their relations to the distribution parameters, which, in turn, are estimated by the MOM, or MLE, or L-MOM methods. For instance, for the Gumbel$_{max}$ distribution with parameters α and β, one knows that $\mu = \beta + 0.5772\alpha$ and $\sigma = \pi\alpha/\sqrt{6}$. The parameters α and β can be estimated either by the MOM, or MLE, or L-MOM methods, yielding three pairs of different estimates, which then should be employed to provide distinct estimates of μ and σ. This is also valid for a three-parameter distribution, as the population coefficient of skewness is related to the parameter estimates. Therefore, Eq. (8.22), as applied to the estimator x_T^D, of the quantile X_T^D, becomes

$$x_T^D = \hat{\mu}_{EM}^D + K_T^D \hat{\sigma}_{EM}^D \tag{8.23}$$

where $\hat{\mu}_{EM}^D$ and $\hat{\mu}_{EM}^D$ respectively denote the mean and the standard deviation, as estimated from the relations of these quantities to the D distribution parameters, which in turn are estimated by the estimation method EM. Equation (8.23) is the essence of quantile estimation with frequency factors. The variation of these with the return period is detailed in the next subsections, for each of the probability distributions most-often used in Statistical Hydrology.

8.3.3.1   Normal Distribution

If X is normally distributed $[X \sim N(\mu,\sigma)]$, so that $D = N$, the frequency factor K_T^N is given by the standard normal variate z_T as corresponding to the return period $T = 1/[1 - \Phi(z_T)]$. The standard normal variate is obtained either from the readings of Table 5.1, or from Eq. (5.17), or from the Excel built-in function NORM.S. INV(.).

8.3.3.2   Lognormal Distribution (Two-Parameter)

The lognormal distribution of $[X \sim LN(\mu_Y, \sigma_Y)]$ corresponds to the normal distribution of $Y = \ln(X)$. As such, in the logarithmic space, the frequency factor K_T^N is given by the standard normal variate z_T that corresponds to the return period $T = 1/[1 - \Phi(z_T)]$. Thus, the quantiles x_T^{LN} can be calculated as

$$x_T^{LN} = \exp\left(\hat{\mu}_{Y,EM}^{LN} + K_T^N \hat{\sigma}_{Y,EM}^{LN}\right) \tag{8.24}$$

If decimal (or common) logarithms are used, such that $Y = \log_{10}(X)$, then Eq. (8.24) must be rewritten as

$$x_T^{LN} = 10^{\hat{\mu}_{Y,EM}^{LN} + K_T^N \hat{\sigma}_{Y,EM}^{LN}} \tag{8.25}$$

The frequency factor for the lognormal distribution can also be computed without prior logarithmic transformation of X. In such a case, Kite (1988) proved that the frequency factor K_T^{LN} is given by

$$K_T^{LN} = \frac{\exp\left[z_T \hat{\sigma}_{Y,EM}^{LN} - \left(\hat{\sigma}_{Y,EM}^{LN}\right)^2 / 2\right] - 1}{\sqrt{\exp\left[\left(\hat{\sigma}_{Y,EM}^{LN}\right)^2\right] - 1}} \tag{8.26}$$

The quantiles x_T^{LN} are then calculated as

$$x_T^{LN} = \hat{\mu}_{X,EM}^{LN} + K_T^{LN} \hat{\sigma}_{X,EM}^{LN} \tag{8.27}$$

where $\hat{\mu}_{X,EM}^{LN}$ and $\hat{\sigma}_{X,EM}^{LN}$ are the estimated mean and standard deviation of the original variable X.

8.3.3.3   Lognormal Distribution (Three-Parameter)

Recall from Sect. 5.3 that the three-parameter lognormal (LN3 or LNO3) distribution corresponds to the normal distribution of $Y = \ln(X - a)$ where a denotes the lower bound of X. Thus, in the logarithmic space, the frequency factor K_T^N is given by the standard normal variate z_T that corresponds to the return period $T = 1/[1 - \Phi(z_T)]$. Then, the quantiles x_T^{LN3} can be calculated as

$$x_T^{LN3} = a + \exp\left(\hat{\mu}_{Y,EM}^{LN3} + K_T^N \hat{\sigma}_{Y,EM}^{LN3}\right) \tag{8.28}$$

Equations (8.26) and (8.27) remain valid for the LN3, recalling that in such a case $Y = \ln(X - a)$.

Table 8.7 Coefficients of Eq. (8.28), as given in Hoshi and Burges (1981)

Coefficient	Z			
	G	$1/A$	B	H^3
a_0	−0.003852050	0.001994470	0.990562000	−0.003651640
a_1	1.004260000	0.484890000	0.031964700	0.017592400
a_2	0.006512070	0.023093500	−0.027423100	−0.012183500
a_3	−0.014916600	−0.015243500	0.007774050	0.007826000
a_4	0.001639450	0.001605970	−0.000571184	−0.000777686
a_5	−0.0000583804	−0.000055869	0.0000142077	0.0000257310

8.3.3.4 Pearson Type III Distribution

The frequency factor for the Pearson type III distribution of $[X \sim \text{P-III}(\alpha,\beta,\xi)]$ can be approximated by the following expression (Wilson and Hilferty 1931):

$$K_T^{\text{P-III}} = \frac{2}{\hat{\gamma}_{\text{EM}}} \left\{ \left[\frac{\hat{\gamma}_{\text{EM}}}{6} \left(z_T - \frac{\hat{\gamma}_{\text{EM}}}{6} \right) + 1 \right]^3 - 1 \right\} \quad \text{for } 0 \le \hat{\gamma}_{\text{EM}} \le 1 \qquad (8.29)$$

where $\hat{\gamma}_{\text{EM}}$ denotes the estimated coefficient of skewness and z_T is the standard normal variate that corresponds to the return period $T = 1/[1 - \Phi(z_T)]$.

For positive values of $\hat{\gamma}_{\text{EM}}$ larger than 1, the modified Wilson-Hilferty approximation, as proposed by Kirby (1972) and given by Eq. (8.30), should be used.

$$K_T^{\text{P-III}} = A \left\{ \text{Max} \left[H, 1 - \left(\frac{G}{6} \right)^2 + z_T \frac{G}{6} \right]^3 - B \right\} \quad \text{for } 0.25 \le \hat{\gamma}_{\text{EM}} \le 9.75 \qquad (8.30)$$

where A, B, G, and H are functions of $\hat{\gamma}_{\text{EM}}$. Hoshi and Burges (1981) provide polynomial approximations for $1/A$, B, G, and H^3, which have the general form

$$Z \approx a_0 + a_1 \hat{\gamma}_{\text{EM}} + a_2 \hat{\gamma}_{\text{EM}}^2 + a_3 \hat{\gamma}_{\text{EM}}^3 + a_4 \hat{\gamma}_{\text{EM}}^4 + a_5 \hat{\gamma}_{\text{EM}}^5 \qquad (8.31)$$

where Z is either $1/A$, or B, or G, or H^3, and a_i, $i = 0, \ldots, 5$, are coefficients given in Table 8.7.

Then, the quantiles from P-III(α,β,ξ) are computed through the following equation:

$$x_T^{\text{P-III}} = \hat{\xi}_{\text{EM}} + \hat{\alpha}_{\text{EM}} \hat{\beta}_{\text{EM}} + K_T^{\text{P-III}} \hat{\alpha}_{\text{EM}} \sqrt{\hat{\beta}_{\text{EM}}} \qquad (8.32)$$

8.3.3.5 Log-Pearson Type III Distribution

The Log-Pearson type III distribution of $[X \sim \text{LP-III}(\alpha_Y,\beta_Y,\xi_Y)]$ corresponds to the Pearson type III distribution of $Y = \ln(X)$. As such, in the logarithmic space, the frequency factor $K_T^{\text{P-III}}$ is given by the same methods described in Sect. 8.3.3.4. Thus, the quantiles $x_T^{\text{LP-III}}$ can be calculated as

$$x_T^{\text{LP-III}} = \exp\left(\hat{\xi}_{Y,\text{EM}} + \hat{\alpha}_{Y,\text{EM}}\hat{\beta}_{,\text{EM}} + K_T^{\text{P-III}}\hat{\alpha}_{Y,\text{EM}}\sqrt{\hat{\beta}_{Y,\text{EM}}} \right) \qquad (8.33)$$

8.3.3.6 Exponential Distribution

The frequency factor for the exponential distribution of $[X \sim \text{E}(\theta)]$ is given by

$$K_T^{\text{E}} = \ln(T) - 1 \qquad (8.34)$$

The quantiles from $\text{E}(\theta)$ are computed through the following equation:

$$x_T^E = \hat{\theta}_{\text{EM}} + K_T^{\text{E}}\hat{\theta}_{\text{EM}} \qquad (8.35)$$

8.3.3.7 Gamma Distribution

The frequency factor for the Gamma distribution of $[X \sim \text{Ga}(\theta,\eta)]$ can be calculated with the methods described for the Pearson type III, since $K_T^{\text{Ga}} = K_T^{\text{P-III}}$. Then, the quantiles from $\text{Ga}(\theta,\eta)$ are given by

$$x_T^{\text{Ga}} = \hat{\eta}_{\text{EM}}\hat{\theta}_{\text{EM}} + K_T^{\text{Ga}}\hat{\theta}_{\text{EM}}\sqrt{\hat{\eta}_{\text{EM}}} \qquad (8.36)$$

8.3.3.8 GEV Distribution

The frequency factor for the GEV distribution of $[X \sim \text{GEV}(\beta,\alpha,\kappa)]$ is

$$K_T^{\text{GEV}} = \frac{\hat{\kappa}_E\Gamma(1 + \hat{\kappa}_{\text{EM}}) - [-\ln(1 - 1/T)]^{\hat{\kappa}_{\text{EM}}}}{|\hat{\kappa}_{\text{EM}}|\sqrt{\Gamma(1 + 2\hat{\kappa}_{\text{EM}}) - \Gamma^2(1 + \hat{\kappa}_{\text{EM}})}} \qquad (8.37)$$

Then, the quantiles from $\text{GEV}(\beta,\alpha,\kappa)$ are calculated as

$$x_T^{GEV} = \hat{\beta}_{EM} + \frac{\hat{\alpha}_{EM}}{\hat{\kappa}_{EM}}[1 - \Gamma(1 + \hat{\kappa}_{EM})] + K_T^{GEV}\frac{\hat{\alpha}_{EM}}{\hat{\kappa}_{EM}}\sqrt{\Gamma(1 + 2\hat{\kappa}_{EM}) - \Gamma^2(1 + \hat{\kappa}_{EM})}$$

$$(8.38)$$

8.3.3.9   Gumbel$_{max}$ Distribution

Many hydrology textbooks, following Gumbel (1958), recommend a method to calculate the frequency factors for the Gumbel$_{max}$ distribution, as dependent on the sample size. Using the results from a Monte Carlo experiment, Lettenmaier and Burges (1982) showed that the frequency factors, calculated as a function of the sample size, are misleading and should be avoided. They recommend that frequency factors for the Gumbel$_{max}$ distribution should be calculated on the basis of an infinite sample size. As such, it can be shown that the frequency factor for the Gumbel$_{max}$ distribution of $[X \sim G(\beta,\alpha)$ or $X \sim Gu_{max}(\beta,\alpha)]$ is

$$K_T^G = -\left\{0.45 + 0.7797\ln\left[-\ln\left(1 - \frac{1}{T}\right)\right]\right\}$$

$$(8.39)$$

Then, the quantiles from $Gu_{max}(\beta,\alpha)$ are calculated as

$$x_T^G = \hat{\beta}_{EM} + 0.5772\hat{\alpha}_{EM} + K_T^G\frac{\pi\hat{\alpha}_{EM}}{\sqrt{6}}$$

$$(8.40)$$

8.3.3.10   GLO (Generalized Logistic) Distribution

The generalized logistic (GLO) distribution was introduced in Sect. 5.9.1 of Chap. 5, as a particular case of the four-parameter Kappa distribution. More specifically, the GLO distribution is defined by the parameters ξ, of location, α, of scale, and κ, of shape, with density function given by

$$f_X(x) = \frac{1}{\alpha}\frac{\left[1 - \kappa\left(\frac{x-\xi}{\alpha}\right)\right]^{\left(\frac{1}{\kappa}-1\right)}}{\left\{1 + \left[1 - \kappa\left(\frac{x-\xi}{\alpha}\right)\right]^{\frac{1}{\kappa}}\right\}^2}$$

$$(8.41)$$

The GLO variate is defined in the range $\xi + \alpha/\kappa \leq x < \infty$ for $\kappa < 0$ and $-\infty \leq x < \xi + \alpha/\kappa$ for $k > 0$. The GLO CDF is given by

$$F_X(x) = \left\{ 1 + \left[1 - \kappa\left(\frac{x - \xi}{\alpha}\right) \right]^{\frac{1}{\kappa}} \right\}^{-1} \tag{8.42}$$

The mean and variance of a GLO variate are respectively given by

$$E[X] = \xi + \frac{\alpha}{\kappa}[1 - \Gamma(1 + \kappa)\Gamma(1 - \kappa)] \tag{8.43}$$

and

$$\text{Var}[X] = \frac{\alpha^2}{\kappa^2}\left[\Gamma(1 + 2\kappa)\Gamma(1 - 2\kappa) - \Gamma^2(1 + \kappa)\Gamma^2(1 - \kappa) \right] \tag{8.44}$$

Analogous to the GEV, the coefficient of skewness of the GLO distribution depends only on the shape parameter κ (see Rao and Hamed 2000). The function for calculating the quantile X_T, of return period T, for the GLO distribution is given by

$$X_T = \xi + \frac{\alpha}{\kappa}[1 - (T - 1)^{-\kappa}] \tag{8.45}$$

Following the publication of the *Flood Estimation Handbook* (IH 1999), the GLO distribution, with parameters fitted by L-MOM, became standard for flood frequency analysis in the UK. Estimating the GLO parameters using the MOM and MLE methods is complicated and the interested reader should consult Rao and Hamed (2000) for details. The L-MOM estimates for the GLO parameters are calculated as

$$\hat{\kappa} = -t_3 \tag{8.46}$$

$$\hat{\alpha} = \frac{l_2}{\Gamma(1 + \hat{\kappa})\Gamma(1 - \hat{\kappa})} \tag{8.47}$$

$$\hat{\xi} = l_1 + \frac{l_2 - \hat{\alpha}}{\hat{\kappa}} \tag{8.48}$$

where l_1 and l_2 are the sample L-moments and t_3 is the sample L-skewness.

The frequency factor for the GLO distribution of $[X \sim \text{GLO}(\xi, \alpha, k)]$ is given by

$$K_T^{\text{GLO}} = \frac{\hat{\kappa}_{\text{EM}}\Gamma(1 + \hat{\kappa}_{\text{EM}})\Gamma(1 - \hat{\kappa}_{\text{EM}}) - (T - 1)^{\hat{\kappa}_{\text{EM}}}}{|\hat{\kappa}_{\text{EM}}|\sqrt{\Gamma(1 + 2\hat{\kappa}_{\text{EM}})\Gamma(1 - 2\hat{\kappa}_{\text{EM}}) - \Gamma^2(1 + \hat{\kappa}_{\text{EM}})\Gamma^2(1 - \hat{\kappa}_{\text{EM}})}} \tag{8.49}$$

The sampling variance of quantiles from $\text{GLO}(\xi, \alpha, \kappa)$, with parameters estimated with the L-MOM method, was studied by Kjeldsen and Jones (2004). Rao and Hamed (2000) provide procedures to calculate the standard errors of quantiles for MOM and MLE estimation methods. The reader interested in details on confidence intervals for quantiles from GLO should consult the referred publications.

8.3.4   Assessing the Uncertainties of Quantile Estimates

As previously mentioned in this chapter, the uncertainties of the T-year quantile can be ascribed part to an incorrect choice of the parent distribution, which cannot be objectively assessed, and part to parameter estimation errors. If one assumes no error has been made in choosing the parent distribution, the uncertainties of quantile estimates can be partly evaluated by the asymptotic method described in Sect. 6.7. The assumption of asymptotically normal quantiles, implied by Eq. (6.26), together with Eqs. (6.27) and (6.28), respectively valid for two-parameter and three-parameter distributions, yields approximate confidence intervals for the T-year quantile. As the partial derivatives and the covariances, in Eqs. (6.27) and (6.28), respectively depend on the specific quantile function and on the method used to estimate the distribution parameters, calculations of approximate confidence intervals for quantiles are tedious and are usually performed with the aid of computer software. The reader interested in evaluating the terms of Eqs. (6.27) and (6.28), as applied to the distributions and estimation methods most-currently used in Statistical Hydrology, is referred to Kite (1988), Rao and Hamed (2000), and Kjeldsen and Jones (2004). Table 8.8 summarizes the results for frequency factors and standard errors for some distributions used in Statistical Hydrology. Alternatively, one can employ the computer-intensive method outlined in Sect. 6.8 or the resampling techniques described in Hall et al. (2004), for assessing standard errors and confidence intervals for quantiles. Also, Stedinger et al. (1993) provide equations to calculate exact and approximate confidence intervals for quantiles from some popular distributions used in Statistical Hydrology. The interval for the T-year quantile at a fixed confidence level $(1-\alpha)$, regardless of the method employed to estimate it, is expected to grow wider as the sample size decreases, and/or as T increases, and/or as the number of estimated parameters increases. It is worth noting, however, that as the eventual errors associated with the incorrect choice of the parent distribution are not objectively accounted for, confidence intervals for quantiles, as estimated for distinct distributions, are not directly comparable. For a given distribution, assuming it adequately describes the parent distribution, the calculated intervals for the T-year quantile at a fixed confidence level $(1-\alpha)$ are usually narrower for MLE estimates, as compared to L-MOM and MOM estimates. Example 8.3 illustrates a complete frequency analysis of flood flow records of the Lehigh River at Stoddartsville.

Example 8.3 Perform a complete frequency analysis of the annual peak discharges of the Lehigh River at Stoddartsville (USGS gauging station 01447500) recorded from the water year 1941/42 to 2013/14, listed in Table 7.1. Consider the following distributions as possible candidate models: one-parameter exponential (E), two-parameter lognormal (LNO2), Gamma (Ga), Gumbel$_{max}$ (Gu), three-parameter lognormal (LNO3), generalized extreme value (GEV), Pearson type III (P III), log-Pearson type III (LP-III), and generalized logistic (GLO).

Table 8.8 Frequency factors and standard errors for some probability distributions used in hydrology

Dist	Frequency factor $K_T = (x_T - \mu)/\sigma^a$	Standard error S_T — MOM	MLE	L-MOM
N	z_T [from $N(0,1)$]	$\sqrt{1 + \frac{z_T^2}{2}}\,\frac{\hat\sigma x}{\sqrt{N}}$	$\sqrt{1 + \frac{z_T^2}{2}}\,\frac{\hat\sigma x}{\sqrt{N}}$	$\sqrt{1 + 0.5113 z_T^2}\,\frac{\hat\sigma x}{\sqrt{N}}$
LNO 2	$\left[\exp(z_T\hat\sigma_w - \hat\sigma_w^2/2) - 1\right]/B$ with $w = \ln(x)$ and $B = \sqrt{\exp(\hat\sigma_w^2) - 1}$	$\sqrt{1 + (B^3 + 3B)K + C\frac{K_T^2}{4}}\,\frac{\hat\sigma x}{\sqrt{N}}$ with $C = B^8 + 6B^6 + 15B^4 + 16B^2 + 2$	$\frac{\ln(B^2+1)(1+BK_T)^2(1+z_T^2/2)}{B^2}\,\frac{\hat\sigma x}{\sqrt{N}}$	NA
LNO 3	$\left[\exp(z_T\hat\sigma_w - \hat\sigma_w^2/2) - 1\right]/B$ with $w = \ln(x - a)$	No explicit or simple expression. See Rao and Hamed (2000)	No explicit or simple expression. See Rao and Hamed (2000)	NA
E	$\ln(T) - 1$	$\sqrt{\hat\theta^2(1 + 2K_T + 2K_T^2)/N}$	$\sqrt{\hat\theta^2(1 + 2K_T + NK_T^2/(N-1))}/N$	$\sqrt{\hat\theta^2(1 + 2K_T + 4K_T^2/3)/N}$
Ga	$\frac{2}{\gamma}\left\{\left[\frac{\gamma}{6}\left(z_T - \frac{\gamma}{6}\right) + 1\right]^3 - 1\right\}$ for $0 < \gamma < 2$; γ from relations to estimated parameters	No explicit or simple expression. See Rao and Hamed (2000)	No explicit or simple expression. See Rao and Hamed (2000)	No explicit or simple expression. See Rao and Hamed (2000)
P III	Same as Gamma	No explicit or simple expression. See Rao and Hamed (2000)	No explicit or simple expression. See Rao and Hamed (2000)	No explicit or simple expression. See Rao and Hamed (2000)
LP-III	Same as Gamma with $x_T = \exp\left[\ln\mu_{\ln(X)} + K_T\sigma_{\ln(X)}\right]$	No explicit or simple expression. See Rao and Hamed (2000)	No explicit or simple expression. See Rao and Hamed (2000)	No explicit or simple expression. See Rao and Hamed (2000)
Gu_{max}	$-0.45 - 0.7797\ln\left[-\ln(1 - \frac{1}{T})\right]$	$\frac{a}{\sqrt{N}}\sqrt{(1.15894 + 0.19187Y + 1.1Y^2)}$ with $Y = -\ln[-\ln(1 - 1/T)]$	$\frac{a}{\sqrt{N}}\sqrt{(1.1087 + 0.5140Y + 0.6079Y^2)}$ with $Y = -\ln[-\ln(1 - 1/T)]$	$\frac{a}{\sqrt{N}}\sqrt{(1.1128 + 0.4574Y + 0.8046Y^2)}$ with $Y = -\ln[-\ln(1 - 1/T)]$
GEV	$\frac{\hat\kappa\,\Gamma(1+\hat\kappa) - [-\ln(1-1/T)]^{\hat\kappa}}{|\hat\kappa|\sqrt{\Gamma(1+2\hat\kappa) - \Gamma^2(1+\hat\kappa)}}$	No explicit or simple expression. See Rao and Hamed (2000)	No explicit or simple expression. See Rao and Hamed (2000)	No explicit or simple expression. See Rao and Hamed (2000)

[a] μ and σ are the mean and variance obtained from the relations with the parameters, as estimated by MOM, or MLE, or L-MOM

Table 8.9 Sample statistics of the annual peak flows of the Lehigh River at Stoddartsville

	X	$\ln(X)$	L-moments		
			l_r or t_r	X	$\ln(X)$
Sample size	73	73	l_1	102.69	4.297
Maximum	903	6.8061	l_2	46.41	0.415
Minimum	14.0	2.6403	l_3	23.71	0.041
Mean	102.69	4.2970	l_4	18.20	0.075
Standard deviation	121.99	0.7536	t_2	0.452	0.097
Coefficient of variation	1.1881	0.1754	t_3	0.511	0.098
Coefficient of skewness	4.5030	0.6437	t_4	0.392	0.181
Coefficient of kurtosis	28.905	4.1346			

Solution Let X denote the flood peak discharges. Table 8.9 lists the descriptive statistics, L-moments, and L-moment ratios of the sample of X and of $\ln(X)$. When compared to the usual descriptive statistics of flood flow samples, the coefficients of variation and skewness of X are relatively high.

The next step is to detect and identify possible outliers in the sample, for which the test of Grubbs and Beck, described in Sect. 7.6, is employed. At the 5 % significance level, the lower bound for detecting low outliers, as calculated with Eq. (7.47), is 6.88 m^3/s, whereas the upper bound for high outliers, as calculated with Eq. (7.46), is 785 m^3/s. As a result, no low outliers are detected and one high outlier is identified: the flood peak discharge 903 m^3/s, which occurred in the 1954/55 water year. According to USGS (1956), the 1955 flood was due to a rare combination of very heavy storms associated with the passing of hurricanes Connie (August 12–13) and Diane (August 18–19) over the northeastern USA, with record-breaking peak discharges consistently recorded at other gauging stations located in the region. Therefore, despite being a high outlier, the discharge 903 m^3/s is consistent and should be retained in the sample, since, as a record-breaking flood-peak flow it will certainly affect model choice and parameter estimation. Unfortunately, no historical information, that could possibly extend the period of time over which the 1955 flood would remain as a maximum, was available for the present analysis.

The third step is to test sample data for randomness, independence, homogeneity, and stationarity. The solution to Example 7.6 has applied the nonparametric tests of the turning points, Wald–Wolfowitz, Mann–Whitney, and Spearman's Rho, and no empirical evidence was found to rule out the hypotheses of randomness, independence, homogeneity, and stationarity (for monotonic trends), respectively.

Table 8.10 lists the estimates for the parameters of all candidate distribution, as resulting from the application of MOM, L-MOM, and MLE estimation methods. In some cases, no results for parameter estimates are provided, either because the employed algorithm did not converge to a satisfactory solution or values were out of bounds.

For the fourth step of goodness-of-fit tests, a common estimation method should be selected, such that the decisions of rejecting (or not rejecting) the null hypotheses, as corresponding to the nine candidate distributions, can be made on a

Table 8.10 Parameter estimates for the candidate probability distributions of X

Distribution	Estimation method								
	MOM			L-MOM			MLE		
	Position	Scale	Shape or LB	Position	Scale	Shape or LB	Position	Scale	Shape or LB
E	–	102.69	–	–	102.69	–	–	102.69	–
LNO2	4.1916	0.9382	–	4.2708	0.8495	–	4.2970	0.7484	–
Ga	–	144.94	0.7084	–	79.660	1.2890	–	62.580	1.6400
Gu	47.780	95.120	–	66.950	64.040	–	65.550	51.110	–
LNO3	4.3395	0.8795	–10.223	3.7689	1.1204	21.519	4.1358	0.8654	8.1244
GEV	48.900	62.081	–0.2289	54.127	33.767	–0.4710	54.827	34.968	–0.4542
P III	48.500	274.68	0.1973	34.349	170.50	0.4008	NS[a]	NS	NS
LP-III	NS	NS	NS	1.8043	0.2225	11.204	1.1112	0.1735	18.363
GLO	79.210	52.096	–0.2538	68.406	28.894	–0.5110	NS	NS	NS

[a]*NS* no satisfactory solution found, *LB* lower bound

Table 8.11 PPCC test statistics and related decisions, at the 5 % significance level

Parent distribution under H_0	Calculated test statistic r_{calc}	Calculated test statistic $r_{crit,}$ $\alpha=0.05$	Decision on H_0
E	0.9039	0.9720 (equation)	Reject
LNO2	0.9481	0.9833 (equation)	Reject
Ga	0.8876	0.9343 (equation)	Reject
Gu	0.8463	0.9720 (equation)	Reject
LNO3	0.9774	0.9833 (table)	Reject
GEV	0.9871	<0.910 (inferred from table)	Not reject
P III	0.9425	0.9343 (equation)	Not reject
LP-III	0.9759	0.9792 (equation)	Reject
GLO	0.9901	<0.910 (inferred from chart)	Not reject

Table 8.12 PPCC test statistics and the related decisions, at the 1 % significance level

Parent distribution under H_0	Calculated test statistic r_{calc}	Calculated test statistic $r_{crit,}$ $\alpha=0.05$	Decision on H_0
LNO3	0.9774	0.9755 (table)	Not reject
GEV	0.9871	<0.817 (inferred from table)	Not reject
P III	0.9425	0.8924 (equation)	Not reject
LP-III	0.9759	0.9662 (equation)	Not reject
GLO	0.9901	<0.910 (inferred from chart)	Not reject

reasonably even basis. From Table 8.10, the MOM and L-MOM estimation methods are the ones that provide results for all candidate distributions, since the algorithm for solving the MLE equations failed to find solutions for the P III and GLO models. The qualities of nearly unbiased and robust estimates of higher order moments and moment-ratios, that are characteristics of L-MOM estimation, appear as appealing to select it as a common basis to perform the GoF tests. Among these and, again, for the sake of evenness, the PPCC Gof test, following comments in Sect. 7.5.5, is an adequate option. Table 8.11 presents the PPCC test statistics and the related decisions, at the 5 % significance level, concerning the different parent distributions. Critical values of the tests statistics were obtained either from tables, or from approximating equations, or from charts, as appropriate (see Sect. 7.5.4).

From Table 8.11, the decisions concerning the two-parameter distributions seem unquestionable as the calculated test statistics are well below their respective critical values, at the 5 % significance level. However, the decisions to reject the LNO3 and LP-III distributions are not particularly compelling, as the calculated test statistics are very close to their respective critical values. Since the critical test statistics are approximate values, a second chance is given for the LNO3 and LP-III distributions, by lowering the significance level to 1 %. Table 8.12 summarizes the PPCC test statistics and the related decisions, at the 1 % significance level. The

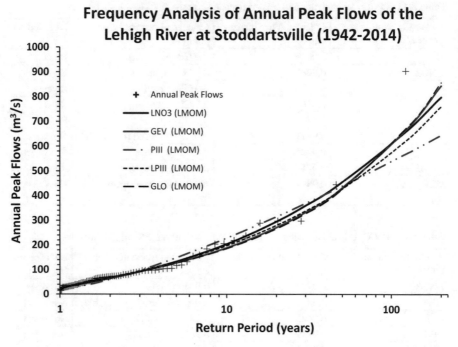

Fig. 8.13 Empirical and theoretical distributions of the annual peak flows of the Lehigh River at Stoddartsville. Parameters estimated by L-MOM

decision, at the 1 % significance level, is that all three-parameter candidate models should be retained for the next steps of analysis.

The empirical and candidate distributions, as plotted on exponential probability paper, are depicted in the chart of Fig. 8.13. The plotting positions used in this chart were calculated with the Cunnane formula, for the empirical distribution, whereas, for the theoretical distributions, the appropriate formulae were used, following the recommendations given in Table 7.14. From the chart of Fig. 8.13, with the exception of the P III model which should be dismissed from the analysis, all three-parameter distributions fit the lower and middle portions of the empirical distributions reasonably well. However, the upper tail of the empirical distribution, as shaped by the tendency imposed by the record-breaking flood discharge 903 m^3/ s, is only approximated by some of the candidate models being analyzed. On the other hand, it is worthwhile remembering, from the discussions in Sect. 8.2.2, that the sampling uncertainties of the plotting position associated with this record-breaking flood discharge, do not warrant the dismissal of any of the fitting models, namely, the LNO3, GEV, LP-III, and GLO. These models should be retained for the next step of the analysis.

The fifth step concerns choosing the model that most adequately represents the parent distribution. As seen in Sect. 8.3.2, the plotting of the $\gamma \times \kappa$ estimates for the sample of 73 annual peak flows of the Lehigh River at Stoddartsville onto the

Table 8.13 AIC and BIC scores for the candidate distributions

Candidate distribution	AIC	BIC	Choice
LNO3	795.89	802.76	3rd
GEV	793.73	800.60	1st
LP-III	794.56	801.43	2nd

conventional moment-ratio diagram of Fig. 8.11 has not led to any conclusion as regards the shape of the parent distribution. The same exercise on the L-moment-ratio diagram of Fig. 8.12 offers some help by delimiting a region on the $\tau_3 \times \tau_4$ plane in which the lines that correspond to candidate distributions GLO, GEV, LNO, and GPA are located. Of these, the GPA model is more frequently used in the context of the peaks-over-threshold (POT) representation for floods, and, as such, shall not be included among the candidates for modeling the annual maximum floods of the Lehigh River. The reason for this will be made clear in Sect. 8.4, where the POT approach is described. The $\tau_3 \times \tau_4$ sample estimates for $\ln(X)$ are also pinpointed on the diagram of Fig. 8.12 and the point location seems not to rule out the plausibility of the P III model in the logarithmic space, which is equivalent to accepting the LP–III as a credible model for X. Thus, analysis of the L-moment-ratio diagram, in spite of not offering an effective tool for discriminating among the candidate models, has been useful in confirming the LNO3, GEV, LP–III, and GLO as promising candidate distributions.

Application of AIC and BIC scores are restricted to the distributions for which MLE parameter estimates are available, namely, the LNO3, GEV, and LP–III. Their respective AIC and BIC scores are displayed in Table 8.13. These scores are congruent in pointing out the GEV, LP–III, and LNO3 distributions as first, second, and third choices, respectively. However, model selection through AIC and BIC scores is limited to MLE-estimated parameters, a fact that has excluded not only the GLO distribution from the comparative analysis but all other distributions with parameters estimated through MOM and L-MOM methods.

The lack of an overall objective criterion leaves the model selection open to some degree of subjectivity. In fact, when a common method of parameter estimation, such as the L-MOM, is adopted for the LNO3, GEV, LP–III, and GLO candidate distributions, a look at the chart of Fig. 8.13 shows that any of the four models fits the sample data accurately and would be a perfect choice for estimating, for instance, the 100-year flood, with relatively small differences among estimates. Figure 8.14a, for parameters estimated with MOM, leads to about the same conclusion; note that the LP–III distribution was not plotted on the chart of Fig. 8.14a because the MOM estimate for parameter α is negative (see Sect. 5.8.2). Figure 8.14b, however, leads to a different conclusion, as the GEV distribution, for parameters estimated with MLE, yields relatively higher flood quantiles for large return periods and appears to better fit the upper tail of the empirical distribution. Overall, taking into account the fittings of the center and upper tail of the empirical distribution, the location of $\tau_3 \times \tau_4$ estimates on the L-moment ratio diagram, the AIC and BIC scores, and the consistency of parameter and quantile estimates, it seems coherent to proceed with the analysis by defining the LNO3,

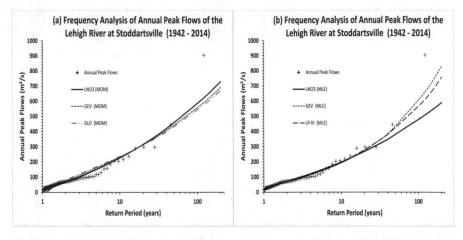

Fig. 8.14 Empirical and theoretical distributions of the annual peak flows of the Lehigh River at Stoddartsville. Parameters estimated by (**a**) MOM and (**b**) MLE

Table 8.14 Estimates of GEV, LP-III and LNO3 parameters, quantiles, and standard errors

Distribution/estimation method →	GEV/MLE	LP-III/MLE	LNO3/L-MOM
Location parameter	54.8271	1.1112	3.7689
Scale parameter	34.9678	0.1735	1.1204
Shape	−0.4542	18.3625	21.5190
50-year quantile	431	405	454
100-year quantile	600	532	609
200-year quantile	831	689	798
50-year std. error	118.90	98.09	141.8
100-year std. error	202.39	154.72	234.8
100-year std. error	334.02	235.18	371.6

with L-MOM estimates, and the GEV and the LP-III, both with MLE estimates, as the set of models to choose from.

The sixth step relates to deciding on the most adequate combination of model and estimation method, on the basis of a trade-off solution between accuracy and quantile estimation uncertainty. Table 8.14 displays the estimates of the GEV, LP-III, and LNO3 parameters and quantiles for the 50-year, 100-year, and 200-year return periods, and their respective standard-errors of estimates, as obtained through the indicated estimation method. The standard errors of estimates (S_T) for MLE were obtained by using the asymptotic method, described in Sect. 6.7, whereas the Monte Carlo simulation method of Sect. 6.8, with 1000 synthetic samples, was employed to provide the estimates of S_T for L-MOM. As expected, the standard errors of MLE estimates are smaller and, thus, reflect the superiority of maximum-likelihood estimates, as far as the quantile estimation uncertainty is

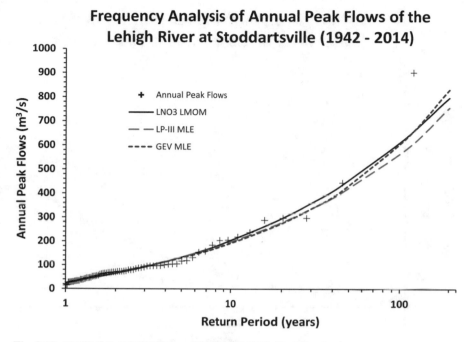

Fig. 8.15   GEV/MLE, LP-III/MLE, and LNO3/L-MOM fitted distributions

concerned. However, the trade-off requires a further examination in terms of accuracy.

Figure 8.15 depicts the GEV/MLE, LP-III/MLE, and LNO3/L-MOM fitted distributions, superimposed over the empirical distribution of the annual peak discharges of the Lehigh River at Stoddartsville. In this chart, the plotting positions used to define the empirical distribution were calculated with the Cunnane formula. A look at the graph of Fig. 8.15 shows that the LNO3 curve, as estimated with L-MOM parameters, is the distribution that best fits the empirical quantiles up to the return period of 50 years. The fit of the LP-III/MLE distribution is slightly inferior, in spite of yielding the smaller standard errors of estimates. As compared to LNO3/L-MOM, the GEV curve, estimated with MLE parameters, provides a worse fit in the middle portion of the empirical distribution. As regards the upper-tail, the LP-III/MLE distribution fails to capture the tendency imposed by the two largest sample records and starts biasing its quantile estimates toward smaller values from the 40-year return period on. In this domain, as compared to the LP-III, the LNO3/L-MOM and GEV/MLE distributions perform better with a slight superiority of the latter, as it approaches the record-breaking flood discharge 903 m^3/s more closely than the LNO3/-MOM does. However, taking into account the uncertainty on the true return period associated with the 903 m^3/s flood discharge, both the GEV/MLE and LNO3/L-MOM would be equally sensible choices, particularly if one considers their respective standard errors of estimates which are roughly equivalent figures.

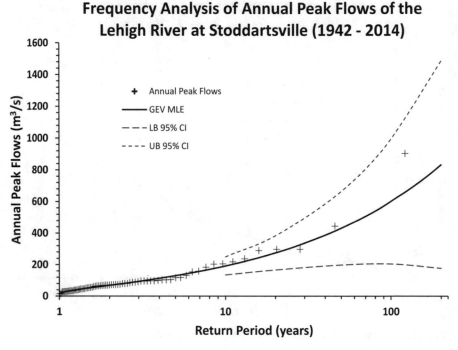

Fig. 8.16 LNO3 adopted distribution with 95 % confidence bands

However, in order to be coherent with the more precise quantiles and the best fit over the empirical upper tail, the GEV distribution, with MLE parameter estimates, is selected as the adopted model for the annual peak discharges of the Lehigh River at Stoddartsville. Figure 8.16 depicts the MLE-estimated GEV distribution, with the 95 % confidence bands calculated with the method described in Sect. 6.7.2.

8.4   Peaks-Over-Threshold (POT) Approach for the Analysis of Partial Duration Series

8.4.1   Theoretical Background

Many hydrological and hydrometeorological variables fluctuate in time such that they constitute alternating sequences of large and small (or even null) values. The large values are usually much above the mean and are clustered in short periods of time, followed by spells of relatively smaller and/or null values. Such a temporal pattern confers to these variables the characteristic configuration of a succession of exceedances over a specified threshold. The number and magnitude of such exceedances can be modeled by a bivariate stochastic process. For the sake of

Representation of a Bivariate Stochastic Process

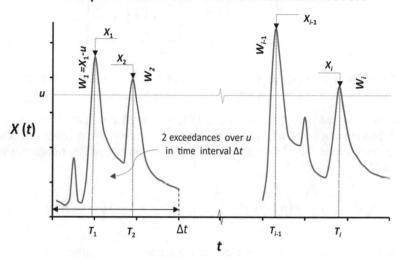

Fig. 8.17 Representation of a bivariate stochastic process

clarity, consider the chart of Fig. 8.17 as depicting part of the time evolution of a hydrologic variable denoted by X, where the exceedances over a threshold u are marked and identified. In such a scheme, for the ith cluster of values of X that have exceeded u, the corresponding maximum value is denoted as X_i, such that $X_i = u + W_i$, where W_i represents the ith exceedance that occurred at time T_i. Such a representation shapes the bivariate stochastic process $\{T_i, X_i; i = 1, 2, \ldots\}$, the modeling of which, in the context of floods and other hydrologic phenomena, has been pioneered by the works of Todorovic and Zelenhasic (1970), Gupta et al. (1976), Todorovic (1978), and North (1980). Other important references in this topic area are Taesombut and Yevjevich (1978), Smith (1984), Rosbjerg (1984), and Van Montfort and Witter (1986).

Under general conditions, the events $\{T_i, X_i; i = 1, 2, \ldots\}$ can be represented as an ample class of nonhomogeneous compound Poisson stochastic processes. For that, two premises are needed:

1. The number of exceedances over the threshold u, denoted as $N_{\alpha\beta}$, that have occurred in the time interval $[\alpha,\beta]$ is a discrete random variable, whose mass function is assumed as a Poisson distribution with time-dependent intensity $\lambda(t)$. As such, the probability that n exceedances have occurred in $[\alpha,\beta]$ is given by

$$P(N_{\alpha\beta} = n) = \frac{\left[\int_{\alpha}^{\beta} \lambda(t)dt\right]^n \exp\left[-\int_{\alpha}^{\beta} \lambda(t)dt\right]}{n!} \qquad (8.50)$$

2. $\{X_i\}$ denotes a sequence of mutually independent random variables, whose probability distributions depend on the occurrence time T_i.

Assume that the time interval $[\alpha,\beta]$ can be partitioned into k_0 subintervals, within which the distribution of $\{X_i\}$ does not depend on the time. Denoting the number of occurrences within the jth subinterval as N_j and the corresponding maximum value of X as M_j, one can write

$$P(M_j \leq x) = P(N_j = 0) + \sum_{n=1}^{\infty} P\left[\bigcap_{i=1}^{n} (X_{i,j} \leq x) \cap (N_j = n)\right] \qquad (8.51)$$

where $X_{i,j}$ represents the ith occurrence that has exceeded u within the jth subinterval, and $\cap$ indicates the joint occurrence or intersection of indicated events. From the condition of mutual independence of events, it follows that

$$P(M_j \leq x) = P(N_j = 0) + \sum_{n=1}^{\infty} P(N_j = n)\left[H_{u,j}(x)\right]^n \qquad (8.52)$$

In Eq. (8.52), $H_{u,j}$ is the probability distribution function of the exceedances of X over the threshold u, inside the jth subinterval. Substituting the appropriate terms of Eq. (8.52) by the Poisson mass function given in Eq. (8.50), it follows that

$$P(M_j \leq x) = \exp\left\{-\left[1 - H_{u,j}(x)\right]\int_j \lambda(t)\,dt\right\} \qquad (8.53)$$

According to North (1980), the probability distribution of the maximum of X, over the time interval $[\alpha,\beta]$, denoted by $M_{\alpha\beta}$, can be derived by equating

$$P(M_{\alpha\beta} \leq x) = P\left(\bigcap_{j=1}^{k_0} M_j \leq x\right) \qquad (8.54)$$

or, from the condition given by premise 2,

$$P(M_{\alpha\beta} \leq x) = \prod_{j=1}^{k_0} P(M_j \leq x) \qquad (8.55)$$

where Π indicates the product of individual terms. Combining Eqs. (8.55) and (8.53), it follows that

$$P\left(M_{\alpha\beta} \leq x\right) = \exp\left\{-\sum_{j=1}^{k_0}\left[1 - H_{u,j}(x)\right]\int_j \lambda(t)\,dt\right\} \qquad (8.56)$$

As $k_0 \to \infty$, Eq. (8.56) becomes

$$P\left(M_{\alpha\beta} \leq x\right) = \exp\left\{-\int_\alpha^\beta\left[1 - H_u(x/t)\right]\lambda(t)\,dt\right\} \qquad (8.57)$$

Equation (8.57) allows the calculation of the probability of the maximum $M_{\alpha\beta}$ within any time interval $[\alpha,\beta]$ to be done. In general, the interest is focused on obtaining the probability distribution of the annual maxima, denoted by $F_a(x)$, by fixing the time interval bounds as $\alpha=0$ and $\beta=1$, such that they respectively represent the beginning and the end of the year. As such, Eq. (8.57) becomes

$$F_a(x) = \exp\left\{-\int_0^1 \lambda(t)\left[1 - H_u(x/t)\right]dt\right\} \qquad (8.58)$$

In Eq. (8.58), the probability distribution of the exceedances of X over the threshold u, denoted by $H_u(x/t)$, depends on time. In instances, however, when empirical evidence indicates that $H_u(x/t)$ does not depend on time, Eq. (8.58) can be substantially simplified and becomes

$$F_a(x) = \exp\left\{-\left[1 - H_u(x)\right]\int_0^1 \lambda(t)\,dt\right\} = \exp\{-\nu\left[1 - H_u(x)\right]\} \qquad (8.59)$$

where ν denotes the mean annual number of exceedances. Equation (8.59) represents the general mathematical formalism for the frequency analysis of the so-called partial duration series, under the POT representation. In order to be solved, Eq. (8.59) requires the estimation of ν and $H_u(x)$. For a given sample of flood records, for instance, ν is usually estimated by the mean annual number of flow exceedances over u. The distribution $H_u(x)$ refers to the magnitudes of the exceedances $W = X - u$ and can be estimated by an appropriate model fitted to the sample exceedances (see solutions to Examples 8.7 and 8.8).

8.4.2 Constraints in Applying the POT Approach

In the theoretical foundations of POT modeling approach, given in the previous subsection, Eq. (8.59) is built upon the assumptions that exceedances over the threshold u are mutually independent and that the number of occurrences is a Poisson variate. These are fundamental assumptions that need to be verified prior to the application of the POT modeling approach to practical situations. A brief discussion on these issues is addressed in the subsections that follow.

8.4.2.1 Mutual (or Serial) Independence of Exceedances

The mutual (or serial) independence of the exceedances over u is a basic assumption and its empirical verification should precede applications of the POT modeling approach. For X being a continuous stochastic process, it is expected that the serial dependence between successive exceedances decreases as the threshold u is raised or, equivalently, decreases as the mean annual number of exceedances ν decreases. In fact, a high threshold would make the mean annual number of exceedances decrease, as the time span between successive events becomes larger. As a result, successive exceedances tend to be phenomenologically more disconnected to each other and, thus, statistically independent. Taesombut and Yevjevich (1978) analyzed the variation of the correlation coefficient between exceedances, as lagged by a unit time interval, in correspondence with the mean annual number of exceedances, over 17 flow gauging stations in the USA. They concluded that the serial correlation coefficient grows with $\hat{\nu}$ and is kept inside the 95 % confidence bands as long as $\hat{\nu} \leq 4,5$. Similar conclusions were obtained by Madsen et al. (1993) for partial duration series of rainfall, recorded at a number of gauging stations in Denmark.

Notwithstanding the difficulties of setting general rules for selecting the threshold u for the POT modeling approach, the inherent physical characteristics of hydrologic phenomena together with empirical studies may suggest conditions under which the mutual independence assumption can hold. As regards the selection of mutually independent rainfall events, for instance, in general, successive occurrences should be separated in time by a significant period with no rainfall within it. For daily rainfall, such a period should be of one or two dry days, whereas for subdaily rainfall, a period of 6 h is usually prescribed.

In regard to flows, setting general rules becomes more difficult, since the catchment actions of retaining or releasing runoff, at time-variable rates, depend on the catchment size, shape, soil, antecedent soil moisture condition, and topographic features, and on the characteristics of the rainfall episode as well. In this regard, analysis of past rainfall-runoff events, as summarized into hyetographs and hydrographs, can prove useful. In general, for flood hydrographs, the flow exceedances should be selected in such a manner that two successive occurrences be separated in time by a long interval of receding flows. This time interval should be sufficiently long to successfully pick up consecutive independent flood exceedances that have been produced by distinct rainfall episodes. WRC (1981) suggests that, in order to select flow exceedances, the flood-peak discharges of two successive hydrographs must be separated by a time interval, in days, equal to or larger than the sum of 5 and the natural logarithm of the catchment drainage area in square miles. Cunnane (1979) suggests that flows within two successive independent flood peaks should recede to at least 2/3 of the first peak discharge. Similarly, WRC (1981) changes the proportion to 75 % of the smallest of the peak discharges pertaining to two successive hydrographs, in addition to the previous criterion based on the catchment drainage area. The reader interested in more details on threshold selection for POT modeling should consult Lang et al. (1999) and Bernardara et al. (2014).

8.4.2.2   Distribution of the Number of Mutually
Independent Exceedances

For hydrological variables, the assumption that the number of exceedances over a sufficiently threshold is distributed as a Poisson variate encounters justifications from both the empirical and theoretical viewpoints. From the empirical standpoint, there are many results that confirm such an assumption for high thresholds (e.g., Todorovic 1978, Taesombut and Yevjevich 1978, Correia 1983, Rosbjerg and Madsen 1992, and Madsen et al. 1993). The theoretical justifications are given in Cramér and Leadbetter (1967), Kirby (1969), and Leadbetter et al. (1983). In particular, Cramér and Leadbetter (1967, p. 256) demonstrate that if a stochastic process is Gaussian, then, under general conditions, it can be stated that the number of exceedances over a high threshold u converges to a Poisson process as $u \to \infty$. In this regard, Todorovic (1978) argues that there is no reason to believe that the Cramér-Leadbetter results would not hold if the process is not Gaussian. Later on, Leadbetter et al. (1983, p. 282) extended the previous results for non–Gaussian processes.

Following the theoretical justifications already given, from the standpoint of practical applications, the question of how high the threshold should be such that the number of mutually independent exceedances can be treated as a Poisson variate remains unanswered. Langbein (1949) suggests the practical criterion of setting the threshold u such that no more than 2 or 3 annual exceedances are selected ($\hat{\nu} \leq 3$). Taesombut and Yevjevich (1978) argue that, in order for the number of exceedances be a Poisson variate, the ratio between the mean and the variance of the annual number of exceedances should be approximately equal to 1. In this context, recall from Sect. 4.2, that the mean and the variance of a Poisson variate are equal. Additional results by Taesombut and Yevjevich (1978) show that, as compared to the analysis of annual maximum series, the frequency analysis of partial duration series provides smaller errors of estimate for Gumbel quantiles only when $\nu \geq 1.65$. They conclude by recommending the use of partial duration series for $\nu \geq 1.95$. Cunnane (1973), in turn, has no reservations in recommending the use of partial duration series for samples with less than 10 years of records.

In spite of the difficulty of proposing a general criterion, experience suggests that specifying a value of ν comprised between 2 and 3 seems to be sufficient to benefit from the advantages of the POT modeling approach, and, at the same time, warrant the serial independence among exceedances and, in many cases, the assumption of Poisson-distributed number of occurrences. However, such a recommendation should be subjected to an appropriate statistical test, in order to check for its adequacy. The test proposed by Cunnane (1979) is very often used in such a context and is based on the approximation of a Poisson variate by a normal variate. If the number of exceedances that occurred in the kth year, denoted as m_k, follows a normal distribution with mean $\hat{\nu}$ and standard deviation $\hat{\nu}$, then the statistics

$$R = \sum_{k=1}^{N} \frac{(m_k - \hat{\nu})^2}{\hat{\nu}} \tag{8.60}$$

follows a Chi-Square distribution with $\eta = N - 1$ degrees of freedom, where N denotes the number of years of records. This is a two-sided test and is valid for $N > 5$. As such, the null hypothesis that the number of independent exceedances follows a Poisson distribution should be rejected, at the significance level α, if

$$R = \sum_{k=1}^{N} \frac{(m_k - \hat{\nu})^2}{\hat{\nu}} < \chi^2_{\frac{\alpha}{2},\eta} \text{ or if } R = \sum_{k=1}^{N} \frac{(m_k - \hat{\nu})^2}{\hat{\nu}} > \chi^2_{1-\frac{\alpha}{2},\eta} \tag{8.61}$$

Example 8.8, later on in this chapter, illustrates the application of the Cunnane test.

8.4.3 Distribution of the Magnitudes of Mutually Independent Exceedances

Let $\{X_1, X_2, \ldots\}$ be a sequence of IID variables, with a common distribution $F_X(x)$. One is most interested in comprehending the probabilistic behavior of the values of $X_{(i)}$, from the occurrences $X_{(1)}, X_{(2)}, \ldots, X_{(i)}$ that have exceeded the threshold u, as referred to in Fig. 8.17. Such a probabilistic behavior can be described by the following conditional distribution:

$$P(X > u + w | X > u) = \frac{1 - F_X(u + w)}{1 - F_X(u)} \tag{8.62}$$

where w denotes the magnitudes of the exceedances of X over u. Application of the asymptotic extreme-value theory to the distribution $F_X(x)$ states that, for large values of N (see Sect. 5.7.2), the distribution of the maximum $Y = \max\{X_1, X_2, \ldots\}$ tends to one of the three limiting forms, which can be condensed into the general equation

$$\lim_{N \to \infty} F_X{}^N(y) = \exp\left\{-\left[1 - \kappa\left(\frac{y - \beta}{\alpha}\right)\right]^{1/\kappa}\right\} \tag{8.63}$$

The resulting limit of Eq. (8.63) is the expression of the CDF of the GEV distribution with parameters α, β, and κ, of location, scale, and shape, respectively. Taking the natural logarithms of both sides of Eq. (8.63) leads to

$$N \ln[F_X(y)] = -\left[1 - \kappa\left(\frac{y - \beta}{\alpha}\right)\right]^{1/\kappa} \tag{8.64}$$

Following Coles (2001), for large values of y, the expansion of $\ln[F_X(y)]$ into a Taylor series results in the following approximate relation:

$$\ln[F_X(y)] \approx -[1 - F_X(y)] \tag{8.65}$$

Replacing the above approximate relation into Eq. (8.64) it follows that, for positive large values of $y = u + w$,

$$[1 - F_X(u + w)] \approx \frac{1}{N}\left[1 - \kappa\left(\frac{u + w - \beta}{\alpha}\right)\right]^{1/\kappa} \tag{8.66}$$

At the point $y = u$,

$$[1 - F_X(u)] \approx \frac{1}{N}\left[1 - \kappa\left(\frac{u - \beta}{\alpha}\right)\right]^{1/\kappa} \tag{8.67}$$

Replacing both Eqs. (8.66) and (8.67) into Eq. (8.62), it follows that

$$P(X > u + w \mid X > u) = \frac{1 - F_X(u + w)}{1 - F_X(u)} = \left[\frac{1 - \kappa\left(\frac{u + w - \beta}{\alpha}\right)}{1 - \kappa\left(\frac{u - \beta}{\alpha}\right)}\right]^{1/\kappa} \tag{8.68}$$

By adding and subtracting $\kappa(u - \beta)/\alpha$ to the numerator of the right-hand side of Eq. (8.68), it can be rewritten as

$$P(X > u + w | X > u) = \left[\frac{1 - \kappa\left(\frac{u-\beta}{\alpha}\right) + \kappa\left(\frac{u-\beta}{\alpha}\right) - \kappa\left(\frac{u+w-\beta}{\alpha}\right)}{1 - \kappa\left(\frac{u-\beta}{\alpha}\right)}\right]^{1/\kappa}$$

$$P(X > u + w | X > u) = \left[1 - \frac{\kappa\left(\frac{u+w-\beta-u+\beta}{\alpha}\right)}{1 - \kappa\left(\frac{u-\beta}{\alpha}\right)}\right]^{1/\kappa} = \left[1 - \frac{\kappa w}{\alpha - \kappa(u-\beta)}\right]^{1/\kappa}$$

(8.69)

Now, if $\alpha' = \alpha - \kappa(u - \beta)$ denotes a parameter, the result expressed by Eq. (8.69) signifies that the exceedances, as denoted by w and conditioned to $X > u$, follow a generalized Pareto (GPA) distribution, with location parameter equal to zero, and shape and scale parameters respectively given by κ and $\alpha' = \alpha - \kappa(u - \beta)$ (see solutions to Examples 5.11, 5.5, and 8.8). Designating the conditional CDF of w by $H_W(w|X > u)$, it follows that

$$H_W(w|X > u) = 1 - \left[1 - \frac{\kappa w}{\alpha'}\right]^{1/\kappa} \text{ for } w > 0, \kappa \neq 0 \text{ and } 1 - \frac{\kappa w}{\alpha'} > 0 \quad (8.70)$$

When $\kappa \to 0$, the limit of $H_W(w/X > u)$ will tend to

$$H_W(w|X > u) = 1 - \exp\left(-\frac{w}{\alpha'}\right) \text{ for } w > 0 \quad (8.71)$$

which is the CDF of an exponential variate with scale parameter α'.

The results given by Eqs. (8.70) and (8.71), as expressed in a more formal manner, by contextualizing the GPA as the limiting distribution of the exceedances as $u \to \infty$, are due to Pickands (1975). In less formal terms, these results state that if the distribution of annual maxima is the GEV, then the exceedances over a high threshold u are distributed as a GPA. Further, the GPA parameters can be uniquely determined by the GEV parameters and vice-versa. In particular, the shape parameter κ is identical for both distributions and the GPA scale parameter α' depends on the threshold value u and is related to the GEV parameters, β of location and α of scale, through $\alpha' = \alpha - \kappa(u - \beta)$.

Example 8.4 [adapted from Coles (2001)]—Assume the parent distribution of the initial variables X is the exponential with parameters of location $\beta = 0$ and of scale $\alpha = 1$, with CDF $F_X(x) = 1 - \exp(-x)$, for $x > 0$. If $w = x-u$ denote the exceedances of X over a high threshold u, determine the conditional probability $P(X > u + w|X > u)$.

Solution From Eq. (8.62), $P(X > u + w | X > u) = \frac{1 - F_X(u+w)}{1 - F_X(u)} = \frac{\exp[-(u+w)]}{\exp(-u)} = \exp(-w)$, for $w > 0$. From the asymptotic extreme-value theory, it is known that the maxima of IID exponential variables X are distributed according to a Gumbel distribution, which is one of the three limiting forms. According to Eq. (8.71), the distribution of the exceedances is the particular form of the GPA given by $H_W(w|X > u) = 1 - \exp(-w)$ for $w > 0$, with scale parameter $\alpha' = \alpha - \kappa(u - \beta) = 1 - 0 \times (u - 0) = 1$, which coincides with the complement, with respect to 1, of the conditional probability $P(X > u + w|X > u)$.

Example 8.5 [adapted from Coles (2001)]—Solve Example 8.4 for the case where X follows a Fréchet distribution with parameters $\kappa = -1$, of shape, $\alpha = 1$, of scale, and $\beta = 1$, of location, such that $F_X(x) = \exp(-1/x)$.

Solution Equation (8.62) gives $P(X > u + w | X > u) = \frac{1-F_X(u+w)}{1-F_X(u)} = \frac{1-\exp[-(1/u+w)]}{1-\exp(-1/u)}$ $\approx \left(1 + \frac{w}{u}\right)$, for $w > 0$. From the asymptotic extreme-value theory, it is known that the maxima of IID Fréchet variables X are distributed according to a Fréchet, which is one of the three limiting forms (GEV with $\kappa < 0$). According to Eq. (8.70), the distribution of the exceedances is given by the GPA $H_W(w | X > u) = 1 - \left[1 - \frac{\kappa w}{\alpha'}\right]^{1/\kappa} = 1 - \left[1 + \frac{w}{u}\right]^{-1}$ for $w > 0$, with scale parameter $\alpha' = \alpha - \kappa(u - \beta) = 1 + 1 \times (u - 1) = u$, which coincides with the complement, with respect to 1, of the conditional probability $P(X > u + w | X > u)$.

Example 8.6 [adapted from Coles (2001)]—Solve Example 8.4 for the case where X follows a Uniform distribution, for $0 \leq X \leq 1$, with parameters $\alpha = 1$, of scale, and $\beta = 0$, of location, such that $F_X(x) = x$.

Solution Equation (8.62) gives $P(X > u + w | X > u) = \frac{1-F_X(u+w)}{1-F_X(u)} = \frac{1-(u+w)}{1-u} = 1 - \frac{w}{1-u}$, for $w > 0$ and $w \leq 1 - u$. From the asymptotic extreme-value theory, it is known that the maxima of IID variables X with an upper bound are distributed according to a Weibull, which is one of the three limiting forms (GEV with $\kappa > 0$). According to Eq. (8.70), the distribution of the exceedances is given by the GPA $H_W(w | X > u) = 1 - \left[1 - \frac{\kappa w}{\alpha'}\right]^{1/\kappa} = 1 - \left[1 - \frac{w}{1-u}\right]$ for $w > 0$, with scale parameter $\alpha' = \alpha - \kappa(u - \beta) = 1 - 1 \times (u - 0) = 1 - u$, which coincides with the complement, with respect to 1, of the conditional probability $P(X > u + w | X > u)$.

8.4.4 Selecting the Threshold u Within the Framework of GPA-Distributed Exceedances

Coles (2001), based on the results of Eqs. (8.70) and (8.71), suggests the following logical framework for modeling the exceedances above a threshold:

- Data are realizations of the sequence of IID variables $X_1, X_2, \ldots$, with common parent CDF $F_X(x)$, which pertains to a domain of attraction for maxima or, in other words, the limit of $[F_X(x)]^N$ converges to one of the three asymptotic forms of extreme values, as $N \to \infty$;
- The extreme events are identified by defining a sufficiently high threshold u and are grouped into the sequence $X_{(1)}, X_{(2)}, \ldots, X_{(k)}$ of independent peak values of X that have exceeded u;
- The exceedances $X_{(j)}$ over u are designated as $W_j = X_{(j)} - u$ for $j = 1, 2, \ldots, k$;

- From Eqs. (8.70) and (8.71), the exceedances W_j are viewed as independent realizations of the random variable W, whose probability distribution is the generalized Pareto (GPA); and
- When enough data on W_j exist, the inference consists of fitting the GPA to the sample, followed by the conventional steps of goodness-of-fit checking and model extrapolation.

Coles (2001) points out that modeling the exceedances W_j differs from modeling annual maxima as the former requires the specification of an appropriate threshold u that should guide the characterization of which data are extreme, in addition to fulfill the requirements for applying Eqs. (8.70) and (8.71). However, at this point, the usual difficulty of making inferences from short samples emerges, which requires the trade-off between biasing estimates and inflating their respective variances. In fact, if a too low threshold is chosen, there will be a large number of exceedances W_j available for inference, but they would likely violate the asymptotic underlying assumptions on which Eqs. (8.70) and (8.71) are based. On the other hand, if a too high threshold is prescribed, there will be too few exceedances W_j available for inference, which will certainly increase the variance of estimates. The usual solution is to choose the smallest threshold such that the GPA is a plausible approximation of the empirical distribution of exceedances.

One of such methods to implement the adequate choice of the threshold is of an exploratory nature and makes use of the property $\alpha' = \alpha - \kappa(u - \beta)$, of linearity between the GPA scale parameter and the threshold u. In fact, the expected value of the exceedances, conditioned on a threshold u_0 and distributed as a GPA with parameters κ and α_{u_0}, provided that $\kappa > -1$, is given by

$$E\left[X - u_0 | X > u_0\right] = \frac{\alpha_{u_0}}{1 + \kappa} \tag{8.72}$$

In fact, when $\kappa \le -1$, the expected value of the exceedances either will tend to infinity or will result in a negative value, making Eq. (8.72) useless. On the other hand, for $\kappa > -1$, assume now the threshold is raised from u_0 to $u > u_0$. If Eq. (8.72) holds for u_0, it should also hold for $u > u_0$, provided the scale parameter is adjusted by using the relation $\alpha_u = \alpha_{u_0} - \kappa u$, with $\beta = 0$. In such a case, Eq. (8.72) becomes

$$E\left[X - u | X > u\right] = \frac{\alpha_{u_0} - \kappa u}{1 + \kappa} \tag{8.73}$$

Therefore, assuming the exceedances follow a GPA for $u > u_0$, the expected value $E\left[X - u | X > u\right]$ must be a linear function of u, with slope coefficient $-\kappa/(1 + \kappa)$. When $\kappa = 0$, the mean exceedances are constant for increasing values of u. When $\kappa > 0$, the mean exceedances decrease with increasing u, with negative slope coefficient equal to $-\kappa/(1 + \kappa)$. Finally, when $-1 < \kappa < 0$, the mean exceedances increase with increasing u, with a positive slope coefficient equal to $-\kappa/(1 + \kappa)$. Such a property of the expected value of exceedances can be employed to build an exploratory method to investigate (1) the correct choice of the threshold

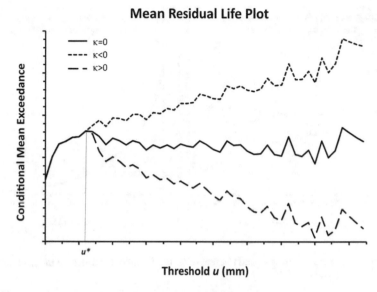

Fig. 8.18 Hypothetical examples of mean residual life plots

u above which the GPA distributional hypothesis holds; and (2) the nullity or the sign of the shape parameter κ and, thus, the weight and shape of the upper tail of $F_X(x)$, among the exponential, polynomial or upper-bounded types.

The described exploratory method is referred to as the *mean residual life plot* and consists of graphing on abscissae the varying values of the threshold u and on ordinates the respective mean exceedances. Figure 8.18 depicts a hypothetical example of the mean residual life plot, where the first point to note is the location of u^*, the apparently correct choice for the threshold above which the exceedances seem to have an approximately stable linear tendency. After choosing the threshold, one should note the slope of the linear tendency: if a relatively constant mean exceedance evolves over the varying values of u, then $\kappa = 0$; if the slope is negative, then $\kappa > 0$, and the angular coefficient is $-\kappa/(1+\kappa)$; and if otherwise, then $\kappa < 0$, and the slope coefficient is $-\kappa/(1+\kappa)$.

In practical situations, however, the use and interpretation of mean residual life plots are not that simple. Coles (2001), Silva et al. (2014), and Bernardara et al. (2014) show examples of how difficult interpreting the information contained in mean residual life plots is. The main issue is related to the few data points that are made available for inference as the threshold is raised to a level where a stable tendency can be discernible. There are cases where inference is very unreliable or meaningless. The practical solution is to gradually lower the threshold to a level such that the mean residual life plot can be employed for inference, with some degree of confidence (Coles 2001, Ghosh and Resnick 2010). In this context, by assuming the mean exceedances estimators as normally distributed, their corresponding $(100-\alpha)$ % confidence intervals can be approximated as $\left[\overline{w}_u - z_{1-\alpha/2}\right.$

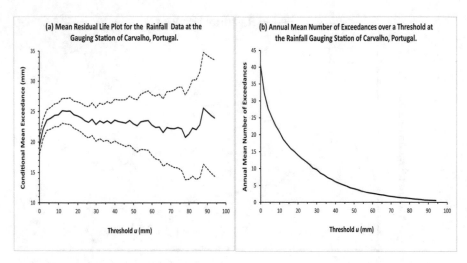

Fig. 8.19 Mean residual life plot and number of daily rainfall exceedances at the gauging station of Carvalho, in Portugal

$s_{w_u}/\sqrt{n_u}; \overline{w}_u + z_{1-\alpha/2}s_{w_u}/\sqrt{n_u}]$, where $\overline{w}_u$ and s_{w_u} respectively denote the mean and standard deviation of the n_u exceedances over u. Example 8.7, adapted from a case study reported in Silva et al. (2012), illustrates an application of the mean residual life plot for choosing the threshold level u to be employed in the modeling of a partial duration series of the extreme daily rainfall depths recorded at the gauging station of Carvalho, located in the Douro River catchment, in Portugal.

Example 8.7 The rainfall gauging station of Carvalho is located at the town of Celorico de Basto, in the district of Braga, in northern Portugal, and has records of daily rainfall depths for 27 water years (Oct 1st–Sep 30th), from 1960/1961 to 1986/87, available from http://snirh.apambiente.pt. Silva et al. (2012) employed the referred data to perform a frequency analysis of the related partial duration series. Initially, they identified all non-zero daily rainfall depths whose successive occurrences were separated by at least 1 day of no rainfall. Following that, the estimates of $\overline{w}_u$, s_{w_u}, n_u and respective 95 % confidence intervals were calculated for threshold values u varying from 0 to 94. The mean residual life plot alongside the 95 % confidence bands are shown in Fig. 8.19a, whereas Fig. 8.19b depicts the number of exceedances as the threshold increases. Note in Fig. 8.19b that the number of exceedances drops to less than 5 points as u rises above 50 mm.

The mean residual life plot of Fig. 8.19a shows that the mean exceedances develop as a concave downward curve up to the threshold $u = 25$ mm, and from that point on up to $u = 75$ mm, it shows an approximately linear and constant development. For larger values of u, the mean exceedances seem erratic, as resulting from the small number of occurrences. For such example, it seem reasonable to admit a constant linear development from the threshold $u = 25$ mm on, which leads to the conclusion of an exponential upper tail ($\kappa = 0$), with CDF given by Eq. (8.71). The second requirement for threshold selection, that is the Poisson-distributed annual number of exceedances (see Sect. 8.4.2), led to $u = 47$ mm, with 123 exceedances above that threshold, so that $\hat{\nu} = 123/27 = 4.56$. The mean magnitude of the exceedances over 47 mm is estimated as 23.394 mm, which yields $\hat{\alpha}' = 23.394$. With these estimates in Eq. (8.59), the exponential distribution of exceedances is given by $H_W(w|X > u) = 1 - \exp[-(x - 47)/23.394]$, which leads to a Gumbel model for the annual maximum daily rainfall depths with CDF expressed as $F_a(x) = \exp\{-\nu[1 - H_W(w)]\} = \exp\{-4.56\exp[-(x - 47)/23.394]\}$.

8.4.5   The Poisson–Pareto Model

The Poisson–Pareto model combines two distributional assumptions: the annual number of exceedances is a discrete variable that follows a Poisson distribution, whereas the magnitude of exceedances follows a generalized Pareto. The derivation of the Poisson–Pareto models starts from Eq. (8.59) by replacing $H_u(x)$ by the CDF of the GPA, which, in its general form, is given by

$$H_u(x) = 1 - \exp(-y)$$
$$\text{where}$$
$$y = -\frac{1}{\kappa}\ln\left[1 - \frac{\kappa(x - \xi)}{\alpha}\right] \text{ for } \kappa \neq 0 \qquad (8.74)$$
$$y = \frac{x - \xi}{\alpha} \text{ for } \kappa = 0$$

and ξ, α, and κ are parameters of location, scale, and shape, respectively. The domains of X are $\xi \leq x \leq \xi + \frac{\alpha}{\kappa}$, for $\kappa > 0$, and $\xi \leq x < \infty$, for $\kappa \leq 0$. For simplicity, the notations $F_a(x)$ and $H_u(x)$ are hereafter replaced by $F(x)$ and $H(x)$, respectively, and Eq. (8.59) is rewritten as

$$F(x) = \exp\{-\nu[1 - H(x)]\} \qquad (8.75)$$

Taking the logarithms of both sides of Eq. (8.75), one can write

$$\ln(F(x)) = -\nu[1 - H(x)] \qquad (8.76)$$

Then,

$$H(x) = 1 + \frac{1}{\nu}\ln(F(x)) \qquad (8.77)$$

By equating the expressions of $H(x)$, as in Eqs. (8.74) and (8.77), one obtains

Table 8.15 Partial duration series of 2-h duration rainfall depths (mm) over 39 mm reduced from the records at the gauging station of Entre Rios de Minas, in Brazil

Water year	Y_i = rainfall (mm)	Water year	Y_i = rainfall (mm)	Water year	Y_i = rainfall (mm)
1973/74	51.1	1978/79	40.2	1981/82	48.3
1974/75	39.6	1978/79	41.0	1981/82	57.2
1974/75	40.0	1978/79	61.1	1982/83	43.3
1974/75	40.5	1979/80	39.2	1982/83	53.1
1975/76	47.4	1979/80	48.4	1984/85	48.6
1976/77	39.4	1980/81	40.8	1984/85	63.1
1976/77	44.5	1980/81	43.6	1984/85	73.4
1977/78	55.6	1980/81	44.3	1985/86	41.2
1977/78	80.0	1980/81	64.1	–	–

$$y = -\ln\left\{-\ln[F(x)]^{1/\nu}\right\} \tag{8.78}$$

For $\kappa = 0$, the GPA standard variate is $y = (x - \xi)/\alpha$, which in Eq. (8.78) yields

$$x = \xi - \alpha\ln\left\{-\ln[F(x)]^{1/\nu}\right\} \tag{8.79}$$

Equation (8.79) is the general quantile function of the Poisson–Pareto model in terms of $F(x)$, for $\kappa = 0$. In terms of the return period T in years, it is expressed as

$$x = \xi - \alpha\left\{\ln\left(\frac{1}{\nu}\right) + \ln\left\{-\ln\left[1 - \frac{1}{T}\right]\right\}\right\} \tag{8.80}$$

For $\kappa \neq 0$, the GPA standard variate is $y = -\ln[1 - \kappa(x - \xi)/\alpha]/\kappa$, which in Eq. (8.78), after algebraic manipulation, yields

$$x = \xi + \frac{\alpha}{\kappa}\left\{1 - \left[-\frac{1}{\nu}\ln[F(x)]\right]^{\kappa}\right\} = \xi + \frac{\alpha}{\kappa}\left\{1 - \left[-\frac{1}{\nu}\ln\left[1 - \frac{1}{T}\right]\right]^{\kappa}\right\} \tag{8.81}$$

Equation (8.81) is the general quantile function of the Poisson–Pareto model, for $\kappa \neq 0$. Example 8.8 illustrates a simple application of the Poisson–Pareto model.

Example 8.8 Fit the Poisson–Pareto model to the partial duration series of rainfall depths of 2-h duration, recorded at the gauging station of Entre Rios de Minas (code 02044007), located in the state of Minas Gerais, in southeastern Brazil, as listed in Table 8.15. The period of available records spans for 13 water years, from 1973/74 to 1985/86. The 26 rainfall depths listed in Table 8.15 are those which have exceeded the threshold of 39 mm over the 13 years of records.

Table 8.16 Elements for calculating the test statistic of the Cunnane test for the partial duration series of 2-h duration rainfall depths at the station of Entre Rios de Minas

Water year	/74	/75	/76	/77	/78	/79	/80	/81	/82	/83	/84	/85	/86	Total
m_k	1	3	1	2	2	3	2	4	2	2	0	3	1	26
R	0.5	0.5	0.5	0	0	0.5	0	2	0	0	2	0.5	0.5	7

Table 8.17 Annual maximum 2-h duration rainfall depths (Poisson–Pareto model)

T (years)	2	5	10	20	30	50	75	100
Quantile estimates (mm)	50.0	62.3	70.8	79.2	84.1	90.4	95.5	99.2

Solution For the threshold $u = 39$ mm, 26 data points were selected from the 13 years of records, which gives the estimate $\hat{\nu} = 2$ for the annual mean number of exceedances. In order to test the null hypothesis that the annual number of exceedances follows a Poisson distribution, the Cunnane test should be used. The elements for calculating the test statistic R, as in Eq. (8.60), are given in Table 8.16. As a two-sided test, the calculated test statistic R should be compared to $\chi^2_{\alpha/2,\eta}$ and $\chi^2_{1-\alpha/2,\eta}$ with 12 degrees of freedom ($\eta = N - 1 = 12$), at the significance level $100\alpha = 5\%$. Readings from the table of Appendix 3 yield $\chi^2_{0.025,12} = 4.404$ and $\chi^2_{0.975,12} = 23.337$. As the calculated statistic $R = 7$ is comprised between the critical values of the test statistic, the null hypothesis that the annual number of exceedances follows a Poisson distribution should not be rejected.

In order to estimate the parameters of the generalized Pareto distribution, recall from the solution to Example 5.11, that the equations for MOM estimates are $\hat{\alpha} = \overline{X}/2\left(\overline{X}^2/S_X^2 + 1\right)$ and $\hat{\kappa} = 1/2\left(\overline{X}^2/s_X^2 - 1\right)$, with $X_i = Y_i - \xi$ denoting the exceedances of the 2-h rainfall depths (Y_i) over $\xi = 39$ mm. With $\overline{X} = 10.577$ and $S_X = 11.063$, the parameter estimates are $\hat{\alpha} = 10.122$ and $\hat{\kappa} = -0.04299$. Equation (8.81), with parameters estimated by the MOM methods, should be used to estimate the quantiles as $x(F) = 39 - 10.122/0.04299\left\{ 1 - [-\ln[F(x)]/2]^{-0.04299}\right\}$. The annual maximum 2-h duration rainfall depths for different return periods, as estimated with the Poisson–Pareto model, are given in Table 8.17.

8.5 Derived Flood Frequency Analysis and the GRADEX Method

The perception that rainfall excess is in most cases the dominating process at the origin of large floods and that rainfall data are usually more abundant and more readily regionalized than streamflow data, has long motivated the development of methods to derive flood probability distributions from rainfall distributions or, at least, to incorporate the hydrometeorological information into flood frequency analyses. Eagleson (1972) pioneered the so-called *derived distribution* approach, by deducing the theoretical peak discharge probability distribution from a given rainfall distribution, through a kinematic wave model. This general approach has been followed by many researchers (e.g., Musik 1993, Raines and Valdes 1993, Iacobellis and Fiorentino 2000, De Michele and Salvadori 2002, among others). In this context, Gaume (2006) presents some theoretical results concerning the asymptotic properties of flood peak distributions and provides a general framework for the analysis of the different procedures based on the derived distribution approach. According to Gaume (2006), despite the numerous developments since Eagleson's original work, the procedures based on the derived distribution approach are still not frequently used in an operational context. Gaume (2006) explains that such a fact may be due either to the more complex mathematics many of these procedures require or by their oversimplified conceptualization of the rainfall-runoff transformation.

Probably with the same motivations that inspired the derived distribution approach and with the empirical support given by the results from the station-year sampling experiment performed by Hershfield and Kohler (1960), Guillot and Duband (1967) introduced the so-called GRADEX (*GRADient of EXtreme values*) method for flood frequency analysis, which has since been widely used in France and other countries, as the reference procedure to construct design floods for dam spillways and other engineering works. The GRADEX method requires simple assumptions on the relationship between rainfall and flood volumes under extreme conditions, as well as on the probability distribution of rainfall volumes, which should exhibit, at least asymptotically, an exponential-like upper tail. Further developments brought forth similar methods, such as those described by Cayla (1993), Margoum et al. (1994), and Naghettini et al. (1996). These are termed GRADEX-type methods since they all share the same basic assumptions introduced by Guillot and Duband (1967). The principles of the GRADEX method are presented next.

The GRADEX method was developed by engineers of the French power company EDF (*Electricité de France*) and was first described by Guillot and Duband (1967). The method's main goal is to provide estimates of low frequency flood volumes, as derived from the upper tail of the probability distribution of rainfall volumes, estimated from local and/or regional rainfall data. In order to accomplish it, the GRADEX method makes use of two fundamental assumptions. The first refers to the relationship between rainfall and flood volumes, of a given duration, as rainfall equals or exceeds the prevailing catchment capacity of storing surface and subsurface water. Under these conditions, it is assumed that any rainfall volume increment tends to yield equal increment in flood volume. The second assumption concerns the upper tail of the probability distribution of the random variable P (the rainfall volume over a given duration), which is supposed to be an exponentially decreasing function of the form

$$1 - F(p) = \exp\left(-\frac{p - K}{a}\right) \tag{8.82}$$

where p represents a large quantile, and the positive constants K and a denote the location and scale parameters, respectively. As shown later in this section, the combination of these assumptions leads to the conclusion that the upper tail of the probability distribution of flood volumes, for the same duration as that specified for rainfall volumes, is also of the exponential-type with the same scale parameter a, the GRADEX parameter, previously fitted to rainfall data. The next paragraphs contain a review of the GRADEX method and the mathematical proofs for its main assertions.

Denote by p_i the average rainfall volume, of duration d, over a given catchment, associated to the ith hypothetical event abstracted from the sample of rainfall data. Duration d is specified as an integer number of days (or hours) large enough to characterize the rising and recession limbs of the catchment flood hydrograph, and may be fixed, for instance, as the average time base (or in some cases as the average time of concentration), estimated from flow data. Suppose x_i represents the flood volume accumulated over the same duration d, associated with the p_i event. Suppose further that the generic pairs $(p_i, x_i; \forall i)$ are expressed in the same measuring units, for instance, mm or $(m^3/s).day$. Under these conditions, the variable R, the i-th occurrence of it being calculated as $r_i = (p_i - x_i)$, represents the remaining volume to attain the catchment total saturation or the catchment water retention volume. Figure 8.20 depicts a schematic diagram of hypothetical realizations of variables X and P: the pairs (p_i, x_i) should all lie below the bisecting line $x = p$, with the exception of some occasional events, which have been supposedly affected by snowmelt or by relatively high baseflow volumes.

The variable R depends upon many related factors, such as the antecedent soil moisture conditions, the subsurface water storage, and the rainfall distribution over time and space. From the GRADEX method standpoint, R is considered a random variable with cumulative probability distribution function, conditioned on P, denoted by $H_{R/P}(r/p)$. Figure 8.20 illustrates the curves that describe the relationship between X and P as corresponding to some hypothetical R quantiles. The first assumption of the GRADEX method states that all curves relating X and P tend to be asymptotically parallel to the bisecting line $x = p$, as the rainfall volume, over the duration d, approaches a value p_0, large enough to exceed the current maximum capacities of absorbing and storing water, in the catchment. The quantity p_0 will depend mainly on the catchment's soil and relief characteristics.

According to Guillot and Duband (1967), the typical p_0 values are associated with return periods of the order of 10 to 25 years, in relatively impermeable and steep catchments, or to return periods of 50 years, in more permeable catchments with moderate reliefs. The relative position of each asymptote depends on the initial conditions prevailing in the catchment. Accordingly, depending on the value taken by the variable R, the curves will be parallel to $x = p$ more rapidly in wet soil than in dry soil. Likewise, the probability distribution function of R, conditioned on P, will

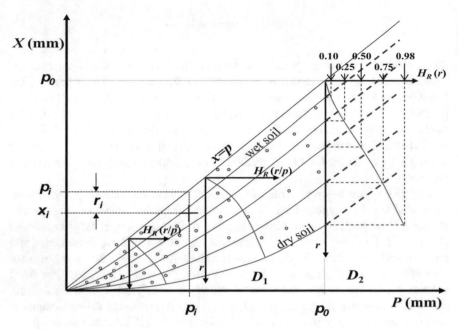

Fig. 8.20 Schematic chart of the relationship between hypothetical rainfall volumes (p_i) and flood volumes (x_i), under the assumptions of the GRADEX method

tend to have a stable shape, constant variance, and a decreasing dependence on rainfall volumes, as P approaches the high threshold value p_0.

In Fig. 8.20, the plane PX may be divided into two domains:

- D_1, containing the points $p < p_0$ and $x \leq p$, in which the probability distribution of R is conditioned on P; and
- D_2, containing the points $p \geq p_0$ and $x \leq p$, where the curves relating X to P are parallel to the bisecting line $x = p$.

Assume that $f_P(p)$, $g_X(x)$ and $h_R(r)$ denote the marginal probability density functions of variables P, X, and R, respectively, whereas $h_{R/(X+R)}[r/(x+r)]$ represents the density of R, conditioned on P, as expressed as $P = X + R$. From the definition of conditional density function, the joint density of $(X + R)$ and R may be written in the form $f_{X+R}(x + r)h_{R/(X+R)}[r/(x+r)]$, the integration of which, over the domain of R, results in the marginal density of X. Formally,

$$g_X(x) = \int_0^\infty f_{X+R}(x + r)h_{R/(X+R)}[r/(x+r)]dr \qquad (8.83)$$

In the domain D_2, or for $x + r > p_0$, once admitted as true the hypothesis that R no longer depends on P, the conditional density $h_{R/X+R}[r/(x+r)]$ becomes the marginal $h_R(r)$ and Eq. (8.83) may be rewritten as

$$g_X(x) = \int_0^\infty f_{X+R}\,(x+r)\,h_R(r)\,dr \qquad (8.84)$$

The second assumption of the GRADEX method refers to the upper tail of the cumulative distribution function $F_P(p)$, or $F_{X+R}(x+r)$, which is assumed to tend asymptotically to an exponential tail. Formally,

$$1 - F_{X+R}\,(x+r) \xrightarrow[x+r\to\infty]{} \exp\left(-\frac{x+r-K}{a}\right) \text{ with } a > 0 \text{ and } x+r > K$$

$$(8.85)$$

where the location parameter K is positive and the scale parameter a is referred to as the GRADEX rainfall parameter. In these terms, the density $f_P(p)$ becomes

$$f_{X+R}(x+r) = \frac{1}{a}\exp\left(-\frac{x+r-K}{a}\right) = f_P(x)\exp\left(-\frac{r}{a}\right) \qquad (8.86)$$

In order to be true, Eq. (8.86) requires the additional condition that $x > K$, which, in practical terms, is easily fulfilled in domain D_2. Replacing Eqs. (8.84) into (8.86), it follows that, in domain D_2 $(x+r > p_0)$ and for $x > K$,

$$g_X(x) = f_P\,(x) \int_0^\infty \exp\left(-\frac{r}{a}\right) h_R(r)\,dr \qquad (8.87)$$

In Eq. (8.87), because r and a are both positive, the definite integral is also a positive constant less than 1. Assuming, for mathematical simplicity, this constant as equal to $\exp(-r_0/a)$, it follows that, for a sufficiently large value x^*, is valid to write

$$g_X\left(x^*\right) = f_P\left(x^* + r_0\right) \qquad (8.88)$$

Therefore, in domain D_2, the density $g_X(x^*)$ can be deduced from $f_P(p^*)$ by a simple translation of the quantity r_0, along the variate axis, which is also valid for the cumulative distributions $G_X(x^*)$ and $F_P(p^*)$. In other terms, it is valid to state that for both x^* and $p^* = x^* + r_0$, the same exceedance probability (or the same return period) can be assigned. Still in this context, it is worth noting that the definite integral, as in Eq. (8.87), represents the expected value of $\exp(-r/a)$, or $E[\exp(-r/a)]$. As a result, the *translation distance* r_0 may be formally written as

Table 8.18 Annual maximum 5-day total rainfall depths (mm), denoted as $P_{5,i}$, and 5-day mean flood flows (m³/s), denoted as $X_{5,i}$, for Example 8.9

Year	$P_{5,i}$	$X_{5,i}$	Year	$P_{5,i}$	$X_{5,i}$	Year	$P_{5,i}$	$X_{5,i}$	Year	$P_{5,i}$	$X_{5,i}$
1939	188		1959	167		1979	136		1999	205	257
1940	96.2		1960	114		1980	213		2000	125	152
1941	147		1961	101		1981	112	120	2001	203	257
1942	83.4		1962	136		1982	98.8	103	2002	141	171
1943	189		1963	146		1983	138	166	2003	183	226
1944	127		1964	198		1984	168	204	2004	93.1	109
1945	142		1965	153		1985	176	215	2005	116	123
1946	238		1966	294		1986	157	187	2006	102	105
1947	164		1967	182		1987	114	122	2007	137	165
1948	179		1968	179		1988	114	122	2008	127	153
1949	110		1969	120		1989	163	196	2009	96.4	104
1950	172		1970	133		1990	265	391	2010	142	171
1951	178		1971	135		1991	150	181	2011	155	187
1952	159		1972	158		1992	148	180	2012	101	103
1953	148		1973	133		1993	244	351	2013	150	179
1954	161		1974	132		1994	149	182	2014	127	152
1955	149		1975	226		1995	216	271	2015	105	108
1956	180		1976	159		1996	118	140	–	–	–
1957	143		1977	146		1997	228	307	–	–	–
1958	96.0		1978	131		1998	192	240	–	–	–

$$r_0 = -a \ln \left\{ E \left[\exp \left(-\frac{r}{a} \right) \right] \right\} \tag{8.89}$$

which, to be evaluated, one would need to fully specify $h_{R/P}(r)$ and $h_R(r)$ distributions (see Naghettini et al. 2012).

In practical applications of the GRADEX method for flood frequency analysis, Guillot and Duband (1967) recommend using the empirical distribution of flood volumes up to a return period within the interval 10–25 years, for relatively impermeable watersheds, and up to 50 years for those with higher infiltration capacity. From that point on, the cumulative distributions of rainfall and flood volumes will be separated by a constant distance r_0, along the variate axis. Equivalently, in domain D_2, the two distributions will plot as straight lines on Gumbel or exponential probability papers, both with slope equal to a (the rainfall GRADEX parameter) and separated by the translation distance r_0, for a given return period. Example 8.9 shows an application of the GRADEX method.

Example 8.9 Table 8.18 displays the annual maximum 5-day total rainfall depths (mm) and annual maximum 5-day mean discharges (m³/s), observed in a catchment of drainage area 950 km², with moderate relief and soils of low permeability. The 5-day duration corresponds to the average time base of observed flood

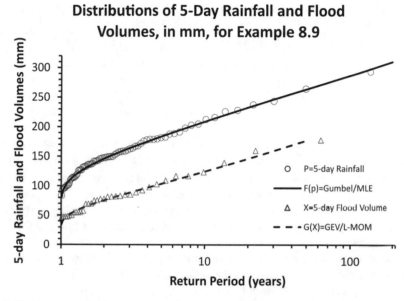

Fig. 8.21 Frequency distributions of 5-day rainfall and flood volumes, in mm, for the data given in Table 8.18

hydrographs in the catchment. The 5-day total rainfall depths are average quantities over the catchment area. Employ the GRADEX method to estimate the frequency curve of annual maximum 5-day mean discharges up to the 200-year return period. Plot on the same chart the frequency curves of annual maximum 5-day total rainfall depths (mm) and annual maximum 5-day mean discharges (m³/s).

Solution The sizes of the samples of rainfall and flow data are 77 and 35 respectively. In order to apply the GRADEX method, the upper-tail of the maximum rainfall distribution must be of the exponential type. Examples of distributions that exhibit such a type of upper tail are exponential, Gamma, normal, and the $\text{Gumbel}_{\max}$. Figure 8.21 depicts the empirical distribution of the annual maximum 5-day total rainfall depths (mm) and the $\text{Gumbel}_{\max}$ distribution with parameters estimated by the maximum-likelihood method, on exponential probability paper, and plotting positions calculated with the Gringorten plotting positions. The parameters estimates are $\hat{\alpha} = 33.39$ and $\hat{\beta} = 134.00$ and it is clear from Fig. 8.21 that the $\text{Gumbel}_{\max}$ fits the sample of rainfall data well, showing plausible exponential upper-tail behavior. The rainfall GRADEX parameter is thus $a = \hat{\alpha} = 33.39$ mm. The 5-day mean flood flows (m³/s) have been transformed into 5-day flood volumes, expressed in mm, and are also plotted on the chart of Fig. 8.21. The transformation factor is given by $F = (5 \text{ days} \times 86,400 \text{ s} \times X_{5,i})/(\text{DA} \times 10^3)$ where $X_{5,i}$ denotes the 5-day mean flood flows, in m³/s, and DA = 950, the drainage area in km².

Table 8.19 Maximum 5-day flood volumes (mm) and 5-day mean discharges (m³/s), for return periods 25–200 years, as estimated with the GRADEX method

Return period (years)	5-day flood volume (mm)	5-day mean discharge (m³/s)
25	153.8	338.2
50	176.9	389.1
100	200.0	440.0
200	223.2	490.9

On the chart of Fig. 8.21, the GEV distribution, with L-MOM parameter estimates $\hat{\alpha} = 22.39$, $\hat{\beta} = 67.11$, and $\hat{\kappa} = -0.115$, is plotted up to the 50-year return period, showing a good fit to the sample of 5-day flood volumes. It is worth noting that any good-fitting distribution could be employed for this purpose, since the GRADEX goal is to extrapolate the 5-day flood volumes frequency curve, guided by the exponential upper tail of 5-day rainfall depths as fitted to the larger sample of rainfall data.

In order to perform such an extrapolation, one needs to know where it should start. According to the recommendations for applying the GRADEX method, for relatively impermeable catchments, the extrapolation should start at a point x_0 with a return period between 10 and 25 years. Let the return period associated with x_0 be $T = 25$ years such that $G(x_0) = 0.96$. In order for the upper tail of 5-day flood volumes to have the same shape of its 5-day rainfall homologous, from the x_0 point on, the location parameter K, in Eq. (8.82), must be known. This can be established by working on Eq. (8.82), where the location parameter K can be written in explicit form as $K = x_0 + a\ln[1 - G(x_0)]$, where a is the GRADEX parameter ($a = 33.39$ mm), and x_0 is the 5-day flood volume quantile of return period $T = 25$ years [or $G(x_0) = 0.96$], which, according to the fitted GEV, is $x_0 = 153.8$ mm. As such, the location parameter is $K = 46.32$. Then, from the point $x_0 = 153.8$ mm on, the x quantiles are calculated with $x_T = K - a\ln(T) = 46.32 - 33.39\ln(T)$ and are given in Table 8.19. Figure 8.22 shows the frequency distribution of 5-day flood volumes extrapolated with the GRADEX method, together with the distribution of the 5-day rainfall depths.

Exercises

1. Construct a Fréchet (Log-Gumbel) probability paper and plot the empirical distribution of the annual flood peaks of the Lehigh River at Stoddartsville given in Table 7.1 and the theoretical distribution found in the solution to Example 8.3.

2. Download the annual total rainfall depths observed at the Radcliffe Meteorological Station, from http://www.geog.ox.ac.uk/research/climate/rms/, and plot the corresponding empirical distribution on lognormal probability paper. Compare the plot with the chart of Fig. 8.7 and decide on the distribution that best fits the sample data.

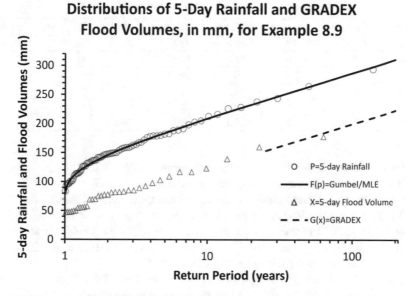

Fig. 8.22 Frequency distributions of 5-day rainfall and flood volumes, with GRADEX extrapolation, for the data given in Table 8.18

3. Perform a complete frequency analysis of the annual maximum mean daily discharges (m³/s) of the Shokotsu River at Utsutsu Bridge, in Hokkaido, Japan, listed in Table 1.3. Use similar procedures as those employed in Example 8.3.
4. Perform a complete frequency analysis of the annual maximum daily rainfall (mm) observed at the gauging station of Kochi, Shikoku, Japan, listed in Table 7.21. Use similar procedures as those employed in Example 8.3.
5. Considering the solution to Example 8.3, what is the probability that the high outlier discharge 903 m³/s (a) will occur at least once in the next 200 years? (b) exactly twice in the next 200 years?
6. Table 2.7 lists the Q_7 flows for the Dore River at Saint-Gervais-sous-Meymont, in France, for the period 1920–2014. Fit a two-parameter Weibull$_{min}$ distribution to these data, using the method of L-moments. Remember that if X follows a two-parameter Weibull$_{min}$, then $Y = -\ln(X)$ is distributed as a Gumbel$_{max}$ distribution, making the Gumbel$_{max}$ estimation procedures and GoF tests also valid for the two-parameter Weibull$_{min}$. As such, if $Y = +\ln(X)$ has L-moments $\lambda_{1,\ln X}$ and $\lambda_{2,\ln X}$, then the parameters of the corresponding two-parameter Weibull$_{min}$ distribution, parameterized as in Eq. (5.85), are $\alpha = \ln(2)/\lambda_{2,\ln(X)}$ and $\beta = \exp[\lambda_{1,\ln(X)} + 0.5772/\alpha]$. Perform a frequency analysis of the Q_7 flows for the Dore River at Saint-Gervais-sous-Meymont, using the two-parameter Weibull$_{min}$, and compare the empirical and theoretical distributions on Gumbel probability paper. Use the Gringorten plotting position formula.

Table 8.20 3-h rainfall depths larger than 44.5 mm (*P*) at Pium-í, in Brazil

Water year	P(mm)	Water year	P(mm)	Water year	P(mm)	Water year	P(mm)
75/76	70.2	78/79	47.6	81/82	53	83/84	46.6
75/76	50	79/80	49.8	82/83	47.9	84/85	72.2
76/77	47.2	79/80	46	82/83	59.4	84/85	46.4
77/78	52	79/80	46.8	82/83	50.2	85/86	48.4
77/78	47.6	80/81	50.6	82/83	53.4		
77/78	47.4	81/82	44.1	82/83	59.4		

7. Derive the frequency factors for the two-parameter Weibull$_{min}$ distribution, as a function of the skewness coefficient and the return period.

8. Fit the Poisson–Pareto model to the partial duration series of rainfall depths, of 3-h duration, observed at the gauging station of Pium-í, code 02045012, listed in Table 8.20. This station is located in the state of Minas Gerais, in southeastern Brazil. The 3-h rainfall depths listed in Table 8.20 refer to the 22 largest values that have occurred from 1975/76 to 1985/86.

9. Table 2.6 lists the 205 independent peak discharges of the Greenbrier River at Alderson (West Virginia, USA) that exceeded the threshold 17,000 cubic feet per second (CFS), in 72 years of continuous flow records, from 1896 to 1967.

   (a) Select the highest possible value of the annual mean number of flood peak discharges such that they can be modeled as a Poisson variate. Check your choices through the Cunnane test;

   (b) After choosing the highest possible value of the annual mean number of flood peak discharges, fit the GPA distribution to the exceedances over the corresponding threshold; and

   (c) Estimate the annual flood quantiles for return periods from 2 to 500 years through the Poisson–Pareto model.

10. Tables 8.21 and 8.22 refer respectively to the maximum rainfall depths and maximum mean discharges, both of 2-day duration, that have been abstracted from the rainfall and flow records of a catchment of 6,520 km^2 of drainage area. This is a steep mountainous catchment with a low average infiltration capacity and time of concentration on the order of 2 days. Employ the GRADEX method to estimate the frequency curve of flood volumes of 2-day duration. Assume a flood wall is being designed to protect this site against flooding and further that the design-flood should have a return period of 500 years. Indicate how the previous results obtained with the GRADEX method can possibly be used to size the flood wall such that to protect this site against the 500-year flood.

Table 8.21 Maximum rainfall depths of 2-day duration (mm)

2-day rainfall (mm)	2-day rainfall (mm)	2-day rainfall (mm)	2-day rainfall (mm)
59	41	79	43
49	42	38	47
100	65	40	44
82	78	52	50
112	80	53	58
62	30	54	60
28	45	57	66
98	47	54	63
92	58	56	75
66	67	57	73
61	35	77	61
55	34	70	45
51	48	39	60
50	52	42	46
30	87	48	75

Table 8.22 Mean 2-day flood flows (m^3/s)

Mean 2d flood flow (m^3/s)	Mean 2d flood flow (m^3/s)	Mean 2d flood flow (m^3/s)	Mean 2d flood flow (m^3/s)
250	405	215	803
428	395	377	255
585	577	340	275
265	470	628	640
285	692	300	742
440	450	381	1063
535	412	475	400
980	408	235	363
1440	170	292	1005
548	260	910	360
1053	840	490	355
150	565	700	
457	202	1270	

References

Akaike H (1974) A new look at the statistical model identification. IEEE Trans Autom Control 19 (6):716–723

ASCE (1996) Hydrology handbook, 2nd edn, chapter 8—floods. ASCE Manuals and Reports on Engineering Practice No. 28. American Society of Civil Engineers, New York

Bayliss AC, Reed DW (2001) The use of historical data in flood frequency estimation. Report to Ministry of Agriculture, Fisheries and Food. Centre for Ecology and Hydrology, Wallingford, UK

Benito G, Thorndycraft VR (2004) Systematic, palaeoflood and historical data for the improvement of flood risk estimation: methodological guidelines. Centro de Ciencias Medioambientales, Madrid

Benjamin JR, Cornell CA (1970) Probability, statistics, and decision for civil engineers. McGraw-Hill, New York

Benson MA (1960) Characteristics of frequency curves based on a theoretical 1,000-year record. In: Dalrymple T (1960) Flood frequency analysis, manual of hydrology: part 3, flood-flow techniques, Geological Survey Water-Supply Paper 1543-A. United States Government Printing Office, Washington

Bernardara P, Mazas F, Kergadallan X, Hamm L (2014) A two-step framework for over-threshold modelling of environmental extremes. Nat Hazards Earth Syst Sci 14:635–647

Bobée B, Rasmussen P (1995) Recent advances in flood frequency analysis. US National Report to IUGG 1991-1994. Rev Geophys. 33 Suppl

Botero BA, Francés F (2010) Estimation of high return period flood quantiles using additional non-systematic information with upper-bounded statistical models. Hydrol Earth Syst Sci 14:2617–2628

Boughton WC (1980) A frequency distribution for annual floods. Water Resour Res 16:347–354

Calenda G, Mancini CP, Volpi E (2009) Selection of the probabilistic model of extreme floods: the case of the River Tiber in Rome. J Hydrol 371:1–11

Cayla O (1993) Probabilistic calculation of design floods: SPEED. In: Proceedings of the International Symposium on Engineering Hydrology, American Society of Civil Engineers, San Francisco, pp 647–652

Chow VT (1954) The log-probability law and its engineering applications. Proc Am Soc Civil Eng 80:1–25

Chow VT (1964) Statistical and probability analysis of hydrologic data. Section 8, part I—frequency analysis. In: Handbook of applied hydrology. McGraw-Hill, New York

Coles S (2001) An introduction to statistical modeling of extreme values. Springer, New York

Coles S, Pericchi LR, Sisson S (2003) A fully probabilistic approach to extreme rainfall modeling. J Hydrol 273:35–50

Correia FN (1983) Métodos de análise e determinação de caudais de cheia (Tese de Concurso para Investigador Auxiliar do LNEC). Laboratório Nacional de Engenharia Civil, Lisbon

Cramér H, Leadbetter MR (1967) Stationary and related stochastic processes. Wiley, New York

Cunnane C (1973) A particular comparison of annual maximum and partial duration series methods of flood frequency prediction. J Hydrol 18:257–271

Cunnane C (1978) Unbiased plotting positions—a review. J Hydrol 37:205–222

Cunnane C (1979) A note on the Poisson assumption in partial duration series models. Water Resour Res 15(2):489–494

De Michele C, Salvadori G (2002) On the derived flood frequency distribution: analytical formulation and the influence of antecedent soil moisture condition. J Hydrol 262:245–258

Eagleson PS (1972) Dynamics of flood frequency. Water Resour Res 8(4):878–898

Fernandes W, Naghettini M, Loschi RH (2010) A Bayesian approach for estimating extreme flood probabilities with upper-bounded distribution functions. Stochastic Environ Res Risk Assess 24:1127–1143

Gaume E (2006) On the asymptotic behavior of flood peak distributions. Hydrol Earth Syst Sci 10:233–243

Ghosh S, Resnick SI (2010) A discussion on mean excess plots. Stoch Process Appl 120:1492–1517.

Guillot P, Duband D (1967) La méthode du GRADEX pour le calcul de la probabilité des crues à partir des pluies. In: Floods and their computation—proceedings of the Leningrad symposium, IASH Publication 84, pp 560–569

Gumbel EJ (1958) Statistics of extremes. Columbia University Press, New York

Gupta VK, Duckstein L, Peebles RW (1976) On the joint distribution of the largest flood and its occurrence time. Water Resour Res 12(2):295–304

Haan CT (1977) Statistical methods in hydrology. The Iowa University Press, Ames (IA)

Hall MJ, van den Boogard HFP, Fernando RC, Mynet AE (2004) The construction of confidence intervals for frequency analysis using resampling techniques. Hydrol Earth Syst Sci 8 (2):235–246

Hershfield DH, Kohler MA (1960) An empirical appraisal of the Gumbel extreme value procedure. J Geophys Res 65(6):1737–1746

Hirsch RM (1987) Probability plotting position formulas for flood records with historical information. J Hydrol 96:185–199

Hirsch RM, Stedinger JR (1986) Plotting positions for historical floods and their precision. Water Resour Res 23(4):715–727

Horton RE (1936) Hydrologic conditions as affecting the results of the application of methods of frequency analysis to flood records. U. S. Geological Survey Water Supply Paper 771:433–449

Hoshi K, Burges SJ (1981) Approximate estimation of the derivative of a standard gamma quantile for use in confidence interval estimate. J Hydrol 53:317–325

Hosking JR, Wallis JR (1997) Regional frequency analysis—an approach based on L-moments. Cambridge University Press, Cambridge

House PK, Webb RH, Baker VR, Levish DR (eds) (2002) Ancient floods, modern hazards—principles and applications of paleoflood hydrology. American Geophysical Union, Washington

Iacobellis V, Fiorentino M (2000) Derived distribution of floods based on the concept of partial area coverage with a climatic appeal. Water Resour Res 36(2):469–482

IH (1999) The flood estimation handbook. Institute of Hydrology, Wallingford, UK

Juncosa ML (1949) The asymptotic behavior of the minimum in a sequence of random variables. Duke Math J 16(4):609–618

Kidson R, Richards KS (2005) Flood frequency analysis: assumptions and alternatives. Prog Phys Geogr 29(3):392–410

Kirby W (1969) On the random occurrence of major floods. Water Resour Res 5(4):778–784

Kirby W (1972) Computer-oriented Wilson-Hilferty transformation that preserves the first three moments and the lower bound of the Pearson type 3 distribution. Water Resour Res 8 (5):1251–1254

Kite GW (1988) Frequency and risk analysis in hydrology. Water Resources Publications, Fort Collins, CO

Kjeldsen T, Jones DA (2004) Sampling variance of flood quantiles from the generalised logistic distribution estimated using the method of L-moments. Hydrol Earth Syst Sci 8(2):183–190

Klemeš V (1987) Hydrological and engineering relevance of flood frequency analysis. Proceedings of the International Symposium on Flood Frequency and Risk Analysis—Regional Flood Frequency Analysis, Baton Rouge, LA, pp 1–18. D. Reidel Publishing Company, Boston

Kottegoda NT, Rosso R (1997) Statistics, probability, and reliability for civil and environmental engineers. McGraw-Hill, New York

Laio F, Di Baldassare G, Montanari A (2009) Model selection techniques for the frequency analysis of hydrological extremes. Water Resour Res 45, W07416

Lang M, Ouarda TBMJ, Bobée B (1999) Towards operational guidelines for over-threshold modeling. J Hydrol 225:103–117

Langbein WB (1949) Annual floods and partial-duration floods series. In: Transactions of the American Geophysical Union 30(6)

Laursen EM (1983) Comment on "Paleohydrology of southwestern Texas" by Kochel R C, Baker VR, Patton PC. Water Resour Res 19:1339

Leadbetter MR, Lindgren G, Rootzén H (1983) Extremes and related properties of random sequences and processes. Springer, New York

Lettenmaier DP, Burges SJ (1982) Gumbel's extreme value I distribution: a new look. ASCE J Hydr Div 108:502–514

Madsen H, Rosbjerg D, Harremoes P (1993) Application of the partial duration series approach in the analysis of extreme rainfalls in extreme hydrological events: precipitation, floods and droughts. Proceedings of the Yokohama Symposium, IAHS Publication 213, pp 257–266

Margoum M, Oberlin G, Lang M, Weingartner R (1994) Estimation des crues rares et extrêmes: principes du modele Agregee. Hydrol Continentale 9(1):85–100

Musik I (1993) Derived, physically based distribution of flood probabilities. In: extreme hydrological events: precipitation, floods, and droughts, Proceedings of Yokohama Symposium, IASH Publication 213, pp 183–188

Naghettini M, Gontijo NT, Portela MM (2012) Investigation on the properties of the relationship between rare and extreme rainfall and flood volumes, under some distributional restrictions. Stochastic Environ Res Risk Assess 26:859–872

Naghettini M, Potter KW, Illangasekare T (1996) Estimating the upper-tail of flood-peak frequency distributions using hydrometeorological information. Water Resour Res 32 (6):1729–1740

NERC (1975) Flood studies report, vol 1. National Environmental Research Council, London

North M (1980) Time-dependent stochastic model of floods. ASCE J Hydraulics Div 106 (5):717–731

Papalexiou SM, Koutsoyiannis D (2013) The battle of extreme distributions: a global survey on the extreme daily rainfall. Water Resour Res 49(1):187–201

Perichi LR, Rodríguez-Iturbe I (1985) On the statistical analysis of floods. In: Atkinson AC, Feinberg SE (eds) A celebration of statistics. Springer, New York, pp 511–541

Pickands J (1975) Statistical inference using extreme order statistics. The Annals of Statistics 3 (1):119–131

Potter KW (1987) Research on flood frequency analysis: 1983–1986. Rev Geophys 26(3):113–118

Raines T, Valdes J (1993) Estimation of flood frequency in ungauged catchments. J Hydraul Eng 119(10):1138–1154

Rao AR, Hamed KH (2000) Flood frequency analysis. CRC Press, Boca Raton, FL

Rosbjerg D (1984) Estimation in partial duration series with independent and dependent peak values. J Hydrol 76:183–195

Rosbjerg D, Madsen H (1992) On the choice of threshold level in partial duration series, Proceedings of the Nordic Hydrological Conference, Alta, Norway, NHP Report 30:604–615

Serinaldi F, Kilsby CG (2014) Rainfall extremes: toward reconciliation after the battle of distributions. Water Resour Res 50:336–352

Silva AT, Portela MM, Naghettini M (2012) Aplicação da técnica de séries de duração parcial à constituição de amostras de variáveis hidrológicas aleatórias. In: 11° Congresso da Água, 2012, Porto, Portugal. Comunicações do 11° Congresso da Água (CD-ROM). Associação Portuguesa de Recursos Hídricos, Lisbon

Silva AT, Portela MM, Naghettini M (2014) On peaks-over-threshold modeling of floods with zero-inflated Poisson arrivals under stationarity and nonstationarity. Stochastic Environ Res Risk Assess 28(6):1587–1599

Smith RL (1984) Threshold models for sample extremes. In: Oliveira JT (ed) Statistical extremes and applications. Reidel, Hingham, MA, pp 621–638

Stedinger JR, Cohn TA (1986) Flood frequency analysis with historical and paleoflood information. Water Resour Res 22(5):785–794

Stedinger JR, Vogel RM, Foufoula-Georgiou E (1993) Frequency analysis of extreme events. Chapter 18. In: Maidment DR (ed) Handbook of hydrology. McGraw-Hill, New York

Taesombut V, Yevjevich V (1978) Use of partial flood series for estimating distributions of maximum annual flood peak, Hydrology Paper 82. Colorado State University, Fort Collins, CO

Todorovic P (1978) Stochastic models of floods. Water Resour Res 14(2):345–356

Todorovic P, Zelenhasic E (1970) A stochastic model for flood analysis. Water Resour Res 6 (6):411–424

USGS (1956) Floods of August 1955 in the northeastern states. USGS Circular 377, Washington. http://pubs.usgs.gov/circ/1956/0377/report.pdf. Accessed 11 Nov 2015

Van Montfort MAJ, Witter JV (1986) The generalized Pareto distribution applied to rainfall depths. Hydrol Sci J 31(2):151–162

Vogel RM, Fennessey NM (1993) L moment diagrams should replace product moment diagrams. Water Resour Res 29(6):1745–1752

Watt WE, Lathem KW, Neill CR, Richards TL, Roussele J (1988) The hydrology of floods in Canada: a guide to planning and design. National Research Council of Canada, Ottawa

Wilson EB, Hilferty MM (1931) The distribution of chi-square. Proc Natl Acad Sci U S A 17:684–688

WMO (1986) Manual for estimation of probable maximum precipitation. Operational Hydrologic Report No. 1, WMO No. 332, 2nd ed., World Meteorological Organization, Geneva

WMO (1994) Guide to hydrological applications. WMO No. 168, 5th edn. World Meteorological Organization, Geneva

WRC (1981) Guidelines for determining flood flow frequency, Bulletin 17B. United States Water Resources Council-Hydrology Committee, Washington

Yevjevich V (1968) Misconceptions in hydrology and their consequences. Water Resour Res 4 (2):225–232

Yevjevich V, Harmancioglu NB (1987) Some reflections on the future of hydrology. Proceedings of the Rome Symposium—Water for the Future: Hydrology in Perspective, IAHS Publ. 164, pp 405–414. International Association of Hydrological Sciences, Wallingford, UK

Chapter 9
Correlation and Regression

Veber Costa

9.1 Correlation

Correlation analysis comprises statistical methods used to evaluate whether two or more random variables are related through some functional form and the degree of association between them. Correlation analysis is one of the most utilized techniques for assessing the statistical dependence among variables or their covariation, and can be a useful tool for indicating the kind and the strength of association between random quantities. Qualitative indications on the association of two variables are readily visualized on a scatterplot. Multiple patterns or no pattern at all may arise from these plots, and, in the former case, can provide evidence of the most appropriate functional form, as given by measures of correlation. If on a given scatterplot, a variable Y systematically increases or decreases as the second variable X increases, the two variables are associated through a monotonic correlation. Otherwise, the correlation is said to be non-monotonic. Figure 9.1 illustrates both types of correlation.

The strength of the association between two variables is usually expressed by correlation coefficients. Such coefficients, hereafter denoted generally as ρ, are dimensionless quantities, which lie in the range $-1 \leq \rho \leq 1$. If the two variables have the same trend of variation or, in other words, if one increases as the other increases, then ρ will be positive. On the other hand, if one of the variables decreases as the other increases then ρ will be negative. Finally, if $\rho = 0$, then either the two variables are independent, in a statistical sense, or the functional form

V. Costa (✉)
Universidade Federal de Minas Gerais, Belo Horizonte, Minas Gerais, Brazil
e-mail: veberc@gmail.com

© Springer International Publishing Switzerland 2017
M. Naghettini (ed.), *Fundamentals of Statistical Hydrology*,
DOI 10.1007/978-3-319-43561-9_9

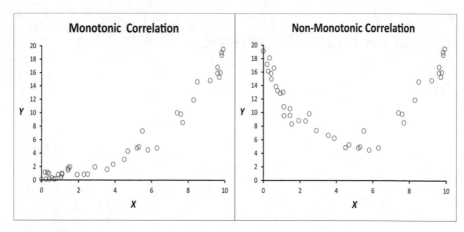

Fig. 9.1 Monotonic and non-monotonic correlations

of the association is not correctly described by the correlation coefficient in use. It is then intuitive to conclude that the measure of the correlation must also take into account the form of the relationship between the variables being studied.

Correlation in a data set may be either linear or nonlinear. Nonlinear associations can be represented by exponential, piecewise linear or power functional forms (Helsel and Hirsch 2002). As such, correlation coefficients used for expressing these kinds of association must be able to measure these particular monotonic relationships. Examples of such coefficients are the Kendall's τ and the Spearman's ρ (see Sect. 7.4.4). These are rank-based statistics, which evaluate the presence of discordant points, as specified by opposite tendencies of variation, in the sample. These coefficients are not discussed here, even though they may be important in some hydrological studies (see Sect. 7.4.4). The interested reader should consult Helsel and Hirsch (2002) for details on rank-based coefficients. If the association between variables is linear, the most common correlation coefficient is the Pearson's r. This coefficient is presented with details in Sect. 9.1.1.

It has to be noted that highly correlated variables do not necessarily have a cause–effect relationship. In fact, correlation only measures the joint tendency of variation of two variables. Obviously, there will be cases in which theoretical evidence of causal associations hold. A typical example in hydrological sciences is the relationship between precipitation and runoff. In these cases, the correlation coefficients may be used as indicators of such a cause–effect situation. However, there are many cases where, even if an underlying cause is present, the description of the phenomenon cannot be addressed as a causal process. These are evident, for instance, for strongly correlated mean monthly discharges in nearby catchments. In this case, a discharge change in one of the catchments is not the cause of a discharge alteration in the other. Concurrent changes may be rather related to common physical or climatological factors at both catchments (Naghettini and Pinto 2007).

9.1.1 Pearson's r Correlation Coefficient

If two random variables, X and Y, are linearly related, the degree of linear association may be expressed through the Pearson correlation coefficient, which is given by

$$\rho_{X,Y} = \frac{\sigma_{X,Y}}{\sigma_X \sigma_Y} \tag{9.1}$$

where $\sigma_{X,Y}$ denotes the covariance between X and Y, whereas σ_X and σ_Y correspond respectively to the standard deviations of X and Y. It is worth noting that $\rho_{X,Y}$, as any other correlation measure, is a dimensionless quantity, which can range from -1, when correlation is negative, to $+1$, when correlation is positive. If $\rho_{X,Y} = \pm 1$, the relationship between X and Y is perfectly linear. On the other hand, if $\rho_{X,Y} = 0$, two possibilities may arise: either (1) X and Y are statistically independent; or (2) the functional form that expresses the dependence of these two random variables is not linear. Figure 9.2 depicts positive and negative linear associations.

The most usual estimator for $\rho_{X,Y}$ is the sample correlation coefficient, which is given by

$$s_{X,Y} = \frac{\sum\limits_{i=1}^{N} (x_i - \bar{x})(y_i - \bar{y})}{N - 1} \tag{9.2}$$

in which x_i and y_i denote the concurrent observations of X and Y, $\bar{x}$ and $\bar{y}$ correspond to their sample means, and N is the sample size.

The Pearson correlation coefficient can thus be estimated as

$$r = \frac{s_{X,Y}}{s_X s_Y} \tag{9.3}$$

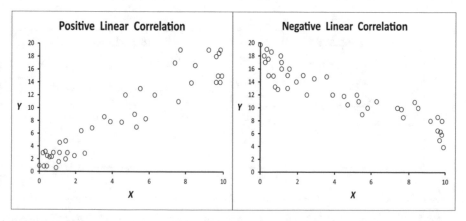

Fig. 9.2 Positive and negative linear correlations

where

$$s_X = \sqrt{\frac{\sum\limits_{i=1}^{N} (x_i - \bar{x})^2}{N - 1}} \tag{9.4}$$

and

$$s_Y = \sqrt{\frac{\sum\limits_{i=1}^{N} (y_i - \bar{y})^2}{N - 1}} \tag{9.5}$$

Figure 9.3 depicts some forms of association between X and Y with the respective estimates of the Pearson correlation coefficient. One may notice that even for an evident nonlinear plot, such as in Fig. 9.3d, a relative high value of the Pearson coefficient is obtained. This fact stresses the importance of visually examining the association between the variables, by means of scatterplots.

It is worth stressing that one must be cautious about the occurrence of spurious correlations. Figure 9.4 provides an interesting example of such a situation.

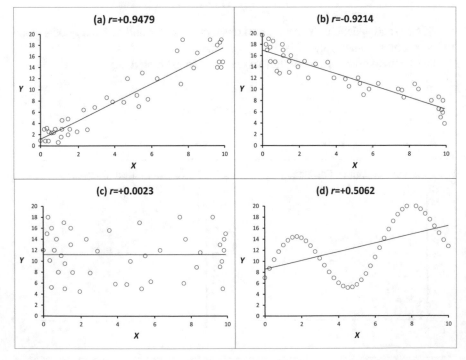

Fig. 9.3 Association between X and Y and the estimate of the Pearson correlation coefficient

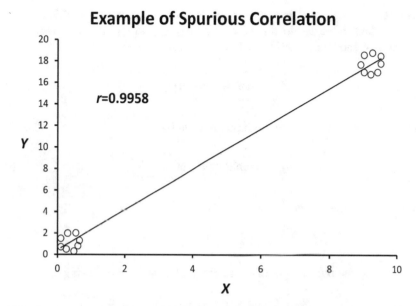

Fig. 9.4 Spurious correlation due to unbalanced distribution of X along its subdomain (adapted from Haan 2002)

In effect, although a high value of r is obtained for the complete sample, one observes that a linear relationship does not hold for the clustered pairs (X, Y) themselves. Thus, the value of r may be attributed to the unbalanced distribution of X along its subdomain instead of an actual linear correlation. Another situation where a spurious correlation may occur is that when the random variables have common denominators. In fact, even when no linear relationship appears to exist when X and Y alone are considered, a linear plot may result if both of them are divided by some variable Z. Finally, if one of the variables is multiplied by the other, a linear relationship may appear despite a nonlinear association being observed between X and Y.

As far as hypothesis testing is concerned, it is often useful to evaluate whether the Pearson's correlation coefficient is null. In this case, the null and alternative hypotheses to test are $H_0 : \rho = 0$ and $H_1 : \rho \neq 0$, respectively. The related test statistic is expressed as

$$t_0 = \frac{r\sqrt{N-2}}{\sqrt{1-r^2}} \tag{9.6}$$

for which the null distribution is a Student's t, with $\nu = N - 2$ degrees of freedom. The null hypothesis is rejected if $|t| > t_{\alpha/2, N-2}$, where α corresponds to the level of significance of the test.

Another possible test is whether the Pearson's correlation coefficient equals some constant value ρ_0. In this case, $H_0 : \rho = \rho_0$ and $H_1 : \rho \neq \rho_0$. According to Montgomery and Peck (1992), for $N \geq 25$, the statistic

$$Z = a\tanh(r) = \frac{1}{2}\ln\left(\frac{1+r}{1-r}\right) \tag{9.7}$$

follows a Normal distribution with mean given by

$$\mu_Z = a\tanh(\rho) = \frac{1}{2}\ln\left(\frac{1+\rho}{1-\rho}\right) \tag{9.8}$$

and variance expressed as

$$\sigma_Z^2 = (N-3)^{-1} \tag{9.9}$$

For testing the null hypothesis, one has to calculate the following statistic

$$Z_0 = [a\tanh(r) - a\tanh(\rho_0)]\sqrt{n-3} \tag{9.10}$$

which follows a standard Normal distribution. The null hypothesis should be rejected if $|Z_0| > Z_{\alpha/2}$, where α corresponds to the level of significance of the test.

It is also possible to construct $100(1-\alpha)\%$ confidence intervals for ρ using the transformation given by Eq. (9.7). Such an interval is

$$\tanh\left[a\tanh(r) - \frac{z_{\alpha/2}}{\sqrt{N-3}}\right] \leq \rho \leq \tanh\left[a\tanh(r) + \frac{z_{\alpha/2}}{\sqrt{N-3}}\right] \tag{9.11}$$

where $\tanh(u) = \frac{(e^u - e^{-u})}{(e^u + e^{-u})}$, r is the sample Pearson correlation coefficient, $z_{\alpha/2}$ corresponds to the standard Normal variate associated with the confidence level $(1-\alpha)$ and N is the sample size.

9.1.2 Serial Correlation

In many hydrologic applications, correlation may also exist between successive observations of the same random variable in a time series. If inference procedures are required when such a condition holds, the effects of the serial correlation (or autocorrelation) must be taken into account, since a correlated series of size N provides less information than an independent series of the same size. According to Haan (2002), this results from the fact that part of the information contained in a given observation is actually already known from the previous observation, through serial correlation.

If the observations of a series are equally spaced in time and the underlying stochastic process is stationary, an estimator of the population serial correlation coefficient is given by

$$r(k) = \frac{\sum\limits_{i=1}^{N-k} x_i x_{i+k} - \frac{\sum\limits_{i=1}^{N-k} x_i \sum\limits_{i=1}^{N-k} x_{i+k}}{(N-k)}}{\left(\sqrt{\sum\limits_{i=1}^{N-k} x_i^2 - \frac{\left(\sum\limits_{i=1}^{N-k} x_i\right)^2}{(N-k)}}\right)\left(\sqrt{\sum\limits_{i=1}^{N-k} x_{i+k}^2 - \frac{\left(\sum\limits_{i=1}^{N-k} x_{i+k}\right)^2}{(N-k)}}\right)} \tag{9.12}$$

where k denotes the lag or the number of time intervals that set apart successive observations. From Eq. (9.12), it is clear that $r(0) = 1$.

A test of significance for the serial correlation coefficient $r(k)$, for stationary and normally distributed time series, was proposed by Anderson (1942). If these assumptions hold, then

$$r(k) = \frac{\sum\limits_{i=1}^{N-k} x_i x_{i+k} - N\bar{x}}{(N-1)\ s_X^2} \tag{9.13}$$

is normally distributed, with mean $\frac{-1}{(N-1)}$ and variance $\frac{(N-2)}{(N-1)^2}$ if $\rho(k) = 0$. Confidence bounds at $(1-\alpha)$ can be estimated as

$$l_b = \frac{-1 - z_{1-\alpha/2}\sqrt{N-2}}{(N-1)} \tag{9.14}$$

and

$$u_b = \frac{-1 + z_{1-\alpha/2}\sqrt{N-2}}{(N-1)} \tag{9.15}$$

where $z_{\alpha/2}$ denotes the standard Normal variate, l_b and u_b denote the lower and upper bounds, respectively. If $r(k)$ is located outside the range given by l_b and u_b, then the null hypothesis $H_0 : \rho(k) = 0$ should be rejected, at the significance level α.

Matalas and Langbein (1962) suggest that, for inferential procedures with autocorrelated series, a number of effective observations, lesser than the actual sample size, must be used. Such a number is given by

$$N_e = \frac{N}{\frac{1+\rho(1)}{1-\rho(1)} - 2\rho(1)\frac{1-\rho^N(1)}{n[1-\rho(1)]^2}} \tag{9.16}$$

in which $\rho(1)$ denotes the autocorrelation of lag 1 and N is the sample size. If $r(1) = 0$, $N_e = N$, whereas if $r(1) > 0$, $N_e < N$.

9.2   Regression Analysis

Regression analysis refers to a collection of statistical techniques commonly used for modeling the association between one or more independent variables, hereafter denoted explanatory variables X, and a dependent variable, which is expressed as a functional form of X and a set of parameters or coefficients, and is referred to as the response variable Y. Regression analysis has become a widespread used tool for data description, estimation and prediction in many fields of science, such as physics, economics, engineering, and, of special interest for this book, hydrology, due to the simplicity of its application framework and the appeal of its rigorous theoretical foundation. The term "regression" is attributed to the English statistician Francis Galton (1822–1911), who, in studies concerning changes in human heights between consecutive generations, notice that such quantities were moving back or "regressing" towards the population mean value.

Regression models are denoted simple when a single explanatory variable is utilized for describing the behavior of the response variable. If two or more explanatory variables take part in the analysis, a multiple regression model is being constructed. As for the parameters, regression models may be grouped into two categories: if the response variable is expressed as a linear combination of the regression coefficients, the model is termed linear; otherwise, the model is nonlinear. In some situations, the term "linear" may be applied even when the (X, Y) plot is not a straight line. For instance, Wadsworth (1990) notes that polynomial regression models, although expressing a nonlinear relationship between X and Y, may still be defined as linear, in a statistical estimation sense, as long as the response variable remains a linear function of the regression coefficients.

Regression analysis usually comprises a two-step procedure. The first one refers to the construction of the regression model. In short, it involves prescribing a functional form between the response and the explanatory variables and estimating the numerical values of the regression coefficients such that an adequate fit to the data is obtained. Several techniques for fitting a regression model are available, from simple (and almost always subjective) visual adjustments, to analytical procedures with formal mathematical background, such as the least squares and the maximum likelihood estimation methods. Once the model is specified and the parameters are estimated, an adequacy check must be performed on the regression equation in order to ascertain the suitability of the fit. This is the basis for the second step: in the light of the intended use of the regression model, a full evaluation must be carried out on whether the assumptions from which the model was derived hold after inference. In this sense, one must assess if the prescribed functional form is appropriate for modeling the processes being studied, if the explanatory variables are, in fact, able to explain some of the variability of the response counterpart, and also if the regression residuals behave as assumed a priori.

In relation to the hydrological sciences, regression analysis finds interesting applications in regionalization studies and in estimating rating curves for a given gauging station. In the first case, the main objective is to derive relationships between hydrologic random variables, such as the mean annual discharges or the mean annual total rainfalls, as observed in different catchments located in a given geographic region, and the physical and/or climatological attributes of the catchments, in order to estimate related quantiles at ungauged sites within this region. The application of regression methods to the regional analysis of hydrologic variables is detailed in Chap. 10. As for the second case, one is asked to identify an adequate model that expresses the relation between discharges and stages at a river cross section. Since discharge measurements are costly and time demanding, such a relation is sought as a means to provide flow estimates from stage measurements.

It is worth mentioning that, although intended to express the association between explanatory and response variables, regression models do not necessarily imply a cause–effect relation. In fact, even when a strong empirical association exists, as evidenced, for instance, by a high value of the Pearson correlation coefficient for linear relationships, it is incorrect to assume from such an evidence alone that a given explanatory variable is the cause of a given observed response. Take as an example the association of the seasonal mean temperatures in New York City and the seasonal number of basketball points scored by the NY Knicks in home games, over the years. These variables are expected to show a significant negative correlation coefficient, but they obviously do not exhibit a cause–effect relation. In fact, both are ruled by the seasons: the temperatures, as naturally governed by the weather fluctuations, and the basketball season schedule, as set out by the American National Basketball Association.

In the subsections that follow, methods for estimating parameters and evaluation of goodness-of-fit are addressed for linear regression models, encompassing both simple and multiple variants of analysis. In addition, a discussion on the practical problems associated to limitations, misconceptions, and/or misuse of regression analysis is presented.

9.2.1 Simple Linear Regression

A simple linear regression model relates the response variable Y to a given explanatory variable X through a straight-line equation, such as

$$Y = \beta_0 + \beta_1 X \tag{9.17}$$

where β_0 is the intercept and β_1 is the slope of the unknown population regression line. It has to be noted, however, that deviations between data points and the theoretical straight line are likely to occur. Such deviations may be the result of the natural variability associated to the process in study, which might be interpreted as a noise component, or may arise from measurement errors or from the requirement of additional explanatory variables for describing the response counterpart. An alternative for accounting for these differences is adding an error term ε to the model given by Eq. (9.17). The errors ε are unobserved random variables, independent from X, which are intended to express the inability of the regression equation to fit the data in a perfect fashion. The simple linear regression model can, thus, be expressed as

$$Y = \beta_0 + \beta_1 X + \varepsilon \tag{9.18}$$

in which the regression coefficients have the same meaning as in Eq. (9.17).

Due to the error term, the response variable Y is necessarily a random variable, whose behavior is described by a probability distribution. This distribution is conditioned on X, or, in other words, on each particular value of the explanatory variable. For a fixed value of X, the possible values of Y should follow a probability distribution. Assuming that the mean value of the error term is null and resorting to the mathematical properties of the expected value discussed in Chap. 3, it is clear that the mean value of the conditional distribution of Y is a linear function of X, and lies on the regression line. In formal terms,

$$E(Y|X) = \beta_0 + \beta_1 X \tag{9.19}$$

Based on a similar rationale, and noting that X is a nonrandom quantity as it is fixed at a point, the conditional variance of the response variable can be expressed as

$$\mathrm{Var}(Y|X) = \mathrm{Var}(\beta_0 + \beta_1 X + \varepsilon) = \sigma^2 \tag{9.20}$$

Equation (9.20) encompasses the idea that the conditional variance of Y is not dependent on the explanatory variable, i.e., the variance of the given response y_i, as related to the explanatory sample point x_i, is the same as the variance of a response y_j, associated to x_j, for every $i \neq j$. This situation is termed homoscedasticity and is illustrated in Fig. 9.5. For this particular figure, the error model is assumed to be the Normal distribution, with $\mu(\varepsilon) = 0$ and $\mathrm{Var}(\varepsilon) = \sigma_\varepsilon^2$. Such a prescription is not required, a priori, for constructing the regression model. However, as seen throughout this chapter, the assumption of normally distributed errors is at the foundations of inference procedures regarding interval estimation and hypothesis tests on the regression coefficients. Also depicted in Fig. 9.5, is the conditional mean values of Y, as given by the regression line.

Parameters β_0 and β_1, however, are unknown quantities and have to be estimated on the basis of the information gathered by the sample of concurrent pairs $\{(x_1, y_1), \ldots, (x_N, y_N)\}$. One of the most common techniques for this purpose is the least squares method, which consists of estimating the regression line, as expressed by its coefficients, for which the sum of squared deviations between each observation and its predicted value attains a minimum value. Such a criterion is a geometry-based procedure, which, as opposed to maximum likelihood estimation methods, does not involve prior assumptions on the behavior of the errors, and has become a standard approach for fitting regression models since it is sign independent and, according to Haan (2002), circumvents the modeling difficulties that may arise

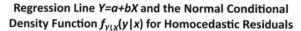

Regression Line *Y=a+bX* and the Normal Conditional Density Function $f_{Y|X}(y|x)$ for Homocedastic Residuals

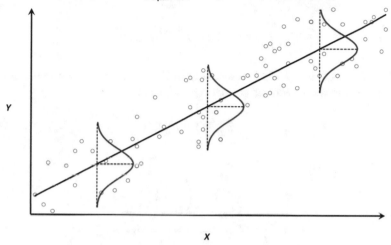

Fig. 9.5 Regression line and the conditional density function $f_{YX}(y|x)$

from applying absolute values in optimization problems. The least squares method is frequently attributed to Legendre (1805) and Gauss (1809), although the latter dates the first developments of the technique to as early as 1795.

To derive the least squares estimators $\hat{\beta}_0$ and $\hat{\beta}_1$, let the linear model in Eq. (9.18) express the regression line for the response variable Y with respect to X. The target function of the least squares criterion is written as

$$M(\beta_0, \beta_1) = \sum_{i=1}^{N} \varepsilon_i^2 = \sum_{i=1}^{N} (y_i - \beta_0 - \beta_1 x_i)^2 \tag{9.21}$$

In order to minimize the function above, the partial derivatives of $M(\beta_0, \beta_1)$, with respect to the regression parameters, as estimated by the least square coefficients, $\hat{\beta}_0$ and $\hat{\beta}_1$, must be null. Thus

$$\frac{\partial M}{\partial \beta_0}\bigg|_{\hat{\beta}_0}, \hat{\beta}_1 = -2\sum_{i=1}^{N}(y_i - \hat{\beta}_0 - \hat{\beta}_1 x_i) = 0$$

$$\frac{\partial M}{\partial \beta_1}\bigg|_{\hat{\beta}_0}, \hat{\beta}_1 = -2\sum_{i=1}^{N}(y_i - \hat{\beta}_0 - \hat{\beta}_1 x_i)x_i = 0 \tag{9.22}$$

This system of equations, after algebraic manipulations, results in

$$\sum_{i=1}^{N} y_i = N\hat{\underline{\beta}}_0 + \hat{\underline{\beta}}_1 \sum_{i=1}^{N} x_i$$

$$\sum_{i=1}^{N} y_i x_i = \hat{\underline{\beta}}_0 \sum_{i=1}^{N} x_i + \hat{\underline{\beta}}_1 \sum_{i=1}^{N} x_i^2 \qquad (9.23)$$

The system of Eq. (9.23) is referred to as the *least-squares normal equations*. The solutions to the system of normal equations are

$$\hat{\underline{\beta}}_0 = \bar{y} - \hat{\underline{\beta}}_1 \bar{x} \qquad (9.24)$$

and

$$\hat{\underline{\beta}}_1 = \frac{\displaystyle\sum_{i=1}^{N} y_i x_i - \frac{\left(\sum_{i=1}^{N} y_i\right)\left(\sum_{i=1}^{N} x_i\right)}{N}}{\displaystyle\sum_{i=1}^{N} x_i^2 - \frac{\left(\sum_{i=1}^{N} x_i\right)^2}{N}} \qquad (9.25)$$

where $\bar{y}$ and $\bar{x}$ denote, respectively, the sample mean of the observations of Y and the sample mean of the explanatory variable X, and N is the sample size. Estimates of the intercept $\hat{\beta}_0$ and the slope $\hat{\beta}_1$ may be obtained by applying Eqs. (9.24) and (9.25) to the sample points. Thus, the fitted simple linear model can be expressed as

$$\hat{y} = \hat{\beta}_0 + \hat{\beta}_1 x \qquad (9.26)$$

where $\hat{y}$ is the point estimate of the mean of Y for a particular value x of the explanatory variable X. One can notice from Eq. (9.24) that the regression line always contains the point $(\bar{x}, \bar{y})$, which corresponds to the *centroid* of the data. For each response sample point y_i, the regression error or regression residual is given by

$$e_i = y_i - \hat{y}_i \qquad (9.27)$$

The system of Eq. (9.22) warrants that the sum of the regression residuals must be zero, and the same holds for the sum of the residuals weighted by the explanatory variable. In practical applications, rounding errors might entail small non-null values. Finally, it is possible to demonstrate that the sum of the observed sample points of Y always equals the sum of the fitted ones.

The regression line obtained from the least squares method is, for practical purposes, a mere estimate of the true population relationship between the response and the explanatory variables. In this sense, no particular set of regression coefficients, as estimated from finite samples, will match the population parameters exactly, and, thus, at most only one point of the two regression lines,

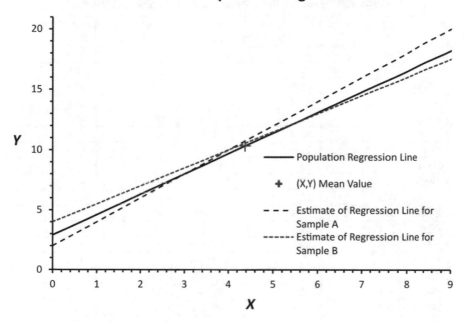

Fig. 9.6 Population and estimated regression lines

the intersection point between the sample-based line estimate and the population true line, will be actually coincident. This situation is depicted in Fig. 9.6. As a limiting case, the sample and population regression lines will be parallel if $\hat{\beta}_1 = \beta_1$ and $\hat{\beta}_0 \neq \beta_0$.

Example 9.1 Table 9.1 presents concurrent observations of annual total rainfall (mm) and annual mean daily discharges (m^3/s) in the Paraopeba River catchment, for the water years 1941/42 to 1998/99. Assuming that a linear functional form holds, (a) estimate the regression coefficients using the least squares method for the annual rainfall depth as explanatory variable; and (b) compute the regression residuals and check if their sum is null.

Solution

(a) Figure 9.7 displays the scatterplot of the two variables: the annual rainfall depth, as the explanatory variable, and the annual mean daily flow, as the response variable. One can notice that some linear association between these variables may be assumed. Thus, for estimating the regression coefficients, one must calculate the sample mean of both variables and apply Eqs. (9.24) and (9.25). The results are summarized in Table 9.2.

The regression coefficients estimates are $\hat{\beta}_1 = \frac{7251946.9 - 4966.4 \times 81953/58}{118823051 - 81953^2/58} =$ 0.07753 and $\hat{\beta}_0 = 85.6 - 0.07753 \times 1413 = -23.917$. Then, the estimated equation for the regression line is $Y = 0.0775X - 23.917$.

Table 9.1 Annual mean flows and annual total rainfall depths (Gauging stations: 4080001 for discharges and 01944004 for rainfall, in the Paraopeba river catchment, in Brazil)

Water year	Annual rainfall depth (mm)	Annual mean daily flow (m³/s)	Water year	Annual rainfall depth (mm)	Annual mean daily flow (m³/s)
1941/42	1249	91.9	1970/71	1013	34.5
1942/43	1319	145	1971/72	1531	80.0
1943/44	1191	90.6	1972/73	1487	97.3
1944/45	1440	89.9	1973/74	1395	86.8
1945/46	1251	79.0	1974/75	1090	67.6
1946/47	1507	90.0	1975/76	1311	54.6
1947/48	1363	72.6	1976/77	1291	88.1
1948/49	1814	135	1977/78	1273	73.6
1949/50	1322	82.7	1978/79	2027	134
1950/51	1338	112	1979/80	1697	104
1951/52	1327	95.3	1980/81	1341	80.7
1952/53	1301	59.5	1981/82	1764	109
1953/54	1138	53.0	1982/83	1786	148
1954/55	1121	52.6	1983/84	1728	92.9
1955/56	1454	62.3	1984/85	1880	134
1956/57	1648	85.6	1985/86	1429	88.2
1957/58	1294	67.8	1986/87	1412	79.4
1958/59	883	52.5	1987/88	1606	79.5
1959/60	1601	64.6	1988/89	1290	58.3
1960/61	1487	122	1989/90	1451	64.7
1961/62	1347	64.8	1990/91	1447	105
1962/63	1250	63.5	1991/92	1581	99.5
1963/64	1298	54.2	1992/93	1642	95.7
1964/65	1673	113	1993/94	1341	86.1
1965/66	1452	110	1994/95	1359	71.8
1966/67	1169	102	1995/96	1503	86.2
1967/68	1189	74.2	1996/97	1927	127
1968/69	1220	56.4	1997/98	1236	66.3
1969/70	1306	72.6	1998/99	1163	59.0

(b) The sum of the residuals equals 1.4. This value is close to zero and the difference may be attributed to rounding errors.

As mentioned earlier, a noteworthy aspect of the least squares method is that no assumptions regarding the probabilistic behavior of the errors is required for deriving the estimators of the regression coefficients. In fact, if the objective is merely a prediction of a value of Y, it suffices that the linear form of the model be correct (Helsel and Hirsch 2002). However, as interest lies in other inference problems, additional assumptions become necessary. For instance, it can be easily shown that the least squares estimators are unbiased (Montgomery and Peck 1992). If the assumption of uncorrelated errors, with $E(\varepsilon) = 0$ and

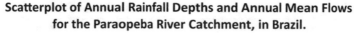

Scatterplot of Annual Rainfall Depths and Annual Mean Flows for the Paraopeba River Catchment, in Brazil.

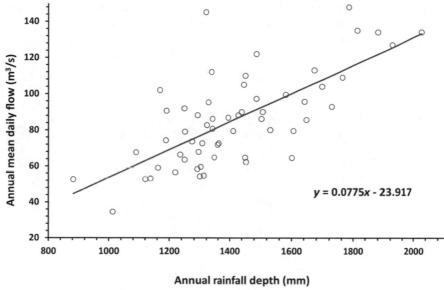

$y = 0.0775x - 23.917$

Annual rainfall depth (mm)

Fig. 9.7 Scatterplot of annual mean daily flow and annual total depth

constant variance $\mathrm{Var}(\varepsilon) = \sigma_\varepsilon^2$, holds, such estimators also have the least variance, as compared to other unbiased estimators which may be obtained from linear combinations of y_i. This result is known as the *Gauss-Markov* theorem. In addition, if the residuals follow a Normal distribution, with $\mu(\varepsilon) = 0$ and $\mathrm{Var}(\varepsilon) = \sigma_\varepsilon^2$, one is able to extend the statistical inference to interval estimation and hypothesis testing through standard parametric approaches, as well as performing a thorough checking on the adequacy of the regression model, on the basis of the behavior of the residuals after the fit. This assumption also enables one to derive an unbiased (yet model-dependent) estimator for σ_ε^2, given in Graybill (1961) as

$$\hat{\sigma}_\varepsilon^2 = s_e^2 = \frac{\sum_{i=1}^{N}(y_i - \hat{y}_i)^2}{N - 2} \qquad (9.28)$$

The quantity s_e^2 in Eq. (9.28) is referred to as *residual mean square*, whereas its square root is usually denoted *standard error of regression* or *standard error of the estimate*. The latter accounts for the uncertainties associated with the inference of the regression model from a finite sample. In addition of being an accuracy measure, providing an estimate to σ_ε^2, the statistic s_e^2 enables one to obtain estimates of the variances of the regression coefficients, which are required for constructing confidence intervals and testing hypotheses on these

Table 9.2 Calculations for the regression coefficients of Example **9.1**

Annual rainfall depth (X)	Annual mean daily flow (Y)	$X \times Y$	X^2	Estimated annual mean daily flow	Residuals
1249	91.9	114,783.1	1,560,001	72.9	19.0
1319	145	191,255	1,739,761	78.3	66.7
1191	90.6	107,904.6	1,418,481	68.4	22.2
1440	89.9	129,456	2,073,600	87.7	2.2
1251	79	98,829	1,565,001	73.0	6.0
1507	90	135,630	2,271,049	92.9	−2.9
1363	72.6	98,953.8	1,857,769	81.7	−9.1
1814	135	244,890	3,290,596	116.7	18.3
1322	82.7	109,329.4	1,747,684	78.5	4.2
1338	112	149,856	1,790,244	79.8	32.2
1327	95.3	126,463.1	1,760,929	78.9	16.4
1301	59.5	77,409.5	1,692,601	76.9	−17.4
1138	53	60,314	1,295,044	64.3	−11.3
1121	52.6	58,964.6	1,256,641	63.0	−10.4
1454	62.3	90,584.2	2,114,116	88.8	−26.5
1648	85.6	141,068.8	2,715,904	103.8	−18.2
1294	67.8	87,733.2	1,674,436	76.4	−8.6
883	52.5	46,357.5	779,689	44.5	8.0
1601	64.6	103,424.6	2,563,201	100.2	−35.6
1487	122	181,414	2,211,169	91.3	30.7
1347	64.8	87,285.6	1,814,409	80.5	−15.7
1250	63.5	79,375	1,562,500	73.0	−9.5
1298	54.2	70,351.6	1,684,804	76.7	−22.5
1673	113	189,049	2,798,929	105.8	7.2
1452	110	159,720	2,108,304	88.6	21.4
1169	102	119,238	1,366,561	66.7	35.3
1189	74.2	88,223.8	1,413,721	68.2	6.0
1220	56.4	68,808	1,488,400	70.6	−14.2
1306	72.6	94,815.6	1,705,636	77.3	−4.7
1013	34.5	34,948.5	1,026,169	54.6	−20.1
1531	80	122,480	2,343,961	94.8	−14.8
1487	97.3	144,685.1	2,211,169	91.3	6.0
1395	86.8	121,086	1,946,025	84.2	2.6
1090	67.6	73,684	1,188,100	60.6	7.0
1311	54.6	71,580.6	1,718,721	77.7	−23.1
1291	88.1	113,737.1	1,666,681	76.1	12.0
1273	73.6	93,692.8	1,620,529	74.8	−1.2
2027	134	271,618	4,108,729	133.2	0.8
1697	104	176,488	2,879,809	107.6	−3.6
1341	80.7	108,218.7	1,798,281	80.0	0.7
1764	109	192,276	3,111,696	112.8	−3.8

(continued)

Table 9.2 (continued)

Annual rainfall depth (X)	Annual mean daily flow (Y)	$X \times Y$	X^2	Estimated annual mean daily flow	Residuals
1786	148	264,328	3,189,796	114.5	33.5
1728	92.9	160,531.2	2,985,984	110.0	−17.1
1880	134	251,920	3,534,400	121.8	12.2
1429	88.2	126,037.8	2,042,041	86.8	1.4
1412	79.4	112,112.8	1,993,744	85.5	−6.1
1606	79.5	127,677	2,579,236	100.6	−21.1
1290	58.3	75,207	1,664,100	76.1	−17.8
1451	64.7	93,879.7	2,105,401	88.6	−23.9
1447	105	151,935	2,093,809	88.2	16.8
1581	99.5	157,309.5	2,499,561	98.6	0.9
1642	95.7	157,139.4	2,696,164	103.4	−7.7
1341	86.1	115,460.1	1,798,281	80.0	6.1
1359	71.8	97,576.2	1,846,881	81.4	−9.6
1503	86.2	129,558.6	2,259,009	92.6	−6.4
1927	127	244,729	3,713,329	125.5	1.5
1236	66.3	81,946.8	1,527,696	71.9	−5.6
1163	59	68,617	1,352,569	66.2	−7.2
Mean					
1413	85.6			85.6	0.0
Sum					
81,953	4966.4	7,251,946.9	118,823,051	4965.0	1.4

quantities. The estimators of the variances of the least squares coefficients are respectively (Montgomery and Peck 1992):

$$\mathrm{Var}(\widehat{\beta_0}) = s_e^2 \left(\frac{1}{N} + \frac{\bar{x}^2}{\sum\limits_{i=1}^{N} (x_i - \bar{x})^2} \right) \tag{9.29}$$

and

$$\mathrm{Var}(\widehat{\beta_1}) = \frac{s_e^2}{\sum\limits_{i=1}^{N} (x_i - \bar{x})^2} \tag{9.30}$$

9.2.2 Coefficient of Determination in Simple Linear Regression

After estimating the regression coefficients, it is necessary to evaluate how well the regression model describes the observed sample points. In other words, one has to resort to an objective goodness-of-fit measure in order to assess the ability of the regression model in simulating the response variable. One of the most common statistics utilized for this purpose is the coefficient of determination r^2. Such a quantity expresses the proportion of the total variance of the response variable Y which is accounted for by the regression linear equation, without relying on any assumption regarding the probabilistic behavior of the error term.

In general terms, the variability of the response variable may be quantified as a sum of squares. In order to derive such a sum, one may express a given observation y_i as

$$y_i = \bar{y} + \hat{y}_i - \bar{y} + y_i - \hat{y}_i \tag{9.31}$$

By rearranging the terms, one obtains

$$(y_i - \hat{y}_i) = (y_i - \bar{y}) - (\hat{y}_i - \bar{y})$$

which, after algebraic manipulation, yields

$$\sum_{i=1}^{n} (y_i - \bar{y})^2 = \sum_{i=1}^{n} (y_i - \hat{y}_i)^2 + \sum_{i=1}^{n} (\hat{y}_i - \bar{y})^2 \tag{9.32}$$

The term on the left side of Eq. (9.32) accounts for total sum of squares of the response variable Y. The terms on the right side correspond to the sum of squares of the residuals of regression and the sum of squares due to the regression model itself, respectively. On the basis of the referred equation, the estimate of the coefficient of determination, as expressed as the ratio of the sum of squares due to the regression to the total sum of squares, is given by

$$r^2 = \frac{\sum_{i=1}^{N} (\hat{y}_i - \bar{y})^2}{\sum_{i=1}^{N} (y_i - \bar{y})^2} \tag{9.33}$$

The coefficient of determination ranges from 0 to 1. One can notice from Eqs. (9.32) and (9.33) that the larger the spread of the observations y_i around the regression line, the less the total variance is explained by the model and, thus, the smaller is the coefficient of determination. In a limiting case, if the regression line does not account for any of the variability of Y, the coefficient of determination will be zero. On the other hand, if the regression model fits the data in a perfect fashion, the sum of residuals squares becomes null and the coefficient of determination equals 1. It has to be noted, however, that large values of r^2 do not imply that the linear functional form prescribed by the model is correct, since such a quantity merely express a geometric description of the scatter of the residuals around the regression model.

9.2.3 Interval Estimation in Simple Linear Regression

The regression line, as estimated from a particular sample, is not necessarily coincident with the population line and expresses one of the infinite possibilities of regression equations that may be obtained from samples of the same size extracted from the population. Due to this fact, it becomes important to provide a measure of accuracy of the estimates of the regression coefficients, by constructing confidence intervals on β_0 and β_1. It is also desirable to evaluate the overall performance of the regression model, by constructing confidence intervals for the regression line itself. These intervals are intended to describe the behavior of the mean response $E(y|x)$ of a given regression model, for observed and unobserved values of the explanatory variable, along its entire domain. Finally, if interest lies in predicting future values of Y, as related to a given value of X through the regression model, a prediction interval must be considered. Here, the term prediction is utilized because a future value of the response variable is itself a random variable, associated to a probability distribution, as opposed to unknown yet fixed parameters of the population under study, such as the mean response of Y. The differences between these two approaches are addressed later in this subsection.

Regarding the regression parameters, it can be shown that, if the errors are IID and follow a Normal distribution, the sampling distributions of the pivot functions $\dfrac{\hat{\beta}_0 - \beta_0}{\sqrt{s_e^2 \left[1/N + \bar{x}^2 \Big/ \sum\limits_{i=1}^{N}(x_i - \bar{x})^2 \right]}}$ and $\dfrac{\hat{\beta}_1 - \beta_1}{\sqrt{s_e^2 \Big/ \sum\limits_{i=1}^{N}(x_i - \bar{x})^2}}$, are, both, Student's t variates, with $\nu = N - 2$ degrees of freedom. Thus, the $100(1 - \alpha)\%$ confidence intervals for β_0 and β_1 are, respectively

$$
\hat{\beta}_0 - t_{\alpha/2, N-2} \sqrt{ s_e^2 \left(\frac{1}{N} + \frac{\bar{x}^2}{\sum\limits_{i=1}^{N}(x_i - \bar{x})^2} \right) } \leq \beta_0
$$

$$
\leq \hat{\beta}_0 + t_{\alpha/2, N-2} \sqrt{ s_e^2 \left(\frac{1}{N} + \frac{\bar{x}^2}{\sum\limits_{i=1}^{N}(x_i - \bar{x})^2} \right) } \tag{9.34}
$$

and

$$
\hat{\beta}_1 - t_{\alpha/2, N-2} \sqrt{ \frac{s_e^2}{\sum\limits_{i=1}^{N}(x_i - \bar{x})^2} } \leq \beta_1 \leq \hat{\beta}_1 + t_{\alpha/2, N-2} \sqrt{ \frac{s_e^2}{\sum\limits_{i=1}^{N}(x_i - \bar{x})^2} } \tag{9.35}
$$

As for σ_ε^2, if the assumption of uncorrelated normal errors holds, it can be shown that the sampling distribution of the pivot function $(N-2)\frac{s_e^2}{\sigma_\varepsilon^2}$ follows a Chi-Square distribution, with $\nu = N - 2$ degrees of freedom. Thus, the $100(1 - \alpha)\%$ confidence interval of σ_ε^2 is

$$\frac{(N-2)s_e^2}{\chi_{\alpha/2,N-2}^2} \le \sigma_\varepsilon^2 \le \frac{(N-2)s_e^2}{\chi_{1-\alpha/2,N-2}^2} \tag{9.36}$$

It is worth noting that the confidence intervals for the regression coefficients and the variance of the errors do not depend on any particular value of the explanatory variable and, as a result, their widths are constant throughout the subdomain comprised by its bounds. However, if the inference is to be performed on the mean response of Y, further discussion is required, since $E(Y|X)$ is a linear function of X and, thus, the variance of a given estimate of the mean response $\hat{y}_0$ will depend on the particular point x_0 for which the mean response is sought. In this sense, one may intuitively expect that the width of the confidence interval on the mean response will vary within the observation range of X.

By expressing $\hat{y}_0$ through the estimated regression line, one may estimate the variance of $E[\hat{y}|x]$ as

$$\mathrm{Var}(\hat{y}_0) = \mathrm{Var}(\hat{\beta}_0 + \hat{\beta}_1 x_0) \tag{9.37}$$

However, from Eq. (9.24), $\hat{\beta}_0 = \bar{y} + \hat{\beta}_1 \bar{x}$. Thus,

$$\mathrm{Var}(\hat{y}_0) = \mathrm{Var}(\bar{y} + \hat{\beta}_1(x_0 - \bar{x})) \tag{9.38}$$

which, after algebraic manipulations, results in

$$\mathrm{Var}(\hat{y}_0) = s_e^2 \left(\frac{1}{N} + \frac{(x_0 - \bar{x})^2}{\sum\limits_{i=1}^{N}(x_i - \bar{x})^2} \right) \tag{9.39}$$

It is possible to demonstrate that the sampling distribution of the pivot function

$(y_0 - E(y|x_0)) / \sqrt{s_e^2 \left[1/N + (x_0 - \bar{x})^2 / \sum\limits_{i=1}^{N}(x_i - \bar{x})^2 \right]}$ is a Student's t variate, with $\nu = N - 2$ degrees of freedom. Thus, the $100(1 - \alpha)\%$ confidence interval on the mean response of Y is

$$\hat{y}_0 - t_{\alpha/2,N-2} \sqrt{s_e^2 \left(\frac{1}{N} + \frac{(x_0 - \bar{x})^2}{\sum\limits_{i=1}^{N} (x_i - \bar{x})^2} \right)} \leq y_0 \leq \hat{y}_0$$

(9.40)

$$+ t_{\alpha/2,N-2} \sqrt{s_e^2 \left(\frac{1}{N} + \frac{(x_0 - \bar{x})^2}{\sum\limits_{i=1}^{N} (x_i - \bar{x})^2} \right)}$$

From Eq. 9.40, it is possible to conclude that the closer x_0 is from the sample mean value $\bar{x}$, the narrower is the confidence interval, with a minimum value attained at $x_0 = \bar{x}$, and the distance between the two lines increases as x_0 goes farther from $\bar{x}$. This situation is illustrated in Fig. 9.8.

As for the prediction of a future value of Y, one must consider an additional source of uncertainty, since, unlike the mean response, which is a fixed parameter, a future value of Y is a random variable. In this sense, if a future value for the response variable is to be predicted, a new random component, which arises from estimating Y by means of the regression model, has to be included in the analysis. Thus, the definitions of confidence and prediction intervals comprise a major distinction: while the former provides a range for $E(Y|X)$, the latter provides a range for Y itself, and, thus, the unexplained variability of the observations has to be considered in order to account for the total uncertainty. Since a given future observation is independent of the regression model, the variance of Y may be

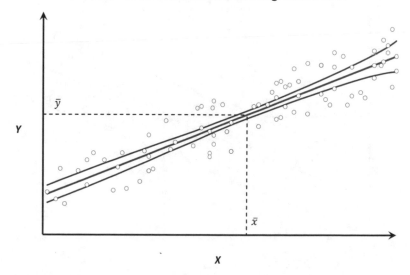

Fig. 9.8 Confidence intervals for the mean response

estimated by s_e^2 and the variance of the predicted error may be expressed as (Montgomery and Peck 1992),

$$\text{Var}(\hat{y}) = s_e^2 \left(1 + \frac{1}{N} + \frac{(x - \bar{x})^2}{\sum\limits_{i=1}^{N} (x_i - \bar{x})^2} \right) \tag{9.41}$$

and the $100(1 - \alpha)\%$ confidence interval for the predicted value of Y is given by

$$\hat{y} - t_{\alpha/2, N-2} \sqrt{s_e^2 \left(1 + \frac{1}{N} + \frac{(x - \bar{x})^2}{\sum\limits_{i=1}^{N} (x_i - \bar{x})^2} \right)} \leq Y \leq \hat{y}$$

$$+ t_{\alpha/2, N-2} \sqrt{s_e^2 \left(1 + \frac{1}{N} + \frac{(x - \bar{x})^2}{\sum\limits_{i=1}^{N} (x_i - \bar{x})^2} \right)} \tag{9.42}$$

where t denotes the Student's t variate with $\nu = N - 2$ degrees of freedom. Equation 9.41 shows that the predicted intervals are wider than the confidence intervals on the regression line due to the additional term in the variance of $\hat{y}$. This is illustrated in Fig. 9.9.

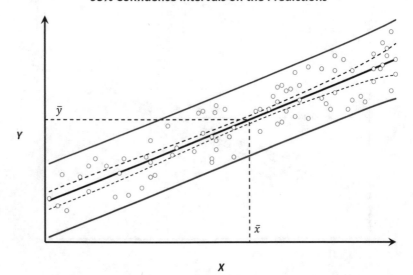

95% Confidence Intervals on the Predictions

Fig. 9.9 Prediction intervals for a given y

9.2.4 Hypothesis Testing in Simple Linear Regression

Hypothesis testing, as related to simple linear regression models, usually have two different points to be emphasized: (1) testing the significance of the regression; and (2) testing particular values of the regression coefficients. Insights on the linear functional form of the regression equation might be provided by the scatterplot of concurrent observations of sample values of variables X and Y, or through the assessment of the Pearson correlation coefficient estimate. These are, however, indicative procedures, since the former does not provide an objective measure of the linear dependence between the variables whereas the latter may be strongly affected by the particular characteristics of the sample, such as the ones described in Sect. 9.1.1, and, thus, might entail misleading conclusions regarding the linear association of X and Y. A more rigorous approach for checking the assumption of linearity requires the assessment of the significance of the regression equation by testing the hypothesis of a null value of the slope β_1. In this case, the hypotheses are $H_0 : \beta_1 = 0$ and $H_1 : \beta_1 \neq 0$.

The test statistic t is constructed by comparing the slope of the estimated and the true regression line, as normalized by the standard error on the estimate of β_1, given by Eq. (9.30). If the null hypothesis holds, then $\beta_1 = 0$ and it follows that

$$ t = \hat{\beta}_1 \Big/ \sqrt{ s_e^2 \Big/ \sum_{i=1}^{N} (x_i - \bar{x})^2 } \qquad (9.43) $$

From Sect. 9.2.3, if the errors are IID normally distributed variates, the sampling distribution of t, as in Eq. (9.43), is a Student's t, with $\nu = N - 2$ degrees of freedom. Thus, the null hypothesis should be rejected if $|t| > t_{\alpha/2, N-2}$, where α denotes the significance level of the test. If H_0 is not rejected, two lines of reasoning might arise: (1) no linear relationship between the variables holds; or (2) although some kind of linear relationship might exist, the explanatory variable X has little value in explaining the variation of Y, and, thus, from a statistical point of view, the mean value of the observations $\bar{y}$ provides a better estimate for the response variable, as compared to the linear model, for all possible values of X. On the other hand, if H_0 is to be rejected, one might infer that: (1) the straight line is a suitable regression model; or (2) the explanatory variable X is useful in explaining the variation of Y, albeit other functional forms, distinct from the linear model, might provide better fits to the data.

An alternative approach for testing the significance of regression is based on Eq. (9.32), which expresses the regression sum of squares. The rationale behind this approach is testing a ratio of the mean squares due to regression to the ones due to residuals. For simplicity, let Eq. (9.32) be expressed as

$$ SS_T = SS_R + SS_E \qquad (9.44) $$

Table 9.3 Analysis of variance for regression of Y on X

Source of variation	Sum of squares	Degrees of freedom
Regression	$\sum_{i=1}^{N} (\hat{y}_i - \bar{y})^2$	1
Residual	$\sum_{i=1}^{N} (y_i - \hat{y}_i)^2$	$N - 2$
Total	$\sum_{i=1}^{N} (y_i - \bar{y})^2$	$N - 1$

where $\mathrm{SS_T}$ refers to the total sum of squares $\sum_{i=1}^{N} (y_i - \bar{y})^2$, whereas $\mathrm{SS_R}$ and SS_E denote, respectively, the sum of squares due to regression, $\sum_{i=1}^{N} (\hat{y}_i - \bar{y})^2$, and the residual sum of squares, $\sum_{i=1}^{N} (y_i - \hat{y}_i)^2$. The quantity $\mathrm{SS_T}$ has $N-1$ degrees of freedom, since one is lost in the estimation of $\bar{y}$. As $\mathrm{SS_R}$ is fully specified by a single parameter, $\hat{\beta}_1$ (see Montgomery and Peck 1992), such a quantity has one degree of freedom. Finally, as two degrees of freedom have been lost in estimating $\hat{\beta}_0$ and $\hat{\beta}_1$, the quantity SS_E has $N-2$ of them. These considerations are summarized in Table 9.3. The procedure for testing the significance of regression on the basis of the sum of squares is termed ANOVA, which is an acronym for analysis-of-variance.

By calculating the ratio of the sum of squares due to regression and residual sum of squares, both divided by their respective degrees of freedom, one may construct the following test statistic

$$F_0 = \frac{\mathrm{SS_R}}{\mathrm{SS_E}/(N-2)} \tag{9.45}$$

which follows a Snedecor's F distribution with $v_1 = 1$ and $v_2 = N - 2$ degrees of freedom. The null hypothesis is rejected if $F_0 > F_{\alpha,1,N-2}$. This test is sometimes referred to as the F test for regression.

Hypothesis testing may be extended for particular values of the regression coefficients. The procedure follows the same reasoning utilized on the test for null slope: the test statistic is obtained on the basis of a comparison between the regression coefficient estimate and the constant value of interest, as normalized by the correspondent estimate of the standard error, which are given by Eqs. (9.29) and (9.30). Thus, for testing the hypothesis that the estimate $\hat{\beta}_0$ equals a given value β_{01}, one has $H_0 : \hat{\beta}_0 = \beta_{01}$ and $H_1 : \hat{\beta}_0 \neq \beta_{01}$. The test statistic is given by

$$t = (\hat{\beta}_0 - \beta_{01}) \Big/ \sqrt{s_e^2 \left[1/N + \bar{x}^2 \Big/ \sum_{i=1}^{n} (x_i - \bar{x})^2 \right]} \tag{9.46}$$

The null hypothesis should be rejected if $|t| > t_{\alpha/2,N-2}$, where t refers to the value of the Student's variate t, with $v = N - 2$ degrees of freedom, and α denotes the level of significance of the test.

As for the slope of the straight-line model, the hypotheses to be tested are $H_0 :$ $\hat{\beta}_1 = \beta_{11}$ against $H_1 : \hat{\beta}_1 \neq \beta_{11}$. The test statistic, which derives from a generalization of Eq. (9.43) for non-null values of β_1, is expressed as

$$t = (\hat{\beta}_1 - \beta_{11})/\sqrt{s_e^2/\sum_{i=1}^{N}(x_i - \bar{x})^2} \qquad (9.47)$$

Again, the null hypothesis should be rejected if $|t| > t_{\alpha/2,N-2}$, where t refers to the value of the Student's t variate, with degrees of freedom, and α denotes the level of significance of the test.

9.2.5 Regression Diagnostics with Residual Plots

After estimating the regression line equation, it becomes necessary to assess if the regression model is appropriate for describing the behavior of the response variable Y. In this context, one must ensure that the linear functional form prescribed to the model is the correct one. This can be achieved by evaluating the significance of regression by means of t and F tests. In addition, it is important to evaluate the ability of the regression model in describing the behavior of the response variable, in terms of explaining its variation. A standard choice for this is the coefficient of determination r^2. Although these statistics provide useful information regarding regression results, they alone are not sufficient to ascertain the model adequacy. Anscombe (1973) provides examples of four graphs with the same values of statistically significant regression coefficients, standard errors of regression, and coefficients of determination. These graphs have been redrawn and are displayed in Fig. 9.10.

Figure 9.10a depicts a reasonable model, for which a linear relationship between explanatory and response variables seems to exist and all points are approximately evenly scattered around the regression line. This is a case of an adequate regression model. Figure 9.10b, on the other hand, shows that X and Y are related through a clearly nonlinear functional form, albeit hypothesis tests on the model slope have resulted in not rejecting the null hypothesis, $H_0 : \beta_1 = 0$. This situation highlights the inadequacy of the proposed linear model despite the fact that the estimated slope is not deemed a result of chance, by an objective test. Figure 9.10c illustrates the effect of a single outlier in the slope of the regression line. In effect, the regression equation is strongly affected by this atypical point, which moves the regression line towards the outlier, reducing the predictive ability of the model. Finally, Fig. 9.10d shows the influence of a single value of the explanatory variable located well

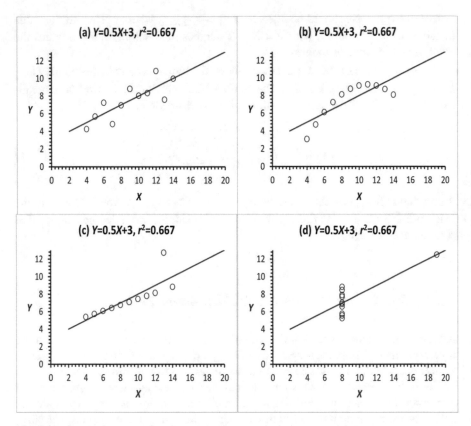

Fig. 9.10 Graphs for regression diagnostics with residual plots (adapted from Anscombe 1973)

beyond the range of the remaining data. Although no apparent relation between X and Y is verified in the cluster of points, the extreme value of X entails the estimation of a misleading regression line. These charts make it clear that evaluating a regression model's adequacy solely on the basis of summary statistics is an inappropriate and unsafe expedient.

Residuals plots are a commonly utilized technique for addressing this problem. In effect, as the regression model is constructed under specific assumptions, the behavior of the residuals provides a full picture on whether or not such assumptions are violated after the fit. In general, three types of plots are of interest: (1) residuals versus fitted values, for linearity, independence and homoscedasticity; (2) residuals versus time, if regression is performed on time series; and (3) normality of residuals.

Regarding the first type, a plot of residuals versus fitted values (or the explanatory variable), such as the one depicted in Fig. 9.11, can provide useful information on the assumed (or assessed) linear function form of the regression model. If such a plot is contained in a horizontal band and the residuals are randomly scattered (Fig. 9.11a), then, no obvious inadequacies on the linear association

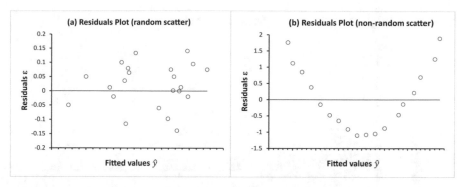

Fig. 9.11 Plot of residuals versus fitted values

assumption are detected. However, if the plot appears to follow some pattern (Fig. 9.11b), it is likely that, although a linear association between X and Y might exist, it does not provide the best fit. In this case, a transformation on X might be a convenient alternative for achieving a linear relationship. In fact, as the errors are assumed to be independent of the explanatory variable X, stretching or shrinking the X axis will not affect the behavior of the residuals regarding the homoscedasticity and normality assumptions and, thus, the only change in the model will be the functional form of the association between explanatory and response variables. As a reference, if the residuals are U-shaped and the association between X and Y is positive, the transformation $X' = X^2$ will yield a linear model. Conversely, if the residuals look like an inverted U and the association between X and Y is still positive, one may use the transformations $X' = \sqrt{X}$ or $X' = \ln(X)$.

The plot residuals versus fitted values is also an adequate tool for checking the assumption of independence of the residuals. In effect, if such a plot appears to assume some functional form, then strong evidence exists that the residuals are correlated. This is particularly clear if regression models are applied to autocorrelated time series and may be easily visualized in a plot of the second type, residuals versus time. Correlated errors do not affect the unbiasedness property of the least squares estimators. However, if a dependence relationship between the residuals exists, the sampling variance of the estimators of β_0 and β_1 will be larger, sometimes at excessive levels, than the ones estimated with Eqs. (9.29) and (9.30), and, as a result, the actual significance of all hypothesis tests previously discussed will be unknown (Haan 2002).

As for the constant variance assumption, the behavior of the residuals may be evaluated by plotting the residuals versus fitted values or the residuals versus explanatory variable X, as exemplified in Fig. 9.12. If these plots show increasing or decreasing variance, then the homoscedasticity assumption is violated. In general, increasing variance is related to right-skewed distribution of the residuals, whereas the opposite skewness property is associated to decreasing variance. Due to this fact, the usual approach for dealing with heteroscedastic residuals is to apply transformations to the response variable Y or to both X and Y. It has to be stressed

Residuals Plot

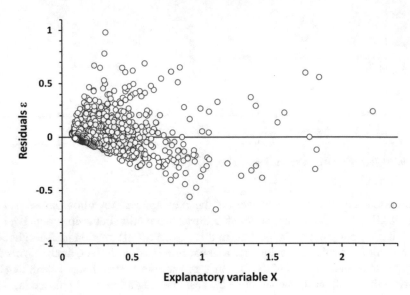

Fig. 9.12 Plot residuals versus explanatory variables

Table 9.4 Transformations for linearization of functions (adapted from Yevjevich 1964)

Type of function	Abscissa	Ordinate	Equation in linear form
$Y = \beta_0 + \beta_1 X$	X	Y	$Y = \beta_0 + \beta_1 X$
$Y = \beta_0 e^{\beta_1 X}$	X	$\ln(Y)$	$\ln(Y) = \ln(\beta_0) + \beta_1 X$
$Y = \beta_0 X^{\beta_1}$	$\ln(X)$	$\ln(Y)$	$\ln(Y) = \ln(\beta_0) + \beta_1 \ln(X)$
$Y = \beta_0 + \beta_1 X + \beta_2 X^2$	$X - x_0$	$\dfrac{Y - y_0}{X - x_0}$	$\left[\dfrac{Y - y_0}{X - x_0}\right] = \beta_1 + 2\beta_1 x_0 + \beta_2 (X - x_0)$
$Y = \beta_0 + \dfrac{\beta_1}{X}$	$\dfrac{1}{X}$	Y	$Y = \beta_0 + \beta_1 \left(\frac{1}{X}\right)$
$Y = \dfrac{\beta_0}{\beta_1 + \beta_2 X}$	X	$\dfrac{1}{Y}$	$\dfrac{1}{Y} = \dfrac{\beta_1}{\beta_0} + \dfrac{\beta_2}{\beta_0} X$
$Y = \dfrac{X}{\beta_0 + \beta_1 X}$	X	$\dfrac{X}{Y}$	$\frac{X}{Y} = \beta_0 + \beta_1 X$

that, when using transformations of variables, the transformed values must be used if interval estimation is to be performed, and, only after obtaining the confidence or prediction bounds in the transformed space, are the endpoints of intervals to be computed for the original units. Table 9.4 presents a collection of transformations for linearizing functions of frequent use in Statistical Hydrology.

The last assumption to be checked is the normal distribution of the residuals. One of the simplest alternatives for this purpose is to plot the residuals, after

Normal Probability Plot

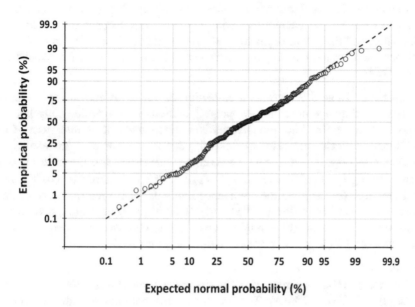

Fig. 9.13   Normal probability plot

ranking them in ascending order, against the theoretical expected normal values, on a normal probability paper (see Sect. 8.2.1). If the referred assumption holds, the points will plot as a straight line. In checking the functional form of the plot, emphasis should be given to its central portion, since, even for approximately normally distributed residuals, departures from the straight line are expected on the extremes due to the observation of outliers (see Sect. 8.2.2 and Montgomery and Peck 1992). The construction of probability papers was described in Sect. 8.2.1. The interested reader is also referred Chambers et al. (1983) for further details on this topic. Figure 9.13 presents a typical normal probability plot.

Another framework for evaluating the normality of the errors is based on the construction of box-plots and frequency histograms. These descriptive tools are useful in demonstrating the conditions of symmetry and spread of the empirical distribution of the errors. As a practical reference, for the Normal distribution, 95 % of the values of the residuals must lie in the range of 2 standard deviations around the mean. If such a condition holds and the histogram is approximately symmetric, then one should be able to assume that the errors are Normal variates.

Wadsworth (1990) states that violating the normality assumption is not as critical as violating the other ones, and may be relaxed up to some limit, since the Student's t distribution, extensively employed for inference procedures in regression models, is robust with respect to moderate departures from a Normal distribution. However, if the distribution of the residuals is too skewed, then the procedures of interval estimation and hypothesis testing previously discussed are clearly inappropriate. According to Helsel and Hirsch (2002), the main consequences of a pronounced non-normality of regression residuals are that confidence and prediction intervals will be excessively wide and may erroneously be considered as symmetric, and the power of the tests will be too reduced, resulting in explanatory variables and regression slopes which may be falsely deemed as statistically insignificant.

Example 9.2 Table 9.5 presents a collection of values of annual maximum discharges and drainage areas for 22 gauging stations in a hydrologically homogeneous region of the São Francisco River catchment, in southeastern Brazil. (a) Estimate a linear regression model using the drainage areas as the explanatory variable. Consider that the annual maximum discharges and drainage areas are related through a model such as $Q = \beta_0 A^{\beta_1}$. (b) Calculate the coefficient of determination. (c) Estimate the confidence intervals for the regression coefficients considering a confidence level of 95 %. (d) Test the significance of regression at the significance level 5 %. (e) Using residuals plots, evaluate the adequacy of the regression model (adapted from Naghettini and Pinto 2007).

Solution

(a) From Table 9.4, one may infer that the linear form of a function $Q = \beta_0 A^{\beta_1}$ is $\ln(Q) = \ln(\beta_0) + \beta_1 \ln(A)$. The fourth and fifth columns of Table 9.5 present, respectively, the natural logarithms of the explanatory variable A and the response variable Q, whereas the sixth column contains the product of these two variables in logarithmic space. With Eqs. (9.24) and (9.25), one obtains $\hat{\beta}_1 = \frac{22 \times 1098.0773 - 182.6197 \times 128.8740}{22 \times 1548.2565 - (182.6197)^2}$, which gives $\hat{\beta}_1 = 0.8751$, and $\ln(\beta_0) = 5.8579 - 0.8751 \times 8.3009 = -1.4062$. Thus, the linear regression model, as estimated from the samples of $\ln(A)$ and $\ln(Q)$, may be expressed as $\ln(Q) = -1.4062 + 0.8751\ln(A)$. The estimates $\ln(\hat{Q})$ and the regression residuals are given in the eighth and ninth columns of Table 9.5, respectively. Figure 9.14 depicts the scatter of the sample points, in logarithmic space, around the regression line. The regression model in arithmetic space is given as $\hat{Q} = 0.2451 A^{0.8751}$.

(b) After estimating the regression line, one is able to calculate the fraction of the variance that is explained by the regression model, i.e., estimate the coefficient of determination r^2. Table 9.6 presents the sum of squares due to regression and the residual sum of squares. By using the values given in the table, $r^2 = 24.7726/25.0529 = 0.989$, or, in other words, 98.9 % of the total variance of the response variable is explained by the regression model.

(c) The standard error of regression is $s_e = \sqrt{\frac{\sum_{i=1}^{N}(y_i - \hat{y}_i)}{N-2}} = \sqrt{\frac{0.2803}{20}} = 0.1184$. The variances of the regression coefficients are respectively given as

$$\text{Var}(\hat{\beta}_0) = s_e^2 \left(\frac{1}{N} + \frac{\bar{x}^2}{\sum_{i=1}^{N}(x_i - \bar{x})^2} \right) = 0.1184^2 \left(\frac{1}{22} + \frac{8.3009^2}{32.3488} \right) = 0.0305 \quad \text{and}$$

Table 9.5 Calculations of the regression coefficients of a transformed power model, by linearly relating the logarithms of drainage areas and of annual maximum discharges

Gauging station #	Drainage area (km²)	Q (m³/s)	$\ln(A)$	$\ln(Q)$	$\ln(A) \times \ln(Q)$	$[\ln(A)]^2$	$\widehat{\ln(Q)}$	Residuals
1	269.1	31.2	5.5951	3.4404	19.2494	31.3050	3.4901	−0.0496
2	481.3	49.7	6.1765	3.9060	24.1254	38.1490	3.9988	−0.0928
3	1195.8	100.2	7.0866	4.6072	32.6490	50.2195	4.7953	−0.1881
4	1055.0	109.7	6.9613	4.6977	32.7024	48.4596	4.6856	0.0121
5	1801.7	154.3	7.4965	5.0389	37.7740	56.1973	5.1540	−0.1151
6	1725.7	172.8	7.4534	5.1521	38.4009	55.5530	5.1163	0.0359
7	1930.5	199.1	7.5655	5.2938	40.0505	57.2373	5.2144	0.0794
8	2000.2	202.2	7.6010	5.3093	40.3557	57.7752	5.2454	0.0638
9	1558.0	207.2	7.3512	5.3337	39.2088	54.0395	5.0268	0.3069
10	2504.1	263.8	7.8257	5.5752	43.6297	61.2413	5.4421	0.1331
11	5426.3	483.3	8.5990	6.1806	53.1474	73.9430	6.1188	0.0618
12	7378.3	539.4	8.9063	6.2905	56.0247	79.3222	6.3877	−0.0972
13	9939.4	671.4	9.2043	6.5094	59.9139	84.7184	6.6484	−0.1391
14	8734.0	690.1	9.0750	6.5368	59.3217	82.3552	6.5353	0.0015
15	8085.6	694.0	8.9978	6.5425	58.8681	80.9611	6.4678	0.0747
16	8986.9	742.8	9.1035	6.6104	60.1782	82.8741	6.5603	0.0501
17	11,302.2	753.5	9.3328	6.6247	61.8270	87.1003	6.7609	−0.1362
18	10,711.6	823.3	9.2791	6.7133	62.2935	86.1014	6.7139	−0.0006
19	13,881.8	889.4	9.5383	6.7905	64.7705	90.9798	6.9408	−0.1502
20	14,180.1	1032.4	9.5596	6.9396	66.3402	91.3859	6.9594	−0.0198
21	16,721.9	1336.9	9.7245	7.1981	69.9978	94.5654	7.1037	0.0944
22	26,553.0	1964.8	10.1869	7.5831	77.2487	103.7729	7.5084	0.0748
Mean			8.3009	5.8579				
Sum			182.6197	128.8740	1098.0773	1548.2565		

Table 9.6 Analysis of variance for regression of ln(Q) on ln(A)

Source of variation	Sum of squares	Degrees of freedom
Regression	24.7726	1
Residuals	0.2803	20
Total	25.0529	21

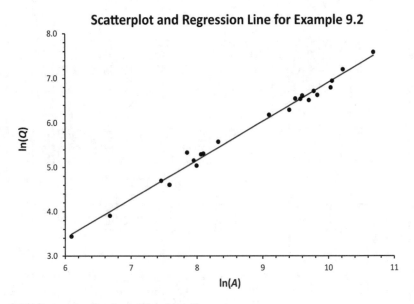

Fig. 9.14 Regression line for ln(Q) in logarithm space

$\text{Var}\left(\hat{\beta}_1\right) = \dfrac{s_e^2}{\sum\limits_{i=1}^{N}(x_i-\bar{x})^2} = \dfrac{0.1184^2}{32.3488} = 0.0004$. The value of the Student's t, for the

95 % confidence level and 21 degrees of freedom is $t_{0.975,21} = 2.086$. Then, the confidence intervals for β_0 and β_1 are, respectively, $-1.7705 \le \beta_0 \le -0.0420$ and $0.8317 \le \beta_1 \le 0.9185$.

(d) The test statistic is $t = \dfrac{0.8751}{\sqrt{0.0004}} = 42.0373$. As $|t| > t_{\alpha/2,N-2} = 2.086$, the null hypothesis should be rejected. Thus, a linear relationship exists between ln(A) and ln(Q) or, in other terms, ln(A) has value in explaining ln(Q).

(e) Figure 9.15a shows that the residuals are approximately contained in a horizontal band throughout the fitted values and are randomly scattered, with no apparent pattern of variation. In turn, Fig. 9.15b shows that the residuals plot as a straight-line on Normal probability paper. It is possible to conclude that the model is adequate since it does not violate any of the assumptions assumed a priori.

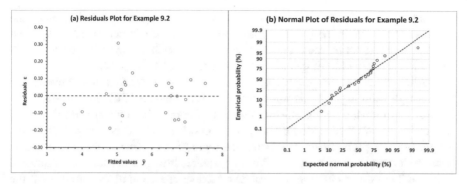

Fig. 9.15 Residuals plots: (**a**) residuals versus fitted values; (**b**) normal probability plot

9.2.6 Some Remarks on Simple Linear Regression Models

Simple linear regression models are widely employed in hydrological sciences as they provide a convenient framework for estimating or predicting the behavior of a given variable on the basis of a simple straight-line equation. There are some situations in which most or even all the information regarding a hydrological variable, at a particular cross-section of a stream, is obtained from a regression model. Therefore, such models play an important role in providing the hydrologist or the engineer with objective indications and guidelines for the design of hydraulic structures, when little or no measured data are available. However, because the method is quick and easy to apply, the technique has been misused, frequently resulting in unreliable results in the statistical modeling of response hydrologic variables. This subsection deals with some problems associated with misusing simple linear regression models.

As a first indication, one must evaluate if the estimates of β_0 and β_1 are reasonable, both in sign and magnitude, and if, for any reasonable value of X, the model responds with unrealistic or physically impossible values of Y, such as, for instance, negative discharges for a given water level, from a rating curve (Helsel and Hirsch 2002). In addition, regression models are constructed for a given range of the subdomain of the explanatory variable. The behavior of the response variable for values of X which are not contained in that subdomain is not known and might not be linear. Furthermore, confidence intervals for values of the explanatory variable beyond the available sample may become too large and lose their physical meaning. Therefore, extrapolation in regression models is not a reliable procedure and must be avoided.

The presence of outliers is also a concern in constructing regression models since these values may greatly disturb the least squares fit. Outliers might be the result of measurement errors, and, if these points are used with no proper corrections, the estimate of the intercept may result in values with no physical meaning and the residual mean square will overestimate the variance of the errors σ_ε^2. On the other hand, the measured value of the hydrologic variable deemed as an outlier may be correct and, thus, it will represent useful information on the behavior of the response variable. In such a case, no correcting action is necessary and robust methods for estimating the regression coefficients, such as the weighted or generalized least squares (Stedinger and Tasker 1985), may be required.

Finally, points of the explanatory variable that are located well beyond the remaining x sample points significantly affect the behavior of the slope coefficient. In such situations, deleting unusual points is recommended when a least squares model is constructed (Montgomery and Peck 1992). Other approaches for addressing this problem are estimating the regression coefficient by means of less sensitive methods as compared to the least squares fitting or introducing additional explanatory variables in order to explain a broader spectrum of the variation of the response variable. The use of multiple explanatory variables is discussed in the next section.

9.3   Multiple Linear Regression

As opposed to simple linear regression models, which describe the linear functional relationship between a single explanatory variable X and the response variable Y, multiple linear regression models comprise the use of a collection of explanatory variables for describing the behavior of Y. In formal terms

$$Y = \beta_0 + \beta_1 X_1 + \ldots + \beta_k X_k + \varepsilon \tag{9.48}$$

where the vector $X_1, \ldots, X_k$ denotes the k explanatory variables, $\beta_0, \ldots, \beta_k$ are the regression coefficients and ε is a random error term which accounts for the differences between the regression model and the response sample points. Each of the regression coefficients, with the exception of the intercept β_0, expresses the change in Y due to a unit change in a given X_i when the other X_j, $i \neq j$, are held constant.

In analogy to the simple linear regression model, it is possible to conclude that, due to the random nature of the error term, the response variable Y is also a random variable distributed according to a model $f_{Y|X_1, \ldots X_k}(y \mid x_1, \ldots, x_k)$, conditioned on $x_1, \ldots, x_k$, whose expected value is expressed as a linear function of the x_i 's and may be written as

$$y = \beta_0 + \sum_{j=1}^{k} \beta_j x_{ij} \tag{9.49}$$

whereas the variance is held constant and equal to σ^2, if all x_i are nonrandom constant quantities.

Parameter estimation in multiple linear regression may also be based on the least squares methods, provided that the number of response sample points y_i is larger than the number of regression coefficients k to be estimated. If such a condition holds, the least square function, which has to be minimized, may be expressed as

$$M(\beta_0, \ldots, \beta_k) = \sum_{i=1}^{N} \varepsilon_i^2 = \sum_{i=1}^{N} \left(y_i - \beta_0 - \sum_{j=1}^{k} \beta_j x_{ij} \right)^2 \qquad (9.50)$$

By taking the first partial derivatives of M with respect to the regression coefficients $\beta_0, \ldots, \beta_k$, as evaluated by the least squares estimators $\hat{\underline{\beta}}_j$, and setting their values to zero, one obtains

$$\left. \frac{\partial M}{\partial \beta_0} \right|_{\hat{\underline{\beta}}_0, \ldots, \hat{\underline{\beta}}_k} = -2 \sum_{i=1}^{N} \left(y_i - \hat{\underline{\beta}}_0 - \sum_{j=1}^{k} \hat{\underline{\beta}}_j x_{ij} \right) = 0$$

$$\left. \frac{\partial M}{\partial \beta_j} \right|_{\hat{\underline{\beta}}_0, \ldots, \hat{\underline{\beta}}_k} = -2 \sum_{i=1}^{N} \left(y_i - \hat{\underline{\beta}}_0 - \sum_{j=1}^{k} \hat{\underline{\beta}}_j x_{ij} \right) x_{ij} = 0 \qquad (9.51)$$

which, after algebraic manipulations, yield the *least squares normal equations*

$$n\hat{\underline{\beta}}_0 + \hat{\underline{\beta}}_1 \sum_{i=1}^{N} x_{i1} + \ldots + \hat{\underline{\beta}}_k \sum_{i=1}^{N} x_{ik} = \sum_{i=1}^{N} y_i$$

$$\hat{\underline{\beta}}_0 \sum_{i=1}^{N} x_{i1} + \hat{\underline{\beta}}_1 \sum_{i=1}^{N} x_{i1}^2 + \ldots + \hat{\underline{\beta}}_k \sum_{i=1}^{N} x_{ik}^2 = \sum_{i=1}^{N} y_i x_{i1}$$

$$\vdots \qquad \vdots \qquad \vdots \qquad \vdots \qquad (9.52)$$

$$\hat{\underline{\beta}}_0 \sum_{i=1}^{N} x_{i1} + \hat{\underline{\beta}}_1 \sum_{i=1}^{N} x_{i1}^2 + \ldots + \hat{\underline{\beta}}_k \sum_{i=1}^{N} x_{ik}^2 = \sum_{i=1}^{N} y_i x_{ik}$$

The least squares estimates of the regression coefficient, $\hat{\beta}_0, \ldots, \hat{\beta}_k$, are obtained by solving the normal equations system with sample points composed of concurrent observations of the explanatory variables X_i, $i = 1, \ldots, k$, and the response variable Y. As with the simple linear regression model, these are sample-based estimates of the unknown population regression coefficients and, due to this fact, as long as one is dealing with finite samples, no particular collection of estimates is expected to match exactly the population's true values.

For mathematical convenience, it is usual to express multiple regression models in matrix notation. Thus, a given model may be written as

$$\mathbf{y} = \mathbf{X}\boldsymbol{\beta} + \boldsymbol{\varepsilon} \qquad (9.53)$$

where $\mathbf{y}$ is the $N \times 1$ vector of observations of the response variable, $\mathbf{y} = \begin{bmatrix} y_1 \\ \vdots \\ y_N \end{bmatrix}$,

$\mathbf{X}$ is an $N \times k$ matrix of explanatory variables, $\mathbf{X} = \begin{bmatrix} 1 & x_{11} & \cdots & x_{1k} \\ \vdots & \vdots & \ddots & \vdots \\ 1 & x_{N1} & \cdots & x_{Nk} \end{bmatrix}$, $\boldsymbol{\beta}$ is the

$N \times 1$ vector of regression coefficients, $\boldsymbol{\beta} = \begin{bmatrix} \beta_1 \\ \vdots \\ \beta_k \end{bmatrix}$, and $\boldsymbol{\varepsilon}$ is an $N \times 1$ vector of errors,

$\boldsymbol{\varepsilon} = \begin{bmatrix} \varepsilon_1 \\ \vdots \\ \varepsilon_N \end{bmatrix}$. The notation in bold for the matrix model is intended to make a

distinction between vector or matrices, and scalar quantities.

Expressing the least squares criterion in matrix notation yields

$$M(\boldsymbol{\beta}) = \sum_{i=1}^{N} \varepsilon^2 = \boldsymbol{\varepsilon}' \boldsymbol{\varepsilon} = (\mathbf{y} - \mathbf{X}\boldsymbol{\beta})^T (\mathbf{y} - \mathbf{X}\boldsymbol{\beta}) \tag{9.54}$$

where the superscript T denotes transposed vector/matrix. Algebraic manipulations allow one to write $M(\boldsymbol{\beta})$ as follows

$$M(\beta) = \mathbf{y}^T \mathbf{y} - 2\boldsymbol{\beta}^T \mathbf{X}^T \mathbf{y} + \boldsymbol{\beta}^T \mathbf{X}^T \mathbf{X}\boldsymbol{\beta} \tag{9.55}$$

By taking the first partial derivative of Eq. (9.55) with respect to the vector $\boldsymbol{\beta}$, as evaluated by the least squares estimators vector $\underline{\hat{\boldsymbol{\beta}}}$, setting its value to zero and simplifying the remainder terms, one obtains

$$\mathbf{X}^T \mathbf{X} \underline{\hat{\boldsymbol{\beta}}} = \mathbf{X}^T \mathbf{y} \tag{9.56}$$

The vector of regression coefficients estimates $\hat{\boldsymbol{\beta}}$ is then obtained by multiplying both sides of Eq. (9.56) by $\left(\mathbf{X}^T \mathbf{X}\right)^{-1}$. Thus,

$$\underline{\hat{\boldsymbol{\beta}}} = \left(\mathbf{X}^T \mathbf{X}\right)^{-1} \mathbf{X}^T \mathbf{y} \tag{9.57}$$

A solution to $\underline{\hat{\boldsymbol{\beta}}}$ exists if and only if $\left(\mathbf{X}^T \mathbf{X}\right)^{-1}$ exists. A requirement for this condition to hold is that the explanatory variables be linearly independent. If so, it can be easily shown that the least squares estimators, as summarized by the vector $\underline{\hat{\boldsymbol{\beta}}}$, are unbiased (Montgomery and Peck 1992). In addition, if the errors are independent, it follows from the Gauss-Markov theorem that such estimators also have the minimum variance, as compared to other possible estimators obtained by linear combinations of y_i. As for the variance of the regression coefficient estimates

$\hat{\boldsymbol{\beta}}$, one may express the covariance matrix as

$$\text{Cov}(\hat{\boldsymbol{\beta}}) = \sigma_\varepsilon^2 (\mathbf{X}^{\mathbf{T}}\mathbf{X})^{-1} \tag{9.58}$$

which is a $p \times p$, $p = k + 1$, symmetric matrix in which the diagonal elements express the variance of $\hat{\beta}_i$ and the off-diagonal elements correspond to the covariance between $\hat{\beta}_i$ and $\hat{\beta}_j$, $i \neq j$. An unbiased (yet model-dependent) estimator for σ_ε^2 is given by

$$s_e^2 = \frac{\text{SS}_\text{E}}{n - k - 1} \tag{9.59}$$

where SS_E corresponds to the residual sum of squares, and $N-k-1$ provides the number of degrees of freedom. Similarly to simple regression models, the quantity s_e^2 is usually referred to as *residual mean square*. The positive square root of s_e^2 is often denoted standard error of multiple linear regression.

Analogously to simple regression models, a goodness-of-fit measure on how much of the variance in Y is explained by the multiple-regression model can be derived. This statistic is termed multiple coefficient of determination R^2 and may be estimated from the ratio of the sum of squares due to regression SS_R to the total sum of squares SS_y. The sum of squares and the degrees of freedom for a linear multiple-regression model are presented in Table 9.7.

The multiple coefficient of determination may be expressed as

$$R^2 = \frac{\text{SS}_\text{R}}{\text{SS}_\text{y}} \tag{9.60}$$

Although still expressing a goodness-of-fit measure, such as in the simple linear regression case, the interpretation of the multiple coefficient of determination requires caution. First, it has to be stressed that large values of R^2 do not mean that the most suitable collection of explanatory variables has been used for constructing the regression model. In effect, the contribution of two strictly opposite variables, as evaluated in the physical description of the phenomenon, might be the same in the regression model. In addition, no indication on the dependence between explanatory variables is provided by the value of the multiple coefficient of determination alone. Such a dependence relationship, which is usually termed

Table 9.7 Analysis of variance in linear multiple regression

Source of variation	Sum of squares	Degrees of freedom	Mean square
Regression	$SS_R = \hat{\boldsymbol{\beta}}^T \mathbf{X}^T \mathbf{y} - n\bar{y}^2$	k	$MS_R = SS_R/k$
Residual	$SS_E = \mathbf{y}^T \mathbf{y} - \hat{\boldsymbol{\beta}}^T \mathbf{X}^T \mathbf{y}$	$n - k - 1$	$MS_E = s_e^2 = SS_E/n - k - 1$
Total	$SS_y = \mathbf{y}^T \mathbf{y} - n\bar{y}^2$	$n - 1$	

multicollinearity, is discussed in Sect. 9.3.3. Multicollinearity might entail an increase in R^2 whilst no additional information is provided by the inclusion of a new explanatory variable that is correlated to another one already considered in the regression model.

Finally, comparing different models, with a distinct number of explanatory variables, is an inappropriate procedure, since the addition of explanatory variables always leads to an increase of the sum of squares due to regression. In this case, the adjusted multiple coefficient of determination, $R^2_{adjusted}$, which can be expressed as

$$R^2_{adjusted} = 1 - \left(1 - R^2\right) \left(\frac{N-1}{N-k-1}\right)$$ (9.61)

is usually preferred to R^2, as it accounts not only for the change in the sum of squares due to regression, but also for the change in the number of explanatory variables.

In addition to measuring the goodness-of-fit of a regression model by means of the coefficient of determination, one may be interested in evaluating the contribution that each of the explanatory variables adds to explaining the variance of Y. Such a quantity is expressed by the coefficient of partial determination, which is given by

$$R^2_{partial} = \frac{SS_E(X_j) - SS_E(X_1, \ldots, X_j, \ldots, X_k)}{SS_E(X_j)}$$ (9.62)

where X_j refers to the a given explanatory variable to be included in the model, $SS_E(X_1, \ldots, X_j, \ldots, X_k)$ is the residuals sum of squares with all explanatory variables and $SS_E(X_j)$ is the residuals sum of squares due to alone X_j. It is worth noting that the calculations of the coefficient of partial determination may be performed directly from an ANOVA table, by comparing the residuals sum of squares before and after the inclusion of X_j.

9.3.1 Interval Estimation in Multiple Linear Regression

As with simple linear regression models, the construction of confidence intervals on the regression coefficients and on the regression line itself might be required for the multiple regression homologous. If the errors ε_i are uncorrelated and follow a Normal distribution, with $\mu(\varepsilon) = 0$ and $Var(\varepsilon) = \sigma_e^2$, then the response variable Y is also normally distributed, with $\mu_i = \beta_0 + \sum_{j=1} \beta_j x_{ij}$ and $Var = \sigma^2$ and standard parametric inference procedures, such as the ones described in Chap. 6, are applicable for this purpose.

As for the regression coefficients, it is easy to demonstrate that, since each y_i follow a Normal distribution and each regression coefficient estimator is a linear combination of the response variable sample points, the marginal distributions of all $\hat{\beta}_j$'s are also Normal, with $\mu = \beta_j$ and $\mathrm{Var} = \sigma_e^2 C_{jj}$, where C_{jj} is the j-th diagonal element of the matrix $(X'X)^{-1}$ (Montgomery and Peck 1992). Given this, it is possible to show that the sampling distribution of the pivot function $\frac{\hat{\beta}_j - \beta_j}{s_e(\hat{\beta}_j)}$, where $s_e(\hat{\beta}_j) = \sqrt{s_e^2 C_{jj}}$ is usually referred to as the standard error of the regression coefficient β_j, is a Student's t, with $\nu = N - p$ degrees of freedom. Thus, the 100 $(1 - \alpha)\%$ confidence intervals of each β_j are given by

$$\hat{\beta}_j - t_{\alpha/2, N-p} s_e(\hat{\beta}_j) \le \beta_j \le \hat{\beta}_j + t_{\alpha/2, N\ p} s_e(\hat{\beta}_j) \tag{9.63}$$

In addition to constructing confidence intervals on the regression coefficients, it is also desirable to evaluate the accuracy of the regression model, in terms of the confidence interval on its mean response y_m, at a given point x_m. An estimate $\hat{y}_m$ for the mean response, as related to x_m, may be obtained by

$$\hat{y}_m = \mathbf{x_m}^T \hat{\boldsymbol{\beta}} \tag{9.64}$$

where $\mathbf{x_m} = \begin{bmatrix} 1 \\ x_{m1} \\ \vdots \\ x_{mk} \end{bmatrix}$ is a $1 \times P$ vector which encompasses the explanatory variables,

and $\hat{\boldsymbol{\beta}}$ is the vector composed by the regression coefficient estimates. The variance of $\hat{y}_m$ is estimated with the following expression:

$$\mathrm{Var}(\hat{y}_m) = s_e^2 \mathbf{x_m}^T (\mathbf{X^T X})^{-1} \mathbf{x_m} \tag{9.65}$$

As with the simple linear regression models, it is possible to show that the sampling distribution of the pivot function $\dfrac{\hat{y}_m - E(y_m|x_m)}{\sqrt{s_e^2 \mathbf{x_m}^T (X^T X)^{-1} x_m}}$ is a Student's t variate, with $\nu = N - p$ degrees of freedom. Thus, the $100(1 - \alpha)\%$ confidence interval of the mean response y_m, at the point x_m, is given by

$$\hat{y}_m - t_{\alpha/2, N-p} \sqrt{s_e^2 \mathbf{x_m}^T (\mathbf{X^T X})^{-1} \mathbf{x_m}} \le y_m \le \hat{y}_m + t_{\alpha/2, N-p} \sqrt{s_e^2 \mathbf{x_m}^T (\mathbf{X^T X})^{-1} \mathbf{x_m}} \tag{9.66}$$

Finally, if interest lies on constructing the interval for a predicted value of the response variable, as evaluated in a particular point x_m, one must account for the uncertainties arising from both response variable and the regression model by adding the estimate of the variance of the errors s_e^2 to the estimate of the variance resulting from the fitted model, $s_e^2 \mathbf{x_m}^T \left(\mathbf{X^T X} \right)^{-1} \mathbf{x_m}$. Thus, the $100(1 - \alpha)\%$ confidence interval predicted value for the response y_m, at the point x_m, is given by

$$
\hat{y}_m - t_{\alpha/2, N-p} \sqrt{ s_e^2 \left(1 + \mathbf{x_m}^T \left(\mathbf{X^T X} \right)^{-1} \mathbf{x_m} \right) } \leq y_m \leq \hat{y}_m
$$
$$
+ t_{\alpha/2, N-p} \sqrt{ s_e^2 \left(1 + \mathbf{x_m}^T \left(\mathbf{X^T X} \right)^{-1} \mathbf{x_m} \right) }
$$

(9.67)

9.3.2 Hypothesis Testing in Multiple Linear Regression

An important hypothesis test in the context of multiple linear regression models is that of evaluating if a linear relationship between the response variable and any of the explanatory counterparts exists. In this case, the test assesses the significance of regression and the hypotheses to test are $H_0 : \beta_1 = \dots = \beta_k = 0$ and $H_1 : \beta_j \neq 0$, for, at least, one j. This test is known as the overall F-test for regression and is based on evaluating the ratio between two measures of variance, namely, the mean square due to regression, SS_R, and the residual sum of squares, SS_E, which, as discussed in Chaps. 5 and 7, follows a Snedecor's F distribution, with $\nu_1 = k$, where k denotes the number of explanatory variables, and $\nu_2 = N - k - 1$ degrees of freedom. In fact, if H_0 is true, it is possible to demonstrate that $SS_R/\sigma^2 \sim \chi^2_k$ and $SS_E/\sigma^2 \sim \chi^2_{N-k-1}$, yield the referred null distribution. The test statistic may be then expressed as

$$
F_0 = \frac{SS_R/k}{SS_E/(N - k - 1)}
$$

(9.68)

The null hypothesis is rejected if $F_0 > F(\alpha, k, N - k - 1)$, where F refers to the value of the Snedecor's F, with k and $N-k-1$ degrees of freedom, at the significance level α. It is worth noting that rejecting the null hypothesis implies that at least one of the explanatory variables has value in explaining the response variable. Not rejecting H_0, on the other hand, might be associated to one of the following lines of reasoning: (1) the relationship between any of the explanatory variables and response counterpart is not linear; or (2) even that some kind of linear relationship exists, none of the explanatory variables has value in explaining the variation of Y on the basis of this functional form.

Hypothesis testing in multiple linear regression may also be performed on individual explanatory variables of the regression model in order to assess their significance in explaining Y. The main objective of such tests is to assess if the inclusion or dismissal of a given explanatory variables entails a significantly better fit to the regression model. In effect, the inclusion of a new explanatory variable always comprises an increase in the sum of squares due to regression, with a concurrent decrease in the residual sum of squares. However, the addition of a variable x_i might imply an increase in the residual mean square due to the loss of one degree of freedom and, as a result, the variance of the fitted values also increases. Thus, one has to assess if the benefits of increasing the regression sum of squares due to the inclusion of X_j outweigh the loss of accuracy in estimating Y.

The hypotheses for testing the significance of an individual response variable are $H_0 : \beta_j = 0$ against $H_1 : \beta_j \neq 0$. The test statistic is expressed as

$$t_0 = \frac{\hat{\beta}_j}{s_e\left(\hat{\beta}_j\right)} \tag{9.69}$$

The null hypothesis should be rejected if $|t| > t_{\alpha/2,N-k-1}$, where t refers to the value of the Student's t, with $N-k-1$ degrees of freedom, at the significance level α. Not rejecting H_0 implies that X_j must not be included into the model. According to Montgomery and Peck (1992), this is a marginal test since $\hat{\beta}_j$ is dependent on all k explanatory variables employed in the construction of the regression model.

An alternative approach to the previous test is to evaluate the contribution of a given explanatory variable with respect to the change in the sum of squares due to regression, provided that all the other explanatory variables have been included in the model. This method is termed *extra-sum-of-squares*. The value of SS_R, as resulting from a given explanatory variable X_j to be included the model, can be calculated as

$$SS_R\left(x_j | x_1, \ldots, x_{j-1}, x_{j+1}, \ldots, x_k\right) = SS_R\left(x_1, \ldots, x_j, \ldots, x_k\right)$$
$$- SS_R\left(x_1, \ldots, x_{j-1}, x_{j+1}, \ldots, x_k\right) \tag{9.70}$$

where the first term on the right side corresponds to the sum of squares with all k explanatory variables included in the model and the second term is the sum of squares computed when X_j is excluded from the model. Equation (9.70) encompasses the idea that $SS_R\left(x_j | x_1, \ldots, x_{j-1}, x_{j+1}, \ldots, x_k\right)$ expresses the contribution of each possible X_j with respect to the full model, which includes all k variables. Again, the hypotheses to test are $H_0 : \beta_j = 0$ and $H_1 : \beta_j \neq 0$. The test statistic is given by

$$F_P = \frac{SS_R\left(x_j | x_1, \ldots, x_{j-1}, x_{j+1}, \ldots, x_k\right)}{SS_E/(N - k - 1)} \tag{9.71}$$

The null hypothesis should be rejected if $F_P > F(\alpha, 1, N - k - 1)$, where F refers to the value of the Snedecor's F, with $\nu_1 = 1$ and $\nu_2 = N - k - 1$ degrees of freedom, at the significance level α. Not rejecting H_0 implies that X_j does not contribute significantly to the regression model and, thus, must not be included in the analysis. The test described by Eq. (9.71) is termed the partial F-test. Partial F-tests are useful for model building since they allow the assessment of the significance of including or dismissing an explanatory variable to the regression model, one at a time. By constructing a regression model based on this rationale, only the explanatory variables that provide a significant increase in R^2 (or $R^2_{adjusted}$) are retained and, as a result, the models obtained are parsimonious and present less uncertainty.

Two different approaches may be utilized for model building through partial F-tests. In the first approach, termed forward stepwise multiple regression, the independent explanatory variable with the highest partial correlation, with respect to the response counterpart, is initially included in the regression model. After that, the remaining independent variables are included, one by one, on the basis of the highest partial correlations with Y, and the procedure is repeated up to the point when all significant variables have been included. At each step, the increment in the value of R^2 due to a new explanatory variable is tested through

$$F_c = \frac{\left(1 - R^2_{k-1}\right)(N - k - 1)}{\left(1 - R^2_{k}\right)(N - k - 2)} \tag{9.72}$$

in which N is the number of observations and k is the number of explanatory variables included in the model. If $F_c > F_{\alpha, N-k-1, N-k-2}$, where F is the value of the Snedecor's F with $\nu_1 = N - k - 1$ and $\nu_2 = N - k - 2$ degrees of freedom and α denotes the level of significance of the test, then the inclusion of the X_k is statistically significant.

In the second approach, denoted by backward stepwise multiple regression, all independent explanatory variables are initially considered in the model. Then, one by one, explanatory variables are eliminated until a significant difference in the value of R^2 is obtained.

The *extra-sum-of-squares* method may be extended to test whether a subset of explanatory variables is statistically significant for the regression model by partitioning the regression coefficients vector into two groups: $\boldsymbol{\beta}_1$, a $p \times 1, p = k + 1$, vector with the variables already included in the model, and $\boldsymbol{\beta}_2$, an $r \times 1, r = N - p$, vector which encompasses the subset to be tested. Thus, the hypotheses to be tested are $H_0 : \beta_2 = 0$ and $H_1 : \beta_2 \neq 0$. The test statistic is given by

$$F_P = \frac{SS_R(x_r, \ldots, x_k | x_1, \ldots, x_{r-1})/r}{SS_E/(N - p)} \tag{9.73}$$

The null hypothesis should be rejected if $F_P > F(\alpha, r, N - p)$, where F refers to the value of the Snedecor's F, with $\nu_1 = r$ and $\nu_2 = N - p$ degrees of freedom and α denotes the level of significance of the test. Not rejecting H_0 implies that $\boldsymbol{\beta}_2$ does not contribute significantly to the regression model and, thus, must not be included in the analysis.

The R software environment for statistical computing provides a comprehensive support for multiple regression analysis and testing. Details can be found in https://cran.r-project.org/doc/contrib/Ricci-refcard-regression.pdf. [Accessed 13th April 2016].

9.3.3 Multicollinearity

In many multiple regression models, situations arise when two explanatory variables are correlated. Particularly for hydrological studies, some level of correlation is almost always detected for the most frequently employed explanatory variables, such as the drainage area of the catchment and the length of its main river. The correlations between explanatory variables or linear combinations of these variables are usually termed multicollinearity, although, for practical purposes, the term should strictly apply for the highest levels of correlation in a collection of X_j's.

When two explanatory variables are highly correlated, they are expected to bring very similar information in explaining the variation of the response variable. Thus, if these two variables are simultaneously included in a regression model, the effects on the response variable will be partitioned between them. The immediate consequence of this fact is that the regression coefficients might not make physical sense, in terms of sign or magnitude. This can induce erroneous interpretations regarding the contributions of individual explanatory variables in explaining Y. Furthermore, when two highly correlated X's are used in the model, the variance of the regression coefficients might be extremely increased and, as a result, they might test statistically insignificant even when the overall regression points to a significant linear relationship between Y and the X's.

In addition to providing improper estimates of the regression coefficients, one must note that, given that both explanatory variables provide approximately the same information to the regression model, their concurrent use will entail only a small decrease in the residuals sum of squares and, therefore, the increase in the coefficient of multiple determination is marginal. This indicates to the analyst that a more complex model will not necessarily perform better in explaining the variation of Y than a more parsimonious one.

According to Haan (2002), evidences of multicollinearity may be identified by:

- High values of correlations between variables in the correlation matrix X;
- Regression coefficients with no physical sense;
- Regression coefficients of important explanatory variables tested as statistically insignificant; and
- Expressive changes in the values of regression coefficients as a result of the inclusion or deletion of an explanatory variable from the model.

An objective index for detecting multicollinearity is the Variance Inflation Factor (VIF), which is formally expressed as

$$\text{VIF} = \frac{1}{1 - R_i^2} \tag{9.74}$$

where R_i^2 corresponds to the multiple coefficient of determination between the explanatory variable X_i and all the others X_j's in the regression equation. If R_i^2 is zero, then X_i is linearly independent of the remaining explanatory variables in the model. On the other hand, if R_i^2 is equal to 1, then VIF is not defined. It is clear then that large values of VIF imply multicollinearity.

As a practical reference for dealing with multicollinearity, one must not include a given explanatory variable in a regression model if the index VIF is larger than 5, which corresponds to $R_i^2 = 0.80$. When this situation holds, one of the variables must be discarded from the regression model. Another approach, suggested by Naghettini and Pinto (2007), is to construct a correlation matrix between the explanatory variables and, provided that a given X_j is selected for estimating the regression equation, all other explanatory variables with Pearson correlation coefficient higher than 0.85, as related to the one initially selected, should be eliminated from the analysis.

Example 9.3 Table 9.8 presents a set of explanatory variables from which one intends to develop a multiple regression model for explaining the variation of the mean annual runoff in 13 small catchments in the American State of Kentucky. Estimate a regression equation using a forward stepwise multiple regression (adapted from Haan 2002).

Solution The first step for constructing the multiple-regression model refers to estimating the correlation matrix for the collection of explanatory variables. Such a matrix is presented in Table 9.9.

From Table 9.9, one may infer that the explanatory variable with the largest linear correlation with the mean annual runoff is the catchment drainage area. Using only this variable in the regression model, one obtains the ANOVA figures given in Table 9.10.

Table 9.8 Explanatory variables for estimating a multiple regression model

Catchment	Runoff	Pr	A	S	L	P	d_i	R_s	SF	R_r
1	441.45	1127.00	5.66	50	3.81	12.69	1.46	0.38	3.48	63.25
2	371.35	1119.89	6.48	7	4.08	12.24	1.97	0.48	6.07	10.48
3	393.19	1047.75	14.41	19	4.98	18.58	3.38	0.57	5.91	14.67
4	373.89	1155.70	3.97	6	2.94	8.50	1.50	0.49	9.91	12.95
5	466.60	1170.69	13.18	16	6.62	18.16	2.61	0.39	8.45	12.95
6	432.05	1247.65	5.48	26	3.07	9.42	2.26	0.71	4.79	43.82
7	462.28	1118.36	13.67	7	7.57	20.14	2.08	0.27	2.41	8.38
8	481.33	1237.23	19.12	11	6.78	19.73	3.76	0.52	3.07	13.72
9	354.08	1128.52	5.38	5	3.20	10.90	1.90	0.53	12.19	7.62
10	473.46	1212.09	9.96	18	3.36	15.79	2.64	0.6	7.88	21.91
11	438.15	1228.85	1.72	21	1.84	6.29	0.99	0.48	7.65	67.06
12	443.99	1244.60	2.18	23	2.03	6.06	1.33	0.61	9.04	57.15
13	334.26	1194.56	4.40	5	3.09	8.30	1.58	0.52	5.96	7.43

where Runoff—mean annual runoff (mm), Pr—mean annual precipitation (mm), A—drainage area (km^2), S—average land slope (%), L—axial length (km), P—perimeter (km), d_i—diameter of the largest circle that can be drawn within the catchment (km), R_s—shape factor (dimensionless), SF—stream frequency—ratio of number of streams in the catchment to total drainage area (km^2), R_r—relief ratio—ratio of the total relief to the largest dimension of the catchment (m/km)

Table 9.9 Correlation matrix for runoff and the explanatory variables

	Runoff	Pr	A	S	L	P	d_i	R_s	SF	R_r
Runoff	1.00									
Pr	0.40	1.00								
A	0.47	−0.25	1.00							
S	0.41	0.08	−0.17	1.00						
L	0.42	−0.34	0.90	−0.21	1.00					
P	0.46	−0.41	0.96	−0.10	0.92	1.00				
d_i	0.33	−0.15	0.91	−0.16	0.67	0.81	1.00			
R_s	−0.15	0.45	−0.25	0.05	−0.58	−0.41	0.15	1.00		
SF	−0.40	0.04	−0.48	−0.30	−0.53	−0.48	−0.32	0.29	1.00	
R_r	0.35	0.42	−0.52	0.80	−0.54	−0.51	−0.50	0.18	−0.08	1.00

Table 9.10 ANOVA table for explanatory variable A

Source of variation	Sum of squares	Degrees of freedom	R^2	$R^2_{adjusted}$	F
Regression	6428.10	1	0.222	0.151	3.130
Residual	22,566.10	11			
Total	28,994.21	12			

Table 9.11 ANOVA table for explanatory variables A and S

Source of variation	Sum of squares	Degrees of freedom	R^2	$R^2_{adjusted}$	F
Regression	13,596.20	2	0.469	0.363	4.415
Residual	15,398.01	10			
Total	28,994.21	12			

The drainage area explains 22.2 % of the runoff total variance. The overall F-test also indicates that the regression model is not significant at the 5 % level of significance, since $F = 3.130 < F_0(0.05, 1, 11) = 4.84$. When adding a new explanatory variable to model, one must be cautious of the occurrence of multicollinearity. By resorting to the criterion proposed by Naghettini and Pinto (2007), all remaining explanatory variables with values of correlation higher than 0.85 with respect to the drainage area are eliminated from the analysis. Thus, by returning to Table 9.9, the variables L, P and d_i must not be included in the model. The explanatory variable with the highest correlation to runoff, after the previous elimination, is S. By including it in the multiple regression model, one obtains the following the ANOVA table presented in Table 9.11:

Again the overall F-test pointed to significant regression model, since $F = 4.415 > F_0(0.05, 2, 10) = 4.10$. The partial F-test also shows that including S in the regression entails a significant improvement to the model, as $F_p = 5.121 > F_0(0.05, 1, 10) = 4.96$. By adding the explanatory variable S to the model, the explained variance corresponds to 46.9 % of the total variance of A. The next explanatory variable to be included is Pr. The corresponding ANOVA is presented in Table 9.12.

Table 9.12 ANOVA table for explanatory variables A, S, and P_r

Source of variation	Sum of squares	Degrees of freedom	R^2	$R^2_{adjusted}$	F
Regression	20,824.909	3	0.718	0.624	7.648
Residual	8169.2961	9			
Total	28,994.205	12			

The overall F-test pointed to a significant regression model because $F = 7.648 > F_0(0.05, 3, 9) = 3.863$. The partial F-test also shows that including Pr in the regression significantly improves the model, since $F_p = 9.733 > F_0(0.05, 1, 9) = 5.117$. The model is now able to explain 71.8 % of the runoff total variance. From this point on, no further explanatory variable is deemed significant for the regression model. Then, the final regression equation can be written as RUNOFF $= -141.984 + 6.190A + 1.907S + 0.410$Pr.

Exercises

1. Referring to the data of annual rainfall depth and annual mean daily flow given in Table 9.1, estimate the Pearson's r correlation coefficient. Test the hypothesis that r is null at the significance level 5 %.
2. From the data given in Table 9.1, test the null hypothesis that $\rho = 0.85$ at the significance level 5 %. Construct the 95 % confidence interval for ρ.
3. Table 9.13 displays 45 observations of the mean annual discharge of the Spray River, near Banff, in Canada. Calculate the serial correlation coefficient for lags 1 and 2. Test the hypothesis that the lag-one serial correlation is null (adapted from Haan 2002).
4. Assuming that the data of mean annual runoff given in Table 9.13 are normally distributed, calculate the number of independent observations that are required for providing the same amount of information as the one given by the 45 correlated observations.
5. Prove that, if the errors are independent, the least squares estimators are unbiased and have the minimum variance.
6. Prove that the correlation coefficient in a simple linear regression model, as expressed by the square root of the coefficient of determination, equals the correlation coefficient between Y and $\hat{Y}$.
7. Derive the normal equations for the regression model $Y = \beta_0 + \beta_1 X + \beta_2 X^2$.
8. Derive the normal equation for a linear model with null intercept. Construct the confidence interval at point $x = 0$. Comment on the obtained results.
9. Table 9.14 provides the values of drainage area A and long-term mean flow Q for a collection of gauging stations in the São Francisco River catchment, in southeastern Brazil. (a) Estimate a simple linear regression model using the drainage area as the explanatory variable. (b) Estimate the coefficient of determination r^2. (c) Estimate the confidence intervals on the mean response of Q (adapted from Naghettini and Pinto 2007).

Table 9.13 Mean annual discharge in the Spray River near Banff, Canada

Water year	Mean annual discharge (m³/s)	Water year	Mean annual discharge (m³/s)	Water year	Mean annual discharge (m³/s)	Water year	Mean annual discharge (m³/s)	Water year	Mean annual discharge (m³/s)
1910–1911	16.5	1919–1920	11.4	1928–1929	13.4	1937–1938	17.7	1946–1947	17.5
1911–1912	14.4	1920–1921	20.0	1929–1930	9.3	1938–1939	12.4	1947–1948	14.5
1912–1913	14.6	1921–1922	12.1	1930–1931	15.7	1939–1940	5.5	1948–1949	12.3
1913–1914	10.9	1922–1923	14.7	1931–1932	10.4	1940–1941	14.3	1949–1950	3.3
1914–1915	14.6	1923–1924	11.4	1932–1933	14.8	1941–1942	16.8	1950–1951	15.9
1915–1916	10.0	1924–1925	10.3	1933–1934	4.6	1942–1943	16.1	1951–1952	16.4
1916–1917	11.8	1925–1926	3.7	1934–1935	15.0	1943–1944	17.9	1952–1953	11.4
1917–1918	14.5	1926–1927	14.4	1935–1936	15.8	1944–1945	6.1	1953–1954	14.6
1918–1919	15.7	1927–1928	12.7	1936–1937	12.5	1945–1946	17.1	1954–1955	13.5

Table 9.14 Drainage areas A and long-term mean flow Q for 22 gauging stations of the São Francisco River catchment, for Exercise 9

Gauging station #	A (km^2)	Q (m^3/s)	Gauging station #	A (km^2)	Q (m^3/s)
1	83.9	1.3	12	3727.4	65.3
2	188.3	2.3	13	4142.9	75.0
3	279.4	4.2	14	4874.2	77.2
4	481.3	7.3	15	5235.0	77.5
5	675.7	8.2	16	5414.2	86.8
6	769.7	8.5	17	5680.4	85.7
7	875.8	18.9	18	8734.0	128
8	964.2	18.3	19	10,191.5	152
9	1206.9	19.3	20	13,881.8	224
10	1743.5	34.2	21	14,180.1	241
11	2242.4	40.9	22	29,366.2	455

Table 9.15 Concurrent measurements of stage H (m) and discharge Q (m^3/s) for the Cumberland River at Cumberland Falls, for Exercise 11

H	Q	H	Q	H	Q	H	Q	H	Q
4.7	1.7	4.1	1.4	3.8	1.2	3.6	1.1	3.3	0.9
4.3	1.5	4.4	1.6	3.8	1.2	3.5	1.1	3.1	0.8
4.5	1.6	4.0	1.3	3.9	1.3	3.4	1.0	2.9	0.7
4.3	1.5	3.8	1.2	2.7	0.6	3.1	0.9	2.7	0.6
4.3	1.5	3.7	1.2	2.5	0.6	3.3	1.0	2.4	0.5

Table 9.16 Concurrent stages H and discharges Q at a river cross section for Exercise 12

H (m)	Q (m^3/s)	H (m)	Q (m^3/s)	H (m)	Q (m^3/s)
0.00	20	2.70	300	5.84	1260
0.80	40	4.07	680	7.19	1920
1.19	90	4.73	990	8.21	2540
1.56	120	4.87	990	8.84	2840
1.91	170	5.84	1260	9.64	3320
2.36	240	7.19	1920	–	–

10. Solve Exercise 9 using the model $Q = \beta_0 A^{\beta_1}$. Discuss the results of both models.

11. Table 9.15 displays concurrent measurements of stage H and discharge Q for the Cumberland River at Cumberland Falls, Kentucky, in USA. Derive the rating curve for this river cross section employing the regression models $Q = \beta_0 + \beta_1 H$ and $Q = \beta_0 H^{\beta_1}$. Which of the two models best fits the data? (adapted from Haan 2002).

12. Concurrent measurements of stage H and discharge Q, at a given river cross section, are displayed in Table 9.16. Figure 9.16 provides a scheme illustrating the reference marks (RM) and the staff gauges, as referenced to the gauge

River Control Section at Gauging Station with Staff Gauges

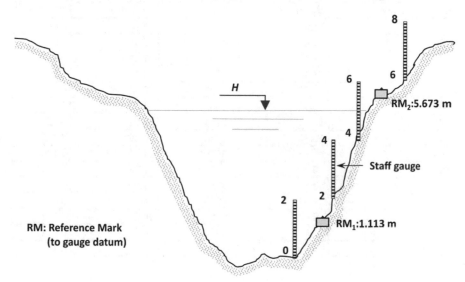

Fig. 9.16 River control section at the gauging station for Exercise 12

datum, at this particular cross section. (a) Estimate the rating curve with the regression models $Q = \beta_0(H - h_0)^{\beta_1}$ and $Q = \beta_0 + \beta_1 H + \beta_2 H^2$. (b) Determine the best-fit model by means of the residual variance, which is given by Eq. (9.28). (c) A bridge will be built at this site, which is located about 500 m downstream of a reservoir. A design guideline imposes that a peak discharge of 5200 m³/s must flow under the bridge deck. What is the minimum elevation of the bridge deck, as referenced to an establish datum, provided that the RM-2 is at the elevation 731.229 m above sea level (adapted from Naghettini and Pinto 2007).

13. Table 9.17 provides the values of the annual minimum 7-day mean flow (Q_7) recorded at a number of gauging stations in the Paraopeba River basin, in the State of Minas Gerais, Brazil, alongside the respective values of potential explanatory variables for a regression model, namely, the catchment drainage area A, the main stream equivalent slope S and the drainage density DD. Estimate a multiple linear regression model using the forward stepwise regression. At each step, calculate the values of R^2 and R^2 adjusted, and perform the overall and partial F-tests (adapted from Naghettini and Pinto 2007).

Table 9.17 Potential explanatory variables for estimating a multiple regression model

Gauging station #	Q_7 (m^3/s)	A (km^2)	S (m/km)	DD (km^{-2})
1	2.60	461	2.96	0.098
2	1.49	291	3.94	0.079
3	1.43	244	7.20	0.119
4	3.44	579	3.18	0.102
5	1.37	293	2.44	0.123
6	2.53	486	1.25	0.136
7	15.12	2465	1.81	0.121
8	16.21	2760	1.59	0.137
9	21.16	3939	1.21	0.134
10	30.26	5414	1.08	0.018
11	28.53	5680	1.00	0.141
12	1.33	273	4.52	0.064
13	0.43	84	10.27	0.131
14	39.12	8734	0.66	0.143
15	45.00	10,192	0.60	0.133

References

Anderson RL (1942) Distribution of the serial correlation coefficient. Ann Math Stat 13:1–13

Anscombe FA (1973) Graphs in statistical analysis. Am Stat 27:17–21

Chambers J, William C, Beat K, Paul T (1983) Graphical methods for data analysis. Wadsworth Publishing, Stamford, CT, USA

Gauss JCF (1809) Theoria Motus Corporum Coelestium in sectionibus conicis solem ambientium (Theory of the motion of heavenly bodies moving about the sun in conic sections). English translation by C. H. Davis (1963). Dover, New York

Graybill FA (1961) An introduction to linear statistical models. McGraw-Hill, New York

Haan CT (2002) Statistical methods in hydrology, 2nd edn. Iowa State Press, Ames, IA, USA

Helsel DR, Hirsch RM (2002) Statistical methods in water resources. USGS Techniques of Water-Resources Investigations, Book 4, Hydrologic Analysis and Interpretation, Chapter A-3. United States Geological Survey, Reston, VA. http://pubs.usgs.gov/twri/twri4a3/pdf/twri4a3-new.pdf. Accessed 1 Feb 2016

Legendre AM (1805) Nouvelles méthodes pour la détermination des orbites des comètes, Firmin Didot, Paris. https://ia802607.us.archive.org/5/items/nouvellesmthode00legegoog/nouvellesmthode00legegoog.pdf. Accessed 11 Apr 2016

Matalas NC, Langbein WB (1962) Information content of the mean. J Geophys Inform Res 67 (9):3441–3448

Montgomery DC, Peck EA (1992) Introduction to linear regression analysis. John Wiley, New York, NY, USA

Naghettini M, Pinto EJA (2007) Hidrologia estatística. CPRM, Belo Horizonte, Brazil

Stedinger JR, Tasker GD (1985) Regional hydrologic regression, 1. Ordinary, weighted and generalized least squares compared. Water Resour Res 21(9):1421–1432

Wadsworth HM (1990) Handbook of statistical methods for engineers and scientists. McGrall-Hill, New York

Yevjevich VM (1964) Section 8-II statistical and probability analysis of hydrological data. Part II. Regression and correlation analysis. In: Chow VT (ed) Handbook of applied hydrology. McGraw-Hill, New York

Chapter 10
Regional Frequency Analysis of Hydrologic Variables

Mauro Naghettini and Eber José de Andrade Pinto

10.1 Introduction

As mentioned in Chap. 8 of this book, at-site frequency analysis of hydrologic variables, in spite of having developed a vast number of models and inference methods in the pursuit of reliable estimates of parameters and quantiles, comes up against practical difficulties imposed by the usually short samples of hydrologic records. In this context, the use of regional frequency analysis has emerged as an attempt to overcome the insufficient description of how a random quantity varies over the period of records with its broader characterization in space, by pooling data from samples of different sizes, collected at distinct sites across a geographic region. In one important variant of regional frequency analysis, the frequency distributions of all gauging stations located inside a *hydrologically homogeneous region* are assumed identical to all sites apart from a site-specific scaling factor, the so-called *index-flood*. Other variants, in turn, use multiple regression methods to represent the relationships between the quantiles or the distribution parameters and the so-called catchment attributes, such as the basin physiographic and soil characteristics alongside climate variables and other inputs. The methods of regional frequency analysis of hydrologic variables can be used either to provide quantile estimates at ungauged sites or to improve quantile estimates at poorly gauged sites.

The general principles that guide regional frequency analysis based on the index-flood approach were formally introduced by Dalrymple (1960). The term

M. Naghettini (✉)
Universidade Federal de Minas Gerais, Belo Horizonte, Minas Gerais, Brazil
e-mail: mauronag@superig.com.br

E.J.d.A. Pinto
CPRM Serviço Geológico do Brasil, Universidade Federal de Minas Gerais, Belo Horizonte, Minas Gerais, Brazil
e-mail: eber.andrade@cprm.gov.br

© Springer International Publishing Switzerland 2017
M. Naghettini (ed.), *Fundamentals of Statistical Hydrology*,
DOI 10.1007/978-3-319-43561-9_10

index-flood, as referring to the scaling factor, was originally used by Dalrymple (1960) in the context of indexing the annual flood series at each site by its respective mean annual flood, as a means of pooling scaled data from multiple sites. As other applications of regional frequency analysis continued using it to designate the scaling factor, the term index–flood became standard use in applying regional techniques to any kind of data and not only to flood data These include rainfall data, such as those used in deriving regional IDF (Intensity–Duration–Frequency) relations for heavy storms of sub-daily durations over an area, and low flow data.

The basic idea of index-flood-based approaches, hereinafter referred to as IFB, is to define a regional frequency curve (or a regional growth curve) that is common to all scaled data recorded at all sites within a homogeneous region. Then, the scaled quantile $\hat{x}_T$ of return period T, as estimated with the regional growth curve, is multiplied by the at-site index-flood $\hat{\mu}_j$ to obtain the estimate of the T-year quantile $\hat{X}_{T,j} = \hat{\mu}_j \hat{x}_T$ for site j. The underlying assumption is that within a homogeneous region, data at different sites follow the same parent distribution with a common shape parameter, but the scale parameter is site-specific and depends on the catchment attributes. For an ungauged site, located in the homogeneous region, it is assumed that the at-site probability distribution can be fully estimated by borrowing its shape from the regional growth curve, while its scale parameter is regressed against the catchment attributes.

A comprehensive application of an IFB approach led to the publication in 1975 of the *Flood Studies Report*, which is a set of five volumes describing the methods for estimating flood-related quantities over nine geographical regions of Great Britain and Ireland (NERC 1975). As regarding the estimation of annual maximum flood flows in those countries, the report recommended the adoption of the GEV (generalized extreme value) distribution, with an index-flood procedure. Later, in 1999, this publication was superseded by the *Flood Estimation Handbook* (IH 1999), which recommends a different model, the GLO (Generalized Logistic) distribution, with estimation based on a unified index-flood-based method for regional frequency analysis that makes use of L-moments, as introduced by Hosking and Wallis (1997). This unified method is described in detail in Sect. 10.4.

The early methods of regional frequency analysis based on the index-flood approach have faced conceptual difficulties in objectively defining a homogeneous region, in which the frequency curves at all sites can be approximated by a regional curve. Such difficulties have led to the development of multiple regression equations for relating quantiles to the catchment attributes (Riggs 1973). As such, the T-year quantile $X_{T,j}$ for site j is directly related to the catchment attributes through a multiple regression equation, using either the method of Ordinary Least Squares (OLS), described in Chap. 9, or variants of it. This regression-based method for quantiles is referred hereinafter as RBQ.

A concurrent application of regression-based methods for regional frequency analysis of hydrologic variables employs regression equations to represent the relationship between the set of parameters Θ_j that describe the probability distribution at site j and the catchment attributes. For this application of regression-based methods, in order to ensure consistency of parameter estimates, the geographic area should be divided into a convenient number of homogeneous regions, inside which data recorded at different sites share the same analytical form of the probability distribution with site-specific parameters. This regression-based method for distribution parameters is referred hereinafter as RBP.

Advanced applications of the multiple regression methods for regional frequency analysis employ the Generalized Least Squares (GLS), introduced by Stedinger and Tasker (1985). GLS regression is an extension of OLS that takes into account the sample size at each gauging station, and the variances and cross correlations of the data records collected at different sites. A full description of advanced regional frequency analysis using GLS is beyond the scope of this introductory textbook; the reader interested in the topic should consult Stedinger and Tasker (1985), Kroll and Stedinger (1998), Reis et al. (2005), and Griffis and Stedinger (2007).

The catchment attributes that have been used in regional frequency analysis of hydrologic variables include (a) physiographic catchment characteristics, such as drainage area, length and slope of the main stream, average basin slope, drainage density, stream order and shape indexes; (b) soil properties, such as the average infiltration capacity and soil moisture deficit; (c) land use patterns as fractions of the total basin area occupied by pastures, forests, agricultural and urban lands; (d) geographical coordinates and elevation of the gauging station and coordinates of the catchment's centroid; and (e) meteorological and climatic inputs, such as the prevailing direction of incoming storms over the basin, the mean annual number of days below a threshold temperature and the mean annual rainfall depth over the catchment.

The catchment attributes are the K potential explanatory variables, denoted as $M_{j,k}, j = 1, \ldots, N; k = 1, \ldots, K$, that can be used to explain the variation of either the index-flood scaling factors μ_j or the T-year quantiles $X_{T,j}$ or the sets of parameters Θ_j, for sites $j = 1, 2, \ldots, N$ across a homogeneous region, in the cases of μ_j and Θ_j, or across a geographic area, in the case of $X_{T,j}$. The regression models that have been used in regional frequency analysis of hydrologic variables are usually of the linear-log type, such as the regression equation $Z_j = \alpha + \beta_1 \ln(M_{j,1}) + \beta_2 \ln(M_{j,2}) + \ldots + \beta_K \ln(M_{j,K}) + \varepsilon$, for a fixed site j, where Z_j can be either μ_j or $X_{T,j}$ or Θ_j, α and β_k are regression coefficients, and ε denotes the residuals. The log–log type of regression can also be used for such a purpose.

One notion that seems to permeate all methods of regional frequency analysis is that the catchment attributes, such as the physiographic characteristics and other inputs, as aggregated at the basin scale, are assumed sufficient to characterize the dominant processes that govern the probability distributions of the variable of interest, at different sites. For the IFB method this notion is oriented towards the delineation of a homogeneous region, in the sense that it should exhibit a growth curve common to all sites within it. RBP is also oriented towards the definition of a homogeneous region, within which the site-specific probability distributions should have a common analytical form, whereas for the RBQ method the notion is related to finding the appropriate sets of gauging stations and explanatory variables that account for distinguishable hydrologic responses. Referring to the RBQ method, Riggs (1973) points out that the *extent of a region encompassed by a regional analysis should be limited to that in which the same* (explanatory) *variables are considered effective throughout*. The next section discusses the concept of hydrologically homogeneous regions, as fundamentally required by the IFB and RBP methods of regional frequency analysis, but which is also relevant to the RBQ method.

10.2   Hydrologically Homogeneous Regions

An important step in regional frequency analysis is the definition of hydrologically homogeneous regions, which is a fundamental assumption of IFB and RBP methods. For the IFB method, a homogeneous region is formed by a group of gauging stations whose standardized frequency distributions of the hydrologic variable of interest are similar. The same applies to the RBP method, by considering that the non-standardized distributions can be represented by a common analytical form with site-specific parameters. Notwithstanding its importance, the delineation of hydrologically homogeneous regions involves subjectivity and is still a matter of debate and a topic of current research (e.g., Ilorme and Griffis 2013, Gado and Nguyen 2016).

Several techniques have been proposed to delineate homogeneous regions but none has established an absolutely objective criterion or a consensual solution to the problem. In this context, Bobée and Rasmussen (1995) argue that the methods of regional frequency analysis, in general, and the techniques of delineating homogeneous regions, in particular, are built upon assumptions that are difficult to check with mathematical rigor. The very notion that catchment attributes are capable of summarizing the processes that govern the probability distributions of the variable of interest, at different sites, is an example of such an assumption. However, acknowledging its shortcomings does not diminish the value of regional frequency analysis, since building models of practical usefulness for water resources engineering is the very essence of Statistical Hydrology.

The first source of controversy concerning the delineation of homogeneous regions is related to the type of information that should support it. A distinction is made between site statistics and site characteristics. The former may refer, for instance, to measures of dispersion and skewness, as estimated from the data samples, which, in turn, are the object themselves of the regional frequency analysis. On the other hand, the site characteristics are, in principle, deterministic quantities and are not estimated from the data samples. Clear examples of such quantities are the latitude, longitude, and elevation of a gauging station. It is also thought to be reasonable to include variable quantities that are not highly correlated to sample data, such as the month with the highest frequency of flood peaks over some threshold, or the mean annual total rainfall depth over the catchment area, or the average base-flow index (BFI) at a given site.

Hosking and Wallis (1997) recommend that the identification of homogeneous regions should be made in two consecutive steps: the first, consisting of a preliminary grouping of gauging stations, based only on site characteristics, followed by the second step, of validation, based on site statistics. Hosking and Wallis (1997) proposed a test based on at-site statistics, to be formally described in Sect. 10.4.2, that serves the purpose of validating (or not) the preliminarily defined regions. Furthermore, they point out that using the same data, to form and test regions, compromises the integrity of the test. The main techniques that have been used for delineating homogeneous regions are categorized and briefly described in the next subsection.

10.2.1   Categories of Techniques for Delineating Homogeneous Regions

The following are the main categories of techniques for delineating homogeneous regions, as viewed by Hosking and Wallis (1997) and complemented by a non-exhaustive list of recent works.

10.2.1.1   Geographic Convenience

In the category of geographic convenience, one can find the many attempts to identify homogeneous regions by forming groups of gauging stations according to subjective criteria and/or convenience arguments, such as site proximity, or the boundaries of state or provincial administration, or other similar arbitrary principles. Amongst the many studies that made use of the geographic convenience criteria are the regional flood frequency analyses of Great Britain and Ireland (NERC 1975) and of Australia (IEA 1987).

10.2.1.2   Subjective Grouping

In this category lie all techniques that group gauging stations according to similarities found in local attributes, such as climate classification, topographic features, and the conformation of isohyetal maps. As these attributes are only approximate in nature, grouping sites using such a criterion certainly entails an amount of subjectivity. As an example, Schaefer (1990) employed comparable depths of annual total rainfall to define homogeneous regions for annual maximum daily rainfall depths in the American State of Washington. Analogously, Pinto and Naghettini (1999) combined contour, Köppen climate classification, and isohyetal maps to identify homogeneous regions for annual maximum daily rainfall depths over the 90,000-km^2 area of the upper São Francisco River catchment, in southeastern Brazil. The results from subjective grouping experiences can be used as preliminary ones and be (or not be) validated later on by an objective criterion such as the heterogeneity measure to be described in Sect. 10.4.2.

10.2.1.3 Objective Grouping

In this case, the regions are formed by assigning sites to one of two groups such that a prescribed site characteristic (or statistic) does or does not exceed a previously specified threshold value. Such a threshold should be specified in such a manner as to minimize some within-group heterogeneity criterion. For instance, Wiltshire (1986) prescribed the within-group variation of the sample coefficient of variation, while Pearson (1991) proposed the within-group variation of the sample L-moment ratios t_2 and t_3, of dispersion and skewness. In the sequence, the groups are further subdivided, in an iterative way, until the desired criterion of homogeneity is met. Hosking and Wallis (1997) note that regrouping sites in such a dichotomous iterative process does not always reach an optimal final solution. They also point out that the within-group heterogeneity statistics can possibly be affected by the eventual cross-correlation among sites.

10.2.1.4 Grouping with Cluster Analysis

Cluster analysis is a method from multivariate statistics designed to find classifications within a large collection of data. After assigning to each gauging station a vector of at-site attributes or characteristics, cluster analysis groups sites based on a statistical distance measuring the dissimilarity among their respective vector of attributes. Cluster analysis has been successfully used to delineate homogeneous regions for regional frequency analysis (e.g., Burn 1989, 1997, Hosking and Wallis 1997, Castellarin et al. 2001, Rao and Srinivas 2006). Hosking and Wallis (1997) consider cluster analysis as the most practical method and recommend its use for grouping sites and delineating preliminary homogeneous regions for regional frequency analysis. As a recommended method, a more detailed description of cluster analysis is provided in Sect. 10.2.2.

10.2.1.5 Other Approaches

In addition to the mentioned techniques, other approaches have been employed for delineating homogeneous regions for regional frequency analysis of hydrologic variables. The following are examples of these approaches, as identified by their core subjects, with related references for further reading: (a) analysis of regression residuals (Tasker 1982); (b) principal components analysis (Nathan and McMahon

1990); (c) factorial analysis (White 1975); (d) canonical correlation (Cavadias 1990, Ribeiro Correa et al. 1995, Ouarda et al. 2001); (e) the ROI (Region Of Influence) approach (Burn 1990); (f) analysis of the shapes of probability density functions (Gingras and Adamowski 1993); and (g) the combined approach introduced by Ilorme and Griffis (2013).

10.2.2   Notions on Cluster Analysis

The term cluster analysis was introduced by Tryon (1939) in the context of behavioral psychology and, nowadays, refers to a large number of different algorithms designed to group similar objects or individuals into homogeneous clusters based on multivariate data. According to Rao and Srinivas (2008), clustering algorithms can be classified into hierarchical and partitional. The former type provides a nested sequence of partitions, which can be done in an agglomerative manner or in a divisive manner. Partitional clustering algorithms, in turn, are developed to recover the natural grouping embedded in the data through a single partition, usually considering the prototype, such as the cluster centroid, as representative of the cluster. The reader is referred to Rao and Srinivas (2008) for an in-depth treatment of clustering algorithms as applied to the regional analysis of hydrologic variables.

Essentially, hierarchical clustering refers to the sequential agglomeration (or division) of individuals or clusters into increasingly larger (or smaller) groups or partitions of individuals according to some criterion, distance, or measure of dissimilarity. An individual or a site may have several influential attributes or characteristics, which can be organized into a vector of attributes $[A_1, A_2, \ldots, A_L]$. The measures or distances of dissimilarity between two individuals should be representative of their reciprocal variation in an L-dimensional space. The most frequently used measure of dissimilarity is the generalized *Euclidean distance*, which is nothing more than the geometric distance in L dimensions. For instance, the Euclidean distance between two individuals i and j, in an L-dimensional space, is given by

$$d_{ij} = \sqrt{\sum_{l=1}^{L} \left(A_{il} - A_{jl}\right)^2} \tag{10.1}$$

To facilitate the understanding of cluster analysis, consider the simplest method of agglomerating individuals into clusters, which is known as the *nearest neighbor* method. For a given set of N attribute vectors, the agglomerative hierarchical clustering begins with the calculation of the Euclidean distances between an individual and all other individuals, one by one. Initially, there are as many clusters as there are individuals, which are referred to as singleton clusters. The first cluster

Table 10.1 Euclidean distances of site-specific attributes at 10 water-quality monitoring stations along the Blackwater River, in England (from data of Kottegoda and Rosso 1997)

Station	1	2	3	4	5	6	7	8	9	10
1	0.00	8.33	6.95	5.95	5.53	4.95	4.70	7.44	7.20	6.93
2	8.33	0.00	1.37	2.38	2.94	3.47	3.93	3.53	3.55	3.56
3	6.95	**1.37**	0.00	1.01	1.64	2.13	2.64	3.16	3.09	2.99
4	5.95	2.38	**1.01**	0.00	0.87	1.20	1.79	3.26	3.12	2.95
5	5.53	2.94	1.64	**0.87**	0.00	**0.58**	1.00	2.74	2.55	2.34
6	4.95	3.47	2.13	1.20	**0.58**	0.00	**0.63**	3.14	2.94	2.69
7	**4.70**	3.93	2.64	1.79	1.00	0.63	0.00	2.98	2.75	2.49
8	7.44	3.53	3.16	3.26	2.74	3.14	2.98	0.00	**0.24**	0.51
9	7.20	3.55	3.09	3.12	2.55	2.94	2.75	**0.24**	0.00	**0.27**
10	6.93	3.56	2.99	2.95	2.34	2.69	2.49	0.51	0.27	0.00

is formed by the pair of the nearest neighbors or, in other terms, the pair of individuals (or singleton clusters) that have the shortest Euclidean distance. This provides $(N-2)$ singleton clusters and 1 cluster with two attribute vectors. In the nearest neighbor approach, the distance between two clusters is taken as the distance between the closest pair of attribute vectors, each of which is contained in one of the two clusters. Then, the two closest clusters are identified and merged, and such a process of clustering continues until the desired number of partitions is reached.

The divisive hierarchical clustering, in turn, begins with a single cluster containing all the N attribute vectors. The vector with the greatest dissimilarity to the other vectors of the cluster is identified and placed into a splinter group. The original cluster is now divided into two clusters and the same divisive process is applied to the largest cluster. The algorithm terminates when the desired number of partitions is reached (Rao and Srinivas 2008). For example, consider the Euclidean distances displayed in Table 10.1, calculated on the basis of two site-specific attributes that are influential on water quality data measured at 10 monitoring stations along the Blackwater River, in England, as originally published in Kottegoda and Rosso (1997).

The hierarchical agglomerative nearest neighbor algorithm begins by grouping stations 8 and 9, which have the shortest Euclidean distance (0.24), into the first cluster. The second cluster is formed by the first cluster (8-9) and station 10, as indicated by the Euclidean distance (0.27) of the closest pair of attribute vectors, one in the first cluster (9) and the other in singleton cluster 10. The next clusters successively group stations 5 and 6, cluster (5-6) and station 7, station 4 and cluster (5-6-7), station 3 and cluster (4-5-6-7), station 2 and cluster (3-4-5-6-7), clusters (2-3-4-5-6-7) and (8-9-10), and, finally, station 1 and cluster (2-3-4-5-6-7-8-9-10).

The hierarchical clustering process, both agglomerative and divisive, can be graphically represented as a tree diagram, referred to as *dendrogram*, illustrating the similar structure of the attribute vectors and how the clusters are formed in the sequential steps of the process. Figure 10.1 depicts the dendrogram of the

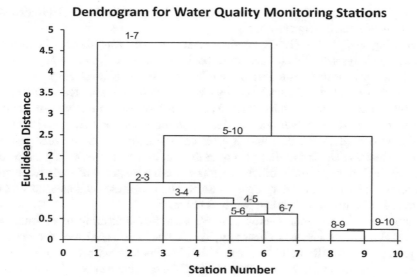

Fig. 10.1 Dendrogram for the example of water-quality monitoring stations along the Blackwater River, in England, from data published in Kottegoda and Rosso (1997)

hierarchical clustering of the 10 water-quality monitoring stations of the Blackwater River, according to the Euclidean distances shown in Table 10.1. The monitoring stations are identified on the horizontal axis and their respective Euclidean distances, as corresponding to the linkages of the clustering process, are plotted on the vertical axis.

In the dendrogram of Fig. 10.1, if only two clusters are to be considered, then the first would be formed by station number 1 and the second by all other nine stations. If three clusters are considered, the previous second cluster would be partitioned into two new clusters: one with stations 8, 9, and 10, and the other with the remaining stations. Now, if six clusters are considered, then stations 1 to 4 would form four different clusters, while the six other stations would form two distinct clusters, one with stations 5,6, and 7, and the other with stations 8,9, and 10, as shown in Fig. 10.1.

As mentioned earlier in this subsection, initially, there are as many clusters as there are individuals and there are no ambiguities in calculating the Euclidean distances d between individuals. However, as one or more clusters are formed, there arises the question as to how the Euclidean distance between clusters should be calculated. In other words, it is necessary to define the linkage criterion that determines the distance between clusters as a function of the pairwise distances between individuals. In the example of the water-quality monitoring stations, where the single linkage criterion (or nearest neighbor) was employed, the distance between two clusters is taken as the distance between the closest pair of individuals, each of which is contained in one of the two clusters. According to Hosking and Wallis (1997), the single-linkage criterion is prone to form a small number of large clusters, with a few small outlying clusters, and is less likely to yield adequate regions for regional frequency analysis.

Hosking and Wallis (1997) highlight that, for achieving success in regional frequency analysis, the linkage criterion should be such that clusters of approximate equal size are formed. Amongst the agglomerative hierarchical clustering algorithms, Ward's method, introduced by Ward (1963), tends to form clusters of roughly equal number of individuals. Ward's method is an iterative process that employs the analysis of variance to determine the distances between clusters. It begins with singleton clusters and at each step it attempts to unite every possible pair of clusters. The union that results in the smallest increase of the sum of the squared deviations of the attribute vectors from the centroid of their respective clusters defines the actual merging of clusters. The mathematical details on Ward's method can be found in Rao and Srinivas (2008).

From the class of non-hierarchical or partitional clustering algorithms, the method, known as k-means clustering, stands out as a useful tool for regional frequency analysis (Rao and Srinivas 2008). The motivation behind this method is that the analyst may have prior indications of the suitable number of clusters to be considered in some geographic region. Then, the k-means method provides the k most distinctive clusters among the possible grouping alternatives. The first step of the method initiates with the formation of k clusters by picking individuals and forming groups. Then the individuals are iteratively moved from one to other clusters (1) to minimize the within-cluster variability, as given by the sum of the squared deviations of the individual distances to the cluster centroid; and (2) to maximize the variability of the k clusters' centroids. This logic is analogous to performing a reverse analysis of variance, in the sense that, by testing the null hypothesis that the group means are different from each other, the ANOVA confronts the within-group variability with the between-group variability. In general, the results from the k-means method should be examined from the perspective of how distinct are the means of the k clusters. Details on the k-means method can be found in Hartigan (1975) and a Fortran computer code for its implementation is described in Hartigan and Wong (1979).

Rao and Srinivas (2008) provide a performance assessment of a number of clustering algorithms and an in-depth account of their main features and applications in regional frequency analysis of hydrologic variables. Computer software for clustering methods can be found at http://bonsai.hgc.jp/~mdehoon/software/cluster/software.htm, as open source, or in R through https://cran.r-project.org/web/views/Cluster.html. Both URLs were accessed on April 8th, 2016.

As applied to the delineation of preliminary homogeneous regions for regional frequency analysis of hydrologic variables, cluster analysis requires some specific considerations. Hosking and Wallis (1997) recommend paying attention to the following points:

(a) Many algorithms for agglomerative hierarchical clustering employ the reciprocal of the Euclidean distance as a measure of similarity. In such a case, it is a good practice to standardize the elements of the attribute vectors, by dividing them by their respective ranges or standard deviations, such that they have comparable magnitudes. However, such a standardization procedure tends to assign equal weights to the different attributes, which has the undesirable effect of hiding the relative greater or lesser influence a given attribute might have on the regional frequency curve.

(b) Some clustering methods, as typified by the k-means algorithm, require the prior specification of the number of clusters, the correct figure of which no one knows a priori. In practical cases, a balance must be pursued between too large and too small regions, respectively, with too many or too few gauging stations. For the index-flood-based methods of regional frequency analysis, there is little advantage in employing too large regions, as little gain in accuracy is obtained with 20 or more gauging stations within a region (Hosking and Wallis 1997). In their words, *there is no compelling reason to amalgamate large regions whose estimated regional frequency distributions are similar.*

(c) The results from cluster analysis should be considered as preliminary. In general, some subjective adjustments arc needed to improve the regions' physical integrity and coherence, as well as to reduce the heterogeneity measure H, to be described in Sect. 10.4.2. Examples of these adjustments are: (1) moving one or more gauging stations from one region to another; (2) deleting a gauging stations or a few gauging stations from the data set; (3) subdividing the region; (4) dismissing the region and moving its gauging stations to other regions; (5) merging regions and redefining clusters; and (6) obtaining more data and redefining regions.

10.3   Example Applications of the Methods of Regional Frequency Analysis

10.3.1   RBQ: A Regression Method for Estimating the T-year Quantile

The Regression-Based Quantile (RBQ) method does not fundamentally require the partition of the geographic area into homogenous regions. Its first step consists of performing an at-site frequency analysis of the records available at each gauging station, according to the procedures outlined in Chap. 8, leading to site-specific quantile estimates for selected values of the return period. It is worthwhile mentioning that, although the method does not require a common probabilistic model be chosen for all sites, consistency and coherence should be exercised in selecting the site-specific distributions. The second step is to fix the target return period T, for which regional quantile estimates are needed, and organize a vector containing the estimated quantiles $\hat{X}_{T,j}, j = 1, 2, \ldots, N$ for the N existing gauging stations across the geographic region of interest, and a $N \times K$ matrix $\mathbf{M}$ containing the K physiographic catchment characteristics and other inputs for each of the N sites. The third step is to use the methods described in Chap. 9 to fit and evaluate a multiple regression model between the estimated quantiles and the catchment physiographic characteristics and other inputs. Example 10.1 illustrates an application of the RBQ method.

Example 10.1 The Paraopeba river basin, located in the State of Minas Gerais (MG), in southeastern Brazil, is an important source of water supply for the state's capital city of Belo Horizonte and its metropolitan area (BHMA), with approximately 5.76 million inhabitants. The total drainage area of the catchment is 13643 km^2, which is located between parallels 18°45′ and 21° South, and meridians 43°30′ and 45°00′ West. Elevation varies in the range of 620 and 1600 m above sea level. The mean annual rainfall depth varies between 1700 mm, at high elevations, and 1250 mm at the catchment outlet. For this part of Brazil, there is a clearly defined wet season from October to March, followed by a dry season from April to September. On average, 55–60 % of annual rainfall is concentrated in the months of November to January. On the other hand, the months of June, July and August account for less than 5 % of the annual rainfall. Figure 10.2 depicts the location, isohyetal, and hypsometric maps for the Paraopeba river basin. Appendix 6 contains a table with the annual minimum 7-day mean flows, in m^3/s, denoted herein as Q_7, at 11 gauging stations in the Paraopeba river basin. The code, name, river, and associated catchment attributes for each gauging station are listed in Table 10.2. In Table 10.2, Area = drainage area (km^2); P_{mean} = mean annual rainfall over the

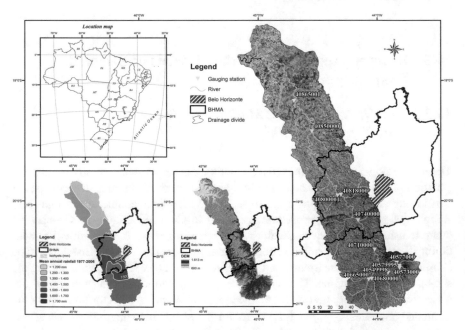

Fig. 10.2 Location, isohyetal, and hypsometric maps of the Paraopeba river basin, in Brazil. Locations of gauging stations are indicated as triangles next to their codes

Table 10.2 Catchment attributes for the gauging stations of Example 10.1

Code	Name of station	River	A (km^2)	P_{mean} (m)	S_{equiv} (m/km)	L (km)	J (km^2)
40549998	São Brás do Suaçuí Montante	Paraopeba	461.4	1.400	2.69	52	0.098
40573000	Joaquim Murtinho	Bananeiras	291.1	1.462	3.94	32.7	0.079
40577000	Ponte Jubileu	Soledade	244	1.466	7.20	18.3	0.119
40579995	Congonhas Linigrafo	Maranhão	578.5	1.464	3.18	41.6	0.102
40680000	Entre Rios de Minas	Brumado	486	1.369	1.25	47.3	0.136
40710000	Belo Vale	Paraopeba	2760	1.408	1.59	118.9	0.137
40740000	Alberto Flores	Paraopeba	3939	1.422	1.21	187.4	0.134
40800001	Ponte Nova do Paraopeba	Paraopeba	5680	1.449	1.00	236.33	0.141
40818000	Juatuba	Serra Azul	273	1.531	4.52	40	0.066
40850000	Ponte da Taquara	Paraopeba	8734	1.434	0.66	346.3	0.143
40865001	Porto do Mesquita (CEMIG)	Paraopeba	10192	1.414	0.60	419.83	0.133

catchment (m); S_{equiv} = equivalent stream slope (m/km); L = length of main stream channel (km); and J = number of stream junctions per km^2. The gauging stations are also located on the map of Fig. 10.2. On the basis of these data, perform a regional frequency analysis of the Q_7 flows for the return period $T = 10$ years, using the RBQ method.

Solution The candidate distributions used to model the Q_7 low flows, listed in the table of Appendix 6, were the two-parameter Gumbel$_{min}$ and the three-parameter Weibull$_{min}$, which were fitted to the samples using the MOM method. As previously mentioned, the RBQ method does not require that a common probabilistic model be chosen for all sites within a region. However, by examining the 11 plots (not shown here) of the empirical, Gumbel$_{min}$, and 3-p Weibull$_{min}$ distributions, on an exponential probability paper with Gringorten plotting positions, the best-fit model over all gauging stations was the 3-p Weibull$_{min}$. As an opportunity to describe a procedure to fit the 3-p Weibull$_{min}$ to a sample, by the MOM estimation method, other than that outlined in Sect. 5.7.2, remember that the Weibull$_{min}$ CDF can be written as

$$F_Y(y) = 1 - \exp\left[-\left(\frac{y - \xi}{\beta - \xi}\right)^{\alpha}\right], \text{ for } y > \xi; \beta \geq 0 \text{ and } \alpha > 0 \qquad (10.2)$$

The quantile, as a function of the return period T, is given by

$$x(T) = \xi + \left\{(\beta - \xi) \times \left[-\ln\left(1 - \frac{1}{T}\right)\right]^{\frac{1}{\alpha}}\right\} \qquad (10.3)$$

According to Kite (1988), parameter α can be estimated as

$$\hat{\alpha} = \frac{1}{C_0 + C_1\gamma + C_2\gamma^2 + C_3\gamma^3 + C_4\gamma^4} \tag{10.4}$$

where γ, the population coefficient of skewness, should be estimated by the sample coefficient of skewness g, as calculated with Eq. (2.13), for $-1.02 \leq g \leq 2$, and the polynomial coefficients are $C_0 = 0.2777757913$, $C_1 = 0.3132617714$, $C_2 = 0.0575670910$, $C_3 = -0.0013038566$, and $C_4 = -0.0081523408$. In the sequence, parameter β is estimated as $\hat{\beta} = \bar{x} + s_X A(\hat{\alpha})$, where the functions $A(\hat{\alpha})$ and $B(\hat{\alpha})$ are given by

$$A(\hat{\alpha}) = \left[1 - \Gamma\left(1 + \frac{1}{\hat{\alpha}}\right)\right] \times B(\hat{\alpha}) \tag{10.5}$$

and

$$B(\hat{\alpha}) = \left[\Gamma\left(1 + \frac{2}{\hat{\alpha}}\right) - \Gamma^2\left(1 + \frac{1}{\hat{\alpha}}\right)\right]^{-1/2} \tag{10.6}$$

Finally, the location parameter ξ is estimated as $\hat{\varepsilon} = \hat{\beta} - s_X B(\hat{\alpha})$. The MOM estimates of the three-parameter Weibull$_{min}$ distribution, alongside the estimated annual minimum 7-day flow of return period $T = 10$ years, denoted as $Q_{7,10}$, for each of the 11 gauging stations, are displayed in Table 10.3.

The next step in the RBQ method is to define, for a fixed return period ($T = 10$ years, in this case), the regression equation between the quantiles $Q_{7,10}$, as in Table 10.3, and the catchment attributes, listed in Table 10.2. The matrix of correlation coefficients among the response and explanatory variables are given in Table 10.4. From the values listed in Table 10.4, it is clear that the variables A and L are highly correlated and should not be considered in possible regression models, in order to reduce the risk of eventual collinearity.

The regression models used to solve this example are of the log–log type. Thus, following the logarithmic transformation of the concerned variables, various regression models were tested, by combining the predictor variables shown in Table 10.2. The significances of predictor variables and of regression equations were evaluated through F-tests. Remember that the partial F-test checks the significance of predictor variables that are added or deleted from the regression equation, whereas the overall F-test checks the significance of the entire regression model. The overall quality of each regression model was further assessed by the sequential examination of the residuals, the standard error of estimates and the coefficient of determination, followed by the consistency analyses of signs and magnitudes of the regression coefficients and the relative importance of the predictors, as assessed by the standardized partial regression coefficients. After all these analyses, the final regression model is given by $Q_{7,10,j} = 0.0047A_j^{0.9629}$, which is valid in the interval $244 \text{ km} \leq A_j \leq 10192 \text{ km}$. The coefficient of

Table 10.3 Estimates of parameters and $Q_{7,10}$ from the 3-p Weibull$_{min}$ distribution

Station	40549998	40573000	40577000	40579995	40680000	40710000	40740000	40800001	40818000	40850000	40865001
$\hat{\alpha}$	3.2926	3.1938	3.6569	3.3357	3.9337	3.4417	4.1393	3.3822	3.5736	3.5654	2.6903
$\hat{\beta}$	2.7396	1.6172	1.5392	4.0013	2.3535	17.4100	21.6187	31.3039	1.4530	42.5004	44.3578
$\hat{\xi}$	0.8542	0.6035	0.3837	0.9098	0.3668	5.5824	7.1853	4.0250	0.2456	8.4619	21.5915
$Q_{7,10}$ (m^3/s)	1.81	1.11	1.01	2.48	1.49	11.7	15.6	18.0	0.889	26.6	31.5

Table 10.4 Matrix of correlation coefficients for Example 10.1

	A (km^2)	P_{mean} (m)	S_{equiv} (m/km)	L (km)	J (junctions/ km^2)	$Q_{7,10}$ (m^3/s)
A (km^2)	1					
P_{mean} (m)	−0.22716	1				
S_{equiv} (m/km)	−0.675	0.600687	1			
L (km)	0.997753	−0.24112	−0.69617	1		
J (junctions/km^2)	0.624234	−0.65707	−0.61808	0.609301	1	
$Q_{7,10}$ (m^3/s)	0.993399	−0.25256	−0.69904	0.992116	0.65669	1

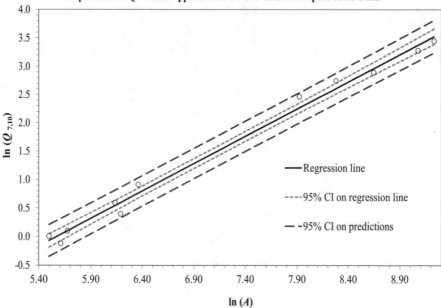

Fig. 10.3 Log–log regression model of $Q_{7,10}$ and drainage area (A) for Example 10.1

determination is 0.9917 and the estimated standard deviation of the transformed variables is 0.138. Figure 10.3 depicts a chart, in logarithmic space, with the scatterplot, the fitted regression equation, and the 95 % confidence bands on the regression line and on the predicted values.

10.3.2 RBP: A Regression Method for Estimating the Distribution Parameters

In order for the Regression-Based Parameter (RBP) method to be applied, a single parametric form should be selected as to represent the probability distributions of the variable of interest at a number of gauging stations inside a homogeneous region. In addition to the methods for delineating homogeneous regions, outlined in Sect. 10.2, an expedite way of grouping gauging stations with a common parametric form consists of plotting the empirical distributions of the data recorded at all stations, as standardized by their respective mean, onto a single probability paper, looking for similarities among them. Then, gauging stations can be moved from one region to another, regions can be merged or dismissed, and new plots are prepared to help make the decision about grouping of sites into homogeneous regions.

Once this step is completed, a number of probability models are fitted to the data of each station and the methods of selecting a distribution function, outlined in Chap. 8, are employed to choose the models that fit the data. The model which seemingly fits data at all sites or in most of them, should be adopted for the entire region. At this point, for a site j within the region, one would have $\hat{\Theta}_j = (\hat{\theta}_1, \hat{\theta}_2, \ldots, \hat{\theta}_P)$ parameters, estimated with a common estimation method, for a P-parameter distribution. Now, for each set of parameter estimates $\hat{\theta}_p, p = 1, 2, \ldots, P$, the next step is to organize a vector containing the estimated parameters $\hat{\theta}_{p,j}, j = 1, 2, \ldots, N$ for the N existing gauging stations across the homogeneous region, and a $N \times K$ matrix $\mathbf{M}$ containing the K catchment attributes for the N sites. The final step is to use the methods described in Chap. 9 in order to fit and evaluate P multiple regression models between the estimated parameters and the catchment attributes. Example 10.2 illustrates an application of the RBP method.

Example 10.2 Appendix 7 lists the annual maximum flows of 7 gauging stations located in the upper Paraopeba River basin, in southeastern Brazil. These stations are located on the map of Fig. 10.2 and their respective catchment attributes are given in Table 10.5. On the basis of these data, perform a preliminary regional frequency analysis of the annual maximum flows, using the RBP method. In Table 10.5, Area = drainage area (km²); P_{mean} = mean annual rainfall over the

Table 10.5 Catchment attributes for the gauging stations of Example 10.2

Code	Name of station	River	A (km²)	P_{mean} (m)	S_{equiv} (m/km)	L (km)	J (km²)
40549998	São Brás do Suaçui Montante	Paraopeba	461.4	1.400	2.69	52	0.098
40573000	Joaquim Murtinho	Bananeiras	291.1	1.462	3.94	32.7	0.079
40577000	Ponte Jubileu	Soledade	244	1.466	7.2	18.3	0.119
40579995	Congonhas Linigrafo	Maranhão	578.5	1.464	3.18	41.6	0.102
40665000	Usina João Ribeiro	Camapuã	293.3	1.373	2.44	45.7	0.123
40710000	Belo Vale	Paraopeba	2760	1.408	1.59	118.9	0.137
40740000	Alberto Flores	Paraopeba	3939	1.422	1.21	187.4	0.134

Table 10.6 Estimates of site statistics for the data samples of Example 10.2

Stations	40549998	40573000	40577000	40579995	40665000	40710000	40740000
Mean (m³/s)	60.9	31.5	29.7	78.2	30.0	351.6	437.1
SD (m³/s)	24.0	10.6	9.2	35.7	10.3	149.0	202.8
CV	0.39	0.34	0.31	0.46	0.34	0.42	0.46

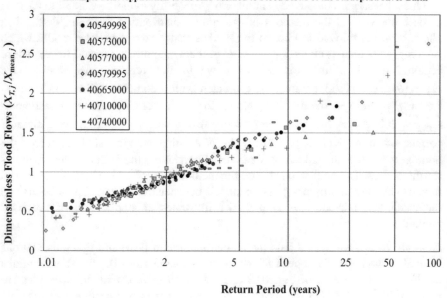

Gumbel Probability Plot of Dimensionless Annual Maximum Flows
Example 10.2: Application of the RBP Method to Flood Flows of the Paraopeba River Basin

Fig. 10.4 Dimensionless empirical distributions for Example 10.2

catchment (m); S_{equiv} = equivalent stream slope (m/km); L = length of main stream channel (km); and J = number of stream junctions per km².

Solution The first step in the RBP method consists of checking whether or not the group of gauging stations forms a homogenous region, inside which a common parametric form can be adopted as a model for the annual maximum flows, with site-specific parameters. Table 10.6 displays some site statistics for the data recorded at the gauging stations, where the sample coefficients of variation show a slight variation across the region, which can be interpreted as a first indication of homogeneity. The coefficients of skewness were not used to compare the site statistics, as some samples are too short to yield meaningful estimates.

In cases of grouping sites with short samples, such as this, a useful tool to analyze homogeneity is to plot the empirical distributions of scaled data onto a single probability paper. These are shown in the chart of Fig. 10.4, on Gumbel probability paper, where the Gringorten plotting positions were used to graph the

Table 10.7 MOM estimates of the parameters of the Gumbel$_{max}$ distribution

Code	Gauging station name	River	$\hat{\alpha}$	$\hat{\beta}$
40549998	São Brás do Suaçui Montante	Paraopeba	18.69	50.07
40573000	Joaquim Murtinho	Bananeiras	8.24	26.71
40577000	Ponte Jubileu	Soledade	7.21	25.53
40579995	Congonhas Linigrafo	Maranhão	27.83	62.13
40665000	Usina João Ribeiro	Camapuã	8.05	25.31
40710000	Belo Vale	Paraopeba	116.15	284.61
40740000	Alberto Flores	Paraopeba	158.13	345.81

Table 10.8 Matrix of correlation coefficients for Example 10.2

	A (km²)	P_{mean} (m)	S_{equiv} (m/km)	L (km)	J (junctions/ km²)	α	β
A (km²)	1.000						
P_{mean} (m)	−0.202	1.000					
S_{equiv} (m/km)	−0.635	0.627	1.000				
L (km)	0.984	−0.306	−0.715	1.000			
J (junctions/km²)	0.688	−0.437	−0.323	0.651	1.000		
α	0.999	−0.193	−0.643	0.978	0.687	1.000	
β	0.995	−0.209	−0.641	0.967	0.699	0.997	1.000

data recorded at the seven gauging stations, as scaled by their respective average flood flows. The dimensionless flood flows of the upper Paraopeba River basin seem to align in a common tendency, thus indicating the plausibility of a single homogeneous region.

Assuming a single homogenous region for the seven gauging stations, the next step is to select a common parametric form as a model for the regional annual maximum flows. However, given the limited number of samples available for this case study, an option was made to restrict model selection to distributions with no more than two parameters, namely, the exponential, Gumbel$_{max}$, two-parameter lognormal, and Gamma. After performing GoF tests and plotting the empirical distributions (not shown here), under different distributional hypotheses, the model that seems to best fit the regional data is the Gumbel$_{max}$. Table 10.7 lists the MOM estimates of the location ($\hat{\alpha}$) and scale ($\hat{\beta}$) parameters of the Gumbel$_{max}$ distribution, for each gauging station.

The next step in the RBP method consists of modeling the spatial variability of parameters α and β through a regression equation with the catchment attributes as explanatory variables. Table 10.8 displays the matrix of the simple correlation coefficients between the parameters and the predictor variables. Since variables A and L are highly correlated, they were not considered into the regression models, so as to reduce the risk of collinearity.

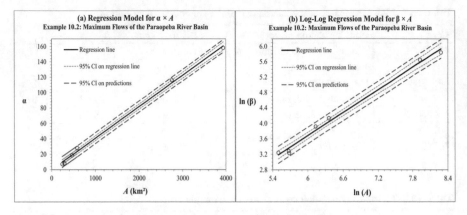

Fig. 10.5 Regression lines and confidence intervals for Example 10.2

The regression models used to solve this example were of the linear and log–log types. Thus, following the logarithmic transformation of the concerned variables, when applicable, various regression models were tested, by combining the predictor variables shown in Table 10.5. The significances of predictor variables and of the total regression equation were evaluated through F-tests. The overall quality of each regression model was further assessed by the sequential examination of the residuals, the standard error of estimates and the coefficient of determination, followed by the consistency analyses of signs and magnitudes of the regression coefficients and the relative importance of the predictors, as assessed by the standardized partial regression coefficients. After these analyses, the following regression equations were estimated: $\hat{\alpha}_j = 0.0408A_j$ and $\hat{\beta}_j = 0.1050A_j^{0.9896}$, for A_j in km^2. Panels a and b of Fig. 10.5 depict charts with the scatterplots, the fitted regression equations, and the 95 % confidence bands on the regression lines and on the predicted values, for parameters α and β, respectively. Substituting the regression equations for α and β into the quantile function of the Gumbel$_{\text{max}}$ distribution, it follows that

$$\hat{X}_{T,j} = \hat{\beta}_j - \hat{\alpha}_j \left\{ \ln\left[-\ln\left(1-\frac{1}{T}\right)\right] \right\} = 0.1050A_j^{0.9896}$$

$$- 0.0408A_j \left\{ \ln\left[-\ln\left(1-\frac{1}{T}\right)\right] \right\}$$

$$(10.7)$$

which is valid for 244 km$^2 \le A_j \le 3940$ km^2 and can be used estimate the annual maximum flows, associated with the T-year return period, for an ungauged site j within the region.

10.3.3   IFB: An Index-Flood-Based Method for Regional Frequency Analysis

The term index-flood was introduced by Dalrymple (1960) in the context of regional frequency analysis of flood flows. It refers to a scaling factor to make dimensionless the data gathered at distinct locations within a homogeneous region, such that they can be jointly analyzed as a sample of regional data. Despite the reference to floods, the term index-flood is in widespread use in regional frequency analysis of any type of data.

Consider the case in which one seeks to regionalize the frequencies of a generic random quantity X, whose variability in time and space has been sampled at N gauging stations across a geographical area. The observations indexed in time by i, recorded at the gauging station j, form a sample of size n_j and are denoted as $X_{i,j}$, $i = 1, \ldots, n_j; j = 1, \ldots, N$. If F, for $0 < F < 1$, denotes the frequency distributions of X at gauging station j, then, the quantile function at this site is written as $X_j(F)$. The basic assumption of the IFB approach is that the group of gauging stations determines a homogeneous region, inside which the frequency distributions at all N sites are identical apart from a site-specific scaling factor, termed index-flood. Formally,

$$X_j(F) = \mu_j \; x(F), j = 1, \; \ldots \; , N \tag{10.8}$$

where μ_j is the index-flood or the scaling factor for site j and $x(F)$ is the dimensionless quantile function or the regional growth curve (Hosking and Wallis 1997), which is common to all sites within the region.

The index-flood μ_j for site j is usually estimated by the corresponding sample mean $\overline{X}_j$, but other central tendency measures of the sample data $\{X_{1,j}, X_{2,j}, \; \ldots \; , X_{n_j,j}\}$, such as the median, can also be used for such a purpose. The dimensionless or scaled data $x_{i,j} = X_{i,j}/\overline{X}_j$, $i = 1, \; \ldots \; , n_j; j = 1, \; \ldots \; , N$ form the empirical basis for estimating the regional growth curve $x(F)$. Analogously to the at-site frequency analysis (see Sect. 8.1), estimation of $x(F)$ can be analytical or graphical. In the former case, the regional growth curve results from fitting a parametric form to empirical scaled data. In the latter case, the empirical regional growth curve can be estimated by finding the median curve, among the empirical distributions of scaled data, as plotted onto a single probability paper.

The inherent assumptions of the IFB approach are

(a) The data observed at any site within the region are identically distributed;
(b) The data observed at any site within the region are not serially correlated;
(c) The data observed at distinct sites within the region are statistically independent;
(d) The frequency distributions at all N sites are identical apart from the scaling factor; and
(e) The analytical form of the regional growth curve has been correctly specified.

Hosking and Wallis (1997) argue that assumptions (a) and (b) are plausible for many kinds of hydrological variables, especially those related to annual maxima. However, they argue that it appears implausible that assumptions (c), (d) and (e) hold for hydrological, meteorological, and environmental data. For example, it is known that frontal systems and severe droughts are events that can extend over large geographic areas and, as such, related data collected at distinct gauging stations located in these areas are likely to show high cross-correlation coefficients. Finally, Hosking and Wallis (1997) point out that the last two assumptions will never be exactly valid in practice and, at best, they may be only approximated in some favorable cases. Despite these shortcomings, the careful grouping of gauging stations into a homogeneous region and the wise choice of a robust probability model for the regional growth curve are factors that can compensate the departures from the assumptions inherent to IFB approaches found in practical cases.

The IFB approach for regional frequency analysis can be summarized into the following sequential steps:

(a) Screening the data for inconsistencies.

   As with at-site frequency analysis, screening the data for gross errors and inconsistencies, followed by performing statistical tests for randomness, independence, homogeneity, and stationarity of data are requirements for achieving a successful regional frequency analysis. The overall directions given in Sect. 8.3.1, in the context of at-site frequency analysis, remain valid for regional frequency analysis. Then, the data samples of the variable to be regionalized are organized and no more than a little missing data can eventually be filled in using appropriate methods, such as regression equations between data of neighboring gauging stations.

(b) Scaling the data.

   This step consists of scaling the elements $X_{i,j}, i = 1, \ldots, n_j; j = 1, \ldots, N$ by their respective sample index-flood factors $\overline{X}_j$, thus forming the dimensionless elements $x_{i,j} = X_{i,j}/\overline{X}_j, i = 1, \ldots, n_j; j = 1, \ldots, N$. In early applications of the IFB approach, such as in Dalrymple (1960), the samples were required to have a common period of records. However, Hosking and Wallis (1997) argue that if the samples are homogeneous and representative (see Sect. 7.4), such a requirement of a common period of record is no longer necessary, provided that inference procedures take into account the different sample sizes recorded at the distinct gauging stations.

(c) Defining the at-site empirical distributions.

   The site-specific empirical distribution is defined using the same methods of graphical frequency analysis outlined in Sect. 8.2. However, for regional frequency analysis, empirical distributions are based on scaled data. In the original work of Dalrymple (1960) and in the British *Flood Studies Report* (NERC 1975) empirical distributions were defined on Gumbel probability paper.

(d) Delineating the homogenous regions.

The concept of a homogeneous region was given in Sect. 10.2, where a distinction was made between site statistics and site characteristics, with reference to the site-specific information that should guide the appropriate grouping of gauging stations. When site statistics are employed, a useful graphical procedure consists of searching for similar tendencies eventually shown by the site-specific empirical distributions of scaled data on a single probability paper. A group of gauging stations with similar empirical distributions curves is likely to form a homogeneous region. When site characteristics are employed, the methods of cluster analysis described in Sect. 10.2.2 should be used to delineate homogeneous regions.

(e) Estimating the regional growth curve.

When an empirical regional curve is sought, the median of the at-site empirical distributions may be a reasonable estimate. When an analyticaly defined regional growth curve is sought, the dimensionless data $x_{i,j} = X_{i,j}/\overline{X}_j$, $i = 1,$ $\dots, n_j; j = 1, \dots, N$ should be used to estimate the function $x(F)$. Assuming the analytical form of $x(F)$ is known, then, the parameters $\theta_1, \theta_2, \dots, \theta_P$, that define F, are dependent upon the population measures of location, dispersion, skewness, and kurtosis, and should be estimated from the scaled data. One possible manner to perform such an estimation is to fit F to each of the N samples, thus obtaining a vector of parameter estimates, denoted as $\hat{\theta}_{p,j}$, $p = 1, \dots, P$, for each site j, with sample size n_j, using one of the estimation method chosen from MOM, LMOM, or MLE. The regional estimate of the pth parameter, denoted as $\hat{\theta}_p^R$ can be found by weighting the site-specific parameter estimates by their respective sample sizes. Formally,

$$\hat{\theta}_p^R = \frac{\sum_{j=1}^{N} n_j \hat{\theta}_p^{(j)}}{\sum_{j=1}^{N} n_j} \tag{10.9}$$

The P regional parameter estimates for the homogeneous region allow the estimation of the regional growth $\hat{x}(F) = x\left(F; \hat{\theta}_1^R, \dots, \hat{\theta}_P^R\right)$. The choice of the probability model F should be oriented by the same general guidelines given in Sect. 8.3.2.

(f) Regression analysis.

Analogously to the RBQ and RBP methods of regional frequency analysis, in the IFB approach, regression methods aim to explain the spatial variation of the index-flood factors $\hat{\mu}_j = \overline{X}_j, j = 1, \dots, N$ from the K potential explanatory variables, as given by the catchment attributes $M_{j,k}, j = 1, \dots, N; k = 1, \dots, K$. The methods for selecting explanatory variables and testing regression models, covered in Sect. 9.4, are then used to fit a parsimonious regression equation, which is usually of the linear-log or log–log types.

(g) Estimating quantiles.

The quantiles of return period T at a site j is estimated as $\hat{X}_{T,i} = \hat{\mu}_j \hat{x}_T$, where $\hat{\mu}_j$ is the index-flood or the scaling factor for site j and $\hat{x}_T$ is the dimensionless quantile function from the regional growth curve. The index-flood $\hat{\mu}_j$, at the site j, inside the homogeneous region, is given by $\overline{X}_j$, in the case of a gauging station, or estimated from the regression model of $\hat{\mu}_j$ against the catchment attributes, in the case of an ungauged site.

Example 10.3 illustrates an application of an IFB method.

Example 10.3 Solve Example 10.2, using the IFB method.

Solution After screening data for inconsistencies and testing them for randomness, independence, homogeneity, and stationarity, they were scaled by their respective sample averages. Figure 10.4 depicts the empirical distributions for the seven gauging stations, plotted on Gumbel probability paper, with plotting positions calculated through the Gringorten formula. The same arguments given in the solution of Example 10.4 apply, first, for grouping the stations into a single homogeneous region, and, second, for choosing the Gumbel$_{max}$ distribution as the regional model for the annual maximum flows of the upper Paraopeba River catchment. As for estimating the regional growth curve, graphical and analytical methods can be used, either by drawing the median line through the empirical distributions or by weighting the at-site distribution parameters by their respective sample sizes, as in Eq. (10.9), to estimate the parameters of the regional growth curve. The latter approach was selected for the solution of this example. The MOM estimates for the Gumbel$_{max}$ distributions, valid for the scaled data of each gauging station, are given in Table 10.9. The regional parameter estimates are listed in the last row of Table 10.9.

The inverse function of the Gumbel$_{max}$ distribution gives the regional growth curve as

$$x(T) = \frac{X_{T,j}}{\overline{X}_j} = 0.819 - 0.314\left\{\ln\left[-\ln\left(1 - \frac{1}{T}\right)\right]\right\} \qquad (10.10)$$

Table 10.9 MOM estimates of Gumbel$_{max}$ distribution for each gauging station

Station code	Mean of scaled data	Standard deviation	Sample size	$\hat{\alpha}$	$\hat{\beta}$
40549998	1	0.498	32	0.307	0.823
40573000	1	0.336	15	0.262	0.849
40577000	1	0.311	20	0.243	0.860
40579995	1	0.456	47	0.356	0.795
40665000	1	0.345	30	0.269	0.845
40710000	1	0.424	25	0.330	0.809
40740000	1	0.464	28	0.362	0.791
Regional parameter estimates				0.314	0.819

Table 10.10   Regional dimensionless quantile estimates for Example 10.3

T (years)	1.01	2	5	10	20	25	50	75	100
Regional quantile estimate	0.339	0.934	1.289	1.525	1.751	1.822	2.043	2.171	2.262

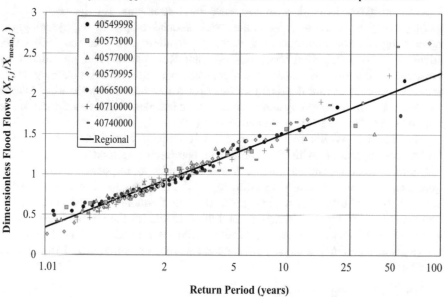

Fig. 10.6   Dimensionless quantiles and regional growth curve for Example 10.3

where $X_{T,j}$ denotes the X quantile of return period T at a site j inside the homogeneous region and $\overline{X}_j$ represents the corresponding index-flood. Table 10.10 presents the regional dimensionless quantiles as calculated with Eq. (10.10). Figure 10.6 depicts a plot of the regional growth curve superimposed over the site-specific empirical distributions.

A regression study aimed to explain the spatial variation of the index-flood factors $\overline{X}_j, j = 1, \ldots, 7$, as given in Table 10.6, from the 5 potential explanatory variables, as given by the catchment attributes $M_{j,k}, j = 1, \ldots, 7; k = 1, \ldots, 5$ of Table 10.5, was performed in an analogous way to Examples 10.1 and 10.2. At the end of the analyses, the final regression equation is written as $\overline{X}_j = 0.1167 A_j$, for $244 \text{ km}^2 \leq A_j \leq 3940 \text{ km}^2$. Thus, by combining Eq. (10.10) with the regression equation for the index-flood, the T-year return period flood discharge for an ungauged site indexed by j, inside the homogeneous region, can be estimated as $X_{T,j} = 0.1167 A_j \{0.819 - 0.314 [\ln(-\ln(1 - \frac{1}{T}))]\}$.

10.4   A Unified IFB Method for Regional Frequency Analysis with L-Moments

The entailed subjectivities at the various stages of regional frequency analysis together with the emergence of more robust inference methods, such as the Probability Weighted Moments (see Sect. 6.5), introduced by Greenwood et al. (1979), have led researcher J. R. M. Hosking, from the IBM Thomas J. Watson Research Center, and Professor J. R. Wallis, from Yale University, to propose a unified index-flood-based method, with inference by L-moments, for the regional frequency analysis of random variables, with special emphasis on hydrological, meteorological, and environmental variables. In their review of the 1991–1994 advances in flood frequency analysis, Bobée and Rasmussen (1995) considered the Hosking–Wallis method as the most relevant contribution for obtaining more reliable estimates of rare floods. In this section, a summary of the Hosking–Wallis method is given, followed by some worked out examples. A full description of the method can be found in Hosking and Wallis (1997).

The Hosking–Wallis method for regional frequency analysis combines the index-flood approach with the L-moment estimation method (see Sect. 6.5). Estimation with L-moments is extended not only to regional parameter and quantiles but is also intended to build regional statistics capable of reducing the subjectivities entailed at critical stages of regional frequency analysis, such as the delineation of homogeneous regions and the selection of the regional probability distribution. As such, the Hosking–Wallis method is viewed as a unified approach for regional frequency analysis. Its sequential steps are summarized in the next paragraphs.

- Step 1: Screening of the Data

    As a necessary step that should precede any statistical analysis, data collected at different gauging stations across a geographic area need to be examined in order to detect and correct gross and systematic errors that may eventually be found in data samples, and further tested for randomness, independence, homogeneity, and stationarity, according to the guidelines given in Sect. 8.3.1. In addition to these guidelines for consistency analysis, Hosking and Wallis (1997) suggest the use of an auxiliary L-moment-based statistic, termed *discordancy measure*, to be described in Sect. 10.4.1, which is based on comparing the statistical descriptors of data from a group of gauging stations with the data descriptors of each specific site.

- Step 2: Identification of Homogeneous Regions

    As with any index-flood-based approach, the Hosking–Wallis method assumes that gauging stations should be grouped into homogeneous regions inside which the probability distributions of the variable being regionalized are identical apart from a site-specific scaling factor. As mentioned earlier in this chapter, in order to group gauging stations into homogenous regions, Hosking and Wallis (1997) suggest a two-step approach. First, they recommend using cluster analysis, based on site characteristics only and preferably on Ward's linkage algorithm, to group gauging stations into preliminary regions (see Sect. 10.2.2). Then, Hosking and Wallis (1997) suggest the use of an auxiliary L-moment-based statistic, termed *heterogeneity measure*, to test whether or not the preliminary regions, defined by cluster analysis, are homogeneous. The heterogeneity measure is based on the difference between the within-group variability of site statistics and the variability expected from similar data as simulated from a hypothetical homogeneous group. The heterogeneity measure will be formally described in Sect. 10.4.2.

- Step 3: Choosing the Appropriate Regional Frequency Distribution

  After having screened regional data for discordancy and grouped gauging stations into a homogenous region, the next step is to choose an appropriate probability distribution to model the frequencies of the variable being regionalized. In order to do so, Hosking and Wallis (1997) suggest the use of an auxiliary L-moment-based statistic, termed *goodness-of-fit measure*, which is formally described in Sect. 10.4.3. The goodness-of-fit measure is built upon the comparison of observed regional descriptors with those that would have been yielded by random samples simulated from a hypothetical regional parent distribution.

- Step 4: Estimating the Regional Frequency Distribution

  Having identified the regional model $F_x(x|\theta_1, \theta_2, \ldots, \theta_P)$ that best fits data, the next step is to calculate the estimates of its regional parameters $\left(\hat{\theta}_1^R, \ldots, \hat{\theta}_P^R \right)$ and of its dimensionless quantiles $\hat{x}(F) = x\left(F; \hat{\theta}_1^R, \ldots, \hat{\theta}_P^R \right)$. The parameter estimates $\hat{\theta}_p^{(j)}$, $p = 1, \ldots, P$ for site j are weighted by its respective sample size, as in Eq. (10.9), in order to estimate the regional parameters. With these, calculations of the regional growth curve $\hat{x}(F) = x\left(F; \hat{\theta}_1^R, \ldots, \hat{\theta}_P^R \right)$ and the site-specific frequency distributions, as given by $\overset{\approx}{X}_j(F) = \mu_j \hat{x}(F)$, are easy and direct.

  Hosking (1996) presents a number of computer routines, coded in Fortran 77, designed to implement the four steps of the Hosking–Wallis unified method for regional frequency analysis. The source codes for these routines are publically available from the Statlib repository for statistical computing at http://lib.stat.cmu.edu/general/lmoments [accessed Dec., 18th 2014] or at ftp://rcom.univie.ac.at/mirrors/lib.stat.cmu.edu/general/lmoments [accessed April, 15th 2016]. Versions of these routines in R are available. One is the lmom package at https://cran.r-project.org/web/packages/lmom/index.html and another is the nsRFA package at https://cran.r-project.org/web/packages/nsRFA/index.html.

10.4.1   Screening of the Data

In addition to the traditional techniques of consistency analysis of hydrologic data, Hosking and Wallis (1997) suggest the comparison of the sample L-moment ratios of different gauging stations as a criterion to identify *discordant* data. They posit that the L-moment ratios are capable of conveying errors, outliers and heterogeneities that may eventually be present in a data sample, and propose a summary statistic, the discordancy measure, for the purpose of comparing L-moment ratios for a given site with the average L-moment ratios for a group of sites.

For a group of gauging stations, the discordancy measure aims to identify the samples that show statistical descriptors too discrepant from the group average descriptors. The discordancy measure is expressed as a summary statistic of three L-moment ratios, namely, the L-CV (or τ), the L-Skewness (or τ_3), and the L-Kurtosis (or τ_4). In the three-dimensional space of variation of these L-moment ratios, the idea is to label as discordant the samples whose estimates $\{\hat{\tau} = t, \hat{\tau}_3 = t_3, \hat{\tau}_4 = t_4\}$, represented as points in space, depart too much from the center where most points are concentrated in. In order to visualize what the discordancy measure means, consider the plane defined by the domains of variation of L-CV and L-Skewness for the data from a group of gauging stations, as depicted in the schematic chart of Fig. 10.7. The point marked with a cross indicates the location of the group average L-ratios, around which concentrical ellipses are drawn. The semi-major and semi-minor axes of the ellipses are chosen to give the best fit to the data, as determined by the sample covariance matrix of the L-moment ratios. The discordant points or discordant data samples are those located outside the perimeter of the outermost ellipse.

The sample L-moment ratios for a site indexed as j, denoted as (t^j t_3^j t_4^j), are the L-moment analogues to the conventional coefficients of variation, skewness, and kurtosis, respectively (see Sect. 6.5). Consider (t^j t_3^j t_4^j) as a point in a 3-dimension space and let u_j denote a 3×1 vector given as

Fig. 10.7 Definition sketch for the measure of discordancy

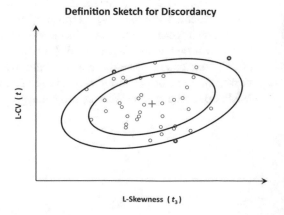

$$u_j = \left(t^j \ t_3^j \ t_4^j \right)^{\mathrm{T}} \tag{10.11}$$

where the superscript T represents vector transposition. Further, let $\bar{u}$ be a 3×1 vector containing the group mean L-moment ratios, calculated as

$$\bar{u} = \frac{\displaystyle\sum_{j=1}^{N} u_j}{N} = \left(t^R \ t_3^R \ t_4^R \right)^{\mathrm{T}} \tag{10.12}$$

where N is the number of gauging stations across the region R.

Given the sample covariance matrix S, calculated as

$$S = (N-1)^{-1} \sum_{j=1}^{N} \left(u_j - \bar{u} \right) \left(u_j - \bar{u} \right)^{\mathrm{T}} \tag{10.13}$$

Hosking and Wallis (1997) define the measure of discordancy for site j, denoted as D_j, by the expression

$$D_j = \frac{N}{3(N-1)} \left(u_j - \bar{u} \right)^{\mathrm{T}} S^{-1} \left(u_j - \bar{u} \right) \tag{10.14}$$

The data for a site j are discordant from the regional data if D_j exceeds a critical value D_{crit}, which depends on N, the number of sites within region R.

In a previous work, Hosking and Wallis (1993) suggested $D_{\mathrm{crit}} = 3$ as a criterion to decide whether a site is discordant from a group of sites. In such a case, if a sample yields $D_j \geq 3$, then it might contain gross and/or systematic errors, and/or outliers that make it discordant from the other samples in the region. Later on, Hosking and Wallis (1997) presented a table for the critical values D_{crit} as a function of the number of sites N within a region. These are listed in Table 10.11.

Hosking and Wallis (1997) point out that, for very small groups of sites, the measure of discordancy is not very informative. In effect, for $N < 3$, the covariance matrix S is singular and D_j cannot be calculated. For $N=4$, $D_j=1$ and, for $N=5$ or $N=6$, the values of D_j, as indicated in Table 10.11, are close to the statistic algebraic bound, as defined by $D_j \leq (N-1)/3$. As a result, the authors suggest the use of the discordancy measure D_j only for $N \geq 7$.

Table 10.11 Critical values D_{crit} for the measure of discordancy D_j

N	D_{crit}	N	D_{crit}
5	1.333	11	2.632
6	1.648	12	2.757
7	1.917	13	2.869
8	2.140	14	2.971
9	2.329	>15	3
10	2.491		

Adapted from Hosking and Wallis (1997)

As regards the correct use of the discordancy measure, Hosking and Wallis (1997) make the following recommendations:

(a) The regional consistency analysis should begin with the calculation of the values of D_j for all sites, regardless of any consideration of regional homogeneity. Sites flagged as discordant should, then, be subjected to a thorough examination, using statistical tests, double-mass curves and comparison with neighbor sites, looking for eventual errors and inconsistencies.
(b) Later on, when the homogeneous regions have been at least preliminarily identified, the discordancy measure should be recalculated for each gauging station within the regions. If a site is flagged as discordant, the possibility of moving it to another homogenous region should be considered.
(c) Throughout the regional consistency analysis, it should be kept in mind that the L-moment ratios are quantities that may differ by chance from one site to a group of sites with similar hydrological characteristics. A past extreme event, for instance, might have affected only part of the gauging stations within a homogeneous region, but it would be likely to affect any of the sites in the future. In such a hypothetical case, the wise decision would be to treat the whole group of gauging stations as a homogeneous region, even if some sites are flagged as discordant.

10.4.2 Identification of Homogeneous Regions

The identification of homogenous regions can be performed either based on site characteristics or on site statistics. Hosking and Wallis (1997) recommend that procedures based on site statistics be used to confirm the preliminary grouping based on site characteristics. In particular, they propose the use of the measure of heterogeneity, based on sample L-moment ratios, whose description is given in this subsection.

As implied by the definition of a homogeneous region, all sites within it should have the same population L-moment ratios. However, the sample estimates of these ratios will show variability as a result of sampling. Hosking and Wallis (1997) argue that it is natural to ask the question whether the between-site variability of the sample L-moment ratios for the group of sites being analyzed is compatible with the one that would be expected from a homogeneous region. This is the essence of the rationale employed to conceive the heterogeneity measure.

The meaning of the heterogeneity measure can be visualized in the L–moment ratio plots of Fig. 10.8. Although other statistics could be employed, in the hypothetical diagrams of Fig. 10.8, the L-CV × L-Skewness plot of panel (a) depicts the sample estimates of these quantities from observed data, whereas panel (b) shows these same quantities as calculated from repeated simulation of a homogeneous region with sites having the same sample sizes as the observed ones. In a diagram such as the ones depicted in Fig. 10.8, a possibly heterogeneous region is expected to show a larger dispersion for the sample L-CVs, for instance,

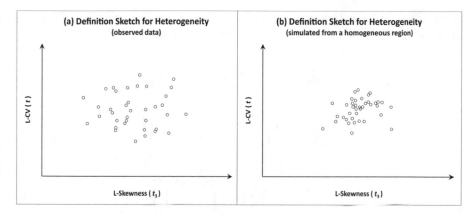

Fig. 10.8 Definition sketch for the measure of heterogeneity

than a homogeneous region would do. The heterogeneity measure seeks to put in quantitative terms the relative difference between the observed and simulated dispersions, as scaled by the standard deviation of the results from simulations.

In order to simulate samples and estimate the L-moment ratios of a homogeneous region, it is necessary to specify a parent probability distribution from which the samples are drawn. Hosking and Wallis (1997) recommend the use of the four-parameter Kappa distribution (see Sect. 5.9.1) for such a purpose arguing that a prior commitment to any two- or three-parameter distribution should be avoided, in order not to bias the results from simulation. Recall that the GEV, GPA, and GLO three-parameter distributions are all special cases of the Kappa distribution. Its L-moments can be set to match the group average L-CV, L-Skewness, and L-Kurtosis, and, thus, it can represent homogeneous regions from many distributions of hydrologic variables.

Consider a region with N gauging stations, each one of which is indexed as j, with sample size n_j, and sample L-moment ratios designated as t^j, t_3^j and t_4^j. In addition, consider that t^R, t_3^R and t_4^R represent the regional average L-moment ratios, as estimated by weighting the site specific estimates by their respective sample sizes, as in Eq. (10.9). Hosking and Wallis (1997) propose that the *measure of heterogeneity*, denoted as H, should be preferably based on the dispersions of the L-CV estimates for actual and simulated samples. First, the weighted standard deviation of the L-CV estimates for the actual samples is calculated as

$$
V = \left[\frac{\sum\limits_{i=1}^{N} n_i \left(t^i - t^R \right)^2}{\sum\limits_{i=1}^{N} n_i} \right]^{\frac{1}{2}}
\tag{10.15}
$$

As mentioned earlier, Hosking and Wallis (1997) suggest that the Kappa distribution should be used to simulate homogeneous regions. Remember that this distribution is defined by the parameters ξ, α, k, and h, with density, cumulative distribution, and quantile functions respectively given by

$$f_X(x) = \frac{1}{\alpha}\left[i - \frac{\kappa(x - \xi)}{\alpha}\right]^{\frac{1}{\kappa}-1}[F(x)]^{1-h} \tag{10.16}$$

$$F_X(x) = \left\{1 - h\left[1 - \frac{\kappa(x-\xi)}{\alpha}\right]^{\frac{1}{\kappa}}\right\}^{\frac{1}{h}} \tag{10.17}$$

$$x(F) = \xi + \frac{\alpha}{\kappa}\left[1 - \left(\frac{1 - F^h}{h}\right)^{\kappa}\right] \tag{10.18}$$

If $\kappa > 0$, x has an upper bound at $\xi + \alpha/\kappa$; if $\kappa \leq 0$, x has no upper bound; if $h > 0$, x has a lower bound at $\xi + \alpha(1 - h^{-\kappa})/\kappa$; if $h \leq 0$ and $\kappa < 0$, the x lower bound is at $\xi + \alpha/\kappa$; and, finally, x has no lower bound if $h \leq 0$ and $\kappa \geq 0$. The L-moments and L-moment ratios of the Kappa distribution are defined for $h \geq 0$ and $\kappa > -1$, and for $h < 0$ and $-1 < \kappa < -1/h$, and given by the following expressions:

$$\lambda_1 = \xi + \frac{\alpha(1 - g_1)}{\kappa} \tag{10.19}$$

$$\lambda_2 = \frac{\alpha(g_1 - g_2)}{\kappa} \tag{10.20}$$

$$\tau_3 = \frac{(-g_1 + 3g_2 - 2g_3)}{g_1 - g_2} \tag{10.21}$$

$$\tau_4 = \frac{(-g_1 + 6g_2 - 10g_3 + 5g_4)}{g_1 - g_2} \tag{10.22}$$

where

$$g_r = \begin{cases} \dfrac{r\Gamma(1 + \kappa)\Gamma\left(\dfrac{r}{h}\right)}{h^{1+\kappa}\Gamma\left(1 + \kappa + \dfrac{r}{h}\right)} & \text{for } h > 0 \\[4mm] \dfrac{r\Gamma(1 + \kappa)\Gamma\left(-\kappa - \dfrac{r}{h}\right)}{(-h)^{1+\kappa}\Gamma\left(1 - \dfrac{r}{h}\right)} & \text{for } h < 0 \end{cases} \tag{10.23}$$

and $\Gamma(.)$ denotes the Gamma function.

The population parameter values of the Kappa distribution are determined to replicate the regional L-moment ratios $(1, t^R, t_3^R, t_4^R)$. Then, a large number N_{SIM} of homogeneous regions, say $N_{\text{SIM}} = 500$, are simulated, assuming no serial correlation or cross-correlation, with samples having the same sizes as the actual samples. Suppose now that the statistics V_l for $l = 1, 2, \ldots, N_{\text{SIM}}$ are calculated for all homogeneous regions through Eq. (10.15). The mean of V yields the value to be expected from a large number of homogeneous regions and can be estimated as

$$\hat{\mu}_V = \frac{\sum\limits_{l=1}^{N_{\text{SIM}}} V_l}{N_{\text{SIM}}} \tag{10.24}$$

The measure of heterogeneity H establishes a relative number for comparing the observed dispersion V and the average dispersion that would be expected from a large number of homogeneous regions. Formally, it is given as

$$H = \frac{(V - \hat{\mu}_V)}{\hat{\sigma}_V} \tag{10.25}$$

where $\hat{\sigma}_V$ is the estimated standard deviation of the N_{SIM} values of V_l, as in

$$\hat{\sigma}_V = \sqrt{\frac{\sum\limits_{l=1}^{N_{\text{SIM}}} (V_l - \hat{\mu}_V)^2}{N_{\text{SIM}} - 1}} \tag{10.26}$$

If H is too large, relatively to what would be expected, the region is probably heterogeneous. According to the significance test proposed by Hosking and Wallis (1997), if $H < 1$, the regions is regarded as "acceptably homogeneous"; if $1 \leq H < 2$, as "possibly heterogeneous"; and, finally, if $H \geq 2$, as "definitely heterogeneous".

As mentioned earlier in this chapter, some ad hoc adjustments, such as breaking up a region or regrouping sites into different regions, may be necessary to make the heterogeneity measure conform to the suggested bounds. In some cases, the apparent heterogeneity may be due to a small number of atypical sites, which can be assigned to another region where they appear to have a more typical behavior. However, Hosking and Wallis (1997) advise that, in those cases, data should be carefully examined and hydrologic arguments should take precedence over the statistical ones. If no physical reason is found to redefine the regions, the wise decision would be that of retaining the sites in the originally proposed regions. They exemplify such a situation by hypothesizing a combination of extreme meteorological events that are plausible to occur at any point inside a region, but that actually has been observed only at some gauging stations, over the period of records. The potential benefits of regional analysis can be attained if, in such a situation, physical arguments allow grouping all sites within the homogeneous region, such that the regional growth curve, as a summary of data observed at multiple sites, can provide better estimates of the likelihood of future extreme events.

The measure of heterogeneity is built as a significance test of the null hypothesis that the region is homogeneous. However, Hosking and Wallis (1997) argue that one should not interpret the H criterion strictly as a hypothesis tests, since the assumptions of Kappa-distributed variables and lack of serial correlation and cross-correlation would have to hold. In fact, even if a test of exact homogeneity could be devised, it would be of little practical interest since even a moderately heterogeneous region would be capable of yielding quantile estimates of sufficient accuracy.

The criteria $H = 1$ and $H = 2$, although arbitrary, are useful indicators of heterogeneity. If these bounds could be interpreted in the context of a true significance test and assuming that V is normally distributed, the criterion to reject the null hypothesis of homogeneity, at a significance level $\alpha = 10\%$, would be a value of the test statistic larger than $H = 1.28$. In such a context, the arbitrary criterion of $H = 1$ would appear as too rigorous. However, as previously argued, the idea is not to interpret the criteria as a formal significance test. Hosking and Wallis (1997) report simulation results that show a sufficiently heterogeneous region, where quantile estimates are 20–40 % less accurate than for a homogeneous region, will on average yield a value of H close to 1, which is then seen as a threshold value of whether a worthwhile increase in the accuracy of quantile estimates could be attained by redefining the regions. In an analogous way, $H = 2$ is seen as the point at which redefining the regions is regarded as beneficial.

In some case, H might take negative values. These indicate that there is less dispersion among the sample L-CV estimates than would be expected from a homogeneous region with independent at-site distributions (Hosking and Wallis 1997). The most frequent cause of such negative values of H is the existence of a positive cross-correlation between data at distinct sites. If highly negative values, such as $H < -2$, are observed, then a probable cause would be either high cross-correlation among the at-site frequency distributions or an unusually low dispersion of the sample L-CV estimates. In such cases, Hosking and Wallis (1997) recommend further examination of the data.

10.4.3 Choosing the Regional Frequency Distribution

As seen in previous chapters, there are many parametric probability distributions that can be used to model hydrologic data. The fitting of a particular distribution to empirical data depends on its ability to approximate the most relevant sample descriptors. However, a successful frequency analysis does not depend only on choosing the distribution that fits the data well, but also on obtaining quantile estimates from a probability model that are likely to occur in the future. In other words, what is being sought is a robust probability model that has both the abilities of describing the observed data and predicting future occurrences of the variable, even if it the true probability distribution may differ from the originally proposed model.

The guidelines given in Chap. 8 for at-site frequency analysis remain valid for regional frequency analysis. However, in the context of regional analysis, the potential benefit comes from the relative more reliable estimation of distributions with more than two parameters, resulting from the principle of substituting space for time. In this regard, Hosking and Wallis (1997) posit that regional parameters can be estimated more reliably than would be possible using only data at a single site and recommend the use of distributions with more than two parameters as a regional model because they can yield less biased quantile estimates in the tails.

Goodness-of-fit tests and other statistical tools, such as those described in Chap. 8, are amenable to be adapted for regional frequency analysis. Plots on probability paper, quantile–quantile plots, GoF tests, and moment-ratio diagrams are useful tools that can be used also in the context of regional frequency analysis. However, for the purpose of choosing an appropriate regional probability distribution, Hosking and Wallis (1997) proposed using a *goodness-of-fit measure,* denoted as Z, which is based on comparing the theoretical L-moment ratios τ_3 and τ_4 of the candidate distribution with the regional sample estimates of these same quantities. The goodness-of-fit measure is described next.

Within a homogeneous region, the site L-moment ratios fluctuate around their regional average values, thus accounting for sampling variability. In most cases, the probability distributions fitted to the data replicate the regional average mean and L-CV. Therefore, the goodness-of-fit of a given candidate distribution to the regional data should necessarily be assessed by how well the distribution's L-Skewness and L-Kurtosis match their homologous regional average ratios, as estimated from observed data. For instance, suppose the three-parameter GEV is a candidate to model the regional data of a number of sites within a homogeneous region and that the fitted distribution yields exactly unbiased estimates for the L-Skewness and L-Kurtosis. When fitted to the regional data, the distribution will match the regional average L-Skewness. Therefore, one can judge the goodness-of-fit of such a candidate distribution by measuring the distance between the L-Kurtosis of the fitted GEV, or τ_4^{GEV}, and the regional average L-Kurtosis, or t_4^R, as illustrated in the diagram of Fig. 10.9.

In order to assess the significance of the difference between τ_4^{GEV} and t_4^R, one should take into account the sampling variability of t_4^R. Analogous to what has been done for the heterogeneity measure, the standard deviation of t_4^R, denoted as σ_4, can be evaluated through simulation of a large number of homogeneous regions, with the same number of sites and record lengths as those of the observed data, all distributed as a GEV. In such a situation, the goodness-of-fit measure of the GEV model would be written as

$$Z^{GEV} = \frac{\left(t_4^R - \tau_4^{GEV}\right)}{\sigma_4} \tag{10.27}$$

As regards the general calculation of Z, to obtain correct values of σ_4, it is necessary to simulate a separate set of regions specifically for each candidate

Fig. 10.9 Definition sketch
for the goodness-of-fit
measure Z

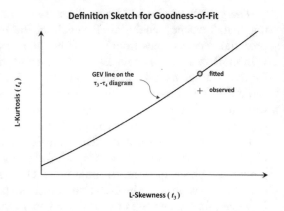

distribution. However, in practice, Hosking and Wallis (1997) argue that it should be sufficient to assume that σ_4 is the same for each candidate three-parameter distribution. They justify such an assumption by positing that, as each fitted distribution has the same L-Skewness, the candidate models are likely to resemble each other to a large extent. As such, Hosking and Wallis (1997) argue that it is then reasonable to assume that a regional Kappa distribution also has a σ_4 close to those of the candidate models. Therefore, σ_4 can be determined by repeated simulation of a large number of homogeneous regions from a Kappa population, similarly to what has been used to calculate the heterogeneity measure.

The statistics used to calculate Z, as described, assume there is no bias in the estimation of L-moment ratios from the samples. Hosking and Wallis (1997) note that such an assumption is valid for t_3 but it is not for t_4, when sample sizes are smaller than 20 or when the population L-Skewness is equal to or larger than 0.4. They advance a solution to such a problem by introducing the term B_4 to correct the bias of t_4. In the example illustrated in Fig. 10.9, the distribution τ_4^{GEV} should be compared to $t_4^R - B_4$ and not to t_4^R as before. The bias correction term B_4 can be determined using the same simulations used to calculate σ_4.

The goodness-of-fit measure Z, as previously described, refers to three-parameter distributions only. Although it is theoretically possible to build similar procedures to two-parameter distributions, they have fixed values for the population τ_3 and τ_4 and, thus, a different and complicated approach would need to be used to estimate σ_4. Despite suggesting some plausible methods to tackle such an issue, Hosking and Wallis (1997) conclude by not recommending the use of the goodness-of-fit measure for two-parameter distributions.

In order to formally define the goodness-of-fit measure Z, consider a homogeneous region with N gauging stations, each one of which is indexed as j, with sample size n_j, and sample L-moment ratios designated as t^j, t_3^j and t_4^j. In addition, consider that t^R, t_3^R and t_4^R represent the regional average L-moment ratios, as estimated by weighting the site specific estimates by their respective sample sizes, as in Eq. (10.9). Finally, consider the following set of candidate three-parameter distributions: generalized logistic (GLO), generalized extreme value (GEV), generalized Pareto (GPA), lognormal (LNO), and Pearson type III (PIII). Each of these distributions is then fitted to the regional average L-moment ratios $(1, t^R, t_3^R, t_4^R)$. Let τ_4^{DIST} denote the L-Kurtosis of the fitted candidate distribution, where DIST is any model from the set $\{GLO, GEV, GPA, LNO, PIII\}$.

In the sequence, a four-parameter Kappa distribution is fitted to the regional average L-moment ratios $(1, t^R, t_3^R, t_4^R)$. The next step is to proceed with the simulation of a large number N_{SIM} of homogeneous regions, with the same number of sites and record lengths as those of the observed data, all distributed as the fitted Kappa model. The simulation of these regions should follow exactly the same sequence as that employed to enable the calculation of the heterogeneity measure (see Sect. 10.4.2).

Next, the regional average L-moment ratios t_3^m and t_4^m for the mth simulated region, with $m = 1, 2, \ldots, N_{SIM}$, are calculated. With these, the bias of t_4^R is estimated as

$$B_4 = \frac{\sum\limits_{m=1}^{N_{SIM}} \left(t_4^m - t_4^R \right)}{N_{SIM}} \tag{10.28}$$

whereas the standard deviation of t_4^R is given by

$$\sigma_4 = \sqrt{\frac{\sum\limits_{m=1}^{N_{SIM}} \left(t_4^m - t_4^R \right)^2 - N_{SIM} B_4^2}{N_{SIM} - 1}} \tag{10.29}$$

Finally, the goodness-of-fit measure Z for each candidate distribution is given by the expression

$$Z^{DIST} = \frac{\tau_4^{DIST} - t_4^R + B_4}{\sigma_4} \tag{10.30}$$

The closer to zero the Z^{DIST}, the better DIST fits the regional data. Hosking and Wallis (1997) suggest the criterion $|Z^{DIST}| \le 1.64$ as reasonable to accept DIST as a fitting regional model.

The goodness-of-fit measure Z is specified as a significance test, under the premises that the region is homogeneous and no cross-correlation between sites is observed. In such conditions, Z has approximately a standard normal distribution. The criterion $|Z^{DIST}| \le 1.64$ corresponds to not rejecting the null hypothesis that regional data were drawn from the hypothesized candidate model, at a significance level of 10%. However, the necessary premises to approximate the distribution of Z as a standard normal are unlikely to hold in practical cases, which is particularly true if serial correlation or cross-correlation is present. In such cases, the variability of t_4^R tends to increase and, as the Kappa region is simulated assuming no correlation, the estimate of σ_4 results in being too small and Z too large, thus leading to a false indication of poor fitting. Hence, the criterion $\left|Z^{DIST}\right| \leq 1.64$ is seen as an indicator of good fit and not as a formal significance test (Hosking and Wallis 1997).

In cases in which the criterion $\left|Z^{DIST}\right| \leq 1.64$ leads to more than one fitting regional model, Hosking and Wallis (1997) recommend comparing the resulting dimensionless quantile curves (or regional growth curves). If these yield quantile estimates of comparable magnitudes, any of the fitting models can be chosen. Otherwise, the search for a more robust model should continue. In such cases, models with more than three parameters, such as the Kappa and the Wakeby distributions, may be chosen, since they are more robust to misspecification of the regional growth curve. The same recommendation may be applied to cases in which no three-parameter candidate satisfies the criterion $\left|Z^{DIST}\right| \leq 1.64$ or in cases of "possibly heterogeneous" or "definitely heterogenous" regions.

Analogous to what has been done in Chap. 8, the goodness-of-fit analysis should be complemented by plotting the regional average values (t_3^R, t_4^R) on an L–moment ratio diagram, such as the one depicted in Fig. 10.10. The expressions used to define the $\tau_3 \times \tau_4$ relations on the L-moment ratio diagram are given by Eqs. (8.12)–(8.17). Hosking and Wallis (1997) suggest that in case where the point defined by the regional estimates (t_3^R, t_4^R) falls above the curve of the GLO distribution, no two- or three-parameter distribution will fit the data, and one should resort to a more general candidate, such as the Kappa or the Wakeby distribution.

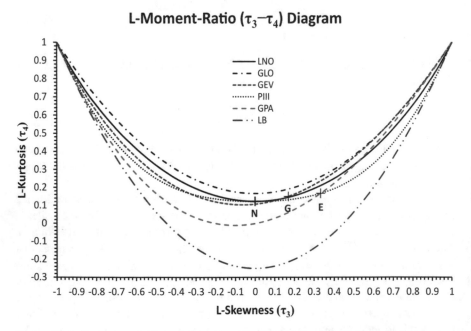

Fig. 10.10 L-Skewness × L-Kurtosis Diagram. (*N* normal, *G* Gumbel, *E* exponential)

The final recommendation relates to the analysis of a large geographic area, which can be subdivided into several homogeneous regions. Hosking and Wallis (1997) argue that if a given candidate distribution fits the data for all or most of the regions, then it would be reasonable to employ it for all regions, even though it may not be selected as the regional model for each region individually.

10.4.4 Estimating the Regional Frequency Distribution by the Method of L-Moments

The goal is to fit a probability distribution function to the data, as scaled by the site specific index-flood factor, recorded at a number of gauging stations inside an approximately homogeneous region. The distribution fitting is performed by the method of regional L-moments which consists of equating the population L-moment ratios to their corresponding regional sample estimates. These are calculated by averaging the L-moment ratios for all sites, as weighted by their respective sample sizes, as in Eq. (10.9). If the index-flood at each site is given by the mean sample value of the data, the regional weighted mean value of scaled data (or the regional L-moment of order 1) must be $\ell_1^R = 1$. As a result, the sample L-moment ratios t, t_3, and t_4 should be identical, regardless of whether they refer to the original data $\left(X_{i,j}, i = 1, \ldots, n_j; j = 1, \ldots, N\right)$ or to the scaled data $\left(x_{i,j} = x_{i,j}/\ell_1^j, i = 1, \ldots, n_j; j = 1, \ldots, N\right)$.

To fit a probability model F to the regional data, the distribution's L-moment quantities $\lambda_1, \tau, \tau_3, \tau_4, \ldots$ are set equal to the regional estimates $1, t^R, t_3^R, t_4^R, \ldots$. If the F distribution is described by P parameters θ_p, $p = 1, \ldots, P$, then equating the appropriate population and sample L-moment quantities would result in a system of P equations and P unknowns, whose solutions are the regional parameter estimates $\hat{\theta}_p$, $p = 1, \ldots, P$. With these, the regional growth curve can be estimated by the inverse function of F, as $\hat{x}(F) = x(F; \hat{\theta}_1, \ldots, \hat{\theta}_p)$. In turn, the quantiles at a given site j, located inside the homogeneous region, can be estimated by multiplying the regional quantile estimate $\hat{x}(F)$ by the site index-flood $\hat{\mu}_j$, or formally, through $\hat{X}_j(F) = \ell_1^j \hat{x}(F)$.

The estimators of τ_r for samples of moderate to large sizes exhibit very small biases. Hosking (1990) employed the asymptotic theory to calculate biases for large samples: for the Gumbel distribution, the asymptotic bias of t_3 is $0.19n^{-1}$, whereas for the Normal distribution, the asymptotic bias of t_4 is $0.03n^{-1}$, where n is the sample size. For small sample sizes, the biases of L-moment ratio estimators can be evaluated through simulation. Hosking and Wallis (1997) note that, for a gamut of distributions and $n \geq 20$, the bias of t can be considered as negligible and, for $n \cong 20$, the biases related to t_3 and t_4 are relatively small and definitely much smaller than the conventional estimators of skewness and kurtosis.

As with any statistical procedure, the results yielded by regional frequency analysis are inherently uncertain. The uncertainties of parameters and quantiles are usually quantified by the construction of confidence intervals, as outlined in Sects. 6.6 and 6.7 of Chap. 6, assuming that all assumptions related to the statistical model hold. Such a rationale can equally be employed to construct confidence intervals for parameters and quantiles yielded by regional frequency analysis with L-moments, provided that the required assumptions hold. In addition to the large-sample approximation by the Normal distribution, building conventional confidence intervals for regional frequency analysis would require that (a) the region is rigorously homogeneous; (b) the regional probability model is correctly specified; (c) no serial correlation and cross-correlation are present in the regional data. Hosking and Wallis (1997) argue that such confidence intervals would be of limited utility because no one could ascertain that all these assumptions hold in real-world cases. As an alternative to assess the uncertainties of regional estimates, Hosking and Wallis (1997) propose an approach, based on Monte Carlo simulations, which allows for possible regional heterogeneity, cross-correlation, misspecification of the regional model, or some combination thereof. The description of such a Monte Carlo-based approach is beyond the scope of this chapter and the interested reader is referred to Hosking and Wallis (1997) for details.

10.4.5   General Comments on the Hosking–Wallis Method for Regional Analysis

Based on the many applications of the unified method for regional frequency analysis using L-moments, Hosking and Wallis (1997) draw the following general conclusions:

- Even in regions with a moderate degree of heterogeneity, presence of cross-correlation in data, and misspecification of the regional probability model, the results from regional frequency analysis are more reliable than those from at-site frequency analysis.
- Regional frequency analysis is particularly valuable for the estimation of very small or very large quantiles.
- In regions with a large number N of gauging stations, errors in quantile estimates decrease fairly slowly as a function of N. Thus, there is little gain in accuracy from enlarging regions to encompass more than 20 gauging stations.
- As compared to at-site estimation, regional frequency analysis is less valuable when longer records are available. On the other hand, heterogeneity is easier to detect when record lengths are large. As such, regions should contain fewer sites when their record lengths are large.
- The use of two-parameter distributions is not recommended in the Hosking–Wallis method of regional frequency analysis. Use of such distributions is beneficial only in the cases where there is sufficient confidence that the sample

regional L-Skewness and L-Kurtosis are close to those of the distribution. Otherwise, large biases in quantile estimates may result.

- The errors of quantile estimates resulting from the misspecification of the regional frequency model are important only far into the tails of the distribution ($F < 0.1$ or $F > 0.99$).
- Robust distributions, such as the Kappa and Wakeby, yield quantile estimates that are reasonably accurate over a wide range of at-site frequency distributions within a region.
- Heterogeneity introduces bias into estimates for sites that are not typical of the region.
- Cross-correlation between sites of a region increases the variability of estimates but has little effect on their bias. A small degree of cross-correlation should not be a concern in regional frequency analysis.
- For extreme quantiles ($F \geq 0.999$), the advantage of using regional frequency analysis over at-site frequency analysis is greater. For extreme quantiles, heterogeneity is less important as a source of errors, whereas misspecification of the regional probability model becomes more important.

Example 10.4 Solve Example 10.2, using the Hosking–Wallis method.

Solution To solve this example, the Fortran 77 routines written by Hosking (1996) and available from ftp://rcom.univie.ac.at/mirrors/lib.stat.cmu.edu/general/lmoments [accessed April, 15th 2016] were compiled and run for the data set of annual maximum flows of the 7 gauging stations (Appendix 7), as scaled by their respective average flood flows. Considering the group of stations as a single one, the first step of the Hosking–Wallis method consists of calculating the discordancy measure $D_j; j = 1, \ldots, 7$, with Eq. (10.14), for each gauging station. The results are listed in Table 10.12. For $N = 7$, Table 10.11 returns the critical value $D_{crit} = 1.917$, which is greater than the calculated $D_j; j = 1, \ldots, 7$, thus allowing the conclusion that no discordant data sample was found among the group of gauging stations.

The second step refers to the identification of homogeneous regions. In the solutions to Examples 10.2 and 10.3, grouping the gauging stations into a single homogeneous region was supported by the comparison of site characteristics, site statistics and the empirical probability distributions of scaled data. Considering the group of stations as a single preliminary homogenous region, the measure of heterogeneity H was calculated as $H = -0.42$, using the procedures described in Sect. 10.4.2 and the mentioned Fortran 77 computer program. The measure H resulted in a slightly negative value, possibly indicating a small degree of positive correlation among sites, which, according to the general conclusions of Sect. 10.4.5, should not be a serious concern for the continuation of the regional

Table 10.12 Discordancy measures for Example 10.4

Station	40549998	40573000	40577000	40579995	40665000	40710000	40740000
Discordancy measure	0.59	0.64	1.45	1.67	0.75	0.8	1.11

Table 10.13 Goodness-of-fit measures for Example 10.4

DIST	GLO	GEV	LNO	PIII	GPA
Z^{DIST}	1.69	**0.44**	**0.21**	**−0.31**	−2.36

Table 10.14 Sample mean and L-moment ratios of scaled data for Example 10.4

Gauging station	Mean (l_1)	L-CV (t_2)	L-Skewness (t_3)	L-Kurtosis (t_4)
40549998	1	0.2147	0.2680	0.1297
40573000	1	0.1952	0.1389	−0.0006
40577000	1	0.1823	0.0134	0.0222
40579995	1	0.2489	0.1752	0.1479
40665000	1	0.1926	0.2268	0.0843
40710000	1	0.2284	0.1414	0.2304
40740000	1	0.2352	0.2706	0.3001
Regional average	**1**	**0.2194**	**0.1882**	**0.1433**

frequency analysis. The selection of the regional probability distribution is then performed with the calculation of the goodness-of-fit measures Z^{DIST}, for each of the following three-parameter models: generalized logistic (GLO), generalized extreme value (GEV), generalized Pareto (GPA), lognormal (LNO), and Pearson type III (PIII). The results are listed in Table 10.13. The criterion $\left| Z^{DIST} \right| \leq 1.64$ allows the selection of the GEV, LNO, and PIII as plausible regional models, whose measures of goodness-of-fit are highlighted in bold typeface in Table 10.13.

In addition to the measures of goodness-of-fit, plotting the regional average L-moment ratios (t_3^R, t_4^R) on the L-moment ratio diagram can prove useful in selecting the regional model. For the scaled data of each gauging station, the unbiased sample estimates of the PWM β_r are calculated with Eq. (6.13) and the results are then used in Eqs. (6.16)–(6.22) to yield the estimates of the first four L-moments and of the L-moment ratios t, t_3, and t_4. These are given in Table 10.14, together with the corresponding regional averages, calculated by weighting the site estimates by their respective sample size, as in Eq. (10.9). Figure 10.11 depicts the L-moment ratio diagram with the regional average estimates (t_3^R, t_4^R) marked with a cross. The relative location of (t_3^R, t_4^R) on the diagram appears to confirm the selection of the GEV, LNO, and PIII as plausible regional models.

The L-moment ratios of Table 10.14 were, then, employed to estimate the parameters of the candidate models GEV, LNO, and PIII. These estimates are given in Table 10.15.

With the parameter estimates for the three candidate models, the dimensionless quantiles can be estimated through their respective inverse functions, when applicable, or, otherwise, through the method of frequency factors, as outlined in Sect. 8.3.3. The dimensionless quantile estimates for the three candidate regional models, for selected values of the return period, are given in Table 10.16. All three candidate models yield relatively congruent quantile estimates, with moderately larger values of GEV estimates, for large return periods. In comparing the large

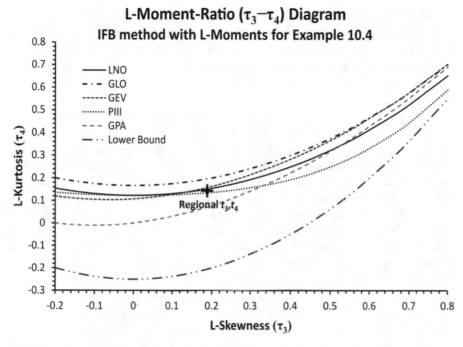

Fig. 10.11 L-moment ratio diagram for Example 10.4

Table 10.15 Parameter estimates of the candidate regional models for Example 10.4

Candidate regional distribution	Location	Scale	Shape
Generalized extreme value (GEV)	0.813	0.308	−0.028
Lognormal (LNO)	0.926	0.365	−0.388
Pearson type III (PIII	1	0.405	1.14

Table 10.16 Dimensionless quantile estimates for Example 10.4

	Return period (years)					
Candidate regional distribution	1.01	2.00	10	20	100	1000
Generalized extreme value (GEV)	0.353	0.927	1.529	1.768	2.327	3.163
Lognormal (LNO)	0.367	0.926	1.533	1.767	2.307	3.108
Pearson type III (PIII	0.397	0.925	1.543	1.769	2.260	2.915

quantile estimates from the three candidate models, the more conservative estimates favor the GEV, which is then selected as the regional model.

The general equation to estimate the regional T-year GEV dimensionless quantiles can be written as $\hat{x}(T) = \hat{\xi} + \frac{\hat{\alpha}}{\hat{k}}\left\{1 - \left[-\ln\left(1 - \frac{1}{T}\right)\right]^{\hat{k}}\right\} = 0.813 - \frac{0.308}{0.028}\left\{1 - \left[-\ln\left(1 - \frac{1}{T}\right)\right]^{-0.028}\right\}$. For an ungauged site, indexed as j, located inside the homogeneous region, it is necessary to estimate the site-specific index-flood $\hat{\mu}_j$

to obtain the quantiles $\hat{X}_j(T) = \hat{\mu}_j \hat{x}(T)$ at that particular location. The same regression model obtained in the solution to Example 10.3 can be used to such an end. Therefore, the general equation for estimating the quantiles for an ungauged site j, inside the homogeneous region, can be written as $\hat{X}_{T,j} = 0.1167 A_j$ $\left\{ 0.813 - \frac{0.308}{0.028}\left[1 - \left[-\ln\left(1 - \frac{1}{T}\right)\right]^{-0.028}\right] \right\}$.

Example 10.5 The intensity–duration–frequency (IDF) relationship of heavy rainfalls is certainly among the hydrologic tools most utilized by engineers to design storm sewers, culverts, retention/detention basins, and other structures of storm water management systems. An IDF relationship is a statistical summary of rainfall events, estimated on the basis of records of intensities abstracted from rainfall depths of sub-daily durations, observed at a recording rainfall gauging station.

An IDF relationship is, in fact, a family of curves which can be expressed in the general form $i_{d,T} = a(T)/b(d)$, where $i_{d,T}$ denotes the rainfall intensity (or rate), in mm/h or in/h, d the rainfall duration, in h, T the return period, in years, $a(T)$ is a nonlinear function of T, defined by the probability distribution function of the maximum rainfall intensities, and $b(d)$ is a function of the duration d only. In general, the functions $a(T)$ and $b(d)$ may be respectively written as $a(T) = c + \lambda \ln(T)$ [or as $a(T) = \alpha T^\beta$] and as $b(d) = (d + \theta)^\eta$, where $c, \lambda, \alpha, \beta, \theta$, and η represent parameters, with $\theta \geq 0$ and $0 < \eta < 1$ (Koutsoyiannis et al. 1998). As an example, the hypothetical IDF equation $i_{d,T} = \{33.4 - 7.56\ln[-\ln(1 - 1/T)]\}/(d^{0.40})$ is depicted in Fig. 10.12 as a family of curves. In general, the shorter the duration, the more intense the rainfall, whereas the rarer the rainfall, the higher its rate.

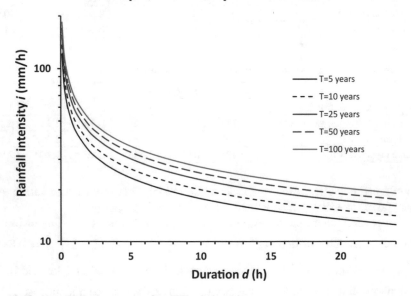

Example of a Family of IDF Curves

Fig. 10.12 Hypothetical example of an IDF relationship

To derive the IDF equation for a given gauging station, it is necessary to estimate the parameters c, λ, α, β, θ, and η, as appropriate, for the functions $a(T)$ and $b(d)$. The conventional approach consists of four steps and assumes that $a(T) = \alpha T^{\beta}$ (Raudkivi 1979). The first step refers to the frequency analysis of the rainfall intensities for each duration d, with $d = 5$ and 10 min, 0.5, 1, 2, 4, 8, 12 and 24 h, following the general guidelines outlined in Chap. 8. The second step relates to the estimation of the rainfall intensities for each duration d and return periods $T = 2$, 5, 10, 25, 50, and 100 years, with the probability distribution fitted in the first step. In the third step, the general IDF equation $i_{d,T} = a(T)/b(d)$ is written as $\ln(i_{d,T}) = [\ln(\alpha) + \beta\ln(T)] - \eta\ln(d + \theta)$ which is a linear relation between the transformed variables $\ln(i_{d,T})$ and $\ln(d + \theta)$, with η representing the slope of the line, β the spacing of the lines for the various return periods, and α the vertical position of the lines as a set (Raudkivi 1979). The fourth and final step consists of estimating the values of parameters c, λ, α, β, θ, and η, either graphically or through a two-stage least squares approach, first θ and η, and then α and β, using the regression methods covered in Chap. 9. Koutsoyiannis et al. (1998) present a comprehensive framework for the mathematical derivation of IDF relationships and propose additional methods for the robust estimation of parameters c, λ, α, β, θ, and η. The reader interested in this general framework and related estimation methods should consult the cited reference.

For a site of interest, a recording rainfall gauging station operating for a sufficiently long time period is generally capable of yielding a reliable estimate of the IDF relationship. In other locations, however, these recording stations may either not exist or have too few records to allow a reliable estimation of the IDF relationship. In parallel, as the interest expands over large areas, short-duration rainfalls may show significant geographical variability, particularly in mountainous regions, as precipitation usually undergoes orographic intensification. These two aspects of short-duration rainfall, the usually scarce records and their spatial variability, seem to have motivated some applications of regional frequency analysis to IDF estimation. Two examples of these are reported in Davis and Naghettini (2000) and Gerold and Watkins (2005), where the Hosking–Wallis method was used to estimate the regional IDF relationships for the Brazilian State of Rio de Janeiro and for the American State of Michigan, respectively. The implicit idea in both applications was that of deriving a relationship of the type $i_{j,d,T} = \mu_{j,d} x_d(T)$, where $i_{j,d,T}$ denotes the rainfall intensity at a site indexed as j, located within a homogeneous region, $\mu_{j,d}$ represents the index-flood, which depends on the duration d and on site-specific attributes, and $x_d(T)$ is the regional growth curve for the duration d.

Table 10.17 displays a list of six recording rainfall gauging stations in the *Serra dos Órgãos* region, in the Brazilian State of Rio de Janeiro, as shown in the location map of Fig. 10.13, together with the hypsometric and isohyetal maps. The region has a sharp relief, with steep slopes and elevations between 650 and 2200 m above sea level. The mean annual rainfall depth varies between 2900 mm, at high elevations, and 1300 mm, at low elevations. The annual maximum rainfall

Table 10.17 Recording rainfall gauging stations in the Serra dos Órgãos region, Brazil

Code	Gauging station	Operator	N (years)	Mean rainfall depth (mm)	Elevation (m)
02243235	Andorinhas	SERLA	20	2462	79.97
02242092	Apolinário	SERLA	19	2869	719.20
02242096	Faz. Sto. Amaro	SERLA	20	2619	211.89
02242070	Nova Friburgo	INMET	18	1390	842.38
02242098	Posto Garrafão	SERLA	20	2953	641.54
02242093	Quizanga	SERLA	21	1839	13.96

Source: Davis and Naghettini (2000)

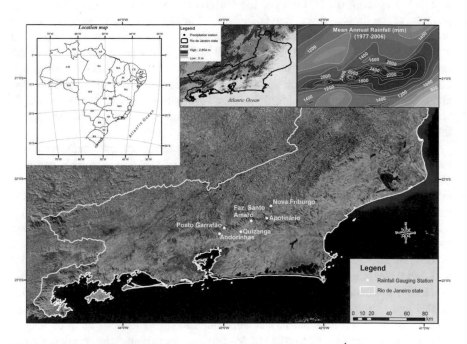

Fig. 10.13 Location, hypsometric, and isohyetal maps of the Serra dos Órgãos region, in Brazil, for the solution of Example 10.5

intensities (mm/h) of sub-daily durations, abstracted from the data recorded at the six gauging stations, are given in part (a) of the table in Appendix 8; the annual maximum quantities were abstracted on the basis of the region's water year, from October to September. Apply the Hosking–Wallis method to these data and estimate the regional IDF relationship for the durations 1, 2, 3, 4, 8, 14, and 24 h.

Solution As mentioned earlier, the regional estimation of rainfall IDF relationships, through an index-flood-based approach, such as the Hosking–Wallis method, implies the use of an equation of the type $i_{j,d,T} = \mu_{j,d}x_d(T)$, relating the rainfall intensity at the site j, to the index-flood $\mu_{j,d}$ and the regional growth curve $x_{j,d}(T)$, for the duration d. As with any index-flood-based approach, the implicit assumption is that of a homogeneous region with respect to the shape of $x_{j,d}(T)$, which is assumed identical to all sites within the region apart from a site-specific scaling factor $\mu_{j,d}$. As such, the first step refers to checking the homogeneity requirement for the region formed by the six gauging stations. This was first performed by plotting the site-specific dimensionless empirical curves on probability paper, for each duration d, as depicted in the panels of Fig. 10.14. The sample mean of each series was used as the scaling factor and the Weibull plotting position formula was employed to calculate the empirical frequencies. In addition to the graphical analysis, the heterogeneity measures were calculated, using the procedures outlined in Sect. 10.4.2. The heterogeneity measures were calculated on the basis of the L-CV, are denoted as H and given in Table 10.18. The results show that the heterogeneity measures are close to 1, which indicate an "acceptably homogeneous" region. The graphs of Fig. 10.14 seem to corroborate such an indication.

After grouping the six sites into an acceptably homogeneous region, the next step consists of choosing a regional growth curve from the set of three-parameter candidate models, through the use of the goodness-of-fit measure Z, as outlined in Sect. 10.4.3. The L-moments and the sample L-moments ratios for the dimensionless rainfall data of each site, for the various durations d, are given in part (b) of the table in Appendix 8. The regional L-moments and L-moment ratios, as estimated with Eq. (10.9), are shown in Table 10.19, and allow the estimation of the parameters of the fitted distributions.

Table 10.20 presents the goodness-of-fit measures Z for the candidate models GLO, LNO, GEV, P III, and GPA, estimated for the various durations d, according to the procedures given in Sect. 10.4.3. The criterion $|Z| \leq 1.64$ is met by the LNO, GEV, P III, and GPA distributions, all durations considered. Among these, the GPA distribution is most often used in the peaks-over-threshold (POT) representation for hydrological extremes, as mentioned and justified in Sect. 8.4.3, and shall not be included among the candidates for modeling the regional IDF relationship, thus leaving the LNO, GEV, and P-III as candidate models. Figure 10.15 depicts the regional (t_3^R, t_4^R), estimated for the various durations d, pinpointed on the L-moment-ratio diagram of the candidate distributions. Although the adoption of a single candidate model for all durations considered is not a requirement, the chart of Fig. 10.15 seems to indicate an overall superior fit by the P III distribution, which is thus chosen as a common regional parametric form, with varying parameters for $d = 1, 2, 3, 4, 8, 14,$ and 24 h. The regional estimates of the Pearson type III parameters, for the durations considered, are displayed in Table 10.21.

The dimensionless quantiles of the rainfall rates, for durations $d = 1, 2, 3, 4, 8,$ 14, and 24 h, as estimated with the corresponding regional Pearson type III parameters can be estimated using the frequency factors, as described in Sect. 8.3.3, with the Eq. (8.32). The calculated quantiles are displayed in Table 10.22 and the corresponding quantile curves $x_d(T)$ are plotted on the charts of Fig. 10.14.

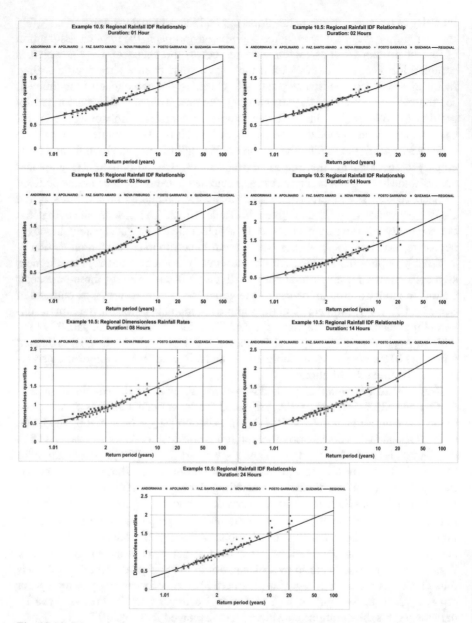

Fig. 10.14 Dimensionless empirical distributions for the rainfall data of Example 10.5

Table 10.18 Heterogeneity measures for the rainfall data of Example 10.5

Duration →	1 h	2 h	3 h	4 h	8 h	14 h	24 h
H	−1.08	−1.34	−1.15	−0.32	−0.13	−0.44	−1.05

Table 10.19 Regional L-moments and L-moment ratios for the data of Example 10.5

Duration	l_1	L-CV(t)	L-Skewness (t_3)	L-Kurtosis (t_4)
1 h	1	0.1253	0.2459	0.1371
2 h	1	0.1293	0.2336	0.1367
3 h	1	0.1564	0.2148	0.1100
4 h	1	0.1707	0.2536	0.1365
8 h	1	0.1911	0.2636	0.1757
14 h	1	0.2038	0.2511	0.1522
24 h	1	0.1885	0.1852	0.1256

Table 10.20 Goodness-of-fit measures (Z) for the data of Example 10.5

Candidate distribution	1 h	2 h	3 h	4 h	8 h	14 h	24 h
Generalized logistic (GLO)	1.95	1.85	2.59	2.03	0.94[a]	1.52[a]	1.86
Lognormal (LNO)	0.80[a]	0.70[a]	1.36[a]	0.88[a]	−0.11[a]	0.41[a]	0.66[a]
Generalized extreme value (GEV)	1.15[a]	1.02[a]	1.64[a]	1.26[a]	0.26[a]	0.77[a]	0.83[a]
Pearson type III (P III)	0.16[a]	0.11[a]	0.81[a]	0.21[a]	−0.76[a]	−0.23[a]	0.25[a]
Generalized Pareto (GPA)	−0.80[a]	−0.99[a]	−0.58[a]	−0.66[a]	−1.45[a]	−1.10[a]	−1.45[a]

[a] $|Z| \leq 1.64$

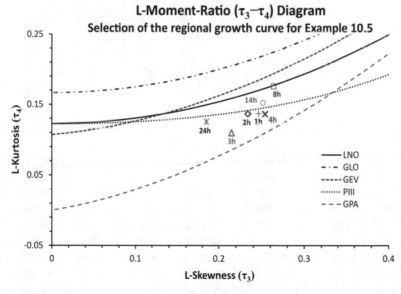

Fig. 10.15 L-moment-ratio diagram of candidate models, with the (t_3^R, t_4^R) estimates for the durations $d = 1, 2, 3, 4, 8, 14,$ and 24 h, for the rainfall data of Example 10.5

   The final step refers to the regression analysis between the scaling factor $\mu_{j,d}$, taken as the mean rainfall intensity of duration d at site j, and the site-specific physical and climatic characteristics, which are the possible explanatory variables taken, in this example, as the site's mean annual rainfall (MAR) depth and

Table 10.21 Regional estimates of Pearson type III parameters

Duration	P III regional parameters		
	Shape (β)	Scale (α)	Location (ξ)
1 h	1.824	0.176	0.679
2 h	2.018	0.172	0.653
3 h	2.374	0.190	0.550
4 h	1.718	0.248	0.574
8 h	1.592	0.152	0.538
14 h	1.750	0.293	0.487
24 h	3.177	0.195	0.381

Table 10.22 Regional estimates of Pearson type III dimensionless quantiles $x_d(T)$

Return period T (years)	Duration d						
	1 h	2 h	3 h	4 h	8 h	14 h	24 h
2	0.944	0.945	0.939	0.921	0.908	0.906	0.936
5	1.165	1.172	1.210	1.223	1.248	1.267	1.258
10	1.317	1.326	1.391	1.433	1.487	1.517	1.465
20	1.464	1.473	1.562	1.635	1.718	1.757	1.658
25	1.510	1.520	1.615	1.699	1.791	1.833	1.719
50	1.651	1.661	1.778	1.895	2.015	2.066	1.900
75	1.733	1.742	1.872	2.008	2.145	2.200	2.004
100	1.790	1.799	1.937	2.087	2.237	2.295	2.076
125	1.834	1.843	1.988	2.149	2.308	2.368	2.131
150	1.870	1.879	2.029	2.199	2.365	2.427	2.177
200	1.927	1.936	2.093	2.277	2.456	2.521	2.247

respective elevation above sea level. These site-specific quantities are given in Table 10.23 for all six gauging stations in the region.

The models tested in the regression analysis were of the log–log type, which has required the logarithmic transformation of the values given in Table 10.23. The significances of including predictor variables and of the complete regression equation were evaluated through the partial F and total F tests, respectively. The overall quality of the regression models was further examined through the analyses of the residuals, the standard errors of estimates, the adjusted coefficients of determination, the signs and magnitudes of the regression coefficients, and the standardized partial regression coefficients. The final regression model is given by $\hat{\mu}_{j,d} = 29.5d^{-0.7238}\text{MAR}_j^{0.0868}$, for $1\,\text{h} \leq d \leq 24\text{h}$, where MAR_j denotes the mean annual rainfall at site j, in mm, and $\hat{\mu}_{j,d}$ is the index-flood in mm/h. For an ungauged site j within the region, MAR_j can be estimated from the isohyetal map shown in Fig. 10.13 and then used with the regression model to yield the estimate of $\hat{\mu}_{j,d}$ for the desired duration d. Finally, the product $\hat{i}_{j,d,T} = \hat{\mu}_{j,d}\hat{x}_d(T)$, with $\hat{x}_d(T)$ taken from Table 10.22, yields the rainfall intensity $\hat{i}_{j,d,T}$ of duration d, for the desired return period T.

Table 10.23 Values of the index-flood $\mu_{j,d}$, for the rainfall duration d, mean annual rainfall depth (MAR), and elevation for the gauging stations of Example 10.5

| Gauging station | Index-flood $\mu_{j,d}$ | | | | | | | Attribute | |
	1 h (mm/h)	2 h (mm/h)	3 h (mm/h)	4 h (mm/h)	8 h (mm/h)	14 h (mm/h)	24 h (mm/h)	MAR (mm)	Elevation (m)
Andorinhas	63.75	43.61	32.09	26.12	15.37	9.47	6.6	2462	79.97
Apolinário	53.77	33.2	25.33	20.58	12.26	7.95	5.57	2869	719.20
Faz. Sto. Amaro	54.2	33.74	25.72	20.99	12.6	8.02	5.56	2619	211.89
Nova Friburgo	52.24	33.13	25.48	20.82	12.13	7.94	5.37	1390	842.38
Posto Garrafão	55.29	34.26	27.19	22.09	13.28	8.7	5.84	2953	641.54
Quizanga	54.2	34.67	25.14	20.13	12.2	8.04	5.5	1839	13.96

MAR mean annual rainfall

Exercises

1. Solve Example 10.1 using an IFB approach with estimation by the method of moments.
2. Solve Example 10.1 using an IFB approach with estimation by the method of regional L-moments, considering as candidate models the Gumbel$_{min}$ and two-parameter Weibull$_{min}$ distributions. The Fortran routines written by Hosking (1996) cannot be used in such a context because they do not encompass candidate models for low flows. As for the L-moment estimation of the Weibull$_{min}$ parameters α and β, recall from Exercise 6 of Chap. 8 that $\alpha = \ln(2)$ $/\lambda_{2,\ln(X)}$ and $\beta = \exp[\lambda_{1,\ln(X)} + 0.5772/\alpha]$.
3. Solve Example 10.2 using the RBQ method, for $T = 5$, 25, and 100 years.
4. On January 11th and 12th, 2011 heavy rains fell over the mountainous areas of the Brazilian State of Rio de Janeiro, known as *Serra dos Órgãos* (see its location on the map in Fig. 10.13), triggering one of the most catastrophic natural disasters in the history of Brazil. Important towns such as Nova Friburgo, Teresópolis, and Petrópolis were hit by heavy rains, which fell on the already saturated steep slopes causing floods, landslides, and massive rock avalanches. More than 900 people were killed, thousands were made homeless and there was huge damage to property as a result of this natural disaster. Figure 10.16 depicts the chart of the cumulative rainfall depths recorded at the gauging station located in the town of Nova Friburgo, during the 24 h, from 7 a.m. Jan 11th to 7 a.m. Jan

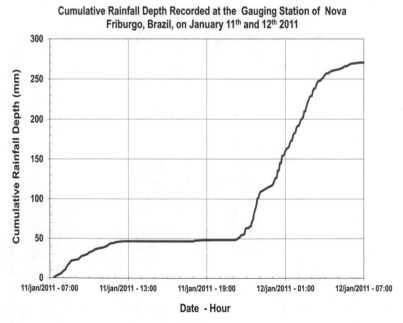

Fig. 10.16 Cumulative rainfall depths recorded at the gauging station of Nova Friburgo, Brazil, on 11th and 12th January 2011

12th, 2011. Based on the results given in the solution to Example 10.5, estimate the return periods associated with the maximum rainfall intensities for the durations $d = 1, 2, 3, 4, 8$, and 24 h, as abstracted from the chart of Fig. 10.16.

5. Extend the solution to Example 10.5 to the subhourly durations $d = 5, 10, 15, 30$, and 45 min. The rainfall data and the estimates of related L-moment ratios are given in the tables of Appendix 8.

6. Appendix 9 presents maps, catchment attributes and low flow data for five gauging stations located in the upper Velhas river basin, in southeastern Brazil. Low flow data refer to annual minimum mean flows for the durations $D = 1$ day, 3 days, 5 days, and 7 days. Employ an index–flood-based approach, with estimation through L-moments, to propose a regional model to estimate the T-year annual minimum mean flow of duration D at an ungauged site within the region. The Fortran routines written by Hosking (1996) cannot be used in such a context because they do not encompass candidate models for low flows. Consider the two-parameter Gumbel$_{min}$ and Weibull$_{min}$ as candidate distributions. As for the L-moment estimation of the Weibull$_{min}$ parameters α and β, recall from Exercise 6 of Chap. 8 that $\alpha = \ln(2)/\lambda_{2,\ln(X)}$ and $\beta = \exp[\lambda_{1,\ln(X)} + 0.5772/\alpha]$.

7. Table A10.1 of Appendix 10 lists the geographic coordinates and elevations of 92 rainfall gauging stations in the upper São Francisco river basin, located in southeastern Brazil, whereas Figs. A10.1 and A10.2 depict the location and isohyetal maps, respectively. The annual maximum daily rainfall depths recorded at the 92 gauging stations are given in Table A10.2. Use the Hosking–Wallis method to perform a complete regional frequency analysis of annual maximum daily rainfall depths over the region.

References

Bobée B, Rasmussen P (1995) Recent advances in flood frequency analysis. US National Report to IUGG 1991-1994. Rev Geophys. 33 Suppl

Burn DH (1989) Cluster analysis as applied to regional flood frequency. J Water Resour Plann Manag 115:567–582

Burn DH (1990) Evaluation of regional flood frequency-analysis with a region of influence approach. Water Resour Res 26(10):2257–2265

Burn DH (1997) Catchment similarity for regional flood frequency analysis using seasonality measures. J Hydrol 202:212–230

Castellarin A, Burn DH, Brath A (2001) Assessing the effectiveness of hydrological similarity measures for flood frequency analysis. J Hydrol 241:270–285

Cavadias GS (1990) The canonical correlation approach to regional flood estimation. In: Beran M, Brilly M, Becker A, Bonacci O (eds) Regionalization in hydrology, IAHS Publication 191, IAHS, Wallingford, UK, pp 171–178

Dalrymple T (1960) Flood-frequency analyses, Manual of hydrology: part.3. Flood-flow techniques, Geological Survey Water Supply Paper 1543-A. U.S. Government Printing Office, Washington

Davis EG, Naghettini M (2000) Regional analysis of intensity-duration-frequency of heavy storms over the Brazilian State of Rio de Janeiro. In: 2000 Joint conference on water resources engineering and planning and management, 2000, Minneapolis, USA. American Society of Civil Engineers, Reston, VA

Gado TA, Nguyen VTV (2016) Comparison of homogeneous region delineation approaches for regional flood frequency analysis at ungauged sites. J Hydrolog Eng 21(3):04015068

Gerold LA, Watkins DW Jr (2005) Short duration rainfall frequency analysis in Michigan using scale-invariance assumptions. J Hydrolog Eng 10(6):450–457

Gingras D, Adamowski K (1993) Homogeneous region delineation based on annual flood generation mechanisms. Hydrol Sci J 38(1):103–121

Greenwood JA, Landwehr JM, Matalas NC, Wallis JR (1979) Probability weighted moments: definition and relation to parameters expressible in inverse form. Water Resour Res 15 (5):1049–1054

Griffis VW, Stedinger JR (2007) The use of GLS regression in regional hydrologic analysis. J Hydrol 344:82–95

Hartigan JA (1975) Clustering algorithms. Wiley, New York

Hartigan JA, Wong MA (1979) Algorithm AS 136: a K-means clustering algorithms

Hosking JRM (1990) L-moments: analysis and estimation of distributions using linear combinations of order statistics. J Roy Stat Soc B 52(2):105–124

Hosking JRM (1996) Fortran routines for use with the method of L-moments, Version 3. Research Report RC 20525, IBM Research Division, Yorktown Heights, NY

Hosking JRM, Wallis JR (1993) Some statistics useful in regional frequency analysis. Water Resour Res 29(1):271–281

Hosking JRM, Wallis JR (1997) Regional frequency analysis—an approach based on L-moments. Cambridge University Press, Cambridge

IEA (1987) Australian rainfall and runoff: a guide to flood estimation, v. 1. Canberra: Institution of Engineers Australia, Canberra

IH (1999) The flood estimation handbook. Institute of Hydrology, Wallingford, UK

Ilorme F, Griffis VW (2013) A novel approach for delineation of hydrologically homogeneous regions and the classification of ungauged sites for design flood estimation. J Hydrol 492:151–162

Kite GW (1988) Frequency and risk analysis in hydrology. Water Resources Publications, Fort Collins (CO)

Koutsoyiannis D, Kozonis D, Manetas A (1998) A mathematical framework for studying rainfall intensity-duration-frequency relationships. J Hydrol 206(1):118–135

Kottegoda NT, Rosso R (1997) Statistics, probability, and reliability for civil and environmental engineers. McGraw-Hill, New York

Kroll CN, Stedinger JR (1998) Regional hydrologic analysis: ordinary and generalized least squares revisited. Water Resour Res 34(1):121–128

Ouarda TBMJ, Girard C, Cavadias GS, Bobée B (2001) Regional flood frequency estimation with canonical correlation analysis. J Hydrol 254:157–173

Pearson CP (1991) Regional flood frequency for small New Zealand basins 2: flood frequency groups. J Hydrol 30:53–64

Pinto EJA, Naghettini M (1999) Definição de regiões homogêneas e regionalização de frequência das precipitações diárias máximas anuais da bacia do alto rio São Francisco. Proceedings of the 13th Brazilian Symposium of Water Resources (CDROM), Belo Horizonte

Reis DS, Stedinger JR, Martins ES (2005) Bayesian generalized least squares regression with application to log Pearson type III regional skew estimation. Water Resour Res 41, W10419

Riggs HC (1973) Regional analyses of streamflow characteristics. Hydrologic analysis and investigations, Book 4, Chapter 3. Techniques of water resources investigations of the United States Geological Survey. U.S. Government Printing Office, Alexandria, VA

Nathan RJ, McMahon T (1990) Identification of homogeneous regions for the purpose of regionalisation. J Hydrol 121:217–238

NERC (1975) Flood studies report, vol 1–5. National Environmental Research Council, London

NRC (1988) Estimating probabilities of extreme floods, methods and recommended research. National Academy Press, Washington

Rao AR, Srinivas VV (2006) Regionalization of watersheds by hybrid-cluster analysis. J Hydrol 318:37–56

Rao AR, Srinivas VV (2008) Regionalization of watersheds: an approach based on cluster analysis. Springer, New York

Raudkivi AJ (1979) Hydrology—an advanced introduction to hydrological processes and modelling. Pergamon Press, Oxford

Ribeiro Correa J, Cavadias GS, Clement B, Rousselle J (1995) Identification of hydrological neighborhoods using canonical correlation analysis. J Hydrol 173:71–89

Schaefer MC (1990) Regional analyses of precipitation annual maxima in Washington State. Water Resour Res 26(1):119–131

Stedinger JR, Tasker GD (1985) Regional hydrologic analysis, 1, ordinary, weighted, and generalized least squares compared. Water Resour Res 21(9):1421–1432

Tasker GD (1982) Simplified testing of hydrologic regression regions. J Hydraul Div ASCE 108 (10):1218–1222

Tryon RC (1939) Cluster analysis. Edwards Brothers, Ann Arbor, MI

Ward JH (1963) Hierarchical grouping to optimize an objective function. J Am Stat Assoc 58:236–244

White EL (1975) Factor analysis of drainage basin properties: classification of flood behavior in terms of basin geomorphology. Water Resour Bull 11(4):676–687

Wiltshire SE (1986) Identification of homogeneous regions for flood frequency analysis. J Hydrol 84:287–302

Chapter 11
Introduction to Bayesian Analysis of Hydrologic Variables

Wilson Fernandes and Artur Tiago Silva

11.1 Historical Background and Basic Concepts

According to Brooks (2003) and McGrayne (2011), Bayesian methods date back to 1763, when Welsh amateur mathematician Richard Price (1723–1791) presented the theorem developed by the English philosopher, statistician, and Presbyterian minister Thomas Bayes (1702–1761) at a session of the Royal Society in London. The underlying mathematical concepts of Bayes' theorem were further developed during the nineteenth century as they stirred the interest of renowned mathematicians such as Pierre-Simon Laplace (1749–1827) and Carl Friedrich Gauss (1777–1855), and of important statisticians such as Karl Pearson (1857–1936). By the early twentieth century, use of Bayesian methods declined due, in part, to the opposition of prestigious statisticians Ronald Fisher (1890–1962) and Jerzy Neyman (1894–1981), who had philosophical objections to the degree of subjectivity that they attributed to the Bayesian approach. Nevertheless, prominent statisticians such as Harold Jeffreys (1891–1989), Leonard Savage (1917–1971), Dennis Lindley (1923–2013) and Bruno de Finetti (1906–1985), and others, continued to advocate in favor of Bayesian methods by developing them and prescribing them as a valid alternative to overcome the shortcomings of the frequentist approach.

In the late 1980s, there was a resurgence of Bayesian methods in the research landscape of statistics, due mainly to the fast computational developments of that decade and the increasing need of describing complex phenomena, for which

W. Fernandes (✉)
Universidade Federal de Minas Gerais, Belo Horizonte, Minas Gerais, Brazil
e-mail: wilson@ehr.ufmg.br

A.T. Silva
CERIS, Instituto Superior Técnico, Universidade de Lisboa, Lisbon, Portugal
e-mail: artur.tiago.silva@tecnico.ulisboa.pt

© Springer International Publishing Switzerland 2017
M. Naghettini (ed.), *Fundamentals of Statistical Hydrology*,
DOI 10.1007/978-3-319-43561-9_11

conventional frequentist methods did not offer satisfactory solutions (Brooks 2003). With increased scientific acceptance of the Bayesian approach, further computational developments and more flexible means of inferences followed. Nowadays it is undisputable that the Bayesian approach is a powerful logical framework for the statistical analysis of random variables, with potential usefulness for solving problems of great complexity in various fields of knowledge, including hydrology and water resources engineering.

The main difference between the Bayesian and frequentist paradigms relates to how they view the parameters of a given probabilistic model. Frequentists believe that parameters have fixed, albeit unknown, true values, which can be estimated by maximizing the likelihood function, for example, as described in Sect. 6.4 of Chap. 6. Bayesian statisticians, on the other hand, believe that parameters have their own probability distribution, which summarize their knowledge, or ignorance, about those parameters. It should be noted that Bayesians also defend that there is a real value (a point) for a parameter, but since it is not possible to determine that value with certainty, they prefer to use a probability distribution to reflect the lack of knowledge about the true value of the parameter. Therefore, as the knowledge of the parameter increases, the variance of the distribution of the parameter decreases. Ultimately, at least in theory, the total knowledge about that parameter would result in a distribution supported on a single point, with a probability equal to one.

Therefore, from the Bayesian perspective, a random quantity can be an unknown quantity that can vary and take different values (e.g., a random variable), or it can simply be a fixed quantity about which there is little or no available information (e.g., a parameter). Uncertainty about those random quantities are described by probability distributions, which reflect the subjective knowledge acquired by the expert when evaluating the probabilities of occurrence of certain events related to the problem at hand.

In addition to the information provided by observed data, which is also considered by the classical school of statistics, Bayesian analysis considers other sources of information to solve inference problems. Formally, suppose that θ is the parameter of interest, and can take values within the parameter space Θ. Let Ω be the available prior information about that parameter. Based on Ω, the uncertainty of θ can be summarized by a probability distribution with PDF $\pi(\theta|\Omega)$, which is called the prior density function or the prior distribution, and describes the state of knowledge about the random quantity, prior to looking at the observed data. If Θ is a finite set, then, it is an inference, in itself, as it represents the possible values taken by θ. At this point, it is worth noting that the prior distribution does not describe the random variability of the parameter but rather the degree of knowledge about its true value.

In general, Ω does not contain all the relevant information about the parameter. In fact, the prior distribution is not a complete inference about θ, unless, of course, the analyst has full knowledge about the parameter, which does not occur in most real situations. If the information contained in Ω is not sufficient, further information about the parameter should be collected. Suppose that the random variable X, which is related to θ, can be observed or sampled; prior to sampling X and assuming that the current value of θ is known, the uncertainty about the amount X can be summarized by the likelihood function $f(x|\theta, \Omega)$. It should be noted that the likelihood function provides the probability of a particular sample value x of X occurring, assuming that θ is the true value of the parameter. After performing the experiment, the prior knowledge about θ should be updated using the new information x. The usual mathematical tool for performing this update is Bayes' theorem. Looking back at Eq. (3.8) and taking as reference the definition of a probability distribution, the *posterior* PDF, which summarizes the updated knowledge about θ is given by

$$\pi(\theta|x, \Omega) = \frac{f(x|\theta, \Omega)\pi(\theta|\Omega)}{f(x|\Omega)}, \tag{11.1}$$

where the prior predictive density, $f(x|\Omega)$, is given by

$$f(x|\Omega) = \int_{\Theta} f(x|\theta, \Omega)\pi(\theta|\Omega)d\theta, \tag{11.2}$$

The posterior density, calculated through Eq. (11.1), describes the uncertainty about θ after taking the data into account, that is, $\pi(\theta|x, \Omega)$ is the posterior inference about θ, according to which it is possible to appraise the variability of θ.

As implied by Eqs. (11.1) and (11.2), the set Ω is present in every step of calculation. Therefore, for the sake of simplicity, this symbol will be suppressed in forthcoming equations. Another relevant fact in this context concerns the denominator of Eq. (11.1), which is expanded in Eq. (11.2): since the integration of Eq. (11.2) is carried out over the whole parameter space, the prior predictive distribution is actually a constant, and as such, it has the role of normalizing the right-hand side of Eq. (11.1). Therein arises another fairly common way of representing Bayes' theorem, as written as

$$\pi(\theta|x) \propto f(x|\theta)\pi(\theta), \tag{11.3}$$

or, alternatively,

$$\text{posterior density} \propto \text{likelihood} \times \text{prior density}. \tag{11.4}$$

According to Ang and Tang (2007), Bayesian analysis is particularly suited for engineering problems, in which the available information is limited and often a subjective decision is required. In the case of parameter estimation, the engineer has, in some cases, some prior knowledge about the quantity on which inference is carried out. In general, it is possible to establish, with some degree of belief, which outcomes are more probable than others, even in the absence of any observational experiment concerning the variable of interest.

Fig. 11.1 Prior probability mass function of variable θ (adapt. Ang and Tang, 2007)

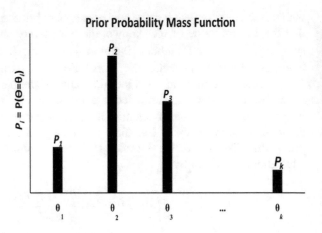

A hydrologist, for example, supported by his/her professional experience or having knowledge on the variation of past flood stages in river reaches neighboring a study site, even without direct monitoring, can make subjective preliminary evaluations (or elicitations) of the probability of the water level exceeding some threshold value. This can be a rather vague assessment, such as "not very likely" or "very likely," or a more informative and quantified assessment derived from data observed at nearby gauging stations. Even such rather subjective information can provide important elements for the analysis and be considered as part of a logical and systematic analysis framework through Bayes' theorem.

A simple demonstration of how Bayes' theorem can be employed to update current expert knowledge makes use of discrete random variables. Assume that a given variable θ can only take values from the discrete set θ_i, $i = 1, 2, \ldots, k$, with respective probabilities $p_i = P(\theta = \theta_i)$. Assume further that after inferring the values of p_i, new information ε is gathered by some data collecting experiment. In such a case, the values of p_i should be updated in the light of the new information ε.

The values of p_i, prior to obtaining the new information ε, provide the prior distribution of θ, which is assumed to have already been elicited and summarized in the form of the mass function depicted in Fig. 11.1.

Equation (11.1), as applied to a discrete variable, may be rewritten as

$$P(\Theta = \theta_i | \varepsilon) = \frac{P(\varepsilon | \theta = \theta_i) P(\theta = \theta_i)}{\sum_{i=1}^{k} P(\varepsilon | \theta = \theta_i) P(\theta = \theta_i)}, \quad i = 1, 2, \cdots, k, \qquad (11.5)$$

where,

- $P(\varepsilon | \theta = \theta_i)$ denotes the likelihood or conditional probability of observing ε, given that θ_i is true;

- $P(\theta = \theta_i)$ represents the prior mass of θ, that is, the knowledge about θ *before* ε is observed; and
- $P(\Theta = \theta_i|\varepsilon)$ is the posterior mass of θ, that is, the knowledge about θ *after* taking ε into account.

The denominator of Eq. (11.5) is the normalizing or proportionality constant, likewise to the previously mentioned general case.

The expected value of Θ can be a Bayesian estimator of θ, defined as

$$\hat{\theta} = E(\Theta|\varepsilon) = \sum_{i=1}^{k} \theta_i P(\Theta = \theta_i|\varepsilon) \qquad i = 1, 2, \cdots, k \qquad (11.6)$$

Equation (11.6) shows that, unlike classical parameter estimation, both the observed data, taken into account via the likelihood function, and the prior information, be it subjective or not, are taken into account by the logical structure of Bayes' theorem. Example 11.1 illustrates these concepts.

Example 11.1 A large number of extreme events in a certain region may indicate the need for developing a warning system for floods and emergency plans against flooding. With the purpose of evaluating the severity of rainfall events over that region, suppose a meteorologist has classified the sub-hourly rainfall events with intensities of over 10 mm/h as extreme. Those with intensities lower than 1 mm/h were discarded. Assume that the annual proportion of extreme rainfall events as related to the total number of events can only take the discrete values $\theta = \{0.0, 0.25, 0.50, 0.75, \text{ and } 1.0\}$. This is, of course, a gross simplification, since a proportion can vary continuously between 0 and 1. Based on his/her knowledge of the regional climate, the meteorologist has evaluated the probabilities respectively associated with the proportions θ_i, $i = 1, \ldots, 5$, which are summarized in the chart of Fig. 11.2.

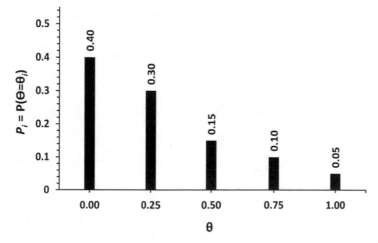

Fig. 11.2 Prior knowledge of the meteorologist on the probability mass function of the annual proportions of extreme rainfall events

Solution In the chart of Fig. 11.2, θ is the annual proportion of extreme rainfall events as related to the total number of events. For instance, the annual probability that none of the rainfall events is extreme is 0.40. Thus, based exclusively on the meteorologist's previous knowledge, the average ratio of events is given by

$$\hat{\theta}' = E(\Theta|\varepsilon) = 0.0 \times 0.40 + 0.25 \times 0.30 + 0.50 \times 0.15 + 0.75 \times 0.10 + 1.0 \times 0.05$$
$$= 0.275$$

Suppose a rainfall gauging station has been installed at the site and that, 1 year after the start of rainfall monitoring, none of observed events could be classified as extreme. Then, the meteorologist can use the new information to update his/her prior belief using Bayes' theorem in the form of Eq. (11.5), or

$$P(\theta = 0.0|\varepsilon = 0.0) = \frac{P(\varepsilon=0.0|\theta=0.0)P(\theta=0.0)}{\sum_{i=1}^{k} P(\varepsilon=0.0|\theta=\theta_i)P(\theta=\theta_i)}, \text{ which, with the new data, yields}$$

$$P(\theta=0.0|\varepsilon=0.0) = \frac{1.0 \times 0.40}{1.0 \times 0.40 + 0.75 \times 0.30 + 0.50 \times 0.15 + 0.25 \times 0.10 + 0.0 \times 0.05}$$
$$= 0.552$$

In the previous calculations, $P(\varepsilon = 0.0|\theta = \theta_i)$ refers to the probability that no extreme events happen, within a universe where $100\theta_i\%$ of events are extreme. Thus, $P(\varepsilon = 0.0|\theta = 0)$ refers to the probability of no extreme events happening, within a universe where 0 % of the events are extreme, which, of course, is 100 %. The remaining posterior probabilities are obtained in a likewise manner, as follows:

$$P(\theta=0.25|\varepsilon=0.0) = \frac{0.75 \times 0.30}{1.0 \times 0.40 + 0.75 \times 0.30 + 0.50 \times 0.15 + 0.25 \times 0.10 + 0.0 \times 0.05}$$
$$= 0.310$$

$$P(\theta=0.50|\varepsilon=0.0) = \frac{0.50 \times 0.15}{1.0 \times 0.40 + 0.75 \times 0.30 + 0.50 \times 0.15 + 0.25 \times 0.10 + 0.0 \times 0.05}$$
$$= 0.103$$

$$P(\theta=0.75|\varepsilon=0.0) = \frac{0.25 \times 0.10}{1.0 \times 0.40 + 0.75 \times 0.30 + 0.50 \times 0.15 + 0.25 \times 0.10 + 0.0 \times 0.05}$$
$$= 0.034$$

$$P(\theta=1.00|\varepsilon = 0.0) = \frac{0.00 \times 0.05}{1.0 \times 0.40 + 0.75 \times 0.30 + 0.50 \times 0.15 + 0.25 \times 0.10 + 0.0 \times 0.05}$$
$$= 0.000$$

Figure 11.3 shows the comparison between the prior and the posterior mass functions. It is evident how the data, of 1 year of records among which no extreme

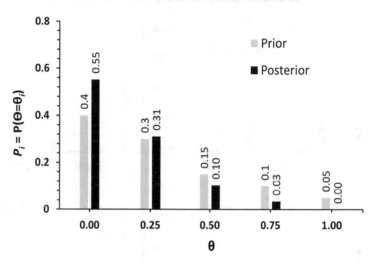

Fig. 11.3 Comparison between prior and posterior mass functions

event was observed, adjust the prior belief since the new evidence suggests that the expected proportion of extreme events is lower than initially thought.

The annual mean proportion of extreme events is given by:

$$\hat{\theta}'' = E(\Theta|\varepsilon) = 0.0 \times 0.552 + 0.25 \times 0.310 + 0.50 \times 0.103 + 0.75 \times 0.034 + 1.0 \times 0.00$$
$$= 0.155$$

It should be noted that classical inference could hardly be used in this case, since the sample size is 1, which would result in $\hat{\theta} = 0$. Bayesian analysis, on the other hand, can be applied even when information is scarce. Suppose, now, that after a second year of rainfall gauging, the same behavior as the first year took place, that is, no extreme rainfall event was observed. This additional information can then be used to update the knowledge about θ through the same procedure described earlier. In such a case, the prior information for the year 2 is now the posterior mass function of year 1. Bayes' theorem can thus be used to progressively update estimates in light of newly acquired information. Figure 11.4 illustrates such a process of updating estimates, by hypothesizing the recurrence of the observed data, as in year 1 with $\varepsilon = 0$, over the next 1, 5, and 10 years.

As shown in Fig. 11.4, after a few years of not a single occurrence of an extreme event, the evidence of no extreme events becomes stronger. As $n \to \infty$, the Bayesian estimation will converge to the frequentist one $[P(\varepsilon = 0.0|\theta = 0) = 1]$. Example 11.1 illustrates the advantages of Bayesian inference. Nevertheless, it also reveals one of its major drawbacks: the subjectivity involved in eliciting prior

Prior and Posterior Distributions for Different Values of *n*

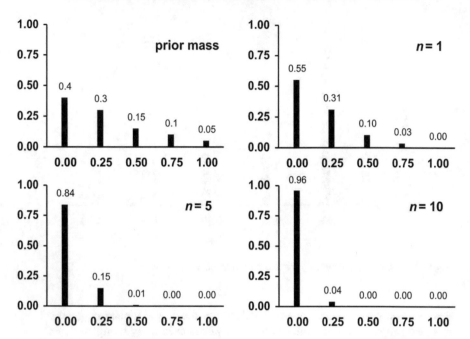

Fig. 11.4 Updating the prior mass functions after 1, 5, and 10 years in which no extreme rainfall event was observed

information. If another meteorologist had been consulted, there might have been a different prior probability proposal for Fig. 11.2, thus leading to different results. Therefore the decision will inevitably depend on how skilled or insightful is the source of previous knowledge elicited in the form of the prior distribution.

Bayesian inference does not necessarily entail subjectivity. Prior information may have sources other than expert judgement. Looking back at Example 11.1, the prior distribution of the ratio of extreme events can be objectively obtained through analysis of data from a rain gauging station. However, when the analyzed data are the basis for eliciting the prior distribution, they may not be used to calculate the likelihood function, i.e., each piece of information should contribute to only one of the terms of Bayes' theorem.

In Example 11.1 there was no mention of the parametric distribution of the variable under analysis. Nevertheless, in many practical situations it is possible to elicit a mathematical model to characterize the probability distribution of the quantity of interest. Such is the case in the recurrence time intervals of floods which are modeled by the geometric distribution (see Sect. 4.1.2), or the probability of occurrence of y floods with exceedance probability θ in N years, which is modeled by the binomial distribution (see Sect. 4.1.1). Bayes' theorem may be applied directly, although that requires other steps and different calculations than those presented so far, which depend on the chosen model. In the next paragraphs, some of the steps required for a general application of Bayes' theorem are presented, taking the binomial distribution as a reference.

Assume Y is a binomial variate or, for short, $Y \sim B(N, \theta)$. Unlike the classical statistical analysis, the parameter θ is not considered as a fixed value but rather as a random variable that can be modeled by a probability distribution. Since θ can take values in the continuous range $[0, 1]$, it is plausible to assume that its distribution should be left- and right-bounded, and, as such, the Beta distribution is an appropriate candidate to model θ, i.e., $\theta \sim \text{Be}(a, b)$. Assuming that the prior distribution of θ is fully established (e.g., that the a and b values are given) and, having observed the event $Y = y$, Bayes' theorem provides the solution for the posterior distribution of θ as follows

$$\pi(\theta|y) = \frac{p(y|\theta)\pi(\theta)}{\int_0^1 p(y|\theta)\pi(\theta)d\theta} \tag{11.7}$$

where

- $\pi(\theta|y)$ is the posterior density of θ after taking y into consideration;

- $p(y|\theta)$ is the likelihood of y for a given θ, that is, $\binom{N}{y}\theta^y(1-\theta)^{N-y}$;

- $\pi(\theta)$ is the prior density of θ, that is, the $\text{Be}(a, b)$ probability density function;

- $\int_0^1 p(y|\theta)\pi(\theta)d\theta$ is the normalizing or proportionality constant, hitherto represented by $p(y)$. It should be noted that $p(y)$ depends only upon y and is, thereby, a constant with respect to the parameter θ.

Under this setting, one can write

$$\pi(\theta|y) = \frac{\binom{N}{y}\theta^y(1-\theta)^{N-y}\frac{\Gamma(a+b)}{\Gamma(a)\Gamma(b)}\theta^{a-1}(1-\theta)^{b-1}}{p(y)} \tag{11.8}$$

After algebraic manipulation and grouping common terms, one obtains

$$\pi(\theta|y) = \frac{1}{p(y)}\binom{N}{y}\frac{\Gamma(a+b)}{\Gamma(a)\Gamma(b)}\theta^{a+y-1}(1-\theta)^{b+N-y-1} \tag{11.9}$$

or

$$\pi(\theta|y) = c(N, y, a, b)\theta^{a+y-1}(1-\theta)^{b+N-y-1} \tag{11.10}$$

Note, in Eq. (11.10), that the term $\theta^{a+y-1}(1-\theta)^{b+N-y-1}$ is the kernel of the $\text{Be}(a+y, b+N-y)$ density function. Since the posterior density must integrate to 1 over its domain, the independent function $c(N, y, a, b)$ must be

$$c(N, y, a, b) = \frac{\Gamma(a + b + N)}{\Gamma(a + y)\Gamma(b + N - y)} \tag{11.11}$$

As opposed to Example 11.1, here it is possible to evaluate the posterior distribution analytically. Furthermore, any inference about θ, after taking the data point into account, can be carried out using $\pi(\theta|y)$. For example, the posterior mean is given by

$$E[\theta|y] = \frac{a + y}{a + b + N} \tag{11.12}$$

which can be rearranged as

$$E[\theta|y] = \frac{a + b}{a + b + N} \left(\frac{a}{a + b}\right) + \frac{N}{a + b + N} \left(\frac{y}{N}\right) \tag{11.13}$$

For the sake of clarity, Eq. (11.12) can be rewritten as

$$E[\theta|y] = \frac{a + b}{a + b + N} \times \{\text{prior mean of } \theta\} + \frac{n}{a + b + N} \times \{\text{data average}\} \tag{11.14}$$

Equation (11.13) shows that the posterior distribution is a balance between prior and observed information. As in Example 11.1, as the sample increases, prior information or prior knowledge becomes less relevant when estimating θ and inference results should converge to those obtained through the frequentist approach. In this case,

$$\lim_{N \to \infty} E[\theta|y] = \frac{y}{N} \tag{11.15}$$

Example 11.2 Consider a situation similar to that shown in Example 4.2. Suppose, then, that the probability p that the discharge Q_0 will be exceeded in any given year is uncertain. An engineer believes that p has mean 0.25 and variance 0.01. Note that, in Example 4.2, p was not considered to be a random variable. Furthermore, it is believed that p follows a Beta distribution, i.e., $p \sim \text{Be}(a, b)$. Parameters a and b may be estimated by the method of moments as $\hat{a} = \bar{p}\left(\frac{\bar{p}}{S_p^2} - 1\right) = 4.4375$ and $\hat{b} = (1 - \bar{p})\left(\frac{\bar{p}}{S_p^2} - 1\right) = 13.3125$, where $\bar{p} = 0.25$ and $S_p^2 = 0.01$. In a sense, the variance of p, as denoted by S_p^2, measures the degree of belief of the engineer as to the value of p. Assume that after 10 years of observations, no flow exceeding Q_0 was recorded. (a) What is the updated estimate of p in light of the new information? (b) What is the posterior probability that Q_0 will be exceeded twice in the next 5 years?

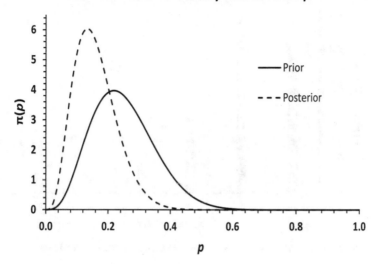

Prior and Posterior Density Functions for p

Fig. 11.5 Prior and posterior densities of p

Solution

(a) As previously shown, $\pi(p|y) = \mathrm{Be}(a+y, b+N-y) = \mathrm{Be}(A,B)$, with $a = 4.4375$, $b = 13.3125$, $N = 10$ and $y = 0$. Equation (11.11) provides the posterior mean of p as $E[p|y] = \frac{4.4375+0}{4.4375+13.3125+10} = 0.1599$. Since no exceedances were observed in the 10 years of records, the posterior probability of flows in excess of Q_0 occurring is expected to decrease, or, in other words, there is empirical evidence that such events are more exceptional than initially thought. Figure 11.5 illustrates how the data update the distribution of p.

Example 4.2 may be revisited using the posterior distribution for inference about the expected probability and the posterior credibility intervals, which are formally presented in Sect. 11.3.

(b) The question here concerns what happens in the 5 next years, thus departing from the problem posed in Example 4.2. Recall that: (1) the engineer has a prior knowledge about p, which is formulated as $\pi(p)$; (2) no exceedance was recorded over a 10-year period, i.e., $y = 0$; (3) the information about p was updated to $\pi(p|y)$; and (4) he/she needs to evaluate the probability that a certain event $\tilde{y}$ will occur in the next 5 years. Formally,

$$
\begin{aligned}
P(\tilde{Y} = \tilde{y} \mid Y = y) &= \int P(\tilde{Y} = \tilde{y}, Y = y)\, dp \\
&= \int P(\tilde{Y} = \tilde{y}, p \mid Y = y)\, dp \\
&= \int P(\tilde{Y} = \tilde{y} \mid p, Y)\, \pi(p \mid Y)\, dp \\
&= \int \binom{N}{\tilde{y}} p^{\tilde{y}} (1-p)^{N-\tilde{y}} f_{\mathrm{Beta}(A,B)}(p)\, dp
\end{aligned}
$$

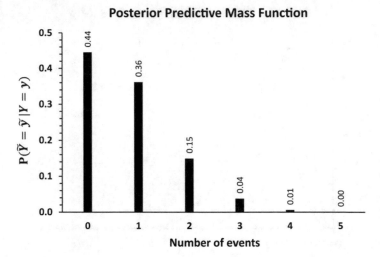

Fig. 11.6 Posterior predictive mass function of the number of events over 5 years

where $N = 5$, $\tilde{y} = 2$, $A = 4.4375$, and $B = 23.3125$. After algebraic manipulation one obtains

$$P\left(\tilde{Y} = \tilde{y} | Y = y\right) = \binom{N}{\tilde{y}} \frac{\Gamma(A+B)}{\Gamma(A)\Gamma(B)} \int p^{\tilde{y}} (1-p)^{N-\tilde{y}} p^{A+1} (1-p)^{B-1} dp.$$ Note that

the integrand in this equation is the numerator of the PDF of the $Be(\tilde{y} + A, N - \tilde{y} + B)$ distribution. Since a density function must integrate to 1, over the domain of the variable, one must obtain

$$P\left(\tilde{Y} = \tilde{y} | Y = y\right) = \binom{N}{\tilde{y}} \frac{\Gamma(A+B)}{\Gamma(A)\Gamma(B)} \frac{\Gamma(\tilde{y}+A)\Gamma(N-\tilde{y}+B)}{\Gamma(N+A+B)} = 0.1494.$$ This is the

posterior predictive estimate of $P(\tilde{y} = 2)$, since it results from the integration over all possible realizations of p. One could further define the probabilities associated with events $\tilde{y} = \{0, 1, 2, 3, 4, \cdots, n\}$ over the next N years. Figure 11.6 illustrates the results for $N = 5$.

11.2   Prior Distributions

11.2.1   Conjugate Priors

In the examples discussed in the previous section, the product *likelihood × prior density* benefited from an important characteristic: its mathematical form was such that, after some algebraic manipulation, a posterior density was obtained which belongs to the same family as the prior density (e.g., *Binomial × Beta → Beta*). Furthermore, in those cases, the proportionality constant (the denominator of Bayes' theorem) was obtained in an indirect way, without requiring integration.

This is the algebraic convenience of using *conjugate* priors, i.e., priors whose combination with a particular likelihood results in a posterior from the same family.

Having a posterior distribution with a known mathematical form facilitates statistical analysis and allows for a complete definition of the posterior behavior of the variable under analysis. However, as several authors point out (e.g., Paulino et al. 2003; Migon and Gamerman 1999) this property is limited to only a few particular combinations of models. As such, conjugate priors are not usually useful in most practical situations. Following is a non-exhaustive list of conjugate priors.

- Normal distribution (known standard deviation σ)

  Notation: $X \sim N(\mu, \sigma)$
  Prior: $\mu \sim N(\varsigma, \tau)$
  Posterior: $\mu|x \sim N(\upsilon(\sigma^2\varsigma + \tau^2 x), \tau\sigma\sqrt{\upsilon})$ with $\upsilon = \frac{1}{\sigma^2 + \tau^2}$

- Normal distribution (known mean μ)

  Notation: $X \sim N(\mu, \sigma)$
  Prior: $\sigma \sim Ga(\alpha, \beta)$
  Posterior: $\sigma|x \sim Ga\left(\alpha + \frac{1}{2}, \beta + \frac{(\mu-x)^2}{2}\right)$

- Gamma distribution

  Notation: $X \sim Ga(\theta, \eta)$
  Prior: $\eta \sim Ga(\alpha, \beta)$
  Posterior: $\eta|x \sim Ga(\alpha + \theta, \beta + x)$

- Poisson distribution

  Notation: $Y \sim P(\nu)$
  Prior: $\nu \sim Ga(\alpha, \beta)$
  Posterior: $\nu|y \sim Ga(\alpha + y, \beta + 1)$

- Binomial distribution

  Notation: $Y \sim B(N, p)$
  Prior: $p \sim Be(\alpha, \beta)$
  Posterior: $p|y \sim Be(\alpha + y, \beta + N - y)$

11.2.2   Non-informative Priors

In some situations there is a complete lack of prior knowledge about a given parameter. It is not straightforward to elicit a prior distribution that reflects total ignorance about such a parameter. In these cases, the so-called non-informative priors or vague priors can be used.

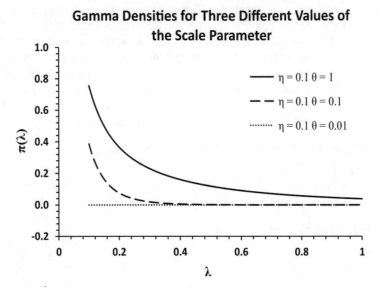

Fig. 11.7 Gamma densities for some values of the scale parameter

A natural impulse for modelers to convey non-information, in the Bayesian sense, is to attribute the same prior density to every possible value of the parameter. In that case, the prior must be a uniform distribution, that is $\pi(\theta) = k$. The problem with that formulation is that when θ has an unbounded domain, the prior distribution is improper, that is, $\int \pi(\theta)\, d\theta = \infty$. Although the use of proper distributions is not mandatory, it is considered to be a good practice in Bayesian analysis. Robert (2007) provides an in-depth discussion about the advantages of using proper prior distributions. A possible alternative to guarantee that $\int \pi(\theta)\, d\theta = 1$ is to use the so-called *vague* priors, which are parametric distributions with large variances such that they are, at least locally, nearly flat. Figure 11.7 illustrates this rationale for a hypothetical parameter λ, which is Gamma-distributed. Note how the Gamma density, with a very small scale parameter, is almost flat. Another option is to use a normal density with a large variance.

Another useful option is to use a Jeffreys prior. A Jeffreys prior distribution has a density defined as

$$\pi(\theta) \propto [I(\theta)]^{1/2} \tag{11.16}$$

where $I(\theta)$ denotes the so-called Fisher information about the parameter θ, as given by

$$I(\theta) = E\left[-\frac{\partial^2 \ln L(x|\theta)}{\partial \theta^2}\right] \tag{11.17}$$

and $L(x|\theta)$ represents the likelihood of x, conditioned on θ. Following is an example of application of a Jeffreys prior.

Example 11.3 Let $X \sim P(\theta)$. Specify the non-informative Jeffreys prior distribution for θ (adapted from Ehlers 2016).

Solution The log-likelihood function of the Poisson distribution can be written as

$\ln L(x|\theta) = -N\theta + \ln(\theta) \sum_{i=1}^{N} x_i - \ln\left(\prod_{i=1}^{N} x_i!\right)$ of which, the second-order derivative is

$$\frac{\partial^2 \ln L(x|\theta)}{\partial \theta^2} = \frac{\partial}{\partial \theta}\left[-N + \frac{\sum_{i=1}^{N} x_i}{\theta}\right] = -\frac{\sum_{i=1}^{N} x_i}{\theta^2}. \text{ Then,}$$

$$I(\theta) = \frac{1}{\theta^2} E\left[\sum_{i=1}^{N} x_i\right] = \frac{N}{\theta} \propto \theta^{-1}.$$

Incidentally, that is the same prior density as the conjugate density of the Poisson model $Ga(\alpha, \beta)$, with $\alpha = 0.5$ and $\beta \to 0$. In general, with a correct specification of parameters, the conjugate prior holds the characteristics of a Jeffreys prior distribution.

11.2.3 Expert Knowledge

Although it is analytically convenient to use conjugate priors or non-informative priors, these solutions do not necessarily assist the modeler in incorporating any existing prior knowledge or belief into the analysis. In most cases, knowledge about a certain quantity does not exist in the form of a particular probabilistic model. Hence the expert must build a prior distribution from whatever input, be it partial or complete, that he/she has. This issue is crucial in Bayesian analysis and while there is no unique way of choosing a prior distribution, the procedures implemented in practice generally involve approximations and subjective determination. Robert (2007) explores in detail the theoretical and practical implications of the choice of prior distributions. In Sect. 11.5, a real-world application is described in detail, which includes some ideas about how to convert prior information into a prior distribution.

In the hydrological literature there are some examples of prior parameter distributions based on expert knowledge. Martins and Stedinger (2000) established the so-called "geophysical prior" for the shape parameter of the GEV distribution, based on past experience gained in previous frequency analyses of hydrologic maxima. The geophysical prior is given by a Beta density in the range $[-0.5, 0.5]$ and defined as

$$\pi(\kappa) = \frac{(0.5 + \kappa)^5 (0.5 - \kappa)^8}{B(6,9)} \tag{11.18}$$

where $B(.)$ is the Beta function. Another example is provided by Coles and Powell (1996) who proposed a method for eliciting a prior distribution for all GEV parameters based on a few "rough" quantile estimates made by expert hydrologists.

11.2.4 Priors Derived from Regional Information

While their "geophysical prior" was elicited based on their expert opinion about a statistically reasonable range of values taken by the GEV shape parameter, Martins and Stedinger (2000) advocate the pursuit of regional information from nearby sites to build a more informative prior for κ. In fact, as regional hydrologic information exists and, in cases, can be abundant, the Bayesian analysis framework provides a theoretically sound setup to formally include it in the statistical inference. There are many examples in the technical literature of prior distributions derived from regional information for frequency analysis of hydrological extremes (see Viglione et al. 2013 and references therein).

A recent example of such an approach is given in Silva et al. (2015). These authors used the Poisson-Pareto model (see Sect. 8.4.5), under a peaks-over-threshold approach, to analyze the frequency of floods in the Itajaí-açu River, in Southern Brazil, at the location of Apiúna. Since POT analysis is difficult to automate, given the subjectivity involved in the selection of the threshold and independent flood peaks, Silva et al. (2015) exploited the duality of the shape parameter of the Generalized Pareto (GPA) distribution for exceedance magnitudes and of the GEV distribution of annual maxima by extensively fitting the GEV distribution to 138 annual maximum flood samples in the region (the 3 southernmost states of Brazil), using maximum likelihood estimation and the resulting estimates of the shape parameter to construct a prior distribution for that parameter. Figure 11.8 shows the location of the 138 gauging stations and the spatial distribution of the estimated values of the GEV shape parameter.

A Normal distribution was fitted to the obtained estimates of κ. Figure 11.9 shows the histogram of the estimates of the GEV shape parameter and the PDF of the fitted Normal distribution. Figure 11.9 also shows, as a reference, the PDF of the geophysical prior elicited by Martins and Stedinger (2000). Silva et al. (2015) found that, while using the geophysical prior is worthwhile in cases where no other prior information about κ is available, in that particular region it did not adequately fit the data.

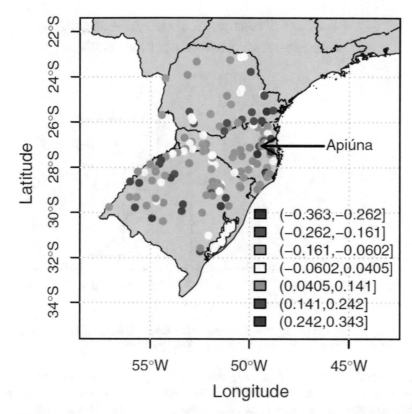

Fig. 11.8 Map of the south of Brazil with the location of the flow gauging stations used to elicit a prior distribution for the GEVshape parameter κ and spatial distribution of κ values (adapted from Silva et al. 2015)

11.3   Bayesian Estimation and Credibility Intervals

Bayesian estimation has its roots in *Decision Theory*. Bernardo and Smith (1994) argue that Bayesian estimation is fundamentally a decision problem. This section briefly explores some essential aspects of statistical estimation under a Bayesian approach. A more detailed description of such aspects can be found in Bernardo and Smith (1994) and Robert (2007).

Under the decision-oriented rationale for Bayesian estimation of a given parameter θ, one has a choice of a loss function, $\ell(\delta, \theta)$, which represents the loss or penalty due to accepting δ as an estimate of θ. The aim is to choose the estimator that minimizes the Bayes risk, denoted as BR and given by

$$\text{BR} = \iint \ell(\delta, \theta) f(x|\theta)\, \pi(\theta)\, dx\, d\theta \tag{11.19}$$

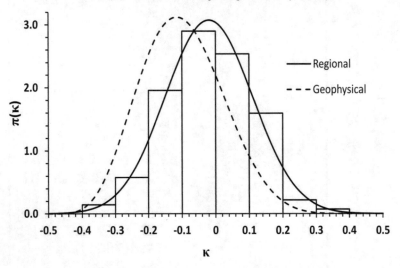

Fig. 11.9 Histogram and normal PDF fitted to the GEV shape parameter estimates, and the PDF of the geophysical prior as elicited by Martins and Stedinger (2000)

The inversion of the order of integration in Eq. (11.19), by virtue of Fubini's theorem (see details in Robert 2007, p. 63), leads to the choice of the estimator δ which minimizes the posterior loss, that is the estimator δ_B of θ such that

$$\delta_B = \min_\delta E\left[\ell(\delta,\theta)\big|x\right] = \min_\delta \int_\Theta \ell(\delta,\theta)\pi\left(\theta\big|x\right)dx \qquad (11.20)$$

The choice of the loss function is subjective and reflects the decision-maker's judgment on the fair penalization for his/her decisions. According to Bernardo and Smith (1994), the main loss functions used in parameter estimation are:

- *Quadratic loss*, in which $\ell(\delta,\theta) = (\delta - \theta)^2$ and the corresponding Bayesian estimator is the posterior expectation $E\left[\pi(\theta|x)\right]$, provided that it exists;
- *Absolute error loss*, in which $\ell(\delta,\theta) = |\delta - \theta|$ and the Bayesian estimator is the posterior median, provided that it exists;
- *Zero-one loss*, in which $\ell(\delta,\theta)=\mathbf{1}_{\delta\neq\theta}$ in which $\mathbf{1}_{(a)}$ is the indicator function and the corresponding Bayesian estimator of θ is the posterior mode.

Robert and Casella (2004) point out two difficulties related to the calculation of δ. The first one is that, in general, the posterior density of θ, $\pi(\theta|x)$, does not have a closed analytical form. The second is that, in most cases, the integration of Eq. (11.20) cannot be done analytically.

Parameter estimation highlights an important distinction between the Bayesian and the frequentist approach to statistical inference: the way in which those two approaches deal with uncertainties regarding the choice of the estimator. In frequentist analysis, this issue is addressed via the repeated sampling principle, and estimator performance based on a single sample is evaluated by the expected behavior of a hypothetical set of replicated samples collected under identical conditions, assuming that such replications are possible. The repeated sampling principle supports, for example, the construction of frequentist confidence intervals, CI (see Sect. 6.6). Under the repeated sampling principle, the confidence level $(1 - \alpha)$ of a CI is seen as the proportion of intervals, constructed on the basis of replicated samples under the exact same conditions as the available one, that contain the true value of the parameter.

The Bayesian paradigm, on the other hand, offers a more natural framework for uncertainty analysis by focusing on the probabilistic problem. In the Bayesian setting, the posterior variance of the parameter provides a direct measure of the uncertainty associated with its estimation. Credibility intervals (or posterior probability intervals) are the Bayesian analogues to the frequentist confidence intervals. The parameter θ, which is considered to be a random object, has $(1-\alpha)$ posterior probability of being within the bounds of the $100(1-\alpha)\%$ credibility interval. Thus the interpretation of the interval is more direct in the Bayesian case: there is a $(1-\alpha)$ probability that the parameter lies within the bounds of the credibility interval. The bounds of credibility intervals are fixed and the parameter estimates are random, whereas in the frequentist approach the bounds of confidence intervals are random and the parameter is a fixed unknown value.

Credibility intervals can be built not only for a parameter but also for any scalar function of the parameters or any other random quantity. Let ω be any random quantity and $p(\omega)$ its probability density function. The $(1-\alpha)$ credibility interval for ω is defined by the bounds (L,U) such that

$$\int_{L}^{U} p(\omega)\, d\omega = 1 - \alpha \tag{11.21}$$

Clearly, there is no unique solution for the credibility interval, even if $p(\omega)$ is unimodal. It is a common practice to adopt the highest probability density (HPD) interval, i.e., the interval $I \subseteq \Omega$, Ω being the domain of ω such that $P(\omega \in I) = 1 - \alpha$ and $p(\omega_1) \geq p(\omega_2)$ for every $\omega_1 \in I$ and $\omega_2 \notin I$ (Bernardo and Smith 1994). Hence the HPD interval is the narrowest interval such that $P(L \leq \omega \leq U) = 1 - \alpha$ and certainly provides a more natural and precise interpretation of probability statements concerning interval estimation of a random quantity.

11.4 Bayesian Calculations

The main difficulty in applying Bayesian methods is the calculation of the proportionality constant, or the prior predictive distribution, given by Eq. (11.2). To make inferences about probabilities, moments, quantiles, or credibility intervals, it is necessary to calculate the expected value of any function of the parameter $h(\theta)$, as weighted by the posterior distribution of the parameter. Formally, one writes

$$E[h(\theta)|x] = \int_{\Theta} h(\theta)\pi(\theta|x)d\theta \qquad (11.22)$$

The function h depends on the intended inference. For point estimation, h can be one of the loss functions presented in Sect. 11.3. In most hydrological applications, however, the object of inference is the prediction itself. If the intention is to formulate probability statements on the future values of the variable X, h can be the distribution of x_{n+1} given θ. In this case, one would have the posterior predictive distribution given by

$$F(x_{n+1}|x) = E[F(x_{n+1}|\theta)|x] = \int_{\Theta} F(x_{n+1}|\theta)\,\pi(\theta|x)\,d\theta \qquad (11.23)$$

The posterior predictive distribution is, therefore, a convenient and direct way of integrating sampling uncertainties in quantile estimation.

The analytical calculation of integrals as in Eq. (11.22) is impossible in most practical situations, especially when the parameter space is multidimensional. However, numerical integration can be carried out using the Markov Chain Monte Carlo (MCMC) algorithms. According to Gilks et al. (1996), such algorithms allow for generating samples with a given probability distribution, such as $\pi(\theta|x)$, through a Markov chain whose limiting distribution is the target distribution. If one can generate a large sample from the posterior distribution of θ, say θ_1, θ_2, ..., θ_m, then the expectation given by Eq. (11.22) may be approximated by Monte Carlo integration. As such,

$$E[h(\theta)|x] \approx \frac{1}{m}\sum_{i=1}^{m} h(\theta_i) \qquad (11.24)$$

In other terms, the population mean of h is estimated by the sample mean of the generated posterior sample.

When the sample $\{\theta_i\}$ is of IID variables, then, by the law of large numbers (see solution to Example 6.3), the approximation of the population mean can be as accurate as possible, requiring that the generated sample size m be increased (Gilks et al. 1996). However, obtaining samples of IID variables with density $\pi(\theta|x)$ is not a trivial task, as pointed out by Gilks et al. (1996), especially when $\pi(\theta|x)$ has a complex shape. In any case, the elements of $\{\theta_i\}$ need not be independent amongst themselves for the approximations to hold. In fact it is required only that the elements of $\{\theta_i\}$ be generated by a process which proportionally explores the whole support of $\pi(\theta|x)$.

To proceed, some definitions are needed. A Markov chain is a stochastic process $\{\theta_t, t \in T, \theta_t \in S\}$, where $T = \{1, 2, \ldots\}$ and S represents the set of possible states of

θ, for which, the conditional distribution of θ_t at any t, given $\theta_{t-1}, \theta_{t-2}, \cdots, \theta_0$, is the same as the distribution of θ_t, given θ_{t-1}, that is

$$P(\theta_{t+1} \in A | \theta_t, \theta_{t-1}, \cdots, \theta_0) = P(\theta_{t+1} \in A | \theta_t), \quad A \subseteq S \qquad (11.25)$$

In other terms, a Markov chain is a stochastic process in which the next state is dependent only upon the previous one. The Markov chains involved in MCMC calculations should generally have the following properties:

- *Irreducibility*, meaning that, regardless of its initial state, the chain is capable of reaching any other state in a finite number of iterations with a nonzero probability;
- *Aperiodicity*, meaning that the chain does not keep oscillating between a set of states in regular cycles; and
- *Recurrence*, meaning that, for every state *I*, the process beginning in *I* will return to that state with probability 1 in a finite number of iterations.

A Markov chain with the aforementioned characteristics is termed *ergodic*. The basic idea of the MCMC sampling algorithm is to obtain a sample with density $\pi(\theta | x)$ by building an ergodic Markov chain with: (1) the same set of states as θ; (2) straightforward simulation; and, (3) the limiting density $\pi(\theta | x)$.

The Metropolis algorithm (Metropolis et al. 1953) is well-suited to generate chains with those requirements. That algorithm was developed in the Los Alamos National Laboratory, in the USA, with the objective of solving problems related to the energetic states of nuclear materials using the calculation capacity of early programmable computers, such as the MANIAC (Mathematical Analyzer, Numerical Integrator and Computer). Although the method gained notoriety in 1953 through the work of physicist Nicholas Metropolis (1915–1999) and his collaborators, the algorithm development had contributions from several other researchers, such as Stanislaw Ulam (1909–1984), John Von Neumann (1903–1957), Enrico Fermi (1901–1954), and Richard Feynman (1918–1988), among others. Metropolis himself admitted that the original ideas were due to Enrico Fermi and dated from 15 years before the date it was first published (Metropolis, 1987). Further details on the history of the development of the Metropolis algorithm can be found in Anderson (1986), Metropolis (1987) and Hitchcock (2003).

The Metropolis algorithm was further generalized by Hastings (1970) into the version widely known and used today. The algorithm uses a reference or jump distribution $g(\theta^* | \theta_t, x)$, from which it is easy to obtain samples of θ through the following steps:

The Metropolis–Hastings algorithm for generating a sample with density $\pi(\theta | x)$:

Initialize θ_0; $t \leftarrow 0$
Repeat {
        Generate $\theta^* \sim g(\theta^* | \theta_t, x)$
        Generate $u \sim$ Uniform(0,1)

$$\text{Calculate } \alpha_{\text{MH}}\left(\theta^*|\theta_t, x\right) = \min\left\{1, \frac{\pi\left(\theta^*|x\right)}{\pi\left(\theta_t|x\right)} \frac{g\left(\theta_t|\theta^*, x\right)}{g\left(\theta^*|\theta_t, x\right)}\right\}$$

If $u \leq \alpha_{\text{MH}}\left(\theta^*|\theta_t, x\right)$

$\qquad \theta_{t+1} \leftarrow \theta^*$

Else

$\qquad \theta_{t+1} \leftarrow \theta_t$

$\quad t \leftarrow (t+1)$

}

An important aspect of the algorithm is that the acceptance rules are calculated using the ratios of posterior densities $\pi(\theta^*|x)/\pi(\theta_t|x)$, thus dismissing the calculation of the normalizing constant. The generalization proposed by Hastings (1970) concerns the properties of the jump distribution $g(\cdot|\cdot)$: the original algorithm, named random-walk Metropolis, required a symmetrical jump distribution, such that $g\left(\theta_i|\theta_j\right) = g\left(\theta_j|\theta_i\right)$. In this case the simpler acceptance rule is

$$\alpha_{\text{RWM}}\left(\theta^*|\theta_t, x\right) = \min\left\{1, \frac{\pi\left(\theta^*|x\right)}{\pi\left(\theta_t|x\right)}\right\} \tag{11.26}$$

Robert and Casella (2004) point out that after a large number of iterations, the resulting Markov chain may eventually reach equilibrium, after a sufficiently large number of iterations, i.e., its distribution converges to the target distribution. After convergence, all the resulting samples have the posterior density, and the expectations expressed by Eq. (11.22) may be approximated by Monte Carlo integration, with the desired precision.

As in most numerical methods, the MCMC samplers require some fine tuning. The choice of the jump distribution, in particular, is a key element for an efficient application of the algorithm. As Gilks et al. (1996) argue, in theory, the Markov chain will converge to its limiting distribution regardless of which jump distribution is specified. However, the numerical efficiency of the algorithm, as conveyed by its convergence rate, is greatly determined by how the jump density g relates to the target density π: convergence will be faster if g and π are similar. Moreover, even when the chain reaches equilibrium, the way it explores the support of the target distribution might be slow, thus requiring a large number of iterations. From the computational perspective, g should be a density that proves to be practical to evaluate at any point and to sample from. Furthermore, it is a good practice to choose a jump distribution with heavier tails than the target, in order to insure an adequate exploration of the support of $\pi(\theta|x)$. Further details on how to choose an adequate jump distribution are given by Gilks et al. (1996).

Another important aspect of adequate MCMC sampling is the choice of starting values: poorly chosen starting values might hinder the convergence of the chains for the first several hundreds or thousands of iterations. This can be controlled by discarding a sufficient number of iterations from the beginning of the chain, which is referred to as the *burn-in* period.

Currently, there are a large number of MCMC algorithms in the technical literature, all of them being special cases of the Metropolis–Hastings algorithm. One of the most popular algorithms is the Gibbs sampler, which is useful for multivariate analyses, and uses the complete conditional posterior distributions as the jump distribution. As a comprehensive exploration of MCMC algorithms is beyond the scope of this chapter, the following references are recommended for further reading: Gilks et al. (1996), Liu (2001), Robert and Casella (2004), and Gamerman and Lopes (2006).

Example 11.4 In order to demonstrate the use of the numerical methods discussed in this section, Example 11.2 is revisited with the random-walk Metropolis algorithm (acceptance rule given by Eq. 11.26). The following R code was used to generate a large sample from the posterior distribution of parameter p:

```
# prior density
pr <- function(theta) dbeta(theta, shape1 = 4.4375, shape2 = 13.3125)

# Likelihood function
ll <- function(theta) dbinom(0,10,theta)

# Unnormalized posterior density (defined for theta between 0 and 1)
unp <-function(theta)  ifelse(theta >=0 && theta <=1,  pr(theta)*ll
(theta),0)

theta_chain <- rep(NA,100000)
theta_chain[1] <- 0.15 #Initialize

# Random-walk Metropolis algorithm
set.seed(123)
for (i in 1:99999) {
    # Jump or proposal density is Normal with standard deviation 0.05
    proposal <- rnorm(1,mean = theta_chain[i],sd = 0.05)
    # Acceptance rule
    U <- runif(1)
    AR <- min(c(1,(unp(proposal)/unp(theta_chain[i]))))
    if (U <= AR) {
    theta_chain[i+1] <- proposal
    } else {
        theta_chain[i+1] <- theta_chain[i]
    }
}
```

The generated sample values are stored in the vector `theta_chain`. These values can be used for inferring any scalar function of the parameter. Figure 11.10 depicts the trace plot of the chain generated with the given code. A trace plot is a

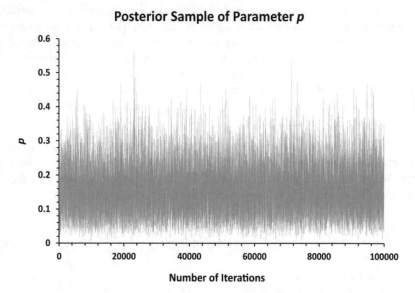

Fig. 11.10 Posterior sample of parameter p generated by the random-walk Metropolis algorithm

basic graphical tool used to detect clear signs of deviant or nonstationary behavior of generated chains, which may indicate a failure of convergence. It can also be used to determine the burn-in period when the starting values are poorly chosen. Visual inspection of a trace plot of a chain with good properties should not detect upward or downward trends or other nonstationary behavior, such as the chain getting stuck in certain regions of the parameter space. It should appear that each element of the plot is randomly sampled from the same target distribution. Figure 11.10, therefore, exemplifies a "good" traceplot.

The sample mean of the chain provides a point estimate for p, $E[p|y] = \frac{1}{N}\sum_{i=1}^{N} p_i = 0.1581$. This estimate is very close to the exact value of p, obtained in Example 11.2, $E[p|y] = 0.1599$. Figure 11.11 compares the exact solution of the posterior PDF of Example 11.2 with the histogram of the MCMC chain. It is clear that two solutions are very similar.

Even in the absence of a thorough monitoring of the MCMC chain convergence (see related procedures in Gilks et al. 1996; Liu 2001), Example 11.4 illustrates the capabilities of MCMC algorithms. In this case, it should be stressed that the posterior sample of the parameter was generated using the un-normalized posterior density, thereby dismissing the calculation of the constant of proportionality in the denominator of Bayes' rule, which, in most practical situations, would reveal impossible to be made.

Posterior Density of Parameter *p*

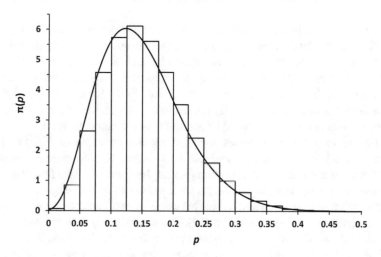

Fig. 11.11 Posterior density of parameter *p*. Exact solution (*continuous line*) and MCMC chain histogram

11.5   Example Application

This section presents an application of the principles of Bayesian analysis to a hydrology-related case study. The research was carried out by Fernandes et al. (2010). Readers interested in details of the case study are referred to this reference. The object of the study is related to the estimation of very rare extreme flood quantiles, with exceedance probabilities ranging from 10^{-6} to 2×10^{-3}, usually required for designing critical hydraulic structures, such as spillways of large dams with high potential flood hazards. The uncertainties associated with such estimates are admittedly very large and are not precisely quantifiable by standard procedures of statistical inference as the available samples of annual maximum floods have typical lengths ranging from 25 to 80 years. To tackle this problem, hydrologists often choose to estimate an upper bound for annual maximum floods, based on current knowledge of hydrological processes under extreme conditions. In this context, the concept of Probable Maximum Flood (PMF) is commonly used in connection to the design of major hydraulic structures (USNRC 1977; ICOLD 1987; FEMA 2004).

   In short, the PMF is the upper bound of potential flooding in a given river section, resulting from a hypothetical rain storm with a critical depth and duration, deemed probable maximum precipitation (PMP), which should be preceded by very severe, but physically plausible hydrological and hydrometeorological conditions (see Sect. 8.3.2). The PMP, in turn, is formally defined by the World Meteorological Organization (WMO 1986) as "the greatest depth of precipitation for a given duration meteorologically possible for a given size storm area at a particular location at a particular time of year, with no allowance made for long-time climatic trends."

Although dam safety guidelines consider "quasi-deterministic" floods, such as the PMF, as a standard design criterion for large hydraulic structures, the estimation of a credible exceedance probability associated with such an extreme flood, in a way that risk-based decisions can be made, is not a trivial task (Dawdy and Lettenmaier 1987; Dubler and Grigg 1996; USBR 2004). Some conceptual changes on how frequency analysis is usually conducted are required, before associating an exceedance probability with the PMF estimate for a given catchment.

There are two major obstacles for merging the concepts of PMF and flood frequency analysis. Firstly, the available information of flow extremes is usually scarce since the available annual maximum samples usually span a few decades and very rarely over more than a century. As such, extrapolation of frequency curves for very rare quantiles, well beyond a credible range of extrapolation, is required, with all the uncertainty it entails (see Sect. 8.3). The second obstacle is related to the fact that many probability distributions used in flood frequency analysis have no upper bound and, therefore, do not accommodate the inherent concept of the PMF. Fernandes et al. (2010) propose the following steps of a workaround procedure:

• Adopting an upper-bounded probability distribution: although the use of such distributions is uncommon and even controversial among hydrologists, their structure is concurrent with limited extreme flood generating physical conditions, that is, they accommodate the notion of an upper bound for a flood in a given catchment.
• Analysis of paleohydrologic proxy data: as discussed in Sect. 8.2.3 and later on in this section, paleoflood data allow for a more comprehensive characterization of rare floods.
• Using the PMF for estimating the upper bound: in the application example described herein, the PMF is not used as a deterministic upper bound but rather informs the elicitation of a prior distribution for that upper bound.
• Using Bayesian inference methods: the Bayesian framework of analysis offers the means to aggregate the information from various sources.

The application example itself is presented in Sect. 11.5.3. The preceding Sect. 11.5.1 and 11.5.2 are necessary for presenting essential concepts and the formalism required to grasp Sect. 11.5.3.

11.5.1 Nonsystematic Flood Data

In recent decades there have been attempts to overcome the lack of data on extreme floods in streamflow records by including proxy information, deemed nonsystematic data, in flood frequency analysis (Stedinger and Cohn 1986; Francés et al. 1994; Francés 2001; Naulet 2002; Viglione et al. 2013). There are two main sources of nonsystematic flood information: historical information, which refers to flood events directly observed or otherwise recorded by humans; and the so-called paleofloods, which correspond to floods that have occurred sometime in the Holocene epoch (approximately in the last 10,000 years) and can be reconstructed using remaining geological and/or botanical physical evidence.

Historical and paleohydrological information generally correspond to censored data which can be either upper-bounded (UB), or lower-bounded (LB), or double-bounded (DB). The type of censoring is determined by detectable limits of water levels of a particular flood during a given span of time. In any case, the exact maximum water level (or discharge) is not known with precision. For UB information, it is known that no flood has ever exceeded a certain level during the considered time span. For LB information, it is known that the flood has exceeded a given level. Finally for DB information, it is known that the maximum flood level is contained within an interval defined by two known bounds (LB,UB). Given some hypotheses, nonsystematic flood data can be incorporated into flood frequency analysis using appropriate statistical methods, thus potentially increasing the reliability of estimated extreme flood quantiles (Francés 2001).

11.5.2   Upper-Bounded Distributions

In the set of probability distributions that are commonly used by hydrologists in flood frequency analysis, some may be upper-bounded, depending on the particular combinations of the numerical values of their parameters. Some of those distributions may have up to 4 or 5 parameters (see Sect. 5.9). Others, such as the generalized extreme value (GEV) distribution is upper-bounded if its shape parameter is positive, which occurs when its skewness coefficient is less than 1.1396. Likewise, the log-Pearson type III (LP3) distribution has an upper limit if its skewness coefficient is negative.

Other upper-bounded probability models that should be mentioned are the Kappa (Hosking and Wallis 1997) and Wakeby distributions. The four-parameter Kappa model may be upper- or lower-bounded for a particular set of values of both its parameters (see Sect. 5.9.1). Analogously, the five-parameter Wakeby model may be upper-bounded for a particular combination of its three shape parameters (see Sect. 5.9.2). These are general distributional forms which can accommodate a number of bounded or unbounded models. For the application described herein, there is the additional requirement that the upper-bound should have an explicit parametric form. Fernandes et al. (2010) proposed the use of the four-parameter lognormal model (LN4) to explicitly incorporate the information provided by PMF estimates into frequency analysis.

The LN4 model branches from the following transformation

$$Y = \ln\left(\frac{X - \varepsilon}{\alpha - X}\right) \tag{11.27}$$

where $\varepsilon \in \mathfrak{R}_+$ denotes the lower bound of X, $\alpha \in \mathfrak{R}_+$ denotes the upper bound of X, and $Y \sim N(\mu_Y, \sigma_Y)$. Takara and Tosa (1999) point out that the LN4 model is not strongly affected by assuming its lower bound is zero. Taking advantage of that observation, the lower bound here is considered as $\varepsilon = 0$, thus rendering the model more parsimonious (one less parameter to estimate and one less prior distribution to elicit).

If $\varepsilon = 0$, the PDF of the variate $X \sim LN4(\mu_Y, \sigma_Y, \alpha)$ is given by

$$f_X(x|\Theta) = \frac{\alpha}{x(\alpha - x)\sigma_Y\sqrt{2\pi}} \exp\left\{-\frac{1}{2\sigma_Y^2}\left[\ln\left(\frac{x}{\alpha - x}\right) - \mu_Y\right]^2\right\} \tag{11.28}$$

with $0 \leq x \leq \alpha$, and the corresponding CDF is given by

$$F_X(x|\Theta) = \Phi\left[\frac{1}{\sigma_Y}\ln\left(\frac{x}{\alpha - x}\right) - \frac{\mu_Y}{\sigma_Y}\right] \tag{11.29}$$

where Φ denotes the CDF of the standard Normal distribution.

Finally, considering the PDF given by Eq. (11.28) and a set of systematic and nonsystematic data, the likelihood function can be constructed in the manner described as follows. Let $X_1, \ldots, X_{\text{Nex}}$ be the sample, of size Nex, of annual maximum floods (systematic data). Upper-bounded censored data are denoted as $\text{UB}_1, \ldots, \text{UB}_{\text{Nub}}$, lower-bounded censored data as $\text{LB}_1, \ldots, \text{LB}_{\text{Nlb}}$, whereas $\text{DB}_1, \ldots, \text{DB}_{\text{Ndb}}$ denote the censored data which is double-bounded within the interval (LB, UB). If the data are IID, then, the likelihood function is give by

$$\begin{aligned}
L(x|\theta) = {}&\prod_i^{\text{Nex}} f_X(\text{EX}_i|\Theta) \times \\
&\prod_i^{\text{Nub}} F_X(\text{UB}_i|\Theta) \times \\
&\prod_i^{\text{Nlb}} \left[1 - F_X(\text{LB}_i|\Theta)\right] \times \\
&\prod_i^{\text{Ndb}} \left[F_X(\text{UR}_i|\Theta) - F_X(\text{LR}_i|\Theta)\right]
\end{aligned} \tag{11.30}$$

11.5.3 Study Site and Data

The application example described here refers to the American River at Folsom Lake, located in the American State of California. This dam site was chosen particularly due to the availability of systematic and nonsystematic flood data. Figure 11.12 shows the schematic location of the study site.

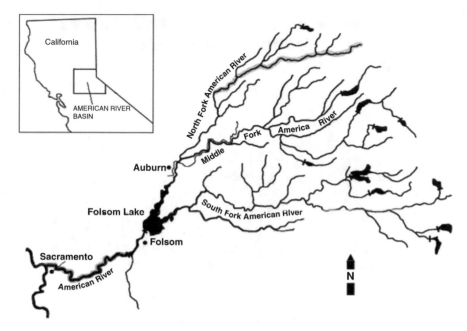

Fig. 11.12 American River basin (adapted from USBR 2002)

According to USBR (2002), the catchment of the American River at Folsom Lake has a drainage area of 4820 km^2 and its flows have been monitored by the US Geological Service since 1905. There are 52 years of systematic data. The annual maximum series passed the tests for randomness, independence and homogeneity and stationarity at the significance level of 5 %, according to the nonparametric hypothesis tests presented in Sect. 7.4.

Regarding the nonsystematic data, studies conducted by USBR (2002) identified the occurrence of 4 distinct paleoflood levels dating back 2000 years before present (BP), all of them being UB censored. Furthermore, there are 5 DB censored floods. The chart in Fig. 11.13 illustrates the data sets used in the case study.

11.5.4 Prior Distribution of Parameters of the LN4 Model

In the Bayesian paradigm, the prior distribution is the mathematical synthesis of the degree of knowledge or belief that some expert has about the quantity of interest. The specification of the distribution is based on personal belief gathered through observation, or experience gained from similar situations, literature review, etc. The prior distribution is elicited before the data are observed.

Amongst the parameters of the LN4, the upper-bound α is perhaps the only one that shows a clear connection to climate and hydrological characteristics of the watershed under extreme hydrometeorological conditions. The hydrological

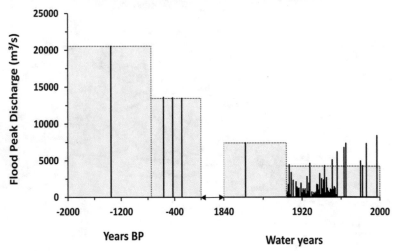

Fig. 11.13 Systematic and nonsystematic data of the American River near Folsom

interpretation of parameters μ_Y and σ_Y (respectively, the mean and standard deviation of the transformed LN4 variate Y) is more complex since it is not straightforward to identify any clear relationship between those parameters and physical characteristics of the catchment. Therefore, eliciting a prior distribution for these parameters is not simple, at least using this approach. A Normal distribution with a large variance was elicited for μ_Y, since it can take any real value, that is $\mu_Y \sim N(1.0, 10^{-6})$. Analogously, since σ_Y can only take positive values, the Gamma distribution $\sigma_Y \sim N(1.0, 10^{-8})$ was adopted as a prior for this parameter. Since these priors are proper, than the posterior parameter distributions should also be proper. Furthermore, it should be mentioned that the parametrization of the Normal and Gamma densities used in this example differ slightly from the parametrizations used elsewhere in this book. As such,

$$f_{\text{Normal}}(z|a,b) = \sqrt{\frac{b}{2\pi}}\exp\left[-\frac{b}{2}(z-a)^2\right] \quad \text{and} \quad f_{\text{Gamma}}(z|a,b) = \frac{b^a z^{a-1}}{\Gamma(a)}\exp(-bz)$$

Unlike the location and scale parameters, the upper bound parameter is directly related to hydrometeorological phenomena in the catchment, thus allowing for a more informative elicitation of a prior distribution. In theory, if there were a set of PMF estimates for the catchment, acquired using the same methodology applied at different points in time, such data might have been used for eliciting a prior distribution for the upper bound parameter. Since that notional data set cannot exist, prior elicitation for that parameter must take a different approach based on

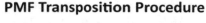

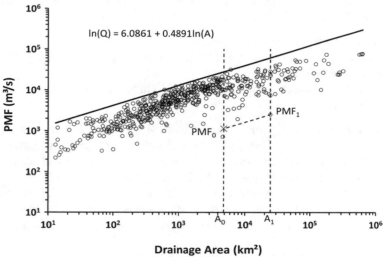

Fig. 11.14 Schematic of the PMF transposition procedure

available information. Fernandes et al. (2010) propose two methods based on a
large set of PMF estimates for different North American catchments. The following
is the detailed presentation of one of such methods.

   The proposed method is based on the transposition of PMF estimates from other
catchments to the catchment of interest. A data set of 561 PMF estimates compiled
by the US Nuclear Regulatory Commission (USNRC 1977) was used, referring to
catchments of varying sizes and characteristics scattered throughout the territory of
the USA. Figure 11.14 shows the applied procedure: rather than transposing the
PMF estimates directly to the study site, the envelope PMF curve was used, as
proposed by USNRC (1977), where A_0 is the drainage area of the American River at
Folsom Lake, resulting in transposing 561 PMF estimates to the catchment area of
the study site, whose frequency analysis can inform a prior distribution for the
upper bound of the LN4 distribution.

   Fernandes (2009) showed that the Gamma distribution is a good candidate for
modeling uncertainties related to the upper bound. Therefore, taking as reference
the 561 PMF estimates transposed to the American River at Folsom Lake, and using
the conventional method of moments, the prior distribution $\alpha \sim \text{Ga}(5.2, 0.00043)$
was elicited for the upper bound.

11.5.5   Posterior Distributions and Further Results

The posterior density of parameters is given by Eq. (11.1). Clearly there is no
analytical solution for such a case, since it would require integration of the product

of the likelihood by the priors over the whole parametric space. For that reason, MCMC algorithms are applied, as discussed in Sect. 11.4. The following questions are posed before an MCMC algorithm is applied:

- What jump distribution should be used?
- Which sampling algorithm should be applied (Metropolis–Hastings, Gibbs sampler, slice sampler...)?
- How does one check whether the Markov chain has converged to the target distribution?

There are no straightforward answers to these questions. It is up to the practitioner to fine-tune the samplers, try different algorithms and jump distributions until a method is found that produces well-converged chains so that usual convergence theorems from Markov theory apply. Fortunately, there are several freely available software packages that provide useful tools for tuning and evaluating MCMC samplers. A thorough presentation of such packages is beyond the scope of this chapter. Gilks et al. (1996) and Albert (2009) are useful references on MCMC software.

In the application described herein, the WinBUGS software was used (Lunn et al. 2000). WinBUGS is a free user-friendly tool that requires minimal programming skills and is available from http://www.mrc-bsu.cam.ac.uk/software/bugs/the-bugs-project-winbugs/ (accessed: 21st February 2016). The following code was used at different stages of analysis of the presented case study:

```
# model with systematic and nonsystematic data
model {              # Likelihood of the LN4 distribution
    #SYSTEMATIC DATA
    for (i in 1:NEX) {
        Z[i] <- abs(x[i]/(alpha-x[i]))
        index[i] <- step(alpha-x[i]) +1
        L[i,1] <- 0
        L[i,2] <- (0.3989*alpha/sigma)*(1/ (x[i]*(alpha-x[i])))
*exp(-0.5*pow((1/sigma)*log(Z[i])-mi/sigma,2))
    }
    #DB DATA
    for (i in NEX+1:NDB+NEX) {
        index[i] <- step(alpha-DBU[i-NEX]) +1
        L[i,1] <- 0
        Zu[i-NEX] <- abs(DBU[i-NEX]/(alpha-DBU[i-NEX]))
        Zl[i-NEX] <- abs(DBL[i-NEX]/(alpha-DBL[i-NEX]))
        L[i,2] <-     phi((1/sigma)*log(Zu[i-NEX])-mi/sigma)-phi
((1/sigma)*log(Zl[i-NEX])-mi/sigma)
    }
    #UB DATA - Level 1
    for (i in NDB+NEX+1:NDB+NEX+NYH[1]) {
        index[i] <- step(alpha-YH[1]) +1
        L[i,1] <- 0
```

```
        Zub1[i-NDB-NEX] < - abs(YH[1]/(alpha-YH[1]))
        L[i,2] < - phi((1/sigma)*log(Zub1[i-NDB-NEX])-mi/sigma)
    }
    #UB DATA - Level 2
    for (i in NDB+NEX+NYH[1]+1:NDB+NEX+NYH[1]+NYH[2]) {
        index[i] <- step(alpha-YH[2])+1
        L[i,1] <- 0
        Zub2[i-NDB-NEX-NYH[1]] < - abs(YH[2]/(alpha-YH[2]))
        L[i,2] <-  phi((1/sigma)*log(Zub2[i-NDB-NEX-NYH[1]])-mi/
sigma)
    }
    #UB DATA- Level 3
    for (i  in  NDB+NEX+NYII[1]+NYH[2]+1:NDB+NEX+NYH[1]+NYH
[2]+NYH[3]) {
        index[i] <- step(alpha-YH[3])+1
        L[i,1] <- 0
        Zub3[i-NDB-NEX-NYH[1]-NYH[2]] <-      abs(YH[3]/(alpha-YH
[3]))
        L[i,2] <-     phi((1/sigma)*log(Zub3[i-NDB-NEX-NYH[1]-NYH
[2]])-mi/sigma)
    }
    #UB DATA- Level 4
    for  (i  in  NDB+NEX+NYH[1]+NYH[2]+NYH[3]+1:NDB+NEX+NYH
[1]+NYH[2]+NYH[3]+NYH[4])
{
        index[i] <- step(alpha-YH[4])+1
        L[i,1] <- 0
        Zub4[i-NDB-NEX-NYH[1]-NYH[2]-NYH[3]] <-         abs(YH
[4]/(alpha-YH[4]))
        L[i,2] <-     phi((1/sigma)*log(Zub4[i-NDB-NEX-NYH[1]-NYH
[2]-NYH[3]])-mi/sigma)
    }
    for (i in 1:NDB+NEX+NYH[1]+NYH[2]+NYH[3]+NYH[4]) {
        dummy[i] <- 0
        dummy[i] ~dgeneric(phi[i])
        phi[i] <- log(L[i,index[i]])
    }
    # PRIOR DENSITIES
    sigma~dgamma(1.0,1.0E-8)
    mi~dnorm(1.0,1.0E-6)
    alpha~dgamma(5.2,0.00043)
}
```

#Systematic and nonsystematic data
```
list(x=c(685,1691,4417,292,3370,2302,2098,1356,1152,1198,
         1911,569,1110,895,1104,396,2818,776,1917,4616,
         691,280,597,467,640,1724,1651,934,3228,309,
         2526,1099,2356,4304,569,2673,1195,790,595,1062,
         974,5097,1053,1407,1206,6201,6796,7362,4955,4304,
         7334,8438),
     NEX=52,
     DBL=c(7419,11327,11327,11327,16990),
     DBU=c(8495,15574,15574,15574,24069),
     NDB=5,
     YH=c(4248,7447,13451,20530),
     NYH=c(44,56,544,1299)
)
```
#Initial values
```
list(  alpha=25000,
       sigma=2,
       mi=2)
```

By applying the WinBUGS code above, a chain of length 600,000 was generated. After a visual analysis of the trace plots of the chain, it was verified that it became stationary after the burn-in period of 100,000 realizations, which were discarded before proceeding with the analysis. Furthermore, the chains were thinned by retaining every 10th value, in order to remove autocorrelation. These actions resulted in a sample of 50,000 posterior parameters. Figure 11.15 shows the prior and posterior densities of the upper bound.

Figure 11.15 shows a clear disparity regarding the lower tails of the prior and posterior distributions. Prior to looking at the data there is no evidence regarding how low the upper bound could possibly be, which provides support for a prior distribution developing throughout the set of all positive real numbers. On the other hand, after taking the data into account, it does not seem logical to say that the upper bound can be lower than the maximum observed flood, so the support was modified for the posterior distribution. Table 11.1 shows some statistics for the prior and posterior distributions, illustrating how the systematic and nonsystematic data significantly reduced the uncertainty about the upper bound, since the posterior coefficient of variation CV is much lower than the prior one.

Several additional analyses can be made using posterior statistics, as shown in Fernandes et al. (2010). Assessment of the quantile curve uses the previously discussed concepts of the Monte Carlo method, particularly Eq. (11.23). The quantile curve together with its respective 95 % HPD credibility intervals are shown in Fig. 11.16.

Concerning the method proposed by Fernandes et al. (2010), some considerations are important to understand the context of the described application. The authors analyzed a wide range of models and approaches, including the following:

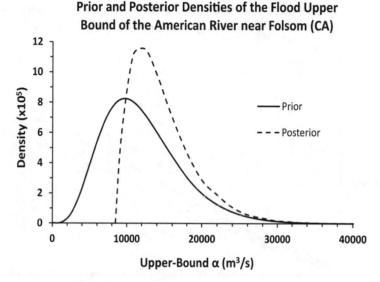

Fig. 11.15 Prior and posterior densities of the upper bound

Table 11.1 Prior and posterior summaries for the upper-bound α

Model	Mean	CV	95 % HPD
Posterior using all data	28,104	0.12	(24,070; 34,980)
Posterior using only systematic data	14,628	0.29	(8609; 22,960)
Prior information	12,018	0.40	

- The likelihood function of the LN4 distribution and of two other upper bounded distributions;
- The likelihood function with and without nonsystematic data;
- Different prior distributions, including non-informative distribution; and
- Parameter estimation under the Bayesian and the classical approaches.

To summarize, results show that the incorporation of nonsystematic data significantly improves estimation of extreme quantiles. They also show that the elicitation of an informative prior distribution is essential when analyzing very rare floods. Furthermore, it was shown that the Bayesian framework is able to combine the objective and subjective aspects of flood frequency analysis.

11.6   Further Reading and Software

This is an entry-level chapter on Bayesian methods, with a particular focus on hydrological applications. For a deeper grasp of Bayesian statistics, interested readers are referred to Bernardo and Smith (1994), Migon and Gamerman (1999),

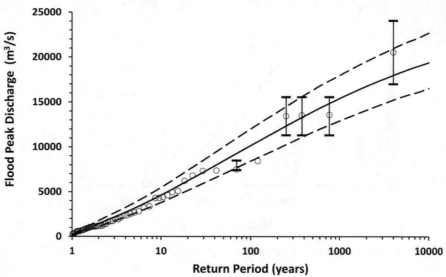

Fig. 11.16 Predictive posterior distribution of flood quantiles (*continuous line*); 95 % credibility intervals (*dashed line*). *Circles* represent systematic data and *circles within bars* represent nonsystematic data

Gelman et al. (2004), Paulino et al. (2003) and Robert (2007). In Chap. 11 of his textbook, entitled "A defense of the Bayesian choice," Robert (2007) makes a point-by-point justification of the Bayesian approach including rebuttals of the most common criticisms of the Bayesian paradigm, which makes for a particularly interesting reading. Renard et al. (2013) provide an up-to-date overview of Bayesian methods for frequency analysis of hydrological extremes with an emphasis on nonstationarity analysis.

There are many software packages available for MCMC. In R, there are the mcmc (http://www.stat.umn.edu/geyer/mcmc/, accessed: 26th March 2016), MCMCpack (Martin et al. 2011) and LaplacesDemon (currently available on https://github.com/ecbrown/LaplacesDemon, accessed: 26th March 2016) packages. Alternatively there are the WinBUGS (Lunn et al. 2000) and JAGS (http://mcmc-jags.sourceforge.net/, accessed: 26th March 2016) software packages.

Exercises

1. Solve Example 11.1 considering that 1 year after the factory operation started, it was verified that all the floods lasted for more than 5 days, that is, $\varepsilon = 1$. Calculate the posterior distribution.

2. Solve Example 11.2 considering, (a) $s_p^2 = 0.1$ and (b) $s_p^2 = 0.001$. Plot the posterior distribution and comment on the uncertainty of the parameter in each case.

3. The daily number of ships docking at a harbor is Poisson-distributed with mean θ, whose prior distribution is Exponential with mean 1. Knowing that in a 5-day stretch the number of arrivals was 3, 5, 4, 3, and 4: (a) determine the posterior distribution of θ; and, (b), obtain the 90 and 05 % credibility intervals for θ (Adapted from Paulino et al. 2003).

4. Show that, for a Normal distribution with known σ, if $\mu_{prior} \sim N(\mu_\mu, \sigma_\mu)$ then
$$\mu_{posterior} \sim N\left(\frac{\mu_\mu(\sigma^2/n)+\bar{x}\sigma_\mu^2}{\sigma^2/n+\sigma_\mu^2}, \sqrt{\frac{\sigma_\mu^2(\sigma^2/n)}{\sigma_\mu^2+(\sigma^2/n)}}\right).$$

5. Two meteorologists, A and B, wish to determine the annual rainfall depth (θ in mm) over an ungauged region. Meteorologist A has the prior belief $\theta \sim N(1850, 30^2)$ while meteorologist B believes that $\theta \sim N(1850, 70^2)$. A rain gauge is installed in a region. After 1 year, the rain gauge registered the cumulative rainfall of $x = 1910$. Find the meteorologists' posterior distributions, considering that the standard error of the annual cumulative rainfall depth is 40 mm and that annual rainfalls are normally distributed, i.e., $X|\theta \sim N(\theta, 40^2)$.

6. Consider a normally distributed variate with mean θ and standard deviation 2. A normal prior for θ, with variance 1, was elicited. What is the minimum size of the sample in order for the posterior standard deviation to be 0.1?

7. Consider that $X \sim Ge(\theta)$. Obtain the Jeffreys prior for θ.

8. Consider the elicitation of the prior distribution for the upper bound as described in Sect. 11.5.4. Suppose further that the upper bound CV $= 0.3$ and that the local PMF is 25,655 m³/s. Elicit the prior distribution of the upper bound based on the gamma distribution and in the following settings: (a) there is very strong evidence that the PMF will be exceeded in the future; (b) there is a very strong evidence that the PMF will not be exceeded in the future; and (c), there is no evidence regarding the probability of the PMF.
   Hint: for the Gamma distribution with parameters ρ_α and β_α, the combination of equations of the method of moments results in $\rho_\alpha = CV^{-2}$. Parameter β_α can be estimated by attributing a non-exceedance probability p to the current PMF estimate, that is, $P(\alpha \leq PMF|\rho_\alpha, \beta_\alpha) = p$.

9. Using the WinBUGS algorithm shown in Sect. 11.5.5, and the prior distributions elicited in Exercise 8, find the posterior distributions of the upper bound, considering that the remaining parameters of the LN4 distribution have non-informative priors.

10. Solve Exercises 8 and 9 for CV $= 0.7$.

11. Consider the sample of annual maximum flows of the Lehigh River at Stoddartsville, listed in Table 7.1. Fit the GEV distribution to the data, with the geophysical prior for the shape parameter, using the following WinBUGS code:

```
# GEV distribution

model {

    for (i in 1:NEX) {
index[i] <- 1-equals(step(x[i]-mi-sigma/(p-0.5)),step((p-0.5)))
index2[i]<-equals(index[i],1)+1
L[i,1]<-0
Z[i] <- index[i]*((p-0.5)/sigma)*(x[i]-mi)
L[i,2]   <-   index[i]*(1/sigma)*pow(1-Z[i],1/(p-0.5)-1)*exp(-pow(1-Z
[i],1/(p-0.5)))
}
for (i in 1:NEX) {
dummy[i] <- 0
dummy[i] ~ dgeneric(phi[i])
# phi(i) = log(likelihood)
phi[i] <- log(L[i,index2[i]])
}

    # priors
    sigma ~ dgamma(1.0,1.0E-8)
    mi ~ dnorm(1.0,1.0E-6)
    p ~ dbeta(6.0,9.0)
    }
```

12. Solve Exercise 11 for the Gumbel distribution, using non-informative priors for both parameters.

References

Albert J (2009) Bayesian computation with R, 2nd edn. Springer, New York
Anderson H (1986) Metropolis, Monte Carlo, and The MANIAC. Los Alamos Science. U.S. Government Printing Office, pp 96–108
Ang A, Tang W (2007) Probability concepts in engineering, 2nd edn. Wiley, Hoboken
Bernardo J, Smith A (1994) Bayesian theory. Wiley, Chichester, UK
Brooks S (2003) Bayesian computation: a statistical revolution. Philos Trans Roy Soc A Math Phys Eng Sci 361(1813):2681–2697
Coles S, Powell E (1996) Bayesian methods in extreme value modelling: a review and new developments. Int Stat Rev 64(1):119
Dawdy D, Lettenmaier D (1987) Initiative for risk-based flood design. J Hydraul Eng 113 (8):1041–1051
Dubler J, Grigg N (1996) Dam safety policy for spillway design floods. J Prof Issues Eng Educ Pract 122(4):163–169
Ehlers R (2016) Inferência Bayesiana (Notas de Aula) Ricardo Ehlers. http://www.icmc.usp.br/~ehlers/bayes/. Accessed 1st Apr 2016

FEMA—Federal Emergency Management Agency (2004) Federal guidelines for dam safety: selecting and accommodating inflow design floods for Dams. The Interagency Committee of the U.S. Department of Homeland Security, Washington

Fernandes W (2009) Método para a estimação de quantis de enchentes extremas com o emprego conjunto de análise bayesiana, de informações não sistemáticas e de distribuições limitadas superiormente. PhD Thesis. Universidade Federal de Minas Gerais, Belo Horizonte, Brazil

Fernandes W, Naghettini M, Loschi R (2010) A Bayesian approach for estimating extreme flood probabilities with upper-bounded distribution functions. Stoch Environ Res Risk Assess 24 (8):1127–1143

Francés F (2001) Incorporating Non-Systematic Information to Flood Frequency Analysis Using the Maximum Likelihood Estimation Method. In: Glade T, Albini P, Francés F (eds) The use of historical data in natural hazard assessments, 1st edn. Springer, Dordrecht, pp 89–99

Frances F, Salas J, Boes D (1994) Flood frequency analysis with systematic and historical or paleoflood data based on the two-parameter general extreme value models. Water Resour Res 30(6):1653–1664

Gamerman D, Lopes H (2006) Markov chain Monte Carlo, 2nd edn. Chapman and Hall/CRC, Boca Raton

Gelman A, Carlin J, Stern H, Rubin D (2004) Bayesian data analysis, 2nd edn. Chapman and Hall/CRC, Boca Raton

Gilks W, Richardson S, Spiegelhalter D (1996) Markov chain Monte Carlo in practice. Chapman & Hall, London

Hastings W (1970) Monte Carlo sampling methods using Markov chains and their applications. Biometrika 57(1):97–109

Hitchcock D (2003) A history of the Metropolis-Hastings algorithm. Am Stat 57(4):254–257

Hosking J, Wallis J (1997) Regional frequency analysis. Cambridge University Press, Cambridge

ICOLD—International Commission of Large Dams (1987) Dam safety guidelines. Bulletin 59. ICOLD (International Congress of Large Dams), Paris

Liu J (2001) Monte Carlo strategies in scientific computing. Springer, New York

Lunn D, Thomas A, Best N, Spiegelhalter D (2000) WinBUGS—a Bayesian modelling framework: concepts, structure, and extensibility. Statist Comput 10:325–337

Martin A, Quinn K, Park J (2011) MCMCpack: Markov Chain Monte Carlo in R. J Stat Software 42(9)

Martins E, Stedinger J (2000) Generalized maximum-likelihood generalized extreme-value quantile estimators for hydrologic data. Water Resour Res 36(3):737–744

McGrayne S (2011) The theory that would not die: how Bayes' rule cracked the enigma code, hunted down Russian submarines, and emerged triumphant from two centuries of controversy. Yale University Press, New Haven, CT

Metropolis N (1987) The beginning of the Monte Carlo Method. Los Alamos Science. U.S. Government Printing Office, Special Issue, 1986-676-104/40022

Metropolis N, Rosenbluth A, Rosenbluth M, Teller A, Teller E (1953) Equation of state calculations by fast computing machines. J Chem Phys 21(6):1087–1092

Migon H, Gamerman D (1999) Statistical inference: an integrated approach. Arnold, London

Naulet R (2002) Utilisation de l'information des crues historiques pour une meilleure prédétermination du risque d'inondation. Application au bassin de l'Ardèche à Vallon Pont-d'Arc et St-Martin d'Ardèche. PhD Thesis. INRS-ETE

Paulino C, Turkman M, Murteira B (2003) Estatística Bayesiana. Fundação Calouste Gulbenkian, Lisboa

Renard B, Sun X, Lang M (2013) Bayesian Methods for Non-stationary Extreme Value Analysis. In: AghaKouchak A, Easterling D, Hsu K, Schubert S, Sorooshian S (eds) Extremes in a changing climate: detection, analysis and uncertainty, 1st edn. Springer, Dordrecht, pp 39–95

Robert C (2007) The Bayesian choice, 2nd edn. Springer, New York

Robert C, Casella G (2004) Monte Carlo statistical methods. Springer, New York

Silva A, Portela M, Naghettini M, Fernandes W (2015) A Bayesian peaks-over-threshold analysis of floods in the Itajaí-açu River under stationarity and nonstationarity. Stoch Environ Res Risk Assess 1:1–20. doi:10.1007/S00477-015-1184-4

Stedinger J, Cohn T (1986) Flood frequency analysis with historical and paleoflood information. Water Resour Res 22(5):785–793

Takara K, Tosa K (1999) Storm and flood frequency analysis using PMP/PMF estimates. In: International Symposium on Floods and Droughts. Nanjing, pp 7–17

USBR—United States Bureau of Reclamation (2002) Flood hazard analysis, Folsom dam. Central Valley Project. USBR, Denver

USBR—United States Bureau of Reclamation (2004) Hydrologic hazard curve estimating procedures. Research Report DSO-04-08. USBR, Denver

USNRC—United States Nuclear Regulatory Commission (1977) Design basis floods for nuclear power plants. Regulatory Guide 1.59. USNRC, Washington

Viglione A, Merz R, Salinas J, Blöschl G (2013) Flood frequency hydrology: 3. A Bayesian analysis. Water Resour Res 49(2):675–692

WMO—World Meteorological Organization (1986) Manual for estimation of probable maximum precipitation. Operational Hydrologic Report No. 1, WMO No. 332, 2nd edn. WMO, Geneva

Chapter 12
Introduction to Nonstationary Analysis and Modeling of Hydrologic Variables

Artur Tiago Silva

12.1 Some Methods for Detecting Changes

The detection of changes in hydrological series plays an important role in the modern management and planning of water resources systems, since it enables a better understanding of the temporal behavior of the underlying hydrological phenomenon. The detection of trends, periodicities, change points or other deterministic components of the hydrological series weakens the assumption of stationarity of the random variable to which the series refers, which in turn, affects the applicability of the standard technical procedures most commonly used in Statistical Hydrology.

There are several kinds of tests and other methods for change detection, each having its own advantages and limitations. These methods should not be applied without considering the specific characteristics of the hydrological variable under analysis or without taking into account their underlying theoretical assumptions. It should be noted that some techniques presented elsewhere in this book are commonly used for the detection of change, such as the Spearman nonparametric test presented in Sect. 7.4.4. Another technique presented in a different chapter is simple linear regression: the methods introduced in Chap. 9, when applied to a hydrological random variable, using time as an independent variable, provide insight into the linear behavior of this variable over time, and thereby detect linear trends.

In this section, two methods for detecting changes are presented: the Mann–Kendall test for monotonic trend and the Pettitt test for change points.

A.T. Silva (✉)
CERIS, Instituto Superior Técnico, Universidade de Lisboa, Lisbon, Portugal
e-mail: artur.tiago.silva@tecnico.ulisboa.pt

© Springer International Publishing Switzerland 2017   537
M. Naghettini (ed.), *Fundamentals of Statistical Hydrology*,
DOI 10.1007/978-3-319-43561-9_12

These methods, in addition to the above mentioned techniques which are presented elsewhere in this book, cover most of the common situations faced by researchers and practitioners in water resources engineering in cases related to the nonstationary behavior of hydrological variables.

12.1.1 Hypothesis Test for a Monotonic Trend

Originally proposed by Mann (1945) and further studied by Kendall (1975), the nonparametric Mann–Kendall test for trend constitutes a widely used approach for detecting monotonic trends in hydrologic time series. This test has the advantage of not assuming a particular form for the distribution of the observed data, while its performance is comparable to that of the hypothesis test of the linear regression slope parameter, presented in Sect. 9.2.4.

Consider the hydrologic series X_t with $t = 1, 2, \ldots, N$. The Mann–Kendall test statistic is given by

$$S = \sum_{i=1}^{N-1} \sum_{j=i+1}^{N} \mathrm{sgn}(X_j - X_i) \tag{12.1}$$

where the sign function $\mathrm{sgn}(y) = 1$ if $y > 0$, $\mathrm{sgn}(y) = -1$ if $y < 0$, and $\mathrm{sgn}(y) = 0$ if $y = 0$. S has a null expected value $E[S] = 0$ and its variance is given by

$$Var[S] = \frac{1}{18}\left[N(N-1)(2N+5) - \sum_{m=1}^{M} t_m(t_m - 1)(2t_m + 5)\right] \tag{12.2}$$

where M is the number of sets of tied groups and t_m is the size of the mth tied group. The standardized test statistic Z, which follows a standard Normal distribution, is computed as

$$Z = \begin{cases} \dfrac{S-1}{\sqrt{Var[S]}} & , \ S > 0 \\[2mm] 0 & , \ S = 0 \\[2mm] \dfrac{S+1}{\sqrt{Var[S]}} & , \ S < 0 \end{cases} \tag{12.3}$$

The Mann–Kendall test has the null hypothesis H_0: {no trend in hydrologic series X_t}. The region of rejection of the standardized test statistic, Z, is dependent on the specification of the significance level α, as well as on the formulation of the alternative hypothesis H_1. If H_1 is {*increasing* monotonic trend in X_t}, it is a right-tailed hypothesis test, with H_0 being rejected if

$Z > z_{1-\alpha}$. If H_1 is {*decreasing* monotonic trend in X_t}, it is a left-tailed hypothesis test, with H_0 being rejected if $Z < z_\alpha$. Alternatively, if H_1 is {monotonic trend in X_t} without specification of the sign of the trend, the hypothesis test is two-tailed and the decision is to reject H_0 if $|Z| > z_{1-\alpha/2}$.

12.1.2 Hypothesis Test for a Change Point

The Pettitt test (Pettitt 1979) considers that a time series of a random variable X_t with $t = 1, 2, \ldots, N$, has a change point at time step τ if the values of X_t for $t = 1$, $2, \ldots, \tau$ have the CDF $F_1(x)$, and the values of X_t for $t = \tau + 1, \tau + 2, \ldots, N$ have the CDF $F_2(x)$ and $F_1(x) \neq F_2(x)$. The application of the Pettitt test does not require the prior specification of the time step at which the change is supposed to have occurred.

The Pettitt test has the null hypothesis H_0: {no change point in hydrologic series X_t}. The procedures to test the hypothesis are as follows. First, consider the matrix elements

$$D_{i,j} = \text{sgn}(X_i - X_j) \tag{12.4}$$

where $\text{sgn}(\cdot)$ is the sign function. The sum of specific submatrices of the matrix D results in the following statistic:

$$U_{t,N} = \sum_{i=1}^{t} \sum_{j=t+1}^{N} D_{i,j} \tag{12.5}$$

The $U_{t,N}$ statistic should be computed for values of t ranging from 1 to N, which can be done using the following iterative formula (Rybski and Neumann 2011):

$$U_{t,N} = U_{t-1,N} + \sum_{j=1}^{N} \text{sgn}(X_t - X_j) \tag{12.6}$$

In order to test H_0 against alternative hypothesis H_1: {change point in X_t}, the test is two-tailed and the Pettitt test statistic is given by

$$K_N = \max_{1 \leq t < N} |U_{t,N}| \tag{12.7}$$

It is also possible to make one-tailed tests of H_0 against alternative hypotheses H_1: {*upward* change in X_t} and H_1: {*downward* change in X_t},

$$K_N^- = - \min_{1 \leq t < N} U_{t,N} \tag{12.8}$$

$$K_N^+ = \max_{1 \leq t < N} U_{t,N} \tag{12.9}$$

such that $K_N = \max\left(K_N^-, K_N^+\right)$. Under H_0, $E\left[D_{i,j}\right] = 0$ and the distribution of $U_{t,N}$ for each t is symmetrical with mean at zero. On one-tailed tests, a high value of K_N^+ is expected when the series exhibits a downward change. In such a situation, the inequality $F_1(x) \leq F_2(x)$ is verified for at least some values of x. Analogously, high values of K_N^- are expected if the series shows an upward change, i.e., $F_1(x) \geq F_2(x)$.

P-values of the Pettit test can be approximated by

$$p \approx \exp\left(-\frac{6K_N^{+2}}{N^3 + N^2}\right) \tag{12.10}$$

for K_N^+ or for K_N^-, and

$$p \approx 2\exp\left(-\frac{6K_N^2}{N^3 + N^2}\right) \tag{12.11}$$

for K_N. These approximations hold for $p \leq 0.5$.

The null hypothesis H_0 should be rejected if $p < \alpha$, at the test significance level α. The supposed change to which the statistical test is applied occurs at the time step corresponding to the maximum (or minimum, depending on which test statistic is used) of $U_{t,N}$.

Example 12.1 Consider the series of annual peak flows of the Lehigh River at Stoddartsville (Table 7.1). (a) Test the hypothesis of that series exhibiting a monotonic trend using the Mann–Kendall test. (b) Use the two-tailed Pettitt test to assess if that series shows any evidence of a change point. Consider the significance level $\alpha = 0.05$.

Solution
(a) The series has length $N = 73$. Equations (12.1), (12.2) and (12.3), provide the following results, respectively, $S = 64$, $V[S] = 44084$, and $Z = 0.300$. At the significance level $\alpha = 0.05$, $z_{1-\alpha/2} = 1.96$. Since $|Z| < z_{1-\alpha/2}$, the null hypothesis (no trend) is not rejected.
(b) The test is two-tailed, so the test statistic is given by Eq. (12.7). Hence $K_N = 285$, which corresponds to the value of $U_{t,N}$ at $t = 60$. To calculate the p-value, Eq. (12.11) is used, thus resulting in $p = 0.581$. Since $p > \alpha$, the null hypothesis (no change point) is not rejected.

Example 12.2 The Três Marias reservoir on the São Francisco upper river catchment, with a drainage area of 50,600 km², in southeastern Brazil, has a total storage volume of 21 hm³ $\left(21 \times 10^9\,\text{m}^3\right)$ and started operation in 1961. Downstream of the dam, at the location of Pirapora, a study is carried out on whether the operation of the reservoir casts a significant regulation on the streamflow. Consider the hydrological series of July mean monthly flows observed at the gauging station of Pirapora, from 1938 to 1992, displayed in Table 12.1. The São Francisco River catchment at Pirapora has a drainage area of $114{,}000\ km^2$ and is located circa 90 km to the north, downstream of the Três Marias dam. July is usually the driest month in this region of Brazil and the July monthly discharges are supposed to be strongly affected by reservoir flow regulation. (a) Apply the nonparametric Mann–Whitney test (see Sect. 7.4.3) to assess whether the subsamples from 1931 to 1961 and from 1962 to 1992 are homogeneous at the 5 % significance level. (b) Apply the two-tailed Pettitt test at the 5 % significance level. Comment on the result of the Pettitt test, particularly in relation to the time position of the change point.

Solution

(a) Homogeneity hypothesis test. The fourth column of Table 12.1 shows the order ranks of the mean monthly flows in July. So $N_1 = 24$, $N_2 = 31$, and $R_1 = 346$. Equations (7.14) and (7.15) give $V_1 = 674$ and $V_2 = 70$, thus, $V = 70$. Equations (7.16) to (7.18) give $E[V] = 372$, $Var[V] = 3472$ and $T = -5.12$. At the significance level $\alpha = 0.05$, $z_{1-\alpha/2} = 1.96$. Since $|T| > z_{1-\alpha/2}$, the null hypothesis is rejected and, as a conclusion, the two subsamples are not homogeneous.

Table 12.1 July mean monthly flows of the São Francisco River at Pirapora, in Brazil, and auxiliary measures for implementing the Mann–Whitney and Pettitt change-point tests

Year	t	X_t	m_t	Subsample	$U_{t,N}$
1938	1	325.00	14	1	−28
1939	2	359.00	18	1	−48
1940	3	248.00	8	1	−88
1941	4	344.00	17	1	−110
1942	5	270.00	9	1	−148
1943	6	478.00	24	1	−156
1944	7	311.00	13	1	−186
1945	8	558.00	34	1	−174
1946	9	422.00	20	1	−190
1947	10	456.00	22	1	−202
1948	11	241.00	7	1	−244
1949	12	517.00	29	1	−241
1950	13	338.00	16	1	−265
1951	14	450.00	21	1	−279
1952	15	487.00	26	1	−282
1953	16	227.00	5	1	−327
1954	17	167.00	3	1	−377
1955	18	153.00	2	1	−429
1956	19	282.00	12	1	−461
1957	20	400.00	19	1	−479

(continued)

Table 12.1 (continued)

Year	t	X_t	m_t	Subsample	$U_{t,N}$
1958	21	273.00	10	1	−515
1959	22	152.00	1	1	−569
1960	23	227.00	5	1	−614
1961	24	278.00	11	1	−648
1962	25	573.00	36	2	-632
1963	26	507.00	28	2	−632
1964	27	337.00	15	2	−658
1965	28	684.00	47	2	−620
1966	29	733.00	51	2	−574
1967	30	622.00	40	2	−550
1968	31	655.00	43	2	−519
1969	32	595.00	37	2	−500
1970	33	525.00	31	2	−493
1971	34	218.00	4	2	−541
1972	35	479.00	25	2	−547
1973	36	487.00	26	2	−550
1974	37	611.00	39	2	−528
1975	38	525.00	31	2	−521
1976	39	853.00	55	2	−467
1977	40	517.00	29	2	−464
1978	41	753.00	53	2	−414
1979	42	678.00	46	2	−378
1980	43	813.00	54	2	−326
1981	44	595.00	37	2	−307
1982	45	750.00	52	2	−259
1983	46	655.00	43	2	−228
1984	47	637.00	42	2	−200
1985	48	666.00	45	2	−166
1986	49	712.00	50	2	−122
1987	50	692.00	49	2	−80
1988	51	623.00	41	2	−54
1989	52	565.00	35	2	−40
1990	53	468.00	23	2	−50
1991	54	547.16	33	2	−40
1992	55	685.39	48	2	

(b) Pettitt change-point test. The fifth column of Table 12.1 shows the $U_{t,N}$ series (Eq. 12.5). From Eq. (12.7) the test statistic is $K_N = 658$, which occurs at $t = 27$, that is, in 1964. This is a two-tailed test, so the p-value is determined by Eq. (12.11) as $p = 4.38 \times 10^{-7}$. Since $p < \alpha$, the null hypothesis is rejected, which suggests that the series exhibits a significant upward change-point in 1964. Figure 12.1 shows the time series X_t and the change-point detected through the Pettitt test.

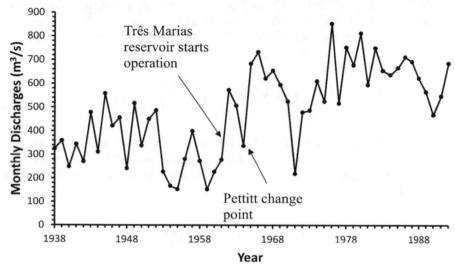

July Mean Monthly Streamflows of the São Francisco River at Pirapora (Brazil) (1938 - 1992)

Fig. 12.1 July mean monthly flows of the São Francisco River at Pirapora: change point identified by the Pettitt test

Note that the Pettitt change-point does not occur when the upstream reservoir begins operation but 3 years after that, in 1964. In fact, the Pettitt change-point signals a sudden change from the statistical viewpoint, and it is not guaranteed to occur in tandem with the real underlying cause of change. It is also noteworthy that due to the great storage capacity of the reservoir (21 billion m^3), the time of filling the reservoir should be taken into account until regular operation begins. However, this period cannot be determined with precision from the available data.

12.2   Kernel Occurrence Rate Estimation

Poisson processes are among the most important stochastic processes, as mentioned in Sect. 4.2. Under the stationarity assumption, the Poisson intensity λ is a constant when the process is homogeneous. If the stationarity assumption is not valid, the Poisson process is nonhomogeneous and $\lambda(t)$ is a non-constant function of time.

The nonparametric kernel estimator was developed by Rosenblatt (1956) and Parzen (1962) as an estimation technique of probability density functions of random variables. Diggle (1985) adapted the technique for smoothing Poisson process data over time.

12.2.1 Formulation

Consider t as ranging from t_0 to t_n, such that, at any instant t, either a "success" or a "failure" occurs and that there were M successes in all. If the times of occurrence of the M successes can be denoted by T_i, with $i = 1, \ldots, M$, the Poisson intensity, as a function of time $\lambda(t)$ can be estimated by

$$\hat{\lambda}(t) = h^{-1} \sum_{i=1}^{M} K\left(\frac{t - T_i}{h}\right) \tag{12.12}$$

where $K(\cdot)$ is a kernel function and h is the bandwidth. The units of $\hat{\lambda}(t)$ are the inverse of the discretization units of t. There are many kernel functions available in the technical literature. One of the most widely used is the Gaussian kernel, expressed as

$$K(y) = \frac{1}{\sqrt{2\pi}} \exp\left(\frac{-y^2}{2}\right) \tag{12.13}$$

12.2.2 Bandwidth Selection

The selection of the bandwidth h determines the bias and variance properties of the estimator: a too small h results in fewer data points that effectively contribute to kernel estimation, which leads to a reduced bias and a high variance, as contrasted to a too large h which leads to an over-smoothing of the estimator, resulting in a small variance and increased bias. In practical terms, the selection of the bandwidth can be seen as a compromise between those two cases. The technical literature lists many optimization procedures and empirical formulae for selecting the optimal bandwidth. One of the most well-known procedures, which is present in many statistical packages, is Silverman's rule of thumb (Silverman 1986, p. 48) defined by

$$h = 0.9 \min\left\{s, \frac{\text{IQR}}{1.34}\right\} M^{-\frac{1}{5}} \tag{12.14}$$

Where s, IQR and M are, respectively, the standard deviation, interquartile range and length of the sample T_i. Equation (12.14) provides values of h that are considered suitable for a wide range of applications.

12.2.3   Pseudodata generation

A direct application of Eq. (12.12) to the series T_i may lead to a boundary bias near the observation limits t_0 and t_n consisting of an underestimation of $\lambda(t)$ due to the nonexistence of data outside the interval $[t_0, t_n]$. This boundary effect may be reduced by generating pseudodata, denoted as pT, outside the range of observed data. A straightforward method to generate pseudotada consists of "reflecting" the observed data near the boundaries: for $t < t_0, pT_i = t_0 - (T_i - t_0)$, covering a range of 3 times h before t_0 and likewise to the right side, for $t > t_0$. Pseudodata generation is equivalent to the extrapolation of the empirical distribution of events near the boundaries (Cowling and Hall 1996). For this reason, the estimation of $\lambda(t)$ near the boundaries should be interpreted with caution. Considering $T_i^\dagger$ as the original point data augmented by the pseudodata and $M^\dagger$ the length of $T_i^\dagger$, Eq. (12.12) can be rewritten as

$$\hat{\lambda}(t) = h^{-1} \sum_{i=1}^{M^\dagger} K\left(\frac{t - T_i^\dagger}{h}\right) \qquad (12.15)$$

Example 12.3 Consider the Poisson process characterized by the occurrence of over-threshold daily rainfalls at São Julião do Tojal, in Portugal. The available data series has a length of 39 water years; the water year in Portugal starts on October 1st. The selected threshold for the peaks-over-threshold sampling is 36 mm. Table 12.2 shows the dates of the occurrences as well as the peak rainfalls. The times of occurrence T_i are relative to $t_0 = 1$, that is the beginning of the time series under analysis. Using the kernel occurrence rate estimator, as formulated in Eq. (12.12), estimate the temporal variation of the Poisson intensity of over-threshold rainfall occurrences (a) with, and (b) without pseudodata generation using the method described in Sect. 12.2.3.

Solution The estimator's bandwidth, according to Silverman's rule of thumb (Eq. 12.14) is $h = 1,566\,\mathrm{d}$. Pseudodata generation uses the method of reflection, which consists of covering a range of $3h$ for $t < t_0$ and $t > t_n$, with the reflection of the occurrences nearing the boundaries t_0 and t_n, respectively. Table 12.3 shows the generated pseudodata points. The kernel occurrence rate estimate without pseudodata uses Eq. (12.12) directly, whereas the estimation with pseudodata uses Eq. (12.15), where $T_i^\dagger$ is obtained by concatenating the T_i values of Table 12.1 with those of Table 12.3.

Figure 12.2 shows the kernel estimates obtained with and without pseudodata. To facilitate the interpretation of the results, $\hat{\lambda}(t)$ was multiplied by 365.25, such that, for a given instant, t, the $\hat{\lambda}(t)$ indicates the estimated number of occurrences above threshold per year. In Fig. 12.2, t takes 512 equidistant values between October 1st, 1955 ($t_0 = 1$) and September 30th, 1994 ($t_n = 14,245$). The chart exemplifies the correction of the boundary bias via the pseudodata generation. The results show that the occurrence rate of over-threshold rainfall events peaked in the mid-1960s, followed by a few decades of lower intensity rainfalls.

Table 12.2 Peaks-over-threshold series and times of occurrences of daily rainfalls exceeding the threshold $u = 36$ mm at São Julião do Tojal, in Portugal

Date	T_i	P (mm)	Date	T_i	P (mm)	Date	T_i	P (mm)
11/4/1955	35	47.0	2/26/1967	4167	55.5	2/9/1979	8533	62.2
12/14/1955	75	37.2	11/26/1967	4440	137.0	10/6/1979	8772	50.5
3/23/1956	175	45.0	2/18/1968	4524	45.5	12/30/1981	9588	78.0
10/12/1956	378	43.1	10/27/1968	4776	45.0	2/15/1982	9635	44.5
3/22/1957	539	38.6	11/6/1968	4786	37.0	11/8/1982	9901	55.5
1/28/1958	851	44.3	11/15/1968	4795	45.5	11/19/1983	10,277	163.7
12/18/1958	1175	37.0	1/10/1969	4851	51.6	10/26/1984	10,619	38.1
3/9/1959	1256	48.5	3/13/1969	4913	53.8	1/5/1985	10,690	41.5
9/11/1959	1442	40.7	10/12/1969	5126	39.0	1/17/1985	10,702	37.0
11/26/1959	1518	38.5	1/3/1970	5209	40.0	4/24/1985	10,799	37.3
1/26/1960	1579	44.5	1/18/1970	5224	54.8	11/5/1985	10,994	46.1
4/3/1960	1647	43.0	5/16/1971	5707	39.5	1/10/1987	11,425	42.5
11/20/1961	2243	92.4	1/2/1972	5938	39.8	2/25/1987	11,471	51.8
3/7/1962	2350	41.9	2/3/1972	5970	44.5	2/22/1988	11,833	65.5
12/11/1962	2629	70.6	12/20/1973	6656	36.2	10/14/1988	12,068	51.0
12/27/1962	2645	46.0	1/30/1976	7427	44.5	4/9/1989	12,245	44.0
2/16/1963	2696	48.8	8/30/1976	7640	42.5	10/13/1989	12,432	38.0
10/29/1963	2951	47.5	9/25/1976	7666	38.5	11/19/1989	12,469	38.5
11/9/1963	2962	38.4	11/11/1976	7713	40.2	12/4/1989	12,484	38.2
11/12/1963	2965	56.0	12/15/1976	7747	66.5	10/14/1990	12,798	39.5
12/14/1963	2997	40.4	10/17/1977	8053	57.0	10/22/1990	12,806	40.3
9/25/1965	3648	39.1	12/22/1977	8119	38.4	2/12/1991	12,919	51.8
10/6/1965	3659	36.2	2/27/1978	8186	38.6	9/17/1993	13,867	37.0
10/26/1965	3679	46.0	5/3/1978	8251	36.4	10/13/1993	13,893	42.3
1/14/1966	3759	37.5	11/7/1978	8439	38.5	10/16/1993	13,896	38.5
2/19/1966	3795	59.5	12/11/1978	8473	38.0	11/1/1993	13,912	68.5
1/10/1967	4120	38.2	12/25/1978	8487	38.2	2/28/1994	14,031	37.2

Table 12.3 Pseudodata obtained with the reflection method

$t < t_0$			$t > t_n$		
−4524	−2965	−1518	14,459	16,058	17,871
−4440	−2962	−1442	14,578	16,245	18,213
−4167	−2951	−1256	14,594	16,422	18,589
−4120	−2696	−1175	14,597	16,657	18,855
−3795	−2645	−851	14,623	17,019	18,902
−3759	−2629	−539	15,571	17,065	
−3679	−2350	−378	15,684	17,496	
−3659	−2243	−175	15,692	17,691	
−3648	−1647	−75	16,006	17,788	
−2997	−1579	−35	16,021	17,800	

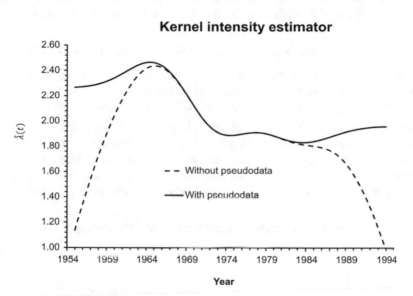

Fig. 12.2 Kernel intensity estimator applied to the occurrence of daily rainfalls exceeding the threshold of 36 mm at São Julião do Tojal

12.2.4   Estimation uncertainty: bootstrap confidence band

Point estimates given by Eq. (12.15) may be difficult to interpret without some measure of uncertainty associated with those estimates. For the purpose of quantifying that uncertainty, a pointwise confidence band around $\hat{\lambda}(t)$ can be constructed using a bootstrap simulation technique. The bootstrap was originally proposed by Efron (1979) as a nonparametric approach for estimating parameter confidence intervals. This technique was later generalized for a wide range of statistical applications, particularly for problems whose solutions by analytical methods were cumbersome. Readers interested in bootstrap techniques are referred to Davison and Hinkley (1997).

For the specific problem of estimating the sampling uncertainty associated with $\lambda(t)$, the procedure, outlined as follows, can be applied.

1. Generate a simulated sample T^* of length $M^{\dagger}$, by random sampling with replacement from the sample $M^{\dagger}$.

2. Calculate $\hat{\lambda}^*(t)$, with Eq. (12.15), using the resampled data and the same bandwidth h.

3. Repeat the resampling-and-estimation procedure until a sufficiently large number of replicates of $\hat{\lambda}^*(t)$ is reached (for example, 2000 replicates).

4. For each time step t the $100(1-\alpha)\%$ confidence interval is defined by the empirical quantiles with a non-exceedance probability $F = \alpha/2$ and $F = 1 - \alpha/2$. For example, supposing 2000 replicates are used, for a confidence level of 90%, the confidence interval is given by the 50th and 1950th order statistics of $\hat{\lambda}^*(t)$.

5. The confidence band is given by such confidence intervals for $t \in [t_0, t_n]$.

The described methodology leads to the estimation of an empirical percentile type confidence band. Other types of bootstrap confidence bands are proposed by Cowling et al. (1996). Figure 12.3 shows the limits of the 90% bootstrap confidence band applied to the case study of Example 12.3, based on 2000 replicates of $\hat{\lambda}^*(t)$. The confidence bands allow the sampling uncertainty to be accounted for in interpreting the results. Figure 12.3 suggests that the increase in extreme rainfall occurrence in the mid-1960s may not be very significant since the confidence band is wide enough to accommodate a constant Poisson intensity during the observation period.

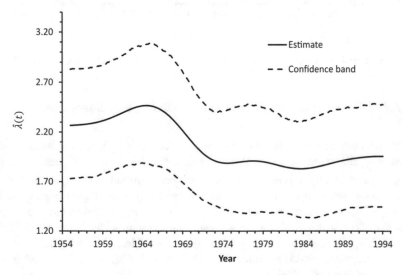

Fig. 12.3 Kernel intensity estimator and respective 90% confidence band applied to the occurrence of daily rainfalls exceeding the threshold of 36 mm at São Julião do Tojal

12.3 Introduction to Generalized Linear Models (GLM)

Regression models are widely applied in statistics to analyze dependent structures between a response or dependent variable and one or more independent variables. Simple and multiple linear regression models are presented in Chap. 9. Those models, as well as other regression models, can find applications in the domain of analysis of nonstationary hydrologic series. In fact, the main question tackled by simple linear regression models is "how does variable Y behave as variable X changes?". This is a central question in nonstationary analysis and modeling of hydrological time series as well.

Nonstationary hydrologic series analysis using regression models considers that there is a relationship between the variable under analysis (dependent variable) and independent variables, such as time and/or covariates (e.g., series of climate indices that evolve over time). In this kind of analysis, the regression residuals should be seen not as an error or imprecision, but as a realization of an independent (but not necessarily) identically distributed random variable. This understanding is illustrated in Fig. 9.5, where the mean of variable Y varies linearly with the dependent variable X and Y conditioned on X follows a Normal distribution.

The application of regression models to the analysis of nonstationary hydrologic series requires the same due attention and careful consideration, as recalled in Chap. 9. For instance, the identification of a linear relationship with time requires special consideration if extrapolation into the far future is needed (see Sect. 9.2.6).

The normal linear regression model is one of the most widely used statistical analysis tools and it is applied in many fields of knowledge and science. Nevertheless, it has its limitations. In many hydrological applications, the variable under analysis is discrete and cannot be modeled by a normal linear model. There are also cases in which the relationship between dependent and independent variables is nonlinear (this subject was addressed in Sect. 9.2.5). Since the 1970s there has been a profusion of generalizations of the normal linear model, driven by two important developments that are addressed here: the systematic use of exponential family distributions and the numerical capability of optimizing likelihood functions, through iterative weighted least squares algorithms.

This section summarizes the mathematical formalism of the generalized linear model in view of the advantages of its use in the context of analysis of nonstationary hydrological variables. For a more formal and comprehensive treatment of generalized linear models, the reader is referred to McCullagh and Nelder (1989), Dobson (2001) and Davison (2003).

12.3.1 Density and Link Functions

In Chap. 9, the three basic aspects of the normal linear model are covered. These are:

1. The mean of the response variable $\mu_{y|x} = E[Y]$ is related to the independent variable x through a linear predictor $\eta = \beta_0 + \beta_1 X$;
2. The density function of the response variable Y, for each value of X, is Normal with mean $\mu_{y|x}$ and a constant variance σ^2; and
3. The mean of Y, for each X, is equal to the linear predictor,.

The generalized linear model (GLM) sustains the first definition, and relaxes the second and third. Regarding the density of the response variable, for a GLM, the response Y may be continuous or discrete, as long as its distribution belongs to the *exponential family of distributions* (not to be confused with the exponential distribution). A distribution belongs to the exponential family if its PDF, or PMF, can be written in the form

$$f_Y(y) = \exp\left\{ \frac{y\theta - b(\theta)}{a(\phi)} + c(y, \phi) \right\} \tag{12.16}$$

where θ is the canonical form of the location parameter, ϕ is the parameter of dispersion, and $a(\cdot)$, $b(\cdot)$ and $c(\cdot, \cdot)$ are known real functions. This family comprises many well-known distributions such as the normal, gamma, binomial, and Poisson distributions.

In a GLM, the linear predictor and the mean of Y are related through a monotonic *link function g*

$$\eta = g\left(\mu_{y|x} \right) \tag{12.17}$$

Therefore, the normal linear model is a special case of the generalized linear model in which Y has a normal density and g is the identity function. Other examples of well-known regression models which are special cases of the GLM are the logistic regression and the Poisson regression. Furthermore, the GLM setup extends the applicability of linear models to nonlinear cases, without the need for the transformations presented in Sect. 9.2.5.

From Eq. (12.16) it follows (McCullagh and Nelder 1989; Davison 2003) that the response variable has mean

$$E[Y] = b'(\theta) = \mu \tag{12.18}$$

and variance

$$\mathrm{Var}[Y] = a(\varphi)b''(\theta) = a(\phi)V(\mu) \tag{12.19}$$

where $'$ denotes differentiation with respect to θ. Hence, the variance of Y is a product of two functions, being that $b''(\theta)$ is called the *variance function* and is usually represented by $V(\mu)$. The variance function depends only on the location parameter θ (and, thus, on the mean μ) and describes the variation of the variance as a function of the mean. The function $a(\varphi)$ depends only on the dispersion parameter φ. It should be noted that, in most practical cases, $a(\varphi) = \varphi/\omega$, where ω is a known constant.

Example 12.4 Consider the discrete random variable $Y \sim P(\lambda)$ which follows the Poisson distribution. Characterize the probability mass function of the Poisson distribution, in the form of Eq. (12.16) and find the parameters of location and dispersion, θ and φ, the mean and variance of Y and the variance function $V(\mu)$.

Solution The Poisson PMF is given by $p_Y(y) = \frac{y^\lambda}{y!}\exp(-\lambda)$. After algebraic manipulation it is possible to rewrite the Poisson PMF as $p_Y(y) = \exp\{y\ln(\lambda) - \lambda - \ln(y!)\}$ which is the form of Eq. (12.16) with $\theta = \ln(\lambda)$, $b(\theta) = e^\theta$, $a(\varphi) = 1$ (with ϕ and ω both equal to 1), and $c(y, \phi) = -\ln(y!)$. In conclusion, $E[Y] = b'(\theta) = e^\theta = \lambda$ and $\mathrm{Var}[Y] = a(\phi)b''(\theta) = e^\theta = \lambda$. The variance function is $V(\mu) = b''(\theta) = e^\theta = \lambda$. Table 12.4 summarizes the previous results and extends them to the normal and binomial distributions.

The link function of a GLM as in Eq. (12.17) defines the relationship between the linear predictor η and the mean of Y, given by μ. The choice of this function should consider the specificities of the case under study. One of the possible choices is the *canonical link function*, which consists of having the linear predictor equal to the location parameter, i.e., $\eta = \theta = b'^{-1}(\mu)$. For instance, if Y is Poisson-distributed, it is known that the mean of Y is always positive. Hence, a logical choice for a link function is the log function, since $g(\mu) = \ln(\mu)$ ensures that μ can only take positive values. Furthermore, because $\theta = \ln(\mu)$, the log link is the canonical link. Choosing the canonical link function has some theoretical advantages as laid out by McCullagh and Nelder (1989, p. 32). However, it is more

Table 12.4 Characterization of some distributions of the exponential family

Distribution	Normal	Binomial	Poisson
Notation	$N(\mu, \sigma)$	$B(N, p)$	$P(\lambda)$
Support	R	$\{0, 1, \ldots, N\}$	N_0
θ	μ	$\ln\left(\dfrac{p}{1-p}\right)$	$\ln(\lambda)$
$a(\varphi)$	σ^2	$1/N$	1
φ	σ^2	1	1
ω	1	N	1
$c(y, \varphi)$	$\dfrac{-1}{2}\left(\dfrac{y^2}{\varphi} + \ln(2\pi\varphi)\right)$	$\ln\left(\dfrac{N}{Ny}\right)$	$-\ln(y!)$
$b(\theta)$	$\theta^2/2$	$\ln\left(1 + e^\theta\right)$	e^θ
$\mu = E[Y]$	θ	$p = \dfrac{e^\theta}{1 + e^\theta}$	$\lambda = e^\theta$
$V(\mu)$	1	$p(1-p)$	λ
$\mathrm{Var}[Y]$	σ^2	$p(1-p)/N$	λ

Table 12.5 Some link functions for generalized linear models

Identity	$g(\mu) = \mu$
Reciprocal	$g(\mu) = \dfrac{1}{\mu}$
Log	$g(\mu) = \ln(\mu)$
Inverse quadratic	$g(\mu) = \dfrac{1}{\mu^2}$
Logit	$g(\mu) = \ln\left(\dfrac{\mu}{1-\mu}\right)$
Probit	$g(\mu) = \Phi^{-1}(\mu)$

important for the choice of the link function to be guided by substantial considerations specific to the problem at hand. Table 12.5 shows some of the more common link functions.

The PDF or PMF of Y is the random component of a GLM, while the link function is the structural or systematic component of a GLM. One can interpret the simple linear regression model as a particular GLM: Y has a Normal density and an identity link function. In Chap. 9, the Normal distribution is presented as the distribution of the "residuals." In the GLM framework, the residuals are the realizations of the random variable around its mean. Notwithstanding the different terminologies applied to the simple linear model and the GLM with Normal density and identity link function, they are mathematically equivalent.

12.3.2 Estimation and Inference

GLM parameters are estimated using the maximum likelihood method. The likelihood function is given by

$$
\begin{aligned}
L(\beta_j|y_i, x_i) &= \prod_{i=1}^{N} f_Y(y_i|\theta_i, \varphi_i, \omega_i) \\
&= \prod_{i=1}^{N} \exp\left\{ \frac{y_i\theta_i - b(\theta_i)}{a(\varphi_i)} + c(y_i, \varphi_i) \right\}
\end{aligned}
\tag{12.20}
$$

One can further define the log-likelihood function $\ell(\beta) = \ln\left(L(\beta)\right)$ as

$$
\ell(\beta_j|y_i, x_i) = \sum_{i=1}^{N} \left\{ \frac{y_i\theta_i - b(\theta_i)}{a(\varphi_i)} + c(y_i, \varphi_i) \right\}
\tag{12.21}
$$

where $\theta_i = \eta_i$, when the link function is canonical, and $\eta_i = \beta_0 + \beta_1 x_i$. The maximum likelihood estimators of the regression parameters $\hat{\beta}_i$ are obtained by solving the following system of equations.

$$\frac{\partial}{\partial \beta_k} \ell(\beta_j | y_i, x_i) = 0 \tag{12.22}$$

The dispersion parameter φ may also be estimated by maximum likelihood, however such an estimation procedure has some practical difficulties (Davison 2003, p. 483). Alternatively, one can expand Eq. (12.19) to

$$
\begin{aligned}
\mathrm{Var}[Y_i] &= a(\phi) b''(\theta_i) \\
&= \frac{\phi}{\omega_i} V(\mu_i)
\end{aligned}
\tag{12.23}
$$

Then, one has

$$\phi = E\left[\frac{\omega_i (Y_i - \mu_i)^2}{V(\mu_i)} \right], i = 1, \ldots, N \tag{12.24}$$

From the previous equation and according to some asymptotic results from mathematical statistics outlined by Turkman and Silva (2000), it is possible to define the following estimator for ϕ

$$\hat{\phi} = \frac{1}{N - q} \sum_{i=1}^{N} \frac{\omega_i (Y_i - \hat{\mu}_i)^2}{V(\hat{\mu}_i)} \tag{12.25}$$

where q is the number of regression parameters. If there is only one dependent variable, $q = 2$.

It is very difficult to solve the system of equations given by Eq. (12.22). To tackle that optimization, the iterative weighted least squares (IWLS) method, which is a variant of the Newton–Raphson algorithm, is recommended in the scientific literature on GLMs. To grasp the IWLS formalism, the works by Dobson (2001) and Davison (2003) are recommended. Furthermore, it should be noted that the IWLS method is implemented in the glm() function in the free statistical software package R (R Core Team 2013).

The regression parameter estimates $\hat{\beta}_k$ benefit from an asymptotic property of maximum likelihood estimators: they are asymptotically Normal with means β_k and variances $\sigma_{\beta_k}^2$, given by the diagonal of the variance-covariance matrix, which was introduced in Sect. 6.7.2. If there is only one dependent variable, one has

$$\mathbf{I} = \begin{bmatrix} \mathrm{Var}(\hat{\beta}_0) & \mathrm{Cov}(\hat{\beta}_0, \hat{\beta}_1) \\ \mathrm{Cov}(\hat{\beta}_1, \hat{\beta}_0) & \mathrm{Var}(\hat{\beta}_1) \end{bmatrix} \tag{12.26}$$

which is given by the inverse of the negated Hessian matrix, $\mathbf{H}$, i.e., the matrix of second-order partial derivatives of the log-likelihood function at the point of maximum likelihood

$$\mathbf{I} = (-\mathbf{H})^{-1} = \left(- \begin{bmatrix} \frac{\partial^2 \ell(\beta_0, \beta_1)}{\partial \beta_0^2} & \frac{\partial^2 \ell(\beta_0, \beta_1)}{\partial \beta_0 \partial \beta_1} \\ \frac{\partial^2 \ell(\beta_0, \beta_1)}{\partial \beta_1 \partial \beta_0} & \frac{\partial^2 \ell(\beta_0, \beta_1)}{\partial \beta_1^2} \end{bmatrix} \Bigg|_{\beta_k = \hat{\beta}_k} \right)^{-1} \tag{12.27}$$

The calculation of the covariance matrix is very complex and is seldom done analytically. However, such a calculation is made numerically within the IWLS method. The numerical solution of the matrix can be obtained through the glm() function in R.

After the regression parameter estimates and corresponding covariance matrix are known, it is possible to construct confidence intervals for those parameters. Furthermore, with such intervals as a basis, one can construct a hypothesis test for the existence of a linear relationship between the link function and the linear predictor. This subject matter is covered in Sect. 12.3.3.1.

Example 12.5 The North Atlantic Oscillation (NAO) is a prominent pattern in climate variability over the northern hemisphere and refers to the redistribution of atmospheric masses between the Arctic and subtropical Atlantic. There are many studies in the technical literature that establish links between the NAO phase and rainfall in Western Europe, particularly during winter in the northern hemisphere. There are also a number of studies on the influence of the NAO on rainfall and river flows in the western Iberian Peninsula during winter months: when the winter NAO is in the negative phase, rainfall and river flows tend to be above normal, and vice versa (e.g. Lorenzo-Lacruz et al. 2011). The NAO is usually characterized by a standardized climatic index which is computed and made available by several organizations, such as the University of East Anglia's Climate Research Unit (Jones et al. 1997, http://www.cru.uea.ac.uk/cru/data/nao/). Consider the case of Example 12.3. Consider the variable Y: "annual number of daily rainfalls exceeding the threshold $u = 36$ mm." Y is a discrete random variable and admittedly follows a Poisson distribution. Using a GLM, analyze the relationship between the dependent variable Y and the independent variable X : "winter NAO index" (the annual November-to-March mean). Both variables are presented in Table 12.6. Estimate GLM regression parameters and the respective covariance matrix, considering the canonical link function of the appropriate model.

Solution The R code for solving this problem is presented in Appendix 11. Y is a Poisson-distributed variable and as such the GLM will have a Poisson probability mass function. In this case, one can write (from Table 12.4) $b(\theta) = e^\theta$. Since the canonical link function is given by $b'^{-1}(\mu)$, $g(\mu) = \ln(\mu)$. This link function suggests that the model is nonlinear. The GLM can then be expressed as

$$\ln\left(\mu_{Y|\text{NAO}}\right) = \beta_0 + \beta_1 \text{NAO}, \quad \text{or,} \quad \text{identically as} \quad \mu_{Y|\text{NAO}} = \exp(\beta_0 + \beta_1 \text{NAO}).$$

Table 12.6 Annual number of threshold exceedances Y and corresponding winter NAO index

HY	Y	NAO	HY	Y	NAO
1955/56	3	−1.068	1975/76	3	0.64
1956/57	2	1.598	1976/77	2	−0.576
1957/58	1	−0.53	1977/78	4	0.338
1958/59	3	0.372	1978/79	4	−0.294
1959/60	3	−0.126	1979/80	1	0.448
1960/61	0	1.892	1980/81	0	0.308
1961/62	2	−0.888	1981/82	2	0.53
1962/63	3	−1.218	1982/83	1	1.942
1963/64	4	−0.642	1983/84	1	0.138
1964/65	1	−0.54	1984/85	4	−0.536
1965/66	4	−0.176	1985/86	1	−0.596
1966/67	2	1.27	1986/87	2	0.952
1967/68	2	−0.196	1987/88	1	−0.056
1968/69	5	−2.068	1988/89	2	1.994
1969/70	3	−0.676	1989/90	3	1.304
1970/71	1	−0.36	1990/91	3	−0.132
1971/72	2	−0.232	1991/92	0	1.68
1972/73	0	1.276	1992/93	1	2.044
1973/74	1	0.342	1993/94	4	1.594
1974/75	0	1.18			

By applying the function glm(), of the statistical software R, one obtains the results $\hat{\beta}_0 = 0.770334$ and $\hat{\beta}_1 = -0.286583$, thus suggesting that the mean of Y decreases as the winter NAO index increases, that is, there tends to be more over-threshold events during the negative phase of the winter NAO. This result is concurrent with the previously mentioned studies. Figure 12.4 shows the resulting curve as well as the scatterplot of the observed data (Table 12.5).

Finally one can obtain the covariance matrix with the command vcov(),

$$
\mathbf{I} = \begin{bmatrix} 1.234574 \times 10^{-2} & 2.876015 \times 10^{-5} \\ 2.876015 \times 10^{-5} & 1.294348 \times 10^{-2} \end{bmatrix}
$$

12.3.3 Model Selection and Evaluation

Generalized linear models constitute a powerful and versatile modeling framework since: (1) they allow for several independent variables; (2) the dependent variable may be described by any probability density function that can be written as in Eq. (12.16); and (3) several link functions can be postulated. The versatility of this modeling methodology requires careful consideration in its application and in selecting the appropriate model for each analysis. Following are three basic and

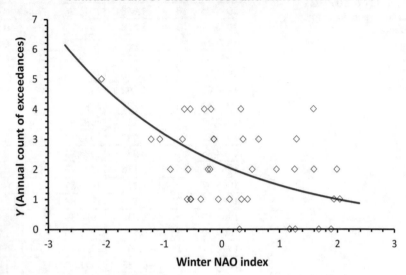

Fig. 12.4 Relation of winter NAO index and annual exceedance counts for Example 12.5

useful tools for statistical modeling using generalized linear models. For a deeper understanding of the selection and evaluation of generalized linear models, the textbook by Davison (2003) is recommended.

12.3.3.1 Hypothesis Tests for Regression Coefficients

As previously mentioned, in Sect. 12.3.2, GLM parameters are estimated by maximum likelihood, since MLE estimates are asymptotically Normal. This result can be used to design a hypothesis test for the regression coefficients β_1, which defines the slope of the linear predictor.

Given the asymptotic normality of the estimator of the regression coefficient β_1, one can define a test with the following hypotheses:

- H_0: {there is no significant linear relationship ($\beta_1 = 0$)}.
- H_1: {there is a significant linear relationship" ($\beta_1 \neq 0$)}.

The null hypothesis H_0 can be rejected at a $100\alpha\%$ significance level if

$$\left|\hat{\beta}_1\right| > z_{1-\alpha/2}\sigma_{\hat{\beta}_1} \tag{12.28}$$

where $\sigma_{\hat{\beta}_1}$ is the standard error of the estimate and $z_{1-\alpha/2}$ is the $1 - \alpha/2$ quantile of the standard Normal distribution.

Example 12.6 Analyze the results of Example 12.5. Apply the hypothesis test for regression coefficients to verify whether or not the linear relationship between variables Y and X is significant.

Solution From the results of Example 12.5, $\hat{\beta}_1 = -0.28658$ and $\sigma^2_{\hat{\beta}_1} = 0.01294$ (second value from the diagonal of the covariance matrix). The standard error of the estimate is $\sigma_{\hat{\beta}_1} = \sqrt{0.01294} = 0.1138$. The limit of the region of rejection is calculated as $z_{1-\alpha/2}\sigma_{\hat{\beta}_1} = 1.96 \times 0.1138 = 0.2230$. Since $|\hat{\beta}_1| > z_{1-\alpha/2}\sigma_{\hat{\beta}_1}$, the decision should be to reject the null hypothesis ($\beta_1 = 0$). Therefore, it can be concluded that the linear relationship modeled in Example 12.5 is significant.

12.3.3.2   Likelihood Ratio Tests

Likelihood ratio tests (LRT) are based on asymptotic results from mathematical statistics (see Casella and Berger 2002; Davison 2003). An LRT is a method for comparing the performance of two competing models fitted to the same sample of the dependent variable: a null model M_0 and a more complex (with more parameters) alternative model M_1, being that M_0 is nested in M_1. By *nested* it is meant that M_1 reduces to M_0 if one or more of its parameters are fixed.

A k-parameter alternative model M_1, with parameter vector θ_{M_1}, can be compared with the null model with $k - q$ parameters, by calculating the test statistic D, also named *deviance statistic*. It is given by

$$D = 2\{\ell(\hat{\theta}_{M_1}) - \ell(\hat{\theta}_{M_0})\} \tag{12.29}$$

where $\ell(\hat{\theta}_{M_1})$ and $\ell(\hat{\theta}_{M_0})$ are the maxima of the log-likelihood function of the alternative and null models, respectively. The deviance statistic follows a chi-square distribution with q degrees of freedom:

$$D \sim \chi^2_q \tag{12.30}$$

where q is the difference in number of parameters between the two models.

The null hypothesis of an LRT $H_0 : \{M = M_0\}$ can be rejected in favor of the alternative hypothesis $H_1 : \{M = M_1\}$, at the $100\alpha\%$ significance level, if

$$D > \chi^2_{1-\alpha,q} \tag{12.31}$$

where $\chi^2_{1-\alpha,q}$ is the $1 - \alpha$ quantile of the chi-square distribution with q degrees of freedom.

The LRT result should be interpreted from the viewpoint of the concept of parsimony: if the deviance statistic D takes a value outside the region of rejection (i.e., $D < \chi^2_{1-\alpha,q}$), the underlying implication is that, notwithstanding the null model's simplicity relative to the alternative model, it is sufficiently adequate to describe the variable under analysis, or, in other words, the "gain" in likelihood achieved by increasing the complexity of M_1 relative to M_0, does not bring about a significant improvement of its capability of modeling the variable.

Example 12.7 Consider Example 12.5. Use an LRT at the 5 % significance level to test whether the fitted model is significantly better than that in which variable Y is considered to be stationary (base model).

Solution (The R code for solving this problem is presented in Appendix 11) The base model M_0 is stationary: Y follows a Poisson distribution with a single parameter which is estimated by the mean. The model fitted in Example 12.5 is the alternative model M_1. M_0 is nested in M_1 because the parameter β_1 is fixed and equal to zero, then the two models are mathematically equal.

The log-likelihood function of the Poisson distribution is $\ell(\lambda) = -\lambda N + \ln(\lambda) \sum_{i=1}^{N} y_i - \ln\left(\prod_{i=1}^{N} y_i!\right)$. The point of maximum likelihood is defined by $\lambda = \hat{\lambda} = \bar{y} = 2.076923$. Then, the maximum of the log-likelihood of model M_0 is $\ell(\hat{\theta}_{M_0}) = -\bar{y}N + \ln(\bar{y}) \sum_{i=1}^{N} y_i - \ln\left(\prod_{i=1}^{N} y_i!\right) = -66.23$.

Using the logLik() function in R, one gets $\ell(\hat{\theta}_{M_1}) = -62.98$. From Eq. (12.29), the result $D = 6.49$ is attained. M_1 has one parameter more than M_0, so $D \sim \chi_1^2$. From Appendix 3, $\chi_{0.95,1}^2 = 3.84$. Since $D > \chi_{0.95,1}^2$, the decision is to reject the null hypothesis, that is, the performance of the more complex alternative model M_1 improves significantly on that of model M_0, hence it can be concluded that the winter NAO index casts a significant influence on variable Y.

12.3.3.3 Akaike Information Criterion (AIC)

The Akaike information criterion (AIC) was introduced by Akaike (1974). It is a widely use method for evaluating statistical models. Founded in information theory, this method does not involve defining hypothesis tests, since there are no null and alternate hypotheses. Rather, it is simply a measure of model performance of a series of candidate models, based on the parsimony viewpoint.

In more general terms than those applied in Sect. 8.3.2, for a given statistical model, the AIC score is given by

$$AIC = 2k - 2\ell(\hat{\theta}) \tag{12.32}$$

where k is the number of parameters and $\ell(\hat{\theta})$ is the maximum of the log-likelihood of the model. The AIC score may be computed for a series of candidate models for the same random sample. According to this criterion, the best model is the one that minimizes the AIC score. AIC rewards the goodness-of-fit of the models but penalizes the increase in complexity (number of parameters).

Example 12.8 Apply AIC to corroborate the results of Example 12.7.

Solution From Example 12.7: model M_0 has one parameter and $\ell(\hat{\theta}_{M_0}) = -66.23$; and model M_1 has 2 parameters and $\ell(\hat{\theta}_{M_1}) = -62.98$. Equation (12.32) yields $\text{AIC}_{M_0} = 134.45$ and $\text{AIC}_{M_1} = 129.96$. Since $\text{AIC}_{M_1} < \text{AIC}_{M_0}$, M_1 is the better model, which corroborates the results of Example 12.7.

12.4 Nonstationary Extreme Value Distribution Models

12.4.1 Theoretical Justification

Extreme-value theory (EVT), which was introduced in Sect. 5.7, provides 3 limiting probability distributions for maximum (or minimum) extreme values, namely the Gumbel, Fréchet, and Weibull distributions. These distributions can be integrated into a single distribution—the generalized extreme value (GEV) distribution (Eq. 5.70). As mentioned in Sect. 5.7.2.3, the GEV distribution has many applications in Statistical Hydrology, such as the statistical modeling of floods or extreme rainfalls, even though the theoretical assumptions that support EVT do not always hold for hydrological variables. In practice, although the theoretical basis of the extremal asymptotic distributions has been specifically developed to analyze extreme value data, frequency analysis using such distributions is carried out analogously as with non-extremal distributions.

As observed by Katz (2013), when confronted with trends in extreme value data, hydrologists tend to abandon analysis frameworks based on EVT, in favor of nonparametric techniques such as the Mann–Kendall test, which is a powerful tool with no distributive restrictions. Nevertheless, it was not developed specifically for extreme values. Another common approach is to conduct inference about trends in extreme value data using a simple linear regression model, which has as one of its central premises that the data is normally distributed. Clarke (2002) points out the inherent methodological inconsistency underlying the application of these approaches to extreme value data: it seems that practitioners of hydrology accept one theory (even if approximate) under stationarity but move away from it under nonstationarity.

In his textbook on EVT, Coles (2001) introduced (based on previous developments by Davison and Smith 1990) nonstationary GEV distribution models, including the particular case of the Gumbel distribution. Coles (2001) argues that it is not possible to deduce a general asymptotic theory of extreme values under nonstationarity, except in a few very specialized forms which are too restrictive to describe nonstationary behavior in real-world applications. However, it is possible to take a pragmatic approach by using existing limiting models for extremes with enhanced estimation procedures, namely regression techniques, which allow modeling the parameters of the GEV or GPA distributions as functions of time. The idea is similar to generalized linear models: the nonstationary GEV model uses link functions applied to the location scale and shape parameters. These link functions are related to linear (or, e.g., polynomial) predictors. Under this approach, at any given time step, the variable is still described by an extreme value distribution, but its parameters are allowed to change over time. Therefore, the statistical model can still be interpreted, although in a contrived manner, in the scope of extreme value theory.

In this subsection, nonstationary models based on limiting distributions for extreme values are presented, namely the GEV and Gumbel (as a special case) distributions, and the generalized Pareto distribution, whose applicability falls in the domain of peaks-over-threshold analysis. It should be noted that the GEV and Gumbel distributions considered here are for maxima. The methodologies described are equally applicable to minima, provided that the negated samples are used.

The presentation of the models is complemented with application examples using the ismev package (Heffernan et al. 2013) in R. That package considers the parametrization of the GEV and GPA distributions used by Coles (2001), which differs from the one adopted in this textbook: the shape parameter of the distributions is, in Coles' parametrization, the symmetrical of the shape parameter used in this textbook.

12.4.2 Nonstationary Model Based on the GEV Distribution

The nonstationary GEV model for frequency analysis consists of fitting that distribution to an observed sample and of estimating one or more of its parameters as a function of time or of a *covariate*. A covariate is a variable that can admittedly exert a time dependence on the hydrological variable under study (e.g. annual maximum flows). Examples of covariates are climate indices (see North Atlantic Oscillation in Example 12.5) and indicators of anthropogenic influence on the catchments.

Consider a series of annual maximum flows X_t that show some signs of changing behavior with time. In order to describe the changing behavior of this extreme variable, it is possible to contemplate the following nonstationary flood frequency model based on the GEV distribution with time-varying parameters β, α, and κ:

$$X_t \sim \text{GEV}(\beta(t), \alpha(t), \kappa(t)) \tag{12.33}$$

where functions $\beta(t)$, $\alpha(t)$, and $\kappa(t)$ define the dependence structure between the model parameters and time.

As shown by Eq. (5.72), the mean of a GEV-distributed variable has a linear dependence on the location parameter β. Then, in order to model a linear temporal trend of a GEV-distributed variable, one can use a nonstationary GEV model with location parameter

$$\beta(t) = \beta_0 + \beta_1 t \tag{12.34}$$

where β_1 determines the rate of change (slope) of the variable. There can be other more convoluted parametric dependence structures for the GEV such as, for example, a 2nd degree polynomial, that is,

$$\beta(t) = \beta_0 + \beta_1 t + \beta_2 t^2 \tag{12.35}$$

or a change point at time t_0

$$\beta(t) = \begin{cases} \beta_1, \text{for } t < t_0 \\ \beta_2, \text{for } t > t_0 \end{cases} \tag{12.36}$$

Nonstationarities may also be introduced in the scale parameter α. This is particularly useful when analyzing changes in variance. A convenient parametrization for a time-changing scale parameter uses the exponential function so as to guarantee that $\alpha(t)$ can only take positive values, that is,

$$\alpha(t) = \exp(\alpha_0 + \alpha_1 t) \tag{12.37}$$

This link is log-linear since it equates to applying a linear relationship to the logarithm of $\alpha(t)$.

For samples of only a few dozen values, as is generally the case in Statistical Hydrology, it is difficult to make a proper estimation of the GEV shape parameter κ even under the stationarity assumption. For that reason, the shape parameter is usually fixed in nonstationary models. Furthermore, to consider a trend in the shape parameter frequently leads to numerical convergence issues when estimating the model parameters. Then, in practice, one should consider $\kappa(t) = \kappa$.

The parameters of the nonstationary GEV model are estimated by maximum likelihood, thus allowing more flexibility for changes in the model structure. The nonstationary GEV model with changing parameters, according to link function of the type of Eqs. (12.34) to (12.37), has the likelihood function

$$L(\theta) = \prod_{t=1}^{N} f_X\left[x_t|\beta(t), \alpha(t), \kappa(t)\right] \tag{12.38}$$

where $f_X(x_t)$ is the probability density function of the GEV, given by Eq. (5.71). Then, the log-likelihood function of the model is obtained as

$$\ell(\theta) = -\sum_{t=1}^{N} \left\{ \ln(\alpha(t)) + \left(1 - \frac{1}{\kappa(t)}\right) \ln\left[1 - \kappa(t)\left(\frac{x_t - \beta(t)}{\alpha(t)}\right)\right] + \left[1 - \kappa(t)\left(\frac{x_t - \beta(t)}{\alpha(t)}\right)\right]^{\frac{1}{\kappa(t)}} \right\}$$
$$\tag{12.39}$$

subject to

$$1 - \kappa(t) \left(\frac{x_t - \beta(t)}{\alpha(t)} \right) > 0 \qquad (12.40)$$

for $t = 1, \ldots, N$.

Parameter estimates are obtained by maximizing Eq. (12.39), which is a complex numerical procedure for which the IWLS method is usually employed. Therefore it is somewhat similar to the GLM, but specifically developed for extreme value data. The function gev.fit() of the package ismev, in R, is one of the free available tools for building and fitting this kind of model.

As in the case of the GLM, parameter estimates of the nonstationary GEV model benefit from the asymptotic properties of maximum likelihood estimators, which enable approximations of the sampling distribution of parameter estimates to the Normal distribution, with mean given by the maximum likelihood estimate and variance given by the diagonal of the covariance matrix **I**. Take as an example the nonstationary GEV model with a linear trend in the location parameter (Eq. 12.34), for which the parameter vector is $\theta = (\beta_0, \beta_1, \alpha, \kappa)^T$, the covariance matrix **I** is

$$\mathbf{I} = \begin{bmatrix} Var(\hat{\beta}_0) & Cov(\hat{\beta}_0, \hat{\beta}_1) & Cov(\hat{\beta}_0, \hat{\alpha}) & Cov(\hat{\beta}_0, \hat{\kappa}) \\ Cov(\hat{\beta}_1, \hat{\beta}_0) & Var(\hat{\beta}_1) & Cov(\hat{\beta}_1, \hat{\alpha}) & Cov(\hat{\beta}_1, \hat{\kappa}) \\ Cov(\hat{\alpha}, \hat{\beta}_0) & Cov(\hat{\alpha}, \hat{\beta}_1) & Var(\hat{\alpha}) & Cov(\hat{\alpha}, \hat{\kappa}) \\ Cov(\hat{\kappa}, \hat{\beta}_0) & Cov(\hat{\kappa}, \hat{\beta}_1) & Cov(\hat{\kappa}, \hat{\alpha}) & Var(\hat{\kappa}) \end{bmatrix} \qquad (12.41)$$

and is given by the inverse of the symmetrical of the respective Hessian matrix of the log likelihood function at the point of maximum likelihood. The Hessian matrix, or matrix of second-order derivatives, is usually obtained by numerical differentiation of the log-likelihood function.

Example 12.9 Consider the series of annual maximum daily rainfalls, at the Pavia rain gauging station, in Portugal (organized by hydrologic year which, in Portugal starts on October 1st), presented in Table 12.7 and in Fig. 12.5. The records range from 1912/13 to 2009/10 with no gaps, adding up to 98 hydrologic years. Fit the following GEV models to the data: (a) a stationary model GEV0; (b) a linear trend in the location parameter GEV1; (c) Determine the quantile with a non-exceedance probability $F = 0.9$ of both models relative to the year 2010/2011.

Solution It is important to recall that the shape parameter returned by the functions of the ismev package is the symmetrical of the shape parameter in the GEV parametrization adopted in this book. Data can be imported from a file using the read.table function, or typed directly on the R console with the following command.

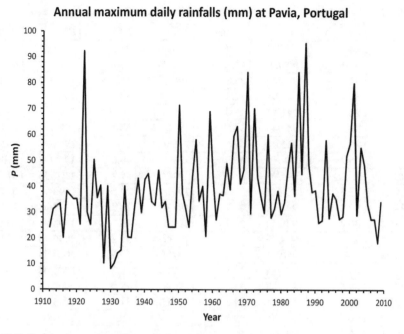

Fig. 12.5 Series of annual maximum daily rainfalls at Pavia

```
> pavia<-c(24.2, 31.3, 32.5, 33.5, 20.2, 38.2, 36.7, 35.2, 35.2,
25.3, 92.3, 30, 25.2, 50.4, 35.7, 40.5, 10.3, 40.2, 8.1, 10.2, 14.2,
15.3, 40.2, 20.4, 20.2, 32.8, 43.2, 29.8, 42.8, 45, 34.2, 32.8,
46.3, 31.9, 34.2, 24.3, 24.3, 24.3, 71.4, 37.4, 31.4, 24.3, 43.8,
58.2, 34.6, 40.2, 20.8, 69, 44, 27.2, 37.2, 36.7, 49, 38.9, 59.6,
63.3, 41.2, 46.6, 84.2, 29.5, 70.2, 43.7, 36.2, 29.8, 60.2, 28,
31.4, 38.4, 29.4, 34, 47, 57, 36.5, 84.2, 45, 95.5, 48.5, 38,
38.6, 26, 27, 58, 27.8, 37.5, 35.2, 27.5, 28.5, 52, 56.8, 80,
29, 55.2, 48.4, 33.2, 27.4, 27.4, 18.2, 34.2)
```

(a) The function used to fit GEV models is gev.fit. Fit the stationary GEV model, GEV0, with the command

```
> GEV0<-gev.fit(pavia)
```

Object GEV0 contains all the information regarding the fitted model including parameter estimates, covariance matrix, maximum log-likelihood, and more. For details on R functions, the help file documentation can be looked up using the command

```
> ?gev.fit
```

Table 12.7 Annual maximum daily rainfalls at Pavia, Portugal (*HY* hydrologic year)

HY	*P* (mm)	HY	*P* (mm)	HY	*P* (mm)	HY	*P* (mm)
1912/13	24.2	1937/38	32.8	1962/63	37.2	1987/88	95.5
1913/14	31.3	1938/39	43.2	1963/64	36.7	1988/89	48.5
1914/15	32.5	1939/40	29.8	1964/65	49.0	1989/90	38.0
1915/16	33.5	1940/41	42.8	1965/66	38.9	1990/91	38.6
1916/17	20.2	1941/42	45.0	1966/67	59.6	1991/92	26.0
1917/18	38.2	1942/43	34.2	1967/68	63.3	1992/93	27.0
1918/19	36.7	1943/44	32.8	1968/69	41.2	1993/94	58.0
1919/20	35.2	1944/45	46.3	1969/70	46.6	1994/95	27.8
1920/21	35.2	1945/46	31.9	1970/71	84.2	1995/96	37.5
1921/22	25.3	1946/47	34.2	1971/72	29.5	1996/97	35.2
1922/23	92.3	1947/48	24.3	1972/73	70.2	1997/98	27.5
1923/24	30.0	1948/49	24.3	1973/74	43.7	1998/99	28.5
1924/25	25.2	1949/50	24.3	1974/75	36.2	1999/00	52.0
1925/26	50.4	1950/51	71.4	1975/76	29.8	2000/01	56.8
1926/27	35.7	1951/52	37.4	1976/77	60.2	2001/02	80.0
1927/28	40.5	1952/53	31.4	1977/78	28.0	2002/03	29.0
1928/29	10.3	1953/54	24.3	1978/79	31.4	2003/04	55.2
1929/30	40.2	1954/55	43.8	1979/80	38.4	2004/05	48.4
1930/31	8.1	1955/56	58.2	1980/81	29.4	2005/06	33.2
1931/32	10.2	1956/57	34.6	1981/82	34.0	2006/07	27.4
1932/33	14.2	1957/58	40.2	1982/83	47.0	2007/08	27.4
1933/34	15.3	1958/59	20.8	1983/84	57.0	2008/09	18.2
1934/35	40.2	1959/60	69.0	1984/85	36.5	2009/10	34.2
1935/36	20.4	1960/61	44.0	1985/86	84.2		
1936/37	20.2	1961/62	27.2	1986/87	45.0		

Maximum likelihood estimates are stored in component $mle. The console returns a vector with the estimated of the location, scale and shape parameters, in the following order $(\beta, \alpha, -\kappa)$:

```
> GEV0$mle
[1] 31.521385926 13.156409604 -0.007569892
```

(b) In order to fit the nonstationary model GEV1, with a linear temporal trend on the location parameter $(\beta(t) = \beta_0 + \beta_1 t)$, it is necessary to create a single column matrix with the variable time, t, taking unit incremental values from 1 to 98 (the length of the sample), being that $t = 1$ corresponds to hydrologic year 1912/13.

```
> time<-matrix(1:98,ncol=1)
```

Subsequently, the function gev.fit is applied with the following arguments

```
> GEV1<-gev.fit(pavia,ydat=t,mul=1)
```

The component \$mle contains the parameter estimates in order $(\beta_0, \beta_1, \alpha, -\kappa)$.

```
> GEV1$mle
[1] 25.57537923 0.12052674 12.37868560 0.02447634
```

(c) In order to estimate the quantile with non-exceedance probability $F = 0.9$, for the hydrologic year 2010/11 ($t = 99$), the GEV quantile function may be applied directly, using Eq. (5.77). Regarding model GEV0, since it is stationary, the quantile function may be applied directly resulting in $x_{0.9,\text{GEV}_0} = 60.88$ mm. Regarding model GEV1, the location parameter for year 2010/11 is determined as $\beta(t = 99) = \beta_0 + \beta_1 \times 99 = 37.5075$. From the application of Eq. (5.77), results $x_{0.9,\text{GEV}_1} = 64.61$ mm.

12.4.3 Nonstationary Model Based on the Gumbel_{max} Distribution

The Gumbel$_{\text{max}}$ (or simply Gumbel) distribution is a limiting case of the GEV distribution when $\kappa \to 0$. Likewise to the GEV, it is possible to specify a nonstationary model for a hydrological variable based on the Gumbel distribution with time-varying parameters β and α, that is,

$$X_t \sim \text{Gum}(\beta(t), \alpha(t)) \tag{12.42}$$

where $\beta(t)$ and $\alpha(t)$ define the dependence structure between the location and scale parameters and time.

The log-likelihood function of the nonstationary Gumbel model is given by

$$\ell(\theta) = -\sum_{t=1}^{N} \left\{ \ln(\alpha(t)) + \left(\frac{x_t - \beta(t)}{\alpha(t)} \right) + \exp\left[-\left(\frac{x_t - \beta(t)}{\alpha(t)} \right) \right] \right\} \tag{12.43}$$

The function gum.fit, of the R package ismev, can be used to fit this model by the IWLS method. This is covered in Example 12.10

Example 12.10 Consider the series of annual maximum daily rainfall at Pavia, in Portugal, shown in Example 12.9. Using the R package "ismev," estimate the parameters of the following Gumbel models: (a) stationary model GUM0; (b) linear trend in the location parameter GUM1; (c) linear trend in the location parameter and log-linear trend in the scale parameter GUM2.

Solution After loading the ismev package and importing the data into R, create a single-column matrix with time t between 1 and 98 (see Example 12.9a).
(a) The stationary model GUM0 is fitted using the function gum.fit, as in

```
> GUM0<-gum.fit(pavia)
```

The parameter estimates in the order (β, α) are given by

```
> GUM0$mle
[1] 31.46559 13.13585
```

(b) The same commands are applied to estimate parameters of the nonstationary model GUM1, with a linear trend on the location parameter $[\beta(t) = \beta_0 + \beta_1 t]$ and parameters in order $(\beta_0, \beta_1, \alpha)$, as in

```
> GUM1<-gum.fit(pavia,ydat=t,mul=1)
> GUM1$mle
[1] 25.7563638 0.1201053 12.4740537
```

(c) The same for model GUM2 with a log-linear trend in the scale parameter $\alpha(t)$ $= \exp(\alpha_0 + \alpha_1 t)$ and parameter vector $(\beta_0, \beta_1, \alpha_0, \alpha_1)$, as in

```
> GUM2<-gum.fit(pavia,ydat=t,mul=1,sigl=1,siglink=exp)
> GUM2$mle
[1] 2.544135e+01 1.262875e-01 2.485020e+00 7.661304e-04
```

12.4.4 Nonstationary Model Based on the Generalized Pareto Distribution

The generalized Pareto (GPA) distribution has its origins in results from EVT, namely in the research by Balkema and de Haan (1974) and Pickands (1975). The GPA distribution is not usually used in frequency analysis of annual maxima, but it is widely applied to peaks-over-threshold data, frequently in combination with the Poisson distribution (see Example 8.8).

Consider a peaks-over-threshold series X_t that shows some signs of changing behavior with time. It is possible to define a nonstationary flood frequency model based on the GPA distribution with scale parameter α and shape parameter κ, both changing with time, as denoted by

$$X_t \sim \text{GPA}(\alpha(t), \kappa(t)) \tag{12.44}$$

As in the case of the GEV model, the shape parameter of the GPA κ defines the shape of the upper tail of the distribution and the precise estimation of this parameter is complex. Likewise, regarding the nonstationary GPA model, it is not usual to allow the shape parameter to vary as a function of time. Therefore, as a rule, the only GPA parameter to be expressed as a function of time is the scale parameter α. Since that parameter can only take positive values, a convenient parametrization for $\alpha(t)$ is

$$\alpha(t) = \exp(\alpha_0 + \alpha_1 t) \tag{12.45}$$

The log-likelihood function of the nonstationary GPA is given by

$$\ell(\theta) = -\sum_{t=1}^{N} \left\{ \ln(\alpha(t)) - \frac{1}{\kappa(t) - 1} \ln\left[1 - \kappa(t)\frac{x_t}{\alpha(t)}\right] \right\} \qquad (12.46)$$

In the R package ismev, the appropriate function for fitting nonstationary GPA models is gpd.fit. The procedures are very similar to the ones described in Examples 12.9 and 12.10.

12.4.5   Model Selection and Diagnostics

The models presented in this subsection are highly versatile since they allow for: (1) having one or two nonstationary parameters; (2) the parameters being dependent on time directly or through a covariate (e.g., climate index); (3) many possible dependence structures between the parameters and time/covariate, i.e., linear, log linear, polynomial, change point, and other dependencies. Therefore, there are several candidate models to each problem in which a nonstationary extreme hydrological variable is present. Model selection under nonstationarity is an important issue, as the consideration of several covariates and possibly convoluted dependence structures can often result in very complex models which fit nicely to the data but may not be parsimonious. The basic aim here is to select a simple model with the capability of explaining much of the data variation.

The logic of model selection for nonstationary extremes is analogous to that of the GLM, whose main tools were presented in Sect. 12.3.3. Relative performances of nested candidate models may be assessed using asymptotic likelihood ratio tests and *AIC* can be used to select the best model from a list of candidates. The models which postulate a linear or log-linear dependence may be evaluated using hypothesis tests of the slope parameter.

The recommended practice for this kind of analysis is to start with a stationary model as a baseline model, with the lowest possible number of parameters, and gradually postulate incrementally complex models, that is, progressively add parameters, and check whether each alternate model has a significantly better performance than the previous one. It is important to be mindful than an LRT is only valid when the 2 models are nested.

Consider the hypothetical scenario in which a trend in the location parameter of a GEV-distributed variable is under analysis, but the parametric dependence structure of the trend is not obvious. One possible approach is to postulate the following models:

- GEV0—"no trend", baseline model, $\beta(t) = \beta$;
- GEV1—"linear trend," $\beta(t) = \beta_0 + \beta_1 t$;
- GEV2—"log-linear trend", $\beta(t) = \exp(\beta_0 + \beta_1 t)$;
- GEV3—"second degree polynomial trend", $\beta(t) = \beta_0 + \beta_1 t + \beta_2 t^2$.

In this scenario, the GEV0 model is nested in any of the other 3 models, since, in any case, it is mathematically equal to having every parameter other than β_0 equal to zero. Obviously, GEV1 is nested in GEV3, so those two models can be compared by means of an LRT. GEV1 is not nested in GEV2 nor is GEV2 in GEV3, hence an LRT may not be used to compare those models. In this kind of situation, it is preferable to compute the *AIC* scores for all models and determine the "best" one, according to that criterion.

Regarding models GEV1 and GEV2, it is also possible to set up a hypothesis test for regression parameters using the same rationale as described in Sect. 12.2.3. For these models, an estimation of the standard error of the slope parameter σ_{β_1} may be obtained by numerical differentiation of the log-likelihood function (see Sect. 12.3.2). One defines the null hypothesis H_0:{there is no trend in the location parameter} (or $\beta_1 = 0$) and alternative hypothesis H_1:{there is a trend in the location parameter} (or $\beta_1 \neq 0$). At the significance level of $100\alpha\,\%$, H_0 may be rejected if

$$\left|\hat{\beta}_1\right| > z_{1-\alpha/2}\sigma_{\beta_1} \tag{12.47}$$

Example 12.11 Consider the models GUM1 and GUM2 from Example 12.10. Determine the significance of the log-linear temporal trend of the scale parameter $\alpha(t)$ of model GUM2. (a) Use an LRT in which GUM1 is the null model; (b) use a hypothesis test for regression parameter α_1 of model GUM2. Consider the significance level of 5 % in both tests.

Solution The R code for solving this problem is presented in Appendix 11. (a) An LRT is employed in which the null model M_0 is GUM1, with 3 parameters, and the alternative model M_1 is GUM2 with 4 parameters. The negated maximum log-likelihood can be obtained on the R console (see Examples 12.9 and 12.10), by calling the component $nllh of the objects generated by the functions gev.fit and gum.fit. As such, $\ell(\hat{\theta}_{M_0}) = -402.5473$ and $\ell(\hat{\theta}_{M_1}) = -402.5087$. From Eq. (12.29), the test statistic $D = 0.0772$ is obtained through Eq. (12.29). The difference in number of parameters of both models is 1, such that $D \sim \chi_1^2$. Since $D < \chi_{0.95,1}^2 = 3.84$ (Appendix 3), the null model is not rejected in favor of the alternative one, at the 5 % significance level. (b) Standard errors of parameters estimates of models fitted using the R functions gev.fit and gum.fit, may be consulted by calling the component $se of the fitted model objects, which returns $\left(\sigma_{\beta_0} = 2.683484, \sigma_{\beta_1} = 0.048396, \sigma_{\alpha_0} = 0.156172, \sigma_{\alpha_1} = 0.002834\right)$. The rejection region of the test is $z_{1-\alpha/2}\sigma_{\alpha_1} = 0.005554$. It is worth remembering that $\hat{\alpha}_1 = 0.0007661$ (Example 12.10). Since $|\hat{\alpha}_1| < z_{1-\alpha/2}\sigma_{\alpha_1}$, the null hypothesis is not rejected, thereby corroborating the results of (a).

The graphical analysis tools presented in Chap. 8 are no longer valid under nonstationarity. Those tools require that the data be identically distributed, but in the nonstationary case, the observations are not homogeneous, since their distribution changes with time. In order to deal with this issue, Coles (2001) suggests the use of modified $Q-Q$ (quantile-quantile) plots to visualize the model fits of nonstationary extreme value models. When the fit is adequate, the scatter of points on the plot should be close to the 1-1 line. In order to apply this technique to a nonstationary model it is first necessary to transform the theoretical quantiles into a standardized and stationary variable.

In case the variable is GEV-distributed, $X_t \sim \text{GEV}(\beta(t), \alpha(t), \kappa(t))$, the standardized variable $\widetilde{X}_t$ is defined by

$$\widetilde{X}_t = \frac{-1}{\hat{\kappa}(t)} \ln\left\{1 - \hat{\kappa}(t)\left[\frac{X_t - \hat{\beta}(t)}{\hat{\alpha}(t)}\right]\right\} \qquad (12.48)$$

and when it's Gumbel-distributed, $X_t \sim \text{GUM}[\beta(t), \alpha(t)]$,

$$\widetilde{X}_t = \frac{X_t - \hat{\beta}(t)}{\hat{\alpha}(t)} \qquad (12.49)$$

The variable resulting from those transformations follows a standardized Gumbel distribution (Gumbel with $\beta = 0$ and $\alpha = 1$), with CDF

$$F_{\widetilde{X}_t}(x) = \exp(-e^{-x}) \qquad (12.50)$$

The previous result enables the making of a standardized Gumbel $Q-Q$ plot. By denoting the order statistics of $\widetilde{x}_t$ as $\widetilde{x}_{(1)}, \ldots, \widetilde{x}_{(N)}$, the $Q-Q$ plot is comprised of the pairs of points

$$\left\{-\ln[-\ln(q_i)], \widetilde{x}_{(i)}, i = 1, \ldots, N\right\} \qquad (12.51)$$

where q_i is the adopted plotting position (the Gringorten plotting position is recommended in the case of Gumbel and GEV models; see Sect. 8.1.2)

A similar technique may be applied when using a nonstationary GPA model. Considering the variable $X_t \sim \text{GPA}(\alpha(t), \kappa(t))$, the distribution used to standardize the variable is the exponential distribution (see Sect. 5.11.4);

$$\widetilde{X}_t = -\frac{1}{\hat{\kappa}(t)} \ln\left\{1 - \hat{\kappa}(t)\left[\frac{X_t}{\hat{\alpha}(t)}\right]\right\} \qquad (12.52)$$

The resulting variable $\widetilde{X}_t$ follows a standardized exponential distribution (exponential with $\theta = 1$), with CDF

$$F_{\widetilde{X}_t}(x) = 1 - \exp(-x) \qquad (12.53)$$

Then, the corresponding $Q-Q$ plot consists of the following pairs of points:

$$\{-\ln(1 - q_i), \tilde{x}_{(i)}, i = 1, \ldots, N\} \tag{12.54}$$

where q_i is the adopted plotting position.

Example 12.12 Consider the 5 stationary and nonstationary models fitted through Examples 12.9 and 12.10. (a) Using AIC, select the best of those 5 models. (b) Check the fit of that model using a $Q-Q$ plot.

Solution
(a) In R, the component \$nllh of the objects fitted in Examples 12.9 and 12.10 returns the negated maximum log likelihood of those models, which are shown in Table 12.8, together with the AIC results.
    The best of the 5 models, according to the AIC scores, is GUM1 (nonstationary Gumbel with a linear trend on the location parameter).
(b) Since $X_t \sim \mathrm{Gum}(\beta(t), \alpha(t))$, the transformation of the theoretical quantiles to the standardized Gumbel distribution uses Eq. (12.49), in which the location parameter has a linear temporal trend $\beta(t) = \beta_0 + \beta_1 t$ and the scale parameter is fixed $\alpha(t) = \alpha$. The $Q-Q$ plot consists of the pairs of points indicated in Eq. (12.51). Table 12.9 shows the necessary calculations to graph the $Q-Q$ plot, which, in turn, is shown in Fig. 12.6.

Table 12.8 Computation of AIC

Model	k (no. parameters)	$\ell(\hat{\theta})$	AIC
GEV0	3	−406.470	818.940
GEV1	4	−402.485	812.970
GUM0	2	−406.477	816.955
GUM1	3	−402.547	811.095
GUM2	4	−402.509	813.017

Table 12.9 Construction of a $Q-Q$ plot of a nonstationary Gumbel model based on the transformation to the standard Gumbel distribution

t	X_t	$\beta(t)$	$\alpha(t)$	q_i (Gringorten)	$\tilde{X}_{(t)}$	$-\ln[-\ln(q_i)]$	$\tilde{x}_{(t)}$
1	24.2	25.87647	12.47405	0.005707	−0.1344	−1.6421	−1.59839
2	31.3	25.99657	12.47405	0.015899	0.425157	−1.42106	−1.53972
3	32.5	26.11668	12.47405	0.026091	0.511728	−1.29368	−1.43967
4	33.5	26.23679	12.47405	0.036282	0.582266	−1.19889	−1.40276
5	20.2	26.35689	12.47405	0.046474	−0.49358	−1.12131	−1.12863
6	38.2	26.47700	12.47405	0.056665	0.939791	−1.05452	−1.05007
⋮	⋮	⋮	⋮	⋮	⋮	⋮	⋮
94	33.2	37.04626	12.47405	0.953526	−0.30834	3.045169	3.48196
95	27.4	37.16637	12.47405	0.963718	−0.78293	3.298009	3.972714
96	27.4	37.28647	12.47405	0.973909	−0.79256	3.632995	4.11714
97	18.2	37.40658	12.47405	0.984101	−1.53972	4.133503	4.859337
98	34.2	37.52668	12.47405	0.994293	−0.26669	5.163149	5.228651

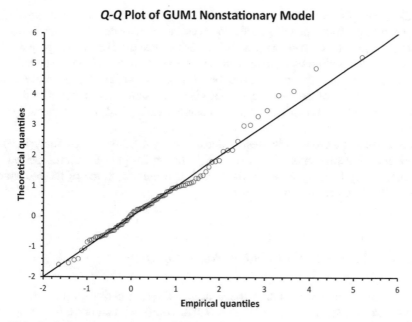

Fig. 12.6 Q–Q plot of nonstationary model GUM1 based on the transformation to the standard Gumbel distribution

12.5 Return Period and Hydrologic Risk in a Nonstationary Context

The concept of return period T of a quantile of a hydrological variable, defined by the inverse of the annual probability of exceedance of that variable, is an important and standard tool for hydrologists, with formal roots in the geometric distribution (see Sect. 4.1.2). Cooley (2013) contends that return periods are created to facilitate interpretation of the rarity of events: the expression "T-year flood" may be more easily interpreted by the general public than "a flood with an annual exceedance probability of $1/T$". The former definition leads to two interpretations of the T-year event:

- The expected waiting times between two events is T years; or
- The expected number of events in a T-year interval is one.

Under the stationarity assumption, both interpretations are correct.

Another notion closely related to return period is the hydrologic risk, defined by the probability that a reference quantile q_T will be exceeded in N years, or, in other words, the probability of occurrence of at least one event larger than q_T in N years. Under the independence and stationarity assumptions, hydrologic risk is given by Eq. (4.15).

The concepts of return period and hydrologic risk are commonly applied in engineering practice and are present in several textbooks on hydrology. However, these concepts do not hold under nonstationarity since the exceedance probabilities of hydrological extremes change from year to year. In any given year, there still exists a one-to-one correspondence between an exceedance probability and a particular quantile of the variable, but the idea of an annual return period is illogical and defeats its purpose as a concept for communicating hydrological hazard under nonstationarity.

As a result of a growing interest in nonstationary flood frequency analysis, some important developments for extending the concept of return period to nonstationarity have appeared in the technical literature. Some of these developments are presented in this section.

12.5.1   Return Period Under Nonstationarity

The first advances in extending the concept of return period to a nonstationary context are due to the work of Wigley (1988, 2009), who showed, in a simplified manner, how to consider nonstationarity when dealing with risk and uncertainty. Olsen et al. (1998) consolidated these original ideas with a rigorous mathematical treatment and defined the return period as the expected waiting time. Formally,

$$T(q_T) = 1 + \sum_{t=1}^{\infty} \prod_{i=1}^{t} F_{i(q_T)} \qquad (12.55)$$

where $F_i(\cdot)$ is the CDF of the variable in year i. Equation (12.55) cannot be written as a geometric series and solving it for q_T is not straightforward. Cooley (2013) shows that, in case $F_i(q_T)$ is monotonically decreasing as $i \to \infty$, it is possible to obtain a bounded estimate of $T(q_T)$ as

$$1 + \sum_{t=1}^{L} \prod_{i=1}^{t} F_i(q_T) < T(q_T) \leq 1 + \sum_{t=1}^{L} \prod_{i=1}^{t} F_i(q_T) + \prod_{i=1}^{L} F_i(q_T) \frac{F_{L+1}(q_T)}{1 - F_{L+1}(q_T)}$$

$$(12.56)$$

with the bounds being of a desired width by choosing a sufficiently large natural number L. Nevertheless, numerical methods must be employed in order to solve the bounds in Eq. (12.56) for q_T. Salas and Obeysekera (2014) built upon the developments of Wigley (1988, 2009) and Olsen et al. (1998), and presented a unified framework for estimating return period and hydrologic risk under nonstationarity.

Parey et al. (2007) and Parey et al. (2010) focused on the interpretation of the return period as the expected number of events in T years being 1 and extended that concept to nonstationarity. Under this interpretation, a T-year flood q_T can be estimated by solving the equation

$$\sum_{i=1}^{T} (1 - F_i(q_T)) = 1 \tag{12.57}$$

where $F_i(\cdot)$ has the same meaning as in Eq. (12.55). Solving Eq. (12.57) also requires numerical methods.

12.5.2 Design Life Level (DLL)

Rootzén and Katz (2013) argue that, for quantifying risk in engineering design, the basic required information consists of (1) the design life period of the hydraulic structure and (2) the probability of occurrence of a hazardous event during that period. These authors propose a new measure of hydrological hazard under nonstationarity: the *Design Life Level*, denoted as *DLL*, which is the quantile with a probability p of being exceeded during the design life period.

To compute the DLL it is necessary to derive the CDF $F_{T_1:T_2}$ of the maximum over the design life period, in which T_1 and T_2 represent the first and the last year of the period, respectively. Formally,

$$F_{T_1:T_2}(x) = P(\max\{X_t, t \in [T_1, T_2]\} \leq x) \tag{12.58}$$

Another way to put it is the probability that every value of X_t must simultaneously be lower than x, or

$$F_{T_1:T_2}(x) = P\left[\bigcap_{t=T_1}^{T_2} (X_t \leq x)\right] \tag{12.59}$$

Under the stationarity assumption (see Sect. 3.3) one would have

$$F_{T_1:T_2}(x) = \prod_{t=T_1}^{T_2} F_t(x) \tag{12.60}$$

The DLL is obtained by numerically inverting Eq. (12.60) for the desired non-exceedance probability $1 - p$. The design life level has a straightforward interpretation and does not imply extrapolations beyond the design life. Obviously, the design life level can also be estimated under stationarity. In that case, Eq. (12.60) is the complementary of the hydrologic risk (Eq. 4.15), with $N = T_2 - T_1 + 1$.

12.6 Further reading

Kundzewicz and Robson (2000, 2004) and Yue et al. (2012) review further methods for the detection of changes in hydrologic series and discuss at length the underlying assumptions and the adequate interpretation of the results of such methods. The presence of serial correlation in hydrologic time series, which is not uncommon in practice, may hinder the detection of trends or change points using the Mann–Kendall and Pettit tests, respectively. Serinaldi and Kilsby (2015b), and references therein, explore the limitations of these tests and suggest pre-whitening procedures designed to remove serial correlation from the data.

Mudelsee (2010) provides an in-depth characterization of the kernel occurrence rate estimation technique, including boundary bias reduction, bandwidth selection and uncertainty analysis via bootstrap techniques. Some examples of application of this technique are Mudelsee et al. (2003, 2004) and Silva et al. (2012).

Finally, it is important to stress that nonstationarity is a property of models and not of the hydrological/hydrometeorological phenomena underlying the time series used in statistical hydrologic analyses. In fact, there is an ongoing debate in the hydrological community on whether the use of nonstationary models is an adequate or even justifiable approach when tackling perceived changes in the statistical properties of hydrologic time series. A review of that debate is beyond the scope of this chapter since it is lengthy and comprises a number of different positions and proposed methodological approaches. Readers interested in such a debate are referred to the following sequence of papers: Milly et al. (2008, 2015), Koutsoyiannis (2011), Lins and Cohn (2011), Stedinger and Griffis (2011), Matalas (2012), Montanari and Koutsoyiannis (2014), Koutsoyiannis and Montanari (2015), Serinaldi and Kilsby (2015a). These works also tend to be very rich in references to up-to-date nonstationary hydrological analyses.

Exercises

1. Solve Example 12.3 (a) with bandwidth values $h = 500$, $h = 1000$ and $h = 2000$. Comment on the results in light of the compromise between variance and bias in estimation.
2. Construct 90 % bootstrap confidence bands for each of the curves obtained in Exercise 2.
3. Show that the probability density function of the Gamma distribution can be written in the form of Eq. (12.16).
4. Solve Example 12.5 considering that $Y \sim \mathrm{Binomial}(N, p)$ and use the AIC to compare the performances of the Poisson and Binomial models.
5. Consider the peaks-over-threshold data of Table 12.2. Fit a nonstationary GPA model with a log-linear dependence between the scale parameter and the winter (November-to-March) NAO index of the corresponding hydrologic year (NAO data shown in Table 12.6).
6. Using a likelihood ratio test, compare the performance of the model estimated in Exercise 5 with that of the corresponding stationary baseline model.

7. Using a hypothesis test for regression coefficients, check the significance of the log-linear relationship of the GPA scale parameter and the winter NAO index of the model estimated in Exercise 5.
8. Consider the series of annual maximum rainfalls at Pavia as listed in Table 12.6. Fit a nonstationary GEV model with a linear trend in the location parameter and a log-linear trend in the scale parameter.
9. Consider the model GUM1 from Example 12.10. Compute the expected waiting time for the exceedance of $q_T = 100\,\text{mm}$, taking the year 2010/11 as reference.
10. Consider the model GUM1 from Example 12.10. Compute the design life level with a non-exceedance probability of $F = 0.9$ and a design life of 50 years, taking the year 2010/11 as reference.

Acknowledgement   The research presented here and the contributions of the author to Chaps. 5 and 11 and Appendix 11 were funded by the Portuguese Science and Technology Foundation (FCT) through the scholarship SFRH/BD/86522/2012. The author wishes to thank Professor Maria Manuela Portela of Instituto Superior Técnico of the University of Lisbon for a thorough review of an earlier version of this book chapter.

References

Akaike H (1974) A new look at the statistical model identification. IEEE Trans Autom Control 19 (6):716–723

Balkema A, de Haan L (1974) Residual life time at great age. Ann Probab 2(5):792–804

Casella G, Berger R (2002) Statistical inference. Thomson Learning, Australia

Clarke R (2002) Estimating trends in data from the Weibull and a generalized extreme value distribution. Water Resour Res 38(6):25.1–25.10

Coles S (2001) An introduction to statistical modeling of extreme values. Springer, London

Cooley D (2013) Return periods and return levels under climate change. In: Extremes in a changing climate. Springer, Dordrecht, pp 97–114

Cowling A, Hall P (1996) On pseudodata methods for removing boundary effects in kernel density estimation. J Roy Stat Soc Series B 58(3):551–563

Cowling A, Hall P, Phillips M (1996) Bootstrap confidence regions for the intensity of a Poisson point process. J Am Stat Assoc 91(436):1516–1524

Davison A (2003) Statistical models. Cambridge University Press, Cambridge

Davison AC, Smith RL (1990) Models for exceedances over high thresholds. J Roy Stat Soc Series B Methodol 393–442

Davison A, Hinkley D (1997) Bootstrap methods and their application. Cambridge University Press, Cambridge

Diggle P (1985) A kernel method for smoothing point process data. Appl Stat 34(2):138

Dobson A (2001) An introduction to generalized linear models. Chapman & Hall/CRC, Boca Raton

Efron B (1979) Bootstrap methods: another look at the jackknife. Ann Stat 7(1):1–26

Heffernan JE, Stephenson A, Gilleland E (2013) Ismev: an introduction to statistical analysis of extreme values. R Package Version 1:39

Jones P, Jonsson T, Wheeler D (1997) Extension to the North Atlantic oscillation using early instrumental pressure observations from Gibraltar and southwest Iceland. Int J Climatol 17 (13):1433–1450

Katz RW (2013) Statistical methods for nonstationary extremes. In: Extremes in a Changing Climate. Springer, Dordrecht, pp 15–37

Kendall M (1975) Rank correlation methods. Griffin, London

Koutsoyiannis D (2011) Hurst-Kolmogorov dynamics and uncertainty. J Am Water Resour Assoc 47(3):481–495

Koutsoyiannis D, Montanari A (2015) Negligent killing of scientific concepts: the stationarity case. Hydrol Sci J 60(7–8):1174–1183

Kundzewicz ZW, Robson AJ (2000) Detecting trend and other changes in hydrological data. World Climate Programme—Water, World Climate Programme Data and Monitoring, WCDMP-45, WMO/TD no. 1013. World Meteorological Organization, Geneva

Kundzewicz ZW, Robson AJ (2004) Change detection in hydrological records—a review of the methodology. Hydrol Sci J 49(1):7–19

Lins H, Cohn T (2011) Stationarity: wanted dead or alive? J Am Water Resour Assoc 47(3):475–480

Lorenzo-Lacruz J, Vicente-Serrano S, López-Moreno J, González-Hidalgo J, Morán-Tejeda E (2011) The response of Iberian rivers to the North Atlantic Oscillation. Hydrol Earth Syst Sci 15(8):2581–2597

Mann H (1945) Nonparametric tests against trend. Econometrica 13(3):245

Matalas N (2012) Comment on the announced death of stationarity. J Water Resour Plann Manag 138(4):311–312

McCullagh P, Nelder J (1989) Generalized linear models. Chapman and Hall, London

Milly PC, Betancourt J, Falkenmark M, Hirsch RM, Kundzewicz ZW, Lettenmaier DP, Stouffer RJ (2008) Climate change. Stationarity is dead: whither water management? Science 319(5863):573–574

Milly PC, Betancourt J, Falkenmark M, Hirsch R, Kundzewicz ZW, Lettenmaier D, Stouffer RJ, Dettinger M, Krysanova V (2015) On critiques of "Stationarity is dead: whither water management?". Water Resour Res 51(9):7785–7789

Montanari A, Koutsoyiannis D (2014) Modeling and mitigating natural hazards: stationarity is immortal! Water Resour Res 50(12):9748–9756

Mudelsee M (2010) Climate time series analysis. Springer, Dordrecht

Mudelsee M, Börngen M, Tetzlaff G, Grünewald U (2003) No upward trends in the occurrence of extreme floods in central Europe. Nature 425(6954):166–169

Mudelsee M, Börngen M, Tetzlaff G, Grünewald U (2004) Extreme floods in central Europe over the past 500 years: role of cyclone pathway "Zugstrasse Vb". J Geophys Res Atmos 109(D23)

Olsen J, Lambert J, Haimes Y (1998) Risk of extreme events under nonstationary conditions. Risk Anal 18(4):497–510

Parey S, Hoang T, Dacunha-Castelle D (2010) Different ways to compute temperature return levels in the climate change context. Environmetrics 21(7–8):698–718

Parey S, Malek F, Laurent C, Dacunha-Castelle D (2007) Trends and climate evolution: statistical approach for very high temperatures in France. Clim Change 81(3–4):331–352

Parzen E (1962) On estimation of a probability density function and mode. Ann Math Stat 33(3):1065–1076

Pettitt AN (1979) A non-parametric approach to the change-point problem. Appl Stat 28(2):126

Pickands J III (1975) Statistical inference using extreme order statistics. Ann Stat 3(1):119–131

Core Team R (2013) R: a language and environment of statistical computing. R Foundation for Statistical Computing, Vienna

Rootzén H, Katz R (2013) Design life level: quantifying risk in a changing climate. Water Resour Res 49(9):5964–5972

Rosenblatt M (1956) Remarks on some nonparametric estimates of a density function. Ann Math Stat 27(3):832–837

Rybski D, Neumann J (2011) A review on the Pettitt test. In: Kropp J, Schellnhuber HJ (eds) In extremis: 202–213. Springer, Dordrecht

Salas J, Obeysekera J (2014) Revisiting the concepts of return period and risk for nonstationary hydrologic extreme events. J Hydrol Eng 19(3):554–568

Serinaldi F, Kilsby C (2015a) Stationarity is undead: uncertainty dominates the distribution of extremes. Adv Water Resour 77:17–36

Serinaldi F, Kilsby C (2015b) The importance of prewhitening in change point analysis under persistence. Stoch Environ Res Risk Assess 30(2):763–777

Silva A, Portela M, Naghettini M (2012) Nonstationarities in the occurrence rates of flood events in Portuguese watersheds. Hydrol Earth Syst Sci 16(1):241–254

Silverman B (1986) Density estimation for statistics and data analysis. Chapman and Hall, London

Stedinger J, Griffis V (2011) Getting from here to where? Flood frequency analysis and climate. J Am Water Resour Assoc 47(3):506–513

Turkman MAA, Silva GL (2000) Modelos lineares generalizados-da teoria à prática. In: VIII Congresso Anual da Sociedade Portuguesa de Estatística, Lisboa

Wigley TML (1988) The effect of changing climate on the frequency of absolute extreme events. Climate Monitor 17:44–55

Wigley TML (2009) The effect of changing climate on the frequency of absolute extreme events. Clim Change 97(1–2):67–76

Yue S, Kundzewicz ZW, Wang L (2012) Detection of changes. In: Kundzewicz ZW (ed) Changes in flood risk in Europe. IAHS Press, Wallingford, UK, pp 387–434

Appendix 1
Mathematics: A Brief Review of Some Important Topics

A1.1 Counting Problems

In some situations, the calculation of probabilities requires counting the number of possible ways of drawing a sample of size k from a set of n elements. The number of sampling possibilities can be easily calculated through the definitions and formulae of combinatorics.

The sampling of the k items may be performed *with replacement*, when it is possible to draw a specific item more than once, or *without replacement*, otherwise. Furthermore, the sequence or order in which the different items are sampled may be an important factor. As a result, the following ways of sampling are possible: with order and with replacement, with order and without replacement, without order and with replacement, without order and without replacement.

In the case of sampling with order and with replacement, the first item should be drawn from the n elements (or possibilities) that constitute the population. Next, this first sampled item is reintegrated to the population and, as previously, the second draw is performed from a set of n items. Based on this rationale, as each drawing is made from n items, the number of possibilities of drawing a sample of k items from a set of size n, with order and with replacement, is n^k.

If the first sampled item is not returned to the population for the next drawing, the number of possibilities for the second item is $(n - 1)$. The third item will then be drawn from $(n - 2)$ possibilities, the fourth from $(n - 3)$, and so forth, until the kth item is sampled. Therefore, the number of possibilities of drawing a sample of size k from a set of n elements is $n(n - 1)(n - 2) \ldots (n - k + 1)$. This expression is equivalent to the definition of permutation in combinatorics. Thus,

$$P_{n,k} = \frac{n!}{(n - k)!} \tag{A1.1}$$

When the order of drawing is not important, sampling without replacement is similar to the previous case, except that the drawn items can be arranged in $k!$

© Springer International Publishing Switzerland 2017
M. Naghettini (ed.), *Fundamentals of Statistical Hydrology*,
DOI 10.1007/978-3-319-43561-9

different ways. In other words, the number of ordered samples, given by Eq. (A1.1), includes $k!$ draws that contain the same elements. Thus, as order is no longer important, the number of possibilities of drawing k items from a sample of size n, without order and without replacement, is $P_{n,k}/k!$. This expression is equivalent to the definition of combination in combinatorics. Thus,

$$C_{n,k} = \binom{n}{k} = \frac{n!}{(n-k)!\ k!} \tag{A1.2}$$

Finally, when the order of drawing is not important and sampling is performed with replacement, the number of possibilities corresponds to sampling, without order and without replacement, of k items from a set of $(n+k-1)$ possibilities. In other words, one can reason that the population has been increased by $(k-1)$ items. Thus, the number of possibilities of drawing k items from a set of n elements, without order and with replacement, is given by

$$C_{n+k-1,k} = \binom{n+k-1}{k} = \frac{(n+k-1)!}{(n-1)!\ k!} \tag{A1.3}$$

The factorial operator, present in many combinatorics equations, can be approximated by Stirling's formula, which is given by

$$n! \cong \frac{\sqrt{2\pi}\, n^{n+1/2}}{e^n} \tag{A1.4}$$

Haan (1977) remarks that, for $n=10$, the approximation error with Stirling's formula is less than 1 % and decreases as n increases.

A1.2   MacLaurin Series

If a function $f(x)$ has continuous derivatives up to the order $(n+1)$, then it can be expanded as follows

$$f(x) = f(a) + f'(a)(x-a) + \frac{f''(a)(x-a)^2}{2!} + \cdots \frac{f^{(n)}(a)(x-a)^n}{n!} + R_n \tag{A1.5}$$

where R_n denotes the *remainder*, after the expansion of $(n+1)$ terms, and is expressed by

$$R_n = \frac{f^{(n+1)}(\tau)(x-a)^{n+1}}{(n+1)!} \qquad a < \tau < x \tag{A1.6}$$

If the expansion given by Eq. (A1.5) converges within a subdomain of x, or, in other words, if $\lim\limits_{n\to\infty} R_n = 0$, then the series expansion is termed *Taylor series* of $f(x)$ around a. If $a=0$, then the expansion is termed *MacLaurin series* and is formally given by

$$f(x) = f(0) + f'(0)x + \frac{f''(0)}{2!}x^2 + \cdots \qquad (A1.7)$$

The MacLaurin series is, therefore, a series expansion in which all terms are non-negative integer powers of the variable being considered. Examples of function expansions by means of MacLaurin series are

$$\cos(x) = 1 - \frac{x^2}{2} + \frac{x^4}{24} - \frac{x^6}{720} - \cdots \qquad -\infty < x < \infty \qquad (A1.8)$$

$$e^x = 1 + x + \frac{x^2}{2} + \frac{x^3}{6} + \frac{x^4}{24} + \cdots \qquad -\infty < x < \infty \qquad (A1.9)$$

$$\ln(1 + x) = x - \frac{x^2}{2} + \frac{x^3}{3} - \frac{x^4}{4} + \cdots \qquad -1 < x < 1 \qquad (A1.10)$$

$$\frac{1}{1 - x} = 1 + x + x^2 + x^3 + x^4 + \cdots \qquad -1 < x < 1 \qquad (A1.11)$$

A1.3   Gamma Function

The Gamma function $\Gamma(z)$ is an extension of the concept of factorial for non-integer numbers. $\Gamma(z)$ is defined, for any real number $z>0$, by the following integral

$$\Gamma(z) = \int\limits_0^\infty x^{z-1} e^{-x} dx \qquad (A1.12)$$

The Gamma function is a continuous function and has continuous derivatives for all orders. When z approaches 0 or $+\infty$, $\Gamma(z)$ tends to $+\infty$. By using integration by parts, it is possible to demonstrate the following property of the Gamma function

$$\Gamma(z + 1) = z\ \Gamma(z) \qquad (A1.13)$$

If z equals a positive integer number n and noting that $\Gamma(1) = 1$, the repeated application of the property given by Eq. (A1.13) leads to

$$\Gamma(n+1) = n! \tag{A1.14}$$

Noteworthy values of the Gamma function are: $\Gamma(2) = \Gamma(1) = 1$ and $\Gamma(0.5) = \sqrt{\pi}$.

The Gamma function may be approximated by several expressions. One of the most efficient, with errors in the range of 2×10^{-10}, is the Lanczos approximation (Lanczoz 1964), which is given by

$$\Gamma(z) = \left[\frac{\sqrt{2\pi}}{z}\left(p_0 + \sum_{i=1}^{8}\frac{p_i}{z+i}\right)\right](z+7,5)^{z+0,5}e^{-(z+7,5)} \tag{A1.15}$$

with

$$p_0 = 0.99999999999980993 \quad p_1 = 676.5203681218851$$

$$p_2 = -1259.1392167224028 \quad p_3 = 771.32342877765313$$

$$p_4 = -176.61502916214059 \quad p_5 = 12.507343278686905$$

$$p_6 = -0.13857109526572012 \, p_7 = 9.9843695780195716 \times 10^{-6}$$

$$p_8 = 1.5056327351493116 \times 10^{-7}$$

A1.4   Beta Function

The Beta function, denoted by $B(z, w)$, is defined, for any positive real number, by the following integral

$$B(z, w) = \int_0^1 x^{z-1}(1-x)^{w-1}dx \tag{A1.16}$$

Cramér (1946) derived the following relationship between the Beta and the Gamma functions

$$B(z, w) = \frac{\Gamma(z)\Gamma(w)}{\Gamma(z+w)} \tag{A1.17}$$

Based on this relationship and using the Lanczos approximation, expressed by Eq. (A1.15), it is possible to evaluate the Beta function for any real numbers z and w.

A1.5 Differentiation Rules and Derivatives for Some Basic Functions

In this section, a list of basic differentiation rules is provided. For all cases, $u = f(x)$ and $v = g(x)$ are differentiable functions of x and c is a constant.

Derivative of a constant: $\frac{dc}{dx} = 0$

Derivative of a constant multiple: $\frac{d(cu)}{dx} = c\frac{du}{dx}$

Derivative of a sum or a difference: $\frac{d(u \pm v)}{dx} = \frac{du}{dx} \pm \frac{dv}{dx}$

Product rule: $\frac{d(uv)}{dx} = v\frac{du}{dx} + u\frac{dv}{dx}$

Quotient rule: $\frac{d}{dx}\left(\frac{u}{v}\right) = \frac{v\frac{du}{dx} - u\frac{dv}{dx}}{v^2}$

Chain rule: $\frac{dy}{dx} = \frac{dy}{du}\frac{du}{dx}$

A1.6 Integration Rules and Indefinite Integrals for Some Basic Functions

Integral of a constant multiple: $\int cf(x)dx = c\int f(x)dx$

Integration of a sum or a difference: $\int [f(x) \pm g(x)]dx = \int f(x)dx \pm \int g(x)dx$

Integral of a chain rule derivative: $\int \frac{d}{dx}\{f[g(x)]\}\frac{d}{dx}[g(x)]dx = f[g(x)] + C$

Table A1.1 Derivatives for some basic functions

Function	Derivative
u^n	$\frac{du^n}{dx} = nu^{n-1}\frac{du}{dx}$
a^u	$\frac{da^u}{dx} = \ln(a)a^u\frac{du}{dx}$
e^u	$\frac{de^u}{dx} = e^u\frac{du}{dx}$
$\log_a u$	$\frac{d}{dx}(\log_a u) = \frac{1}{(\ln a)u}\frac{du}{dx}$
$\ln u$	$\frac{d}{dx}(\ln u) = \frac{1}{u}\frac{du}{dx}$
$\sin u$	$\frac{d}{dx}(\sin u) = \cos u\frac{du}{dx}$
$\cos u$	$\frac{d}{dx}(\cos u) = -\sin u\frac{du}{dx}$
$\tan u$	$\frac{d}{dx}(\tan u) = \sec^2 u\frac{du}{dx}$

Table A1.2 Indefinite integrals for some basic functions

Function	Indefinite integral
x^n, for $n \neq -1$	$\int x^n dx = \frac{x^{n+1}}{n+1} + C$
a^x or $e^{x\ln(a)}$, for $a \neq 1, a > 0$	$\int a^x dx = \frac{a^x}{\ln a} + C$
e^{ax}	$\int e^{ax} = \frac{1}{a} e^{ax} + C$
$\frac{1}{x}$, for $x \neq 0$	$\int \frac{1}{x} dx = \ln\lvert x \rvert + C$
$\sin(ax)$	$\int \sin(ax) dx = -\frac{1}{a} \cos(ax) + C$
$\cos(ax)$	$\int \cos(ax) dx = \frac{1}{a} \sin(ax) + C$
$\tan(ax)$	$\int \tan(ax) x = -\frac{1}{a} \ln(\cos(ax)) + C$

Integral by parts: $\int f(x) \frac{d}{dx}[g(x)] dx = f(x)g(x) - \int \frac{d}{dx}[f(x)]g(x) dx$

Integral by substitution: $\displaystyle\int_{\varphi(a)}^{\varphi(b)} f(x) dx = \int_a^b f[\varphi(t)] \frac{d}{dt}[\varphi(t)] dt$

References

Cramér H (1946) Mathematical Methods of Statistics. Princeton University Press, Princeton.

Haan CT (1977) Statistical Methods in Hydrology. The Iowa University Press, Ames, IA.

Lanczos C (1964) A precision approximation of the Gamma function. J Soc Ind Appl Math Series B Numer Anal 1:86–96.

Appendix 2
Values for the Gamma Function $\Gamma(t)$

$$\Gamma(t) = \int_0^\infty e^{-x} x^{t-1} dx$$

t	$\Gamma(t)$	t	$\Gamma(t)$	t	$\Gamma(t)$	t	$\Gamma(t)$
1.00	1.00000	1.25	0.90640	1.50	0.88623	1.75	0.91906
1.01	0.99433	1.26	0.90440	1.51	0.88659	1.76	0.92137
1.02	0.98884	1.27	0.90250	1.52	0.88704	1.77	0.92376
1.03	0.98355	1.28	0.90072	1.53	0.88757	1.78	0.92623
1.04	0.97844	1.29	0.89904	1.54	0.88818	1.79	0.92877
1.05	0.97350	1.30	0.89747	1.55	0.88887	1.80	0.93138
1.06	0.96874	1.31	0.89600	1.56	0.88964	1.81	0.93408
1.07	0.96415	1.32	0.89464	1.57	0.89049	1.82	0.93685
1.08	0.95973	1.33	0.89338	1.58	0.89142	1.83	0.93969
1.09	0.95546	1.34	0.89222	1.59	0.89243	1.84	0.94261
1.10	0.95135	1.35	0.89115	1.60	0.89352	1.85	0.94561
1.11	0.94739	1.36	0.89018	1.61	0.89468	1.86	0.94869
1.12	0.94359	1.37	0.89931	1.62	0.89592	1.87	0.95184
1.13	0.93993	1.38	0.88854	1.63	0.89724	1.88	0.95507
1.14	0.93642	1.39	0.88785	1.64	0.89864	1.89	0.95838
1.15	0.93304	1.40	0.88726	1.65	0.90012	1.90	0.96177
1.16	0.92980	1.41	0.88676	1.66	0.90167	1.91	0.96523
1.17	0.92670	1.42	0.88636	1.67	0.90330	1.92	0.96878
1.18	0.92373	1.43	0.88604	1.68	0.90500	1.93	0.97240

(continued)

© Springer International Publishing Switzerland 2017
M. Naghettini (ed.), *Fundamentals of Statistical Hydrology*,
DOI 10.1007/978-3-319-43561-9

t	$\Gamma(t)$	t	$\Gamma(t)$	t	$\Gamma(t)$	t	$\Gamma(t)$
1.19	0.92088	1.44	0.88580	1.69	0.90678	1.94	0.97610
1.20	0.91817	1.45	0.88565	1.70	0.90864	1.95	0.97988
1.21	0.91558	1.46	0.88560	1.71	0.91057	1.96	0.98374
1.22	0.91311	1.47	0.88563	1.72	0.91258	1.97	0.98768
1.23	0.91075	1.48	0.88575	1.73	0.91466	1.98	0.99171
1.24	0.90852	1.49	0.88595	1.74	0.91683	1.99	0.99581

- For other t values, employ the recurrence relation $\Gamma(t+1) = t\Gamma(t)$.
- For high values of t, one can also use Stirling's approximation:

$$\Gamma(t) \approx t^t e^{-t} \sqrt{\frac{2\pi}{t}} \left(1 + \frac{1}{12t} + \frac{1}{288t^2} - \frac{139}{51840t^3} - \frac{571}{2488320t^4} + \cdots \right)$$

Appendix 3
$\chi^2_{1-\alpha,\nu}$ Quantiles from Chi-Square Distribution, with ν Degrees of Freedom

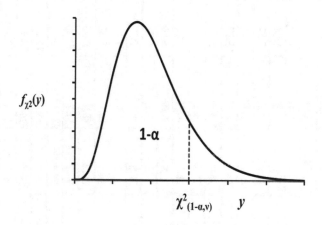

ν	$\chi^2_{0.995,\nu}$	$\chi^2_{0.99,\nu}$	$\chi^2_{0.975,\nu}$	$\chi^2_{0.95,\nu}$	$\chi^2_{0.90,\nu}$	$\chi^2_{0.10,\nu}$	$\chi^2_{0.05,\nu}$	$\chi^2_{0.025,\nu}$	$\chi^2_{0.01,\nu}$	$\chi^2_{0.005,\nu}$
1	7.88	6.63	5.02	3.84	2.71	0.0158	0.0039	0.0010	0.0002	0.0000
2	10.6	9.21	7.38	5.99	4.61	0.211	0.103	0.0506	0.0201	0.0100
3	12.8	11.3	9.35	7.81	6.25	0.584	0.352	0.216	0.115	0.072
4	14.9	13.3	11.1	9.49	7.78	1.06	0.711	0.484	0.297	0.207
5	16.7	15.1	12.8	11.1	9.24	1.61	1.15	0.831	0.554	0.412
6	18.5	16.8	14.4	12.6	10.6	2.20	1.64	1.24	0.872	0.676
7	20.3	18.5	16.0	14.1	12.0	2.83	2.17	1.69	1.24	0.989
8	22.0	20.1	17.5	15.5	13.4	3.49	2.73	2.18	1.65	1.34
9	23.6	21.7	19.0	16.9	14.7	4.17	3.33	2.70	2.09	1.73
10	25.2	23.2	20.5	18.3	16.0	4.87	3.94	3.25	2.56	2.16
11	26.8	24.7	21.9	19.7	17.3	5.58	4.57	3.82	3.05	2.60
12	28.3	26.2	23.3	21.0	18.5	6.30	5.23	4.40	3.57	3.07
13	29.8	27.7	24.7	22.4	19.8	7.04	5.89	5.01	4.11	3.57
14	31.3	29.1	26.1	23.7	21.1	7.79	6.57	5.63	4.66	4.07
15	32.8	30.6	27.5	25.0	22.3	8.55	7.26	6.26	5.23	4.60
16	34.3	32.0	28.8	26.3	23.5	9.31	7.96	6.91	5.81	5.14
17	35.7	33.4	30.2	27.6	24.8	10.1	8.67	7.56	6.41	5.70
18	37.2	34.8	31.5	28.9	26.0	10.9	9.39	8.23	7.01	6.26
19	38.6	36.2	32.9	30.1	27.2	11.7	10.1	8.91	7.63	6.84
20	40.0	37.6	34.2	31.4	28.4	12.4	10.9	9.59	8.26	7.43
21	41.4	38.9	35.5	32.7	29.6	13.2	11.6	10.3	8.90	8.03
22	42.8	40.3	36.8	33.9	30.8	14.0	12.3	11.0	9.54	8.64
23	44.2	41.6	38.1	35.2	32.0	14.8	13.1	11.7	10.2	9.26
24	45.6	43.0	39.4	36.4	33.2	15.7	13.8	12.4	10.9	9.89
25	46.9	44.3	40.6	37.7	34.4	16.5	14.6	13.1	11.5	10.5
26	48.3	45.6	41.9	38.9	35.6	17.3	15.4	13.8	12.2	11.2

27	49.6	47.0	43.2	40.1	36.7	18.1	16.2	14.6	12.9	11.8
28	51.0	48.3	44.5	41.3	37.9	18.9	16.9	15.3	13.6	12.5
29	52.3	49.6	45.7	42.6	39.1	19.8	17.7	16.0	14.3	13.1
30	53.7	50.9	47.0	43.8	40.3	20.6	18.5	16.8	15.0	13.8
40	66.8	63.7	59.3	55.8	51.8	29.1	26.5	24.4	22.2	20.7
50	79.5	76.2	71.4	67.5	63.2	37.7	34.8	32.4	29.7	28.0
60	92.0	88.4	83.3	79.1	74.4	46.5	43.2	40.5	37.5	35.5
70	104.2	100.4	95.0	90.5	85.5	55.3	51.7	48.8	45.4	43.3
80	116.3	112.3	106.6	101.9	96.6	64.3	60.4	57.2	53.5	51.2
90	128.3	124.1	118.1	113.1	107.6	73.3	69.1	65.6	61.8	59.2
100	140.2	135.8	129.6	124.3	118.5	82.4	77.9	74.2	70.1	67.3

Appendix 4
$t_{1-\alpha,\nu}$ Quantiles from Student's t Distribution, with ν Degrees of Freedom

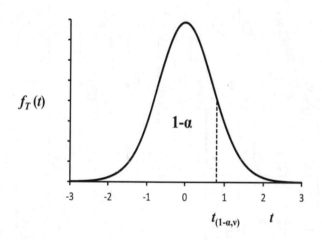

© Springer International Publishing Switzerland 2017
M. Naghettini (ed.), *Fundamentals of Statistical Hydrology*,
DOI 10.1007/978-3-319-43561-9

ν	$t_{0.55,\nu}$	$t_{0.60,\nu}$	$t_{0.70,\nu}$	$t_{0.75,\nu}$	$t_{0.80,\nu}$	$t_{0.90,\nu}$	$t_{0.95,\nu}$	$t_{0.975,\nu}$	$t_{0.99,\nu}$	$t_{0.995,\nu}$
1	0.158	0.325	0.727	1.000	1.376	3.08	6.31	12.71	31.82	63.66
2	0.142	0.289	0.617	0.816	1.061	1.89	2.92	4.30	6.96	9.92
3	0.137	0.277	0.584	0.765	0.978	1.64	2.35	3.18	4.54	5.84
4	0.134	0.271	0.569	0.741	0.941	1.53	2.13	2.78	3.75	4.60
5	0.132	0.267	0.559	0.727	0.920	1.48	2.02	2.57	3.36	4.03
6	0.131	0.265	0.553	0.718	0.906	1.44	1.94	2.45	3.14	3.71
7	0.130	0.263	0.549	0.711	0.896	1.42	1.90	2.36	3.00	3.50
8	0.130	0.262	0.546	0.706	0.889	1.40	1.86	2.31	2.90	3.36
9	0.129	0.261	0.543	0.703	0.883	1.38	1.83	2.26	2.82	3.25
10	0.129	0.260	0.542	0.700	0.879	1.37	1.81	2.23	2.76	3.17
11	0.129	0.260	0.540	0.697	0.876	1.36	1.80	2.20	2.72	3.11
12	0.128	0.259	0.539	0.695	0.873	1.36	1.78	2.18	2.68	3.06
13	0.128	0.259	0.538	0.694	0.870	1.35	1.77	2.16	2.65	3.01
14	0.128	0.258	0.537	0.692	0.868	1.34	1.76	2.14	2.62	2.98
15	0.128	0.258	0.536	0.691	0.866	1.34	1.75	2.13	2.60	2.95
16	0.128	0.258	0.535	0.690	0.865	1.34	1.75	2.12	2.58	2.92
17	0.128	0.257	0.534	0.689	0.863	1.33	1.74	2.11	2.57	2.90
18	0.127	0.257	0.534	0.688	0.862	1.33	1.73	2.10	2.55	2.88
19	0.127	0.257	0.533	0.688	0.861	1.33	1.73	2.09	2.54	2.86
20	0.127	0.257	0.533	0.687	0.860	1.32	1.72	2.09	2.53	2.84
21	0.127	0.257	0.532	0.686	0.859	1.32	1.72	2.08	2.52	2.83
22	0.127	0.256	0.532	0.686	0.858	1.32	1.72	2.07	2.51	2.82
23	0.127	0.256	0.532	0.685	0.858	1.32	1.71	2.07	2.50	2.81
24	0.127	0.256	0.531	0.685	0.857	1.32	1.71	2.06	2.49	2.80
25	0.127	0.256	0.531	0.684	0.856	1.32	1.71	2.06	2.48	2.79
26	0.127	0.256	0.531	0.684	0.856	1.32	1.71	2.06	2.48	2.78

27	2.77	2.47	2.05	1.70	1.31	0.855	0.684	0.531	0.256	0.127
28	2.76	2.47	2.05	1.70	1.31	0.855	0.683	0.530	0.256	0.127
29	2.76	2.46	2.04	1.70	1.31	0.854	0.683	0.530	0.256	0.127
30	2.75	2.46	2.04	1.70	1.31	0.854	0.683	0.530	0.256	0.127
40	2.70	2.42	2.02	1.68	1.30	0.851	0.681	0.529	0.255	0.126
60	2.66	2.39	2.00	1.67	1.30	0.848	0.679	0.527	0.254	0.126
120	2.62	2.36	1.98	1.66	1.29	0.845	0.677	0.526	0.254	0.126
∞	2.58	2.33	1.96	1.645	1.28	0.842	0.674	0.524	0.253	0.126

Appendix 5
Snedecor's F Quantiles, with $\gamma_1=m$ (Degrees of Freedom of the Numerator) and $\gamma_2=n$ (Degrees of Freedom of the Denominator)

© Springer International Publishing Switzerland 2017
M. Naghettini (ed.), *Fundamentals of Statistical Hydrology*,
DOI 10.1007/978-3-319-43561-9

$1-\alpha$	n	$m=1$	$m=2$	$m=3$	$m=4$	$m=5$	$m=6$	$m=7$	$m=8$	$m=9$	$m=10$	$m=12$	$m=15$	$m=20$	$m=30$	$m=60$	$m=120$	$m=\infty$
0.9	1	39.86	49.50	53.59	55.83	57.24	58.20	58.91	59.44	59.86	60.19	60.71	61.22	61.74	62.26	62.79	63.06	63.32
0.95	1	161.45	199.50	215.71	224.58	230.16	233.99	236.77	238.88	240.54	241.88	243.90	245.95	248.02	250.10	252.20	253.25	254.29
0.975	1	647.79	799.48	864.15	899.60	921.83	937.11	948.20	956.64	963.28	968.63	976.72	984.87	993.08	1001	1010	1014	1018
0.99	1	4052	4999	5404	5624	5764	5859	5928	5981	6022	6056	6107	6157	6209	6260	6313	6340	6366
0.995	1	16,212	19,997	21,614	22,501	23,056	23,440	23,715	23,924	24,091	24,222	24,427	24,632	24,837	25,041	25,254	25,358	25,462
0.9	2	8.53	9.00	9.16	9.24	9.29	9.33	9.35	9.37	9.38	9.39	9.41	9.42	9.44	9.46	9.47	9.48	9.49
0.95	2	18.51	19.00	19.16	19.25	19.30	19.33	19.35	19.37	19.38	19.40	19.41	19.43	19.45	19.46	19.48	19.49	19.50
0.975	2	38.51	39.00	39.17	39.25	39.30	39.33	39.36	39.37	39.39	39.40	39.41	39.43	39.45	39.46	39.48	39.49	39.50
0.99	2	98.50	99.00	99.16	99.25	99.30	99.33	99.36	99.38	99.39	99.40	99.42	99.43	99.45	99.47	99.48	99.49	99.50
0.995	2	198.50	199.01	199.16	199.24	199.30	199.33	199.36	199.38	199.39	199.39	199.42	199.43	199.45	199.48	199.48	199.49	199.51
0.9	3	5.54	5.46	5.39	5.34	5.31	5.28	5.27	5.25	5.24	5.23	5.22	5.20	5.18	5.17	5.15	5.14	5.13
0.95	3	10.13	9.55	9.28	9.12	9.01	8.94	8.89	8.85	8.81	8.79	8.74	8.70	8.66	8.62	8.57	8.55	8.53
0.975	3	17.44	16.04	15.44	15.10	14.88	14.73	14.62	14.54	14.47	14.42	14.34	14.25	14.17	14.08	13.99	13.95	13.90
0.99	3	34.12	30.82	29.46	28.71	28.24	27.91	27.67	27.49	27.34	27.23	27.05	26.87	26.69	26.50	26.32	26.22	26.13
0.995	3	55.55	49.80	47.47	46.20	45.39	44.84	44.43	44.13	43.88	43.68	43.39	43.08	42.78	42.47	42.15	41.99	41.83
0.9	4	4.54	4.32	4.19	4.11	4.05	4.01	3.98	3.95	3.94	3.92	3.90	3.87	3.84	3.82	3.79	3.78	3.76
0.95	4	7.71	6.94	6.59	6.39	6.26	6.16	6.09	6.04	6.00	5.96	5.91	5.86	5.80	5.75	5.69	5.66	5.63
0.975	4	12.22	10.65	9.98	9.60	9.36	9.20	9.07	8.98	8.90	8.84	8.75	8.66	8.56	8.46	8.36	8.31	8.26
0.99	4	21.20	18.00	16.69	15.98	15.52	15.21	14.98	14.80	14.66	14.55	14.37	14.20	14.02	13.84	13.65	13.56	13.47
0.995	4	31.33	26.28	24.26	23.15	22.46	21.98	21.62	21.35	21.14	20.97	20.70	20.44	20.17	19.89	19.61	19.47	19.33
0.9	5	4.06	3.78	3.62	3.52	3.45	3.40	3.37	3.34	3.32	3.30	3.27	3.24	3.21	3.17	3.14	3.12	3.11
0.95	5	6.61	5.79	5.41	5.19	5.05	4.95	4.88	4.82	4.77	4.74	4.68	4.62	4.56	4.50	4.43	4.40	4.37
0.975	5	10.01	8.43	7.76	7.39	7.15	6.98	6.85	6.76	6.68	6.62	6.52	6.43	6.33	6.23	6.12	6.07	6.02
0.99	5	16.26	13.27	12.06	11.39	10.97	10.67	10.46	10.29	10.16	10.05	9.89	9.72	9.55	9.38	9.20	9.11	9.02
0.995	5	22.78	18.31	16.53	15.56	14.94	14.51	14.20	13.96	13.77	13.62	13.38	13.15	12.90	12.66	12.40	12.27	12.15
0.9	6	3.78	3.46	3.29	3.18	3.11	3.05	3.01	2.98	2.96	2.94	2.90	2.87	2.84	2.80	2.76	2.74	2.72
0.95	6	5.99	5.14	4.76	4.53	4.39	4.28	4.21	4.15	4.10	4.06	4.00	3.94	3.87	3.81	3.74	3.70	3.67
0.975	6	8.81	7.26	6.60	6.23	5.99	5.82	5.70	5.60	5.52	5.46	5.37	5.27	5.17	5.07	4.96	4.90	4.85
0.99	6	13.75	10.92	9.78	9.15	8.75	8.47	8.26	8.10	7.98	7.87	7.72	7.56	7.40	7.23	7.06	6.97	6.88

1 − α	n																	
0.995	6	8.88	9.00	9.12	9.36	9.59	9.81	10.03	10.25	10.39	10.57	10.79	11.07	11.46	12.03	12.92	14.54	18.63
0.9	7	2.47	2.49	2.51	2.56	2.59	2.63	2.67	2.70	2.72	2.75	2.78	2.83	2.88	2.96	3.07	3.26	3.59
0.95	7	3.23	3.27	3.30	3.38	3.44	3.51	3.57	3.64	3.68	3.73	3.79	3.87	3.97	4.12	4.35	4.74	5.59
0.975	7	4.14	4.20	4.25	4.36	4.47	4.57	4.67	4.76	4.82	4.90	4.99	5.12	5.29	5.52	5.89	6.54	8.07
0.99	7	5.65	5.74	5.82	5.99	6.16	6.31	6.47	6.62	6.72	6.84	6.99	7.19	7.46	7.85	8.45	9.55	12.25
0.995	7	7.08	7.19	7.31	7.53	7.75	7.97	8.18	8.38	8.51	8.68	8.89	9.16	9.52	10.05	10.88	12.40	16.24
0.9	8	2.29	2.32	2.34	2.38	2.42	2.46	2.50	2.54	2.56	2.59	2.62	2.67	2.73	2.81	2.92	3.11	3.46
0.95	8	2.93	2.97	3.01	3.08	3.15	3.22	3.28	3.35	3.39	3.44	3.50	3.58	3.69	3.84	4.07	4.46	5.32
0.975	8	3.67	3.73	3.78	3.89	4.00	4.10	4.20	4.30	4.36	4.43	4.53	4.65	4.82	5.05	5.42	6.06	7.57
0.99	8	4.86	4.95	5.03	5.20	5.36	5.52	5.67	5.81	5.91	6.03	6.18	6.37	6.63	7.01	7.59	8.65	11.26
0.995	8	5.95	6.06	6.18	6.40	6.61	6.81	7.01	7.21	7.34	7.50	7.69	7.95	8.30	8.81	9.60	11.04	14.69
0.9	9	2.16	2.18	2.21	2.25	2.30	2.34	2.38	2.42	2.44	2.47	2.51	2.55	2.61	2.69	2.81	3.01	3.36
0.95	9	2.71	2.75	2.79	2.86	2.94	3.01	3.07	3.14	3.18	3.23	3.29	3.37	3.48	3.63	3.86	4.26	5.12
0.975	9	3.33	3.39	3.45	3.56	3.67	3.77	3.87	3.96	4.03	4.10	4.20	4.32	4.48	4.72	5.08	5.71	7.21
0.99	9	4.31	4.40	4.48	4.65	4.81	4.96	5.11	5.26	5.35	5.47	5.61	5.80	6.06	6.42	6.99	8.02	10.56
0.995	9	5.19	5.30	5.41	5.62	5.83	6.03	6.23	6.42	6.54	6.69	6.88	7.13	7.47	7.96	8.72	10.11	13.61
0.9	10	2.06	2.08	2.11	2.16	2.20	2.24	2.28	2.32	2.35	2.38	2.41	2.46	2.52	2.61	2.73	2.92	3.29
0.95	10	2.54	2.58	2.62	2.70	2.77	2.85	2.91	2.98	3.02	3.07	3.14	3.22	3.33	3.48	3.71	4.10	4.96
0.975	10	3.08	3.14	3.20	3.31	3.42	3.52	3.62	3.72	3.78	3.85	3.95	4.07	4.24	4.47	4.83	5.46	6.94
0.99	10	3.91	4.00	4.08	4.25	4.41	4.56	4.71	4.85	4.94	5.06	5.20	5.39	5.64	5.99	6.55	7.56	10.04
0.995	10	4.64	4.75	4.86	5.07	5.27	5.47	5.66	5.85	5.97	6.12	6.30	6.54	6.87	7.34	8.08	9.43	12.83
0.9	12	1.90	1.93	1.96	2.01	2.06	2.10	2.15	2.19	2.21	2.24	2.28	2.33	2.39	2.48	2.61	2.81	3.18
0.95	12	2.30	2.34	2.38	2.47	2.54	2.52	2.69	2.75	2.80	2.85	2.91	3.00	3.11	3.26	3.49	3.89	4.75
0.975	12	2.73	2.79	2.85	2.96	3.07	3.18	3.28	3.37	3.44	3.51	3.61	3.73	3.89	4.12	4.47	5.10	6.55
0.99	12	3.36	3.45	3.54	3.70	3.86	4.01	4.16	4.30	4.39	4.50	4.64	4.82	5.06	5.41	5.95	6.93	9.33
0.995	12	3.91	4.01	4.12	4.33	4.53	4.72	4.91	5.09	5.20	5.35	5.52	5.76	6.07	6.52	7.23	8.51	11.75
0.9	15	1.76	1.79	1.82	1.87	1.92	1.97	2.02	2.06	2.09	2.12	2.16	2.21	2.27	2.36	2.49	2.70	3.07
0.95	15	2.07	2.11	2.16	2.25	2.33	2.40	2.48	2.54	2.59	2.64	2.71	2.79	2.90	3.06	3.29	3.68	4.54
0.975	15	2.40	2.46	2.52	2.64	2.76	2.86	2.96	3.06	3.12	3.20	3.29	3.41	3.58	3.80	4.15	4.77	6.20

(continued)

$1-\alpha$	n	$m=1$	$m=2$	$m=3$	$m=4$	$m=5$	$m=6$	$m=7$	$m=8$	$m=9$	$m=10$	$m=12$	$m=15$	$m=20$	$m=30$	$m=60$	$m=120$	$m=\infty$
0.99	15	8.68	6.36	5.42	4.89	4.56	4.32	4.14	4.00	3.89	3.80	3.67	3.52	3.37	3.21	3.05	2.96	2.87
0.995	15	10.80	7.70	6.48	5.80	5.37	5.07	4.85	4.67	4.54	4.42	4.25	4.07	3.88	3.69	3.48	3.37	3.26
0.9	20	2.97	2.59	2.38	2.25	2.16	2.09	2.04	2.00	1.96	1.94	1.89	1.84	1.79	1.74	1.68	1.64	1.61
0.95	20	4.35	3.49	3.10	2.87	2.71	2.60	2.51	2.45	2.39	2.35	2.28	2.20	2.12	2.04	1.95	1.90	1.84
0.975	20	5.87	4.46	3.86	3.51	3.29	3.13	3.01	2.91	2.84	2.77	2.68	2.57	2.46	2.35	2.22	2.16	2.09
0.99	20	8.10	5.85	4.94	4.43	4.10	3.87	3.70	3.56	3.46	3.37	3.23	3.09	2.94	2.78	2.61	2.52	2.42
0.995	20	9.94	6.99	5.82	5.17	4.76	4.47	4.26	4.09	3.96	3.85	3.68	3.50	3.32	3.12	2.92	2.81	2.69
0.9	30	2.88	2.49	2.28	2.14	2.05	1.98	1.93	1.88	1.85	1.82	1.77	1.72	1.67	1.61	1.54	1.50	1.46
0.95	30	4.17	3.32	2.92	2.69	2.53	2.42	2.33	2.27	2.21	2.16	2.09	2.01	1.93	1.84	1.74	1.68	1.62
0.975	30	5.57	4.18	3.59	3.25	3.03	2.87	2.75	2.65	2.57	2.51	2.41	2.31	2.20	2.07	1.94	1.87	1.79
0.99	30	7.56	5.39	4.51	4.02	3.70	3.47	3.30	3.17	3.07	2.98	2.84	2.70	2.55	2.39	2.21	2.11	2.01
0.995	30	9.18	6.35	5.24	4.62	4.23	3.95	3.74	3.58	3.45	3.34	3.18	3.01	2.82	2.63	2.42	2.30	2.18
0.9	60	2.79	2.39	2.18	2.04	1.95	1.87	1.82	1.77	1.74	1.71	1.66	1.60	1.54	1.48	1.40	1.35	1.29
0.95	60	4.00	3.15	2.76	2.53	2.37	2.25	2.17	2.10	2.04	1.99	1.92	1.84	1.75	1.65	1.53	1.47	1.39
0.975	60	5.29	3.93	3.34	3.01	2.79	2.63	2.51	2.41	2.33	2.27	2.17	2.06	1.94	1.82	1.67	1.58	1.48
0.99	60	7.08	4.98	4.13	3.65	3.34	3.12	2.95	2.82	2.72	2.63	2.50	2.35	2.20	2.03	1.84	1.73	1.60
0.995	60	8.49	5.79	4.73	4.14	3.76	3.49	3.29	3.13	3.01	2.90	2.74	2.57	2.39	2.19	1.96	1.83	1.69
0.9	120	2.75	2.35	2.13	1.99	1.90	1.82	1.77	1.72	1.68	1.65	1.60	1.55	1.48	1.41	1.32	1.26	1.19
0.95	120	3.92	3.07	2.68	2.45	2.29	2.18	2.09	2.02	1.96	1.91	1.83	1.75	1.66	1.55	1.43	1.35	1.26
0.975	120	5.15	3.80	3.23	2.89	2.67	2.52	2.39	2.30	2.22	2.16	2.05	1.94	1.82	1.69	1.53	1.43	1.31
0.99	120	6.85	4.79	3.95	3.48	3.17	2.96	2.79	2.66	2.56	2.47	2.34	2.19	2.03	1.86	1.66	1.53	1.38
0.995	120	8.18	5.54	4.50	3.92	3.55	3.28	3.09	2.93	2.81	2.71	2.54	2.37	2.19	1.98	1.75	1.61	1.44
0.9	∞	2.71	2.30	2.08	1.95	1.85	1.78	1.72	1.67	1.63	1.60	1.55	1.49	1.42	1.34	1.24	1.17	1.03
0.95	∞	3.84	3.00	2.61	2.37	2.22	2.10	2.01	1.94	1.88	1.83	1.75	1.67	1.57	1.46	1.32	1.22	1.05
0.975	∞	5.03	3.69	3.12	2.79	2.57	2.41	2.29	2.19	2.12	2.05	1.95	1.84	1.71	1.57	1.39	1.27	1.06
0.99	∞	6.64	4.61	3.79	3.32	3.02	2.81	2.64	2.51	2.41	2.32	2.19	2.04	1.88	1.70	1.48	1.33	1.07
0.995	∞	7.89	5.30	4.28	3.72	3.35	3.10	2.90	2.75	2.63	2.52	2.36	2.19	2.00	1.79	1.54	1.37	1.08

Appendix 6
Annual Minimum 7-Day Mean Flows, in m^3/s, (Q_7) at 11 Gauging Stations in the Paraopeba River Basin, in Brazil

© Springer International Publishing Switzerland 2017
M. Naghettini (ed.), *Fundamentals of Statistical Hydrology*,
DOI 10.1007/978-3-319-43561-9

i	Year	40549998	Year	40573000	Year	40577000	Year	40579995	Year	40680000	Year	40710000	Year	40740000	Year	40800001	Year	40818000	Year	40850000	Year	40865001
1	1957	2.171	1945	1.877	1943	1.966	1939	3.930	1940	2.461	1966	18.071	1967	21.600	1938	44.814	1944	1.830	1968	42.17	1978	39.23
2	1958	2.576	1946	1.741	1944	1.601	1940	3.581	1942	2.847	1967	16.771	1968	22.000	1939	35.214	1945	1.716	1969	29.67	1980	41.33
3	1959	1.974	1947	1.991	1945	1.820	1941	5.160	1946	2.687	1968	15.800	1969	16.343	1940	29.900	1946	1.450	1970	36.39	1981	39.96
4	1960	2.271	1949	2.080	1946	1.450	1942	3.886	1948	1.960	1969	12.657	1970	18.100	1941	39.114	1947	1.553	1971	16.00	1982	59.97
5	1961	3.147	1950	1.580	1947	1.807	1944	5.950	1949	2.443	1970	13.757	1971	10.771	1942	37.457	1948	1.280	1972	34.14	1984	42.10
6	1962	2.414	1951	1.510	1948	1.453	1945	4.980	1950	2.250	1971	8.257	1972	19.500	1943	52.143	1949	1.884	1973	43.14	1986	36.60
7	1963	1.704	1952	1.880	1949	1.924	1946	4.054	1954	1.979	1972	14.886	1973	23.886	1944	30.300	1950	1.630	1974	32.44	1987	37.20
8	1965	2.770	1953	1.963	1950	1.600	1947	5.220	1955	1.451	1973	17.686	1976	18.171	1945	39.786	1951	1.463	1975	30.51	1988	36.70
9	1966	3.089	1954	1.310	1951	1.276	1948	2.966	1956	2.104	1974	14.557	1977	19.171	1946	33.086	1952	1.400	1976	31.81	1989	37.50
10	1967	2.751	1955	1.110	1952	1.700	1949	4.500	1958	2.344	1975	12.343	1978	20.314	1947	39.929	1954	0.973	1977	25.57	1990	30.41
11	1968	2.657	1956	1.431	1953	1.649	1950	3.800	1959	1.670	1976	15.129	1980	23.214	1948	25.143	1955	0.805	1978	33.01	1991	48.34
12	1969	2.633	1957	1.237	1954	1.160	1951	3.710	1960	2.130	1977	15.271	1981	23.271	1949	36.014	1956	1.089	1979	61.20	1992	54.63
13	1972	1.854	1958	1.484	1955	0.817	1952	3.089	1962	1.871	1978	15.514	1982	27.243	1950	34.329	1957	1.744	1980	41.61	1993	47.31
14	1973	3.180	1959	1.110	1956	1.250	1953	4.000	1964	1.460	1979	21.800	1984	23.914	1951	31.157	1958	1.559	1981	41.00	1994	34.40
15	1974	2.120	1960	1.230	1957	1.276	1956	2.650	1965	2.714	1980	17.200	1986	21.557	1952	33.072	1959	1.190	1982	46.60		
16	1975	2.000	1961	1.360	1958	1.643	1957	3.190	1966	2.350	1981	20.700	1987	21.071	1953	25.900	1960	0.912	1984	43.59		
17	1976	2.480	1962	1.346	1959	1.250	1958	3.179	1969	1.884	1982	23.300	1988	18.414	1954	18.114	1961	1.080	1985	56.06		
18	1977	2.259	1963	1.160	1960	1.289	1961	3.659	1970	2.397	1984	16.986	1993	25.129	1955	15.271	1962	1.000	1986	34.24		
19	1978	2.194	1964	1.260	1961	1.276	1962	3.620	1971	1.000	1985	21.671	1994	18.814	1956	21.729	1963	0.944	1987	41.34		
20	1981	3.500	1965	1.566	1962	0.969	1963	2.300	1972	2.680	1986	14.257	1995	17.114	1957	26.072	1964	0.944	1988	34.56		
21	1982	3.370			1963	0.978	1964	2.340	1973	2.683	1987	15.400	1996	21.300	1958	30.043	1965	1.550	1989	39.63		
22	1984	2.190			1964	1.080	1965	3.474	1974	2.060	1988	13.943	1997	22.386	1959	19.600			1990	35.63		
23	1985	3.770			1965	1.560	1966	4.580	1975	2.131	1989	16.200	1998	18.929	1960	21.457			1991	43.97		
24	1986	2.221					1968	3.244	1976	2.327	1990	12.929	1999	14.829	1961	28.872			1992	51.54		
25	1987	2.666					1972	3.907	1977	2.764	1992	20.286			1962	26.529			1994	41.79		
26	1988	2.367					1973	5.217	1978	2.909					1963	18.043			1995	42.24		
27	1989	3.009					1974	4.390	1981	2.780					1964	18.314			1996	46.50		
28	1990	2.046					1975	3.639	1982	2.641					1965	35.100			1997	51.34		
29	1992	3.609					1976	4.583	1984	1.891					1966	32.100			1998	26.71		
30	1993	3.227					1977	3.233	1985	2.781					1967	27.272						
31	1994	1.887					1978	2.859	1986	1.140					1968	28.357						

#	Year	Value	Year	Value	Year	Value	Year	Value
32	1995	1.840	1979	3.271	1988	1.454	1969	21.730
33	1996	2.406	1980	2.830	1989	2.679	1970	25.843
34	1997	3.004	1981	2.080	1992	2.553	1971	12.857
35	1999	1.727	1982	3.170	1995	1.466	1972	24.757
36			1983	4.520	1996	2.037	1973	31.157
37			1984	2.279	1997	1.654	1974	24.743
38			1985	2.967	1999	1.669	1975	21.700
39							1976	24.286
40							1978	24.929
41							1979	38.886
42							1980	28.871
43							1982	35.100
44							1984	29.414
45							1985	45.400
46							1986	26.143
47							1987	26.243
48							1988	23.429
49							1989	25.571
50							1990	21.357
51							1991	32.357
52							1992	37.229
53							1993	30.129
54							1994	25.829
55							1995	22.514
56							1996	28.285
57							1997	30.971
58							1998	16.286
59							1999	12.771
60								

Appendix 7
Annual Maximum Flows for 7 Gauging Stations in the Upper Paraopeba River Basin

© Springer International Publishing Switzerland 2017
M. Naghettini (ed.), *Fundamentals of Statistical Hydrology*,
DOI 10.1007/978-3-319-43561-9

WY	40549998	WY	40573000	WY	40577000	WY	40579995	WY	40665000	WY	40710000	WY	40740000
56/57	97.2	46/47	33.0	42/43	43.0	38/39	93.6	38/39	22.6	65/66	457	67/68	315
57/58	42.2	49/50	28.8	43/44	22.2	39/40	85.6	39/40	19.3	66/67	350	68/69	356
58/59	44.1	51/52	39.3	44/45	34.4	40/41	112	43/44	24.5	67/68	220	69/70	255
59/60	51.7	52/53	34.9	45/46	19.9	41/42	48.4	44/45	31.5	68/69	268	70/71	182
60/61	80.3	53/54	20.0	46/47	26.5	42/43	103	45/46	23.5	69/70	190	71/72	474
61/62	56.6	55/56	22.0	47/48	30.9	43/44	62.9	46/47	24.2	70/71	147	72/73	410
62/63	44.0	56/57	49.3	48/49	44.8	44/45	76.7	47/48	22.1	71/72	378	73/74	351
65/66	81	57/58	20.5	49/50	21.8	45/46	41.2	48/49	30.7	72/73	330	76/77	456
66/67	77.8	58/59	23.5	51/52	34.1	46/47	54.1	49/50	19.7	73/74	295	77/78	723
67/68	58.0	59/60	21.0	52/53	28.2	47/48	62.3	50/51	26.2	74/75	207	78/79	457
68/69	56.9	60/61	50.7	53/54	12.9	48/49	104	51/52	25.7	75/76	150	79/80	460
72/73	91.7	61/62	40.6	54/55	39.3	49/50	46.0	52/53	19.5	76/77	350	80/81	432
73/74	53.3	62/63	18.5	55/56	16.5	50/51	206	53/54	24.4	77/78	670	81/82	519
74/75	48.0	63/64	36.1	57/58	22.5	51/52	110	54/55	25.7	78/79	403	82/83	443
75/76	33.1	64/65	33.7	58/59	26.6	52/53	89.0	55/56	24.9	79/80	336	83/84	387
76/77	65.6			59/60	40.4	55/56	41.2	56/57	27.0	80/81	385	84/85	816
77/78	112			60/61	39.3	56/57	112	64/65	21.0	81/82	460	85/86	345
78/79	132			61/62	23.8	57/58	44.1	65/66	18.9	82/83	451	86/87	423
82/83	68.6			63/64	39.3	61/62	75.6	66/67	14.7	83/84	374	87/88	455
83/84	48.5			64/65	27.4	62/63	52.8	72/73	44.4	84/85	785	88/89	222
84/85	50.4					63/64	72.9	73/74	41.0	85/86	287	90/91	715
85/86	38.3					64/65	68.2	74/75	29.1	86/87	322	92/93	300
86/87	36.4					65/66	112	75/76	33.1	87/88	418	93/94	336
87/88	77.0					72/73	77.0	76/77	50.7	88/89	161	94/95	461
88/89	37.2					73/74	111	77/78	44.7	89/90	397	95/96	372
89/90	44.6					74/75	45.5	78/79	39.0			96/97	1133

Year	Value		Year	Value		Year	Value		Year	Value
97/98	205		79/80	46.6		75/76	30.8		92/93	57.9
98/99	235		80/81	42.4		76/77	55.8		93/94	33.0
			83/84	29.5		77/78	148		94/95	63.2
			84/85	52.0		78/79	128		95/96	42.4
						79/80	59.4		96/97	85.7
						80/81	50.9		98/99	39.0
						81/82	94.6			
						82/83	132			
						83/84	88.2			
						84/85	132			
						87/88	54.7			
						88/89	22.1			
						89/90	63.4			
						90/91	19.8			
						91/92	74.6			
						92/93	57.2			
						93/94	55.1			
						94/95	60.3			
						97/98	66.5			
						98/99	84.2			
						99/00	90.2			

Appendix 8a
Rainfall Rates (mm/h) Recorded at Rainfall Gauging Stations in the Serra dos Órgãos Region

© Springer International Publishing Switzerland 2017
M. Naghettini (ed.), *Fundamentals of Statistical Hydrology*,
DOI 10.1007/978-3-319-43561-9

WY	5 min	10 min	15 min	30 min	45 min	1 h
	ANDORINHAS	ANDORINHAS	ANDORINHAS	ANDORINHAS	ANDORINHAS	ANDORINHAS
77/78	156.9	118	99.6	81.6	68.5	61.7
78/79	132	115	102.1	72.8	52.4	47.2
79/80	176.1	147.3	135.7	125.3	110.8	83.6
80/81	162.5	133.6	113.9	92.7	75.5	63
81/82	139.5	115.7	112.8	94.2	80.7	74.9
82/83	160.7	154.5	124.8	92.3	85	77.5
83/84	141.2	130.4	125.6	110.7	99.2	84.4
84/85	141.6	117.8	102.9	80.5	70.7	60.7
85/86	235.2	164.6	116	86.2	73.8	56.9
86/87	178.9	128.5	110.9	85.5	58.8	44.6
87/88	116.6	110.9	97.6	70.7	62.6	59.7
88/89	142.4	127.7	114.2	88	71.6	56.8
89/90	130.8	103.9	86.4	65.4	48	42.5
90/91	128.7	123.7	112	84.5	70	59.2
91/92	209.1	157.4	140.2	107.3	97.1	81.4
92/93	229	171.8	152.7	115.1	114.5	102.8
93/94	177.8	123.9	94.5	72.3	53.5	41.5
94/95	133.4	119.1	100.6	79.6	65.4	60.1
95/96	128.5	113.8	94.2	72	53	47.7
97/98	112	106.2	89.7	76.4	74	68.7

WY	2 h	3 h	4 h	8 h	14 h	24 h
	ANDORINHAS	ANDORINHAS	ANDORINHAS	ANDORINHAS	ANDORINHAS	ANDORINHAS
77/78	47.8	36.9	31.5	15.9	9.1	5.9
78/79	36.9	25.8	22	13.7	8	5.3
79/80	42.3	28.6	23.7	12.4	7.1	6.2

WY	5 min	10 min	15 min	30 min	45 min	1 h
	APOLINARIO	APOLINARIO	APOLINARIO	APOLINARIO	APOLINARIO	APOLINARIO
80/81	38.5	26.2	20.5	10.3	6.1	4.4
81/82	67.2	53.7	45.6	31.7	21.3	13.1
82/83	44.1	31.9	24.6	12.4	7.1	4.8
83/84	43.6	30.4	22.9	11.5	6.6	4.2
84/85	39.5	27.2	20.4	10.3	6.3	4.7
85/86	38.6	30	25	12.6	7.4	5.1
86/87	34.1	24.9	20.1	11.2	6.8	5.8
87/88	45.9	38.6	29.2	24.3	15.2	9.1
88/89	36.6	24.7	18.7	10.1	6.6	7.1
89/90	33.7	22.9	17.2	9.2	5.4	3.7
90/91	38.6	32.8	30.1	20.9	13.3	9.4
91/92	48.4	34.5	26.1	15.7	9.1	5.7
92/93	65.6	43.74	32.8	16.5	9.5	6.8
93/94	30.5	23.6	19.2	10.1	8.1	7.2
94/95	48.4	34.7	27.9	18.3	10.5	7.6
95/96	34.1	22.8	17.2	8.7	5.1	3.6
97/98	57.8	47.9	47.7	31.6	20.8	12.2
WY	5 min	10 min	15 min	30 min	45 min	1 h
	APOLINARIO	APOLINARIO	APOLINARIO	APOLINARIO	APOLINARIO	APOLINARIO
76/77	124.8	112.9	101.5	73	66.2	63
77/78	106	102.9	94.2	71.9	65.1	49.6
78/79	148.2	115.8	108.6	79.6	61.3	48.9
79/80	133.7	104.1	102.5	92.5	89	76.2
80/81	188.8	150.7	125.8	79.3	71.9	58.6
81/82	130.5	130.5	125.2	80.6	62.9	48.8
82/83	165.4	111.1	86.4	61.3	53.6	46.4
83/84	129	104.9	96.9	71.5	66.7	59.6

(continued)

WY	5 min ANDORINHAS	10 min ANDORINHAS	15 min ANDORINHAS	30 min ANDORINHAS	45 min ANDORINHAS	1 h ANDORINHAS
84/85	120.3	112.9	98.6	70	59.6	55
85/86	134.7	129.5	110.3	75.5	64.4	57.5
86/87	144.5	122.5	108.1	75.2	62.7	48.2
87/88	142.4	156.5	127.7	89.5	64.3	66.4
88/89	120.2	95.9	84.4	64.1	54.7	43.1
89/90	126.1	114.3	98.1	64.5	54.8	43
90/91	126	120.1	117	97.6	81.4	74.5
91/92	111.9	107	93.9	66.8	55.7	43.6
92/93	191.6	121	108.6	88.2	68	51.2
93/94	120	118.2	111.7	77.9	62.9	48.1
94/95	113.1	90.9	84.7	60.4	51.4	40

WY	2 h APOLINARIO	3 h APOLINARIO	4 h APOLINARIO	8 h APOLINARIO	14 h APOLINARIO	24 h APOLINARIO
76/77	43.1	40.7	41	22.6	13.2	7.8
77/78	31.5	28.3	22.1	11.2	6.8	4
78/79	31.6	22.6	17.6	12.8	8	4.8
79/80	43.9	29.3	22	11	6.4	4.3
80/81	35.4	25.5	22.8	12	7.5	5.1
81/82	33.2	24.4	19.8	14	8.9	6.4
82/83	27	21.7	16.4	9.2	5.3	3.2
83/84	32.5	24.6	20	11.9	7.3	4.3
84/85	36	31.1	25.4	15.5	9.2	5.4
85/86	44.7	40.6	30.6	15.4	9	5.3
86/87	28.6	19.2	16.2	10.8	7.3	6.8
87/88	38	29.1	23.9	16.1	12.2	8.4
88/89	27.6	20	15.3	9.2	6.1	3.7

	5 min FAZ. SANTO AMARO	10 min FAZ. SANTO AMARO	15 min FAZ. SANTO AMARO	30 min FAZ. SANTO AMARO	45 min FAZ. SANTO AMARO	1 h FAZ. SANTO AMARO
89/90	26.6	17.73	14.1	12.2	11.2	9.9
90/91	37.9	27	20.4	11.5	10.2	7
91/92	27.5	18.8	15.2	9.3	5.4	4.7
92/93	29.8	23.2	18.2	9.4	5.6	6
93/94	31.8	21.2	17.2	11.6	7.3	5.9
94/95	24.1	16.2	12.9	7.2	4.2	2.8
WY	5 min FAZ. SANTO AMARO	10 min FAZ. SANTO AMARO	15 min FAZ. SANTO AMARO	30 min FAZ. SANTO AMARO	45 min FAZ. SANTO AMARO	1 h FAZ. SANTO AMARO
77/78	121	114	102.9	78.3	78.1	58.8
78/79	164.9	123.1	99.6	70.3	58.5	45.9
79/80	128.8	106.5	94.3	66.2	56.2	44.3
80/81	134.7	112.7	103	71.1	60.8	47.1
81/82	142.4	136.8	122.8	82.7	71.1	63
82/83	162.8	116.9	131	107.8	88.4	70.3
83/84	168.5	150.9	137	86.9	68.2	51.5
84/85	126.5	109	100.8	111.8	91.1	69.3
85/86	159.4	125.2	112.5	74.3	63.4	49.7
86/87	109.5	113	95	67.5	57.2	50.2
87/88	135.1	117.1	116.9	83.7	71.9	75.8
88/89	132.4	110.3	104.6	109.3	98.3	83.1
89/90	162.2	143.9	132.1	106.2	71.1	53.3
90/91	123.9	115.3	105.1	78.5	65.3	57.5
91/92	105	103.7	91.5	66.9	61.3	47.3
92/93	105.2	98.8	87.4	61.8	52.7	41.3
93/94	189.3	152.6	114	77.5	65	52.2
94/95	115.4	102.5	84.3	59.8	50.9	39.8

(continued)

WY	5 min	10 min	15 min	30 min	45 min	1 h
	ANDORINHAS	ANDORINHAS	ANDORINHAS	ANDORINHAS	ANDORINHAS	ANDORINHAS
95/96	126.7	106.4	95.5	69.9	57.7	45.5
96/97	112.7	94.8	81.2	57.7	49	38.2
WY	2 h	3 h	4 h	8 h	14 h	24 h
	FAZ. SANTO AMARO	FAZ. SANTO AMARO	FAZ. SANTO AMARO	FAZ. SANTO AMARO	FAZ. SANTO AMARO	FAZ. SANTO AMARO
77/78	45.6	30.4	23.4	11.9	7.8	5.1
78/79	26.3	19	15.8	11.4	9.1	6.6
79/80	27.8	26.4	21.5	16.3	10.1	6.9
80/81	29.3	19.6	14.7	10.3	6.5	3.9
81/82	53.9	36.8	35.3	18.2	11.7	8
82/83	35.4	23.6	17.7	10.7	6.2	3.7
83/84	32.7	32.4	24.7	13	7.6	4.5
84/85	34.8	23.6	17.9	9.1	8.6	5.2
85/86	35.5	25.6	21.5	13.3	8.6	5.1
86/87	35.2	41.8	35.5	25.6	15	9.1
87/88	41.8	29.6	24.7	14.2	8.2	4.8
88/89	44.6	36.2	27.8	15	9.4	6.3
89/90	27.1	20.5	15.4	7.9	5	5.4
90/91	38.3	28.1	27.6	18.5	12.1	8.6
91/92	30.2	24.6	19.1	9.7	5.6	5.2
92/93	29.7	21.3	16.3	9.5	5.7	5.2
93/94	31	23	18.8	11.6	7.5	5.8
94/95	24.5	17	13.1	6.8	4.8	5.1
95/96	27.8	18.7	15.2	11.2	6.5	3.8
96/97	23.3	16.2	13.7	7.8	4.5	3

WY	5 min	10 min	15 min	30 min	45 min	1 h
	NOVA FRIBURGO	NOVA FRIBURGO	NOVA FRIBURGO	NOVA FRIBURGO	NOVA FRIBURGO	NOVA FRIBURGO
71/72	121.4	102.5	95.5	65.4	55.5	42.2
72/73	136	125.4	131.5	92.4	79.9	60.8
73/74	188	129.2	103.4	80.3	66.4	50.7
74/75	232.3	177.2	158.9	120.2	96.3	81.4
75/76	130.6	111.6	104	75.4	60.5	48
76/77	151.7	134.2	109.1	82.6	71	57.2
77/78	136.2	108.8	97.4	64.4	51.9	39.5
78/79	134.6	121.6	120.8	89.7	66.8	61.8
79/80	120	117	113.7	79.6	62.3	49.8
80/81	125.1	113.2	99.3	70.7	60.4	48.1
81/82	109.5	101.7	92.7	63	52.2	42
82/83	220	173.7	116.2	73.7	63.2	50.6
83/84	112	103.8	90.6	67.7	70.1	62.9
84/85	145.3	110.1	106.9	83.6	76	58.9
85/86	104.2	100	89.8	68.7	58.7	44.7
87/88	107.6	98.5	78.2	58.4	52.2	42.5
88/89	130.5	103.7	92.8	71.5	61.7	54.2
89/90	115	106.3	85.2	62.1	55	45.1

WY	2 h	3 h	4 h	8 h	14 h	24 h
	NOVA FRIBURGO	NOVA FRIBURGO	NOVA FRIBURGO	NOVA FRIBURGO	NOVA FRIBURGO	NOVA FRIBURGO
71/72	25	17.2	13.8	8.6	6.1	4.2
72/73	35.5	26	20.4	10.5	7.1	4.6
73/74	29.6	19.8	14.9	7.5	5	3.5
74/75	50.9	41.2	32.8	16.5	9.5	6.4
75/76	24.2	16.4	12.9	7.1	4.3	2.8

(continued)

WY	5 min ANDORINHAS	10 min ANDORINHAS	15 min ANDORINHAS	30 min ANDORINHAS	45 min ANDORINHAS	1 h ANDORINHAS
76/77	40.8	30.9	26.9	13.7	8.4	5.7
77/78	23.6	21.3	17.5	14.6	13.1	7.7
78/79	35.7	39	34.9	21.5	13.9	8.4
79/80	37.5	28.2	21.2	10.6	6.8	4.8
80/81	29.5	21.8	20.5	17.9	10.9	8.1
81/82	26.6	22.9	17.5	14.1	9.9	6.8
82/83	31.7	24	18.1	9.2	5.3	3.6
83/84	39.8	29	21.9	11.1	6.9	4.8
84/85	46.3	32.7	26.3	13.3	8.2	5.7
85/86	29.1	19.5	15.6	8.9	5.6	4.7
87/88	27.2	19.3	18.8	11.3	7.3	5.2
88/89	32.8	26.3	23	12.5	8.4	5.8
89/90	30.5	23.2	17.7	9.5	6.2	3.8

WY	5 min POSTO GARRAFAO	10 min POSTO GARRAFAO	15 min POSTO GARRAFAO	30 min POSTO GARRAFAO	45 min POSTO GARRAFAO	1 h POSTO GARRAFAO
77/78	121	111.4	92.6	63.5	54	44.2
78/79	120.9	116.7	100.6	77.8	64.2	51.3
79/80	125.6	113.5	91	60.3	48.8	40.2
80/81	138.8	143.5	133.7	82.4	71.2	83.2
81/82	108.9	109.5	96.9	78.2	63.9	51.1
82/83	197.2	160.2	135.6	91	73.8	57.2
83/84	126.9	103.9	92.2	65.7	56.7	44
84/85	126.1	101	86.1	61	54.6	43.4
85/86	111.2	105.2	88.4	65.7	60.1	45.6

WY	2 h POSTO GARRAFAO	3 h POSTO GARRAFAO	4 h POSTO GARRAFAO	8 h POSTO GARRAFAO	14 h POSTO GARRAFAO	24 h POSTO GARRAFAO
86/87	105.8	111.3	113.6	75.4	60.3	50.2
87/88	120.5	119.3	118.5	112.8	98.3	73.8
88/89	147.1	129.1	123.5	108.5	74	57.3
89/90	111.7	109.4	104	69.8	59.8	45.2
90/91	141.8	134.5	121.9	116.2	95.1	83.6
91/92	181.3	134.8	108.2	64.8	55.4	43
92/93	145.2	145.8	103.9	82.1	95.2	81.1
93/94	126.6	105.3	95.1	79.1	67.9	57.5
94/95	131.2	104.1	97.4	66	53.2	45.9
95/96	182.1	103.6	101.2	74.2	66.8	53.9
97/98	137.3	103.9	88.3	72.8	67.5	54.1
77/78	33.8	37.6	32.6	18.4	12.3	7.2
78/79	37.1	40.2	32.7	16.4	9.6	8.4
79/80	24.3	20	15.1	9.8	5.7	4.4
80/81	41.8	28.4	21.4	10.8	6.4	3.9
81/82	30.6	24.1	18.4	10.3	6.8	4
82/83	38.7	28.5	23.1	20.5	14.4	8.4
83/84	27.6	18.7	14.3	8.2	5.2	3.4
84/85	26.6	17.8	15.1	8.8	5.1	3.3
85/86	29	24.9	20.2	11.9	7.9	5.1
86/87	28.7	19.4	15.7	14	8.8	5.5
87/88	37.2	39.7	36.5	20.4	11.7	7.1
88/89	31.6	21.1	15.9	8.5	5.6	4.6
89/90	27.5	20.5	16.2	12.8	12.4	8.1

(continued)

WY	5 min ANDORINHAS	10 min ANDORINHAS	15 min ANDORINHAS	30 min ANDORINHAS	45 min ANDORINHAS	1 h ANDORINHAS
90/91	54.4	42.8	37.4	21	13.5	7.9
91/92	23.6	16.7	12.7	7.1	4.5	3.4
92/93	40.9	27.9	21.2	11.5	6.9	4.6
93/94	32.3	22.2	17.5	9.2	7.4	8
94/95	28.8	22.9	17.2	9.3	7.1	5.6
95/96	31.6	25	19	10.8	6.4	4.2
97/98	59	45.4	39.6	25.9	16.3	9.6

WY	5 min QUIZANGA	10 min QUIZANGA	15 min QUIZANGA	30 min QUIZANGA	45 min QUIZANGA	1 h QUIZANGA
76/77	166.9	113.3	96.3	79.1	70.1	62.5
77/78	122.2	110.9	94.5	66.4	56	44.3
78/79	183.7	122.7	111.2	100.1	85.7	68.9
79/80	225.2	169.2	150.1	77.5	58.8	48.6
80/81	117.7	112	107.4	68.1	58.2	44.8
81/82	140.7	114.3	100	77.8	66.8	51.8
82/83	127.1	119	117.1	99.4	71.3	65
83/84	153.3	107.4	99.8	67	55.4	45.4
84/85	109	101.5	89.1	66.3	61.6	55.8
85/86	134.5	117.7	111.7	81.6	59.3	44.9
86/87	122.7	106.6	99.2	77.7	67.5	54.1
87/88	127.9	115	107.8	66.4	62.8	56.6
88/89	175	169.8	162.5	117.7	100.2	81.6
89/90	157	126.4	85.4	64.5	51.9	40.1
90/91	126.8	110.7	103.3	73.4	62.7	52.2
91/92	198.2	154.7	141	86.4	66.4	52

WY	2 h QUIZANGA	3 h QUIZANGA	4 h QUIZANGA	8 h QUIZANGA	14 h QUIZANGA	24 h QUIZANGA
92/93	119	107.2	98.6	103.1	70.6	53.4
93/94	115.4	103	84.4	64.7	58.7	46.3
94/95	100.1	99.5	94.7	66.9	57	45.6
95/96	132.4	118.5	114.2	80.7	89	84.6
97/98	106.4	99.9	83.9	60.1	51.1	39.8
76/77	43.9	29.3	22.8	12.2	15.1	10.2
77/78	31.1	20.9	16.9	9.7	6.4	4.2
78/79	37.1	24.9	18.9	11.4	7.6	6.8
79/80	38.2	26.2	19.8	12.8	7.9	6.3
80/81	25.8	17.2	13.4	7.5	4.9	3.4
81/82	37.9	31.5	27.1	14.9	9	5.4
82/83	43.9	33.3	25.1	13.1	11	7.5
83/84	30.6	21.7	17.7	16.7	9.7	5.9
84/85	45.6	36.7	27.6	16.1	9.5	5.7
85/86	26.4	17.6	18.2	11	7.3	4.8
86/87	28.7	20.3	16.2	9.1	5.3	3.8
87/88	34.4	25.8	26	23	13.5	9.2
88/89	46.8	31.7	23.8	12	7.4	5.2
89/90	25.6	23.5	18	10.6	6.9	4.3
90/91	32.6	27.5	22.4	16.4	10.4	7.2
91/92	28.5	19.7	15.6	10.4	6.5	4.2
92/93	31.6	24.4	19.4	10	6.5	4.7
93/94	32.4	22.6	17.9	10	6.1	5.2
94/95	27.9	19.8	15	8.3	5.2	3.4
95/96	55.1	37.3	28	14.1	8.2	5
97/98	24	16	13	6.9	4.5	3.2

Appendix 8b
L-Moments and L-Moment Ratios of Rainfall Rates (mm/h) Recorded at Rainfall Gauging Stations in the Serra dos Órgãos Region, Located in the State of Rio de Janeiro, Brazil

© Springer International Publishing Switzerland 2017
M. Naghettini (ed.), *Fundamentals of Statistical Hydrology*,
DOI 10.1007/978-3-319-43561-9

	Duration	Gauging station	N	l_1	l_2	t_2	t_3	t_4	t_5
1	5 min	ANDORINHAS	20	156.65	19.4334	0.124056	0.2681	0.1387	−0.0029
2	5 min	APOLINARIO	19	135.64	12.7503	0.094001	0.3066	0.2241	0.0374
3	5 min	FAZ. SANTO AMARO	20	136.32	13.7384	0.100781	0.1522	0.0603	−0.0087
4	5 min	NOVA FRIBURGO	18	140	18.9824	0.135589	0.3953	0.2559	0.1067
5	5 min	POSTO GARRAFAO	20	135.36	13.5558	0.100146	0.2956	0.2025	0.0532
6	5 min	QUIZANGA	21	141.01	17.9824	0.127526	0.285	0.156	0.0049
1	10 min	ANDORINHAS	20	129.19	11.0332	0.085403	0.25	0.118	−0.0471
2	10 min	APOLINARIO	19	116.93	9.1164	0.077965	0.1793	0.2441	0.0697
3	10 min	FAZ. SANTO AMARO	20	117.68	9.1839	0.078041	0.2436	0.1708	−0.027
4	10 min	NOVA FRIBURGO	18	118.81	11.5938	0.097583	0.4313	0.2379	0.1121
5	10 min	POSTO GARRAFAO	20	118.3	9.3	0.078614	0.3404	0.0819	−0.009
6	10 min	QUIZANGA	21	119.01	10.2267	0.085931	0.3995	0.2813	0.1096
1	15 min	ANDORINHAS	20	111.32	9.9389	0.089282	0.1772	0.1399	0.0589
2	15 min	APOLINARIO	19	104.43	7.7655	0.074361	0.0596	0.0798	0.0027
3	15 min	FAZ. SANTO AMARO	20	105.58	9.2403	0.087519	0.1349	0.1076	−0.0354
4	15 min	NOVA FRIBURGO	18	104.78	10.2451	0.097777	0.2548	0.2405	0.1133
5	15 min	POSTO GARRAFAO	20	104.64	8.595	0.082139	0.2224	0.0638	−0.0151
6	15 min	QUIZANGA	21	107.25	11.0876	0.103381	0.307	0.2387	0.1043
1	30 min	ANDORINHAS	20	87.66	9.0834	0.103621	0.2186	0.139	0.0341
2	30 min	APOLINARIO	19	75.76	6.1386	0.081027	0.112	0.1069	0.0357
3	30 min	FAZ. SANTO AMARO	20	79.41	9.4753	0.119321	0.2239	0.0961	−0.0602
4	30 min	NOVA FRIBURGO	18	76.08	7.7686	0.102111	0.2785	0.2176	0.1683
5	30 min	POSTO GARRAFAO	20	78.36	9.0513	0.115509	0.3084	0.1581	0.0551
6	30 min	QUIZANGA	21	78.33	8.331	0.106358	0.2997	0.1272	0.0474
1	45 min	ANDORINHAS	20	74.25	10.7308	0.144523	0.1822	0.1485	0.0323
2	45 min	APOLINARIO	19	64.03	5.0175	0.078362	0.2025	0.2635	0.195

3	45 min	FAZ. SANTO AMARO	20	66.81	7.53	0.112708	0.2283	0.1605	0.0347
4	45 min	NOVA FRIBURGO	18	64.45	6.152	0.095454	0.2635	0.1999	0.1593
5	45 min	POSTO GARRAFAO	20	67.04	7.8147	0.116568	0.271	0.1585	0.0269
6	45 min	QUIZANGA	21	65.77	6.6195	0.100646	0.3122	0.2251	0.0948
1	1 h	ANDORINHAS	20	63.75	9.225	0.144706	0.1432	0.1261	0.0388
2	1 h	APOLINARIO	19	53.77	5.8942	0.109619	0.229	0.1131	-0.0032
3	1 h	FAZ. SANTO AMARO	20	54.2	6.9613	0.128437	0.2391	0.1345	-0.0072
4	1 h	NOVA FRIBURGO	18	52.24	5.6667	0.108474	0.2334	0.1472	0.1233
5	1 h	POSTO GARRAFAO	20	55.29	7.4989	0.135629	0.3275	0.1135	-0.0214
6	1 h	QUIZANGA	21	54.2	6.6452	0.122605	0.2986	0.1856	0.0657
1	2 h	ANDORINHAS	20	43.61	5.5421	0.127083	0.2673	0.1722	0.0538
2	2 h	APOLINARIO	19	33.2	3.5187	0.105985	0.1568	0.0746	-0.0249
3	2 h	FAZ. SANTO AMARO	20	33.74	4.4111	0.130738	0.2259	0.147	0.0545
4	2 h	NOVA FRIBURGO	18	33.13	4.333	0.130788	0.2129	0.1234	0.0519
5	2 h	POSTO GARRAFAO	20	34.26	4.9271	0.143815	0.3197	0.2145	0.0873
6	2 h	QUIZANGA	21	34.67	4.7195	0.136126	0.2143	0.0868	0.019
1	3 h	ANDORINHAS	20	32.09	4.7458	0.14789	0.2782	0.1318	0.0893
2	3 h	APOLINARIO	19	25.33	3.7846	0.149412	0.2166	0.1655	0.0745
3	3 h	FAZ. SANTO AMARO	20	25.72	4.0316	0.15675	0.183	0.1048	0.0284
4	3 h	NOVA FRIBURGO	18	25.48	3.9605	0.155436	0.2212	0.1363	0.0398
5	3 h	POSTO GARRAFAO	20	27.19	5.0237	0.184763	0.2594	0.0594	-0.0606
6	3 h	QUIZANGA	21	25.14	3.6238	0.144145	0.1349	0.0699	-0.0108
1	4 h	ANDORINHAS	20	26.12	4.48	0.171516	0.3133	0.1919	0.1158
2	4 h	APOLINARIO	19	20.58	3.424	0.166375	0.3099	0.2595	0.194
3	4 h	FAZ. SANTO AMARO	20	20.99	3.7176	0.177113	0.2437	0.0927	0.0355
4	4 h	NOVA FRIBURGO	18	20.82	3.3964	0.163132	0.24	0.183	0.0393
5	4 h	POSTO GARRAFAO	20	22.09	4.6563	0.210788	0.3265	0.0776	-0.0907

(continued)

	Duration	Gauging station	N	l_1	l_2	t_2	t_3	t_4	t_5
6	4 h	QUIZANGA	21	20.13	2.7333	0.135782	0.0976	0.0302	−0.0915
1	8 h	ANDORINHAS	20	15.37	3.6132	0.235081	0.3866	0.1738	0.0411
2	8 h	APOLINARIO	19	12.26	1.8193	0.148393	0.2457	0.2651	0.1214
3	8 h	FAZ. SANTO AMARO	20	12.6	2.4289	0.19277	0.2552	0.2107	0.096
4	8 h	NOVA FRIBURGO	18	12.13	2.1556	0.177708	0.1991	0.1406	0.0765
5	8 h	POSTO GARRAFAO	20	13.28	2.9384	0.221265	0.2895	0.0776	−0.0449
6	8 h	QUIZANGA	21	12.2	2.0624	0.169049	0.2014	0.187	0.0661
1	14 h	ANDORINHAS	20	9.47	2.3574	0.248933	0.4397	0.2508	0.0589
2	14 h	APOLINARIO	19	7.95	1.393	0.17522	0.1665	0.1355	0.0056
3	14 h	FAZ. SANTO AMARO	20	8.02	1.5182	0.189302	0.1919	0.1526	0.1104
4	14 h	NOVA FRIBURGO	18	7.94	1.5127	0.190516	0.2166	0.1498	0.0331
5	14 h	POSTO GARRAFAO	20	8.7	1.9768	0.227218	0.246	0.0475	−0.0499
6	14 h	QUIZANGA	21	8.04	1.5238	0.189527	0.2389	0.1747	0.0713
1	24 h	ANDORINHAS	20	6.6	1.4229	0.215591	0.2931	0.1856	0.0733
2	24 h	APOLINARIO	19	5.57	1.0497	0.188456	0.1523	0.1488	0.0397
3	24 h	FAZ. SANTO AMARO	20	5.56	0.9066	0.163058	0.1795	0.2052	−0.0121
4	24 h	NOVA FRIBURGO	18	5.37	0.9373	0.174544	0.1213	0.1036	−0.0296
5	24 h	POSTO GARRAFAO	20	5.84	1.1766	0.201473	0.1164	−0.0635	−0.0344
6	24 h	QUIZANGA	21	5.5	1.0271	0.186745	0.2378	0.1707	0.0802

Appendix 9
Regional Data for Exercise 6 of Chapter 10

Figure A9.1 shows the location, isohyetal, and hypsometric maps of the upper Velhas River basin, located in the State of Minas Gerais, in southeastern Brazil. This 5000-km^2 river catchment is the major source of water supply for the state's capital city of Belo Horizonte and its metropolitan area (BHMA), with approximately 5.76 million inhabitants. The mean annual rainfall depths decreases from 1600 mm, at high elevations, to 1400 mm at the outlet, with a well-defined wet season from October to March and a dry season from April to September. Near the source of the Velhas River, at an average elevation of 1500 m, the terrain is hilly to mountainous, with narrow valleys and streams of low sinuosity.

Table A9.1 lists the code, name, river, and associated catchment attributes of each of five gauging stations in the river basin. In Table A9.1, Area=drainage area (km^2); P_{mean}=mean annual rainfall over the catchment (m); L=length of main stream channel (km); and J=number of stream junctions per km^2, and S_{equiv}=equivalent stream slope (m/km). The gauging stations are also located on the map of Fig. A9.1. Table A9.2 lists the annual minimum mean flows for the durations 1 day, 3 days, 5 days, and 7 days, recorded at the five gauging stations in the river basin.

© Springer International Publishing Switzerland 2017
M. Naghettini (ed.), *Fundamentals of Statistical Hydrology*,
DOI 10.1007/978-3-319-43561-9

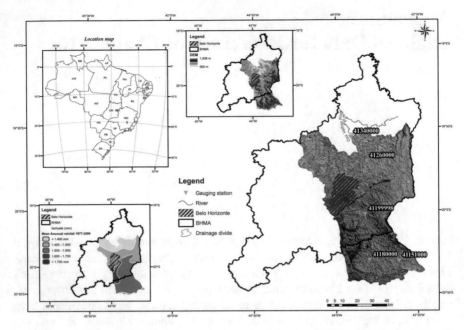

Fig. A9.1 Location, isohyetal, and hypsometric maps of the upper Velhas River basin, in Brazil. Locations of gauging stations are indicated as *triangles* next to their codes

Table A9.1 Gauging stations attributes for regional analysis of low flows

Code	Gauging station	River	Area (km²)	P_{mean} (m)	L (km)	J (Junctions/ Km²)	S_{equiv} (m/km)
41151000	Faz. Água Limpa	Velhas	174.6	1.498	26.15	0.115	8.59
41180000	Itabirito-Linígrafo	Itabirito	330	1.518	47.7	0.252	5.25
41199998	Honório Bicalho—Mont.	Velhas	1698	1.535	90.3	0.212	2.56
41260000	Pinhões	Velhas	3727	1.475	156.8	0.204	1.42
41340000	Ponte Raul Soares	Velhas	4874	1.458	200.3	0.209	1.13

Table A9.2 Annual minimum mean flows (m³/s) of five gauging stations in the upper Velhas river basin, in Brazil, for durations 1 day, 3 days, 5 days, and 7 days

Station 41151000—Faz Água Limpa					Station 41180000—Itabirito Linígrafo				
Year	1 Day	3 Days	5 Days	7 Days	Year	1 Day	3 Days	5 Days	7 Days
1957	1.330	1.363	1.390	1.409	1967	3.700	3.700	3.740	3.797
1958	1.600	1.600	1.612	1.626	1968	4.650	4.650	4.694	4.713
1959	1.230	1.230	1.230	1.230	1969	3.600	3.600	3.640	3.657
1960	1.180	1.180	1.190	1.201	1970	3.750	3.750	3.750	3.750
1961	1.380	1.413	1.420	1.423	1972	2.700	2.700	2.712	2.717
1962	1.180	1.197	1.210	1.230	1973	4.480	4.480	4.480	4.480
1964	0.978	0.995	1.009	1.015	1974	4.390	4.390	4.390	4.390
1965	1.220	1.220	1.240	1.249	1975	4.640	4.640	4.640	4.640
1966	1.450	1.490	1.486	1.510	1976	3.830	3.830	3.830	3.830
1969	1.120	1.153	1.160	1.163	1977	3.860	3.860	3.886	3.916
1970	1.390	1.410	1.426	1.441	1978	4.080	4.130	4.170	4.187
1971	1.220	1.220	1.264	1.270	1979	5.800	5.800	5.800	5.824
1973	1.760	1.760	1.760	1.760	1980	3.680	3.760	3.850	3.944
1974	1.520	1.520	1.520	1.520	1981	4.440	4.487	4.524	4.560
1975	1.310	1.310	1.340	1.386	1982	3.200	3.273	3.310	3.310
1976	1.040	1.057	1.070	1.083	1983	5.520	5.520	5.618	5.684
1977	1.350	1.350	1.350	1.390	1984	4.010	4.010	4.010	4.010
1978	1.320	1.320	1.340	1.349	1985	5.940	5.940	5.940	5.989
1979	1.980	1.980	1.994	2.010	1986	3.020	3.020	3.206	3.230
1980	1.790	1.877	1.994	2.010	1987	4.230	4.280	4.308	4.340
1981	1.730	1.730	1.754	1.764	1988	4.500	4.593	4.668	4.660
1982	2.110	2.180	2.180	2.200	1989	3.600	3.600	3.600	3.600
1983	2.110	2.110	2.110	2.110	1990	2.880	2.913	2.962	2.954
1984	1.610	1.630	1.646	1.670	1991	3.530	3.530	3.530	3.530
1985	2.180	2.227	2.236	2.240	1992	4.380	4.380	4.482	4.500
1986	1.790	1.790	1.790	1.790	1993	4.810	4.810	4.810	4.830
1987	1.440	1.440	1.474	1.496	1994	4.060	4.083	4.102	4.110
1988	1.730	1.813	1.860	1.877	1995	4.350	4.350	4.396	4.426
1989	1.380	1.380	1.380	1.380	1996	3.810	3.810	3.810	3.810
1990	1.280	1.280	1.280	1.294	1997	3.630	3.690	3.794	3.810
1991	1.730	1.750	1.766	1.801	1998	3.520	3.520	3.520	3.520
1992	1.610	1.730	1.754	1.764	1999	2.420	2.703	2.760	2.760
1993	1.670	1.710	1.718	1.721					
1994	1.360	1.470	1.470	1.470					
1995	1.290	1.290	1.290	1.290					
1996	1.420	1.420	1.420	1.430					
1997	1.650	1.650	1.650	1.661					
1998	1.350	1.350	1.350	1.350					
1999	0.898	0.939	0.960	0.977					

Station 41199998—Honório Bicalho

Year	1 Day	3 Days	5 Days	7 Days
1971	7.500	7.950	8.860	8.979
1972	13.100	13.800	14.260	14.529
1973	10.100	11.700	12.240	12.771
1974	9.620	10.240	10.468	10.491
1975	8.280	9.183	9.554	9.981
1976	8.280	9.963	10.654	10.710
1977	10.600	10.867	10.840	10.900
1978	11.300	12.467	12.520	12.671
1979	14.700	15.467	16.160	16.686
1980	14.200	16.567	17.040	17.086
1981	12.700	14.533	14.720	14.714
1982	15.800	16.200	16.380	16.457
1983	14.600	14.933	15.120	15.200
1984	12.500	13.500	13.740	13.743
1985	19.700	19.900	20.060	20.743
1986	18.000	18.400	18.480	18.429
1987	11.500	11.833	12.100	12.357
1988	16.800	17.700	17.980	18.100
1989	12.500	12.667	12.900	12.929
1990	13.500	13.900	14.180	14.457
1991	16.200	17.300	17.420	17.629
1992	16.200	16.600	16.780	16.929
1993	16.800	17.333	17.760	17.871
1994	15.400	16.100	16.580	16.700
1995	10.800	12.133	12.400	12.586
1996	13.700	14.367	14.500	14.714
1997	17.500	17.667	18.500	19.100
1998	10.100	11.033	11.320	11.643
1999	9.620	9.780	10.008	10.106

Station 41260000—Pinhões

Year	1 Day	3 Days	5 Days	7 Days
1980	27.80	27.97	28.12	29.43
1981	24.50	24.67	24.70	24.80
1982	30.50	31.30	31.68	31.76
1983	33.40	33.80	34.24	34.60
1984	24.50	25.37	25.98	26.56
1985	33.90	34.87	36.02	36.83
1986	22.00	22.83	23.84	24.03
1987	16.70	17.60	19.56	19.83
1988	24.00	24.33	25.04	25.64
1989	19.80	20.90	21.62	22.14
1990	21.70	22.37	23.90	24.64
1991	32.20	32.93	32.96	33.06
1992	31.70	31.87	32.10	32.36
1993	25.40	26.77	27.40	27.47
1994	16.90	17.47	20.14	21.39
1995	16.00	16.57	16.94	17.17
1996	23.20	24.03	24.40	24.74
1997	32.90	33.03	33.54	34.04
1998	23.70	23.87	24.08	24.34
1999	19.40	19.53	19.74	19.83

Station 41340000—Ponte Raul Soares

Year	1 Day	3 Days	5 Days	7 Days	Year	1 Day	3 Days	5 Days	7 Days
1938	33.70	35.20	36.52	36.70	1990	30.80	30.97	31.42	31.77
1939	28.30	29.87	31.04	31.60	1991	34.50	34.87	35.16	35.76
1940	22.10	25.00	26.18	26.93	1992	41.10	41.50	41.70	41.77
1941	28.30	31.87	33.62	34.91	1993	32.90	33.63	34.20	34.46
1942	24.20	29.87	31.28	32.74	1994	31.90	32.93	34.10	34.61
1944	30.60	33.10	34.08	34.90	1995	23.80	24.73	24.92	25.21
1947	32.50	36.97	38.38	39.27	1996	27.70	28.93	29.58	30.07
1950	27.10	30.47	31.02	32.97	1997	34.00	35.43	36.58	37.24
1951	18.00	19.13	19.36	20.23	1998	28.00	28.33	28.80	29.14
1952	29.40	31.43	32.84	34.19	1999	20.90	21.57	22.36	23.14

(continued)

Table A9.2 (continued)

Station 41340000—Ponte Raul Soares

Year	1 Day	3 Days	5 Days	7 Days	Year	1 Day	3 Days	5 Days	7 Days
1953	25.40	28.53	29.12	30.07					
1954	17.10	20.60	22.16	22.59					
1955	14.80	16.97	17.94	18.43					
1956	17.50	19.03	20.78	20.81					
1957	31.50	31.50	31.78	31.87					
1958	31.90	33.10	35.00	36.47					
1959	17.50	18.17	18.58	18.86					
1960	21.00	21.93	22.32	23.03					
1961	25.40	28.47	29.08	29.51					
1962	18.00	19.23	20.60	21.31					
1963	17.10	18.83	19.18	19.70					
1964	19.40	21.03	21.04	21.19					
1965	34.90	35.17	35.26	35.23					
1967	22.40	22.67	22.98	23.26					
1968	22.00	23.13	23.54	23.59					
1969	19.10	19.23	19.58	19.80					
1970	24.60	24.73	24.94	25.03					
1971	16.30	16.83	16.94	17.04					
1972	21.30	21.53	21.76	21.80					
1973	19.90	22.97	23.30	23.60					
1974	21.80	21.97	22.50	22.80					
1975	24.40	24.57	24.60	24.97					
1976	16.90	18.73	19.26	19.44					
1977	24.00	24.83	25.20	25.64					
1978	22.00	22.67	24.40	25.43					
1979	33.40	34.13	35.40	36.69					
1981	31.20	31.37	31.84	32.20					
1984	26.90	29.57	31.64	32.37					
1985	40.20	41.40	41.88	42.36					
1986	25.70	26.70	27.00	27.00					
1987	22.30	23.13	23.48	23.43					
1988	25.70	26.20	26.70	27.30					
1989	25.70	26.20	27.00	27.73					

Appendix 10
Data of 92 Rainfall Gauging Stations in the Upper São Francisco River Basin, in Southeastern Brazil, for Regional Frequency Analysis of Annual Maximum Daily Rainfall Depths

© Springer International Publishing Switzerland 2017
M. Naghettini (ed.), *Fundamentals of Statistical Hydrology*,
DOI 10.1007/978-3-319-43561-9

Table A10.1 Geographic coordinates and elevations (m) of rainfall gauging stations

	Code	Gauging station name	Location	Latitude	Longitude	Elevation
1	01844000	CURVELO	Curvelo	18°45′58″ S	44°25′34″ W	608
2	01844001	SANTO HIPÓLITO	Santo Hipólito	18°18′00″ S	44°13′22″ W	510
3	01844010	PONTE DO LICÍNIO (JUSANTE)	Presidente Juscelino	18°40′22″ S	44°11′28″ W	560
4	01845002	FAZENDA SÃO FÉLIX	São Gonçalo do Abaeté	18°27′52″ S	45°38′48″ W	760
5	01845004	LAGOA DO GOUVEIA	Tiros	18°50′29″ S	45°51′05″ W	1035
6	01845008	MORAVANIA	Morada Nova de Minas	18°40′ S	45°21′ W	600
7	01845009	TRÊS MARIAS	Três Marias	18°10′ S	45°18′ W	570
8	01845010	VILA CANASTRÃO	Tiros	18°34′ S	45°43′ W	835
9	01845011	SÃO GONÇALO DO ABAETÉ	São Gonçalo do Abaeté	18°21′ S	45°50′ W	800
10	01845012	ANDREQUICÉ	Três Marias	18°17′ S	45°00′ W	830
11	01845013	S. GONÇALO DO ABAETÉ	São Gonçalo do Abaeté	18°20′37″ S	45°50′12″ W	836
12	01845014	TIROS	Tiros	18°59′59″ S	45°57′58″ W	1030
13	01845026	FAZENDA DAS PEDRAS	Três Marias	18°00′ S	45°06′ W	600
14	01846003	MAJOR PORTO	Patos de Minas	18°42′25″ S	46°02′13″ W	672
15	01943000	MIN. MORRO VELHO	Nova Lima	19°58′45″ S	43°51′00″ W	770
16	01943006	SABARÁ	Sabará	19°53′35″ S	43°48′54″ W	720
17	01943009	VESPASIANO	Vespasiano	19°41′14″ S	43°55′15″ W	676
18	01943010	CAETÉ	Caeté	19°54′00″ S	43°40′03″ W	840
19	01943011	INSTITUTO AGRONÔMICO	Belo Horizonte	19°55′ S	43°54′ W	850
20	01943012	LAGOA SANTA	Lagoa Santa	19°38′ S	43°54′ W	777
21	01943013	CARLOS PRATES	Belo Horizonte	19°54′43″ S	43°57′28″ W	915
22	01943022	CAIXA DE AREIA	Belo Horizonte	19°56′42″ S	43°54′45″ W	950
23	01943053	AVENIDA DO CONTORNO	Belo Horizonte	19°56′04″ S	43°57′07″ W	915
24	01944000	PRUDENTE DE MORAIS - A	Prudente de Morais	19°29′01″ S	44°10′14″ W	732
25	01944003	MATEUS LEME	Mateus Leme	19°59′18″ S	44°25′48″ W	836

(continued)

Table A10.1 (continued)

Code		Gauging station name	Location	Latitude	Longitude	Elevation
26	01944004	PONTE NOVA DO PARAOPEBA	Juatuba	19°57′20″ S	44°18′24″ W	721
27	01944005	BETIM	Betim	19°58′17″ S	44°12′06″ W	832
28	01944007	FAZENDA ESCOLA FLORESTAL	Florestal	19°52′47″ S	44°25′18″ W	745
29	01944009	PEDRO LEOPOLDO	Pedro Leopoldo	19°38′04″ S	44°03′12″ W	698
30	01944010	HORTO FLORESTAL (Paraopeba)	Paraopeba	19°16′05″ S	44°24′06″ W	733
31	01944011	JAGUARUNA (Onça do Pitangui)	Onça de Pitangui	19°43′37″ S	44°48′24″ W	685
32	01944016	SETE LAGOAS	Sete Lagoas	19°28′01″ S	44°15′02″ W	780
33	01944018	CAETANÓPOLIS	Caetanópolis	19°17′33″ S	44°24′40″ W	738
34	01944019	FÁBRICA TECIDOS S. ANTÔNIO	Sete Lagoas	19°28′03″ S	44°14′14″ W	751
35	01944021	VELHO DA TAIPA	Conceição do Pará	19°41′46″ S	44°55′46″ W	585
36	01944023	COMPANHIA INDUSTRIAL B.H..	Pedro Leopoldo	19°36′53″ S	44°02′31″ W	720
37	01944024	FAZENDA VARGEM BONITA	Jequitibá	19°14′14″ S	44°07′23″ W	636
38	01944027	JUATUBA	Mateus Leme	19°57′20″ S	44°20′04″ W	728
39	01944031	PONTE DA TAQUARA	Paraopeba	19°25′23″ S	44°32′54″ W	624
40	01944032	PITANGUI	Pitangui	19°41′04″ S	44°52′44″ W	696
41	01944040	POMPÉU VELHO	Pompéu	19°16′ S	44°49′ W	650
42	01944049	PAPAGAIOS	Papagaios	19°25′42″ S	44°43′11″ W	703
43	01944060	PORTO MESQUITA	Pompéu	19°10′ S	44°40′ W	670
44	01945000	ARAÚJOS	Araújos	19°56′54″ S	45°10′01″ W	813
45	01945002	BARRA DO FUNCHAL	Serra da Saudade	19°23′41″ S	45°53′04″ W	720
46	01945004	ESTAÇÃO ÁLVARO DA SILVEIRA	Bom Despacho	19°45′06″ S	45°07′01″ W	648
47	01945008	BOM DESPACHO	Bom Despacho	19°44′33″ S	45°15′18″ W	750

(continued)

Table A10.1 (continued)

	Code	Gauging station name	Location	Latitude	Longitude	Elevation
48	01945013	MATUTINA	Matutina	19°14′ S	45°58′ W	1100
49	01945014	ENGENHO RIBEIRO	Bom Despacho	19°41′ S	45°23′ W	650
50	01945015	FAZENDA NOVO HORIZONTE	Córrego Danta	19°43′ S	45°56′ W	1050
51	01945016	FAZENDA DA CURVA	Luz	19°58′ S	45°35′ W	650
52	01945017	PORTO PARÁ	Pompéu	19°18′ S	45°05′ W	600
53	01945019	DORES DO INDAIÁ	Dores do Indaiá	19°28′07″ S	45°36′06″ W	692
54	01945035	ABAETÉ	Abaeté	19°09′47″ S	45°26′33″ W	565
55	01946000	TAPIRAÍ	Tapiraí	19°52′46″ S	46°01′58″ W	670
56	01946004	IBIÁ	Ibiá	19°28′32″ S	46°32′33″ W	855
57	01946007	FAZENDA SÃO MATEUS	Ibiá	19°31′03″ S	46°34′22″ W	870
58	01946009	SÃO GOTARDO	São Gotardo	19°18′55″ S	46°02′40″ W	1100
59	01946010	PRATINHA	Pratinha	19°45′05″ S	46°22′43″ W	1150
60	01946011	TAPIRA	Tapira	19°55′37″ S	46°49′31″ W	1120
61	02043002	LAGOA GRANDE	Nova Lima	20°10′45″ S	43°56′34″ W	1350
62	02043004	RIO DO PEIXE	Nova Lima	20°08′16″ S	43°53′33″ W	1097
63	02043013	CONGONHAS	Congonhas	20°31′19″ S	43°49′48″ W	871
64	02043016	RIO ACIMA	Rio Acima	20°05′15″ S	43°47′16″ W	730
65	02043018	CARANDAÍ	Carandaí	20°57′21″ S	43°48′03″ W	1056
66	02043042	REPRESA DAS CODORNAS	Nova Lima	20°09′53″ S	43°53′31″ W	1200
67	02044002	ITAÚNA	Itaúna	20°04′17″ S	44°34′13″ W	859
68	02044003	CARMO CAJURU	Carmo do Cajuru	20°11′32″ S	44°47′37″ W	746
69	02044005	CARMO DA MATA	Carmo da Mata	20°33′28″ S	44°52′03″ W	846
70	02044006	DIVINÓPOLIS	Divinópolis	20°08′13″ S	44°53′31″ W	672
71	02044007	ENTRE RIOS DE MINAS	Entre Rios de Minas	20°39′40″ S	44°04′14″ W	885
72	02044008	MELO FRANCO	Brumadinho	20°11′52″ S	44°07′15″ W	761
73	02044009	FAZENDA CAMPO GRANDE	Passa Tempo	20°37′31″ S	44°26′00″ W	915
74	02044012	IBIRITÉ	Ibirité	20°02′34″ S	44°02′36″ W	1073
75	02044016	FAZENDA BENEDITO CHAVES	Itatiaiuçu	20°10′09″ S	44°30′54″ W	944

(continued)

Table A10.1 (continued)

	Code	Gauging station name	Location	Latitude	Longitude	Elevation
76	02044019	FAZENDA VISTA ALEGRE	Mateus Leme	20°03′05″ S	44°27′06″ W	913
77	02044021	ALTO DA BOA VISTA	Mateus Leme	20°06′20″ S	44°24′04″ W	905
78	02044024	FAZENDA CURRALINHO	Igarapé	20°00′27″ S	44°19′52″ W	754
79	02044026	FAZENDA COQUEIROS	Itaúna	20°07′47″ S	44°28′28″ W	975
80	02044036	ITAGUARA	Itaguara	20°24′ S	44°28′ W	840
81	02044040	USINA JOÃO RIBEIRO	Entre Rios de Minas	20°38′07″ S	44°02′56″ W	850
82	02044046	BONFIM	Bonfim	20°20′ S	44°15′ W	952
83	02045001	BAMBUÍ	Bambuí	20°01′16″ S	45°57′58″ W	654
84	02045002	IGUATAMA	Iguatama	20°10′44″ S	45°42′01″ W	606
85	02045005	LAMOUNIER	Itapecerica	20°28′20″ S	45°02′10″ W	738
86	02045010	ARCOS (COPASA)	Arcos	20°17′41″ S	45°32′34″ W	791
87	02045011	LAGOA DA PRATA	Lagoa da Prata	20°02′12″ S	45°32′07″ W	658
88	02045012	PIUMHI	Piumhi	20°27′31″ S	45°56′38″ W	806
89	02045013	ST° ANTONIO DO MONTE	S. Antonio do Monte	20°05′04″ S	45°17′48″ W	950
90	02045015	FAZENDA OLOS D'ÁGUA	Pimenta	20°26′ S	45°50′ W	810
91	02046007	FAZENDA AJUDAS	Bambuí	20°06′06″ S	46°03′18″ W	705
92	02046009	DELFINÓPOLIS	Delfinópolis	20°20′50″ S	46°50′46″ W	680

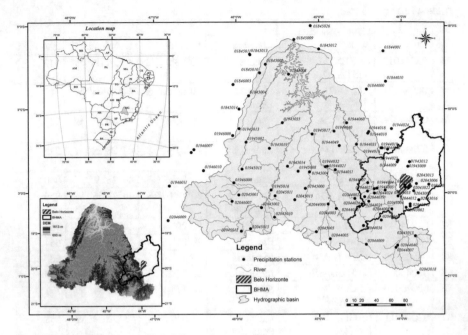

Fig. A10.1 Location map of rainfall gauging stations

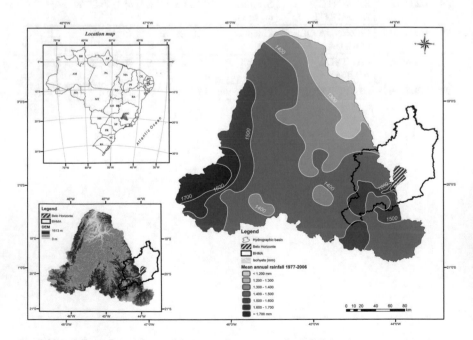

Fig. A10.2 Maps of elevation and isohyets of mean annual rainfall (mm)

Table A10.2 Annual maximum daily rainfall depths (mm) for 92 gauging stations of the upper São Francisco River basin, in Brazil

	Station	N	40/41	41/42	42/43	43/44	44/45	45/46	46/47	47/48	48/49	49/50	50/51	51/52	52/53	53/54	54/55	55/56	56/57	57/58	58/59	59/60	60/61	61/62	62/63	63/64	64/65	65/66	66/67
1	01844000	10		71.4	83.8		101.4	71.8	60.9	85		50	121.2																
2	01844001	22		63	63.3	167.4	58.2				110												92.3		70.8		78.2	63.8	58.2
3	01844010	10																											
4	01845002	11																											
5	01845004	10																											
6	01845008	14																											
7	01845009	17																											
8	01845010	16																											
9	01845011	18																											
10	01845012	18																											
11	01845013	15																					110.4						
12	01845014	15																						94.6	63.2				66.5
13	01845026	13																					127.8		63.2				
14	01846003	19																											
15	01943000	39		59.2	144		71.1	60.9	79.2	103.6	126	54.6	96	68.1	100.8	101.1	80.8		81.3	90.4	56.1	84.8	87.4		73.9	88.4	76.7	76.7	
16	01943006	26		69			70.8	58.8	114	57.4	132.4		110	100	75	118		49		68			72	80	100				
17	01943009	31		62.6	61	86	108.6	54.4	90	79				81.4					83.4	63	54.6	86	70	75.8			79	84	
18	01943010	37		72.8	69.4	77.8	74.2	102.2	93.4	75	117.4	47.2	67.4	76	102.6	87	112.8		80.1	95.7	71.8	102.3	105.5				75.9		
19	01943011	24					72.2	63.2	66.2	73.5		78	90	80	69.8	71.8	88.2	56	116.2	56.8		76	109	133	94	108.4	80	86.8	87.5
20	01943012	20		45	71.4	77	71.6			79	81.5	103	86.4			66.5	64.5	49.6	72	72.5	40.5		94.8		97.7	137.5	78		
21	01943013	13		65	69	94.4	139	56.6	149	85.4		74.6	118.6	89.4	63.8	75	71.4												
22	01943022	16			86																				61	116			
23	01943053	15							161.5	94.4	77	58.5	122		78.8	111.8	85.4	74.4	98.7			69.8	79.8		74		76.7	116.9	
24	01944000	33		55.4	75.2		72.9	65.6	88.7	87.7	81.9	105.9	156.8	63.8	47.6	83.1	101.5	71.7	96	69	56.4	71	83.9	76.4	96.6	110.2	102.8	79.2	62.5
25	01944003	34		48.1	65.4	94.8	64.2	50.8	79.8	122.8	73.8	52.6	120.8	60.4	128.4	80.4	93.8	113.2	122.8	73.4	83.2	73.4	60.2	52.4	85.4	159.4	60.8	80.2	48.0
26	01944004	48		68.8			67.3		70.2	113.2	79.2	51.2	66.4	65.1	115.0	67.3	102.2	54.4	69.3	54.3	36.0	64.2	83.4	64.2	76.4		62.1	78.3	74.3
27	01944005	15					64.2	94.0		104.0	85.0	54.0		84.0	68.1				83.0	51.0			87.0						
28	01944007	34		72.0	84.3	78.2								58.8					60.4	71.5	51.6	130.4	61.8	80.0		153.2	64.4	79.8	70.2
29	01944009	26		72.2	57.2	76	54	64.2	70							72						55	74		80			75	102.2
30	01944010	27		108.6																72.4	71.6	136.8		72.0			67.0	95.0	55.0

(continued)

Table A10.2 (continued)

Station	N	40/41	41/42	42/43	43/44	44/45	45/46	46/47	47/48	48/49	49/50	50/51	51/52	52/53	53/54	54/55	55/56	56/57	57/58	58/59	59/60	60/61	61/62	62/63	63/64	64/65	65/66	66/67
31 01944011	30						74.6					100.2		64.0						55.2		66.2	73.1	75.0	104.0	75.2	58.1	88.8
32 01944016	22										120.5	149.8	68.1			70.1	68.8	99.6	72.5	64.5	98.5	86.4	111.6	82		105	103.5	
33 01944018	18																	71.7	54.5		116.3	72.5	51.6		78.2	76.8		77.8
34 01944019	12																		64.1	63.4	75.8	84.6	91.6	96.3	79.2			70.2
35 01944021	22																								71.0	84.8		
36 01944023	10																						90	90.4	68		70	55.3
37 01944024	21																						80.2	48.2		57.2		
38 01944027	18																											
39 01944031	11																											
40 01944032	15																											
41 01944040	17																											
42 01944049	12																											
43 01944060	16																											
44 01945000	24			74.0	105.6		88.0		50.8	100.5		101.8	51.2	32.1	80.6	74.0	112.0	74.0	100.0	97.0	39.0	69.0		62.2	72.2			60.0
45 01945002	26		59.1					130.4	60.4	48.0	48.2	88.2	118.2	76.3	77.0				66.4									

	Station	68/69	68/69	69/70	70/71	71/72	72/73	73/74	74/75	75/76	76/77	77/78	78/79	79/80	80/81	81/82	82/83	83/84	84/85	85/86	86/87	87/88	88/89	89/90	90/91	91/92	92/93	93/94
1	01844000				60.2		56.3																					
2	01844001		62		66.8	57.2	78.4		51.4						82		123	43.3	59.9	74.6					60			
3	01844010									60.6	86.4	84.2	101	70	39.8	104					67	46.2				121.6		
4	01845002		72.8	135.8	72.6	80.0	90.2	89.0	87.2		71.0				90.4		109.0									59.0		
5	01845004									65.0	68.2	81.0				99.0		93.0	75.3				75.4	78.2	72.1	91.1		
6	01845008									73.2		101.2	90.8	80.2	127.2	140.0	104.8	71.4	151.4	82.0	76.7	66.0	70.3	72.2	65.1	84.3		
7	01845009										80.5	83.2	85.6	85.0	92.6	94.6	87.0	67.0	60.0	57.0	65.6	104.0	53.4	103.0	52.4	78.6	79	
8	01845010									98.4		125.0	77.8	128.0	80.4	84.2	170.0		40.3	103.0	57.6	89.6	70.9	87.9	105.7	80.6		
9	01845011									61.2	75.0	75.2	124.0	166.0	81.8	83.2	92.6	98.8	77.0	74.0	73.6	73.0	80.0	58.0	75.1	95.0	78.4	
10	01845012									62.4	92.4	114.8	109.8	105.8	75.6	66.8	104.6	122.4	70.0	74.0	79.6	73.2	69.8	79.4	82.2	95.2	61.8	
11	01845013				50						73.2	107.8	126.8	88.4	80	88	123	79	78	95						92.3	83.2	
12	01845014		72.6														132.2	83.3			72.1		76	102	78.6		63.2	
13	01845026								62.2						62.3	71.8	104.4	79.6	94.6	68.4	61.2	86.4	63.4	112.0	48.8	78.0	82.8	
14	01846003	61.7	63.7		40.5	74.4	81.6	87.4	72		54.2	146.6	83			98	96	83	75	67	70	71.2	62.2	71		95	75.1	
15	01943000		104.4		55.1		78.3		72	52		119.1	130			140.4	99.1	70.3	110	100	76.2		51	75				
16	01943006				70		75		58.4		70.8	100	72.6	70.6	70.4		87.4			80.4	64.2							
17	01943009	54				65.2	128.9		61.2	32.8		210.8	102.6	77.9		98.2			114	62.5	60				60.8	62	88	
18	01943010	50.7	90								66.9	210.2	92.1	86.5	86.3	123.6	84.6	64.6	80.7	73	83.4	73.6	57.2	97.7	116.2	100.9		
19	01943011	66.4	70																									
20	01943012	108.5																										
21	01943013																											
22	01943022								68.1		67.3		97.5	78.2	83				138.4		112.4	79.2	71	110	124		147.3	
23	01943053										152.9																	
24	01944000	60.4	88.2	83.2	77.8	99.9	70.6	93.9	76.6	55.4																		
25	01944003	58.8	110.4	85.6	78.0	98.6	86.0	61.1	72.0											62.2	49.5	67.5	62					
26	01944004	41.0	101.6		51.4	70.3	81.3	85.3	58.4	66.3		72.8	100.0	73.4	61.8	83.4	93.4	99.0	133.0	101.0	109.0	88.0	99.6	74.0	94.0	99.2	101.6	
27	01944005				55.0	54.6		51.4	40.2																			
28	01944007	69.6	60.0	59.8	82.6	79.0	94.0	80.0	68.0	104.6	75.2	86.4					69.0	69.0		60.0	86.0	166.0	53.3	105.1		78.8		
29	01944009	65	90				60.6		58		134.6	45.2		50.1	69	39.2	74.2				49.5	67.5	62					
30	01944010	57.3	58.5	103.0	80.8	129.2	47.8	71.8	74.0	58.4	97.6	178.0	146.0	84.2	57.3			66.6			55.4	75.2	65.0	67.6				
31	01944011	58.3	91.2	107.4			51.8			88.0	72.5	146.0	65.5	101.3	65.3	69.3	82.8					122.3	74.3	100.3		84.5		
32	01944016	76.7				97.2	75.7	84.8	90.8	60.4	148.5	146.7							90.2	58.2	57.3							

(continued)

#	Station	68/69	69/70	70/71	71/72	72/73	73/74	74/75	75/76	76/77	77/78	678/79	79/80	80/81	81/82	82/83	83/84	84/85	85/86	86/87	87/88	88/89	89/90	90/91	91/92	92/93	93/94
33	01944018	80.3	85.8	69.2	136.2	83.8	74.5	85.3	60.8	64.5	157.5	147.3	75.2														
34	01944019	66.3			95	75																					
35	01944021	57.1	107.0	157.0				51.8	75.0		55.0	123.7		60.2	56.0		72.1	94.3	76.3	59.2	126.3	56.8	71.6	64.6	87.6		
36	01944023	91.6					68	77	45.5		151			60	123												
37	01944024	58.3		73.5	102.5	67.8			90.4	63.3	125	108.0	91.3	86.6	102.8	84.7	52	102.6	57.6	49.2	67.6	55.2	70	200	69	63.2	
38	01944027				69.6	93.0	115.6	57.0	76.0	89.4	70.3		90.4	57.2	83.8		93.4	140.6	85.4	76.8	124.7		95.5				
39	01944031						103.8	71.2									75.6	94.0	54.8			50.8					
40	01944032							66.5		76.8	54.0	130.0	116.0	73.8	91.8	83.0	61.0			109.9	139.9	71.7	67.9	70.8	91.8		
41	01944040								68.0	139.2	67.2	114.8	50.0	80.2	80.3	112.4	64.6	93.0	74.0	40.0	72.0	47.0	60.0	87.0	110.0		
42	01944049								60.1	153.2	111.0	92.2	75.0	54.1	97.0	53.0	93.0		97.6		61.6	100.6	69.8				
43	01944060								49.6	70.2	90.8	80.4	85.0	70.2		75.0	63.0	82.0	75.0	66.2	60.0	111.0	69.3	85.0	75.0		
44	01945000		42.5				78.2	61.6	85.4	70.4									46.0			47.0	66.2				
45	01945002			75.0	75.0	69.0	80.0		56.0	74.2	70.4	80.0	88.0	68.0			93.0	62.2			95.0				78.0		

#	Station	N	40/41	41/42	42/43	43/44	44/45	45/46	46/47	47/48	48/49	49/50	50/51	51/52	52/53	53/54	54/55	55/56	56/57	57/58	58/59	59/60	60/61	61/62	62/63	63/64	64/65	65/66	66/67
46	01945004	15		69.0	52.0	57.2		78.3	98.0				105.4								84.0	50.0							
47	01945008	11																											
48	01945013	17																											
49	01945014	16																											
50	01945015	16																											
51	01945016	15																											
52	01945017	18																											
53	01945019	21																			69.2	72.2		58.2	96.2	126.4	78.8		
54	01945035	10																											
55	01946000	33		75.0	56.0		73.0		144.0	60.0	80.1	96.0	76.1	93.5	58.0	70.2	62.2	73.2				80.0		79.1		63.0	59.1		
56	01946004	34							140.4	78	52	77	64	88	60		52					53.2			105.2		67.6	78	
57	01946007	23																	82.4										67
58	01946009	17																											
59	01946010	14																											
60	01946011	14																											
61	02043002	27			79		130	98.5		96.5	165.1		100.6	71.9	90.2	55.9				92.5	82.6	64.8				108.2	62.2		
62	02043004	27					76.7	100.6	78.5	103.9	111.8	66.5	100.6	66.3	104.6	86.1	159	97.5				73.7		79.5	80.8	92.5			
63	02043013	31		51.6	90.2	49.2	45.4	47.0			108.2	56.8	102.2	55.2	96.6	52.8	110.4	68.2	71.2	60.2	65.8		97.6	57.2	71.4	52.6	102.4	79.6	
64	02043016	10				76.4	62.4			92	97	89.4	84	55.2		58.2	81.2	48											
65	02043018	30	44			66.4				67.4					50.6	75.8		75.6	74.4	50.6				58	70.2	55.2			
66	02043042	11																											
67	02044002	24			51.0		80.2		90.0																				
68	02044003	38		74.8	86.2	70.6	82.2	54.8	76.6	64.4	94.2	106.2	66.6	71.8		72.4					53.0	73.2	129.0	71.0	73.1	91.2	73.2	62.1	91.1
69	02044005	22		95.0	65.0	59.0		69.0	79.3	89.0		100.2		73.1		86.0			82.1				91.1		60.1	117.1			74.0
70	02044006	31		78.4	58.6	92.6	63.3		107.8		57.2	83.1											149.3	57.4		81.9			
71	02044007	30		75.0		105.3	99.0	62.0	62.0	90.0	94.0	70.0			89.0	75.0						95.0	115.0	97.0			80.0		
72	02044008	21			68.0	75.4	78.0	75.0	65.0		118.2		87.3	94.0	43.0	86.3	105.6	80.0		68.2				55.0			62.2	62.2	
73	02044009	30																			93.2	77.0		63.0	87.2				94.0
74	02044012	33						62.5	65.0	66.4		62.5	71.4		64.0	74.8			64.2	107.4		74.0		87.8	80.8	100.0	86.8		77.4
75	02044016	13																											
76	02044019	18																											
77	02044021	18																											
78	02044024	15																											

(continued)

Station	N	40/41	41/42	42/43	43/44	44/45	45/46	46/47	47/48	48/49	49/50	50/51	51/52	52/53	53/54	54/55	55/56	56/57	57/58	58/59	59/60	60/61	61/62	62/63	63/64	64/65	65/66	66/67
79	02044026	17																											
80	02044036	16																											
81	02044040	11																											
82	02044046	14									63.8				61.4														73.8
83	02045001	29			62.6	69.2	140.4	103.4	82.2												65.6					99.8		79.2	
84	02045002	31		61.0	61.4				142.0	70.0		62.4	88.0			55.0	50.0	103.0	71.6	53.4						74.4			
85	02045005	44		62.2	116.4	79.2	78.2	64.8	75.2	74.8	166.8		125.0	68.6	78.6	93.0	58.2	170.0	80.2	93.6	70.2	68.4	83.2		68.6	125.2	98.0		
86	02045010	12																											
87	02045011	17																											
88	02045012	12																											
89	02045013	16																											
90	02045015	16																											
91	02045007	38		50.7	82.4	73.0	110.0	76.0	105.6	81.3	57.2	86.4	69.0	90.5		71.5	45.4		54.1	69.0									
92	02046009	22		83.6	61.4	88.4	54.4	90.2	73.2			55.4	93.2	60.4	108	74	66.2	94	82	74	56	73	99	75	57.6	58.2	84.2		

#	Station	68/69	69/70	70/71	71/72	72/73	73/74	74/75	75/76	76/77	77/78	678/79	79/80	80/81	81/82	82/83	83/84	84/85	85/86	86/87	87/88	88/89	89/90	90/91	91/92	92/93	93/94	94/95
46	01945004						72.4			84.2	74.8	83.2					68.2				67.0			68.8	76.0			
47	01945008								67.8	77.2	68.8					128.0	118.0	85.8	74.0	113.6	105.6	75.0		98.0				
48	01945013								74.4	73.2	81.0	67.0	76.0	90.0	88.3		74.6	62.4	67.6	64.6	67.0	70.0	83.4	89.6	130.0	68.6		
49	01945014									55.6	106.2	100.7	87.0	66.4	78.0	100.2	113.8	76.6	99.8	64.6	73.6	58.4	87.8	71.2	91.4			
50	01945015								71.9		63.0	72.4	56.0	71.0	96.0	135.0	114.0	130.0	99.0	67.2	68.6	56.6	110.8	99.0	87.6			
51	01945016									68.4	91.6	67.8	74.0	66.0	96.2	143.2	46.0	76.2		98.6	60.0	78.0	86.2	195.2	81.2			
52	01945017	110.2				66.0	79.2			60.2	73.2	80.2	56.8	62.6	98.2	146.6	60.6	71.6	80.6	65.0		54.5	93.0	69.1				
53	01945019	129.1	104.6				62.1			64.2	119.2		94.0		74.0			100.9	68.0	76.9		56.4	89.4	70.2	116.0			
54	01945035								108	87.8	79.2		58.7	82.2		96.2		90			86.1					78.4	87.3	
55	01946000	88.4	84.5	77.3			47.3	63.2	126.1	86.4	93.0	61.3	69.4	73.4		81.0	70.1	104.0	73.0									
56	01946004	59.4	94		93.8	72.4	70.2		71	108.2	56		59	59	58	74	91		125.7	120			110	71.7	124	75	71.4	
57	01946007	104.8	62.8	60	71.2	88.6	46.6		61.3		54	85.9	86.2	71.2	53.2	47.8	95.4	78	57.3	122.3			75.9	68.2	37	56.5	66	
58	01946009								67	65.5	103.2	82.6	79.7		95.8	100.3	80	78	72.3			50.5	87	81.5	113	80	95.5	71.3
59	01946010									70		67	80	76.2	95	125	65	62.5	67.3	52			108	136	100	78.3	105.5	
60	01946011				74.3				91	83	59	96.4	87.1	71.6	81.6	56.8	67	68.2		72.1			77	72.8	68.4	72.4	70	
61	02043002			63	54.1		65.2	64.9			105.2	79.2						76.2	56.8				65.1	79.4				
62	02043004						52.6				125								75.4			58.2		67.3	120.2			
63	02043013	85.4	67.6	97.4	68.2		69.8			88.6				54.2							79.0							
64	02043016																											
65	2043018 ANL	51.8	95	73.2		57	61.4	70	80		125.1	109.5	60	57.5	107	95		65	78.2	56	77.2		89					
66	02043042									68.4	97	113.0	92.2	70.1	63	76.2				32.3	108.4		76.4	100	130.1			
67	02044002	70.0	59.0	115.0	89.0		70.4	80.3	60.3	71.4			90.4	80.2	79.4	85.4	60.4	92.4	48.1	85.3	75.4	87.2	87.2	70.4	80.4			
68	02044003	78.5	91.2	91.1			59.3	102.3	84.0	56.0	81.2	133.0	77.2	79.0	91.2	72.0	64.2	72.0	79.4			89.6		80.8				
69	02044005	83.1	82.0		69.0	100.5	62.1	70.0	96.0							79.0												
70	02044006		100.0	72.2	75.0	85.0	64.0	97.3	97.1	67.0	56.0	118.1	62.4	70.1	78.0	73.4	69.3	79.0	63.0	92.0	69.0	60.0	74.0	121.2	52.3			
71	02044007		65.0		81.8	66.8	78.2	43.7	52.1	79.2	107.2	72.4	64.0		75.3	108.3	76.9	69.0	61.1	70.0	67.0		93.0					
72	02044008				92.4	78.2	58.0			78.3		65.1	95.3	82.5		66.3	59.9	74.3										
73	02044009	77.0	78.0	82.0	80.2		66.2			117.0	91.0	76.0	66.6	146.0		79.0	107.8	95.0	82.4	120.0	74.0	60.0	74.0	74.0	113.0			
74	02044012	59.2		57.0	79.3		46.1					105.0	144.4	94.6	96.2	90.4	107.8	117.4	61.2		75.4	79.8	85.4	129.4	71.6			
75	02044016								60.6	53.2	56.0	94.8	55.2	61.6	85.5	97.4			81.4	75.2	59.6	89.6			79.2			
76	02044019		89.2		63.0	100.0	62.0	50.2	70.2	75.2	52.0	64.2	50.0		90.1		82.0	87.2	78.0	95.0	83.0	45.0	97.0					
77	02044021				76.5		105.0	62.6	80.5	64.0	71.6	94.2	65.0	80.2	81.1	111.8	108.6	87.8	107.0	90.0	83.0	80.4	106.4	112.8				

(continued)

#	Station	68/69	68/69	69/70	70/71	71/72	72/73	73/74	74/75	75/76	76/77	77/78	678/79	79/80	80/81	81/82	82/83	83/84	84/85	85/86	86/87	87/88	88/89	89/90	90/91	91/92	92/93	93/94	94/95
78	02044024							95.0	58.0	75.0	60.6	80.0	75.0		74.0	64.2	95.6		113.6	61.2	109.0		70.4	93.8		87.4			
79	02044026								66.2		51.6	89.6	99.0	65.6		64.6	69.4	110.0	85.1	95.2	80.4	101.4	72.4	89.2	69.0	80.2	83.2		
80	02044036									59.0		71.6	82.2	69.0	93.0	149.0	59.0	52.0	147.0	60.4	72.0	110.0	80.0	59.0	130.0	52.0			
81	02044040								65.6	53.0	87.0	123.0	80.2		90.0							72.0	50.5	88.0	97.0	65.0			
82	02044046											74.2	63.4	58.0	69.0	128.4	77.6	73.8	97.4	49.4	85.0	56.0	63.2		75.0	82.6			
83	02045001	54.0	112.2	56.0	50.0	59.2	73.2	61.0		135.0	88.0	71.6	50.0	66.4	51.0	95.0	200.0	146.0	75.0	60.2		75.3	72.0						
84	02045002							56.4	69.2	104.2	65.0	74.2	53.8	86.2	46.8	92.2	165.0	63.0	84.6	50.4	95.2	94.8	72.0		100.5	134.3			
85	02045005	97.2	155.0	100.2	56.0			57.0	67.0	82.6	94.6	76.2	104.0			117.4	139.0	83.4	116.8	53.8	81.6	92.4	60.6	115.0	57.0	117.4			
86	02045010									96.8	86.5	84.3	61.0	75.9		88.0	109.0	67.1	146.2	86.4	108.0	54.1	65.0						78.3
87	02045011									94.2	79.1	87.3	71.3		74.2	97.3	123.4	62.5	69.3	63.5	90.3	63.2		64.5			97.3	112.9	
88	02045012										74.9	68.2	70.8	61.8	50.0	117.5		50.0	123.4	50.4	87.0	64.0	112.0	140.0		84.0			
89	02045013									81.0	57.0	69.8		96.2	54.0	96.0	102.2	75.0	82.0	107.2	123.0	85.0	72.4	63.0	102.0	73.0			
90	02045015									62.2	93.0		84.6	65.0	57.0	87.0	81.2		100.1	45.0	94.0		59.0	31.0	89.3	142.0			
91	02046007	85.0	91.5	76.9		69.2	80.1	52.0	79.0	91.0	57.1	74.2	102.2	70.2	52.1	64.3		70.2	75.3	70.2	127.2	71.2	38.0	67.2	110.0	81.3			
92	02046009																												

Appendix 11
Solutions to Selected Examples in R

A11.1. Introduction

R (R Core Team 2013) is simultaneously an open source software, a programming language, and a powerful environment for statistical analysis. The R software is available for downloading from https://www.r-project.org/ and is easy to install on Windows, MacOS and UNIX platforms.

This appendix contains a collection of guidelines and suggested approaches to tackle some of the statistical hydrology methodologies covered throughout this textbook, by providing the solutions to some of the exercises and worked out examples.

Newcomers to R are invited to complement the reading of this appendix with further exploration of the R environment. A good starting point is to type `help.start()` in the R console which opens an internet browser page with an extensive introductory material. Other helpful resources are the books by Matloff (2009) and Crawley (2012).

One of the main strengths of R is the collaborative nature with which users create and maintain packages of new R functions and other features. There are hundreds of packages available on the Web in several repositories. One of these repositories is CRAN (Comprehensive R Archive Network), which stores packages that have been validated by the R Core Team. A list of CRAN packages is available at https://cran.r-project.org/web/packages/available_packages_by_name.html. CRAN packages can be installed directly from the R console with the command `install.packages()`. For example to install the package lmomco (Asquith 2007) containing function for estimating the parameters of several distributions by the L-moments method, type the command

```
>install.packages('lmomco')
```

© Springer International Publishing Switzerland 2017
M. Naghettini (ed.), *Fundamentals of Statistical Hydrology*,
DOI 10.1007/978-3-319-43561-9

This needs to be done only once. However, to use that particular function in any R session, the package must be called by typing the following library command:

```
>library(lmomco)
```

R has many distribution functions in the base packages (basic packages that come installed with R) and many more are defined in user-created packages in CRAN and other repositories. As a rule, the distribution functions, as prefixed by the letters d, p, q, and r, are used to calculate their respective PDF, CDF, quantiles, and random number generation; e.g., for the Normal distribution, dnorm, pnorm, qnorm, and rnorm. A list of distributions in the base stats packages can be seen at https://stat.ethz.ch/R-manual/R-devel/library/stats/html/Distributions.html. One should always check which parametrization of the distribution considered by these functions is being used, prior to call them. For example, the dispersion parameter used in the normal distribution is the standard deviation and not the variance. This can be checked by typing ?dnorm in the console, which opens a web browser page with a help file for the Normal distribution function.

A11.2   Solutions

The examples were selected so as to provide a broad overview of R resources for Statistical Hydrology. The solutions presented here are not unique as there are many ways to tackle the same problem and there are many alternative functions for the same purpose. Some comments were added (comment character: #) to explain some of the choices. The code files are available online at https://github.com/fshydrology/fsh.

Chapter 2, Exercise 7 The solution to this example uses the package psych (Revelle 2014).

```
x <- c(1.06, 0.74, 1.5, 0.5, 0.89, 1.77, 0.96, 3.31, 0.69, 1.07, 2.73,
1.79, 2.04, 0.73, 0.2, 0.97, 2.66, 0.83, 1.17, 1.47, 1.23, 2.01,
0.86, 0.63, 0.93, 0.72, 1.31, 0.79, 1.47, 0.08, 0.19, 2.31, 0.32,
0.81, 1.79, 1.27, 4.87, 0.87, 2.34, 1.05, 1.12, 0.7, 0.32, 2.24,
0.86, 1.7, 1.05, 1.2, 1.86, 3.03, 1.09, 1.58, 1.77, 2, 0.74,
0.88, 0.82, 3.13, 0.75, 1, 1.86, 2, 1.61, 0.49, 1.16, 0.45, 0.74,
1.38, 1.28, 0.97, 0.85, 1, 1.57, 1.46, 1.32, 1.62, 0.97, 0.91,
1.24, 1.79, 1.68, 1.69, 1.49, 2.31, 0.81, 1.61, 0.88, 1.52, 2.54,
3.11)

# a) Histogram
hx <- hist(x, freq=FALSE)

# b) Frequency polygon
plot(c(hx$breaks[1],hx$mids,tail(hx$breaks,1),0),
```

```
       c(0,hx$density,0,hx$breaks[1]),type='l',
       xlab='x', ylab='Relative frequency')
# c) summary statistics
library(psych) # contains the useful function describe()
describe(x)
```

Example 5.16 The solution to this example uses the packages copula (Yan 2007) and VGAM (Yee 2010).

```
library(copula) # fitting the copula
library(VGAM) # GEV distribution functions

# Data Input
vol <- c(12.42, 13.77, 14.96, 17.66, 14.56, 12.28, 14.92, 21.12, 10.72,
15.79, 7.75, 62.61, 24.47, 12.43, 13.61, 6.43, 9.16, 10.28, 5.6, 14.99,
14.16, 4.89, 3.93, 9.59, 9.24, 10.74, 11.98, 9.48, 19.88, 21.21, 13.83,
13.23, 13.13, 9.47, 12.7, 13.34, 11.03, 15.17, 6.74, 18.16, 18.9,
24.33, 19.26, 15.39, 10.54, 16.15, 6.36, 8.22, 5.08, 16.01, 10.87,
6.54, 22.6, 14.14, 9.06, 12.8, 8.81, 8.35, 11.75, 18.76, 27.03, 26.08,
40.51, 11.45, 18.24, 11.82, 14.69, 26.09, 14.58, 7.94, 6.02) #10^6 m^3

peak <- c(79.29, 87.5, 74.76, 158.57, 55.22, 70.51, 52.95, 205.3,
65.41, 103.07, 33.98, 903.31, 131.67, 38.51, 100.52, 52.39, 97.41,
46.44, 31.71, 62.86, 64.28, 14.02, 15.86, 28.32, 27.18, 47.01, 51.82,
33.41, 90.9, 218.04, 80.42, 60.88, 68.53, 56.35, 84.67, 69.94, 65.98,
38.23, 32.28, 118.65, 104.77, 237.3, 116.67, 95.71, 41.06, 88.63,
29.45, 45.59, 28.06, 104.21, 68.53, 29.45, 203.6, 97.41, 38.23,
72.49, 33.13, 45.59, 154.04, 79.29, 288.83, 104.06, 297.33, 73.06,
96.56, 73.06, 74.19, 297.33, 27.55, 42.76, 95.43) #m3/s

# Distribution parameters
par.vol <- c(10.5777, 4.8896, -0.1799)
par.peak <- c(4.2703, 0.8553)

# Plot of Figure 5.24
plot(vol, peak)

U <- as.vector(pgev(vol, par.vol[1], par.vol[2], -par.vol[3]))
V <- plnorm(peak, meanlog=par.peak[1], sdlog=par.peak[2])

# Plot of Figure 5.25
plot(U, V)

fcop=fitCopula(copula=gumbelCopula(), cbind(U, V))
```

```
# Figure 5.26
layout(t(1:2))
persp(gumbelCopula(fcop@estimate),dCopula,xlim=c(0,1),
     ylim=c(0,1), xlab='u', ylab='\nv',zlab='\nJoint density')
     title('(a)')
persp(gumbelCopula(fcop@estimate),pCopula,xlim=c(0,1),ylim=c(0,1),
      xlab='u', ylab='\nv',zlab='\nC(u,v)')
      title('(b)')

mymvdist <- mvdc(copula=gumbelCopula(fcop@estimate),
     margins=c('gev','lnorm'),
     paramMargins=list(list(location=par.vol[1],scale=par.vol[2],
     shape=(-par.vol[3])),
     list(meanlog=par.peak[1],sdlog=par.peak[2])))

# Figure 5.27
layout(t(1:2))
persp(mymvdist,dMvdc,xlim=c(0,40),ylim=c(0,300)
     , xlab='x', ylab='y',zlab='\n\n Joint PDF')
     title('(a)')

persp(mymvdist,pMvdc,xlim=c(0,40),ylim=c(0,300)
     , xlab='x', ylab='y',zlab='\n Joint CDF')
     title('(b)')
```

Example 6.10 The solution to this example uses the packages lmomco (Asquith 2007), ismev (Heffernan, Stephenson and Gilleland 2013), and VGAM (Yee 2010).

```
x <- c(68.8, 67.3, 70.2, 113.2, 79.2, 61.2, 66.4, 65.1, 115, 67.3,
102.2, 54.4, 69.3, 54.3, 36, 64.2, 83.4, 64.2, 76.4, 159.4, 62.1,
78.3, 74.3, 41, 101.6, 85.6, 51.4, 70.3, 81.3, 85.3, 58.4, 66.3,
91.3, 72.8, 100, 78.4, 61.8, 83.4, 93.4, 99, 133, 101, 109, 88,
99.6, 74, 94, 99.2, 101.6, 76.6, 84.8, 114.4, 95.8, 65.4, 114.8)

library(lmomco) # sample moments and LMOM estimators
library(ismev) # GEV ML estimators
library(VGAM) # GEV functions

# MOM estimators of the GEV
gevMOM=function (mom) {
     m <- mom$moments[1]    # mean
     s <- mom$sd            # standard deviation
     g <- mom$skew          # skewness coefficient
```

```
        if (g == 1.1396) {# subroutine to solve Equation (5.74)
        } else {
            if (g > 1.1396) {
                domain <- c(-0.333,-0.00001)
            } else {
                domain <- c(0.00001,2)
            }
            Kappa <- uniroot(function (x)(g - sign(x)*
            (-gamma(1+3*x)+3*gamma(1+x)*gamma(1+2*x)-2*gamma(1+x)^3)
            /(gamma(1+2*x)-gamma(1+x)^2)^(3/2)),
            interval=domain, tol=1e-10)$root
            }

        Alpha <- s*abs(Kappa)/(sqrt(gamma(1+2*Kappa)-gamma(1+Kappa)
^2))
        Beta <- m-(Alpha/Kappa)*(1-gamma(1+Kappa))
        res <- c(Beta,Alpha,Kappa)
        return(res)
}


# Parameter estimation
parMOM <- gevMOM(pmoms(x)) # MOM
parLMOM <- as.numeric(pargev(lmoms(x))$para) # LMOM
parML <- gev.fit(x,quiet=TRUE)$mle # ML


# Exceedance probabilities of quantile x=150
1-pgev(q=150,location=parMOM[1],scale=parMOM[2],shape=-parMOM[3])
1-pgev(q=150,location=parLMOM[1],scale=parLMOM[2],shape=-parLMOM
[3])
1-pgev(q=150,location=parML[1],scale=parML[2],shape=parML[3])


# Quantile with F=0.99 (T=100)
qgev(p=0.99,location=parMOM[1],scale=parMOM[2],shape=-parMOM[3])
qgev(p=0.99,location=parLMOM[1],scale=parLMOM[2],shape=-parLMOM
[3])
qgev(p=0.99,location=parML[1],scale=parML[2],shape=parML[3])
```

Example 6.15 The solution to this example uses the packages ismev (Heffernan, Stephenson and Gilleland 2013) and VGAM (Yee 2010).

```
x <- c(68.8, 67.3, 70.2, 113.2, 79.2, 61.2, 66.4, 65.1, 115, 67.3,
102.2, 54.4, 69.3, 54.3, 36, 64.2, 83.4, 64.2, 76.4, 159.4, 62.1,
78.3, 74.3, 41, 101.6, 85.6, 51.4, 70.3, 81.3, 85.3, 58.4, 66.3,
```

91.3, 72.8, 100, 78.4, 61.8, 83.4, 93.4, 99, 133, 101, 109, 88,
99.6, 74, 94, 99.2, 101.6, 76.6, 84.8, 114.4, 95.8, 65.4, 114.8)

```
library(ismev)
library(VGAM)

gumfit <- gum.fit(x) # fit Gumbel distribution
parML <- gumfit$mle
vcov <- gumfit$cov # covariance matrix

# gradient of quantile function at F=0.99
h <- c(1,-log(-log(0.99)))

# standard quantile error at F=0.99
SF <- sqrt(t(h)%*%vcov%*%h) #

#bounds of the confidence interval for quantile F=0.99 (T=100)
LB <- qgumbel(0.99,parML[1],parML[2])+qnorm(0.025)*SF
UB <- qgumbel(0.99,parML[1],parML[2])+qnorm(0.975)*SF
```

Example 8.3   The solution of this example focused only on the GEV distribution. In
this example maximum likelihood estimation of the GEV could have used functions
from the ismev package. Instead, the presented code shows how to construct the
log-likelihood function of the GEV using its density function. The same principle
can be applied to any distribution in R.

```
library(lmomco)
library(VGAM)

x <- c(444.57, 79.29, 70, 87.5, 74.76, 158.57, 55.22, 70.51, 52.95,
205.3, 65.41, 103.07, 33.98, 903.31, 131.67, 38.51, 100.52, 52.39,
97.41,
46.44, 31.71, 62.86, 64.28, 14.02, 15.86, 28.32, 27.18, 47.01, 51.82,
33.41, 90.9, 218.04, 80.42, 60.88, 68.53, 56.35, 84.67, 69.94, 65.98,
38.23, 32.28, 118.65, 104.77, 237.3, 116.67, 95.71, 41.06, 88.63,
29.45, 45.59, 28.06, 104.21, 68.53, 29.45, 203.6, 97.41, 38.23,
72.49, 33.13, 45.59, 154.04, 79.29, 288.83, 184.06, 297.33, 73.06,
96.56, 73.06, 74.19, 297.33, 27.55, 42.76, 95.43)

# Parameter estimation
gevMOM <- function (mom) {
      m <- mom$moments[1]   # mean
      s <- mom$sd            # standard deviation
      g <- mom$skew          # skewness coefficient
```

```
        if (g == 1.1396) {# subroutine to solve Equation (5.74)
        } else {
            if (g > 1.1396) {
                domain <- c(-0.333,-0.00001)
            } else {
                domain <- c(0.00001,2)
            }
            Kappa <- uniroot(function (x)(g - sign(x)*
            (-gamma(1+3*x)+3*gamma(1+x)*gamma(1+2*x)-2*gamma(1+x)^3)
            /(gamma(1+2*x)-gamma(1+x)^2)^(3/2)),
            interval=domain, tol=1e-10)$root
        }

    Alpha  <- s*abs(Kappa)/(sqrt(gamma(1+2*Kappa)-gamma(1+Kappa)
^2))
    Beta <- m-(Alpha/Kappa)*(1-gamma(1+Kappa))
    res <- c(Beta,Alpha,Kappa)
    return(res)
}
parMOM <- gevMOM(pmoms(x)) #MOM

parLMOM <- as.numeric(pargev(lmoms(x))$para) #LMOM

max.ll <- optim(par=parMOM,
    fn =  function(pars)  -sum(dgev(x,pars[1],pars[2],-pars[3],
log=TRUE))
,hessian=TRUE)

parML <- max.ll$par  #ML

# PPCC test statistic
cunnanePP <- ppoints(x,a=0.4)
cor(qgev(cunnanePP,parLMOM[1],parLMOM[2],-parLMOM[3]),sort(x))

# AIC and BIC
aic <- max.ll$value*2+6
bic <- 2*max.ll$value+log(length(x))*3

# Quantile estimates
T <- c(1.0001,1.2,1.5,2,5,10,20,50,100,200,300,400,500,1000)
F <- 1-1/T

quantile <- qgev(F,parML[1],parML[2],-parML[3])
```

```
# quantile standard errors
Beta <- parML[1]
Alpha <- parML[2]
Kappa <- parML[3]
r <- rep(0,length(F))
h <- attr(numericDeriv(quote(qgev(F,Beta,Alpha,-Kappa)),
    theta=c("Beta","Alpha","Kappa")),'gradient')
vcov <- solve(max.ll$hessian)
for (k in 1:length(F)) r[k]=t(h[k,])%*%vcov%*%h[k,]
SF <- sqrt(r)
```

Example 9.3

```
mydata <- read.csv(url
('https://github.com/fshydrology/fsh/raw/master/data9.3.csv'),
row.names=1)

cor(mydata) # correlation matrix

step1 <- lm(Runoff ~ A, data=mydata)
summary(step1)
anova(step1)

step2 <- lm(Runoff ~ A + S, data=mydata)
summary(step2)
anova(step2)

step3 <- lm(Runoff ~ A + S + Pr, data=mydata)
summary(step3)
anova(step3)
```

Examples 12.5 and 12.7

```
data12.5 <- data.frame(
Y=c(3, 2, 1, 3, 3, 0, 2, 3, 4, 1, 4, 2, 2, 5, 3, 1, 2, 0, 1, 0,
3, 2, 4, 4, 1, 0, 2, 1, 1, 4, 1, 2, 1, 2, 3, 3, 0, 1, 4),
NAO=c(-1.068, 1.598, -0.53, 0.372, -0.126, 1.892, -0.888, -1.218,
-0.642, -0.54, -0.176, 1.27, -0.196, -2.068, -0.676, -0.36, -0.232,
1.276, 0.342, 1.18, 0.64, -0.576, 0.338, -0.294, 0.448, 0.308,
0.53, 1.942, 0.138, -0.536, -0.596, 0.952, -0.056, 1.994, 1.304,
-0.132, 1.68, 2.044, 1.594))

glm1 <- glm(Y ~ NAO,data=data12.5,family=poisson(link='log'))
coef(glm1) # regression parameters
vcov(glm1) # covariance matrix

glm0 <- glm(Y ~ 1,data=data12.5,family=poisson(link='log'))
```

```
D <- 2*(logLik(glm1)-logLik(glm0)) #likelihood ratio test statistic
qchisq(0.95,1) #critical region


AIC(glm1)
AIC(glm0)
```

Example 12.11 The solution to this example uses the package ismev (Heffernan, Stephenson and Gilleland 2013).

```
library(ismev)


pavia <-c(24.2, 31.3, 32.5, 33.5, 20.2, 38.2, 36.7, 35.2, 35.2, 25.3,
92.3, 30, 25.2, 50.4, 35.7, 40.5, 10.3, 40.2, 8.1, 10.2, 14.2,
15.3, 40.2, 20.4, 20.2, 32.8, 43.2, 29.8, 42.8, 45, 34.2, 32.8,
46.3, 31.9, 34.2, 24.3, 24.3, 24.3, 71.4, 37.4, 31.4, 24.3, 43.8,
58.2, 34.6, 40.2, 20.8, 69, 44, 27.2, 37.2, 36.7, 49, 38.9, 59.6,
63.3, 41.2, 46.6, 84.2, 29.5, 70.2, 43.7, 36.2, 29.8, 60.2, 28,
31.4, 38.4, 29.4, 34, 47, 57, 36.5, 84.2, 45, 95.5, 48.5, 38,
38.6, 26, 27, 58, 27.8, 37.5, 35.2, 27.5, 28.5, 52, 56.8, 80,
29, 55.2, 48.4, 33.2, 27.4, 27.4, 18.2, 34.2)
# time covariate in a single column matrix
t <- matrix(1:length(pavia),ncol=1)


# fitting GEV and Gumbel models
GEV0 <- gev.fit(pavia)
GEV1 <- gev.fit(pavia,ydat=t,mul=1)
GUM0 <- gum.fit(pavia)
GUM1 <- gum.fit(pavia,ydat=t,mul=1)
GUM2 <- gum.fit(pavia,ydat=t,mul=1,sigl=1,siglink=exp)


# maximum log-likelihood of each model
-GEV0$nllh
-GEV1$nllh
-GUM0$nllh
-GUM1$nllh
-GUM2$nllh


# likelihood ratio test statistic
2*(GUM1$nllh-GUM2$nllh)


# standard errors of GUM2 parameters
GUM2$se
```

```
# Rejection region of test
qnorm(0.975)*GUM2$se[4]
```

References

Asquith WH (2007) Lmomco: L-moments, Trimmed L-moments, L-comoments, and Many Distributions. R package version 0.97, 4.

Crawley MJ (2012) The R book. Wiley, New York.

Heffernan JE, Stephenson A, Gilleland E (2013) Ismev: an introduction to statistical analysis of extreme values. R package version 1.39.

Matloff N (2011) The art of R programming: a tour of statistical software design. No Starch Press, San Franscisco.

R Core Team (2013) R: a language and environment of statistical computing. R Foundation for Statistical Computing, Vienna.

Revelle W (2014) Psych: procedures for personality and psychological research, R package version, 1(1). Northwestern University, Evanston.

Yan J (2007) Enjoy the joy of copulas: with a package copula. J Stat Software 21(4):1–21.

Yee TW (2010) The VGAM package for categorical data analysis. J Stat Software 32(10):1–34.

Index

Printed in the United States
By Bookmasters